Probability and Statistics

A Traditional/Technology Approach

First Edition

Raymond Badalian

Educo International Inc.
1-800-963-3826

ACKNOWLEDGEMENTS

Managing Editor	**Dr. Kewal P. Chinda,** Educo International, Inc.
Chief Editors	**George Dekermenjian, Professor of Mathematics,** Los Angeles City College
Editors	**Galust Mardirussian,** Professor of English, Los Angeles City College **Ronald Kendis,** Professor of Mathematics, Los Angeles City College **Kamal Kapoor,** Professor of Statistics, University of Delhi, India
Production Services	Educo International, Inc.
Desktop Publishing	Sunita Verma and her team Educo International, Inc.

Special Thanks

To the following persons for the advises, suggestions, and enconcepts.

i) Roger Wolf, Department Chair, Los Angeles City College, CA

ii) KianKaviani, Assistant Chair, Los Angeles City College, CA

iii) Robert Mardirosian, Professor of Mathematics, Los Angeles City College, CA

iv) Prabha Betene, Professor of Mathematics, LaGuaurdia

v) Steven Castillo, Professor of Mathematics, Los Angeles Valley College, CA

vi) Tooraj Gaurdi, Professor of Mathematics, Antelope Valley College, Lancaster, CA

vii) F. Abebe, Professor of Mathematics, Clark Atlanta University, Atlanta, CA

Special thanks to **Dr. Man M. Sharma,** Professor Emeritus of Mathematics at Clark Atlanta University, for support, continous encouragements throughtout the process of completing this book.

ISBN: 13 digit: 978-1-935168-53-9
ISBN: 10 digit: 1-935168-53-3

Printed at: Replika Press Pvt. Ltd.,

PREFACE

The approach to this book, "**Probability and Statistics**" is based on the practice of teaching statistics to students whose preparation in mathematics is limited to basic algebra. To that end, a pedagogical decision was made to motivate the thinking process behind the practice of using statistics with the help of examples. The use of technology to perform computations is applied throughout the text. Microsoft™ Excel® was chosen as the application and its relevant commands and functions are available to perform at all appropriate places. Each chapter includes a section called "**Extended Topics and Proofs**". This section is optional and presents some theoretical concepts and selected proofs of relevant results that may be used for more advanced students. Each chapter ends with a large array of exercises to enrich student understanding. The majority of these exercises are real-world applications in education, political science, demographics, business, sports, and health.

Certain difficult-to-teach concepts are presented using simulations with the aid of Microsoft Excel®. Two such examples are:

- the limiting process as a binomial distribution approaches a normal distribution
- the properties of sampling distributions

This book is supplemented with an electronic version that can be accessed from the web by faculty as lecture notes and by students as tutorials. The electronic version of the book is completely aligned with this printed textbook and has the advantage of dynamic graphics, pedagogically sound animations, and visually appealing screen designs with attractive color combinations. As a tutorial for students, the electronic version provides an effective tool for self learning that is highly interactive and engaging. The electronic version can also be used for online teaching. The discussion in this book is divided into ten chapters and five appendices including one on brief introduction to Microsoft™ Excel® (Appendix 1), the application used throught the book. The aim is to focus on conceptual understanding and deal with the computational part of the problem solving it with the appropriate use of Microsoft™ Excel®.

Highlights of Contents:

TECHNOLOGY SUPPORT

One of the **most useful features** of this book is the technology support to teachers and students in the form of lecture notes and tutorials, that are embedded with illustrations and examples with step-by-step solutions. **All technology components are created in Macromedia Flash delivered through the web requiring no special installations.** These features are described below in more detail.

A. Multimedia Electronic Lecture Notes (ELN) for Instructors and Tutorials for Students

- Tutorials with examples are embedded with dynamic graphics, animations, and step by step solutions.
- Vivid screen designs and animations make the presentation of topics clearer, as documented by students and instructors in several pilot studies.
- The ELN for instructors are designed to enhance interactivity with students, affording instructors minimal writing and drawing and more emphasis on class discussion.
- The ELN, which can be used in class to supplement conventional or online teaching, are available to students as tutorials from the web for self study, anywhere at any time. Each example in the tutorial has several versions generated by randomized parameters. The student can see the solution to the first version, and then click the 'Next Version' to attempt problems.

B. Homework on the Internet with Embedded Tutorial

- Different types of free response questions, with several variations for each question, are used to create online homework. Instructors can create the homework using a large question bank of free response questions, provided by the system.
- Each student gets a different set of questions, with tutorial type assistance, a solution to the first version, and feedback for other versions, with the option to change the answer if answered incorrectly. Students can complete the homework in multiple sessions.
- Homework is instantly graded by the system and scores are transferred automatically to the grade book.

C. Quizzes/Tests on Internet (Free-Response, or Multiple-Choice or Mixed)

- Responses to test questions may be entered from the keyboard, or through a **small online keypad containing Mathematical symbols needed for answers.**
- After every quiz, students get instant feedback on their score. They can see step-by-step solutions, go to practice mode with an embedded tutorial, or take the quiz again, if allowed by the instructor.
- There will be no loss of work done by a student if during a test or homework session the power goes off, the web connection is lost, or the student closes Test/Assignment window improperly.
- A large question bank of, both Multiple Choice and Free Response questions is available for each objective of the content for web based testing and practice testing.
- A browser based math editor is available to instructors to add their own questions.

- An easy-to-use online test generator, is available to instructors to create online quizzes/tests, or to print multiple versions of class test with answer keys.
- The test generator provides several options to instructors to deliver tests and homework to students in several test modes: with or without feedback, with or without solutions, single or multiple sessions, graded or just for practice.
- Instructors may allow students to take quizzes more than once, from anywhere at any time, and every time the quiz is different, covering similar types of items.
- For every graded test created by an instructor, the system generates an online practice test for students by default. Items for these practice tests are selected by the system and cover the objectives selected by the instructor for the graded test items. The system provides different sets of items for every practice test session.
- Instructors may conduct proctored major tests, password-protected, in local computer labs, or print the tests and manually transfer the scores to the online graded book.
- **Clickers** type feature is already built-in the system with added advantage that students can respond from their computers. Instructors can watch live, as students are responding, and view analysis of responses for each item.

D. Activity and Grade Reports

- Instructors can view/print **several types of student activity reports** at any time. Students can view/print their updated progress reports anytime.
- **Enhanced electronic grade** book (online) is very flexible, where instructors can have any number of grading categories or set-up any grading scales, and add/edit messages to students based on their grades at the time they view their progress. Grade book categories may also include attendance or time spent on tutorial.
- **Electronic attendance marking** is unique and flexible, teacher can mark: a) All present (P) with one click and then change P to A for those not present, to T (for Tardy) for those who come late, or to E (for Exempt) for those who have valid excuse. The system calculates the present percentage and transfers the attendance grades to the grade book.

E. Several other Communication Tools

- **Announcements**
  - Faculty to his/her students.
  - Campus coordinator to Instructors, students of a section/course or to all.
  - Educo to different groups of users.
- **Chat, threaded discussion, group discussions**

All the above features of technology support make the use of this book ideal for teaching and delivering online courses.

F. **Students can purchase the book online through e-commerce of this learning portal.**

To the students:

We sincerely hope that the approaches followed in this book with several options of technology support will help you better learn the concepts provided in this book or its online supplement.

To the instructors:

The authors would appreciate any suggestion for improvement in any component of this package, either for the presentations in the textbook or its support material on the web. Suggestions can be forwarded through **Contact us** link on the login page.

Raymond Badalian

TABLE OF CONTENTS

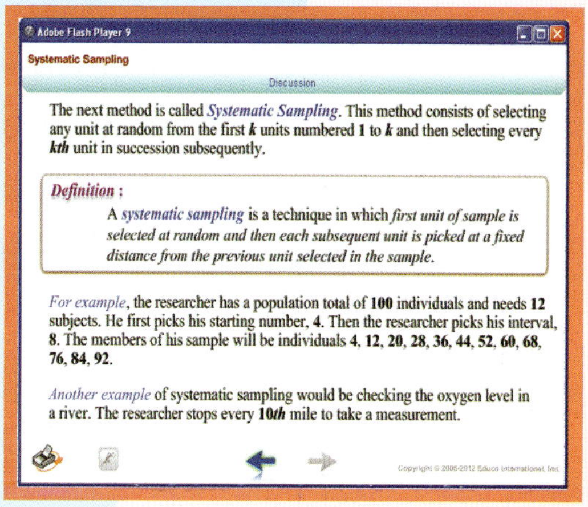

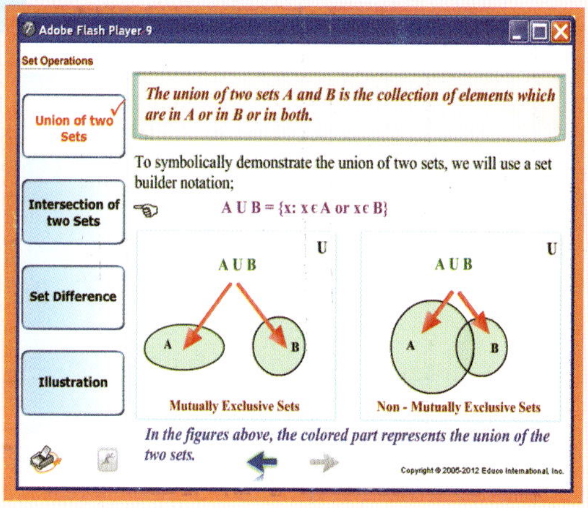

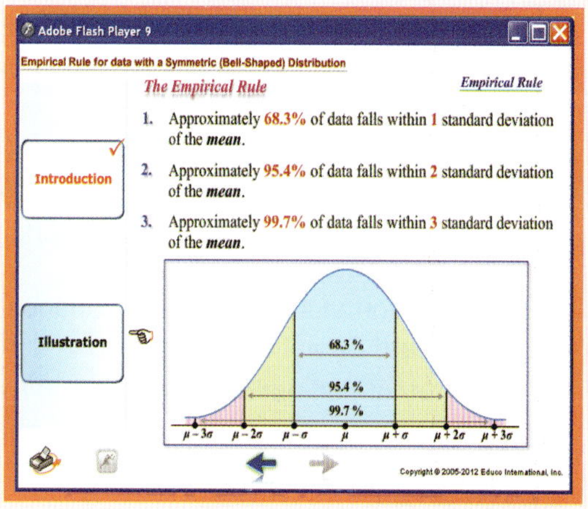

TABLE OF CONTENTS

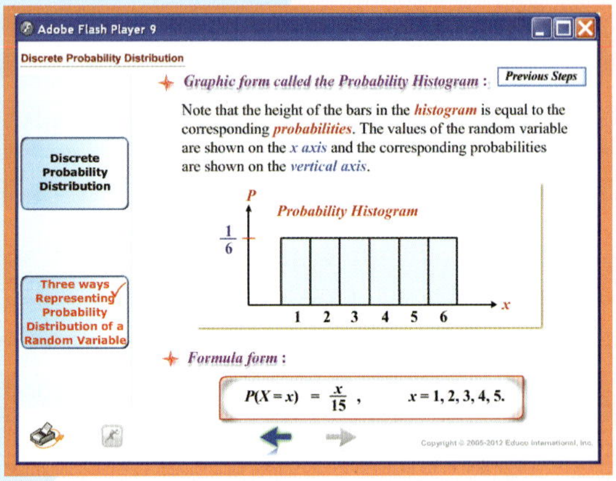

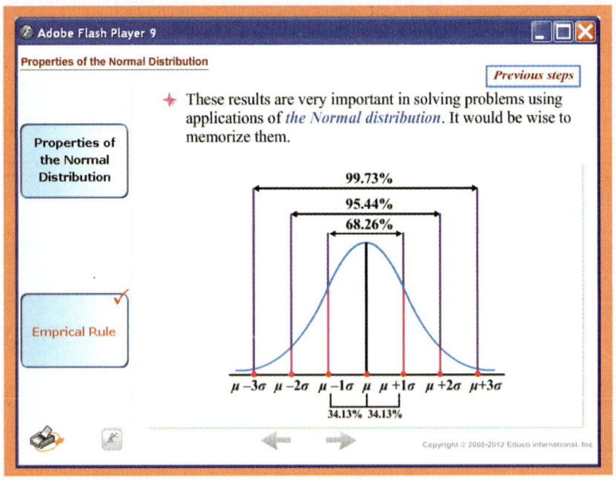

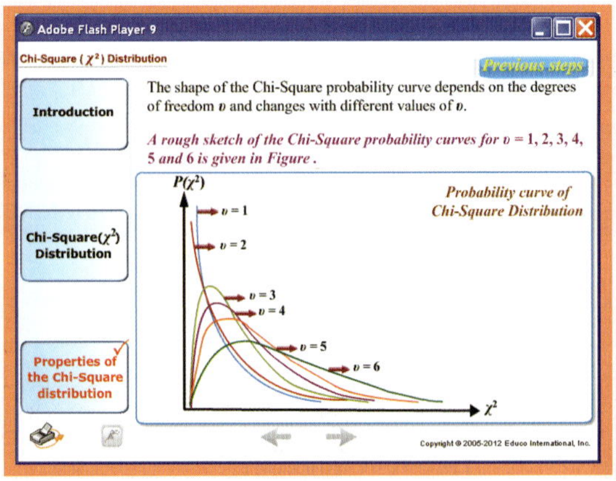

TABLE OF CONTENTS

CHAPTER 7: Hypothesis Testing (One Population)

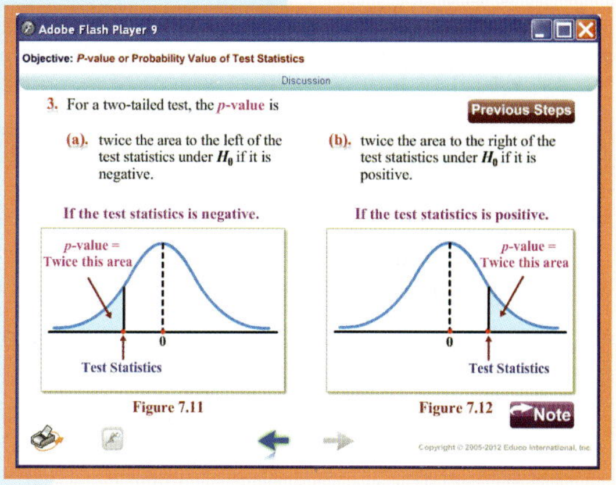

CHAPTER 8: Inferences from Two Populations

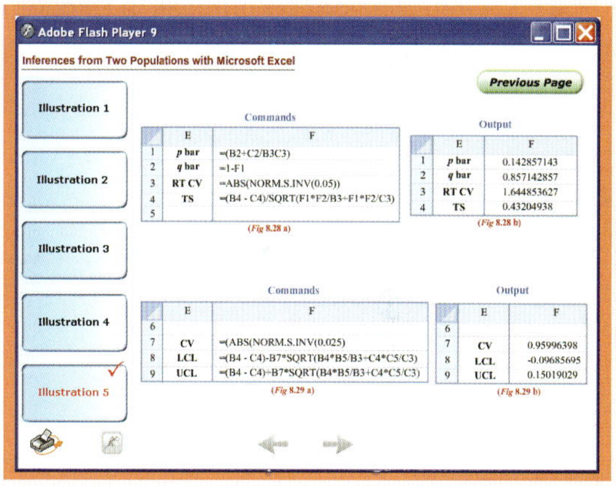

CHAPTER 9: Multinomial Experiments Chi-Square Tests for Categorical Data

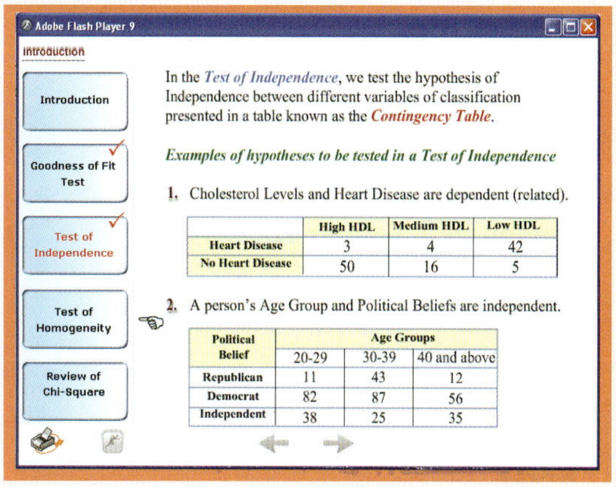

TABLE OF CONTENTS

CHAPTER 10: Correlation and Regression Analysis

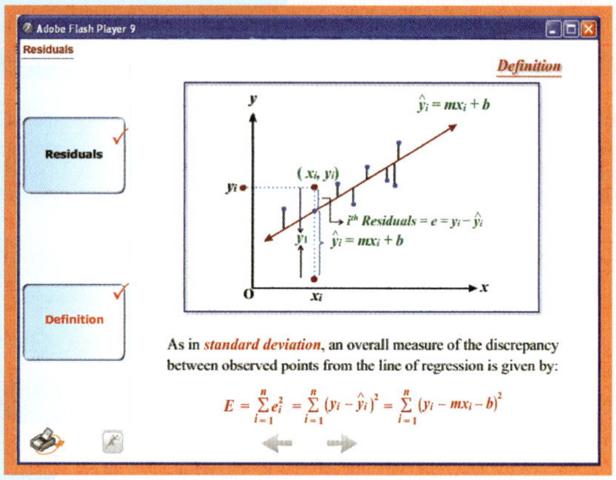

Appendix 1: A Brief Introduction to Microsoft Excel®

Appendix 2: Analysis of Variance

Appendix 3: Non-Parametric Statistics

1

INTRODUCTION TO STATISTICS

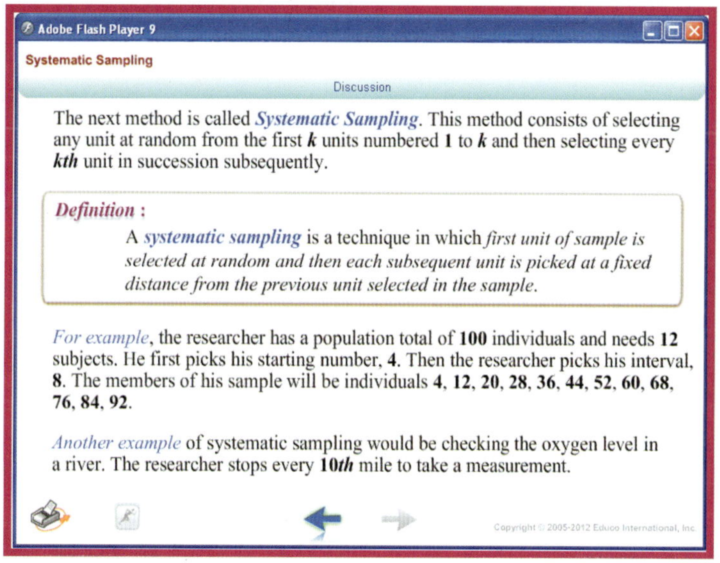

CHAPTER 1

INTRODUCTION TO STATISTICS

Chapter Overview:

1. *Overview of Statistics*
2. *Statistical Data and Design of Experiments*
3. *Sampling Methods*

1.1: OVERVIEW OF STATISTICS

Objectives

After completing this section you will be able to describe:

A. Descriptive Statistics.
B. Inferential Statistics.
C. Important Terms Related to Inferential Statistics.

Students who are studying Social and Natural Sciences such as Psychology, Economics, Sociology, Business, Biology, Computer Science, Biotechnology *etc.* are often surprised to learn how much of the curriculum in their field focuses on understanding and learning "how to analyze data." Because statistics is an essential part of these fields, students need to understand how to use statistics and how to interpret with statistical results. The science of statistics provides students guidelines for describing data and for interpreting patterns of differences and similarities that may occur within certain groups of data or between different groups. Statistics also provides scientifically rigorous techniques for analyzing data and making inferences from numerical results. Statistics has the power to transform data into knowledge that can be used to understand and solve real world problems.

Definition:

Statistics *is the science of collecting, organizing and summarizing data, and drawing conclusions for the population based on those results.*

Alternative Definition:

Statistics *is a collection of methods for decision making in the face of uncertainty on the basis of numerical data and calculated risks.*

The two major branches of statistics are **Descriptive** and **Inferential Statistics**. These branches are not mutually exclusive, but rather built on one another.

A. Descriptive Statistics

Descriptive statistics is a statistical procedure that is used to describe characteristics of data sets. It refers to the measurement of data that is presently occurring within all subjects. This type of statistics involves methods of organizing, picturing, and summarizing information from samples or populations so that we can communicate our results as accurately and completely as possible. For example, one of the first things that we would want to do with our data is to graph them, to calculate averages and to look for extreme scores or oddly shaped distribution of scores. These procedures are called *descriptive statistics* because they are primarily aimed at describing the data. It also allows us to simplify large quantities of data in an understandable and logical manner.

Definition:

The branch of statistics which deals with collecting, organizing, summarizing and graphing a data set is called ***descriptive statistics.***

B. Inferential Statistics

Once we have described our data in detail and are satisfied that we understand what the numbers have to say on a superficial level, we will employ the use of Inferential Statistics. Inferential statistics refers to assessing the likelihood of something happening at some point in the future tests or testing a sample of the population in order to generalize to the entire population. Perhaps one of the simplest inferential tests is used when you want to compare the average performance of two groups on a single measure to see if there is a difference. You might want to know whether sixth-grade boys and girls differ in math test scores or whether a program group differs on the outcome measure from a control group. Most of this book will deal with inferential statistics as it allows us to draw conclusions about a population based on the data in our sample.

Definition:

The branch of statistics which deals with drawing conclusions based on information collected from the sample is called ***inferential statistics***.

C. Important Terms Related to Inferential Statistics

• Population and Sample

Population is the aggregate of objects, animate or inanimate, under study in any statistical investigation. A finite subset of the population that has been selected from it, with the objective of investigating its properties is called a *sample* and the number of units in the sample is known as the *sample size*. Sampling is a tool which enables us to draw conclusions about the characteristics of the population after studying only those objects or items that are included in the sample.

Population:

In most statistical investigations, the target group is known as the population. A population is not a group of people, but a set of responses from a group. For example, if we are interested in the effect that watching Sesame Street has on preschool children's vocabulary scores, the population of interest is **preschool children's vocabulary scores**. In a second case, if we are interested in differences in cooking abilities between men and women, our population of interest is **differences in cooking abilities of men and women**. As you can see, the population of interest is dependent on the research question itself. In most statistical studies, the population can be too large, making it difficult, sometimes even impossible, to try to extract information from the population. Fortunately, statistics has allowed us to draw conclusions about an entire population by studying only a subset (sample) of responses drawn from that population.

Definition:

*A **population** is the entire collection of outcomes, measurements, and counts to be studied.*

Sample:

A sample is a subset or part of the population that is being examined. It is important that the data sample we examine is representative of the population of interest. This means that the sample must possess the same important qualities and characteristics that are in the population. For example, if the population of interest is preschool children's vocabulary scores, our sample responses should be reflective of the wide range of possible vocabulary scores found in the population. If our sample consisted of only the vocabulary scores of preschoolers from higher socio-economic status families who have had extensive experience in high-quality preschools, our sample would not be representative of the entire population; therefore, any conclusions drawn from that sample may be very limited.

Definition:

*A **sample** is a subset of members from an entire population.*

Case Study:

In a recent study of the proportion of college students addicted to nicotine, it was found that out of 13,847 students, a total of 36.2% smoked cigarettes. A researcher randomly selected a group of 86 students and determined that 37.1% were addicted to nicotine. In this particular study, the size of the population was 13,847 students and the size of the sample was 86 students.

Note: As the size of the sample increases, the sample percentage becomes closer to the population percentage. This means that we can reduce the error in estimation by increasing the sample size. It is quite obvious that using samples to draw conclusions about a population is much faster and more cost effective than using the entire population for the study.

• Parameter and Statistic

In statistical research, the measurements obtained can be classified into two categories: **Parameter and Statistic.** One important distinction is that *Statistic* takes into consideration numbers or indices that are provided by our sample (*e.g.*, the average preschool vocabulary score in the sample) that is used to estimate the same values within the population, called *parameters*.

Parameter:

A parameter is a true value in the population of interest. When describing parameters, *the population mean* is most often used. The following are examples of parameters:

- The average weight of females between ages 25 to 30.
- The median income of males in San Francisco.
- The number of abandoned dogs in the city of LosAngeles.

In the previous case study, the value 36.2% describes a characteristic of the population and is called a **Parameter**.

Definition:

*A **parameter** is a numerical measurement describing a certain characteristic of a population.*

Statistic:

A statistic is a numerical quantity that describes characteristics of the sample. A statistic is a number you calculate from your sample data in order to estimate the parameter. The following are examples of statistics:

- The average weight of a random sample of 1,000 females between the ages of 25 to 30.
- The median income of a random sample of 1,000 males in San Francisco.
- The percentage in a random sample of 1,000 abandoned dogs in Los Angeles.

Again, looking at the earlier case study, the value of 37.1% is considered a statistic, because it describes a characteristic of the sample.

Definition:

*A **Statistic** is a numerical measurement describing a certain characteristic of a sample.*

1.2: Statistical Data and Design of Experiments

Objectives

After completing this section you will be able to describe:

A. Quantitative and Qualitative Data.

B. Discrete and Continuous Data.

C. Levels of Measurement.

D. Independent and Dependent Variables.

In simple terms, data is the information that is compiled during any type of experiment or study. Statistics is a science that consists of different models of collecting, summarizing, and graphing measurements from populations or samples. The compilation of this information is called a data set. Some data sets consist of numerical values while others consist of non-numeric measurements.

Definition:

Data is a set of numerical or categorical values of the variable.

A. Quantitative and Qualitative Data

Quantitative data are purely numeric. A quantitative variable has a value or numerical measurement for which operations such as addition or averaging make sense. For example, the number of packages delivered to a particular business each month is a quantitative data. On the other hand, *Qualitative* data deals with descriptions. This type of data can be extremely varied in nature and virtually includes any information that can be captured and that is not numerical in nature. For example, the biological sex of a chimp, the color of a child's eyes, or someone's chosen political party. Luckily, these variables can easily be quantified by assigning numbers to represent the different categories. Once assigned, these categories allow us to use statistical analysis to describe and explain our data.

Definition:

Quantitative Data is a *data set which consists of numbers representing counts or measurements.*

Examples of Quantitative data are:

- The heights of children in a kindergarten class.
- The speed of a train en-route to Chicago.
- The temperature of an erupting volcano.

Definition:

Qualitative Data *is a data set which consists of non-numeric measurements or counts.*

Examples of Qualitative data are:

- The eye-color of a 30 year old male.
- The gender of a patient.
- The label on a soda can.

B. Discrete and Continuous Data

Quantitative data can be further described in terms of discrete or continuous data. It is important to note the differences between these variables because they can be analyzed and presented in many different ways.

• Discrete Data

Discrete data has a distinct countable number of values. In statistics, any data with fractional values are not possible. For example, if we counted the number of books on a bookshelf, we might find that there are 159 books or 160 books. We will never find 159.5 books or 160.3 books on the shelf.

Definition:

Discrete Data *is data that represents distinct counts or is countable.*

Examples of discrete data are:

- Number of students in classrooms (22, 24, 19, 32).
- Number of households in an apartment complex (3, 5, 1, 7).
- Number of left handed students in a class (2,4,7,0).
- Number of delayed flights on a Monday (3,10,4,12).

• Continuous Data

Continuous data takes on any value within the limits of the data. For example, in measuring the distance from home to our college campus, we could find that we are 5.88882 miles away or 6.298 miles away. Continuous data has no gaps in its scale. In this particular case, fractional values are always possible. It is always possible to find values that fall between two points on the scale. Continuous data is often related to physical measurements such as weight, height, and volume.

Definition:

Continuous Data *is data that represents values corresponding to some continuous scale without interruptions.*

Examples of Continuous data are:

- Amount of gasoline in a car (15.5, 8.6, 38.12).
- Temperature in a city at midnight (12.6, 19.8, 26.4, 9.3).
- The range of voltage in a voltmeter.
- Variation of body temperature of a healthy adult in a month.

C. Levels of Measurement

Another characteristic of data is its level of measurement. There are four levels of measurement in statistics:

i) Nominal Measures

ii) Ordinal Measures

iii) Interval Measures

iv) Ratio Measures

i) Nominal Measures:

The Nominal level of measurement applies to data that consists of names, labels, or categories. There are no criteria by which the data can be ordered from smallest to largest. As such, nominal measurements place specific qualities into separate categories. These categories are called mutually exclusive. Any item in one category cannot be in any other category and they never overlap. One category is not "greater" or "better" than any other; therefore, categories are merely different. For example, the sex of a person is a nominal variable. A male is considered one category and a female another. In sports, basketball jerseys would be considered nominal in the sense that a player that has a high jersey number is not necessarily better or worse than a player with a low jersey number. Considering political parties as another example, one can be registered to vote as a Democrat, Republican, or as any other political party, but ultimately cannot be registered to vote as all.

> **Definition:**
>
> *Data at **nominal** level of measurement is categorized by names, labels and categories.*

At this level of measurement data is qualitative only and there is no order. Colors, Models, Brands, *etc.* are examples of such data.

Examples of nominal data:

(a) Color of Laptop used by students at Clark Atlanta University (black, white, gray, purple, pink, blue).

(b) Brand of Cell Phone used by students at Cambridge University (Nokia, Samsung, Sony Ericsson, Apple).

(c) Honors courses taught in leading universities like Cambridge or Oxford (physics, chemistry, statistics, mathematics, history, economics).

(d) Eye color of employees of Educo: (blue, green, brown, black).

ii) Ordinal Measures:

The second type of measurement scale is Ordinal measures. Like a categorical scale, this may involve different qualitative categories, but these particular categories are ranked in some kind of order. However, differences between data values either cannot be determined or are meaningless. For example, runners in a race finish in different places. For example, runners can finish in 1st place, 2nd place, 3rd place, and so on to "last place". These numbers are not truly numeric because adding first place to second place does not equal third place, but they do represent different degrees of quality. In terms of runners, the runner who came in first is certainly "better" than the person who came in third place and the runner who finished third is certainly "better" than the person who came in "last." Similarly, if we designated employees within a manufacturing firm according to their rank (e.g., 1 = executive, 2 = managerial, 3 = administrative, 4 = factory line worker), the numbers have no true numeric meaning ("executive" is not twice as much as "managerial"), but they are in an ordinal (or "ordered") form, so that executive is "higher" than managerial or administrative positions. While the categories in an ordinal scale are ranked, we cannot distinguish the distance or difference between two categories just from their values. While our 1st place runner is the "best," we have no idea from this label how much better she is than the 2nd place, or 3rd place runner. The 2nd place runner could be one second or ten minutes behind the winner; an ordinal scale does not provide this information.

Note: Ordinal data can be treated as nominal, but not vice-versa.

Definition:

*Data at **ordinal** level of measurement can be arranged in order, but their differences are meaningless and cannot be determined.*

In this case, data can be both qualitative and quantitative. Data can be put in order but it can't perform any mathematical operations such as adding, subtracting, multiplying or dividing. Grades (A, B, C, D, F), Rank (sergeant, lieutenant), *etc.* are examples of such data.

Examples of ordinal data:

(a) Freeway traffic on Christmas day (heavy, medium, light).

(b) Working position at Call Center (Customer support levels 1, 2, 3).

(c) Star rating of newly released movies (★, ★★, ★★★, ★★★★, ★★★★★).

(d) Rank of new appointees at Oxford University (Professor, Associate Professor, Assistant Professor).

iii) Interval Measures:

The third type of measurement scale is called an Interval measure. The interval level of measurement applies to data that can be arranged in order. In addition, differences between data values are meaningful. Like ordinal measures, there is logic to the values of the numbers that are used. For example, a 5 is "better" or "worse" than a 3, which is "better" or "worse" than a 1 (depending on the direction of our scale). However, unlike an ordinal scale, the distance between two numbers on an interval scale has a meaning.

Definition:

*Data at **interval** level of measurement can be arranged in order and the differences between entries* are *meaningful, but it doesn't have a starting point zero.*

In this case, data is quantitative only. Data can be placed in order and differences in the data have proper meanings. In addition, there is no natural zero starting point. We can see that zero is not a starting point for year and temperature because there are values before or below 0.

iv) Ratio Measures:

The ratio level of measurement applies to data that can be arranged in order. Ratio scales of measurement are *truly* numeric. A score or a number on a ratio scale represents a true quantity or amount. For example, if we are interested in the number of television sets in people's homes, we would count them. If we are interested in the number of times a person has been married, we would count them too. If Bob has been married four times and Sue has been married twice, we could accurately say that Bob has had twice as many marriages as Sue. Ratio scales have a *true zero point.* A person could have zero television sets, or zero marriages, but cannot have anything less than zero.

Definition:

Data at ***ratio*** *level of measurement can be arranged in order and the differences between entries are meaningful. Data at the ratio level has an absolute zero.*

Ratio level of measurement has the same characteristic as interval level of measurement with an added starting point of zero. Speed, Distance, Marriages, *etc.* are few examples of such data. There is a natural zero starting point for the above examples.

D. Independent and Dependent Variables

In any experiment, the variables of interest are classified as **Independent** and **Dependent**.

- **Independent Variable**

The independent variable is the variable that is systematically manipulated by the experimenter. The experimenter controls the variable and sets different levels of the variable. These levels may reflect exposure to the variable or degree of its intensity. The term "independent" is used because whether or not it occurs is not dependent on any other variable. The experimenter decides how, when, and to what degree the independent variable will occur.

- **Dependent Variable**

When a variable is influenced by a set of independent variables, then it is called dependent variable. For example, Dr. Echo may be interested in the effects of sleep deprivation on young adults' driving performance. He is hypothesizing that individuals who are sleep-deprived will exhibit poorer driving performance. In order to evaluate the validity of his hypothesis, he conducts an experiment where he systematically varies level of sleep deprivation (the independent variable) and then measures how those variations relate to variations in driving performance (the dependent variable).

1.3: Sampling Methods

Objectives

After completing this section you will be able to define:

A. Random Sampling.
B. Convenience Sampling.
C. Systematic Sampling.
D. Stratified Sampling.
E. Cluster Sampling.

There are different sampling methods to select research participants whose responses will make up our data. We list a few for your review:

A. Random Sampling

The best and easiest method is called **Random Sampling**. This is where research participants are chosen from a specified population in such a way that each individual has an equal chance of being selected.

Definition:

A random sampling is a technique in which each member of population has equal chances of being selected in the sample.

For example, suppose George wanted to study the level of job satisfaction among workers in a certain factory. In order to choose a random sample of 50 workers from the total population of 500 workers, he could begin with a list of all of the workers. He uses a table of random numbers or a computer program that generates random numbers to randomly select 50 workers. With a small population of interest, random sampling is not too difficult to work with. However, with a very large population (like preschool children's vocabulary scores), true random sampling may not be possible and a **Convenience Sample** is more likely to be used.

B. Convenience Sampling

Conveniences sampling is often used when the population is too large. This method is often used because the sample is readily available and the responses are easily accessed. As with the example of job satisfaction in a factory, an example of a convenient sample would be to survey only those employees who were in a particular room in the entire factory.

Definition:

*A **convenience sampling** is a technique in which a unit of sample is selected with convenience of persons. In other words, whatever sample is handy is chosen.*

C. Systematic Sampling

The next method is called **Systematic Sampling**. This method consists of selecting any unit at random from the first k units numbered 1 to k and then selecting every k^{th} unit in succession subsequently.

Definition:

*A **systematic sampling** is a technique in which first unit of sample is selected at random and then each subsequent unit is picked at a fixed distance from the previous unit selected in the sample.*

For example, the researcher has a population total of 100 individuals and needs 12 subjects. He first picks his starting number, 4. Then the researcher picks his interval, 8. The members of his sample will be individuals 4, 12, 20, 28, 36, 44, 52, 60, 68, 76, 84, 92.

Another example of systematic sampling would be checking the oxygen level in a river. The researcher stops every 10th mile to take a measurement.

D. Stratified Sampling

Another method is **Stratified Sampling**. We use stratified sampling when there are smaller sub-groups that are to be investigated. It begins with dividing the population into a set of smaller non-overlapping sub-groups (strata), and then randomly a sample is taken from each sub-group. Strata can be natural groupings, such as age ranges or ethnic origins.

Definition:

*A **statified sampling** is a technique in which the population is divided into a finite number of stratas (hetrogenous subgroups) and then units (proportional to size of strats) are picked randomly from each strator.*

For example if a company employs more men than women but it is required to have each group equally represented, we would employ stratified sampling. Two strata are thus created, of men and women, with an equal number in each to sample from.

E. Cluster Sampling

The last method is called **Cluster Sampling**. Cluster sampling is a sampling technique in which the entire population of interest is divided into groups (clusters) and a random sample of these clusters is selected. Each cluster must be mutually exclusive, and the clusters all together must include the entire population. After clusters are selected, then all units within the clusters are selected. This method is widely used when the studied population is spread across a wide area and other sampling methods would be too difficult to implement.

> **Definition:**
>
> *A **cluster sampling** is one where the population is divided into sections and then one of the sections is selected with all members represented in it.*

For example, in order to obtain the percentage of voters favoring a republican candidate for presidency, three suburbs are randomly selected from the entire region and all the qualified voters in these suburbs are questioned.

Chapter 1: General Exercises

1. Use a search engine to find the definition of the following terms. If possible, use each term in an example demonstrating their use in the real world.

- Statistics
- Variable
- Descriptive Statistics
- A Population
- A Sample
- Inferential Statistics
- Qualitative Variables
- Quantitative Variables
- Discrete Variables
- Continuous Variables
- Nominal Measurement
- Ordinal Measurement
- Random Sampling
- Systematic Sampling
- Stratified Sampling
- Cluster Sampling
- Independent Variable (explanatory variable)
- Dependent Variable (outcome variable)

2. **Answer each of the following questions in a sentence or two.**

(a) Why do we use samples instead of the population in statistical studies?

(b) What is the difference between ordinal and nominal measurements?

(c) What is the difference between independent and dependent variables?

(d) Use the internet, newspaper, or an article to find a statistical study, and determine if it's an observational or experimental study?

(e) In the article dealt with in (d) above, identify the dependent and independent variables.

3. Answer in short paragraphs.

(a) What is the difference between descriptive and inferential statistics?

(b) How do sample and population differ?

(c) What is the difference between a parameter and a statistic?

(d) What are population variables called?

(e) How do discrete data differ from continuous data?

(f) How many levels of measurements are there?

(g) What is the difference between a nominal level of measurement and an ordinal level of measurement?

(h) What is the difference between the interval level of measurement and the ratio level of measurement?

(i) What are the variables of interest called?

(j) Why is it important to select the right method of sampling?

(k) How many sampling techniques are there?

(l) What is a random sample and how is it obtained?

(m) How do you obtain a convenience sample?

(n) How do random samples and convenience samples differ?

(o) What is systematic sampling?

(p) What is the difference between stratified sampling and cluster sampling?

(q) When do we use cluster sampling?

(r) How do we obtain a cluster sample?

4. **Determine whether the given statement is true or false.**

(a) Descriptive statistics is a procedure used to describe a set of data.

(b) A sample is a subset of the population.

(c) A parameter is a numerical measurement describing a certain characteristic of a sample.

(d) Data that represents values that correspond to some continuous scale without interruptions is called discrete data.

(e) Data at ordinal level of measurement can be arranged in order but the difference in between is meaningless.

(f) Data at ratio level of measurement can be arranged in order and the difference in between entries is meaningful but it doesn't have a starting point zero.

(g) A systematic sample is one which uses results that are easy to obtain.

(h) Cluster Sampling is widely used when the studied population is spread across a wide area and other sampling methods would be difficult to implement in accessing the selected sample.

5. **Identify the type of units (Sample or Population).**

(a) At a university library there are 281 computers for use. In order to check the operating condition of computers a technician selects 23 for testing.

(b) A Gallup poll of 214 randomly surveyed students found that 7% are having difficulties with on-line testing.

(c) In a study of animal behavior, a researcher randomly selects 4 kangaroos from a zoo which has 32 kangaroos.

(d) During an investigation process of the parking citations issued by Officer Johnson, 57 cases were selected out of total 512 for a closer examination.

(e) A customer service manager in a call center at a bank randomly selects 19 out of 341 calls received in a day.

6. **Identify the level of measurement.**

(a) Temperature, year, account balance.

(b) IQ score, distance, speed.

(c) Color, blood type, gender.

(d) Weight, pressure, age.

(e) Rank, grade, size (small, large …)

7. **Identify the type of data (Qualitative or Quantitative).**

(a) Year, weight, time.

(b) Zip code, religion, vehicle model.

(c) Alcohol concentration, height, voltage.

(d) Mood, fish type, nationality.

(e) Oxygen level, acceleration, amplitude

8. **Identify the sampling method used.**

(a) An insurance agent picks every 17th quote to respond.

(b) A medical researcher selects a group of students by selecting classrooms and selecting students from each class.

(c) A store manager surveys only those employees who showed up to work 15 minutes early.

(d) On a roulette table, in order to select 3 numbers to place a bet on, a gambler closes his eyes and points at 3 different numbers.

(e) On a street, every 5th pedestrian is being surveyed during a 2 hours period to obtain a group of responses for a research study on health care reform.

2

DATA ANALYSIS

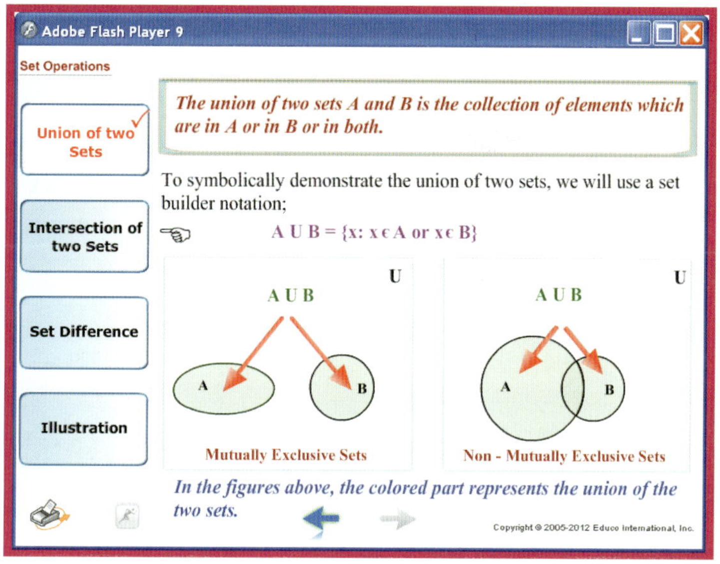

CHAPTER 2

DATA ANALYSIS

Chapter Overview:

2.1: INTRODUCTION

In Chapter 1 you were introduced to the concepts governing descriptive statistics in preparation for analyzing data. You learned how to collect and organize data using concepts such as population, sample, variable, and random sampling. In this chapter we primarily deal with a variety of methods and approaches of analyzing data. At the same time we demonstrate the usefulness of producing visual explanations-tables, charts, and graphs-in order to facilitate the understanding of the analysis of data and the conclusions drawn based on a given analysis. As it is said: a picture speaks louder than a thousand words.

In what follows we will discuss descriptive statistics-numbers used to summarize the data-that consists of measures of central tendency, variation or dispersion of the values around the central tendency, 5-Number Summary, and Box-and-Whisker Plot. Central tendency is concerned with averages (Mean, Median, and Mode) and dispersion refers to the variation or spread of the data values around the central tendency in a distribution. The measures of dispersion that are most commonly used are the range, variance, and standard deviation. The collection of minimum data value, maximum data value, the Median, the first quartile and the third quartile of a data set are called the 5-Number Summary and the statistical chart used to graph the 5-Number Summary is called the Box-and-Whisker Plot. Finally, in the last two sections we will deal with the highly popular and sought after use of Microsoft Excel® to analyze the data. In fact, if you were to Google "data analysis in statistics," what you will get is a large number of sites that deal with the use of the proper software for a given analysis.

2.2: Organization of Quantitative Data

Objectives

After completing this section you will be able to:

A. Make Frequency Distributions.

B. Draw Histogram, Bar Graph and Pie Charts.

C. Understand Shapes of Frequency Distributions.

A fair die has been rolled 40 times, and the outcomes have been recorded in the following table.

4	3	5	1	3	6	5	1
6	6	6	5	6	6	4	3
1	5	3	4	1	4	4	6
6	4	6	5	3	5	4	2
3	3	1	4	1	2	5	4

Table 2.1

A. Frequency Distributions

The data given in the above experiment is called raw or ungrouped data. In raw form, the data is so scattered that even after careful observation, the message remains unclear. The above presentation of the raw data fails to give us any useful information. Our objective in this section will be to express raw data in a meaningful way, from which we can highlight the significant facts and provide more useful information.

Definition of Frequency Table: *Organization of raw data and how often each value or a range of values occurs (frequency) is called a Frequency Distribution or a Frequency Table.*

- **Types of Frequency Distributions**

There are two different types of Frequency Tables:

i) Frequency Distribution Tables **ii)** Grouped Frequency Tables

i) Frequency Distribution Tables, also known as Single value Frequency Tables, are used for moderately small data sets with repeating patterns.

Definition of a Frequency Distribution Table: *Single value Frequency tables are statistical tables that organize raw data into distinct numerical values and their corresponding frequencies (how often each value occurs).*

Construction of Frequency Distributions (Single value Frequency tables):

***Step* 1** Separate and organize the data so that each distinct value is grouped together.

***Step* 2** Draw a 2 column table. Label the first column X and the second column f, where X represents the distinct data values and f represents the corresponding frequencies (number of times each distinct data value occurs).

***Step* 3** Write all distinct values of data in the X-column.

***Step* 4** Write the number of times each distinct data value occurs in the f-column by using counting techniques such as tally marks or spreadsheet commands.

Example 1: In a random sample of 50 families in a suburb, the number of automobiles per family are recorded as follows:

2	0	1	1	3	1	2	0	1	1
0	1	2	2	1	0	0	1	4	5
1	0	1	1	1	2	3	3	2	2
2	3	5	6	1	1	1	2	0	1
0	1	1	2	2	1	3	4	1	5

Organize the data in the form of a frequency distribution.

Solution:

***Step* 1** Make a two-column chart. Label one column X and the other column f (Table 2.2) where X represents the number of automobiles per family and f represents the frequency or number of families.

***Step* 2** Write all distinct values of data in the X-column.

***Step* 3** Record each value in the data by a tally mark next to its value in the table. After four tally marks have been recorded, the fifth mark is drawn crossing the other four (||||) for ease of counting by fives.

***Step* 4** Now add the tally marks and complete the frequency (f) column.

X	*Tally bars*	f
0	卌 |||	8
1	卌 卌 卌 卌	20
2	卌 卌 |	11
3	卌	5
4	||	2
5	|||	3
6	|	1

Table 2.2

This data is discrete and its frequency is a single value corresponding to each value of the variable x.

Example 2: Given the following data set, construct the Frequency Distribution.

1	2	1	1	2	3	2	2	1	6	4
5	4	3	5	3	2	4	4	5	4	5
1	4	2	4	1	2	5	2	4	4	5
5	1	2	4	2	3	6	5	1	5	6
4	4	4	1	4	2	6	6	2	6	6

Solution: We select one of each data value as the representative for each class.

Raw data organized and separated by values	Frequency Count
1 1 1 1 1 1 1 1 1	9
2 2 2 2 2 2 2 2 2 2 2 2	12
3 3 3 3	4
4 4 4 4 4 4 4 4 4 4 4 4 4 4	14
5 5 5 5 5 5 5 5 5	9
6 6 6 6 6 6 6	7

Therefore, the frequency table will look like;

X(number)	1	2	3	4	5	6
f (frequency)	9	12	4	14	9	7

Table 2.3

Note: The sum of the numbers in the frequency column is equal to the sample size. (9 + 12 + 4 + 14 + 9 + 7 = 55)

Try it yourself:

1. Given the following data set, construct the frequency distribution using Tally bars.

3	3	1	1	1	4	1	4	2	4
4	1	2	2	2	1	1	2	3	3
2	2	2	1	2	4	1	3	2	2
1	2	1	1	0	4	2	1	3	4
0	3	0	1	1	1	3	4	4	2

2. The following data represents the favorite colors of the kids of age group 2-5. Construct the frequency distribution by using Tally bars.

Red	Green	Red	Red	Blue	Blue	Blue	Green	Red	Red
Blue	Red	Red	Green	Blue	Red	Green	Blue	Blue	Blue
Green	Red	Red	Green	Red	Green	Blue	Red	Green	Blue
Blue	Red	Red	Blue	Blue	Red	Blue	Blue	Green	Green
Red	Green	Blue	Blue	Green	Green	Green	Blue	Red	Blue

3. In a random sample of 50 college students, the number of W's (withdrawals) for each student is given below.

2	0	1	4	3	1	2	0	6	1
0	1	2	2	3	0	4	1	1	4
1	0	1	4	2	5	4	1	0	4
2	3	5	0	4	1	6	2	1	5
0	2	3	1	1	3	1	3	3	3

Organize the data using a Frequency Distribution Table.

4. As an independent project, a team of Economics students collected data on calories per gram for 50 different kind of chocolates popular among University students. Construct the frequency distribution.

5	4	9	9	8	9	3	8	7	2
9	0	8	5	1	3	5	7	4	8
5	4	9	0	2	9	0	4	9	6
9	6	7	0	0	0	5	4	3	3
2	0	4	3	6	6	1	0	5	5

ii) Grouped Frequency Tables are used for large data sets with no specific pattern.

A large collection of data with many distinct values can be organized using Grouped Frequency Tables.

Definition: ***Grouped Frequency tables*** *organize raw data into classes of equal length and their corresponding frequencies (how many data values fall in each class).*

Classes	Frequency
0 - 10	9
10 - 20	19
20 - 30	18
30 - 40	0
40 - 50	1
50 - 60	1

Table 2.4

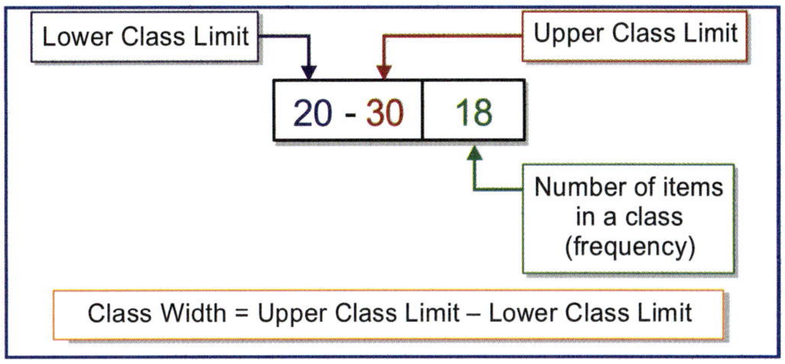

Figure 2.1

Let's take a closer look at one of the rows:

Lower Class Limit: Smallest number in each class (interval)

Upper Class Limit: Largest number in each class (interval)

Class Width: Difference of the Upper and Lower class limits.

Frequency: Number of data values in each class.

Class mark: The midpoint of a class is called the Class Mark and is given by

$$\text{Class Mark} = \frac{\text{Upper class limit} + \text{Lower class limit}}{2}$$

- ## Construction of Grouped Frequency Tables

In order to construct a Grouped Frequency Table, we must first decide on the number of classes. The number of classes is subjective and ultimately depends on the size of the data set and the guidelines established by the stakeholder.

Requirements in constructing Frequency Tables

1. Each class width must be the same.
2. To find frequencies, we will use the lower class limit for counting.
3. Each and every data value must be classified into a class (interval).
4. There must be enough classes to classify all the data.

Procedure:

1. Unless otherwise stated, we will use a subjective process to decide on the number of classes as outlined in the following example.
2. After the number of classes is determined, we must calculate the Class Width.

$$\text{Class width} \approx \frac{\text{largest value - smallest value}}{\text{number of classes selected}}$$

3. Once classes have been constructed, we need to find the frequencies (number of data values in each class). Record each value in the data set with a tally mark next to the class in which it falls. The total number of tally marks in each class gives the frequency for that class.

Example 3: Given the following data set, construct a Grouped Frequency Table with at least 4 classes.

3	11	15	22	28	15	21	8	15	21	27	23
6	12	17	27	5	12	17	24	14	20	25	50
13	48	15	22	5	11	16	15	6	13	17	24
19	25	24	7	7	13	20	25	8	14	20	13

Solution: A rule of thumb is to make sure that the number of classes is a *divisor* of the range.

$$\text{Class width} \approx \frac{\text{largest value} - \text{smallest value}}{\text{number of classes selected}} = \frac{50-3}{?} = \frac{47}{?}$$

Note that 47 cannot be evenly divided into equal intervals, but 50 can. So we can use 50 as the approximate range, and choose 5 as the number of our classes since it divides 50.

Hence, the class width is $\frac{50}{5} = 10$.

Now we can construct our classes.

Classes	Frequency
0 - 10	?
10 - 20	?
20 - 30	?
30 - 40	?
40 - 50	?
50 - 60	?

Table 2.5

To find the frequency for each class, we must count the number of data values that fall in that class, using the lower class limit for categorization. This means that to find the frequency for our first class 0-10, we need to count the number of data values between 0 and 10 (not including 10). For the second class 10-20, we count the number of data values between 10 and 20 (not including 20) and so on.

In general, we can use two types of classes.

i) *Inclusive type*: In this class, we include both lower as well as upper limits during counting of observation in the class. In other words, if $x \in$ class $a - b$, then $a \le x \le b$.

ii) *Exclusive type*: In this class, we include only lower limit of the class when counting of observations in the class. In other words, if $x \in$ class $a - b$, then $a \le x < b$. We will be adopting the exclusive type in this text.

Note: The sum of the class frequencies must be equal to the total number of data values.

The procedure is shown below:

Classes	Separated Data values per class	Frequency Count
0-10	3 5 5 6 6 7 7 8 8	9
10-20	11 11 12 12 13 13 13 13 14 14 15 15 15 15 15 16 17 17 17 19	20
20-30	20 20 20 21 21 22 22 23 24 24 24 25 25 25 27 27 28	17
30-40		0
40-50	48	1
50-60	50	1

Table 2.6

Finally, the Grouped Frequency Table will look like this.

Classes	0 - 10	10 - 20	20 - 30	30 - 40	40 - 50	50 - 60
Frequency	9	20	17	0	1	1

Table 2.7

Try it yourself:

1. The following table displays 210 randomly selected Final Exam Scores.

60	50	49	71	85	8	96	82	56	91	93	88	46	92
58	58	88	75	69	82	67	32	71	47	72	89	77	47
79	28	92	70	4	32	38	93	100	96	100	68	79	68
56	79	71	69	49	93	41	55	68	84	87	85	57	85
80	98	90	56	101	55	0	57	100	97	86	23	82	70
78	45	90	68	6	57	80	88	100	97	100	4	55	64
99	56	68	64	47	88	87	78	87	58	76	44	92	88
55	89	95	85	88	77	44	69	71	88	89	75	73	65
54	60	85	50	16	69	38	81	73	88	94	50	64	52
62	72	95	47	45	81	78	69	96	97	95	3	74	86
58	80	86	88	90	69	95	7	35	60	63	55	95	100
48	68	94	67	5	7	90	22	51	34	0	62	70	62
48	77	98	58	97	22	81	63	71	74	68	55	88	100
81	57	87	92	51	63	60	70	93	90	81	81	88	63
66	48	59	66	46	70	62	24	0	47	0	79	72	66

Construct the Grouped Frequency Distribution for the above data set. Use at least 4 classes.

2. Construct the Grouped Frequency Distribution with at least five classes of the following data.

57	4	52	61	6	68	37	90	46	72
76	29	34	75	84	66	74	16	28	70
11	12	20	11	37	31	12	10	3	36
15	32	12	46	50	94	20	64	73	35
78	59	2	17	34	61	81	62	39	87
35	41	4	34	63	25	91	44	25	55

3. Arrange the data given below into at least 4 classes.

46	29	48	42	45	14	14	13	5	24
33	11	14	31	48	20	28	42	30	48
30	33	16	11	7	9	33	21	49	5
5	32	39	23	16	25	12	33	15	10
29	5	41	5	6	24	14	16	13	16

4. Construct a frequency distribution table with at least 4 classes of IQ scores of Economics students.

80	79	48	71	81	25	54	26	49	58
57	50	12	30	15	42	12	85	50	55
79	81	13	87	67	48	64	57	85	50
90	37	93	76	22	78	84	86	92	10
46	65	16	11	18	53	90	12	16	34

5. The following table represents number of withdrawls per day, for 60 randomly selected days at a particular ATM. Construct a frequency distribution with at least 6 classes.

21	32	8	31	5	21	9	30	42	20
19	10	43	40	38	32	22	40	5	15
6	40	14	26	19	43	19	38	31	25
13	36	21	36	16	32	32	39	28	44
14	38	41	12	9	8	19	5	6	10
10	13	13	31	6	6	10	11	30	42

B. Draw Histogram, Bar-Graph and Pie Charts

Pictures and graphs can reveal a lot more information than tables or words. For this, we want to present the data or the frequency table in a form that can expose more detailed characteristics at a glance. Three such forms are Histogram, Bar-Graph and Pie Chart.

These forms are popular and commonly used devices for pictorial representation of frequency distributions. Histograms and Bar-Graphs are constructed by drawing a series of vertical rectangles on the horizontal axis with equal widths corresponding to the class intervals, and heights corresponding to the frequencies of the classes. Pie Charts are circular graphs representing the frequency of the classes in the form of the sectors of the circle.

• Histograms

A **Histogram** consists of vertical bars or rectangles whose bases represent the classes and whose heights represent the frequencies so that the area of the rectangles are proportional to the frequencies of the corresponding classes. Histograms are primarily used for representing quantitative data.

The Frequency Table for 120 rolls of a fair die is given.

Number on a die	1	2	3	4	5	6
Frequency	23	19	22	19	20	17

Table 2.8

Construction of Histograms:

Step 1 Draw a pair of horizontal and vertical axis.

Step 2 Label all classes on the horizontal axis.

Step 3 Select a suitable scale on the vertical axis that will show all the frequencies.

Step 4 For each class, draw a rectangle having the class interval as its base and the frequency value as its height.

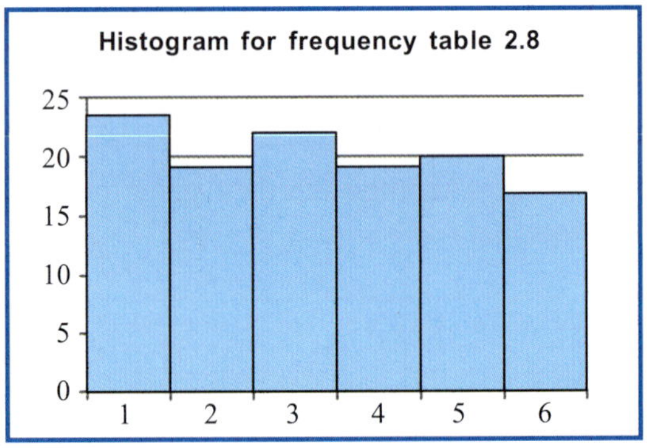

Figure 2.2

Try it yourself:

1. Draw the histogram for the following frequency table.

(a)

Classes	0 - 20	20 - 40	40 - 60	60 - 80	80 - 100
Frequency	10	24	46	20	12

(b)

Classes	0	1	2	3	4	5	6	7	8
Frequency	2	5	6	7	9	5	8	5	3

(c)

Classes	0 - 40	40 - 80	80 - 120	120 - 160	160 - 200
Frequency	21	25	31	23	11

- **Bar Graphs**

Bar graphs are commonly used to display Categorical (Qualitative) data. The following example will illustrate the use of a Bar Graph (Bar-Chart).

Example 4: A car dealership sold 20 Hondas, 10 BMWs, 35 Toyotas, and 18 Cadillacs in the month of December. Construct a bar graph to show the distribution of sales in December.

Model	Frequency
Honda	20
BMW	10
Toyota	35
Cadillac	18

Solution:

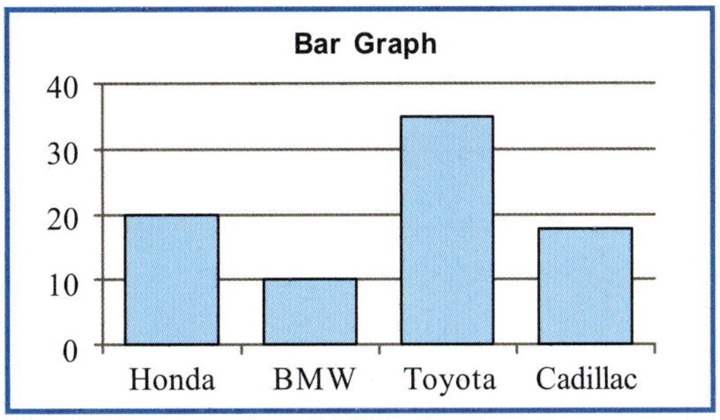

Figure 2.3

- **Pie Charts**

Pie charts are circular graphs that represent the relative frequency of each class with a sector (slice) from the circle.

Construction of a Pie Chart:

***Step* 1** Construct the relative frequency for each category (class).

***Step* 2** Compute the central angle for each class by using the formula

Central Angle = Relative frequency × 360°.

***Step* 3** Start from any radial line and construct the first angle with the use of a protractor.

***Step* 4** With the terminal side of the first angle as the initial side of the next angle, construct the second central angle in the similar way.

***Step* 5** Slice up the pie by using the created central angles and label each slice with the name of the category it represents.

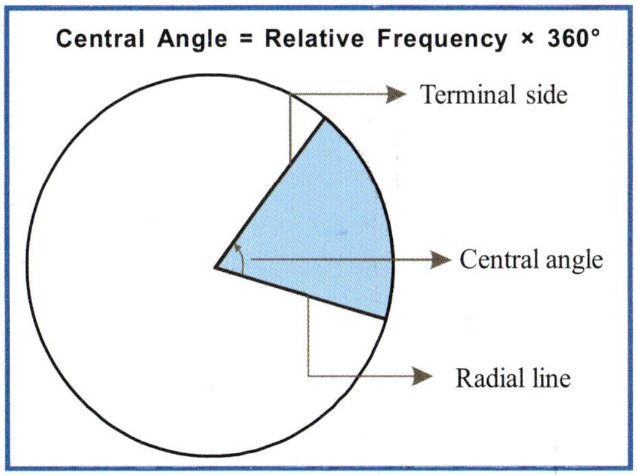

Figure 2.4

Example 5: Given the following frequency table, draw the Pie chart.

Classes	0 - 30	30 - 60	60 - 90	90 - 120
Frequency	10	24	17	8

Solution:

Classes	Frequency	Rel. Frequency	Central Angle	%
0 - 30	10	0.169	$(0.169)(360) = 60.8°$	17
30 - 60	24	0.407	$(0.407)(360) = 146.5°$	41
60 - 90	17	0.288	$(0.288)(360) = 103.7°$	29
90 - 120	8	0.136	$(0.136)(360) = 48.9°$	14

Table 2.9

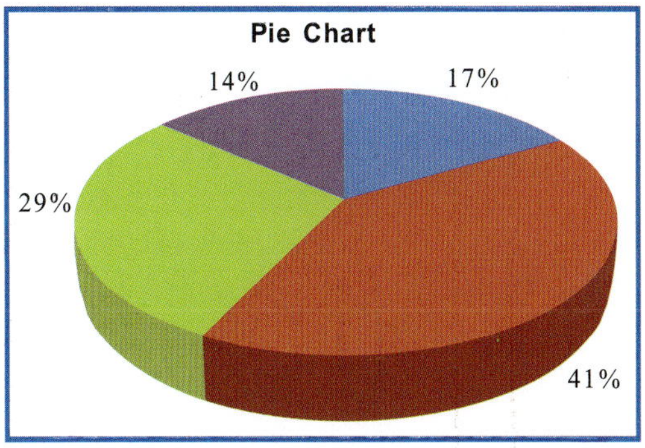

Figure 2.5

Try it yourself:

1. Draw Pie Charts for the data given below:

(a)

Classes	0 - 40	40 - 80	80 - 120	120 - 160	160 - 200
Frequency	54	39	21	37	11

(b)

Classes	Red	Green	Blue	White	Yellow	Brown
Frequency	1	3	4	8	5	3

(c)

Classes	A	B	C	D	E	F
Frequency	7	4	3	9	7	6

C. Shapes of Frequency Distributions

Graphs of Frequency Tables (Distributions) possess different shapes. However, there are three type of distinct shapes that occur in nature

i) Skewed **ii)** Uniform **iii)** Symmetric

i) Skewed Distributions: A Frequency Distribution is said to be skewed if the density of the graph is concentrated in the left or right tail areas.

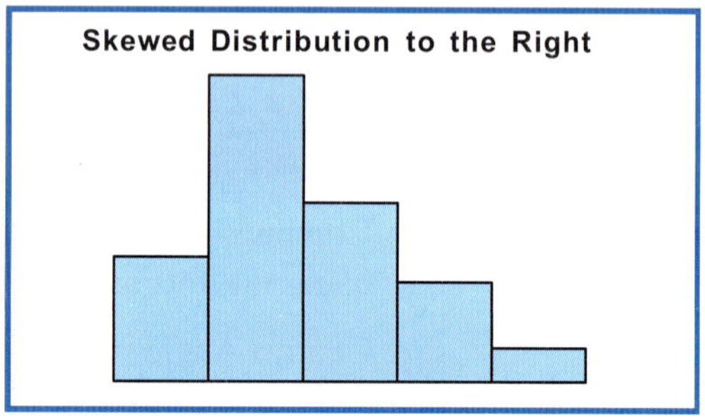

Figure 2.6

ii) Uniform Distributions: A frequency distribution is said to be Uniform when all classes have approximately equal frequencies.

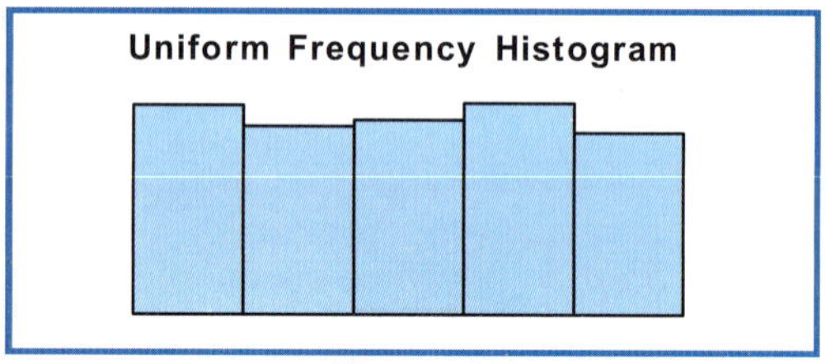

Figure 2.7

iii) Symmetric Distributions: A frequency distribution is said to be Symmetric when the left side of the graph is the mirror image of the right side.

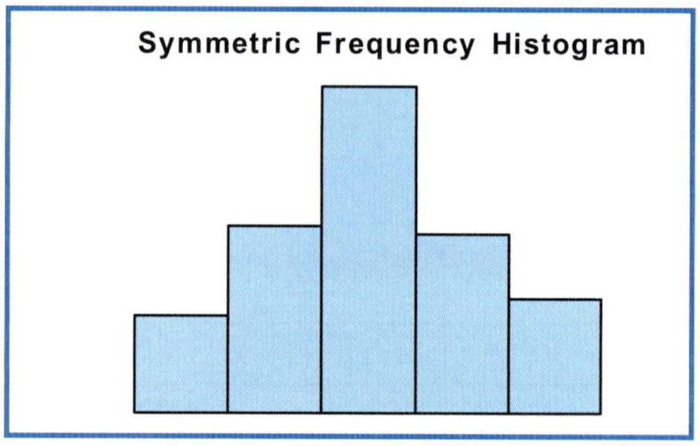

Figure 2.8

2.3: Relative and Cumulative Frequency Distributions and their Histograms

A. Relative Frequency Distributions and their Histograms

In order to generalize the information in a Frequency Distribution relative to the whole population, we need an alternative form of the Frequency Table called Relative Frequency Table. The Relative Frequency Distribution contains the ratios of the frequencies of each class to the sample size. To calculate the Relative Frequencies for each class, we must divide the class frequency by the total frequency.

$$\text{Relative Frequency} = \frac{\text{Class Frequency}}{\text{Total Frequency}} = \frac{f}{n}$$

Objectives

After completing this section you will be able to:

A. Make Relative Frequency Distributions and Draw their Histograms.

B. Make Cumulative Frequency Distributions and Draw their Histograms.

Example 6: Given the following Frequency Table, construct the Relative Frequency Distribution and its histogram.

Classes	0-10	10-20	20-30	30-40	40-50	50-60
Frequency	5	13	2	13	21	12

Solution: In order to calculate the Relative Frequencies, we must divide each class frequency by the total frequency $\left(\sum f = 66\right)$.

Relative Frequency for the first class: $RF_1 = \frac{5}{66} = 0.07576 \approx 0.08$

Relative frequency for the second class: $RF_2 = \frac{13}{66} = 0.19697 \approx 0.20$, and so on.

Classes	Relative Frequency
0-10	$\frac{5}{66} \approx 0.08$
10-20	$\frac{13}{66} \approx 0.2$
20-30	$\frac{2}{66} \approx 0.03$
30-40	$\frac{13}{66} \approx 0.2$
40-50	$\frac{21}{66} \approx 0.32$
50-60	$\frac{12}{66} \approx 0.18$

Table 2.10

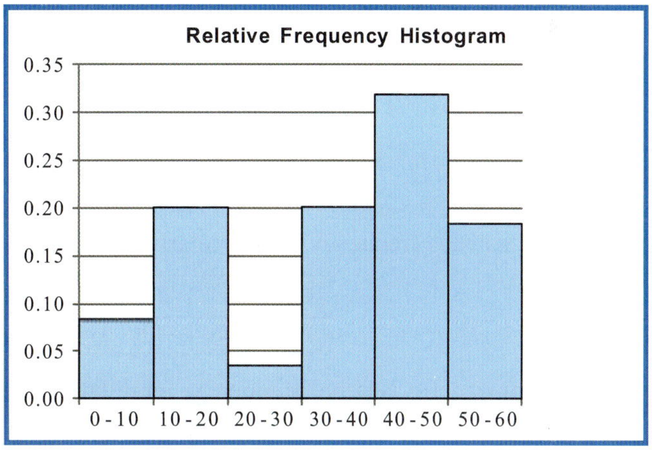

Figure 2.9

B. Cumulative Frequency Distributions and their Graphs

The Cumulative Frequency of a class is the sum of that class frequency and the frequencies of the previous classes. The Cumulative Frequency Distribution shows classes with their corresponding Cumulative Frequencies.

Example 7: Given the following Frequency Table, construct the Cumulative Frequency Distribution and its histogram.

Classes	Frequency
0-10	5
10-20	13
20-30	2
30-40	13
40-50	21
50-60	12

Solution:

Classes	Frequency	Cumulative Frequency
0-10	5	5
10-20	13	13+5
20-30	2	2+13+5
30-40	13	13+2+13+5
40-50	21	21+13+2+13+5
50-60	12	12+21+13+2+13+5

Table 2.11 a

Classes	Cumulative Frequency
0-10	5
10-20	18
20-30	20
30-40	33
40-50	54
50-60	66

Table 2.11 b

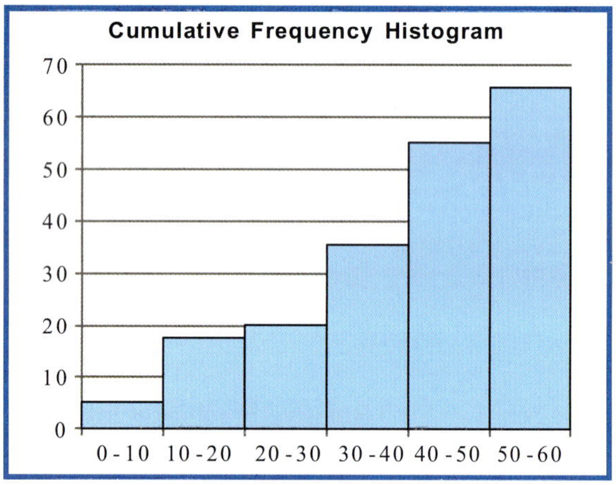

Figure 2.10

Try it yourself:

1. Given the following Grouped Frequency table, construct the Relative and Cumulative Frequency Distributions and their histograms.

Classes	0-40	40-80	80-120	120-160	160-200
Frequency	27	10	21	37	11

2. Construct frequency and a relative frequency histogram for this data.

Classes	0-10	10-20	20-30	30-40	40-50
Frequency	18	9	22	37	25

3. Construct a relative frequency, cumulative frequency and histogram for this data.

(a)

Classes	0-10	10-20	20-30	30-40	40-50
Frequency	12	17	20	8	3

(b)

Classes	30-40	40-50	50-60	60-70	70-80
Frequency	10	17	47	20	9

2.4: Graphs of Frequency Tables

There are several different ways of graphing Frequency Tables. These graphs include Bar-Graphs used to display nominal data and Histograms (discussed above), Frequency Polygons, and Pie Charts.

Objectives

After completing this section you will be able to draw:

A. Frequency Polygons.

B. Relative Frequency Polygons.

C. Cumulative Frequency Polygons (Ogive).

A. Frequency Polygons

Frequency Polygons are line graphs of class midpoints (known as Class Marks) and their corresponding frequencies.

Example 8: Given the following frequency table, draw the Frequency Polygon (Line Graph).

Classes	0-30	30-60	60-90	90-120
Frequency	10	24	17	8

Solution:

Class Mark (midpoint)	Frequency
15	10
45	24
75	17
105	8

Table 2.12

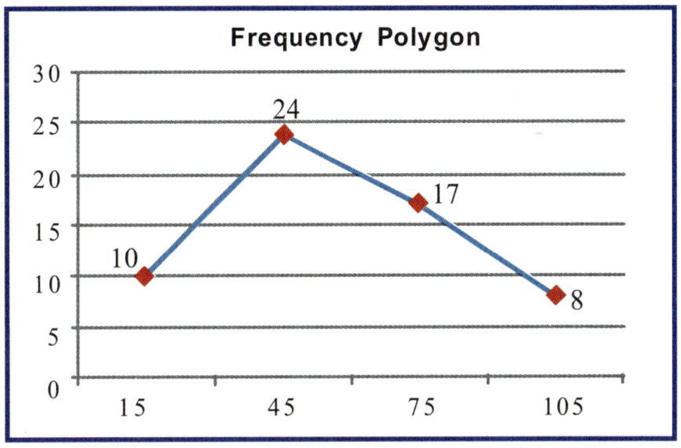

Figure 2.11

Try it yourself:

1. Given the following Grouped Frequency Table, draw the Frequency Polygon (Line Graph).

(a)

Classes	0-40	40-80	80-120	120-160	160-200
Frequency	54	39	21	37	11

(b)

Classes	0-15	15-30	30-45	45-60	60-75	75-90	90-105
Frequency	28	30	23	21	15	35	12

(c)

Classes	10-20	20-30	30-40	40-50	50-60	60-70	70-80
Frequency	27	45	23	35	29	19	28

B. Relative Frequency Polygons

Relative Frequency Polygons are line graphs of class midpoints (known as Class Marks) and their corresponding relative frequencies.

Example 9: Given the following frequency table, draw the Relative Frequency Polygon (Line Graph).

Classes	0-30	30-60	60-90	90-120
Frequency	10	24	17	8

Solution:

Class Mark (midpoint)	Relative Frequency
15	$\frac{10}{59} \approx 0.169$
45	$\frac{24}{59} \approx 0.407$
75	$\frac{17}{59} \approx 0.288$
105	$\frac{8}{59} \approx 0.136$

Table 2.13

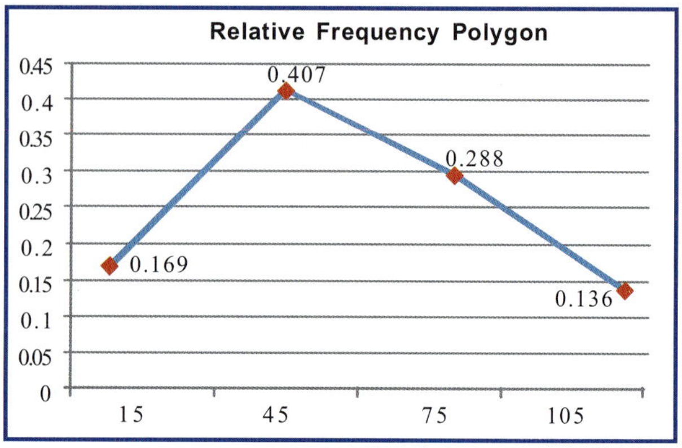

Figure 2.12

Try it yourself:

1. Given the following Grouped Frequency Table, draw the Relative Frequency Polygon (Line Graph).

(a)

Classes	0-40	40-80	80-120	120-160	160-200
Frequency	54	39	21	37	11

(b)

Classes	0-15	15-30	30-45	45-60	60-75	75-90	90-105
Frequency	33	45	32	26	28	40	20

(c)

Classes	10-20	20-30	30-40	40-50	50-60	60-70	70-80
Frequency	28	38	26	30	6	16	21

C. Cumulative Frequency Polygons (Ogive)

Cumulative Frequency Polygons are line graphs of Upper Class limits and their corresponding Cumulative Frequencies.

Example 10: Given the following frequency table, draw the Cumulative Frequency Polygon (Ogive).

Classes	0 - 30	30 - 60	60 - 90	90 - 120
Frequency	10	24	17	8

Solution:

UCL	Cumulative Frequency
30	10
60	34
90	51
120	59

Table 2.14

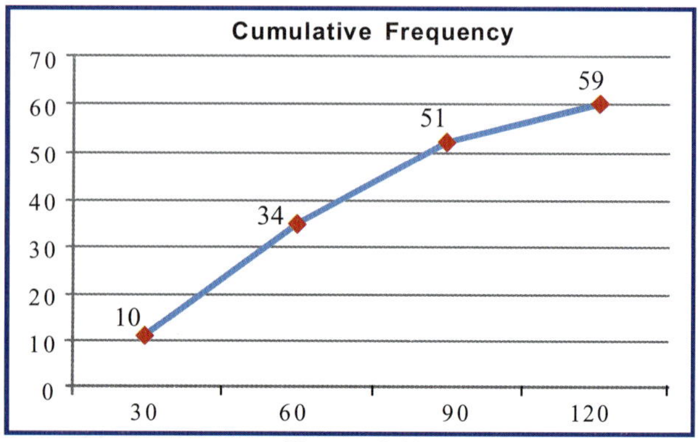

Figure 2.13

Important Rule: When constructing a Cumulative Frequency Polygon, our line graph must start with the lower class limit of the first class and end with the upper class limit of the last class. Therefore, we must always connect the beginning of the line graph to the origin.

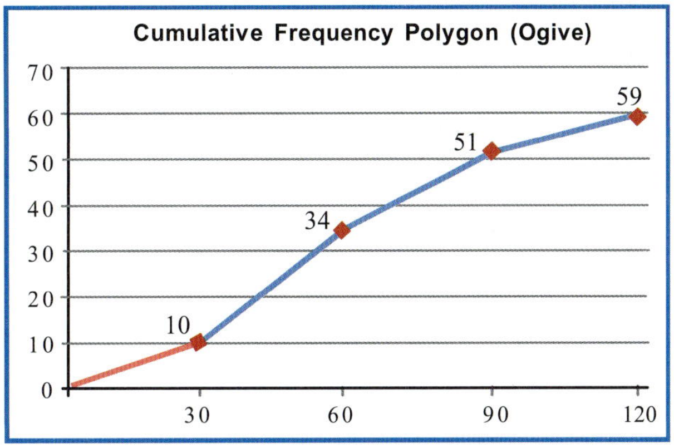

Figure 2.14

Try it yourself:

1. Given the following Grouped Frequency Table, draw the Cumulative Frequency Polygon (Ogive).

(a)

Classes	0-40	40-80	80-120	120-160	160-200
Frequency	54	39	21	37	11

(b)

Classes	0-15	15-30	30-45	45-60	60-75	75-90	90-105
Frequency	24	5	31	38	39	31	32

(c)

Classes	0-5	5-10	10-15	15-20	20-25	25-30	30-35
Frequency	18	14	34	7	27	23	22

2.5: Organizing Data Using Stem and Leaf Plots

Stem and Leaf Plots are an alternative to frequency tables for organizing data graphically. Their advantage over a frequency distribution is that in a Stem and Leaf Plot the actual data values are preserved, and can be recognized by simply looking at the table.

Objectives

After completing this section you will be able to:

A. Define Stem-Leaf Plot.

B. Construct Stem-Leaf Plot.

A. Stem-Leaf Plot

A Stem-Leaf Plot is a 2-column chart, labeled 'stem' and 'leaf', in which each value of data is recorded as part stem and part leaf.

B. Construction of a Stem and Leaf Plot

Separate each number in the data into two parts (stem and leaf). The stem may be the first digit of the number or it may have as many digits as necessary to make the leaf a single digit number.

The following example will illustrate the steps in creating a Stem and Leaf plot.

Example 11: The following data set represents 20 randomly selected scores from a group of college level statistics classes. Organize the above data in a stem and leaf plot.

90	73	30	88	78	43	45	58	63	69
65	64	70	27	42	86	86	38	18	95

Solution: **(a)** We must first arrange the data in an ascending (increasing) order.

18	27	30	38	42	43	45	58	63	64
65	69	70	73	78	86	86	88	90	95

Table 2.15 a

(b) We then arrange the data into separate groups according to their 1st digit.

18		27		30	38		42	43	45		58			
63	64	65	69		70	73	78		86	86	88		90	95

Table 2.15 b

(c) We can now display the results in a stem and leaf plot by using the 1st digits on the stem and the next digits on the leaf.

Stem	Leaf			
1	8			
2	7			
3	0	8		
4	2	3	5	
5	8			
6	3	4	5	9
7	0	3	8	
8	6	6	8	
9	0	5		

Table 2.15 c

Try it yourself:

1. Construct the Stem and Leaf Plot for the following data sets.

(a)

46	65	87	50	85	93	81	117	115
75	88	70	80	95	45	52	100	63
83	7	70	10	90	80	35	95	100
90	50	98	54	65	78	68	68	75

(b)

81	39	10	51	27	8	70	52	15	29
88	45	17	64	62	33	55	32	20	66
17	51	60	80	34	26	3	8	0	76
28	8	36	39	6	80	39	76	77	83
39	41	79	56	49	23	2	38	13	13

(c)

20	43	37	65	77	76	35	70	52	32
50	52	43	6	45	54	75	44	67	85
41	4	35	74	20	21	13	40	76	7
37	19	5	54	82	55	74	83	44	29

2.6: Measures of Central Tendency and Dispersion (Variation)

Objectives

After completing this section you will be able to compute:

A. Mean, Median, and the Mode of Raw Data.

B. Empirical Relation among Mean, Median, and Mode.

C. Measure of Dispersion: Range, Variance, and Standard Deviation.

D. Range of Usual and Unusual Values.

One of the most important objectives of statistical data analysis is to determine various numerical measures which describe the characteristics of a data set. The Mean, Median, Mode, Geometric Mean, and Harmonic Mean are common measures of central tendency. However, in this section we will discuss the three most common and important measures of central tendency: mean, median, and mode.

A. Mean, Median, and the Mode of Raw Data

• Mean

Mean is a measure of the average value of a data set.

Notation:

μ (mu): The symbol used for the mean of the entire population.

$\overline{x}$ (x bar): The symbol used for the mean of a sample from the population.

Formal Definition: *A measure of central tendency is the value in the central part of the distribution of a data set around which the individual observations cluster.*

The formulas for the mean:

$$\textbf{Population Mean } (\mu) = \frac{\sum x_i}{N} \qquad \textbf{Sample Mean } (\overline{x}) = \frac{\sum x_i}{n}$$

Note: N represents the size of the population, and n represents the sample size. The sample is always a subset of the population.

How to compute the Mean

Add all the data values (x values) and divide by the number of data point in the sample or the population.

Example 12: Find mean of the following sample observations

94	33	45	41	22	77	20	11	3	64

Solution: Number of observations = 10

$$\text{Mean} = \frac{\sum x}{n}$$

$$= \frac{\text{Sum of all observations}}{n}$$

$$= \frac{94 + 33 + 45 + 41 + 22 + 77 + 20 + 11 + 3 + 64}{10} = \frac{410}{10} = 41$$

- ## Median

Median is the measure of the center value of a data set when the data is arranged in increasing order.

Notation:

η (eta): The symbol used for the median of the entire population.

$\tilde{x}$ (x tilde): The symbol used for the median of a sample from the population.

Definition: *The median is the middle observation of the distribution when observations have been arranged in an increasing order. The median breaks the data set into two equal parts. Fifty percent of the data is below the median and the rest above the median.*

How to compute the Median

Arrange the data in an increasing order. The median is the middle observation in the ordered data set.

In general, if the number of observations n is odd, then the median is the value of $\left(\frac{n+1}{2}\right)^{\text{th}}$ observation, when the data set has been arranged in an increasing order.

Example 13: Find the median of following observations:

12	25	12	23	70	40	69	33	12

Solution: Arrange the data in an increasing order:

12	12	12	23	25	33	40	69	70

Number of observations = 9

Median is the value of $\left(\frac{9+1}{2}\right)^{\text{th}} = 5^{\text{th}}$ observation, which is 25.

If n, is even, then there are two middle observations, $\left(\frac{n}{2}\right)^{\text{th}}$ and $\left(\frac{n}{2}+1\right)^{\text{th}}$ observations. In this case, the median is computed by the arithmetic mean of these two middle observations.

Example 14: Find the median of the following observations;

70	12	25	12	23	33	69	40

Solution: Arrange the data in an increasing order:

12	12	23	25	33	40	69	70

Number of observations = 8. Since $n = 8$ is even,

Median = Average of $\left(\frac{n}{2}\right)^{\text{th}} = 4^{\text{th}}$ and $\left(\frac{n}{2}+1\right)^{\text{th}} = 5^{\text{th}}$ observations.

The 4th observation is 25 and 5th observation is 33

Therefore, median $= \frac{25+33}{2} = 29$

- **Mode**

The mode of a data set is the value that occurs most frequently.

For the given data set {70, 12, 25, 12, 23, 33, 69, 40}, the mode is 12 because it occurs twice whereas all the other values occur only once.

Example 15: The amounts of contributions (in $) for the Christmas gifts for the secretaries in an office are 40, 30, 50, 30, 100, 50, 21, 100, 50, 50, and 40. Find mean, median and mode of this data set.

Solution: **Mean:** $\frac{\sum x}{n} = \frac{561}{11} = 51$

Median: Arrange the data in an increasing order:

21	30	30	40	40	50	50	50	50	100	100

Number of observations = 11. Since $n = 11$ is odd,

$$\text{Median} = \left(\frac{11+1}{2}\right)^{th} = 6^{th} \text{ observation in the ordered data set} = 50$$

Mode: As the observation 50 is repeated maximum number of times. Therefore, Mode = 50

Round off Rule: The round off rule for the measure of central tendency is to round the final answer to one more decimal place than the original data.

A comparison among the mean, median and the mode in different distributions:

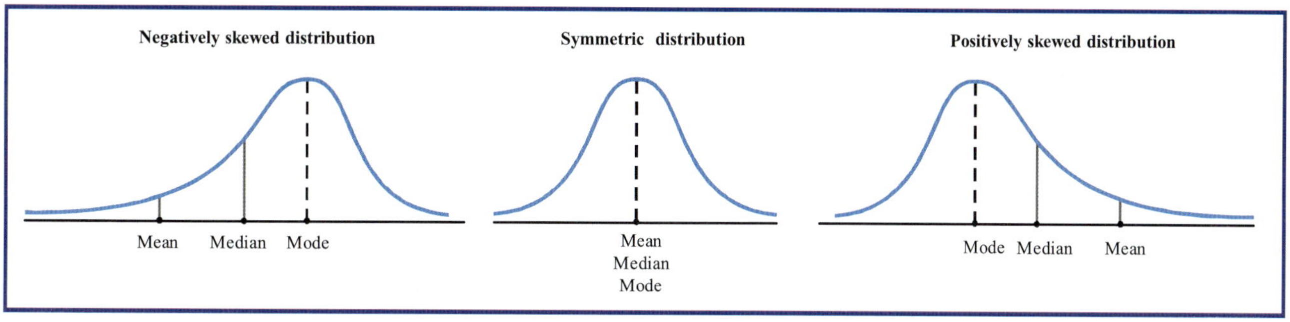

Figure 2.15

B. Empirical Relation among Mean, Median, and Mode

When the distribution of data is Normal (symmetric), the mean, median and the mode are equal.

In skewed or non-symmetric distributions, the *approximate* relationship between the mean, median and the mode are as follows:

$$\text{mean} - \text{mode} = 3\,(\text{mean} - \text{median}) \Rightarrow \text{mode} = 3 \text{ medians} - 2 \text{ mean}$$

Try it yourself:

1. Find mean of the following sample observations.

37	6	20	6	31	30	34	7	34	30

2. Find the median of following observations.

(a)

36	36	23	36	23	5	29	34

(b)

9	10	14	24	12	6	25	19	37	25	23

3. Find the mode of the following observations.

12	25	36	42	0	28	28	3	33	30
10	31	1	0	43	9	5	5	27	3
32	41	4	19	36	14	11	19	15	9
44	24	44	36	23	23	5	5	8	10
11	11	2	41	40	36	40	32	1	37

4. Given the following data set, find the mean, median and the mode. Is the distribution of the data symmetric, skewed to the left, or skewed to the right?

85	15	14	25	69	7	8
15	11	35	8	23	41	52
24	18	6	5	9	50	47
31	3	30	44	65	12	89
75	1	9	6	7	15	100
88	45	92	36	45	76	25

C. Measure of Dispersion: Range, Variance, and Standard Deviation

A measure of center is an essential piece of information in data analysis; however it is not enough to fully describe the nature (shape) of a distribution. Three different populations can have identical means, but the spread of the data in one can be significantly different from the other two.

Note that the mean of the three sets of data are the same. However, the scatter plots of these data sets reveal information about the dispersion of the data which makes them quite different.

	Data Set A	Data Set B	Data Set C
	2	18	65
	15	12	2
	8	18	7
	10	7	3
	7	9	4
	6	10	6
	15	11	10
	2	15	12
	45	10	1
Mean	**12.22**	**12.22**	**12.22**

Table 2.16

Data Set A

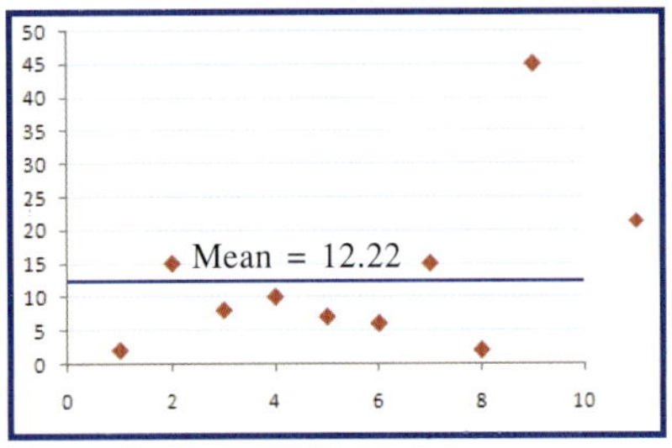

Low Variation

Figure 2.16 a

Data Set B

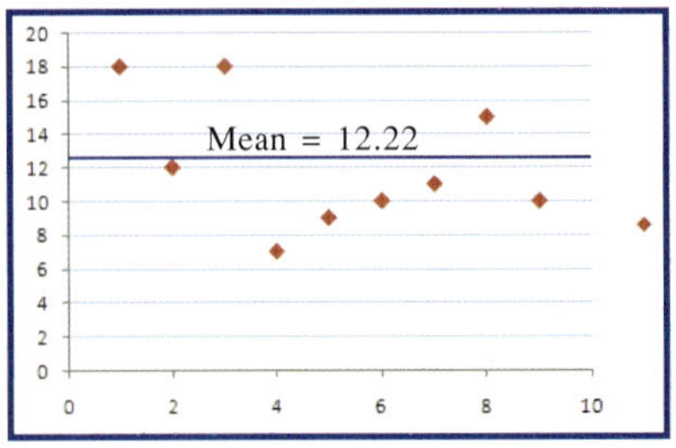

Medium Variation

Figure 2.16 b

Data Set C

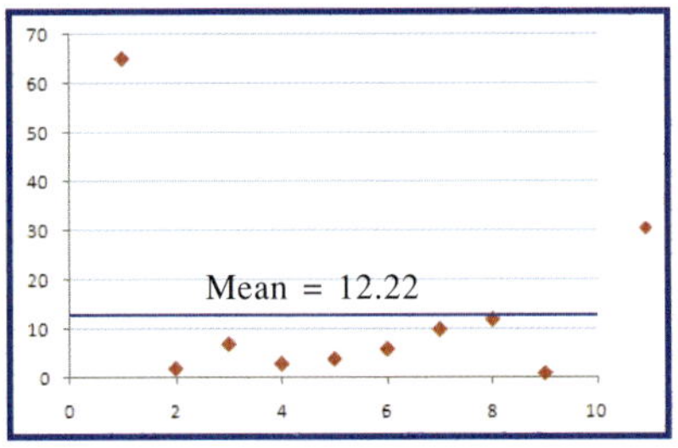

High Variation

Figure 2.16 c

From the Figures 2.16 a, b and c, it is quite clear that the measure of central tendency is not enough to completely describe the behavior of a data set. Populations can have the same mean yet the data could have quite different spread. Therefore, to completely describe the characteristics of data sets, we need to measure the amount of variation in a data set in addition to a measure of central tendency.

We will now quantify dispersion by introducing three measures variation: **Range**, **Variance**, and **Standard Deviation**.

- **Range**

The range measures, the difference between maximum and minimum observations in the data set.

Range = Maximum Value – Minimum Value

Example 16: Given the following 3 data sets, find the range for each one.

Data Set A	Data Set B	Data Set C
2	18	65
15	12	2
8	18	7
10	7	3
7	9	4
6	10	6
15	11	10
2	15	12
45	10	1

Solution: Range for Data Set A:

Range = Max – Min = 45 – 2 = 43

Range for Data Set B:

Range = Max – Min = 18 – 7 = 11

Range for Data Set C:

Range = Max – Min = 65 – 1 = 64

- **Variance and Standard Deviation**

Standard deviation was first suggested by **Karl Pearson** as a measure of variability in **1893**.

Notation:

σ^2: The symbol used for the variance of the entire population.

s^2: The symbol used for the variance of a sample from the population.

Definition:

Variance is the measure of variation of a data set about its central value and is defined as the average of the sum of the squares of the deviation of each data value from the mean.

$$\sigma^2 = \frac{\sum(x-\mu)^2}{N} = \frac{(x_1-\mu)^2 + (x_2-\mu)^2 + \ldots + (x_N-\mu)^2}{N}$$

where x represents each individual value, μ is the population mean, and N is the size of the population. The formula for sample variance is $s^2 = \dfrac{\sum(x-\bar{x})^2}{n-1}$ *where n represents the size of the sample.*

Definition:

Standard Deviation also measures the variation of a data set about its central value in the units as of the units of data values. It is defined as the square root of the variance.

$$\sigma = \sqrt{\sigma^2} = \sqrt{\frac{\sum(x-\mu)^2}{N}}$$

The formula for sample standard deviation is

$$s = \sqrt{\frac{\sum(x-\bar{x})^2}{n-1}} = \sqrt{\frac{n\sum x^2 - \left(\sum x\right)^2}{n(n-1)}}$$

Note: $\sum x^2 = x_1^2 + x_2^2 + \ldots + x_n^2$, $\left(\sum x\right)^2 = (x_1 + x_2 + \ldots + x_n)^2$

Trying to understand variance and standard deviation from the definition is quite difficult. We will give few examples to clarify the procedure of finding the variance and the standard deviation, and use the results to show how important the value of standard deviation is in calculating the range of usual values, z-scores, and variation in data sets.

Example 17: Given the following numbers, find the variance and the standard deviation.

25	18	20	15	19	20	21	23	20	14

(a) Assume that the data set is the population.

(b) Assume that the data set is a sample.

Solution: First, we compute mean of the entire data set:

$$\mu = \frac{25+18+20+15+19+20+21+23+20+14}{10} = 19.5$$

In the table given below, we calculate values corresponding to $x - \mu$ and $(x - \mu)^2$ for each number in data set.

x	25	18	20	15	19	20	21	23	20	14
$(x-\mu)$	5.5	−1.5	0.5	−4.5	−0.5	0.5	1.5	3.5	0.5	−5.5
$(x-\mu)^2$	30.25	2.25	0.25	20.25	0.25	0.25	2.25	12.25	0.25	30.25

Table 2.17

(a) Population Variance:

$$\sigma^2 = \frac{\sum(x-\mu)^2}{N} = \frac{(25-19.5)^2 + (18-19.5)^2 + \ldots + (14-19.5)^2}{10}$$

$$= \frac{30.25 + 2.25 + \ldots + 30.25}{10} = \frac{98.5}{10} = 9.85 \approx 9.9$$

Population Standard Deviation: $\sigma = \sqrt{\sigma^2} = \sqrt{9.85} = 3.138 \approx 3.1$

(b) Sample Mean $\bar{x} = \frac{\sum x}{n} = \frac{195}{10} = 19.5$

Sample Variance $s^2 = \frac{\sum(x-\bar{x})^2}{n-1} = \frac{98.5}{9} = 10.944 \approx 10.9$

Sample Standard Deviation $s = \sqrt{s^2} = \sqrt{10.944} \approx 3.3$

Note: As we can see, the final answers for both the population and the sample are similar. For the sake of simplicity and to avoid confusion throughout the text. We can also use the following formula to calculate the standard deviation.

$$s \approx \sigma \approx \sqrt{\frac{n\sum x^2 - \left(\sum x\right)^2}{n(n-1)}}$$

Procedure for the above formula:

(a) Square each data value: $x_1^2, x_2^2, \ldots, x_n^2$

(b) Add all the squares from (a) and multiply the result by the sample size n: $n \times \sum x^2$.

(c) Square the sum of all the data points $\left(\sum x\right)^2$.

(d) Subtract the result in (c) from the result in (b): $n \sum x^2 - \left(\sum x\right)^2$.

(e) Multiply n by $n - 1$.

(f) Divide the result in part (d) by the results in part (e), and take the square root:

$$\sqrt{\frac{n\sum x^2 - \left(\sum x\right)^2}{n(n-1)}}$$

Notational Summary:

	Population	Sample
Mean	μ	$\overline{x}$
Median	η	x
Standard Deviation	σ	s
Variance	σ^2	s^2

Table 2.18

Try it yourself:

1. Given the following numbers, find the variance and the standard deviation.

39	9	36	19	11	33	36	14	34	34

(a) Assume that the data set is the population.

(b) Assume that the data set is a sample.

2. Given the following raw data, find the range, variance, and the standard deviation.

(a)

21	34	37	16	14	31	14	34	30	28
20	29	28	28	21	10	35	11	21	31

(b)

22	7	26	14	22	30	35	26	33	31
37	8	32	19	32	9	22	17	32	22
25	17	29	7	22	23	8	18	18	18
27	7	27	23	35	20	36	23	9	34
23	15	18	19	18	29	35	38	7	14

(c)

85	15	14	25	69	7	8
15	11	35	8	23	41	52
24	18	6	5	9	50	47
31	3	30	44	65	12	89
75	1	9	6	7	15	100
88	45	92	36	45	76	25

D. Range of Usual and Unusual Values

Once a mean (measure of central tendency) and a standard deviation (measure of the dispersion) are calculated, we can construct the range of usual and unusual observations in an experiment using the following rule of thumb:

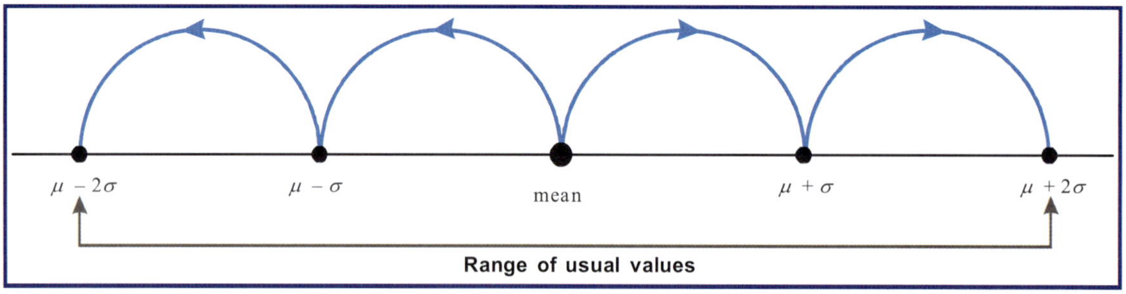

Figure 2.17

The range of usual values is $\mu \pm 2\sigma$. The observations which lie outside $\mu \pm 2\sigma$ are called unusual observations. If μ and σ are not given, we estimate these values from the sample observation. In that case range for usual values become $\bar{x} \pm 2s$ and observations outside $\bar{x} \pm 2s$ are called unusual sample observations.

Note: The rule remains the same for a sample from a population: $\bar{x} \pm 2s$

Example 18: If the average IQ is 100 with a standard deviation of 15, find the range of usual and unusual IQ scores.

Solution: The range of usual values is $\mu \pm 2\sigma = 100 \pm 2 \times 15 \rightarrow (70, 130)$.

Hence, any IQ between 70 to 130 is considered usual. All other IQs below 70 and above 130 are considered unusual.

Try it yourself:

1. Find the range of usual values, using the following data sets.

(a)

23	13	23	16	24
23	15	25	20	25
19	26	22	28	17
28	16	28	29	32

(b)

9	3	23	6	17	9	31	6	17	9
6	31	34	20	11	24	33	27	10	33
19	21	24	7	17	26	23	18	4	19
14	8	16	20	30	6	20	8	19	32
11	31	30	10	27	31	27	3	11	27

(c)

22	82	61	72	48	69	37	94	45	4
89	51	45	96	42	34	86	52	75	6
8	68	19	34	19	77	83	23	84	39

2.7: Mean, Median and Mode of a Frequency Distribution

Calculating the mean, median and mode from a data given in grouped frequency distribution format is quite different from calculating the central tendency for **raw data**. Distinct data values are not available when working with frequency tables. Therefore, we must use an approximating procedure to calculate the above mentioned statistics.

Objectives

After completing this section you will be able to compute:

A. Mean, Median and Mode of a Frequency Distribution.

B. Variance and Standard Deviation of a Frequency Distribution.

A. Mean, Median and Mode of a Frequency Distribution (Grouped Data)

The following example illustrates the procedure for computing mean, median and mode of a Grouped frequency distribution.

Example 19: Compute the **(a)** Mean, **(b)** Median, and **(c)** Mode of the following grouped frequency table.

Classes	10-20	20-30	30-40	40-50	50-60	60-70
Frequency	4	7	15	17	8	3

Solution: In order to calculate the parameters in question, we need to construct a Class Mark (the midpoint of a class) labeled as (x) for each class and use it as the single numerical representative of that particular class.

Classes	**Class Mark (x)**	**Frequency (f)**	$f \cdot x$	**Cumulative Frequency**
10-20	15	4	60	4
20-30	25	7	175	11
30-40	35	15	525	26
40-50	45	17	765	43
50-60	55	8	440	51
60-70	65	3	195	54
	Sums:	**54**	**2160**	

Table 2.19

(a) Mean of the frequency table $\mu = \dfrac{\sum(f \cdot x)}{\sum f} = \dfrac{f_1x_1 + f_2x_2 + \ldots + f_kx_k}{f_1 + f_2 + \ldots + f_k}$

Note: x's in the formula represent the class marks, and $\sum f = n$

$$\mu = \frac{\sum(f \cdot x)}{\sum f} = \frac{2160}{54} = 40.0$$

(b) A class that contains the median is called median class. Now median $= \left(\dfrac{n+1}{2}\right)^{\text{th}}$

observations $= \dfrac{54+1}{2} = 27.5^{\text{th}}$ observation which belongs to class 40 - 50. Therefore, median class in this example is 40-50.

Now, median is calculated by using the formula:

Median $= l + \dfrac{h}{f_m}\left(\dfrac{n+1}{2} - C\right)$, where

l = the lower limit of median class = 40

h = class width = 10

f_m = frequency of median class = 17

C = cumulative frequency of a class preceding the median class = 26

Thus, the median $= 40 + \dfrac{10}{17}(27.5 - 26) = 40 + 0.88 = 40.88$.

(c) The class with largest frequency is 40-50 is called the model class. The mode is the class mark of the model class.

$$\text{Mode} = \frac{40+50}{2} = 45.$$

B. Variance and Standard Deviation of a Frequency Distribution (Grouped Data)

The following example illustrates the procedure in calculating the range, standard deviation, and variance of a grouped frequency distribution.

Variance of a Frequency Table:

$$\sigma^2 = \frac{n\sum(f \cdot x^2) - \left(\sum f \cdot x\right)^2}{n(n-1)} = \frac{n\left(f_1x_1^2 + f_2x_2^2 + \ldots + f_kx_k^2\right) - \left(f_1x_1 + f_2x_2 + \ldots + f_kx_k\right)^2}{n(n-1)}$$

Standard deviation of a Frequency Table:

$$\sigma = \sqrt{\frac{n\sum(f \cdot x^2) - \left(\sum f \cdot x\right)^2}{n(n-1)}} = \sqrt{\frac{n\left(f_1x_1^2 + f_2x_2^2 + \ldots + f_kx_k^2\right) - \left(f_1x_1 + f_2x_2 + \ldots + f_kx_k\right)^2}{n(n-1)}}$$

Example 20: Compute the variance and standard derivation of the following grouped frequency table.

Classes	5-15	15-25	25-35	35-45	45-55	55-65	65-75
Frequency	4	6	9	11	9	8	3

Solution: We calculate class marks for each class and will make the table for computing variance and standard derivation.

Classes	Class Mark (x)	Frequency (f)	$f \cdot x$	$f \cdot x^2$
5 - 15	10	4	40	400
15 - 25	20	6	120	2400
25 - 35	30	9	270	8100
35 - 45	40	11	440	17600
45 - 55	50	9	450	22500
55 - 65	60	8	540	28800
65 - 75	70	3	210	14700
	Sums:	**50**	**2010**	**94500**

Table 2.20

Variance of a frequency table:

$$\sigma^2 = \frac{n\sum(f \cdot x^2) - \left(\sum f \cdot x\right)^2}{n(n-1)} = \frac{50(94500) - (2010)^2}{50(50-1)} = 279.551$$

Standard deviations of a frequency table: $\sigma = \sqrt{\sigma^2} = 16.72$

Try it yourself:

1. Compute the Mean, Median, and Mode of the following grouped frequency table.

(a)

Classes	0-5	5-10	10-15	15-20	20-25	25-30	30-35
Frequency	9	25	10	15	24	24	21

(b)

Classes	0-10	10-20	20-30	30-40	40-50	50-60	60-70	70-80
Frequency	41	51	45	50	49	39	42	15

2. Compute the variance and standard derivation of the following grouped frequency table.

(a)

Classes	0-15	15-30	30-45	45-60	60-75	75-90	90-105
Frequency	33	27	27	11	14	16	28

(b)

Classes	10-20	20-30	30-40	40-50	50-60	60-70	70-80
Frequency	32	38	53	46	50	23	33

3. Find the mean, median, mode, variance, and the standard deviation of the following grouped frequency table.

(a)

Classes	0-10	10-20	20-30	30-40	40-50	50-60	60-70	70-80	80-90	90-100
Frequency	8	9	14	29	11	31	19	17	15	22

(b)

Classes	20-25	25-30	30-35	35-40	40-45	45-50	50-55	55-60
Frequency	6	4	17	49	67	18	19	9

2.8: The Empirical Rule and Measures of Relative Standing

Objectives

After completing this section you will be able to:

- **A.** Apply Chebyshev's Rule.
- **B.** Apply Empirical Rule.
- **C.** Compute Z-Scores.
- **D.** Compute Quartiles.
- **E.** Verify Outliers.
- **F.** Compute Percentiles.

In this section, we will combine the measures of central tendency and variability to create a more precise picture of the distribution of a data set. In addition, measures such as *Z*-Scores, Quartiles, and Percentiles will be introduced to facilitate the comparison of different populations using the same scale.

We will start this section by looking at what percent of the data is near the mean using the standard deviation as a yardstick.

A. Chebyshev's Rule for all Types of Distributions

Chebyshev's rule describes the proportion of the spread (variation) in terms of the standard deviation.

The Chebyshev's Rule: *The proportion of data points that fall within k standard deviations of the mean is at least* $1 - \frac{1}{k^2}$, *given* $k > 1$.

Example 21: Find the proportion of data points that fall within 3 standard deviations of mean.

Solution: Here $k = 3$, then $1 - \frac{1}{3^2} = 0.889 = 88.9\%$, which means that at least 88.9% of the data falls within 3 standard deviations of the mean.

Example 22: The mean score in an Algebra Exam is 72 with a standard deviation of 10. Find the score range for which at least 75% of the grades will fall.

Solution: Using Chebyshev's Theorem, we have $1 - \frac{1}{k^2} = \frac{75}{100} \Rightarrow k = 2$

Therefore, by Chebyshev's Theorem, at least 75% of observation lies in the interval $\mu \pm k\sigma = 72 \pm 2(10)$. Hence $72 - 2(10) = 52$, and $72 + 2(10) = 92$. This means that at least 75% of the students will get grades between 52 to 92.

Example 23: A survey of Math Academia (a tutoring center) revealed that the mean monthly amount of expenditure on office supplies was \$400. The standard deviation was \$50. Use the Chebyshev's theorem, to find the minimum percentage of the data values that will fall between \$250 to \$550.

Solution: Using the formula, $2k\sigma$ = max value – min value, we have

$$2k\sigma = 550 - 250, \text{ where } \sigma = 50$$

$$\Rightarrow \quad k = \frac{300}{2\sigma} = \frac{300}{2 \times 50} = 3$$

$$\Rightarrow \quad \left(1 - \frac{1}{k^2}\right) \times 100\% = \left(1 - \frac{1}{9}\right) \times 100\% = 88.9\%$$

Hence, at least 88.9% of the data will fall between \$250 to \$550. This means that at least 89% of the time the center's supply cost will be from \$250 to \$550.

B. Empirical Rule for Data with a Symmetric (Bell-Shaped) Distribution

Chebyshev's Theorem applies to all types of distributions regardless of their shape. However, when a distribution is approximately Bell-Shaped, the exact distribution of the data around the mean follows a different rule called the **Empirical Rule.**

The Empirical Rule

1. Approximately 68.3% of data falls within 1 standard deviation of the mean.
2. Approximately 95.4% of data falls within 2 standard deviation of the mean.
3. Approximately 99.7% of data falls within 3 standard deviation of the mean.

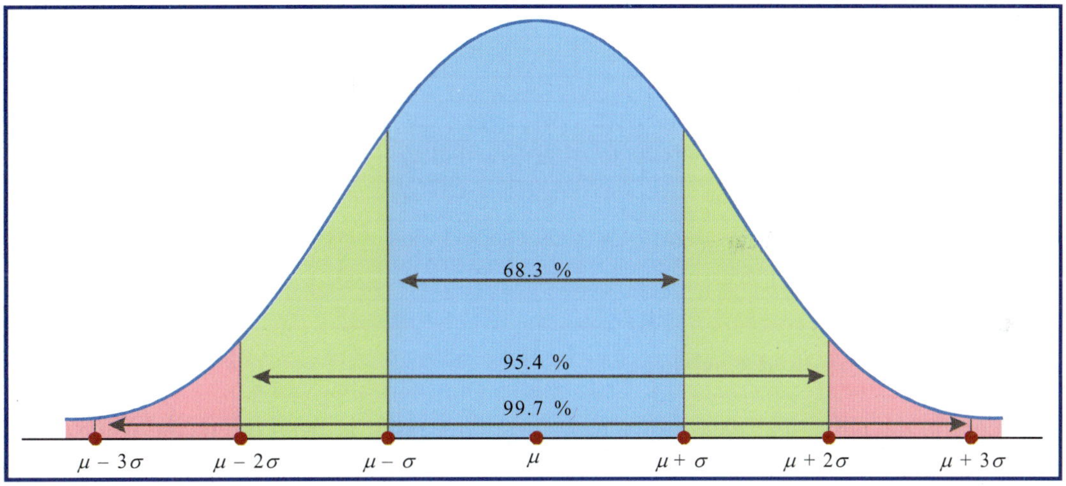

Figure 2.18

Example 24: Verify Empirical Rule for the approximately symmetric data given below.

55	28	68	35	59	57	42	32	35	44
52	47	47	40	46	65	54	68	46	33
37	58	51	64	52	32	57	59	39	41
60	62	56	55	50	54	43	21	36	49

Solution: In this case $\mu = 48.21$ and $\sigma = 11.43$

Number of observations falls within $(\mu - \sigma, \mu + \sigma)$ or (36.78, 59.64) = 25

Number of observations falls within $(\mu - 2\sigma, \mu + 2\sigma)$ or (25.34, 71.08) = 39

Number of observations falls within $(\mu - 3\sigma, \mu + 3\sigma)$ or (13.91, 82.51) = 40

Also, 68.3% of total observation is $= 40 \times 0.683 = 27.32 \approx 27$

95.4% of total observation is $= 40 \times 0.954 = 38.16 \approx 38$

99.7% of total observation is $= 40 \times 0.997 = 39.88 \approx 40$

Try it yourself:

1. The mean and standard deviation of a sample of size 180 are 12 and 2 respectively.
   (a) Use Chebyshev's Rule to determine the least percentage of sample values that lie between 8 and 16.
   (b) Use the Empirical Rule to find the percentage of sample values that lie between 8 and 16.

2. Find the proportion of data points that fall within k standard deviations of mean, if

(a) $k = 3$ **(b)** $k = 1.5$ **(c)** $k = 2$

3. According to Chebyshev's Theorem at least what percentage of data values lie between

(a) $\mu - 2.5\sigma$ and $\mu + 2.5\sigma$ **(b)** $\mu - 2\sigma$ and $\mu + 2\sigma$

(c) $\mu - \sigma$ and $\mu + \sigma$ **(d)** $\mu - 1.2\sigma$ and $\mu + 1.2\sigma$

4. Verify Empirical Rule for the approximately symmetric data given below.

(a)

58	11	48	17	26	11	19	42	4	47
8	46	39	19	17	69	49	36	27	25
0	31	1	64	58	35	47	27	36	62
25	53	52	53	37	22	67	67	1	9
68	15	65	44	45	7	48	22	35	11

(b)

54	34	16	30	24	80	44	46	52	32	67	46
40	43	86	24	19	74	37	37	76	62	87	47
40	17	18	62	21	69	60	44	65	75	13	84
77	50	74	21	24	29	24	56	45	66	67	60
35	71	85	63	21	88	82	11	40	31	63	27

C. *Z*-Scores (Standard Positions)

A z-score is the numerical measure of how far a data point is from the mean, using standard deviation as the scale of measurement. In other words, a z-score provides the position of the data point from mean in term of number of standard deviations. For example, if a point has a z-score of 2.3, it means that it is 2.3 standard deviations away from the mean.

A data point is considered usual if its z-score is in $(-2, 2)$. If the z-score of a point is less than -2 or greater than 2, then it is considered an unusual data value. This is because most data values are expected to be near the mean. Saying that a z-score is between $(-2, 2)$ means that it is within 2 standard deviations of the mean in both directions.

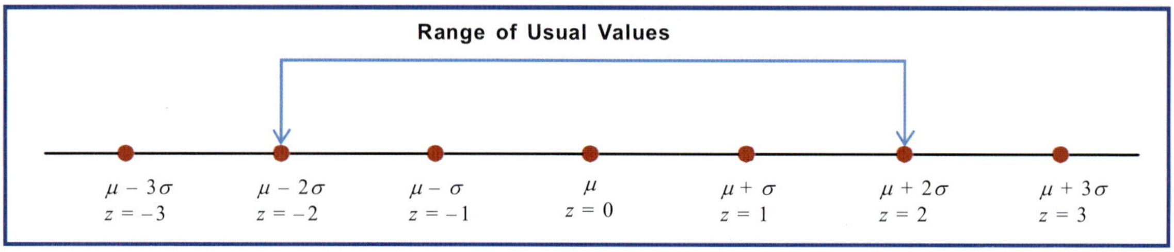

Figure 2.19

The formula to calculate a z-score for a given data point (x) is

$$z = \frac{x - \mu}{\sigma}$$

Example 25: Find the z-score for an IQ of 143, when average of IQ's is 100 and standard deviation is 15. Is it unusual?

Solution: $z = \dfrac{143 - 100}{15} = 2.87$. It is an unusual data value as it lies out side $(-2, 2)$

Example 26: By using the mean and the standard deviation of the above example, what would be an IQ score corresponding to $z = -1.96$

Solution: $z = \dfrac{x - \mu}{\sigma} \Rightarrow x = z\sigma + \mu$ $x = (-1.96)(15) + 100 = 70.6$

Try it yourself:

1. The mean GPA in a local High School is 2.9, with a standard deviation of 0.6. What is the z-score for a 3.8 GPA? Is it a usual GPA?

2. Find the z-score:

   (a) for an observation of 125, when average of IQ's is 110 and standard deviation is 15. Is it unusual?

   (b) for an observation of 113, when average of IQ's is 105 and standard deviation is 20. Is it unusual?

   (c) for an IQ of 90, when average of IQ's is 85 and standard deviation is 1. Is it unusual?

D. Quartiles

To partition the data sets in four equal parts, we need three numbers Q_1, Q_2 and Q_3, called Quartiles. Recall that a median divides a data set into two equal parts. Therefore Q_2 is the median.

To calculate Q_1 and Q_3 follow the instructions given below.

(a) Arrange the data in an increasing order and find the median (Q_2).

(b) The **first quartile** Q_1 is the median of the first 50% of the ordered list, which is to the left of the median including the median itself. 25% of the data falls below Q_1.

(c) The **third quartile** Q_3 is the median of the second 50% of the ordered list, which is to the right of the median including the median itself. 75% of the data falls below Q_3.

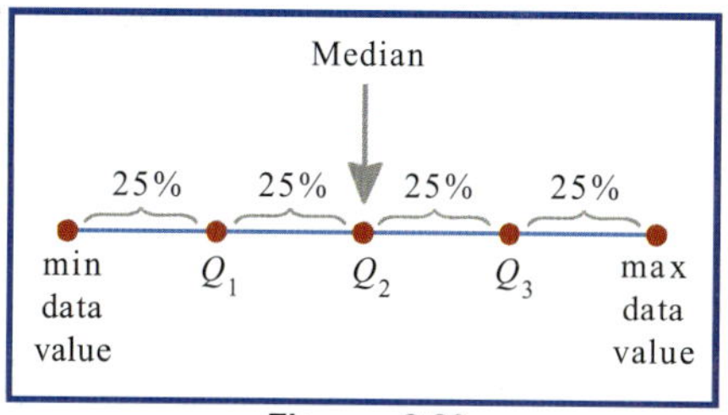

Figure 2.20

Note: There are several ways to calculate quartiles, all with slightly different answers. Throughout the text, we will adopt the above procedure.

Example 27: From the data given below, compute median Q_2, first quartile Q_1 and the third quartile Q_3.

1	3	3	5	6	7	7	10	10	12	14	17	17	21	25	29	30	33	37	40	45

Solution: $n = 21$, and all the data is already arranged in an increased order.

$$\text{Median} = \text{value of } \left(\frac{21+1}{2}\right)^{\text{th}} \text{ observation, in order data set}$$

$$= \text{value of } 11^{\text{th}} \text{ observation} = 14$$

Q_1 = Median of lower 50% observation including median

= Median of first 11 observations: 1, 3, 3, 5, 6, 7, 7, 10, 10, 12, 14

$$= \left(\frac{11+1}{2}\right)^{th} = 6^{th} \text{ observation} = 7$$

Q_3 = Median of upper 50% observations, including median.

= Median of last 11 observations: 14, 17, 17, 21, 25, 29, 30, 33, 37, 40, 45

$$= \text{value of} \left(\frac{11+1}{2}\right)^{th} \text{ observation} = 6^{th} \text{ observation} = 29$$

1	3	3	5	6	7	7	10	10	12	14	17	17	21	25	29	30	33	37	40	45
										Q_2										
1	3	3	5	6	7	7	10	10	12	14										
					Q_1					14	17	17	21	25	29	30	33	37	40	45
															Q_3					

Table 2.21

We can also find the quartiles using rank formulas.

The Rank of $Q_1 = 0.25\ (n + 1)$	(position of 1st quartile)
The Rank of $Q_2 = 0.5\ (n + 1)$	(position of 2nd quartile)
The Rank of $Q_3 = 0.75\ (n + 1)$	(position of 3rd quartile)

To calculate Q_1 for the above example by using rank formulas, we have

$$Q_1 = 0.25\ (n + 1)$$
$$= 0.25\ (21 + 1)$$
$$= 5.5 \approx 6^{th} \text{ position}$$

The 6th position in our data set is 7. Therefore $Q_1 = 7$

E. Outliers

Extreme (unusual or suspicious) values in a data set are called outliers. In most common statistical studies, outliers are usually found and disregarded before the actual statistical study begins. There are many ways to determine outliers. In the following example, we will demonstrate a simple way of finding outliers in a data set.

i) For a given data set, find Q_1 and Q_3

ii) Find the Interquartile Range defined as IQR = $Q_3 - Q_1$

iii) Multiply IQR by 2.

iv) The accepted range of values in a data set should be between $Q_1 - 2 \times \text{IQR}$ and $Q_3 + 2 \times \text{IQR}$

v) Any data value less than $Q_1 - 2 \times \text{IQR}$ or greater than $Q_3 + 2 \times \text{IQR}$ is considered as outlier.

Example 28: Determine the outliers in the following data set.

10	7	9	15	38	90	17	23

Solution: Arrange the data in an increasing order.

7	9	10	15	17	23	38	90

In order to find the outliers, we first calculate Q_1, Q_3 and the interquartile range (IQR).

The median is the average of 15 and 17.

7	9	10	15	17	23	38	90

$\Rightarrow$ Median $= \dfrac{15+17}{2} = 16$

Since the 1st quartile is the median of the bottom 50% of a data set,

7	9	10	15

Q_1 will be the average of 9 and 10.

$$\Rightarrow \quad Q_1 = \frac{9+10}{2} = 9.5$$

Since the 3rd quartile is the median of the top 50% of a data set,

17	23	38	90

Q_3 will be the average of 23 and 38.

$$\Rightarrow \quad Q_3 = \frac{23+38}{2} = 30.5$$

The interquartile range IQR $= Q_3 - Q_1 = 30.5 - 9.5 = 21$

Once the quartiles and interquartile range have been calculated, we are ready to find the range of the outliers.

$$Q_1 - 2 \times \text{IQR} = 9.5 - 2(21) = -32.5 \text{ and } Q_3 + 2 \times \text{IQR} = 30.5 + 2(21) = 72.5$$

We can now conclude that any data value outside −32.5 to 72.5 range is an outlier. An observation 90 as outlier.

Try it yourself:

1. From the data given below, compute quartiles.

(a)

11	10	39	38	16	29	31	42	44	26	25	19

(b)

28	19	29	18	41	25	13	24	42	28
30	10	43	25	22	16	14	28	16	15
19	17	44	44	17	22	42	42	13	21
17	10	16	34	38	28	15	34	36	28
22	40	27	17	31	18	10	42	28	34

2. Determine the outliers (if any) in the following data sets (discrete data).

(a)

11	38	144	16	37	43	17	31	22	29	31	15

(b)

28	18	33	22	32	40	21	44	35	32
11	30	39	22	28	15	16	10	28	13
25	34	24	26	38	37	34	37	30	29
23	35	26	25	14	10	33	44	41	32
28	44	42	29	37	41	14	31	30	18

(c)

3	2	24	38	21	6	7	10
15	46	5	93	7	21	15	46
0	3	56	34	45	9	17	38
4	27	2	25	33	80	0	58
32	20	31	18	1	15	11	67

F. Percentiles

Percentiles are values that divide a given distribution into 100 equal parts with each partition encompassing 1% of the distribution. A data set can be divided into 99 percentiles. Each percentile is a 1% partition of a data set, when the data is arranged in an increasing order from min to max.

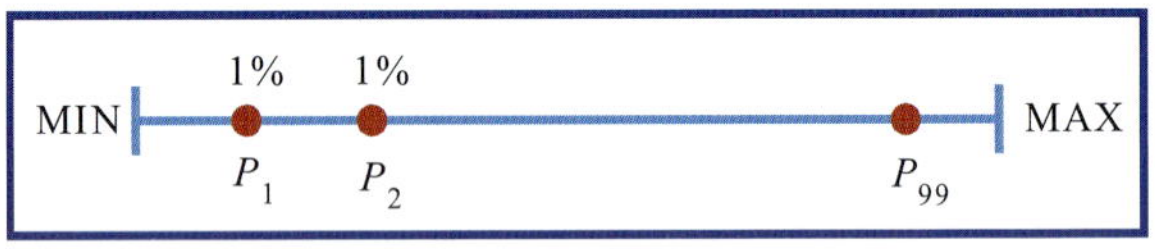

Figure 2.21

Note that P_1 represents the 1st percentile, and P_{99} represents the 99th percentile.

Given that the entire distribution of data in an experiment is represented as a bell shaped graph, P_{24}, the 24th percentile, separates the bottom 24% of the data from the top 76%.

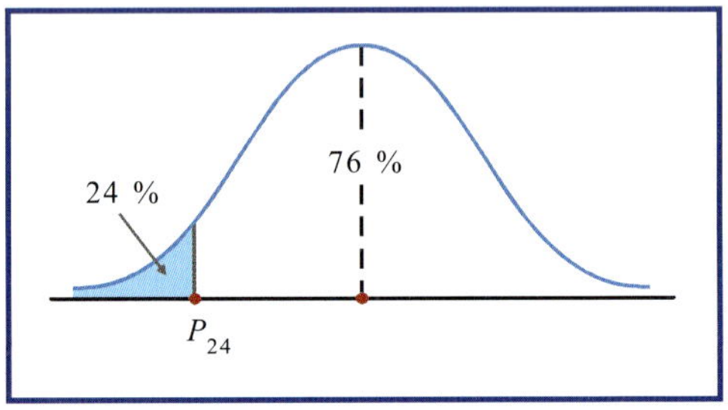

Figure 2.22

We can also use histograms to show P_{24}.

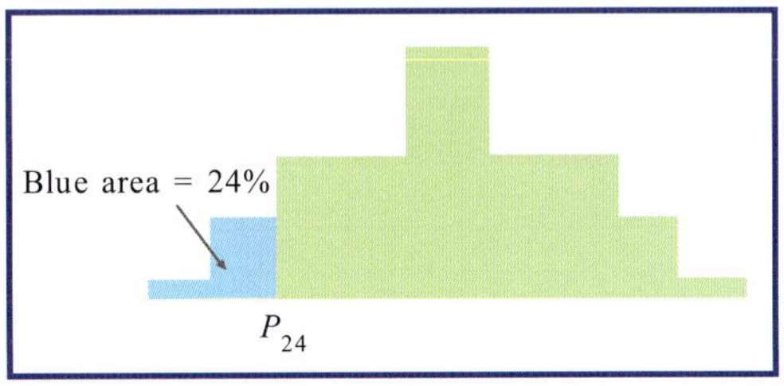

Figure 2.23

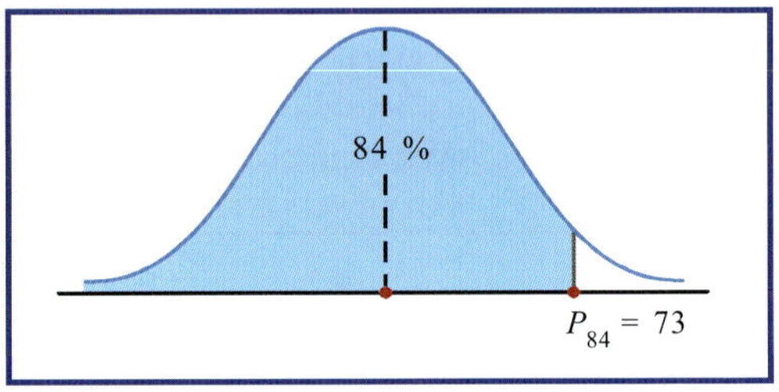

Figure 2.24

If the 84th percentile of a data set is 73, then 84% of the data is less than 73

- ### Percentile of a data point

To find the percentile of a value in a data set arranged in an increasing order, we use the following formula.

$$\text{Percentile of } x = \frac{(\text{number of data values below } x) + 0.5}{\text{Total number of values in the data set}} \times 100\%$$

Note: We must round the final answer up to the nearest whole number.

- **Convert a percentile to a data value**

To convert a percentile to a data value (to find the value in a data set corresponding to a given percentile), we find the position of that data value using the formula $\frac{\text{percentile}}{100} \times n$, where n represents the total number of data points. Once the position of a desired percentile is found, we can then find the data value corresponding to that position by counting the numbers from left to right after the data has been arranged in an increasing order.

Note: The position computed by using the formula $\frac{\text{percentile}}{100} \times n$ must be rounded up to the next higher integer.

Example 29: A statistics professor randomly selects 20 of her students and records their final exam scores.

47	30	68	42	75	89	96	74	73	19
53	88	74	65	59	58	84	96	10	67

(a) Find the percentile for 53.

(b) Find P_{36} (the value that corresponds to the 36th percentile)

Solutions: **(a)** Arrange the data in an increasing order.

10	19	30	42	47	**53**	58	59	65	67
68	73	74	74	75	84	88	89	96	96

The number of data values below 53 are 5. Using the above percentile formula we have

percentile of 53 $= \frac{5 + 0.5}{20} \times 100\% = 27.5\% \Rightarrow 28^{\text{th}}$ percentile. A student with a score of 53 did better than 28% of the class.

(b) The position (rank) of $P_{36} = \frac{36}{100} \times 20 = 7.2 \approx 8$ (by rounding it to next number).

This means that the 36th percentile of the above data set is at the 8th position.

10	19	30	42	47	53	58	**59**	65	67
68	73	74	74	75	84	88	89	96	96

$P_{36} = 59$. This means that 36% of the data values in the above data set are below 59.

Try it yourself:

1. Given the following data set,

3	2	24	38	21	6	7	10
15	46	5	93	7	21	15	46
0	3	56	34	45	9	17	38
4	27	2	25	33	80	0	58
32	20	31	18	1	15	11	67

(a) compute P_{20}.

(b) compute P_{83}.

(c) find the percentile rank for data value 58.

2. Given the following data sets, compute P_{42}, P_{64}, and P_{93}.

(a)

39	34	31	17	26	43	26	36	17	42	37	18

(b)

30	27	34	35	10	33	21	31	40	27
41	44	35	21	33	21	26	15	37	10
32	25	36	24	36	14	43	18	35	44
16	20	21	20	34	11	15	39	12	20
14	29	12	40	13	33	27	39	18	42

2.9: 5-Number Summary and Box-and-Whisker Plot

Objectives

After completing this section you will be able to make:

A. Box-and-Whisker Plot.

In statistics, the collection of the minimum data value, maximum data value, the Median, Q_1 and Q_3 of a data set is called the **5-Number Summary**. The statistical chart used to graph the **5-Number Summary** is called the Boxplot (Box-and-Whisker Plot).

A. Box-and-Whisker Plot

The 5-Number Summary can be represented graphically and is called Box-and-Whisker Plot. A Box-and-Whisker Plot consists of a rectangular box with endpoints at Q_1 and Q_3, a vertical line inside the box is at the median Q_2, two horizontal lines called whiskers extending from Q_3 to the largest value in the data set, and Q_1 to the smallest value in the data set.

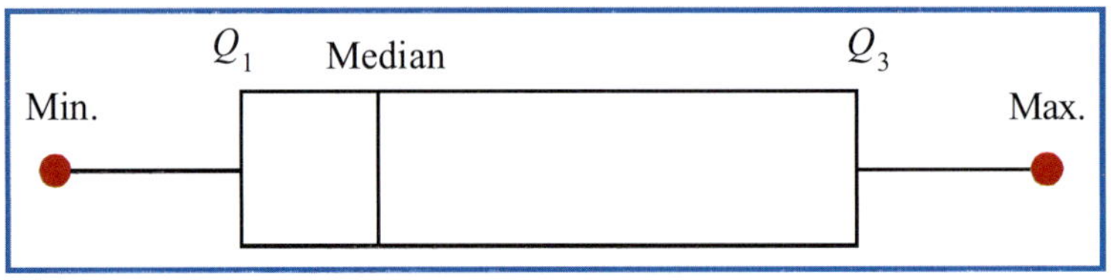

Figure 2.25

The length of the box is the Interquartile Range denoted as IQR, where $IQR = Q_3 - Q_1$. Recall that values that are $2 \times$ IQR, below Q_1 or above Q_3 are outliers.

Example 30: Given the following data set;

65	45	12	74	100	15	74	54	63	15	24	60

Construct the Box-and-Whisker Plot for the 5-Number Summary.

Solution: We first must calculate the 5-Number Summary

min = 12	Q_1 = 19.5	median = 57	Q_3 = 69.5	max = 100

and then construct the Boxplot as shown below:

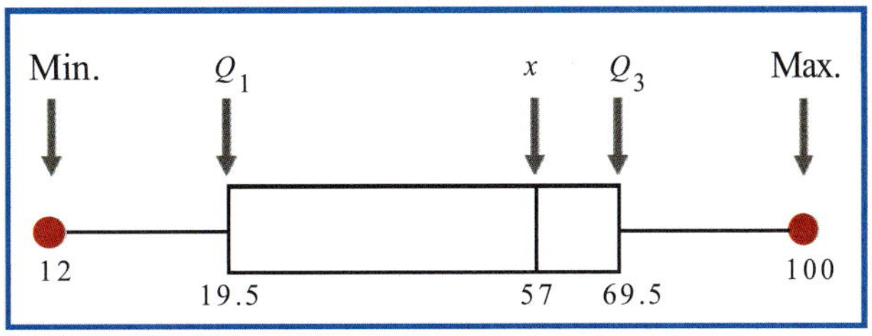

Figure 2.26

Try it yourself:

1. Given the following data sets, compute the 5-Number Summary and construct the Box-and-Whisker Plot.

(a)

3	2	24	38	21	6	7	10
15	46	5	93	7	21	15	46
0	3	56	34	45	9	17	38
4	27	2	25	33	80	0	58
32	20	31	18	1	15	11	67

(b)

34	40	37	15	35	27	12	23	13	35	29	37
28	17	23	10	42	16	26	30	41	18	21	18

(c)

37	18	41	30	23	29	22	44	37	12
31	29	29	35	29	36	14	34	16	35
12	35	44	38	13	29	37	18	41	19
29	11	13	19	42	42	44	18	39	10
37	14	25	34	36	17	28	42	24	23

2.10: Data Analysis with Microsoft Excel®

In this section, we will explain the computations in Data Analysis by using Microsoft Excel® commands, that will enable you to:

Objectives

After completing this section you will be able to carryout the computations in Chapter 2 using Microsoft Excel®.

i) draw Histograms, Pie Charts, and Line Graphs.

ii) computing mean, median and standard deviation.

iii) computing statistical parameters by using the descriptive statistics package.

iv) making 5-number summary.

Illustration 1: Given the following frequency table, draw the Histogram and the Pie Chart.

Classes	0-20	20-40	40-60	60-80	80-100
Frequency	15	8	20	40	23

Solution: To produce the Histogram (Figure 2.28) we highlight the Frequency Distribution Table (Figure 2.27) in Excel® and press the F11 Key. To connect the bars, right click on one of the bars, select the Format Data Series, click on the Option tab and turn the Gap Width to zero.

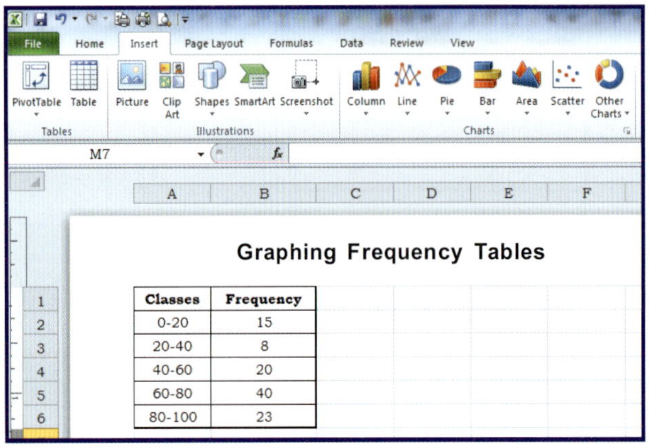

Figure 2.27

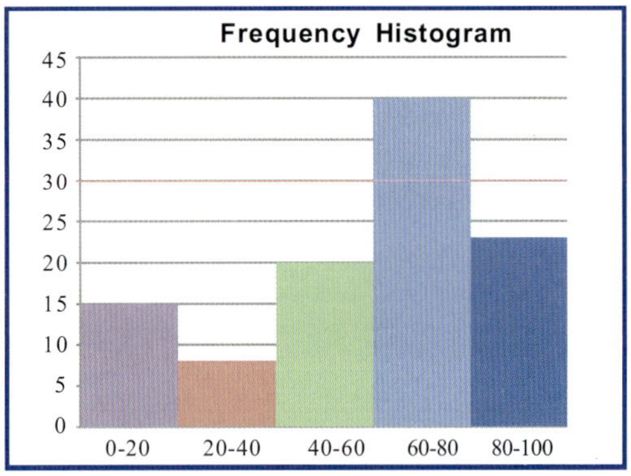

Figure 2.28

We can also produce the chart using the Tabs and Ribbons.

Select the Frequency Table on your spreadsheet. Go to the Insert Tab, and select the Column icon from the Charts Group.

For making a Pie Chart we select the Frequency Table on your spreadsheet.

Go to the Insert Tab, and select the Pie icon from the Charts Group.

For making line graph select the Frequency Table on spreadsheet.

Go to the Insert Tab, and select the 2-D line under Line chart group.

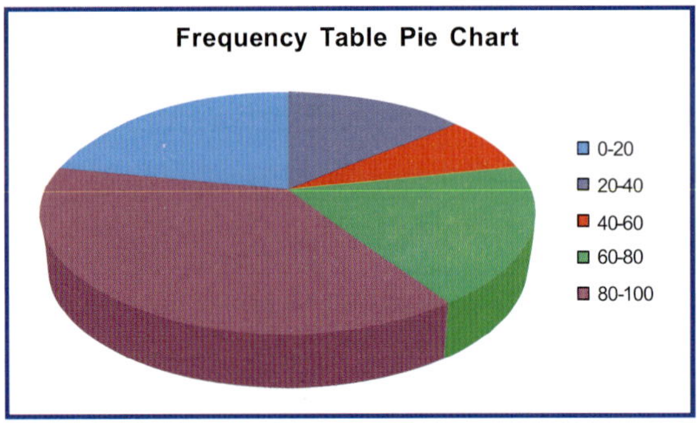

Figure 2.29 a

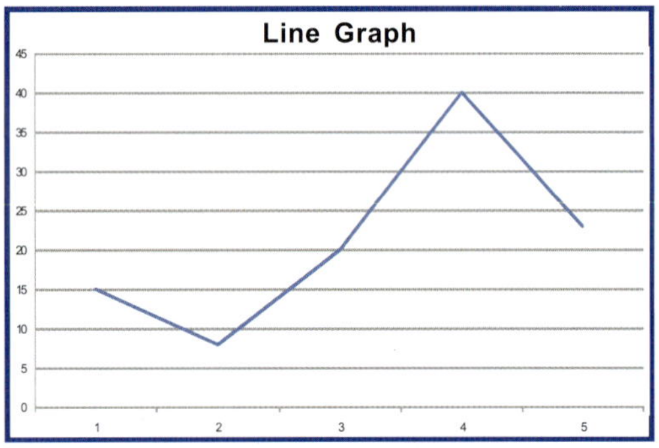

Figure 2.29 b

Illustration 2: Compute mean, median, standard deviation, and variance for the data given below.

25	18	20	15	19	20	21	23	20	14

Solution:

- **Mean:** $$\frac{\sum x}{n} = \frac{195}{10} = 19.5$$

  Excel command: = AVERAGE(Initial_Cell_Reference: Final_Cell_Reference)

- **Median:** Arrange the data in an increasing order. The median is the midpoint of the data set if it is distinct. Otherwise, the median is the mathematical average of the 2 middle values.

14	15	18	19	20	20	20	21	23	25

The median is $\frac{20+20}{2} = 20.$

Excel command: = MEDIAN(Initial_Cell_Reference: Final_Cell_Reference)

- **Standard Deviation:** $\sqrt{\frac{n\sum x^2 - \left(\sum x\right)^2}{n(n-1)}} = \sqrt{\frac{10\,(3901) - 195^2}{10(9)}} \approx 3.31$

  Excel command: = STDEV(Initial_Cell_Reference: Final_Cell_Reference)

- **Variance:** $\sigma^2 = 3.31^2 = 10.96$

  Excel command: = VAR(Initial_Cell_Reference: Final_Cell_Reference)

Command Summary:

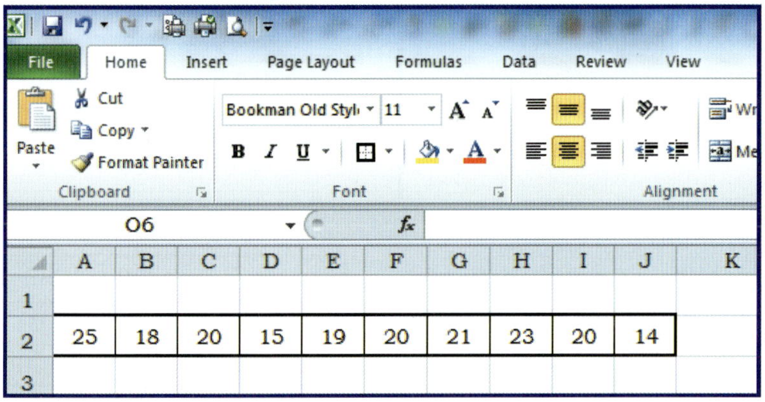

Figure 2.30

mean	=AVERAGE(A2:J2)
median	=MEDIAN(A2:J2)
SD	=STDEV(A2:J2)
var	=VAR(A2:J2)

Table 2.22

Illustration 3: For the Data Set given in Illustration 2, use the data analysis package in excel to calculate the mean, standard deviation, variance, range, min, and max.

25	18	20	15	19	20	21	23	20	14

Solution: ***Step 1*** Go to the Data Tab in Excel®.

Step 2 Click on Data Analysis icon from the Analysis Group.

Step 3 Click on Descriptive Statistics and Press OK. (Figure 2.31)

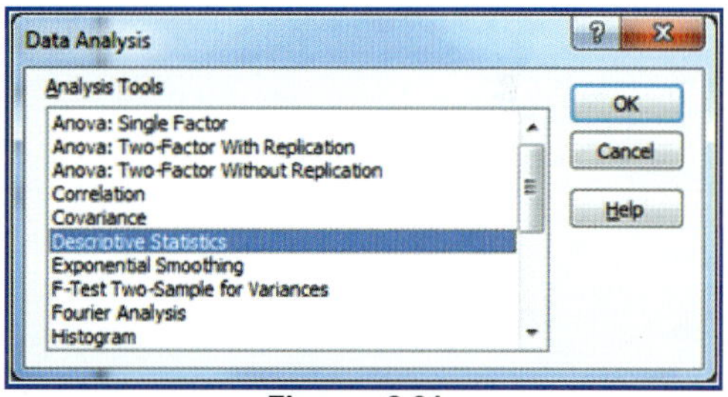

Figure 2.31

The following Dialog Box (Figure 2.32) will appear.

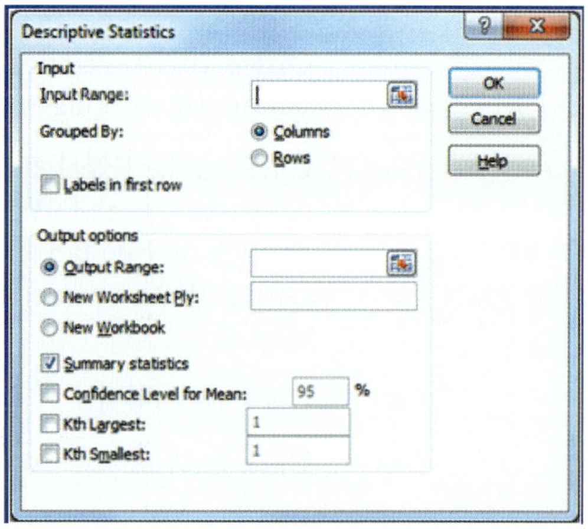

Figure 2.32

Step 4 Click in the Input Range box, and select your data set.

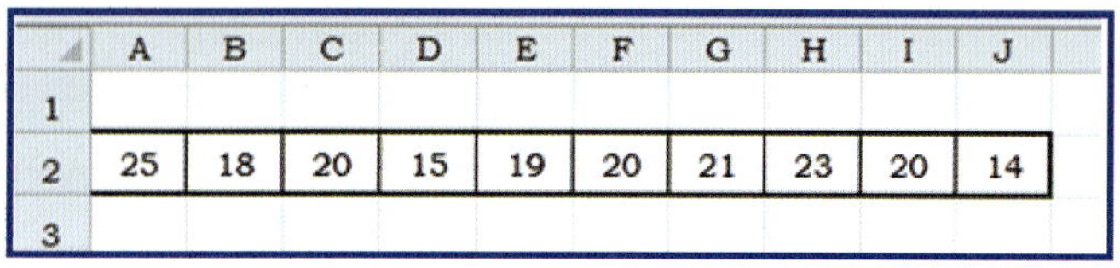

Figure 2.33

Step 5 Select row or column depending on whether your data is typed in a row or column. If your data set has a label, select the box for Labels in the first column. Under the output options, select new workbook, and check the Summary Statistics box. (Figure 2.34)

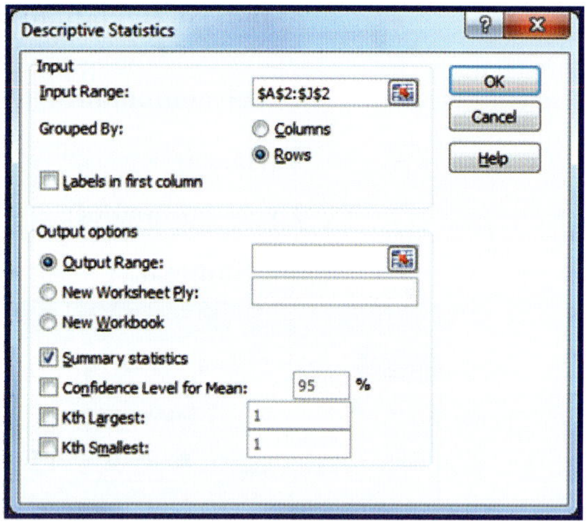

Figure 2.34

The Table 2.23 should be displayed. Here you can see the mean, standard deviation, variance, range, min, and max.

Row 1	
Mean	19.5
Standard Error	1.046156988
Median	20
Mode	20
Standard Deviation	3.308238874
Sample Variance	10.94444444
Kurtosis	0.034918263
Skewness	-0.184127289
Range	11
Minimum	14
Maximum	25
Sum	195
Count	10

Table 2.23

Most of the time, during the initial installation of Microsoft Excel®, the Data Analysis Package is not installed. If you cannot find the Data Analysis icon under the Analysis Group, follow the following steps.

Step 1 Click on the File Tab (if working with Office 2010), or click on the Office button. (if working with Office 2007)

Step 2 Click on the Excel® Option box (if working with Office2007), or click on File tab (if working with Office2010), and select Option from the list.

Step 3 Click on Add-Ins.

Step 4 Click on Analysis ToolPak.

Step 5 Click on OK.

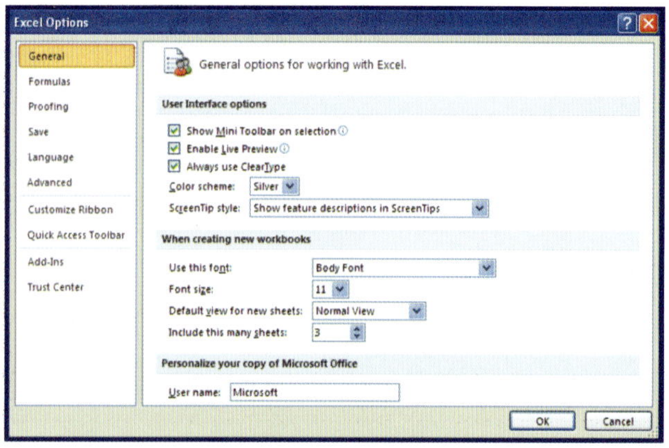

Figure 2.35

Note: Sometimes you may have to restart the program to see the packages you have added.

Illustration 4: Compute the mean, range, standard deviation, and the variance of the following frequency distribution.

x	1	2	3	4	5	6
Frequency	5	8	15	10	7	3

Solution: The solution of the above example using excel is shown in the following tables.

	A	B	C	D	E
1	x	f	x*f	x^2*f	
2	1	5	=A2*B2	=A2^2*B2	
3	2	8			
4	3	15			
5	4	10			
6	5	7			
7	6	3			
8					
9					

Table 2.24 a

	A	B	C	D
1	x	f	x*f	x^2*f
2	1	5	5	5
3	2	8	16	32
4	3	15	45	135
5	4	10	40	160
6	5	7	35	175
7	6	3	18	108
8		=sum(B2:B7)	=sum(C2:C7)	=sum(D2:D7)
9				

Table 2.24 b

	A	B	C	D	E	F	G	H	I
1	x	f	x*f	x^2*f		n	=B8		
2	1	5	5	5		n-1	=B8-1		
3	2	8	16	32		Sum(x*f)	=C8		
4	3	15	45	135		Sum(x^2*f)	=D8		
5	4	10	40	160					
6	5	7	35	175		mean	=G3/G1		
7	6	3	18	108		range	=(MAX(A2:A7)-MIN(A2:A7))		
8		48	159	615		sd	=SQRT((G1*G4-G3^2)/(G1*G2))		
9						variance	=G8^2		

Table 2.24 c

	A	B	C	D	E	F	G
1	x	f	x*f	x^2*f		n	48
2	1	5	5	5		n-1	47
3	2	8	16	32		Sum(x*f)	159
4	3	15	45	135		Sum(x^2*f)	615
5	4	10	40	160			
6	5	7	35	175		mean	3.3125
7	6	3	18	108		range	5
8		48	159	615		sd	1.370762
9						variance	1.878989
10							

Table 2.25 d

To find the above parameters manually, we perform the following calculations:

(a) Mean $= \mu = \dfrac{159}{48} \approx 3.313$

(b) Range = Highest data value – lowest data value = 6 – 1 = 5

(c) Standard deviation $= s = \sqrt{\dfrac{n\sum(f\cdot x)^2 - \left(\sum f\cdot x\right)^2}{n(n-1)}} = \sqrt{\dfrac{48(615)-(159)^2}{48(47)}} = \sqrt{1.88} \approx 1.37$

(d) Variance $= s^2 = \left(\sqrt{1.88}\right)^2 = 1.88$

Illustration 5: Given the following data set, find the median (Q_2), first quartile (Q_1), third quartile (Q_3), min, and the max values. (5-Number Summary)

	A	B	C	D	E	F	G	H	I	J	K	L	M
1													
2		65	45	12	74	100	15	74	54	63	15	24	60

Solution: We will use the Quartile command within Excel® to calculate the 5-Number summary.

Click on f_x on the formula bar. Search for Quartile. Press OK. Figure 2.36 will appear.

In the Array box select the range of your data set from the Excel® worksheet.

Click on the Quart input box. You will see a description of how to find min, max, median, Q_1 and Q_3.

To find the min value type "0" in the quart input box. To find the max value type "4" in the quart input box. To find the median value type "2" in the quart input box. To find the Q_1 type "1" in the quart input box. To find the Q_3 type "3" in the quart input box.

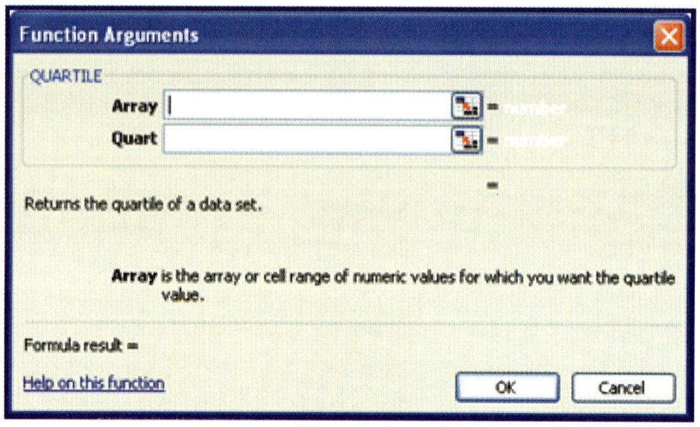

Figure 2.36

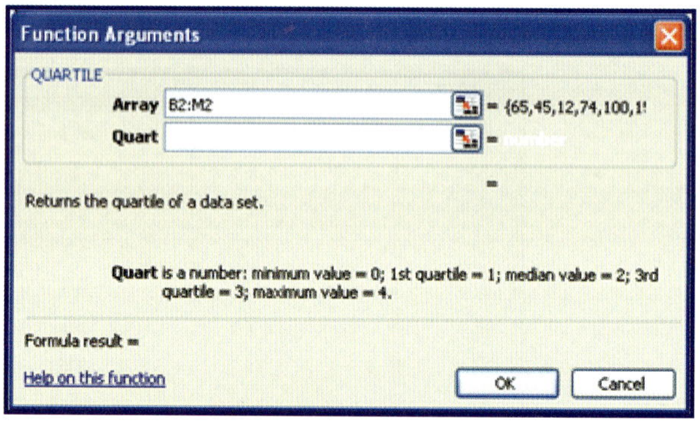

Figure 2.37

For example, typing a 3 in the Quart input box will calculate the value of Q_3 as shown in Figure. 2.38.

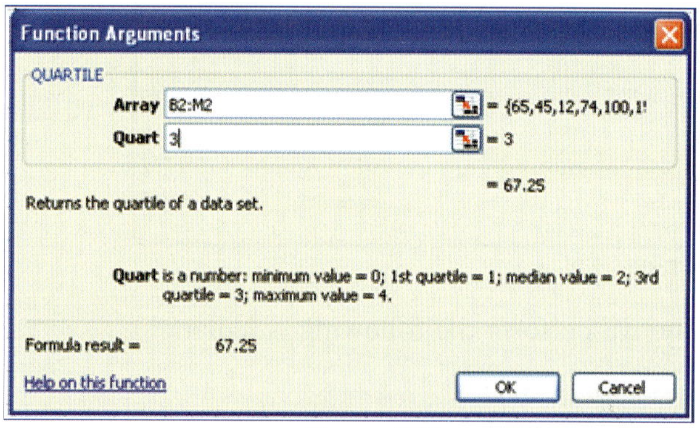

Figure 2.38

Here is the summary of the values.

Statistics	Data Set 1
Q_1	21.75
min	12
median	57
max	100
Q_3	67.25

Table 2.26

You can also calculate the above statistics by using the following Excel® Commands:

min	Q_1	median	Q_3	max
=QUARTILE(B2:M2, 0) or = MIN(B2:M2)	=QUARTILE(B2:M2, 1)	=QUARTILE(B2:M2, 2) or = MEDIAN(B2:M2)	=QUARTILE(B2:M2, 3)	=QUARTILE(B2:M2, 4) or = MAX(B2:M2)

Table 2.27

Illustration 6: Given the data set, find (a) P_{23} and (b) Percentile rank of 30.

	A	B	C	D	E	F	G	H	I	J	K	L	M
1													
2		65	45	12	74	100	15	74	54	63	15	24	60

Solution: **(a)** To find the 23rd Percentile of the data set above, use the command:

= PERCENTILE(B2:M2,0.23) = 19.77 $\Rightarrow P_{23} \approx 20$

(b) To find the Percentile Rank of 30, use the command;

= PERCENTRANK(B2:M2,30) = 0.298 $\Rightarrow$ The number 30 is at the 29.8 $\approx$ 30th percentile.

Illustration 7: Given the following data set, construct a Grouped Frequency Table.

0	21	40	61	82	100	74	12	34	55
0	23	40	62	83	100	74	12	36	55
1	29	41	63	83	17	74	13	37	55
3	29	41	65	84	17	74	69	96	52
4	30	43	66	84	17	75	69	98	15
4	31	45	67	85	18	76	73	98	38
4	32	48	67	86	18	76	68	92	55
6	32	50	67	90	18	76	9	33	73
7	32	50	68	90	38	77	58	57	98
9	32	51	68	92	39	79	17	58	38

Solution: We will use Excel® to count the number of values in the following ranges (classes).

Classes	Frequency
0-20	?
20-40	?
40-60	?
60-80	?
80-100	?
100-120	?

Table 2.28

***Step* 1** Key in the given data set into Excel®, and set up your classes and a column for frequencies as shown in the following figure.

	A	B	C	D	E	F	G	H	I	J	K	L	M
1	0	21	40	61	82	100	74	12	34	55		Class	Frequency
2	0	23	40	62	83	100	74	12	36	55		0-20	
3	1	29	41	63	83	17	74	13	37	55		20-40	
4	3	29	41	65	84	17	74	69	96	52		40-60	
5	4	30	43	66	84	17	75	69	98	15		60-80	
6	4	31	45	67	85	18	76	73	98	38		80-100	
7	4	32	48	67	86	18	76	68	92	55		100-120	
8	6	32	50	67	90	18	76	9	33	73			
9	7	32	50	68	90	38	77	58	57	98			
10	9	32	51	68	92	39	79	17	58	38			

Figure 2.39

***Step* 2** Since in the first class (0-20) we count the values from (0 to 19), and in the second class we count the values from (20 to 39), etc., create a column of the upper class values we use for counting, as in Figure 2.40.

	A	B	C	D	E	F	G	H	I	J	K	L	M	N
1	0	21	40	61	82	100	74	12	34	55		Class	Frequency	Upper class values for counting
2	0	23	40	62	83	100	74	12	36	55				
3	1	29	41	63	83	17	74	13	37	55		0-20		19
4	3	29	41	65	84	17	74	69	96	52		20-40		39
5	4	30	43	66	84	17	75	69	98	15		40-60		59
6	4	31	45	67	85	18	76	73	98	38		60-80		79
7	4	32	48	67	86	18	76	68	92	55		80-100		99
8	6	32	50	67	90	18	76	9	33	73		100-120		119
9	7	32	50	68	90	38	77	58	57	98				
10	9	32	51	68	92	39	79	17	58	38				

Figure 2.40

***Step* 3** Select all the cells in the frequency column, click on the *fx* command line, and type the command shown in the following display.

M3 *fx* =FREQUENCY(A1:J10,N3:N8)

	A	B	C	D	E	F	G	H	I	J	K	L	M	N
1	0	21	40	61	82	100	74	12	34	55		Class	Frequency	Upper class values for counting
2	0	23	40	62	83	100	74	12	36	55				
3	1	29	41	63	83	17	74	13	37	55		0-20	=FREQUENCY(A1:J10,N3:N8)	19
4	3	29	41	65	84	17	74	69	96	52		20-40		39
5	4	30	43	66	84	17	75	69	98	15		40-60		59
6	4	31	45	67	85	18	76	73	98	38		60-80		79
7	4	32	48	67	86	18	76	68	92	55		80-100		99
8	6	32	50	67	90	18	76	9	33	73		100-120		119
9	7	32	50	68	90	38	77	58	57	98				
10	9	32	51	68	92	39	79	17	58	38				

Figure 2.41

Note: A1:J10 is the array of the given sample data and N3:N8 is the array of the upper class values used for counting.

***Step* 4** Press enter. To execute this particular sequential command:

i) Copy cell M3.

ii) Select cell range M3 to M8 and press F2.

iii) Hold down Ctrl+Shift and then press enter.

Excel® will automatically group the 100 individual data points in to your constructed classes as shown below.

	A	B	C	D	E	F	G	H	I	J	K	L	M	N
1	0	21	40	61	82	100	74	12	34	55		Class	Frequency	Upper class values for counting
2	0	23	40	62	83	100	74	12	36	55				
3	1	29	41	63	83	17	74	13	37	55		0-20	22	19
4	3	29	41	65	84	17	74	69	96	52		20-40	18	39
5	4	30	43	66	84	17	75	69	98	15		40-60	18	59
6	4	31	45	67	85	18	76	73	98	38		60-80	25	79
7	4	32	48	67	86	18	76	68	92	55		80-100	15	99
8	6	32	50	67	90	18	76	9	33	73		100-120	2	119
9	7	32	50	68	90	38	77	58	57	98		**Total Frequency**	**100**	
10	9	32	51	68	92	39	79	17	58	38				

Figure 2.42

Here is the Grouped Frequency Table.

Classes	Frequency
0-20	22
20-40	18
40-60	18
60-80	25
80-100	15
100-120	2

Table 2.29

2.11: Extended Topics and Proofs in Data Analysis (Optional)

Objectives

After completing this section you will be able to:

A. Compute other measures of the Average.

B. Measures of Dispersion (Theoretically).

C. Draw Graph of the Cumulative Percentile.

A. Other measures of the Average

i) The Arithmetic Mean or just the mean of a set of "n" numbers is

$$\bar{x} = \frac{x_1 + x_2 + \ldots + x_n}{n} = \frac{1}{n}\sum x_i$$

For simplicity, the above formula can be written

as $\bar{x} = \dfrac{\sum x}{n}$

ii) Geometric Mean (GM) of a set of numbers is the nth root of the product of the numbers.

$$\text{GM} = \sqrt[n]{(x_1)(x_2)\ldots(x_n)}$$

iii) Harmonic Mean "HM" of a set of "n" numbers is the reciprocal of the Arithmetic Mean of the reciprocals of the numbers.

$$\text{HM} = \frac{1}{\dfrac{\sum_{i=1}^{n} \frac{1}{x_i}}{n}} = \frac{1}{\dfrac{\sum \frac{1}{x}}{n}} = \frac{n}{\sum \frac{1}{x}}$$

iv) Root Mean Square (RMS), or the Quadratic Mean, of a set of "n" numbers is defined by

$$\text{RMS} = \sqrt{\overline{x^2}} = \sqrt{\frac{\sum_{i=1}^{n} x_i^2}{n}} = \sqrt{\frac{\sum x^2}{n}}$$

Example 31: On the way to Las Vegas, because of all the cops with radar guns, Raymond could not drive at a constant speed of 100 mph (as he loves to do). He drove 180 miles at 70 mph, 80 miles at 100 mph, and the remaining 20 miles at 50 mph. Find the Arithmetic and Harmonic means of his speeds.

Solution: Time of travel at 70 mph is $= \dfrac{180}{70} = 2.57$ hours.

Time of travel at 100 mph is $= \dfrac{80}{100} = 0.8$ hours.

Time of travel at 50 mph is $= \dfrac{20}{50} = 0.4$ hours.

The true average speed is $= \dfrac{180 + 80 + 20}{2.57 + 0.8 + 0.4} = \dfrac{280}{3.77} = 74.27$ mph

Calculating the average speed using the definition of:

i) Arithmetic Mean: $= \dfrac{70 + 100 + 50}{3} = 73.33$ mph

ii) Harmonic Mean: $= \dfrac{3}{\dfrac{1}{70} + \dfrac{1}{100} + \dfrac{1}{50}} = 67.74$ mph

Example 32: Prove that the sum of the deviation of a data set from its mean is zero.

$$\sum_{i=1}^{n}(x_i - \overline{x}) = \sum(x - \overline{x}) = 0$$

Note: To simplify the notation, we will use $\sum x$ to represent $\sum_{i=1}^{n} x_i$.

Solution: Let d's represent the difference between each x value and the mean $\overline{x}$.

Each individual deviation can be represented by $d_i = x_i - \overline{x}$, $i = 1, 2, \ldots, n$.

We need to show that $\sum_{i=1}^{n} d_i = 0$

The sum of deviations $= \sum d_i$

$$= \sum(x - \overline{x})$$

$$= \sum x - \sum \overline{x}$$

$$= \sum x - n\overline{x}$$

$$= \sum x - \frac{n\sum x}{n}$$

$$= \sum x - \sum x = 0$$

Example 33: The formula for the Geometric Mean of a data set, where each data value x_j has a particular frequency f_j is given by :

$$\text{GM} = \sqrt[N]{x_1^{f_1}\, x_2^{f_2} \ldots x_n^{f_n}}$$

where f_j, $j = 1, 2, \ldots, n$ is the frequency of data value x_j and $\text{N} = f_1 + f_2 + \ldots + f_n$. Derive an expression for $\ln(\text{GM})$.

Solution: **GM** $= \sqrt[N]{x_1^{f_1}\, x_2^{f_2} \ldots x_n^{f_n}}$

After taking the natural log of both sides, we have:

$$\ln(\text{GM}) = \ln\left(\sqrt[N]{x_1^{f_1}\, x_2^{f_2} \ldots x_n^{f_n}}\right)$$

$$= \frac{1}{N}\left[\ln x_1^{f_1} + \ln x_2^{f_2} + \ldots + \ln x_n^{f_n}\right]$$

$$= \frac{1}{N}\left[f_1 \ln x_1 + f_2 \ln x_2 + \ldots + f_n \ln x_n\right]$$

$$= \frac{1}{N}\sum_{i=1}^{n} f_i \ln x_i = \frac{1}{N}\sum f \ln x$$

B. Measures of Dispersion

The formula for Sample Variance is (Measure of average variation) $s^2 = \dfrac{\sum(x_i - \bar{x})^2}{n-1}$.

The above measure is also called the mean square deviation from the mean. Another similar measure for population is Variance and its formula is $\sigma^2 = \dfrac{\sum(x_i - \mu)^2}{N}$.

Example 34: Prove that the shortcut formula for Sample Variance is: $s^2 = \dfrac{n\sum x^2 - \left(\sum x\right)^2}{n(n-1)}$

Solution:

$$s^2 = \frac{\sum(x_i - \bar{x})^2}{n-1}$$

$$= \frac{\sum\left(x_i^2 - 2x_i\bar{x} + \bar{x}^2\right)}{n-1} = \frac{\sum x_i^2 - \sum 2x_i\bar{x} + \sum \bar{x}^2}{n-1} = \frac{\sum x_i^2 - 2\bar{x}\sum x_i + \bar{x}^2 \sum 1}{n-1}$$

$$= \frac{\sum x_i^2 - 2n\bar{x}^2 + \bar{x}^2 n}{n-1} \qquad \left(\textbf{Note } \bar{x} = \frac{\sum x}{n} \Rightarrow \sum x = n\bar{x}\,,\ \sum 1 = n\right)$$

$$= \frac{\sum x_i^2 - n\bar{x}^2}{n-1}$$

Replacing $\bar{x}$ by $\dfrac{\sum x}{n}$ in $\dfrac{\sum x_i^2 - n\bar{x}^2}{n-1}$, we have:

$$s^2 = \frac{\sum x_i^2 - n\left(\dfrac{\sum x_i}{n}\right)^2}{n-1} = \frac{\sum x_i^2 - \dfrac{\left(\sum x_i\right)^2}{n}}{n-1} = \frac{n\sum x_i^2 - \left(\sum x_i\right)^2}{n(n-1)}$$

C. Graph of the Cumulative Percentile

In the example given below, plot cumulative frequencies in percentages.

Example 35: For the following frequency table, construct and plot cumulative frequencies in percentages.

Classes	0-20	20-40	40-60	60-80	80-100
Frequency	5	5	10	29	12

Formula for Cumulative %

$$\text{Cumulative Percent} = \frac{\text{Cumulative Frequency}}{n} \cdot 100\%$$

Solution:

Classes	**Frequency**	**Cumulative Frequency**	**Cumulative Percent**
0-20	5	5	8.2
20-40	5	10	16.39
40-60	10	20	32.79
60-80	29	49	80.33
80-100	12	61	100
Total frequency	61		

Table 2.30 a

UCL	C.F.
0	0
20	8
40	16
60	33
80	80
100	100

Table 2.30 b

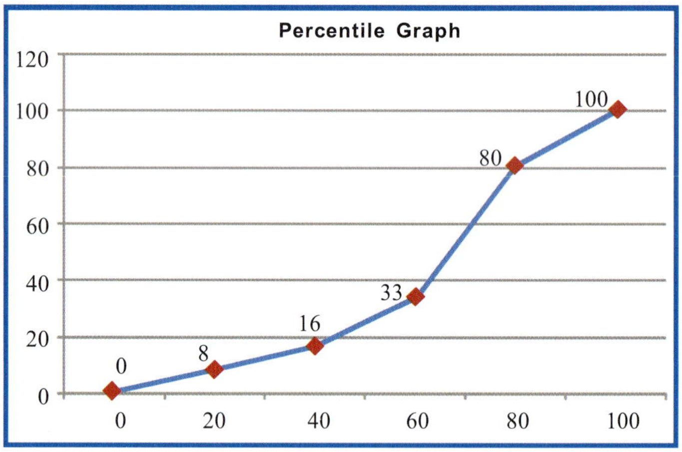

Figure 2.43

Chapter 2: General Exercises

1. Graph the histogram for the following frequency table. Calculate the Relative and Cumulative Frequencies. Graph the Relative and Cumulative Frequency Histograms.

Classes	Frequency
10-20	4
20-30	7
30-40	15
40-50	17
50-60	8
60-70	3

2. Graph the histogram for the following frequency table. Graph the Frequency Polygon, and the Ogive.

Classes	Frequency
0-20	5
20-40	5
40-60	10
60-80	29
80-100	12

3. The following table lists the percentage of sales for different model BMW's. Illustrate the distribution of % frequencies with a Pie Chart. Label the chart.

3 Series	5 Series	6 Series	7 Series	M Series
36%	21%	10%	14%	19%

4. Given the following data set, construct the Frequency Table.

1	4	1	6	3	2	1	2	2	4	2	6	3	5	5	5
1	1	6	6	4	5	3	3	4	5	5	2	2	3	3	2
3	5	2	2	5	5	4	1	2	6	1	3	1	4	1	2
6	1	5	5	6	6	5	6	5	1	2	5	3	5	2	6
6	4	6	3	6	5	2	4	2	6	3	6	6	3	3	4
2	3	1	1	4	6	6	6	2	5	1	2	1	5	1	2
5	4	5	6	3	1	2	6	6	5	3	6	2	6	5	4
2	5	5	1	6	5	2	3	3	3	5	3	4	5	4	4
3	3	1	6	1	5	5	5	1	3	2	2	3	1	3	4
2	6	2	5	3	4	6	3	6	2	1	5	3	2	2	5

5. The following data represents the IQ scores for 60 randomly selected 5th graders. Construct a Frequency Distribution Table with at least 4 classes for the data below and graph the Histogram. Draw the Pie Chart for the same Frequency Table.

145	139	126	122	125	130	35	110	118	118
101	142	134	124	112	109	134	110	80	75
123	94	88	100	136	140	109	65	110	127
106	57	115	133	116	102	126	20	109	136
117	90	80	83	103	114	141	101	122	105
89	102	108	110	128	104	112	68	102	97

6. The data shown below represents the number of miles per gallon that 30 selected SUV's obtained in city driving. Construct the Frequency Distribution, and draw the Histogram.

15	15	18	25	9	8
7	5	10	11	24	20
19	20	14	16	13	12
11	10	8	4	9	10
11	15	17	15	20	17

7. The following data represents the IQ scores for 30 randomly selected 4th graders. Construct the Cumulative Frequency Table and the Pie Chart.

102	93	80	128	111	110	100	106	96	84
143	103	99	134	146	97	82	83	83	156
96	58	72	21	69	120	48	67	105	119

8. Construct the Pie Chart for the following qualitative data.

Military Spending	Social Security	Health Care	Welfare	Education
120	110	60	50	30

(a) Out of 300 eligible voters in a community, 160 are Democrats, 96 are Republicans, and 44 support other parties.

(b) The main items in the US budget for 1989 were in billions of dollars.

9. The following data set lists the time in minutes spent on Facebook in a two hour period for 40 randomly selected people at an internet café.

3	2	24	38	21	6	7	10
15	46	5	93	7	21	15	46
0	3	56	34	45	9	17	38
4	27	2	25	33	80	0	58
32	20	31	18	1	15	11	67

(a) Construct the Frequency Table with at least 4 classes and draw the Histogram and the Pie Chart.

(b) Construct the Relative Frequency Table and its Histogram.

(c) Construct the Cumulative Frequency Table and draw its Histogram.

(d) Draw the Frequency Polygon.

(e) Draw the Relative Frequency Polygon.

(f) Draw the Ogive.

10. The US Census Bureau published the number of businesses in certain fields in the state of California. The table below summarizes this information.

16	54	32	21	9	205	31	311	97	984
85	8543	3234	234	456	43	75	98	312	99
78	31	432	43	234	54	23	4	527	3361
6345	321	4654	4224	9743	432	43	174	318	713
846	81	728	24	434	64	153	483	11	8374

(a) Construct the Frequency Table with at least 4 classes, draw the Histogram and the Pie Chart.

(b) Construct the Relative Frequency Table and its Histogram.

(c) Construct the Cumulative Frequency Table and draw the Ogive.

(d) Draw the Frequency Polygon.

(e) Draw the Relative Frequency Polygon.

11. For the given frequency table

Classes	Frequency
0-40	54
40-80	39
80-120	21
120-160	37
160-200	11

(a) Draw the Frequency Polygon and the Ogive.

(b) Determine whether the distribution is approximately Normal, Uniform or Skewed.

12. Given the following 140 randomly selected Final Exam scores from several Statistics classes, construct the Frequency Table with 5 classes.

42	47	75	59	63	80	6	19	23	95
81	76	20	13	33	54	90	37	60	52
55	9	90	52	63	76	97	81	54	95
14	2	59	31	30	40	9	98	28	71
1	78	78	11	6	54	61	92	40	13
23	43	23	37	49	17	98	100	49	41
90	30	39	78	28	0	87	93	12	11
81	22	69	91	70	33	76	13	100	98
62	87	93	40	68	81	96	91	25	71
35	68	18	15	22	30	71	31	54	24
60	28	66	14	41	29	64	29	63	33
8	89	88	52	41	49	37	37	40	68
31	44	26	21	38	62	77	67	18	8
30	71	29	60	98	70	90	91	17	7

13. Given the following data set

1	4	1	6	3	2	1	2	2	4
2	6	3	5	5	5	1	1	6	6
4	5	3	3	4	5	5	2	2	3
3	2	3	5	2	2	5	5	4	1
2	6	1	3	1	4	1	2	6	1
5	5	6	6	5	6	5	1	2	5
3	5	2	6	6	4	6	3	6	5
2	4	2	6	3	6	6	3	3	4
2	3	1	1	4	6	6	6	2	5
1	2	1	5	1	2	5	4	5	6
3	1	2	6	6	5	3	6	2	6
5	4	2	5	5	1	6	5	2	3
3	3	5	3	4	5	4	4	3	3
1	6	1	5	5	5	1	3	2	2
3	1	3	4	2	6	2	5	3	4
6	3	6	2	1	5	3	2	2	5

(a) Construct the Single Column Frequency Table.

(b) Calculate all Relative Frequencies and Cumulative Frequencies on the same table.

(c) Draw the Histogram for the Frequency Table.

(d) Draw the Line Graph for the Frequency Table (called the Frequency Polygon).

(e) Draw the Relative Frequency Polygon.

(f) Draw the Ogive.

14. The following data represents the amount (in US dollar) of gratuity at Seaside restaurants for 33 customers.

5	20	3	0	25	10	13	14	0	4	23
26	0	5	30	10	50	35	14	16	13	0
2	7	3	19	21	0	21	0	100	45	8

(a) Arrange the above data in at least 4 classes. Find the Relative Frequency, and Cumulative Frequency.

(b) Construct the Stem and Leaf Plot.

15. Construct a Stem and Leaf graph for the following data set.

19	2	21	15	22	34	30	100
5	28	1	10	29	7	22	105
1	9	28	27	15	1	1	115
30	19	16	1	8	37	30	117
40	18	6	7	28	22	32	138
17	12	1	23	39	19	31	133
11	13	36	8	1	7	31	102

16. Given the following data, draw the Stem and Leaf Plot.

39	54	61	23	42	53	39	38	36	44	42	37
44	38	18	51	44	60	17	42	42	12	47	39

17. Construct a Stem and Leaf graph using the following data (administrator's salaries in thousands US dollars) and find the outliers if any.

127	41	110	60	61	64	65	65	111	67	67	69	112
70	72	100	73	132	76	77	78	79	80	81	81	115
83	138	86	87	88	89	91	92	93	95	113	111	95
96	98	99	100	101	102	103	105	106	108	109	153	90
117	80	118	120	121	125	127	129	131	135	139	140	89

18. Given the following frequency distribution, find the mean and standard deviation.

Classes	Frequency
10-20	4
20-30	7
30-40	15
40-50	17
50-60	8
60-70	3

19. Given the following frequency distribution, find the variance.

Classes	Frequency
0-20	5
20-40	5
40-60	10
60-80	29
80-100	12

20. The table below gives the altitude of the moon between 1:05 AM and 5:10 AM in Malibu, CA on a certain day (as collected by the U.S. Naval Observatory). Negative values are highlighted in yellow.

9.8	−7.1	34.2	−0.9	−4.8	29.8	4.6	30.5
10.5	26.4	31.6	23.3	30.3	32.4	13.6	23.4
13.7	24.4	22.2	31.1	−3.4	−5.7	19.6	−0.4
−7.8	32.1	21.8	28.7	24.9	9.2	−4.1	−6.6
20.2	33	28.9	31.5	27.7	21.5	15.8	18.5
−7.5	20.1	32.3	−8.5	−8.4	−5.3	18.3	−7.7

(a) Calculate the mean and the standard deviation of the above data set.

(b) Construct a Frequency Table using at least 4 classes, and then calculate the mean and standard deviation of the frequency table.

(c) Compare and explain the answers from part (a) and (b).

21. The following represents 30 rolls of a fair die.

3	2	6	3	1	5	3	5	4	2
2	2	6	3	5	1	4	3	1	1
6	5	2	4	2	1	5	3	1	4

(a) Set up a frequency table including a column of relative frequencies.

(b) Calculate the mean and the standard deviation of the frequency table.

22. The following table contains 99 randomly selected final exam grades from Prof. Mardirosian's Statistics class.

14	70	59	83	82	64	65	35	70	66	100
80	56	47	88	76	64	68	76	52	65	76
68	98	58	67	77	46	88	98	81	94	70
67	55	66	69	21	49	20	51	69	79	84
79	34	67	52	85	59	14	72	73	68	31
63	72	32	69	72	90	81	89	11	75	68
71	92	32	65	47	71	71	46	18	74	81
57	70	73	27	65	80	73	39	59	99	57
59	42	75	43	74	102	92	28	68	54	68

(a) Construct a Frequency Distribution Table with at least 4 classes. Draw the Histogram representing the distribution of data graphically. Label the x axis as "classes" and y axis as "frequency".

(b) Find the mean and standard deviation of the frequency table from part (a).

(c) Determine the range of usual and unusual grades for Prof. Mardirosian's Statistics classes based on the collected data.

(d) Would it be usual for a student to receive a grade of 96?

23. The following table contains 55 randomly selected salaries (in US dollar) of college graduates with 10 years of work experience.

30,273	70,089	83,006	96,180	117,621	41,500	72,252
84,571	98,116	118,148	59,027	72,717	86,677	99,596
120,788	60,442	73,960	87,008	100,951	121,725	61,896
74,908	88,708	101,951	125,022	64,488	76,140	89,676
102,197	127,491	65,239	77,472	91,755	103,811	129,684
65,954	78,036	92,367	105,013	131,844	66,711	79,767
93,722	106,091	135,460	67,175	80,972	95,751	108,404
139,452	69,057	81,166	113,247	109,956	141,367	

(a) Construct a Frequency Distribution Table with at least 4 classes. Draw the Histogram representing the distribution of data graphically. Label the x axis as "classes" and y axis as "frequency".

(b) Determine the range of usual and unusual salaries based on the collected data.

(c) Would it be usual for a professional to have a $200,000 income?

24. The table below gives the altitude of the sun between 6:10 a.m. and 1:20 p.m. in Los Angeles, CA on a certain day (as collected by the U.S. Naval Observatory).

11.6	14.8	8.4	22.1	6.5	2.7	31.3
4.8	30.6	33.7	33.3	20.7	26.9	12.3
33.3	31.3	29.8	16.3	1.4	6.5	8.2
17.8	23.4	27.9	33.7	32.5	19.3	0.8
13.2	32.9	32	32.5	4.6	9.9	24.6
31.9	3.1	25.8	28.9	33.5	33.5	32.9

(a) Construct a Relative Frequency Histogram for this data using at least five classes.

(b) Calculate the mean and the standard deviation for the frequency table you have constructed.

(c) Using the results in (b), determine the interval for usual and unusual values.

25. The following data (in %) set displays the first test results of 36 students taking a statistics class.

14	43	70	39	67	36	73	0	15
45	80	3	21	58	76	2	77	19
23	19	83	90	7	18	47	71	10
24	0	59	0	0	13	66	41	19

(a) Find the mean and standard deviation of the data, and determine the range of usual values.

(b) What would be the Z-Score for a score of 18?

26. Determine whether the gasoline price of $3.12 (per gallon) in a suburb of a city is usual or unusual where the mean price is $3.05 and the variance is $0.3.

27. At the Grocery store the mean charge for a customer is $51.83 and the variance is $29.32. Is it usual for a customer to pay $118.94 for groceries?

28. A random set of scores in Professor Badalian's Statistics class are given in the following table.

46	65	87	50	85	93	68	81	63
75	88	70	80	95	45	75	52	100
83	7	70	10	90	80	100	35	79
90	50	98	54	65	78	95	68	72

(a) Find the mean, standard deviation, and the range of usual and unusual values.

(b) What is the relative position of a test score of 53.

(c) Find Q_1, Q_3 and the median, and determine the range of outliers.

29. A random sample of 10 students had the following IQ's.

70	120	110	101	88	83	95	98	107	100

(a) Find the mean IQ score $(\bar{x})$.

(b) Verify that the sum of the variations from mean is zero. That is, $\sum(x-\bar{x}) = 0$.

30. Twenty-four applicants for an astronaut position in NASA are sent for a special JAR-FCL 3 Class 2 medical examination. During urine stick analysis they were tested for Leucocytes (white blood cells in a liter of blood $\times 10^9$) and the following results were observed.

4.1	5.3	4.7	9.1	8.7	6.3	7.1	5.9
6.8	9.7	5.5	7.2	8.9	4.8	5.2	7.3
8.6	6.7	5.9	8.1	6.9	6.9	7.3	8.5

(a) Find the mean and standard deviation for the raw data.
(b) Use the raw data to find the Q_1, Q_3 and the median.
(c) Find the range of usual and unusual values.
(d) Find the range of outliers.

31. Raymond's gasoline expense (in US dollar) for 18 randomly selected weeks during Spring 2010 was recorded.

49.32	52.13	25.43	59.33	43.00	69.05	56.95	64.00	64.00
47.32	20.00	69.43	74.54	30.00	17.34	61.32	34.32	40.00

(a) Find the mean, variance, standard deviation, and the range of usual and unusual values.

(b) Use the Range Rule of Thumb to estimate the standard deviation.

(c) What is the relative position of $43?

(d) Find Q_1, Q_3 and the median, and determine the range of outliers.

32. At a Learning Center, before applicants start the preparation course for IELTS (International English Language Testing System) Exam, they must take a practice test to determine their knowledge base on a scale of 1 to 9. The practice test results of twenty-one applicants were as follows.

3.2	1.3	2.5	4.1	2	6.4	1.5
4	1.2	1	1.9	2.1	3.4	2.1
2.3	1.7	1	3.6	7.1	5	1.6

(a) Find the 5-Number-Summary.

(b) Draw the Box Plot.

(c) Find P_{45}.

33. At a Plutonium Mine 18 workers that received an acute dose of radiation were sent to a treatment center. The following table displays the radiation indicator device readings (in rad) for those infected miners.

13	40	21	19	14	38	69	43	73
37	10	11	27	32	32	41	25	16

(a) Find the 5-Number-Summary.
(b) Draw the Box-Plot.
(c) Find P_{58}.

34. Use the following data from a previous exercise and answer the questions below.

49.32	52.13	25.43	59.33	43.00	69.05	56.95	64.00	64.00
47.32	20.00	69.43	74.54	30.00	17.34	61.32	34.32	40.00

(a) Find P_{87}.

(b) Find the percentile for data value $30.00.

35. At an Audi dealership, a financial analyst decided to work on a declined customer application for financing a vehicle. In order to get a feel for the type of credit score that is needed to be approved by this particular financial agency, he looked at twenty four approved applications with following credit scores. What value corresponds to P_{85}?

625	589	685	673	682	699
562	689	701	600	711	654
591	684	667	690	588	703

36. The following data represents the number of years 27 randomly selected criminals were convicted at a courthouse in Los Angeles.

7	0	4	5	4.5	0	2	0	4
1	9	0	25	8	15	0	2.5	0
3.5	6	1.5	2	8	4.5	30	5	5

(a) What percentile corresponds to $x = 6$? **(b)** Find P_{90}.

37. Given the following data;

70	55	60	37	65	26	30	29	91	77	40
88	31	43	57	22	41	86	35	72	53	32
79	26	73	56	39	65	84	23	19	66	56
32	17	38	90	25	53	45	87	19	20	87

(a) Find P_{32}. **(b)** Find P_{79}. **(c)** Find P_5.

(d) Find the mean, median, Q_1, Q_3, variance, standard deviation, range of usual values, outliers, and all the *Z*-Scores (construct a corresponding table for all the *Z*-Scores).

38. At a Korean BBQ Restaurant, the wait times for 32 randomly selected customers was recorded by the manager who had recently received complaints. The following data shows the wait time (in minutes). Find P_{34}.

7	4	9	1	7	4	8	15	23	7	2
8	19	11	6	10	9	12	5	2	14	3
7	9	20	11	5	7	5	5	3	8	

39. The following data represents the time (in minutes) it took for 60 randomly selected customers to get their transactions processed from the time they walked in the bank to the time they left.

18	7	5	25	19	7	8	11	13	12
9	8	7	6	9	10	10	19	27	5
9	19	23	24	21	8	8	17	16	14
12	12	9	3	38	19	16	15	12	10
28	35	36	9	8	8	8	16	15	15
24	20	16	13	15	15	12	10	9	20

(a) Construct a Stem and Leaf Plot.

(b) Calculate the mean, variance, and standard deviation for the raw data.

(c) Calculate the mean, variance, and standard deviation for the grouped frequency table.

(d) Find the range of the usual and unusual values.

(e) Find P_{95}.

(f) Calculate the 5-Number Summary.

40. The National Climatic Data Center collected air pressure in a certain city in January, 2001. The following table gives measurements that some of the stations obtained.

62	36	61	27	61	31	56	61	59	62
60	59	61	62	61	125	0	124	93	5
58	1	120	124	29	5	7	0	40	124
62	124	124	61	55	62	61	33	60	60

Find the mean, median, standard deviation, variance, Q_1, Q_3, P_{26}, P_{95}, and the range of outliers.

41. Given the following data, construct the 5-Number Summary and draw the Box Plot.

20	32	42	26	37	39	26	23	9	15	45	62

42. Given the following data, construct the 5-Number Summary and draw the Box Plot.

85	15	14	25	69	7	8	75	88	7	76
36	45	31	15	11	35	8	23	41	52	1
45	15	25	89	6	30	24	18	6	5	9
50	47	9	92	100	30	44	65	12		

43. Find the mean, standard deviation and the 5-Number Summary for the following data. Draw the Box Plot.

23	13	23	16	24	23	15	25	29	32
20	25	19	26	22	28	17	28	16	28

44. Given that the mean IQ is 100 with a standard deviation of 15, use Chebyshev's Theorem to find the minimum percentage of IQs that will fall between 70 to 130.

45. A sample of adjunct faculty monthly salary has a mean of \$2500 with a standard deviation of \$300. Using Chebyshev's Theorem, find the range in which at least 80% of the salaries will fall.

46. The IQ of the author's college students is normally distributed with a mean of 100 and standard deviation of 15. What percentage of college students have IQs between 70 to 130? **(Use the empirical rule to solve the problem)**

47. At a high school, GPA's are normally distributed with a mean of 2.6 and a standard deviation of 0.5. What percentage of students at the college have a GPA between 2.1 and 3.1?

48. Draw the % frequency bar graph of the following discrete data.

X	1	2	3	4	5	6
Frequency (in %)	18	15	17	16	15	19

Assuming the percentages represent the heights of the bars, calculate the total area under the histogram using areas of rectangles. Use 18 to represent 18%, *etc.* (***Hint:*** You are basically finding areas of rectangles). What does this exercise tell you about the relationship between percentages and areas?

3

COUNTING AND PROBABILITY THEORY

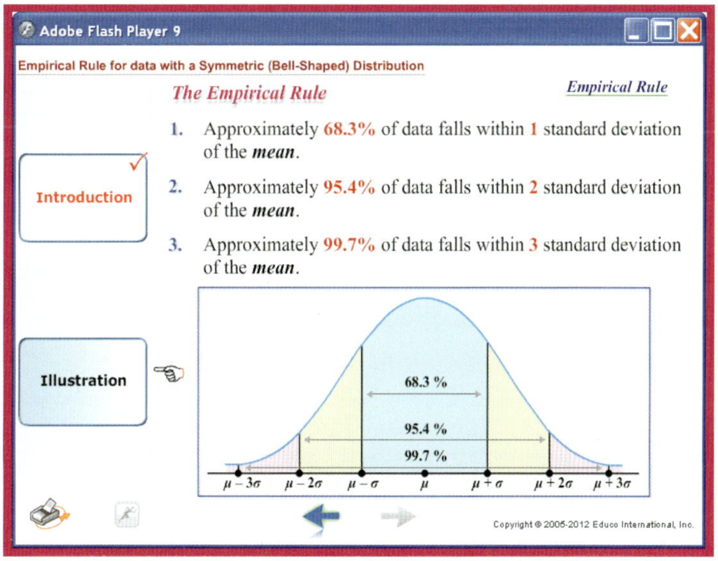

CHAPTER 3

COUNTING AND PROBABILITY THEORY

Chapter Overview:

3.1: INTRODUCTION

The earliest analytical work on the theory of probability arose from gambling problems. It dates back to the 15th century when several works on probability emerged. In 1494 Fra Luca Paccioli authored the first printed book on probability. Inspired by this book, Geromimo Cardano wrote a book on games of chances in 1550.

In 1654, a French nobleman and gambler named Chevalier de Méré created a dice game which he believed would make him money. In his dice game, he would bet even money that he could roll at least one 12 in 24 rolls of two fair dice. Unfortunately, his gambling scheme was a losing proposition. When he began losing money, he contacted his mathematician friend Blaise Pascal (1623-1662) to help him discover the flaw in the game. Pascal determined that in this game of dice, he would lose 51% of the time. Inspired by this game, Pascal began a more detailed study of similar games and concepts. This analysis would lead Pascal to collaborate with another famous French mathematician Pierre de Fermat (1601-1665) and together they would start to lay the foundation of **Modern Probability Theory.**

Several other mathematicians continued the work of Pascal and Fermat in the 17th and 18th centuries. Among them were Christian Huygens (1629-1695), Jacob Bernoulli (1654-1705), Abraham de Moivre (1667-1754) and Pafnuty Lvovich Chebyshev (1821-1894), to mention a few.

Probability is the framework in which the validity of statistical analysis about a population may be examined. To understand statistical methodology, an understanding of the basic laws of probability is imperative. In this chapter we will discuss the basics of Probability Theory, a branch of mathematics that has become a significant part of research in social, natural and hard sciences. We will also introduce Microsoft Excel® commands that will enable us to generate simulations and facilitate the tedium of mathematical calculations.

Since the formal definition of probability is based on the basic concepts from Set Theory, we will start our discussion with an overview of the fundamentals of the theory of sets. We will then proceed to look at combinatorics (counting techniques), and continue the chapter with an Introduction to Probability Theory, Addition Rule, and Multiplication Rule.

3.2: A Brief Review of Set Theory

Objectives

After completing this section you will be able to understand:

A. Important Definition in Set Theory.

B. Set Operations.

A. Important Definition in Set Theory

A set is a well defined collection of objects or elements. Capital letters are used to label sets, and braces are used to display them.

The set of integers between 0 and 10 inclusive is written as $A = \{0, 1, 2, 3, 4, 5, 6, 7, 8, 9, 10\}$. The number of elements in set A is written as $n(A)$. Here $n(A) = 11$.

The set consisting of a triangle, a rectangle, a circle and a parallelogram, all well defined, is written as $E = \{\triangle, \square, \odot, ▱\}$.

- **Universal Set:**

  Collection of all possible well defined elements is called universal set. It is denoted by U.

- **Empty Set:**

  A set with no elements is called the empty set and is labeled as $\varnothing$.

- **Venn Diagrams:**

  Venn diagrams are graphical representation of sets and their operations.

- **Disjoint Sets (Mutually Exclusive sets):**

  Disjoint sets are sets with no common element. In Figure 3.1, sets A and B are disjoint.

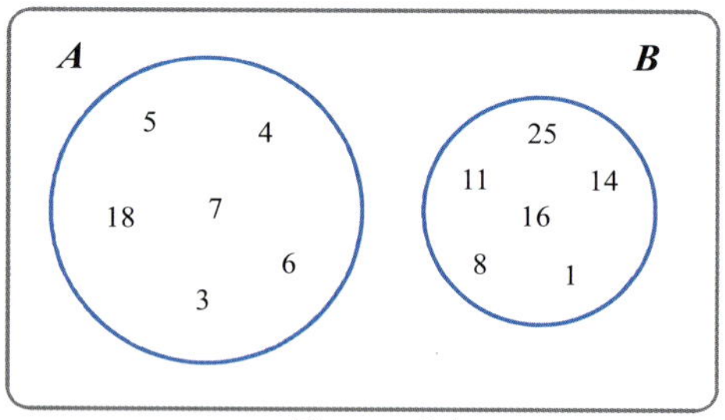

Figure 3.1

Note: We will only deal with finite sets *i.e.*, for any set A, $n(A)$ is finite.

- **Subset:**

  If every element in set A is also an element of set B, then set A is called a subset of set $B (A \subseteq B)$. If $A \subseteq B$ and $A \neq B$, then A is said to be a proper subset of B and is denoted by $A \subset B$. In other words, set B contains at least one element that is not in A.

  If $A \subseteq B$ then $n(A) \leq n(B)$ and if $A \subset B$ then $n(A) < n(B)$.

  (This is true only for finite sets not in general.)

  For example, if $U = \{1, 2, 3, 4, 6, 7, 8, 9, 12, 15, 26\}$, $B = \{1, 2, 3, 6, 7, 8, 12, 15\}$ and $A = \{1, 3, 12\}$, then we can say that $A \subset B$. (Figure 3.2)

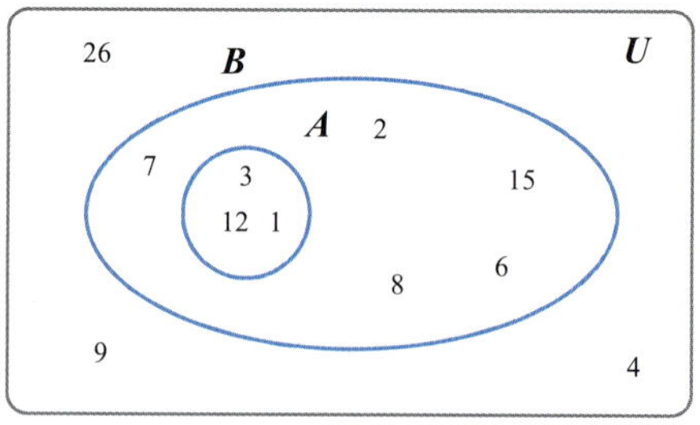

Figure 3.2

Note: Every set is a subset of itself. The empty set is a subset of all sets.

Example 1: Given a set $A = \{1, 2, 3, 4, 5\}$, list all the possible subsets of set A.

Solution:

{1}	{2}	{3}	{4}	{5}	{1,2}	{1,3}	{1,4}
{1,5}	{2,3}	{2,4}	{2,5}	{3,4}	{3,5}	{4,5}	{1,2,3}
{1,2,4}	{1,2,5}	{1,3,4}	{1,3,5}	{1,4,5}	{2,3,4}	{2,3,5}	{2,4,5}
{3,4,5}	{1,2,3,4}	{1,2,3,5}	{1,2,4,5}	{1,3,4,5}	{2,3,4,5}	{1,2,3,4,5}	$\varnothing$

- **Power Set:**

  The set of all subsets of a set A is called the power set of A. The total number of subsets of a set A is calculated by using the formula 2^n, where n is number of elements in A. In the above example, the solution is called the power set of set A.

- **Complement of Set *A*:**

  Complement of set A is the set of all elements in the universal set that are not in A. We denote the complements of a set using bar notation. $\bar{A}$ is called the complement of A.

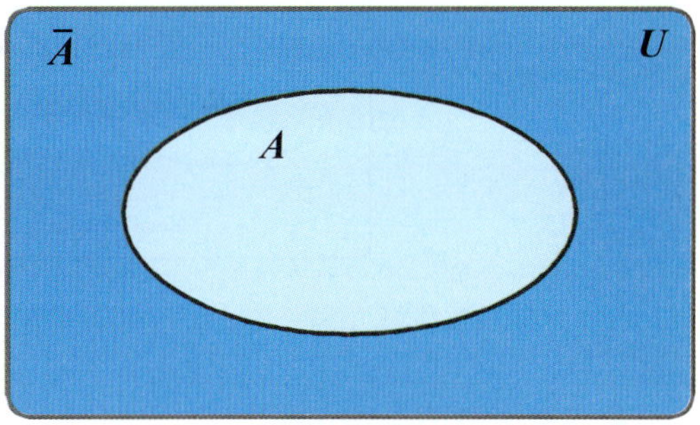

Figure 3.3

Note: i) $\bar{U} = \varnothing$ and $\bar{\varnothing} = U$. Recall that U represents the universal set.

ii) $\bar{A} = U - A$.

iii) $n(\bar{A}) + n(A) = n(U) \Rightarrow n(\bar{A}) = n(U) - n(A)$

Example 2: Given $U = \{1, 2, 3, 4, 5, 36, 58\}$, and $A = \{1, 2, 3, 4, 58\}$, find $\bar{A}$.

Solution: Since we know that, $\bar{A} = U - A$ then $\bar{A} = \{5, 36\}$.

Try it yourself:

1. Find all the subsets for the set $E = \{1, 3, 5, 7\}$.
2. List the elements in the set $\{x \mid x$ is a positive integer less the $8\}$.
3. $U = \{23, 24, 25, 26, 27, 28, 29, 30\}$, $A = \{23, 24, 25, 26\}$. Find $\bar{A}$.

B. Set Operations

- **Union of two Sets:**

  The union of two sets A and B is the collection of elements which are in A or in B or in both. To symbolically demonstrate the union of two sets, we will use a set builder notation;

$$A \cup B = \{x : x \in A \text{ or } x \in B\}$$

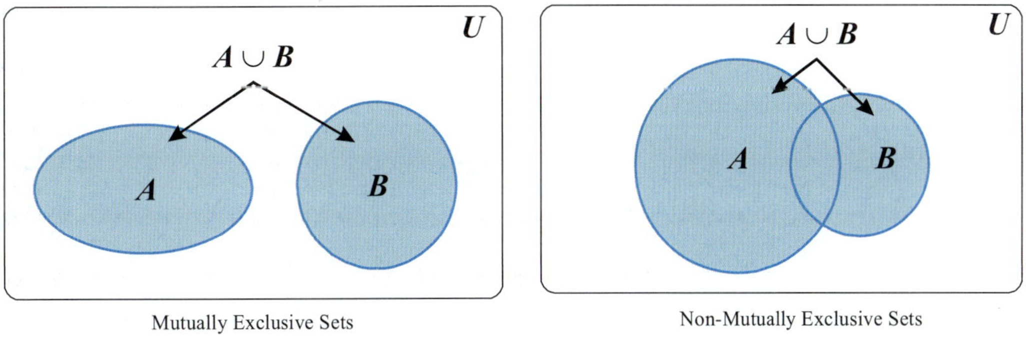

Figure 3.4

Note: i) In the figures above, the color blue represents the union of the two sets.

ii) Here "or" means "either ... or" or "both".

- ## Intersection of two Sets:

The intersection of two sets A and B is the set of elements that are common to both A and B.

$$A \cap B = \{x : x \in A \text{ and } x \in B\}$$

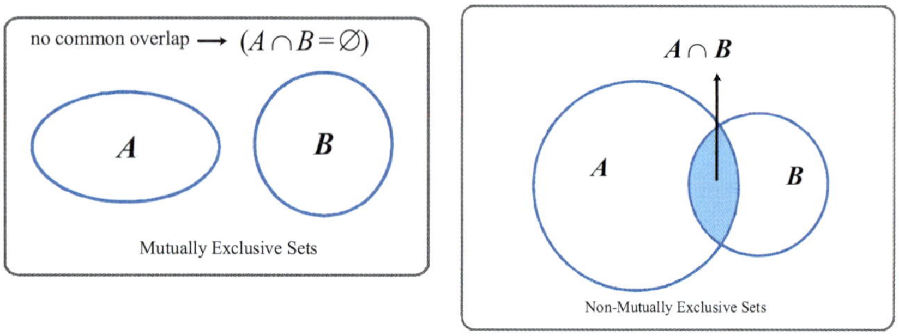

Figure 3.5

- ## Set Difference:

The difference of two sets A and B is written as $A - B$ (or $A \cap B$) and is the set of all elements of A that are not in B.

Example 3: Given $A = \{1, 2, 3\}$ and $B = \{5, 6, 7\}$, find $A \cup B$, and $A \cap B$.

Solution: $A \cup B = \{1, 2, 3, 5, 6, 7\}$

$A \cap B = \varnothing = \{\ \}$

Note that $n(A) + n(B) = 3 + 3 = 6 = n(A \cup B)$. This is not a coincidence; in fact the following formula holds when A and B are mutually exclusive.

$$n(A \cup B) = n(A) + n(B)$$

Example 4: Given $A = \{1, 2, 3, 4, 5\}$ and $B = \{4, 5, 6, 7, 8, 9, 10\}$, find $A \cup B$, and $A \cap B$.

Solution: $A \cup B = \{1, 2, 3, 4, 5, 6, 7, 8, 9, 10\}$

$A \cap B = \{4, 5\}$

Note that $n(A) + n(B) - n(A \cap B) = 5 + 7 - 2 = 10 = n(A \cup B)$

In fact if A and B are not mutually exclusive then

$$n(A \cup B) = n(A) + n(B) - n(A \cap B)$$

Try it yourself:

1. Given the following Venn Diagram, find

(a) $A \cup B$. **(b)** $A \cap B$.

(c) $\overline{A}$. **(d)** $\overline{A} \cap B$.

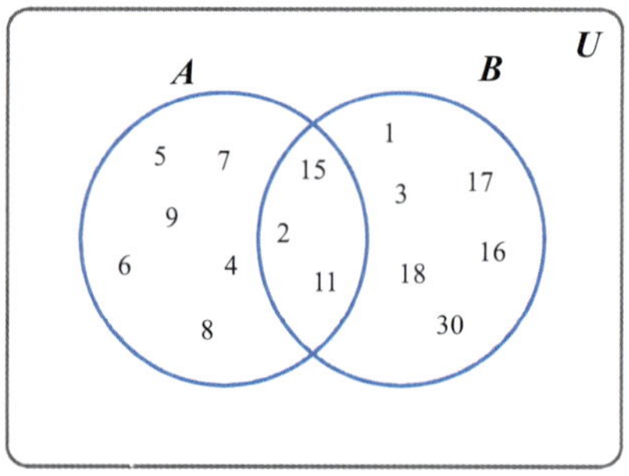

3.3: Counting Techniques (Combinatorics)

Counting techniques are absolutely essential in calculating probabilities. As we will see in the next section, the definition of classical probability is based on the ratio of the number of elements of a set A to the number of elements in the universal set. Therefore, to understand the theory of probability, it is necessary to be familiar with counting techniques known as **Combinatorics**. In this section we will look at the use of Tree-Diagrams to list all possible outcomes of experiments, and fundamentals of **Counting Techniques.**

Objectives

After completing this section you will be able to use:

- **A.** Fundamental Rule of Counting.
- **B.** Tree Diagrams.
- **C.** Factorial Rule.
- **D.** Arrangements and Combinations.

A. Fundamental Rule of Counting (The *MN* Rule)

If one operation can be performed in "m" different ways and another operation can be performed in "n" different ways, then two operations when associated together can be performed in $m \times n$ distinct ways.

Alternative Definition: *If event A has m possible outcomes, and event B has n possible outcomes, then the joint event A and B has $m \times n$ possible outcomes.*

Example 5: There are six routes to travel from A to B and seven routes to travel from B to C. Find the total number of routes one can travel from A to C, via B.

Solution:

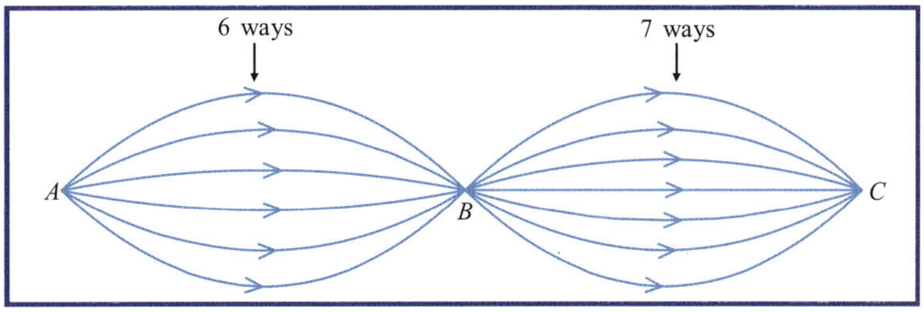

Figure 3.6

Total number of routes from A to C through B $= 6 \times 7 = 42$

Example 6: Suppose you can order a computer in 3 different types with 4 different configurations. How many options are available for you to order a computer?

Solution: In, $3 \times 4 = 12$ different ways you can order a computer.

The $m \times n$ rule can be generalized to more than two operations. Suppose an experiment is performed in k stages with n_1 ways to accomplish the first stage, n_2 ways to accomplish the second stage, and so on. Then the total number of ways to accomplish the experiment is

$n_1 \times n_2 \times n_3 \times \ldots \times n_k$ many ways

Example 7: In how many ways can 5 marbles be distributed among 3 boxes?

Solution: Any one of the 5 marbles can be placed in the first box, any one of the remaining 4 marbles can be placed in the second box, and the 3 remaining marbles can be placed in the third box.

Hence $5 \times 4 \times 3 = 60$ different ways.

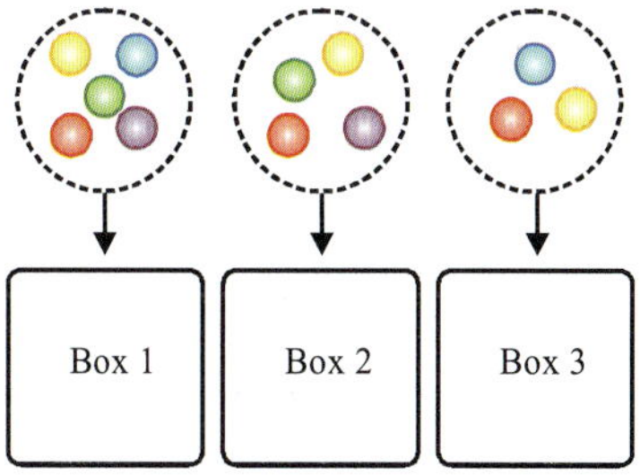

Figure 3.7

Example 8: With 10 consecutive on/off switches, how many different settings are possible?

Solution:

2	2	2	2	2	2	2	2	2	2

There are 10 possible available slots with 2 choices for each one. Therefore, total different possible setting are = $2 \times 2 \times 2 \times 2 \times 2 \times 2 \times 2 \times 2 \times 2 \times 2 = 2^{10} = 1{,}024$.

Example 9: There are 6 seats around a table. In how many ways can 3 friends be seated?

Solution: The first person approaching the table can choose any one of the available six seats. Then the second one can choose any one of the remaining five seats, and the third person can choose any one of the four seats left. According to the ***MN* rule**, the total number of possible configurations are $6 \times 5 \times 4 = 120$ ways.

Example 10: How many different Social Security number combinations are possible if the digits can repeat?

Solution:

10	10	10	10	10	10	10	10	10

Each one of the nine slots can take on 10 different digits. Hence number of different combinations = $10 \times 10 \times 10 \times 10 \times 10 \times 10 \times 10 \times 10 \times 10 = 10^9 = 1{,}000{,}000{,}000$

Example 11: How many gender sequences are possible when a couple plans to have 5 children?

Solution:

2	2	2	2	2

There are 5 children with 2 possible genders. Therefore, using the *MN* Rule, we have = $2 \times 2 \times 2 \times 2 \times 2 = 2^5 = 32$ possible gender sequences.

Example 12: How many different arrangements of 5 children are possible?

Solution: $5 \times 4 \times 3 \times 2 \times 1 = 120$ different arrangements.

Example 13: A table tennis team has 5 members. In national competitions, the coach wants the 1st, 2nd, and 3rd places to go to her team members. In how many ways can the 5 team members take on the 1st, 2nd, and 3rd places?

Solution: $5 \times 4 \times 3 = 60$ ways.

The 1st place can be taken by any one of the 5 members. So there are 5 ways for that. The 2nd place can now be taken in 4 ways and the 3rd in 3 ways. Hence total number of ways $5 \times 4 \times 3 = 60$ ways.

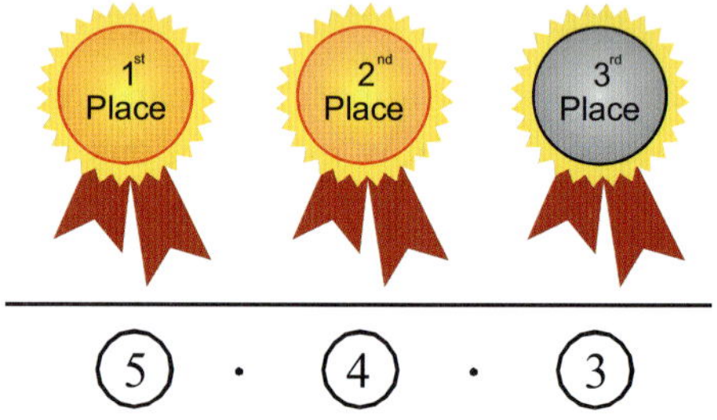

Figure 3.8

Example 14: In how many ways can the above 5 team members be arranged in a van with 5 seats?

Solution: $5 \times 4 \times 3 \times 2 \times 1 = 120$ different arrangements.

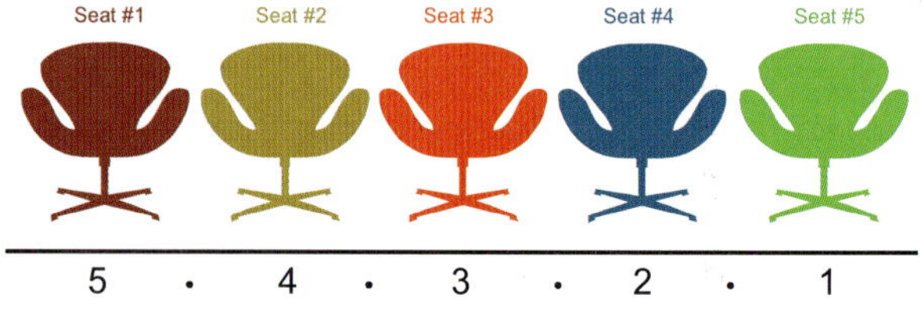

Figure 3.9

Try it yourself:

1. Diners at Fringe Pizza may order a pizza with 3 different crusts, 2 sauces, and 15 different toppings. How many different orders can a customer place?
2. How many different ways can first and second places be awarded in a race with 8 contestants?
3. How many 8 character passwords can be created, if you can use letters and numbers, and **(a)** they cannot be repeated? **(b)** they can be repeated?
4. A yogurt store has six flavors of yogurt. How many ways can four customers order one flavor each?
5. Robert may select a desktop from a group of seven, and a laptop from a group of 5. How many ways can he select a desktop and a laptop?

B. Tree Diagrams

Sometimes calculating the total number of outcomes is not enough, and it is necessary to list all possible outcomes in the experiment. If the number of outcomes in an experiment is small (e.g., flipping a coin or rolling a die), it is quite simple to just list the outcomes. However, as experiments get more involved and complicated, it becomes difficult to list all their possible outcomes by guessing alone. Rather than trying to list all the outcomes in an experiment in a haphazard way, one can simply use tree diagrams to do the job systematically. A **Tree Diagram** is a device used to list all the outcomes of experiments in a systematic way. As we will see in next sections, tree diagrams are also quite useful in determining the chances of complicated events.

Example 15: Suppose a tourist can travel from Germany to France by plane, car, or train, and from France to England by car, boat, or plane. List all possible ways she can travel from Germany to England, by using tree diagram.

Solution:

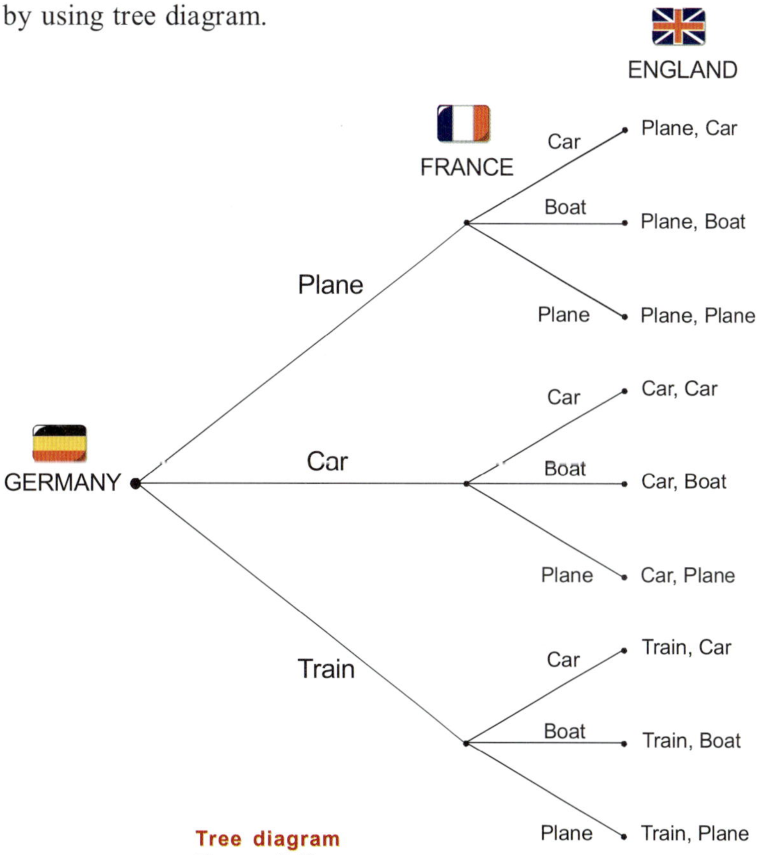

Tree diagram
Figure 3.10

Example 16: A couple plans to have three children.

(a) How many gender sequences are possible?

(b) List all the possible gender sequences using a Tree Diagram.

Solutions: **(a)** Using the *MN* Rule, total number of gender sequences for 3 children is $2 \times 2 \times 2 = 8$.

(b) The tree diagram listing all gender sequences for 3 children is shown in adjoining Figure 3.11.

$$S = \{\text{BBB,BBG,BGB,BGG,GBB,GBG,GGB,GGG}\}$$

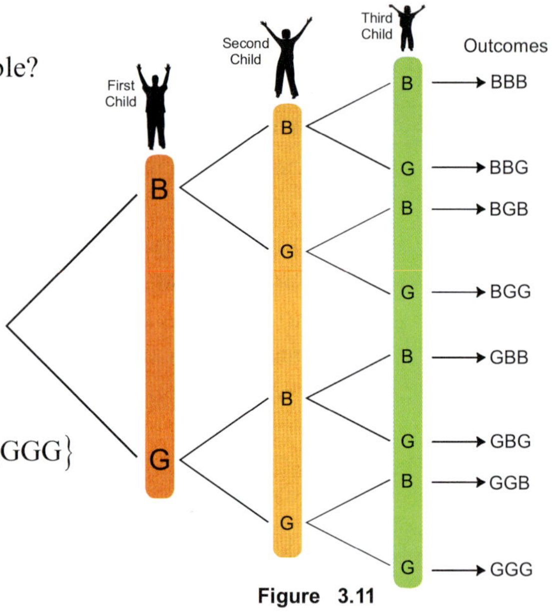

Figure 3.11

Example 17: You have been invited to Raymond's restaurant called International Fusion. There is a pre-set 3 course meal.

(a) How many different choices do you have for your 3-course meal?

(b) Construct the Tree diagram for all your choices.

Solutions: **(a)** Total number of choices $= 4 \times 4 \times 3 = 48$

(b)

Figure 3.12

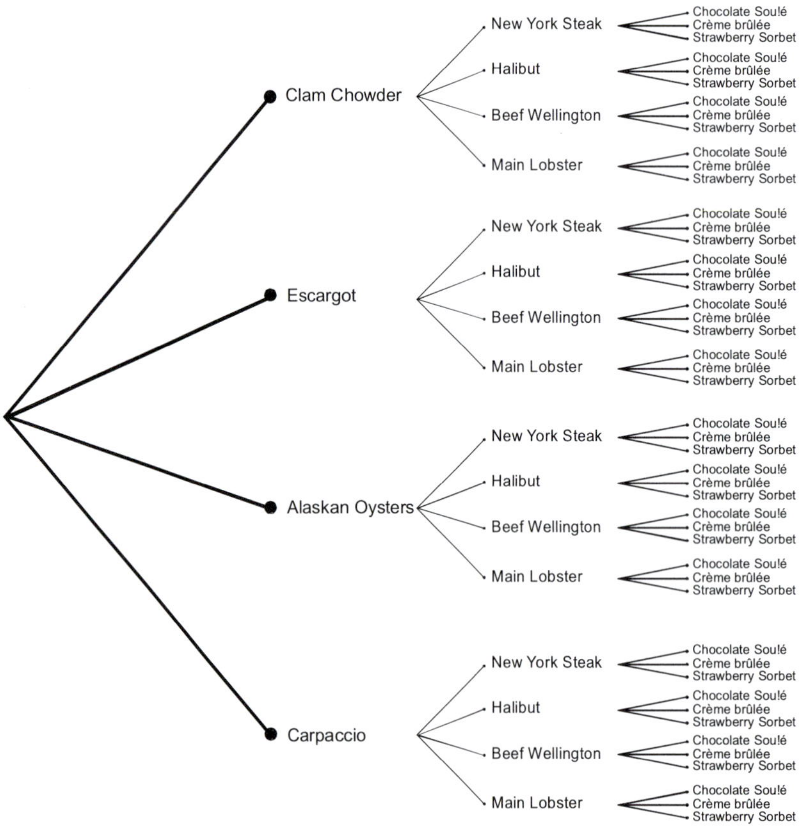

Figure 3.13

Example 18: A BMW dealership offers 5 model cars in 5 colors, 3 transmission options, and 2 packages (Premium and Sports). How many choices does a customer have to buy a car?

Solution:

5	5	3	2

$5 \times 5 \times 3 \times 2 = 150$ *different combinations.*

We can use a tree diagram to list all the possible 150 combinations.

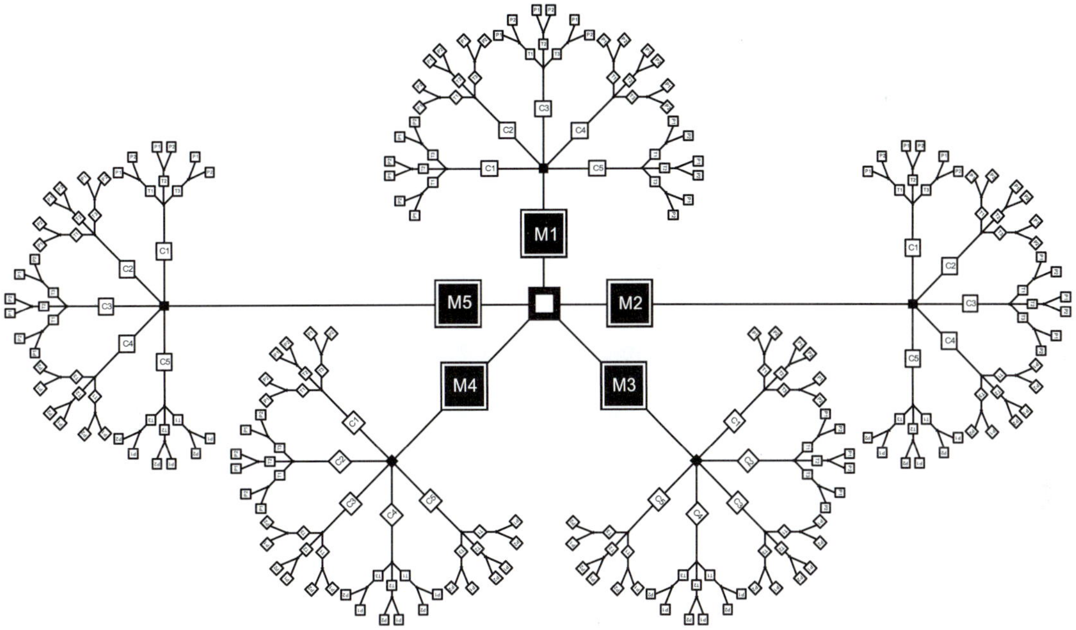

Figure 3.14

Try it yourself:

1. How many different 6 character passwords can be created using letters and digits with repetition?
2. How many different 6 character passwords can be created using letters and digits without repetition?
3. What is the total number of gender sequences when a couple plans to have 8 children?
4. In how many ways can 5 students take on the first, second, and third places in AMATYC (American Mathematical Association of Two-Year Colleges) contest?
5. In how many different ways can a blood donor have his or her blood labeled? Recall that there are 4 blood types, A, B, AB, and O. Blood can also be Rh^+ and Rh^-. Construct the tree Diagram for this experiment.
6. A die is rolled once, then a coin is flipped. List all the outcomes using a tree diagram.
7. A coin is tossed 4 times. List all the outcomes using a tree diagram.

C. Factorial Rule

In the *MN* Rule when the number of items to be arranged (n) is equal to the number of positions (k) (similar to Example 10), we can use a new mathematical function called Factorials to carry out the multiplication process.

The **factorial symbol "!"** represents a specific product of decreasing positive whole numbers. For example, $5! = 5 \times 4 \times 3 \times 2 \times 1 = 120$. In general,

$$n! = n \times (n-1) \times (n-2) \times \ldots \times 3 \times 2 \times 1$$

By special definition $0! = 1$.

Note: The Factorial Rule is a special case of the *MN* Rule.

A collection of "n" different items can be arranged in $n!$ different ways. In other words, the total number of arrangements of "n" objects (items), in "n" positions is $n!$.

Example 19: How many different 6 letter words can be created from the following letters?

"S, W, A, Y, T, B"

Solution: $6! = 6 \times 5 \times 4 \times 3 \times 2 \times 1 = 720$ distinct words.

Example 20: In how many ways can you travel to 7 distinct cities?

Solution: Any of the seven cities can be placed in the first position. The second position can take on any of the remaining six cities, and so on. Therefore, the total number of possible travel arrangements are $7! = 7 \times 6 \times 5 \times 4 \times 3 \times 2 \times 1 = 5{,}040$.

Notice that the factorial command in excel calculates the product much faster than multiplying the numbers one by one as done in a typical *MN* rule.

Example 21: Four cars can be painted in four different colors (red, white, blue, and yellow). How many different color configurations are possible?

Solution: $4! = 24$

Example 22: In how many ways can 100 senators be arranged in the senate?

Solution: The arrangement of 100 senators in the senate is equivalent to the arrangement of 100 distinct items in a row. That is 100!.

Try it yourself:

1. In how many ways can 6 athletes take on 6 positions?
2. A traveler wants to visit 5 cities. Determine the number of different routes he may visit all 5 cities.

D. Arrangements and Combinations of Distinct Objects

So far, we have looked at the *MN* Rule and the Factorial Rule as two fundamental Counting Techniques. Recall that the Factorial Rule was the total number of arrangements of "n" objects into "k" positions, where $n = k$.

What if we want to find the number of possibilities of organizing "n" objects into "k" positions with or without order, where $k < n$. The Mathematical functions that carry out these types of operations are called **Arrangements (Permutation) and Combinations.**

- ## Arrangements (Permutations):

An Arrangement (Permutation) of "n" distinct objects taken "k" at a time is an **ordered** arrangement of "k" objects at a time out of "n" objects.

In Arrangements order is important. Therefore, in the following set the number of arrangements is 3 even though all 3 outcomes are made out of the same letters.

{ABC, CBA, BAC}

The number of arrangements (permutations) of n distinct objects taken k at a time is denoted by A_k^n and its value is calculated by the following formulas:

(a) $A_k^n = n(n-1)(n-2)\ldots(n-k+1)$ For example, $A_3^5 = 5 \times 4 \times 3 = 60$

(b) $A_k^n = \dfrac{n!}{(n-k)!}$ For example, $A_3^5 = \dfrac{5!}{(5-3)!} = \dfrac{120}{2} = 60$

Example 23: How many different 4 digit numbers can be formed from the digits 1, 2, 3, 4, 5, and 6 if

(a) repetitions are not allowed? **(b)** repetitions are allowed?

Solutions: **(a)** If repetitions are not allowed, there are 6 choices for the first digit, 5 choices for the second digit, 4 choices for the third digit, and 3 choices for the fourth digit. Hence, the number of arrangements is $A_4^6 = 6 \times 5 \times 4 \times 3 = 360$.

Note: Order is important.

(b) If repetitions are allowed, there are 6 choices for each of the four digits. Using the *MN* Rule there are $6 \times 6 \times 6 \times 6 = 1{,}296$ four-digit numbers.

Note: Part "b" is not an arrangement so the order is irrelevant.

Example 24: From a group of 10 committee members, a chairperson, a vice chairperson and a secretary should be selected. How many different slates (arrangements) of candidates are possible?

Solution: Because order does count with the slates of candidates, we want the number of arrangements of $r = 3$ people (groups of size 3) selected from the $n = 10$ available members. Therefore,

$$A_k^n = A_3^{10} = \frac{10!}{(10-3)!} = \frac{10!}{7!} = \frac{10 \times 9 \times 8 \times 7!}{7!} = 10 \times 9 \times 8 = 720$$

- ## Special Case for Arrangements:

i) Arrangements of "n" distinct objects taken all at a time

The total number of arrangements of "n" distinct objects taken all at a time is given by:

$$A_n^n = n(n-1)\cdots(4)(3)(2)(1) = n! \text{ or } A_n^n = \frac{n!}{(n-n)!} = n!$$

ii) Circular Arrangements

The total number of arrangements of "n" distinct objects in a circle is $(n-1)!$.

Example 25: How many ways 6 students can be seated on six seats **(a)** in a line and **(b)** around a table?

Solutions: **(a)** All six students can be arranged in a line in 6! = 720 ways.

(b) Since students are to be arranged around a table, therefore we use the formula for circular permutations and required arrangements are $5! = 5 \cdot 4 \cdot 3 \cdot 2 \cdot 1 = 120$ ways

Note: The arrangement rule turns into the factorial rule when $n = k$.

iii) Arrangements of Objects not All Distinct (Repeated Factorials)

The number of arrangements of "n" objects taken all at a time is given by formula

$$\frac{n!}{n_1!\, n_2! \cdots n_k!}$$

when "n_1" objects are alike of one kind, "n_2" objects are alike of second kind, ..., "n_k" objects are alike of kth kind.

Example 26: Find the total number of arrangements (permutations) of the letters of the word ATLANTA taken all at a time.

Solution: Since in ATLANTA there are 7 letters, out of which "A" repeats 3 times, "T" repeats twice and the rest are all different, occurring once, the total number of arrangements is

$$\frac{7!}{3! \times 2!} = \frac{7 \times 6 \times 5 \times 4 \times 3 \times 2 \times 1}{3 \times 2 \times 2} = 420$$

Try it yourself:

1. A statistics author wants to write several books on Probability, Statistics, and Computer Methods in Statistics respectively. If he has 8 assistants to select from, how many ways can he assign a different assistant to each book? (Arrangement)
2. A list of 12 topics for a semester project is given to a class. How many ways can 5 students select different topics?
3. How many distinct "words" can be formed using all the letters of EXERCISES with or without meaning?
4. At the International Conference in Technology in Las Vegas, 16 faculty from LA City College are scheduled to speak. 4 are from the Math Department, 7 are from the English department, and 5 are Computer Science Professors. How many orders of speaking are possible?

- ## Combinations:

A combination of "n" different (distinct) objects taken "k" at a time is a selection of only "k" objects out of "n" objects without any regard to the order.

In Combinations, order is not important. Therefore, in the following set the number of combinations is 1.

{ABC, CBA, BAC}

The number of combinations of n objects taken k at a time is denoted by C_k^n and is calculated by the formulas:

(a) $C_k^n = \dfrac{n!}{k!(n-k)!}$, or **(b)** $C_k^n = \dfrac{n(n-1)(n-2)\cdots(n-k+1)}{1 \cdot 2 \cdot 3 \cdots k}$

Alternative notation for combinations: C_k^n are ${}_nC_k$, $C(n, k)$ and $\binom{n}{k}$.

Example 27: Calculate C_2^6 using formula "a", and C_3^{50} using formula "b".

Solutions: **(a)** $C_2^6 = \frac{6!}{2!4!} = \frac{720}{48} = 15$ **(b)** $C_3^{50} = \frac{50 \times 49 \times 48}{1 \times 2 \times 3} = 19{,}600$

***Remarks*:**

1. Use the formula (b) if ***n*** is large, but ***k*** is not.

2. $C_k^n = C_{n-k}^n$. Use this formula, if both ***n*** and ***k*** are large. For example,

$$C_{98}^{100} = C_{100-98}^{100} = C_2^{100} = \frac{100 \times 99}{1 \times 2} = 4{,}950.$$

By special definition $C_0^n = 1$.

Example 28: A box contains 10 distinct rocks. In how many ways can 3 rocks be drawn from the box?

Solution: Since order is not important, the total number of combinations is

$$C_3^{10} = \frac{10 \times 9 \times 8}{1 \times 2 \times 3} = 120 \text{ ways.}$$

Example 29: How many 5 member FBI special operation teams can be selected from 40 original available agents?

Solution: A five member team can be selected in

$$C_5^{40} = \binom{40}{5} = \frac{40 \times 39 \times 38 \times 37 \times 36}{1 \times 2 \times 3 \times 4 \times 5} = 658{,}008 \text{ ways.}$$

Example 30: Professor Badalian grades homework by choosing 6 out of 10 homework problems. How many different groups of problems of size 6 can he choose from the original 10 problems?

Solution: The total number of combinations of 10 items taken 6 at a time is

$$C_6^{10} = C_{10-6}^{10} = C_4^{10} = \frac{10 \times 9 \times 8 \times 7}{1 \times 2 \times 3 \times 4} = 210$$

Example 31: In a 5 card poker game, what is the total number of possible hands?

Solution:

$$\binom{52}{5} = \frac{52 \times 51 \times 50 \times 49 \times 48}{5 \times 4 \times 3 \times 2 \times 1} = 2{,}598{,}960$$

Example 32: Four cards are drawn from a standard deck of 52 playing cards. What is the number of combinations that two red and two black cards can be selected using the *MN* rule?

Solution:

$$C_2^{26} \times C_2^{26} = \frac{26 \times 25}{1 \times 2} \times \frac{26 \times 25}{1 \times 2} = 105{,}625.$$

Try it yourself:

1. Five cards are selected in series from a standard deck of 52 cards. Find the number of ways this can be accomplished.

2. There are 12 members on the board of directors for a general hospital.

(a) If they must elect a chairperson, first vice chairperson, second vice chairperson, and secretary, how many different slates of candidates are possible?

(b) If they must form an ethics subcommittee of four members, how many different subcommittees are possible?

3. A box contains 5 red and 7 green marbles. In how many ways can two marbles of the same color be selected?

4. A university debate team includes 5 women and 4 men.

(a) In how many ways can all 9 team members be seated in a row of 9 seats?

(b) In how many ways can the team members be seated if men and women were to alternate seats?

(c) In how many ways can the team members be seated if all men and all women are to seat together?

Summary of Counting Rules:

Multiplication Rule: In a sequence of n events, where the first event can occur k_1 ways, the second event can occur k_2 ways, the third events can occur k_3 ways, *etc.*, then the total number of occurrences is: $k_1 \times k_2 \times k_3 \times \ldots \times k_n$

Arrangement Rule: The total number of arrangements (permutations) of n objects taken k at a time (in order) is: $A_k^n = \dfrac{n!}{(n-k)!}$

Combination Rule: The total number of combinations of n objects taken k at a time (without order) is: $C_k^n = \dfrac{n!}{k!(n-k)!}$

3.4: Introduction to Probability

A. Translation of the terminology from Set Theory to Probability Theory

Language of Set Theory	Language of Probability
Elements, Members	Outcomes
Universal Set	Sample Space
Any Subset of the Universal Set	Event
Complement of a set	Complement of an event

Table 3.1

Objectives

After completing this section you will be able to understand:

A. Translation of the terminology from Set Theory to Probability Theory.

B. Theory of Probability.

- **Outcomes:**

Outcomes are the most fundamental building blocks of experiments. When an experiment is conducted, every simple occurrence is called an outcome.

Example 33: List all the outcomes when a fair die is rolled.

Solution: $\{1\},\{2\},\{3\},\{4\},\{5\},\{6\}$

Example 34: List all the possible outcomes when a couple plans to have 3 children.

Solution: We can use a Tree Diagram (see figure 3.15 along side) to help us find all possible gender sequences.

All possible outcomes are:
{BBB, BBG, BGB, GBB, GGG, GGB, GBG, BGG}.

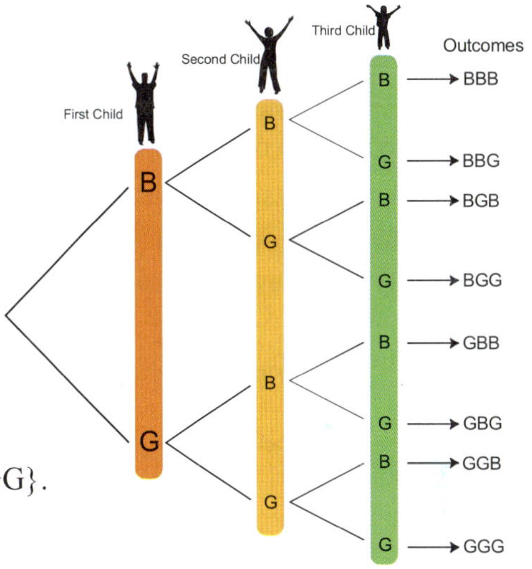

Figure 3.15

- **Sample Space:**

The collection of all possible outcomes in an experiment is called a Sample Space and is labeled by S.

Example 35: Construct the Sample Space for the experiment of rolling a fair die.

Solution: $S = \{1, 2, 3, 4, 5, 6\}$

Example 36: Construct the Sample Space for the experiment of rolling two dice or a single die twice.

Solution:

$$S = \begin{Bmatrix} (1,1) & (1,2) & (1,3) & (1,4) & (1,5) & (1,6) \\ (2,1) & (2,2) & (2,3) & (2,4) & (2,5) & (2,6) \\ (3,1) & (3,2) & (3,3) & (3,4) & (3,5) & (3,6) \\ (4,1) & (4,2) & (4,3) & (4,4) & (4,5) & (4,6) \\ (5,1) & (5,2) & (5,3) & (5,4) & (5,5) & (5,6) \\ (6,1) & (6,2) & (6,3) & (6,4) & (6,5) & (6,6) \end{Bmatrix}$$

Example 37: Construct the Sample Space for the experiment of tossing a coin.

Solution: $S = \{H, T\}$

- ### Events:

All possible subsets of a sample space are called events. An event is a collection of simple outcomes. A simple event is a subset of a sample space that contains a single outcome.

Note: Events are labeled using capital letters.

i) Mutually Exclusive Events or Cases:

Two or more events are said to be mutually exclusive if the occurrence of one excludes the others in the same experiment.

For example, in the toss of a coin the events 'head' and 'tail' are mutually exclusive because if head comes, one can not get tail. Similarly in the throw of a die, the six faces numbered 1, 2, 3, 4, 5 and 6 are mutually exclusive. Thus, *events are said to be mutually exclusive if no two or more of them can happen simultaneously.*

ii) Equally Likely Events or Cases:

The outcomes of an experiment are said to be equally likely or equally probable if none of them is expected to occur in preference to other, that is if each one of them has an equal chance of occurrence.

Thus, in tossing a coin (die), all the outcomes, H, T(the faces 1, 2, 3, 4, 5, 6) are equally likely for a fair coin (die).

Example 38: Given the sample space of rolling a die, list 5 different events and label them.

Solution: $E_1 = \{3\}$, $E_2 = \{1,3,5\}$, $E_3 = \{6,1,5\}$, $E_4 = \{2,4,6\}$, $E_5 = \{1,2,3,4,5,6\}$

B. Theory of Probability

Probability is a number between 0 and 1 inclusive that measures the likelihood or chance of the occurrence of an event. The probability of an event is expressed as $P(E)$, and has the following basic properties:

1. $P(E) \geq 0$ **2.** $P(S) = 1$

3. For any two disjoint events E_1 and E_2, $P(E_1 \cup E_2) = P(E_1) + P(E_2)$.

Round Off Rule for Probability:

Probabilities in this chapter will be expressed using 3 decimal places. When the probability of an outcome or event is too small, it is a common practice to round the decimal to the first non-zero digit after the decimal point.

- ### Empirical Probability:

Empirical Probability stems from the definition of Relative Frequency and relies on the actual experiment to determine the likelihood of outcomes.

$$\text{Probability of an event} = \text{Relative Frequency} = \frac{f}{n}$$

Example 39: A fair die is rolled 20 times, and the following outcomes are recorded.

1	5	3	1	4
5	2	4	4	5
1	1	3	2	6
6	5	3	2	2

Use Empirical probability to find the probability of rolling a 1.

Solution: Let E represent the event of rolling a 1. Then $P(E)=\frac{f}{n}=\frac{4}{20}=\frac{1}{5}=0.2$

Example 40: A coin is tossed 45 times, and the outcomes are recorded in the table below.

H	T	H	H	T	H	H	T	H
T	H	H	T	T	T	T	T	H
T	H	H	T	T	T	H	H	H
H	H	H	H	H	T	T	H	H
T	H	T	H	T	T	H	T	H

Use Empirical probability to find the probability of getting tails.

Solution: Let E represent the event of tossing tails.

$$P(E)=\frac{f}{n}=\frac{20}{45}=0.444$$

Notice that the answer is different from the expected probability of getting tails, which is 0.5.

- **Classical or Priory Probability:**

Classical probability uses sample spaces to determine the numerical probability of an event. Classical probability assumes that all outcomes in the sample space are equally likely (all outcomes have an equal chance of occurring).

Suppose an experiment (trial) has "N" exhaustive, mutually exclusive and equally likely outcomes. Then the probability of the desired event E, denoted by $P(E)$, is given by $P(E)=\frac{M}{N}$, where M is the number of outcomes favorable to E denoted by $n(E)$, and N is the total number of outcomes in the sample space denoted by $n(S)$.

$$P(E)=\frac{n(E)}{n(S)}$$

- **The Law of large Numbers:**

In experiments, as the sample size increases over time, the relative frequency of outcomes or events gets closer to the theoretical (classical) probability value.

Example 41: Find the probability of getting exactly 2 boys, when a couple plans to have 3 children.

Solution: Sample space for the experiment: $S = \{BBB, BBG, BGB, GBB, GGG, GGB, GBG, BGG\}$

The event of getting exactly 2 boys: $E = \{BBG, BGB, GBB\}$

$$P(E) = \frac{n(E)}{n(S)} = \frac{3}{8} = 0.375$$

Example 42: If one card is randomly drawn from a standard deck of playing cards, find the probability of getting a face card.

Solution: Sample space for the experiment: $S = \{52 \text{ cards}\}$

$$S = \left\{ \begin{array}{ccccccccccccc} A\spadesuit & K\spadesuit & Q\spadesuit & J\spadesuit & 10\spadesuit & 9\spadesuit & 8\spadesuit & 7\spadesuit & 6\spadesuit & 5\spadesuit & 4\spadesuit & 3\spadesuit & 2\spadesuit \\ A\heartsuit & K\heartsuit & Q\heartsuit & J\heartsuit & 10\heartsuit & 9\heartsuit & 8\heartsuit & 7\heartsuit & 6\heartsuit & 5\heartsuit & 4\heartsuit & 3\heartsuit & 2\heartsuit \\ A\clubsuit & K\clubsuit & Q\clubsuit & J\clubsuit & 10\clubsuit & 9\clubsuit & 8\clubsuit & 7\clubsuit & 6\clubsuit & 5\clubsuit & 4\clubsuit & 3\clubsuit & 2\clubsuit \\ A\diamondsuit & K\diamondsuit & Q\diamondsuit & J\diamondsuit & 10\diamondsuit & 9\diamondsuit & 8\diamondsuit & 7\diamondsuit & 6\diamondsuit & 5\diamondsuit & 4\diamondsuit & 3\diamondsuit & 2\diamondsuit \end{array} \right\}$$

♥ = Hearts, ♠ = Spades, ♣ = Clubs, ♦ = Diamonds

The events of getting a face card are $E = \{\text{face cards}\}$

$$E = \left\{ \begin{array}{cccc} K\spadesuit & K\heartsuit & K\clubsuit & K\diamondsuit \\ Q\spadesuit & Q\heartsuit & Q\clubsuit & Q\diamondsuit \\ J\spadesuit & J\heartsuit & J\clubsuit & J\diamondsuit \end{array} \right\}$$

$$P(E) = \frac{n(E)}{n(S)} = \frac{12}{52} = 0.231$$

Example 43: If a die is rolled twice, find the probability of getting a sum of 5.

Solution: Sample space for the experiment:

$$S = \left\{ \begin{array}{cccccc} (1,1) & (1,2) & (1,3) & (1,4) & (1,5) & (1,6) \\ (2,1) & (2,2) & (2,3) & (2,4) & (2,5) & (2,6) \\ (3,1) & (3,2) & (3,3) & (3,4) & (3,5) & (3,6) \\ (4,1) & (4,2) & (4,3) & (4,4) & (4,5) & (4,6) \\ (5,1) & (5,2) & (5,3) & (5,4) & (5,5) & (5,6) \\ (6,1) & (6,2) & (6,3) & (6,4) & (6,5) & (6,6) \end{array} \right\}$$

The events of getting a sum of 5 are as follows:

$$E = \{(1,4),(2,3),(3,2),(4,1)\}$$

$$P(E) = \frac{n(E)}{n(S)} = \frac{4}{36} = 0.111$$

Example 44: A shot hits the target 70 times out of 100 fires. What is the probability of hitting the target?

Solution: Let event E represent the hitting the target.

$$P(E)=\frac{n(E)}{n(S)}=\frac{70}{100}=0.7$$

Example 45: There are 15 Republicans and 20 Democrats. If a senator is randomly selected, what is the probability that he or she is Republican?

Solution: Let event E represent a Republican. Then $P(E)=\frac{n(E)}{n(S)}=\frac{15}{15+20}=0.428$

Example 46: A box contains 8 red and 12 white marbles. A marble is drawn from the box at random. Find the probability that

(a) the marble is red. **(b)** the marble is white.

Solutions: Let events A and B represent red and white marbles respectively. Then,

(a) $P(A)=\frac{n(A)}{n(S)}=\frac{8}{8+12}=0.4$ **(b)** $P(B)=\frac{12}{8+12}=0.6$

- **Probability of Complementary Events:**

Given that $\overline{E}$ is the complement of E, then $P(\overline{E})=1-P(E)$

Example 47: $P(E) = 0.84$. Find $P(\overline{E})$.

Solution: $P(\overline{E}) = 1 - P(E) = 1 - 0.84 = 0.16$

Example 48: The probability that a DVD is defective is 0.18. What is the probability that a randomly selected DVD is not defective?

Solution: $P(\text{defective DVD}) = P(A) = 0.18$

$P(\text{non defective DVD}) = P(\overline{A}) = 1 - P(A) = 1 - 0.18 = 0.82$

Example 49: A 4 member FBI team is to be selected from 12 agents. If a 4 member team is to be randomly selected, find the probability of selecting the one that includes the 4 most experienced agents in the original group?

Solution: Let E represent the event in question. Then, $P(E)=\frac{n(E)}{n(S)}=\frac{1}{C_4^{12}}=\frac{1}{495}=0.002$

Example 50: A CD box contains 10 CD's and among these 3 are defective. If six CD's are picked at random, find the probability of selecting two defective CD's.

Solution: The probability formula used for this problem is $P(A) = \frac{M}{N}$ (where A is the event of selecting 2 defective CD's out of 6). It is quite simple to find N. N is the total number of outcomes $C_6^{10} = 210$.

To find M, we will need to use the MN rule – the number of outcomes favorable to draw 2 defective CD's out of 3 defective CD's and 4 non-defective CD's out of the initial 7 non-defective CD's. The Figure 3.16 will facilitate understanding the above theory.

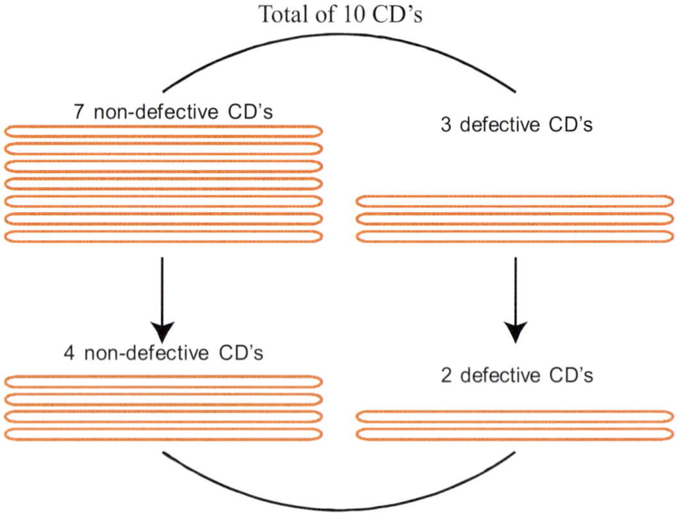

Figure 3.16

Therefore,

$M = C_2^3 \cdot C_4^7 = 105$. Hence

$$P(A) = \frac{M}{N} = \frac{105}{210} = 0.5$$

$$\left(\text{or simply put: } P(A) = \frac{C_2^3 \times C_4^7}{C_6^{10}} = 0.5 \right)$$

Example 51: A taxi cab company has 12 Fords and 8 Chevrolets. If 5 of these taxi cabs are in the shop for repairs, what is the probability that 3 of them are Fords and 2 are Chevrolets?

Solution:

$$P(E) = \frac{m}{n} = \frac{C_3^{12} \cdot C_2^8}{C_5^{20}} = \frac{\frac{12 \cdot 11 \cdot 10}{1 \cdot 2 \cdot 3} \cdot \frac{8 \cdot 7}{1 \cdot 2}}{\frac{20 \cdot 19 \cdot 18 \cdot 17 \cdot 16}{1 \cdot 2 \cdot 3 \cdot 4 \cdot 5}} = \frac{385}{959} = 0.397$$

Example 52: If a single card is drawn at random from a standard 52-card deck, find the probability of

(a) drawing a King.

(b) drawing a Red Card.

(c) drawing a Face Card.

(d) drawing a Club.

Solutions :

(a) $P(K) = \frac{n(K)}{n(S)} = \frac{4}{52} = 0.077$

(b) $P(RC) = \frac{n(RC)}{n(S)} = \frac{26}{52} = 0.5$

(c) $P(FC) = \frac{n(FC)}{n(S)} = \frac{12}{52} = 0.231$

(d) $P(C) = \frac{n(C)}{n(S)} = \frac{13}{52} = 0.25$

Try it yourself:

1. Robert runs a tutoring center. Yesterday he counted 127 people who walked by his center, 58 of whom came in. Of the 58, only 6 signed up for tutoring.
   (a) Estimate the probability that a person who walks by the center will come in.
   (b) Estimate the probability that a person who comes in the center will not sign up.
2. A box contains 8 red and 12 white marbles. A marble is drawn from the box at random. Find the probability that **(a)** the marble is red and **(b)** the marble is white.
3. A tire store sells tires in groups of 4. The store has 28 tires out of which 7 are used. If 4 tires are randomly selected for sale, what is the probability
   (a) that they are all new tires?
   (b) that they are all old tires?
   (c) that out of 4 exactly one tire is used?
4. If a couple has three children, what is the probability that they have two or more girls? First construct the sample space.
5. 3 digits are picked at random from digits 0 to 9. Find the probability that the digits are consecutive digits.

3.5: Addition Rule

We have seen how simple it is to compute the probability of getting a king when a card is randomly drawn from a standard deck of playing cards. We also can quite simply compute the probability of drawing a heart when a card is randomly drawn from a deck.

What if we were asked to calculate the probability of two or more events? For instance, what is the probability of drawing a king or a heart? (Again, "or" means "either … or" or "both".)

Objectives

After completing this section you will be able to use:

A. Addition Rule for Two Events.

B. Addition Rule for three events.

A. Addition Rule for Two Events

The probability of two or more events can be calculated using the addition Rules.

i) For non-mutually exclusive events:

For any two events E and F of an experiment, the probability of occurrence of at least one of the events E or F is given by

$$P(E \cup F) = P(E) + P(F) - P(E \cap F)$$

where $P(E \cap F)$ is the probability of E and F to occur jointly.

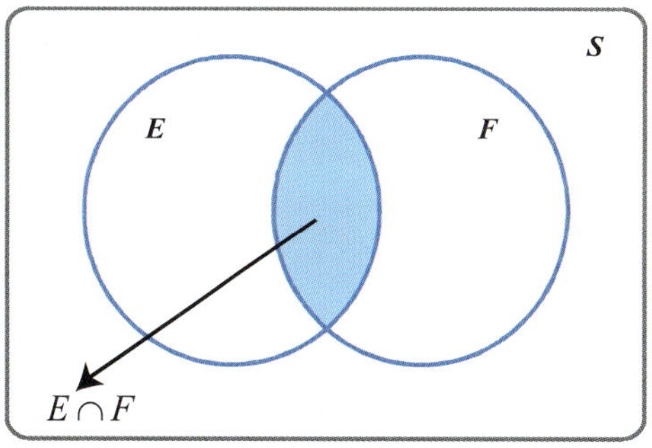

Figure 3.17

Note: Typically, when we compute probabilities of sets, we use the symbols ∪ and ∩; however, when computing the probabilities of outcomes, we use the words "**or**" and "**and**". In this textbook we will use the words and symbols interchangeably for simplicity.

Example 53: A card is drawn at random from an ordinary deck of playing cards. Find the probability that the card is a King or a diamond?

Solution: Since the events in this example are non-mutually exclusive (there is a king of diamond), we will use; $P(K \text{ or } \Diamond) = P(K) + P(\Diamond) - P(K \text{ and } \Diamond) = \frac{4}{52} + \frac{13}{52} - \frac{1}{52} = 0.308$

Example 54: A card is drawn from an ordinary deck of 52 cards. Find the probability that the card is either an ace or a red card.

Solution:

Figure 3.18

Let A be the outcome (event) that the card is an ace and let R be the outcome (event) that the card is a red card. Then we have

$$P(A \cup R) = P(A) + P(R) - P(A \cap R)$$

$$P(A) = P(\text{Drawing an ace}) = \frac{4}{52} \qquad P(R) = P(\text{Drawing a red card}) = \frac{26}{52}$$

$$P(A \cap R) = P(\text{The card is a red ace}) = \frac{2}{52}$$

Therefore,

$$P(A \cup R) = P(A) + P(R) - P(A \cap R)$$

$$= \frac{4}{52} + \frac{26}{52} - \frac{2}{52}$$

$$= \frac{28}{52} = 0.538$$

Example 55: In a residential community, the probability of randomly selecting a lawyer is 68%. The probability of selecting a female is 42%. The probability of selecting a female lawyer is 34%. If a resident is randomly selected from this community, what is the probability that the resident will be a lawyer or a female?

Solution: $P(L \text{ or } F) = P(L) + P(F) - P(F \text{ and } L) = 0.68 + 0.42 - 0.34 = 0.76$

Example 56: A family consists of the parents, two grandparents, 2 sons and 3 daughters. One of them should be chosen to make dinner. Find the probability that it will be one of the grandparents or a female.

Solution: Let G be the event that grandparent are chosen and let F be the event that a female is chosen. Then we have

$$P(G \text{ or } F) = P(G) + P(F) - P(G \text{ and } F)$$

$$P(G) = P(\text{Grandparent are chosen}) = \frac{2}{9}, \qquad P(F) = P(\text{Female is chosen}) = \frac{5}{9}$$

$$P(G \text{ and } F) = P(\text{Grandmother is chosen}) = \frac{1}{9} \Rightarrow P(G \text{ or } F) = \frac{2}{9} + \frac{5}{9} - \frac{1}{9} = \frac{2}{3} = 0.67$$

Example 57: A survey of 2000 college students produced the following results.

	Passed	Failed
Male	573	680
Female	400	347

(a) Find the probability that a randomly selected student will be male or someone that failed.

(b) Find the probability that a randomly selected student will be female or someone that passed.

Solutions: To compute the above probabilities, we need to calculate the row and column total of the above table.

	Passed	Failed	Total
Male	573	680	**1253**
Female	400	347	**747**
Total	**973**	**1027**	**2000**

Table 3.2

(a) $P(\text{Male or someone that failed}) = P(M \text{ or } F)$

$$= P(M) + P(F) - P(M \text{ and } F)$$

$$= \frac{1253}{2000} + \frac{1027}{2000} - \frac{680}{2000} = 0.8$$

(b) $P(\text{Female or someone that passed}) = P(F \text{ or } P)$

$$= P(F) + P(P) - P(F \text{ and } P)$$

$$= \frac{747}{2000} + \frac{973}{2000} - \frac{400}{2000} = 0.66$$

ii) For Mutually Exclusive Events:

If E and F are mutually exclusive events, then the probability of occurrence of E or F is the sum of their individual probabilities.

$$P(E \cup F) = P(E) + P(F)$$

Note: The concepts and formulas in the Addition Rule can easily be verified using Venn-Diagrams from Set Theory.

Example 58: A card is drawn at random from an ordinary deck of playing cards. Find the probability that the card is a King or a Queen.

Solution: Since the events in this example are mutually exclusive (there is no king of queens, at least in a deck of cards), we will use;

$$P(K \text{ or } Q) = P(K) + P(Q) = \frac{4}{52} + \frac{4}{52} = 0.154$$

Example 59: A bag contains 8 white, 7 red and 5 green marbles. A marble is drawn at random. Find the probability that the marble is not white.

Solution: Let ***R*** and ***G*** (mutually exclusive events) denote that the marble is red and green respectively. Then we have

$$\begin{aligned} P(\text{The marble is not white}) &= P(\text{The marble is either red or green}) \\ &= P(R \text{ or } G) \\ &= P(R) + P(G) \\ &= \frac{7}{8+7+5} + \frac{5}{8+7+5} \\ &= \frac{12}{20} = 0.6 \end{aligned}$$

Example 60: A card is drawn from a deck of 52 cards. What is the probability that the card is either a red king or a black card?

Solution: Let K be the event that the card is a red king and B be the event that the card is black. Note that K and B are mutually exclusive events. Then

$$\begin{aligned} P(K \text{ or } B) &= P(K) + P(B) \\ &= \frac{2}{52} + \frac{26}{52} = \frac{28}{52} = 0.538 \end{aligned}$$

B. Addition Rule for Three Events

The addition rule for any three events A, B, and C is
$P(A \cup B \cup C) = P(A) + P(B) + P(C) - P(A \cap B) - P(B \cap C) - P(A \cap C) + P(A \cap B \cap C)$ and if events are mutually exclusive, then $P(A \cup B \cup C) = P(A) + P(B) + P(C)$.

Example 61: A day of the week is selected at random. Find the probability that it is weekday.

Solution:

$$\begin{aligned} P(\text{weekday}) &= P(M \text{ or } T \text{ or } W \text{ or } TH \text{ or } F) \\ &= P(M) + P(T) + P(W) + P(TH) + P(F) \\ &= \frac{1}{7} + \frac{1}{7} + \frac{1}{7} + \frac{1}{7} + \frac{1}{7} \\ &= \frac{5}{7} \approx 0.714 \end{aligned}$$

Try it yourself:

1. A recent survey asked 200 college students if they thought men are better in mathematics than women. The results of the survey are shown below.

Gender	Yes	No	Total
Female	56	44	100
Male	39	61	100
Total	95	105	200

(a) Find the probability that a randomly selected student was a female or responded "yes".

(b) Find the probability that a randomly selected student was a male or responded "no".

2. A card is randomly drawn from an ordinary deck of playing cards. Find the probability of getting

(a) a ten or a jack or a five.

(b) a diamond or a club or a heart.

(c) an ace or diamond or a queen or a seven.

3. It is a sunny day in Los Angeles; A Big Blue bus is headed for the Santa Monica Pier with 47 passengers on board. After careful observation of the passengers, the following data is obtained.

	Eye Color			
	Blue	Black	Green	Total
Male	3	26	1	30
Female	2	11	4	17
Total	5	37	5	47

(a) If a passenger is randomly selected, what is the probability that it is a female or someone that has green eyes?

(b) If a passenger is randomly selected, what is the probability that it is a male or has blue eyes?

3.6: Multiplication Rule

Objectives

After completing this section you will be able to use:

A. Multiplication Rule (Conditional probability).

B. Probability Tree Diagrams.

C. Probability of At Least One Event.

A. Multiplication Rule (Conditional Probability)

- **For Dependent Events:**

The probability of occurrence of two events E and F without replacement is given by

$$\begin{aligned} P(E \cap F) &= P(E)\,P(F\,|\,E) \\ &= P(F)\,P(E\,|\,F) \end{aligned}$$

where $P(F\,|\,E)$ is the conditional probability of an event F, when event E has occurred. The probability $P(E\,|\,F)$ can be similarly defined.

Example 62: Two cards are randomly drawn from a standard deck of playing cards without replacement. Find the probability that the first one is a king and the second one is a Jack.

Solution: $P(K \text{ and } J) = P(K) \cdot P(J \mid K) = \frac{4}{52} \cdot \frac{4}{51} = \frac{4}{663} \approx 0.006$

> **Note:** $P(K) \cdot P(J|K)$ represents the probability of getting a king on the first draw times the probability of getting a jack on the second draw given that a king has already been pulled out.

Example 63: Two cards are randomly drawn from a standard deck of playing cards without replacement. Find the probability that the first one is a king and the second one is also a king.

Solution: $P(K$ on the first draw and K on the second draw$)$

$$= P(K \text{ on the first draw}) \cdot P(K \text{ on the second draw} \mid K \text{ on the first draw})$$

$$= \frac{4}{52} \cdot \frac{3}{51} = \frac{1}{221} \approx 0.005$$

In Multiplication Rule, sometimes the dependency is not mentioned and is left to the reader to realize that the events are dependent, meaning there is no replacement. The following example will demonstrate this fact.

Example 64: In a box of 18 alarm clocks, there are 6 defective ones. If 3 alarm clocks are randomly selected, what is the probability that all of them are defective?

Solution: Let E represent the event that all 3 alarm clocks are defective.

$$P(E) = P\left(1^{\text{st}} \text{ D}\right) \cdot P\left(2^{\text{nd}} \text{ D} | 1^{\text{st}} \text{ D}\right) \cdot P\left(3^{\text{rd}} \text{ D} | 2^{\text{nd}} \text{ D}\right)$$

$$= \frac{6}{18} \cdot \frac{5}{17} \cdot \frac{4}{16} = \frac{5}{204} \approx 0.025$$

- **For Independent Events:**

If two events $\boldsymbol{E}$ and $\boldsymbol{F}$ are independent then $P(E \mid F) = P(E)$. In other words, the occurrence of event F has no bearing on the occurrence of event E. In this case, the multiplication rule reduces to

$$P(E \text{ and } F) = P(E) \cdot P(F)$$

Example 65: Two cards are randomly drawn from a standard deck of playing cards (with replacement.) Find the probability that the first one is a king and the second one is a jack.

Solution: $P(K \text{ and } J) = P(K) \cdot P(J) = \frac{4}{52} \cdot \frac{4}{52} = \frac{1}{169} \approx 0.006$

Example 66: The probability that an alarm clock is defective is 0.034. If 3 alarm clocks are randomly selected, (with replacement) find the probability that all of them are defective.

Solution: Let E represent the event that all 3 alarm clocks are defective.

$$P(E) = P\left(1^{\text{st}} \text{ D}\right) \cdot P\left(2^{\text{nd}} \text{ D}\right) \cdot P\left(3^{\text{rd}} \text{ D}\right)$$

$$= (0.034)(0.034)(0.034)$$

$$= 0.000039304 \approx 0.00004$$

B. Probability Tree Diagrams

For problems below, based on Multiplication Rule, it is helpful to use Probability Tree Diagrams (PTD), which are the same as Tree Diagrams introduced in the combinatorics section, with probabilities for each branch added to the diagram.

Example 67: There are 5 red and 3 white marbles in a bag. If two marbles are randomly selected with replacement (1st marble is selected, put back, and then the 2nd marble is selected), find the probability that there will be one of each color in the selection.

Solution: $P(\text{both colored marbles are selected}) = P[(R \text{ and } W) \text{ or } (W \text{ and } R)]$

$$= P(R \text{ and } W) + P(W \text{ and } R)$$

$$= \frac{5}{8}\cdot\frac{3}{8}+\frac{3}{8}\cdot\frac{5}{8}$$

$$= \frac{30}{64}=\frac{15}{32} \approx 0.469$$

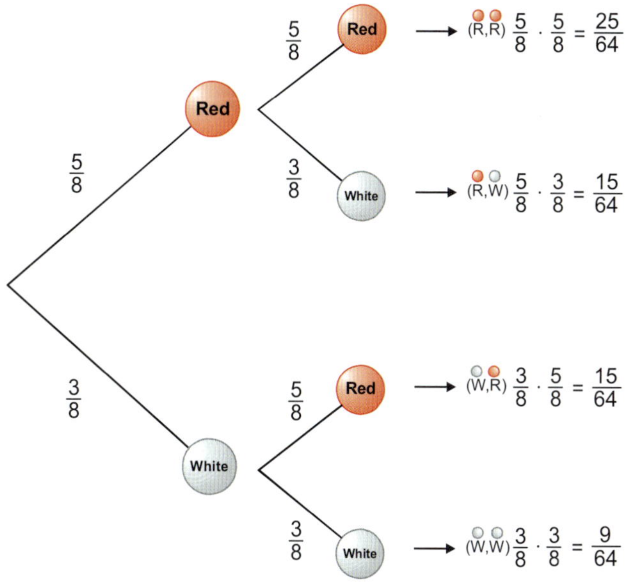

Figure 3.19

Example 68: There are 5 red and 3 white marbles in a bag. If two marbles are randomly selected without replacement (1st marble is selected and taken out, and then the 2nd marble is selected), find the probability that there will be one of each color in the selection.

Solution: Let R and W denote the events that the marble is red and white respectively. Then we have $P(\text{one marble of each color is selected})$

$$= P(R \text{ and } W) + P(W \text{ and } R)$$

$$= \frac{5}{8}\cdot\frac{3}{7}+\frac{3}{8}\cdot\frac{5}{7} = \frac{30}{56} \approx 0.536$$

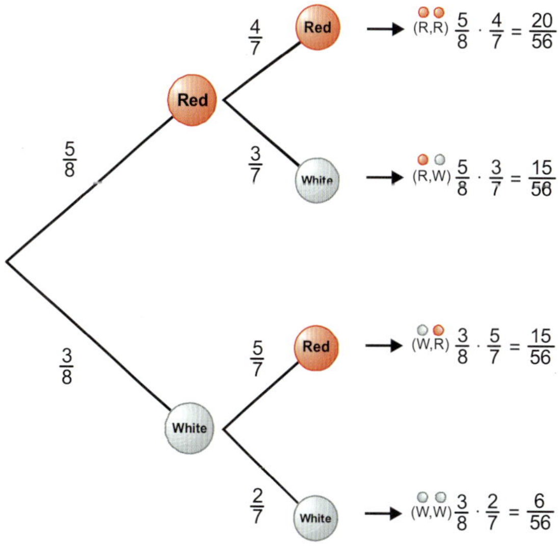

Figure 3.20

C. Probability of at Least One Event

Suppose E is an event. Then we know that $P(E) = 1 - P(\overline{E})$. This formula comes in handy particularly when computing probabilities of events containing outcomes with repetition. For example, if E is the event containing all outcomes with *at least one occurrence of outcome T*, then $\overline{E}$ is the event containing all outcomes with *no occurrence of outcome T.*

$$\begin{aligned} P(\text{at least one occurrence of } T) &= P(E) \\ &= 1 - P(\overline{E}) \\ &= 1 - P(\text{no occurrence of } T). \end{aligned}$$

Example 69: The probability that an alarm clock works properly is 98%. If 4 alarm clocks are set to go off at the same time, find the probability that at least one will work properly.

Solution:

$$P(\text{alarm clock working properly}) = 0.98$$

$$P(\text{alarm clock will not be working properly}) = 1 - 0.98 = 0.02$$

Let A_1 represent the event that the first alarm clock will not work, A_2 represent the event that the second alarm clock will not work, and so on.

$$\begin{aligned} P(\text{at least one will work properly}) &= 1 - P(\text{none will work properly}) \\ &= 1 - \left[P(\overline{A}_1)\cdot P(\overline{A}_2)\cdot P(\overline{A}_3)\cdot P(\overline{A}_4)\right] \\ &= 1 - (0.02)^4 = 0.99999984 \end{aligned}$$

> **Note:** The probability of the alarm clock working properly being 98% is an example of what is called an "iid". "iid" probabilities represent what is known in statistics as Identically and Independently Distributed outcomes. This means that the probability of one outcome does not affect the probability of the other outcome, and they are identical. Any randomly selected alarm clock at any time will have a 98% probability of working properly.

Example 70: A coin is tossed 8 times. Find the probability of getting at least one tail in 8 tosses.

Solution: Let A be the event of getting a tail in single toss of a coin.

$$\Rightarrow P(A) = \frac{1}{2}.$$

P(of getting at least one tail in 8 tosses of a coin)

$= 1 - P$(none of 8 tosses resulted in tail)

$$= 1 - \left(P(\overline{A})\right)^8 = 1 - \left(\frac{1}{2}\right)^8 = 1 - \frac{1}{256} = \frac{255}{256} = 0.996$$

Example 71: A box contains 10 light bulbs out of which 3 are defective. Two bulbs are drawn at random without replacement. Find the probability that

(a) both bulbs are non-defective.

(b) one out of two is defective.

(c) at least one bulb is non-defective.

Solutions: Let's define the events:

E: First bulb is non-defective, F: Second bulb is non-defective.

(a) $P(E \cap F) = P(E) \cdot P(F \mid E) = \frac{7}{10} \cdot \frac{6}{9} = \frac{42}{90} = 0.47$

This part can also be calculated using combination formula as:

$$P(E \cap F) = \frac{C_2^7 \times C_0^3}{C_2^{10}} = 0.47$$

(b) $P\left[(E \cap \bar{F}) \cup (\bar{E} \cap F)\right] = P(E \cap \bar{F}) + P(\bar{E} \cap F)$

$$= P(E) \cdot P(\bar{F} \mid E) + P(\bar{E}) \cdot P(F \mid \bar{E})$$

$$= \frac{7}{10} \cdot \frac{3}{9} + \frac{3}{10} \cdot \frac{7}{9}$$

$$= \frac{42}{90} = 0.47$$

(c) $P(\text{at least one is non-defective}) = 1 - P(\text{both are defective})$

$$= 1 - P(1^{\text{st}} \text{ is defective and } 2^{\text{nd}} \text{ is defective})$$

$$= 1 - P(\bar{E} \cap \bar{F})$$

$$= 1 - P(\bar{E}) \cdot P(\bar{F} \mid \bar{E})$$

$$= 1 - \frac{3}{10} \cdot \frac{2}{9}$$

$$= \frac{14}{15} = 0.93$$

Alternative method **:**

$P(\text{at least one non-defective}) = 1 - P(\text{none is non def(all is defective)})$

$$= 1 - \frac{C_0^7 \cdot C_2^3}{C_2^{10}}$$

$$= 0.93$$

Example 72: Two cards are drawn, without replacement, from an ordinary deck of 52 cards. Find the probability that

(a) both cards are kings. **(b)** one out of two is a king.

(c) at least one card is a king.

Solutions: Let's define the events

E : First card is a king and F : Second card is king

(a) $P(E \cap F) = P(E) \cdot P(F \mid E) = \frac{4}{52} \cdot \frac{3}{51} = \frac{1}{221} \approx 0.005$

Alternative method **:**

$$P(E \cap F) = \frac{C_2^4 \times C_0^{48}}{C_2^{52}} = 0.005$$

(b) $P\left[\left(E \cap \bar{F}\right) \cup \left(\bar{E} \cap F\right)\right] = P\left(E \cap \bar{F}\right) + P\left(\bar{E} \cap F\right)$

$$= \frac{4}{52} \cdot \frac{48}{51} + \frac{48}{52} \cdot \frac{4}{51} = \frac{32}{221} \approx 0.145$$

(c) P(at least one card is a king) $= 1 - P\left(\bar{E} \cap \bar{F}\right)$

$$= 1 - P\left(\bar{E}\right) \cdot P\left(\bar{F} \mid \bar{E}\right)$$

$$= 1 - \frac{48}{52} \cdot \frac{47}{51} = \frac{396}{2652} = 0.15$$

Alternative method **:**

P(at least one card is a king) $= 1 - P\left(\bar{E} \cap \bar{F}\right)$

$$= 1 - \frac{C_0^4 \times C_2^{48}}{C_2^{52}} = 0.15$$

Example 73: Two fair dice are rolled. One is red and the other is white. Calculate

(a) P(3 on red die and even number on white die)

(b) P(3 on red die and even number on white die or vice versa)

Solutions: **(a)** P(3 on red die and even number on white die)

= P(3 on red die) × P(even number on white die)

$$= \frac{1}{6} \times \frac{3}{6} = \frac{1}{12} = 0.083$$

(b) P(3 and even number on both dice in either order)

= P[(3 on red die and even number on white die) or (even number on red die and 3 on white die)]

= P(3 on red die and even number on white die) + P(even number on red die and 3 on white die)

= P(3 on red die) × P(even number on white die) + P(even number on red die) × P(3 on white die)

$$= \frac{1}{6} \times \frac{3}{6} + \frac{3}{6} \times \frac{1}{6} = \frac{1}{6} \approx 0.17$$

Example 74: A box contains 6 red and 4 white marbles (box 1) and another contains 8 red and 2 white marbles (box 2). One marble from each box is drawn at random. Find P(marbles are of different colors).

Solution: P(marbles are of different colors)

= P[(red and white) or (white and red)]

= P(red and white) + P(white and red)

$$= P(\text{red marble from box 1}) \times P(\text{white marble from box 2}) + P(\text{white marble from box 1}) \times P(\text{red marble from box 2})$$

$$= \frac{6}{10} \times \frac{2}{10} + \frac{4}{10} \times \frac{8}{10} = \frac{11}{25} = 0.44$$

Example 75: What is the probability that 3 randomly selected students will all have the same birthday?

Solution: Let E represent the event in question.

$P(E) = P(1^{st}$ one has a B'day$) \times P(2^{nd}$ one has the same B'day$) \times P(3^{rd}$ one has the same B'day$)$

$$= \frac{365}{365} \times \frac{1}{365} \times \frac{1}{365} \approx 0.00000751 = 0.000008 \approx 0.000$$

Example 76: Find probability of getting four cards of same type, when five cards are dealt out of 52 cards.

Solution: E: getting four cards of same type in 5 cards drawn

$$\Rightarrow \quad n(E) = \binom{4}{4}\binom{12}{1} = 12 \quad \text{and } n(S) = \binom{52}{5} = 2{,}598{,}960$$

$$P(E) = \frac{n(E)}{n(S)} = \frac{12}{2598960} = \frac{1}{216580} \approx 0.000005$$

Example 77: Find the probability of getting a full house when 5 cards a dealt.

> **Note:** Full house means we have 3 cards of same type and other 2 also are of same type but different from first.

Solution: There are 13 choices for the card which occurs 3 times, and for each one of the 3 cards there are $C_3^4 = 4$ possibilities. There are 12 choices for the card which occurs 2 times, and for each one of the 2 cards there are $C_2^4 = 6$ possibilities. Therefore total number of possible full houses will be = (13)(4)(12)(6) = 3,744.

The total number of possible 5 cards is $n(S) = C_5^{52} = 2{,}598{,}960$.

$$P(\text{getting a full house}) = \frac{3744}{2598960} = 0.0014 \approx 0.001$$

Try it yourself:

1. Find the probability of getting 3 consecutive kings when 3 cards are drawn without replacement from a shuffled deck of ordinary playing cards.

2. Find the probability of getting 3 consecutive kings when 3 cards are drawn with replacement from a shuffled deck of ordinary playing cards.

3. Three cards are drawn at random from an ordinary deck of 52 playing cards without replacement. Find the probability that

   (a) all 3 are aces. **(b)** all 3 are jacks. **(c)** all 3 are spades.

   (d) 1st one is a King, 2nd one is a Queen, and the 3rd one is a Jack.

4. A bin containing light bulbs has a 16% defect rate. If 4 light bulbs are randomly selected from the bin, find the probability that all the bulbs selected are defective.

5. The probability that a Blu-ray disk is defective is 3%. If 15 Blu-ray disks are randomly selected from the production line, what is the probability that they are all defective?

6. In the 103rd Congress, the Senate consists of 8 women and 92 men. If a lobbyist for the alcohol industry randomly selects three different Senators, what is the probability that they are all men?

7. If 4 people are randomly selected, find the probability that all 4 were born in January.

3.7: Probability with Microsoft Excel®

Objectives

After completing this section you will be able to carryout the computations in Chapter 3 using Microsoft Excel®.

Computing Probabilities by using Microsoft Excel®

The following examples illustrate the use of Microsoft Excel® commands in Probability.

Illustration 1: In how many ways can you travel to 10 distinct cities?

Solution:

= FACT(10)	→ 3628800

Illustration 2: Calculate A_4^6 .

Solution:

Method 1: **Using commands**

= PERMUT(6, 4)	→ 360

Method 2: **Using a Dialog Box from the insert function icon**

1. Go to insert function by clicking "f_x" in the formula bar.
2. Go to categories and select **Statistical.**
3. Scroll down and choose the **Permut** function. Press OK.
4. You should see the display box in Figure 3.22.
5. Type in the total number of objects 6 in the number box, and number of objects in each arrangement 4 in the number_chosen box. Press OK.

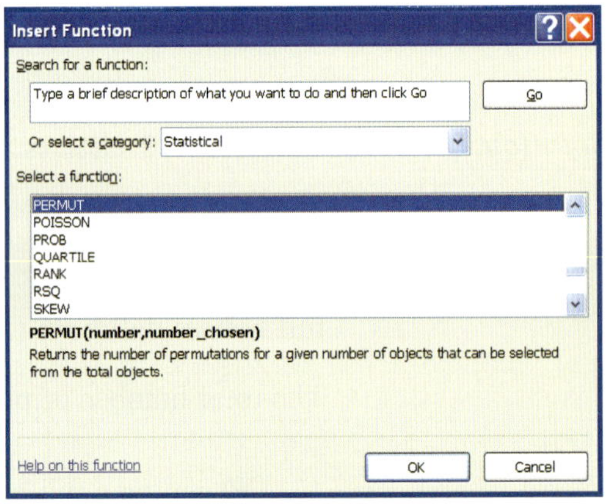

Figure 3.21

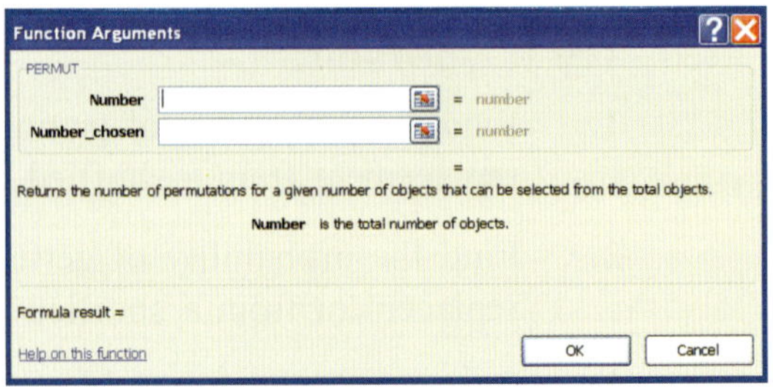

Figure 3.22

Note: The same procedure applies for computing C_4^6 : = COMBIN(6,4) = 15

Illustration 3: Find the total number of arrangements (permutations) of the letters of the word STATISTICS taken all at a time.

Solution:

$\frac{10!}{(3!)(3!)(2!)} = 50400$	= FACT(10)/(FACT(3)*FACT(3)*FACT(2))	50400

Illustration 4: Calculate the probability $P(A) = \frac{C_2^3 \times C_4^7}{C_6^{10}} = 0.5$ in Excel®

Solution:

= COMBIN(3,2)*COMBIN(7,4)/COMBIN(10,6)	0.5

Illustration 5: There are 10 CD's in a box, out of which 3 are defective. 6 CD's are drawn at random. Find the probability that 2 out of 6 are defective.

Solution: *The probability formula used for this problem is* $P(A) = \frac{M}{N}$ (where A is the event of selecting 2 defective CD's out of 6).

Now N represents total number of outcomes $= C_6^{10} = \mathbf{210}$ and outcome favors to event A: $M = C_2^3 \cdot C_4^7 = \mathbf{105}$. Hence $P(A) = \frac{M}{N} = \frac{n(E)}{n(S)} = \frac{\mathbf{105}}{\mathbf{210}} = \mathbf{0.5}$

Using Excel;

M	= COMBIN(7, 4)*COMBIN(3,2)	105
N	= COMBIN(10,6)	210
$p = \frac{M}{N}$	$= \frac{105}{210}$	0.5

Table 3.3

Simply put: $P(A) = \frac{C_2^3 \times C_4^7}{C_6^{10}} = 0.5$

Illustration 6: A taxi cab company has 12 Fords and 8 Chevrolets. If 5 of these taxi cabs are in the shop for repairs, what is the probability that 3 of them are Fords and 2 are Chevrolets?

Solution: $P(E) = \frac{M}{N} = \frac{C_3^{12} \cdot C_2^8}{C_5^{20}} = 0.40$

Excel command: = (COMBIN(12,3)*COMBIN(8,2))/COMBIN(20,5)

3.8: Extended Topics and Proofs in Probability Theory (Optional)

Objectives

After completing this section you will be able to use:

A. Properties of Classical Probability (Axioms).

B. Stirling's Approximation.

A. Properties of Classical Probability

- **Kolmogorov's Axiom of Probability Function:**

i) For all events A, $0 \le P(A) \le 1$

ii) $P(S) = 1$

iii) If $E_1, E_2, \ldots, E_k$ are finite collection of mutually exclusive events, then

$$P(E_1 \cup E_2 \cup \ldots \cup E_k) = P(E_1) + P(E_2) + \ldots + P(E_k)$$

$$= \sum_{i=1}^{k} P(E_i) \qquad \text{(Finite Sum)}$$

If $E_1, E_2, \ldots,$ are countably infinite collection of mutually exclusive events, then

$$P(E_1 \cup E_2 \cup \ldots) = P(E_1) + P(E_2) + \ldots$$

$$= \sum_{i=1}^{\infty} P(E_i) \qquad \text{(Infinite Sum)}$$

- **Theorem 3.1:** If E_1 and E_2 are subsets of S such that $E_1 \subset E_2$, then $P(E_1) \le P(E_2)$.

Proof: Let $E_2 = E_1 \cup (E_2 \cap \bar{E}_1)$.

Since E_1 and $(E_2 \cap \bar{E}_1)$ are mutually exclusive, then

$P(E_2) = P(E_1) + P(E_2 \cap \bar{E}_1)$. (By using property (iii))

But $P(E_2 \cap \bar{E}_1) \ge 0$, therefore $P(E_2) \ge P(E_1)$.

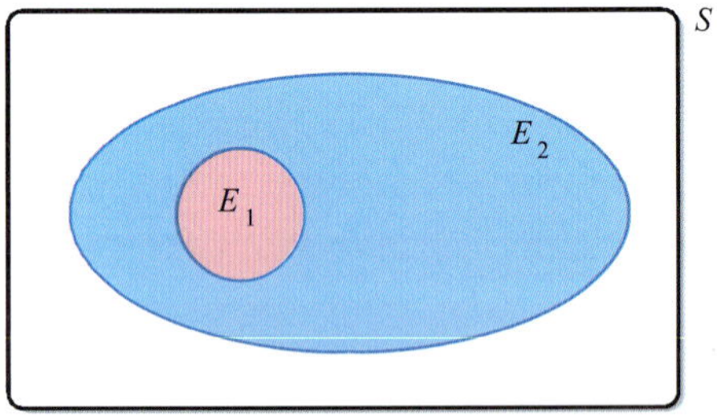

Figure 3.23

- **Theorem 3.2:** For all events A, $0 \le P(A) \le 1$.

Proof: If any given event E is from a sample space $\varnothing \subset E \subset S$, then using Theorem 3.1, we have $P(\varnothing) \le P(E) \le P(S) \Rightarrow 0 \le P(E) \le 1$.

- **Theorem 3.3:** Addition Rule of Probability for two mutually exclusive independent events is

$$P(A \cup B) = P(A) + P(B)$$

Proof: Since A and B are mutually exclusive, then $n(A \cap B) = 0$.

We know that $n(A \cup B) = n(A) + n(B) - n(A \cap B)$ (1)

Dividing equation (1) by $n(S)$ give us $\dfrac{n(A \cup B)}{n(S)} = \dfrac{n(A)}{n(S)} + \dfrac{n(B)}{n(S)} - \dfrac{n(A \cap B)}{n(S)}$ (2)

Given that $n(A \cap B) = 0$, equation (2) reduces to $\dfrac{n(A \cup B)}{n(S)} = \dfrac{n(A)}{n(S)} + \dfrac{n(B)}{n(S)}$ (3)

However, by the definition of classical probability, equation (3) will reduce to $P(A \cup B) = P(A) + P(B)$.

- **Theorem 3.4:** $P(A \text{ and } B) = 0$ if A and B are mutually exclusive.

Proof: If A and B are mutually exclusive events then $n(A \cap B) = 0$

$$\Rightarrow \quad \frac{n(A \cap B)}{n(S)} = 0$$

$$\Rightarrow \quad P(A \cap B) = 0$$

- **Theorem 3.5:** For all $A \subset S$, $P(\bar{A}) = 1 - P(A)$.

Proof: $1 = P(S) = P(A \text{ or } \bar{A}) \Rightarrow P(A \text{ or } \bar{A}) = 1$

Since A and $\bar{A}$ are mutually exclusive, then

$$P(A \text{ or } \bar{A}) = P(A) + P(\bar{A}) = 1 \Rightarrow P(\bar{A}) = 1 - P(A)$$

Alternate Proof: $P(\bar{A}) = \dfrac{n(\bar{A})}{n(S)} = \dfrac{n(S) - n(A)}{n(S)} = \dfrac{n(S)}{n(S)} - \dfrac{n(A)}{n(S)} = 1 - P(A)$.

B. Stirling's Approximation

Evaluating facotrials of sizable numbers can be very cumbersome – we are already in millions when we talk of 10!. A helpful tool to evaluate/approximate factorials of large numbers is Stirling's approximation formula derived by James Stirling, a Scottish mathematician (1692-1770). It states:

If "n" is sufficiently large, then we can use the following approximation formula for computing $n!$,

$$n! \approx \sqrt{2\pi n}\left(\frac{n}{e}\right)^n$$

It is interesting to note that as the value of n becomes larger and larger the percentage difference between the actual value of n! and the approximation of it by the Stirling's formula, i.e. the percentage error, gets smaller and smaller.

Example 78: Using Stirling's approximation find the value of 5! and calculate the percentage error.

Solution: Value of $5! = 5 \times 4 \times 3 \times 2 \times 1 = 120$

Approximation by Stirling formula ($n = 5$) $= \sqrt{2\pi \times 5} \times \left(\dfrac{5}{e}\right)^5 = \sqrt{10\pi} \times \left(\dfrac{5}{e}\right)^5$

$= 5.605 \times (1.84)^5 = 118$

error $= 2$

% error $= \dfrac{2}{120} \times 100 = 1.66\%$

Example 79: Using Stirling's approximation find the value of 10! and calculate the percentage error.

Solution: Value of $10! = 10 \times 9 \times 8 \times 7 \times 6 \times 5 \times 4 \times 3 \times 2 \times 1 = 3628800$

$$\text{Approximation by Stirling formula } (n = 10) = \sqrt{2\pi \times 10} \times \left(\frac{10}{e}\right)^{10} = \sqrt{20\pi} \times \left(\frac{10}{e}\right)^{10}$$

$$= 7.9266 \times 453999.2976 \approx 3598695$$

$$\text{error} = 30105$$

$$\%\text{ error} = \frac{30105}{3628800} \times 100 = 0.8296 = 0.83\%$$

Chapter 3: General Exercises

1. Let $A = \{-3, 4, 5, 6, 7, 24\}$ and $B = \{4, 5, 18, 26, 120\}$. Find **(a)** $A \cup B$ and **(b)** $A \cap B$.

2. $U = \{1, 2, 3, 4, 5, 6, 7, 8, 9, 10\}$ and $A = \{2, 7, 8, 9, 10\}$. Find $\bar{A}$.

3. $U = \{a, b, c, d, e, f, g, h\}$ and $E = \{d, e, f\}$. Find $\bar{E}$.

4. For sets E_1 and E_2, $n(E_1) = 50$, $n(E_2) = 25$ and $n(E_1 \text{ and } E_2) = 15$. Find $n(E_1 \text{ or } E_2)$.

5. Given $U = \{-5, -4, -3, -2, -1, 0, 1, 2, 3, 4, 5, 6, 7, 8, 9, 10\}$, $A = \{1, 2, 3, 4\}$, $B = \{1, 2, 3, 4, 5, 6, 7, 8\}$ and $C = \{-5, -1, 0, 4, 6, 8\}$, find the following sets.

(a) $A \cup B$ **(b)** $A \cap B$ **(c)** $A \cap B \cap C$ **(d)** $B \cap \bar{C}$

(e) $\bar{A} \cup C$ **(f)** $\bar{A} \cup C \cap B$ **(g)** $A \cup \varnothing$ **(h)** $A \cap \varnothing$

6. $A = \{3, 4, 5, 6, 7\}$ and $B = \{6, 7, 8, 9\}$. Draw a Venn diagram showing the location of the elements in A and B.

7. $n(A) = 45$, $n(B) = 30$ and $n(A \cap B) = 15$. Find $n(A \cup B)$.

8. $n(A) = 25$, $n(A \cap B) = 8$, $n(A \cup B) = 32$. Find $n(B)$.

9. Set A contains 35 elements, B contains 26 and their union contains 45. How many elements are there in $A \cap B$?

10. Set A contains 30 elements, B contains 20 and their intersection contains 12. How many elements are there in $A \cup B$?

11. A teenage girl wants to select a 5-digit pin for her social networking account. How many different possible pin numbers are there if

(a) the digits can repeat. **(b)** the digits cannot repeat.

12. A taxicab can choose one of 3 routes to go from city ***A*** to city ***B***; 4 routes to go from city ***B*** to city ***C***; and 2 routes to go from city ***C*** to city ***D***. How many different routes can the taxicab driver choose to go from city ***A*** to city ***D***?

13. How many different (distinct) 13 digit passwords can be created using letters and digits *with repetition*?

14. What is the total number of gender sequences when a couple plans to have 8 children?

15. A major indoor soccer team plays 40 games during a regular season. Each game can result in one of 3 outcomes: a win, a loss, or a tie. Find the total number of possible outcomes for 40 games.

16. How many different three-letter "words" (with or without meaning) can be formed from the letters of UNIVERSAL?

17. Jack Bauer a CTU agent has been ordered by President Palmer to lead a covert operation in Eastern Europe. He has a choice between 10 safe houses and Jack is to give the President his first, second, and third choices. In how many ways can this be done?

18. A club has 30 male and 20 female members. In how many different ways can 2 males and 3 females be selected from the club?

19. A 5 character code must be formed using a letter of the alphabet for each character. How many different codes can be created if characters

    (a) may be repeated within a code? **(b)** may not be repeated within a code?

20. From a committee of 5 men and 4 women, a chair and a secretary has to be selected. Find the number of ways in which this can be done if both are men or both are women.

21. How many three-digit numbers can be formed using the digits $\{1,2,3,4,5,6\}$

    (a) if a digit may be repeated? **(b)** if a digit may not be repeated?

22. How many three-digit numbers can be formed using the digits $\{1, 2, 3, 4, 5\}$ if the first digit is odd and a digit may be repeated within a number?

23. How many 15 digits passwords are possible if

    (a) the digits can repeat. **(b)** the digits cannot repeat.

24. Prof. Badalian has 8 test problems on combinations and 5 on arrangements. How many ways can he select one of each?

25. From 12 people 3 are selected for first, second, and third awards. How many ways can the

    (a) awards be distributed.

    (b) second and third awards be made if award for the first place has already been decided.

26. A test has five true-false questions. In how many different ways can the answers be given to the five questions?

27. New car models come with a choice of three body styles, four interior package options, five different colors and an automatic or standard transmission. Determine the total number of the different types of cars possible.

28. A certain car is available in red, black, silver, white, or blue. It can also be purchased with two or four doors, leather or cloth seats, sunroof or no sunroof, and manual or automatic. How many different combinations are possible for this car?

29. 4 couples attend a movie and are seated in a row of 8 seats. How many different seating arrangements are possible if couples are seated together?

30. Professor Badalian has 7 statistics books that he wishes to arrange side by side. How many different arrangements are possible.

31. In a political election, 6 candidates are seeking nominations. In how many ways can voters rank their first, second, and third choices?

32. A televised talk show includes 5 women and 4 men as panelists.

    (a) In how many ways can the panelists be seated in a row of 9 chairs?

    (b) In how many ways can the panelists be seated if the men and women are to be alternated?

33. Four FBI agents are to be selected from a group of 20 to investigate a case.

    (a) In how many different ways can the agents be selected?

    (b) In how many ways can the group of 4 be selected if a certain agent must be included?

34. A CEO must select 5 division heads for promotion from 15 eligible division heads.

(a) In how many ways can the 5 candidates be chosen?

(b) In how many ways can the 5 promoted candidates be chosen to be placed in 5 different positions?

35. Use a tree diagram to construct the sample space for tossing a coin

(a) 3 times. **(b)** 4 times.

36. For breakfast Nick always selects one item from each of the following: (Orange juice or Grapefruit juice), (Eggs or Hash brown), and (Toast or Bagels). Draw a tree diagram to show all his possible menus.

37. There are 2 freeways from Smallville to Narnia and 4 highways from Narnia to Rivendell. Construct a tree diagram to show all the possible routes from Smallville to Rivendell.

38. Andy may select a small, medium, or large fries and then select one of ketchup, mustard, and mayonnaise for a spread. Draw a tree diagram that shows the different ways these selections may be made.

39. A college administrator is considering evaluating 6 sections: 3 in the Mathematics Department, 2 in the English Department, and 1 in the History Department. Due to budget cuts and lack of personnel, she decides to evaluate 2 sections selected at random with each section equally likely to be selected, but no section selected twice. What is the probability that one Mathematics section and 1 English section are selected.

(***Hint***: Use a tree diagram to display all possible outcomes).

40. Suzanna has 4 necklaces and 3 pairs of earrings. Construct a tree diagram that will show all her choices of wearing both jewelries.

41. Evaluate the following.

(a) $(3!)(5!)$ **(b)** $\frac{5!}{3!}$ **(c)** $\frac{8!}{3!5!}$ **(d)** $\frac{6!}{3!3!}$ **(e)** $\frac{18!}{2!3!2!4!}$

42. In how many ways can a traveling salesman travel to 50 states?

43. How many different numbers can be formed by using all the digits of number 5569885122?

44. How many distinct words (with or without meaning) can be created by using alphabets of words:

(a) FREEDOOMMM, **(b)** PROBABILITY, **(c)** COUNTING, and **(d)** ILLUSTRATE?

45. Evaluate the following arrangements.

(a) A_3^8 **(b)** A_2^5 **(c)** A_5^5 **(d)** A_1^6

46. Evaluate the following combinations.

(a) C_3^8 **(b)** C_2^5 **(c)** C_5^5 **(d)** C_1^6

47. A store selects 5 items from a selection of 10 items to arrange in a display. How many different arrangements are possible?

48. A panel of 6 is selected from 14 people and seated in a conference room. In how many different orders can they be seated?

49. Professor Badalian grades homework by choosing 6 out of 10 homework problems. How many different groups of 6 problems can be chosen from the original 10 problems?

50. A department chair must select 5 faculty members to serve on the hiring committee. 18 faculty members are available.

(a) In how many ways can the 5 faculty members be chosen?

(b) In how many ways can the 5 faculty members be chosen with special task assigned to each one?

51. How many ways can 4 statistics books be selected from a collection of 12 statistics books?

52. A club has 20 sophomores and 35 freshmen. In how many ways can they select a president, vice-president, secretary, and a committee of 6 if the officers must be sophomores and the committee members freshmen?

53. Two people are selected from 5 men and 6 women. In how many ways can the selection be made so that the two people are of the same sex?

54. A group consists of 8 freshmen and 7 sophomores.

(a) In how many ways can one select a committee of 5 from the group?

(b) In how many ways can one select a committee of 3 freshmen and 2 sophomores?

55. In how many ways can a team of 5 boys or 4 girls be selected from a class of 18 boys and 15 girls?

56. George receives a Preferred Customer Award from MFCS, an Office Supplies Company which entitles him to select 3 printers from a list of 18 printers or a computer from a list of 6 computers. In how many ways can he make his selection?

57. Construct the sample space for rolling a die twice. List the outcomes in each of the following events and then find the probability of each event.

(a) Sample space. **(b)** The score is 7. **(c)** The total score is less than 6.

58. If a couple has 4 children, what is the probability that they have two or more girls? Construct the sample space first.

59. Suppose a fair die is rolled. Use the sample space {1, 2, 3, 4, 5, 6} to give the probability of each of the following events.

(a) E_1: the die shows an odd number.

(b) E_2: the die shows a number less then and equal to 6.

(c) E_3: the die shows a 3.

60. If a single playing card is randomly selected from a standard 52-card deck, then find the probability of the following events.

(a) Drawing a Queen. **(b)** Drawing a diamond.

(c) Drawing a face card. **(d)** Drawing a heart or club.

61. Two fair dice are rolled. Find the probability of each of the following events. (***Hint:*** Construct a sample space).

(a) Getting even dots on both dice.

(b) Getting a sum of 5 on both dice.

(c) Getting an even number of dots on the first die and an odd number on the second die.

62. If a fair die is rolled, what is the probability that any number except 4 will come up?

63. Find thc probability of obtaining an cvcn numbcr in onc roll of a fair dic.

64. Given the following probability table,

x	0	1	2	3	4	5
$P(X=x)$	0.237	0.396	0.264	0.088	0.014	0.001

(a) calculate the probability that X is greater than 4.

(b) calculate the probability that X is at most 2.

(c) calculate the probability that X is 3.

(d) calculate the probability that X is at least 3.

65. A study of 350 randomly selected Quantis Airlines flights showed that 235 arrived on time. What is the estimated probability of a Quantis Airline flight arriving late?

66. After survey of a sample of 1600 students at a community college the age of each student is recorded and the results are shown in the frequency distribution (See table 3.4). If you randomly select another student, what is the probability that the student is between 15 and 24?

Student Ages	Frequency
15-25	572
25-35	465
35-45	253
45-55	212
55-65	46
65 and over	52
	$\Sigma f = 1600$

Table 3.4

67. A sculptor wants to arrange his 36 sculptures in a museum for display, 6 sculptures per room. If there are 6 sculptures of women in the collection, what is the probability that all 6 of them will end up in the same room, assuming that the sculptor does not mind?

68. The table below shows the frequency of a college student passing statistics, depending on the number of hours the students study in a week.

Hour Studied/Week	0-2	2-4	4-6	6-8	8-10
Frequency	3	7	9	13	16

Find the probability that a student passes when they

(a) study at least 3 hours. **(b)** spend less than 7 hours studying.

69. In a study of Americans over 55 years of age, it is found that 250 have Arthritis and 2155 do not. If an American over 55 years of age is selected, what is the estimated probability that he/she has Arthritis? Based on that estimated probability is Arthritis a major concern for those over 55 years of age?

70. On a multiple choice test, each question has 5 possible answers. If you make a random guess on the first question, what is the probability that you are correct?

71. Two 6-sided dice are rolled. What is the probability that the sum of the two numbers on the dice will be 4?

72. The probability that Andy will pass his statistics test is 0.90. Find the probability that he will fail his statistics test.

73. Find the probability of correctly answering the first 7 questions on a multiple choice test if random guesses are made and each question has 6 possible answers.

74. A bag contains 6 red ribbons, 7 blue ribbons, and 3 green ribbons. If a ribbon is randomly selected, what is the probability that it is not red?

75. There are 7 cans of Cola and 9 cans of Sprite in a cooler. If a can of soda is randomly selected, find the probability that it is a can of cola.

76. A coin is tossed so that it has a 60% chance of getting landing on heads. If it is tossed 3 times, find the probability of getting

(a) 3 heads. **(b)** 2 heads and 1 tail. **(c)** at least 1 head.

77. On a TV Quiz show a contestant is asked to pick an integer at random from 1 to 100. If the number picked is divisible by 9 or 12, the contestant will win a free trip to Cabo San Lucas. What is the probability that the contestant will win?

78. In a company, the probability that an employee picked at random is an engineer is 0.63, the probability that the employee is male is 0.73, and the probability that the employee is a male engineer is 0.54. Find the probability that the member is male or an engineer.

79. If a card is randomly selected from an ordinary 52-card deck, then find the probability that it will be a black or a face card.

80. A card is randomly drawn from ordinary deck of playing cards. Find the probability of drawing

(a) a black or a number 5 card. **(b)** number on the card is either 4 or 5.

(c) a club or a red card. **(d)** a face or a spade card. **(e)** a diamond or a jack card.

81. A die is rolled. What is the probability that the number of dots on top is either 3 or an even number?

82. Among 100 students at a small college, 45 are Math majors, 30 are Physics majors, and 15 majoring in both (double major). If a student is selected at random, find the probability that she is neither a Math major nor a Physics major.

83. At a store that has 1500 CD's and 980 DVD's in stock, 10 of the CD's and 4 of the DVD's are known to have a scratch on them. If a customer picks a disk at random, what is the probability that it will be

(a) a DVD or scratched? **(b)** a CD or not scratched? **(c)** a DVD or a CD?

84. At BADKS Electronics, all 140 employees were asked about their political affiliation. The employees were grouped by type of work, as Executives or Production Workers. The result is summarized in the table below:

Employee Type	Democrat (D)	Republican (R)	Independent (I)	Totals
Executive (E)	5	34	9	48
Production Worker (PW)	63	21	8	92
Total	68	55	17	140

Suppose an employee is randomly selected from the 140 employees. Find

(a) $P(\text{D or PW})$ **(b)** $P(\text{I or E})$

85. The table below lists the blood types of a group of patients in Hackilumas Memorial Hospitals Cancer program.

		Group				
		A	B	O	AB	Total
Blood Type	+	72	43	21	30	166
	–	61	52	19	7	139
	Total	133	95	40	37	305

If a patient is randomly selected, what is the probability that he/she will be from the AB group or have a Positive Blood Type?

86. A class consists of 45 women and 86 men. If 3 students are randomly selected, what is the probability that all of them are women?

87. In a suburban town, 66% of automobile owners have full coverage auto insurance. What is the probability that 7 automobile owners selected at random all have full coverage auto insurance?

88. A study conducted in a community college revealed that 49% of the graduating students make at least $4000 a month within a year after graduation. Find the probability that all 12 randomly selected graduates will be making at least $4000 a month.

89. Three cards are drawn from an ordinary deck of 52 playing cards without replacement. Find the probabilities of getting

(a) three kings. **(b)** at least 2 kings. **(c)** an ace, a king and a queen respectively.

(d) a club, a spade and a heart respectively. **(e)** three hearts.

90. In Raymond's statistics class, there are 16 men and 14 women. Four students are randomly selected to demonstrate the use of Microsoft Excel®. Find the probability that the group consists of

(a) all men. **(b)** all women. **(c)** one man and three women.

91. In Casino Royal you are dealt three cards successively from a shuffled deck of 52 playing cards. Find the probability that the first card is a jack, the second card is a ten, and the third card is a ten

(a) without replacement. **(b)** with replacement.

92. Two cards are drawn from a standard deck of playing cards *without replacement*. Find the probability that the first card is a spade and the second card is red. (Use a probability tree diagram to answer this question.)

93. Four people are randomly selected from 20 faculty and 15 students. What is the probability that they are all students?

94. You have a box containing 6 red candles, 5 green candles, and 1 gold candle. 2 candles are drawn at random *without replacement*. Find the probability that at least one is red.

95. A sample of 5 different laptop computers is randomly selected from a group containing 40 that are defective and 25 that have no defects. What is the probability that all 5 of the laptop computers selected are defective?

96. A company manufactures 10,000 flash drives daily, with a known 3% defect rate. The company quality control policy for shipments requires for a package of flash drives to completely be defect free before shipping. If a package containing 20 flash drives is randomly selected and tested, what is the probability that it will satisfy the company's quality control requirement?

97. In a group of 120 lawyers there are 80 Republicans and 40 Democrats. Find the probability of selecting 10 Republicans from the entire group.

98. There are 120 lawyers and the probability that a lawyer is Republican is 74%. Find the probability of selecting 10 Republicans from the entire group.

99. Approximately 10% of us are left-handed. If 4 people are selected at random, find the probability that all of them are left-handed?

100. There are 15 juniors and 20 seniors in the Table Tennis team at GGG College.

(a) Find the probability that if Coach Raymond randomly picks 3 members for the team, they will all be seniors.

(b) Find the probability that if Coach Raymond randomly picks 5 members for the team, there will be 2 juniors and 3 seniors.

101. A box contains 10 light bulbs out of which 3 are defective. Two bulbs are drawn at random, without replacement. Find the probability that

(a) both bulbs are non-defective.

(b) one out of two is defective.

(c) at least one bulb is non-defective.

102. An electronics company uses a procedure called approval sampling. A sample of electronics is randomly selected without replacement, and the entire batch is rejected if there is at least one defective. The company has just manufactured 5000 Blu-ray disks and 4% are defective. If 15 of the Blu-ray disks are selected and tested, what is the probability that the entire batch will be rejected?

103. Two fair coins were tossed, and it is known that at least one was a tail. Find the probability that both were tails.

In exercise 104-107, evaluate the mathematical expressions using Microsoft Excel ®.

104. $\dfrac{20!}{(3!)(2!)(5!)(4!)(6!)}$

105. $\binom{10}{2}\binom{9}{3}\binom{6}{4}\binom{5}{2}$

106. $\dfrac{C_3^{16}C_5^{8}}{C_8^{24}}$

107. $\dfrac{A_8^{10}\,A_4^{6}}{A_2^{5}\,A_4^{7}}$

108. How many 7 character license plates are possible with 2 digits (0 to 9), 3 letters (A to Z), and 2 symbols from the list ($ ☼ ☻ ♀ ♂ ♠ ♥ ♣ ♦ ☺ ♪) ?

109. In how many ways can 2 men, 3 women, and 2 children be selected from 8 men, 15 women, and 5 children?

110. At the International Technology Conference in Las Vegas, George plans to attend presentations given by several 3D companies. His schedule allows him to attend the 10:00, 11:00, and 2:00 o'clock presentations. He can choose from 8 companies at 10:00, 12 at 11:00, and 5 at 2:00. How many ways can he select the three presentations?

111. The ID on student parking stickers consists of 3 letters excluding P and X, followed by 5 digits. How many ID's are possible?

112. In a sample of 1000 people, 120 are left-handed. If two persons are selected at random from the sample, find the probability of the following events

(a) Both people are left-handed.

(b) At least one of the two people is left-handed.

(c) Neither person is left-handed.

(d) Two of the events from (a), (b), and (c) are complementary events. Which two are they?

113. A carton contains 12 eggs, 3 of which are cracked. If we randomly select 5 of the eggs for hard-boiling, what is the probability of the following events?

(a) 3 of the cracked eggs are selected.

(b) None of the cracked eggs are selected.

(c) All of the cracked eggs are selected.

114. Construct a sample space for rolling 2 dice at the same time and answer the following.

(a) Write the sample space.

(b) P(both show the same number)?

(c) P(sum of the dots is less than 5)?

(d) P(dots are both even)?

115. A coin is tossed 10 times. What is the probability of getting at least one tail?

116. A freshman college student experiences difficulties with malfunctioning alarm clocks. Instead of using one alarm clock, he decides to use four alarm clocks. What is the probability that at least one of the alarm clocks will work correctly if each individual alarm clock has a 99% chance of working correctly?

117. If a couple plans on having 4 children, what is the probability that they will have at least two boys?

118. A box contains 2 blue and 3 green marbles. Two marbles are drawn at random without replacement.

(a) Construct a probability Tree diagram for the experiment.

(b) What is the probability of getting at least one green marble?

(c) What is the probability of getting one blue and one green marble?

(d) What is the probability of two of the same color?

119. There are 6 red marbles out of 10 in the box. 7 marbles are drawn. What is the probability that 4 marbles out of these 7 are red?

120. Find the probability that 4 randomly selected people have the same Birthday.

121. A committee of 5 senators is to be selected from 10 Democrats and 14 Republicans. Find the probability that the committee will consist of

(a) 3 Republicans and 2 democrats. **(b)** all republicans.

122. You have a box containing 4 red, 3 green and 1 white marble. Two marbles are randomly drawn without replacement find the probability of getting

(a) 0 green marbles when 2 marbles are randomly selected.

(b) 1 green marble when 2 marbles are randomly selected.

(c) 2 green marbles when 2 marbles are randomly selected.

123. Nine high school baseball all-stars are to be selected to play in a baseball charity game. The players will be selected from a list of 27 athletes. If the players are selected randomly, what is the probability that the 9 tallest players will be selected?

124. Raymond bought 10 stocks of which 4 are communication stocks and 6 technology stocks. If 3 stocks are randomly selected, what is the probability 2 are communication stocks and 1 is a technology stock?

125. Bella has 15 stocks in her portfolio. Five are communication stock, and the remaining ten are transportation stocks. She selects 4 stocks at random, as an anniversary gift for her daughter and son-in-law. Find the probability that among the 4 randomly selected stocks

(a) three are of communication stocks.

(b) none are of communication stocks.

(c) all are of transportation stocks.

(d) there are two transportation and two communication stocks.

126. A student at a university needs to take 18 courses in order to complete his major. He can only take 3 courses per quarter. He decides to first schedule the class he thinks is the hardest, then the one of less difficulty, and the easiest one last. Let's assume that all classes are available. What is the probability that next quarter he has the same schedule as his best friend who is also in the same major?

127. Two dice are rolled. Find the probability of rolling a pair of 2's or getting a sum of the dots that is at least 5.

128. A person playing single deck blackjack has a king in his hand and he is alone at the table. What is the probability that the next card dealt to him will be an ace or a face card?

129. A federal messenger delivery route must include stops in five cities.

(a) How many different routes are possible?

(b) If the route is randomly selected, what is the probability that the cities will be arranged in alphabetical order?

130. A messenger service dispatcher sends a delivery van to seven different locations. If the order in which the deliveries are made is randomly determined, find the probability that the resulting route is one of the two shortest possible routes. (The shortest routes are the same route in two opposite directions.)

131. The adjacent figure describes the distribution of Republicans and Democrats in 5 interest groups. Assuming that a member is randomly selected out of the entire population in the figure, find the indicated probabilities when

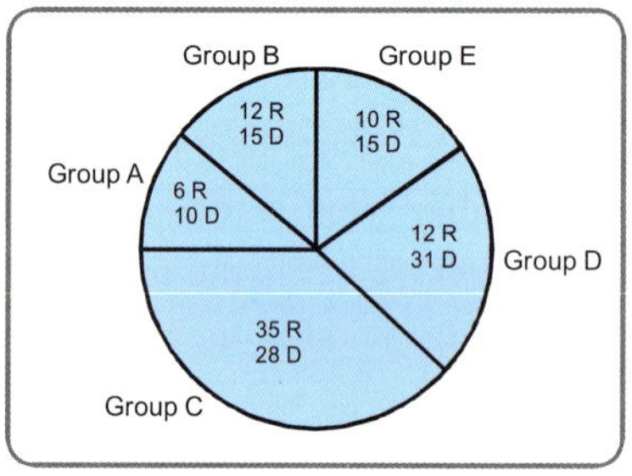

(a) P(Republican).

(b) P(Democrat or Group C).

(c) P(not Group E).

(d) P(Group A or Democrat).

(e) P(Republican or Group B).

132. A quality control manager wishes to inspect 6 manufacturing plants for the end of the year report. 3 of the plants are in New York, 2 are in Los Angeles, and 1 is in Florida. He randomly selects 2 plants for inspection. Assuming that all selections are equally likely and no plant can be selected twice, construct a probability tree diagram to help determine the probability that 1 Florida plant and 1 New York plant are selected.

133. How many different hands are possible in a 5 card poker game? Also find the probability that a particular player will get

(a) a diamond flush when 5 cards are dealt.

(b) a flush of any suite.

(c) a full house.

(d) any two-pair.

(e) a Royal Flush of hearts.

(f) a Royal Flush of any suite.

134. Approximate using Stirling's formula and calculate the percentage error.

(a) 6! **(b)** 8! **(c)** 12!

4

DISCRETE PROBABILITY DISTRIBUTIONS

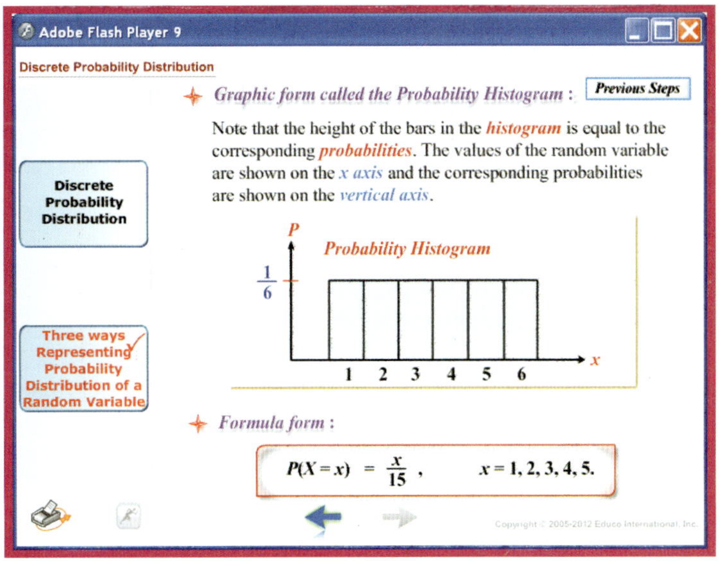

Chapter 4

DISCRETE PROBABILITY DISTRIBUTIONS

Chapter Overview:

1. *Introduction*
2. *Random Variables*
3. *Discrete Probability Distributions*
4. *Mean, Variance and Standard Deviation*
5. *Binomial Distribution*
6. *Poisson Distribution*
7. *Discrete Probability Distributions with Microsoft Excel®*
8. *Extended Topics and Proofs in Discrete Probability Distribution (Optional)*

4.1: Introduction

A probability distribution is a table, a formula, or a graph composed of a measurable characteristic called a *random variable* and the corresponding probability for each value that such a variable takes. Probability distributions are theoretical models that help us calculate the expected outcomes of events in a given experiment and the probabilities associated with those events.

Probability distributions can be discrete or continuous. In this chapter, we will focus on discrete probability distributions and their characteristics. Continuous probability distributions will be discussed in the next chapter.

4.2: Random Variables

> **Objectives**
>
> After completing this section you will be able to describe:
>
> **A.** Discrete random variables.
>
> **B.** Continuous random variables.

Random Variable (X) is a rule that assigns a number to every outcome in the experiment.

The nature of a probability distribution is determined by the type of random variable it models. There are two types of random variables: *Discrete random variables*, and *continuous random variables*.

A. Discrete Random Variables

Probabilistic variables that are distinct and can be counted (countable) are called **Discrete Random Variables**.

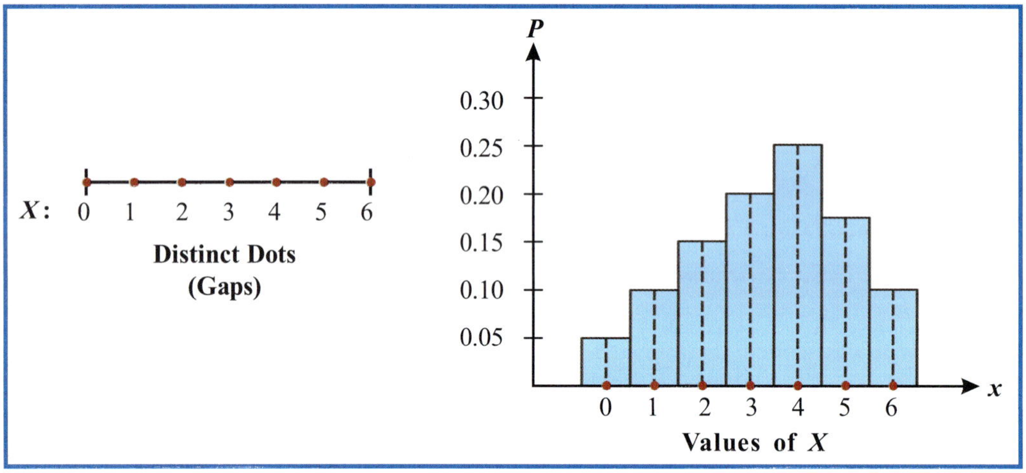

Figure 4.1

Examples of Discrete Random Variables

1. In a coin flip experiment, we may choose a random variable that assigns numbers 0 and 1 to the outcomes as follows:

   0 to H (heads).

   1 to T (tails).

   Note: The choice of assigning these specific values are arbitrary.

2. In a game of Roulette with a \$5 initial bet on a number, we need to choose a random variable that assigns {–\$5, \$175} to the outcomes {losing, winning}.

   Note: Here the rules of the game determine the random variable. Namely, losing the game results in a loss of the initial bet and winning the game results in a net gain of 35 times the original bet. Hence, the assignment of values is not arbitrary.

3. If we are interested in classifying random samples of 10 students based on the number of students that are left handed, we will choose a random variable that assigns each random sample with the number of students that are left handed. Therefore, this random variable can take any one of values {0, 1, 2, 3, 4, 5, 6, 7, 8, 9, 10}.

4. If we are interested in classifying families with 5 children based on the number of girls in each family, we will choose a random variable that assigns each family with the number of girls. Therefore, this random variable will take any one of values {0, 1, 2, 3, 4, 5}.

5. Consider a large number of airports each serving several departing flights per day. If we are interested in classifying airports based on the number of delayed flights per day, we would choose a random variable that assigns each airport with one of the values $\{0, 1, 2, 3, 4, 5, 6, 7, \ldots, n\}$.

B. Continuous Random Variables

A random variable is called a **continuous random variable** if its values assume real numbers in some interval.

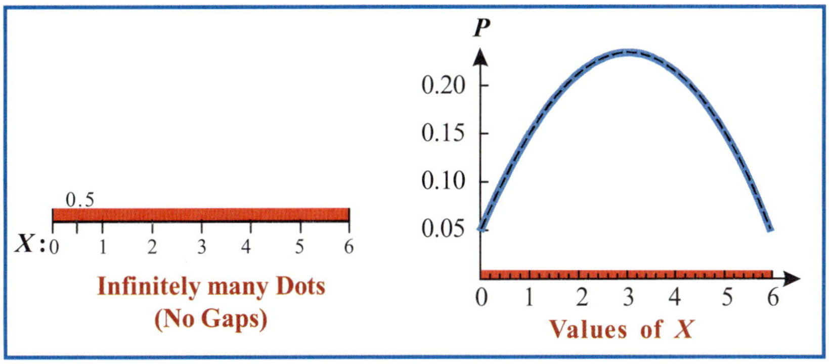

Figure 4.2

Examples of Continuous Random Variables

1. In an experiment of reading the voltage of electrical circuits using a voltmeter, the voltage reading is a continuous random variable.
2. In an experiment monitoring the body temperature of a child in a given day, the body temperature is a continuous random variable.
3. In a statistics exam, the time it takes to finish the test is a continuous random variable.

Note: The values that a continuous random variable can take are not countable.

Continuous random variable will be studied in detail in the next chapter.

- ## Discrete Random Variables

Consider the experiment of rolling two dice. The sample space is as follows:

$$\left\{\begin{matrix} (1,1) & (1,2) & (1,3) & (1,4) & (1,5) & (1,6) \\ (2,1) & (2,2) & (2,3) & (2,4) & (2,5) & (2,6) \\ (3,1) & (3,2) & (3,3) & (3,4) & (3,5) & (3,6) \\ (4,1) & (4,2) & (4,3) & (4,4) & (4,5) & (4,6) \\ (5,1) & (5,2) & (5,3) & (5,4) & (5,5) & (5,6) \\ (6,1) & (6,2) & (6,3) & (6,4) & (6,5) & (6,6) \end{matrix}\right\}$$

When an experiment is performed, there are many possible outcomes. For instance, in the above experiment every simple outcome is presented and it can easily be determined that the probability for each simple outcome is $\frac{1}{36}$.

However, in most statistical studies, while performing an experiment, we may only be interested in a particular aspect of the outcomes of the experiment (an event), rather than every outcome in the sample space. For example, in the above experiment, we may be interested in the sum of the dots when 2 dice are rolled. Let, the random variable (X) represent the sum of the dots when two dice are rolled. Now X can take any one of the values 2, 3, 4, 5, 6, 7, 8, 9, 10, 11, and 12 as displayed in the Figure 4.3.

Output		(1, 1)	(1, 2)	(1, 3)	(1, 4)	(1, 5)	(1, 6)
X	→	2	3	4	5	6	7
		(2, 1)	(2, 2)	(2, 3)	(2, 4)	(2, 5)	(2, 6)
		3	4	5	6	7	8
		(3, 1)	(3, 2)	(3, 3)	(3, 4)	(3, 5)	(3, 6)
		4	5	6	7	8	9
		(4, 1)	(4, 2)	(4, 3)	(4, 4)	(4, 5)	(4, 6)
		5	6	7	8	9	10
		(5, 1)	(5, 2)	(5, 3)	(5, 4)	(5, 5)	(5, 6)
		6	7	8	9	10	11
		(6, 1)	(6, 2)	(6, 3)	(6, 4)	(6, 5)	(6, 6)
		7	8	9	10	11	12

Figure 4.3

Notes:

1. A random variable is denoted by capital letters X, Y, Z *etc.*, and the particular values which the random variable takes are denoted by lower case letters x, y, z *etc.*
2. In the study of random variables, it is important to distinguish between a random variable and its values. For example, in three tosses of a fair coin, the number of tails is the random variable which can take any one of the four values 0, 1, 2, or 3.

Example 1: Ten senators are randomly selected from a group of 10 democrats and 20 republicans. Let a random variable X denote the number of democrats selected. What are the possible values that X can take?

Solution: In this example X is a discrete random variable which can take any one of following possible values 0, 1, 2, 3, 4, 5, 6, 7, 8, 9 or 10.

Example 2: A coin is tossed three times. Write the sample space of the experiment and the values of the random variable X, where

(a) X is the number of tails,

(b) X is the difference between the number of heads and the number of tails.

Solutions: The sample space (S) of the experiment is

$$S = \{\text{HHH, HHT, HTH, THH, HTT, THT, TTH, TTT}\}.$$

(a) X = Number of tails

Outcomes	X
HHH	0
HHT	1
HTH	1
THH	1
HTT	2
THT	2
TTH	2
TTT	3

$\Rightarrow X = \{0, 1, 2, 3\}$

Table 4.1

(b) X = number of heads – number of tails

Outcomes	X
HHH	$3 - 0 = 3$
HHT	$2 - 1 = 1$
HTH	$2 - 1 = 1$
THH	$2 - 1 = 1$
HTT	$1 - 2 = -1$
THT	$1 - 2 = -1$
TTH	$1 - 2 = -1$
TTT	$0 - 3 = -3$

$\Rightarrow X = \{-3, -1, 1, 3\}$

Table 4.2

Example 3: A box contains three defective and seven good transistors. A transistor is selected at random. If it is defective, it is discarded and a new transistor is tried. The process continues until a good transistor is selected. Let the random variable X represent the number of draws required until a good transistor is selected. Draw a tree diagram for the experiment and write the numerical values of the random variable X.

Solution: Let D represent defective transistors and G represent good transistors.

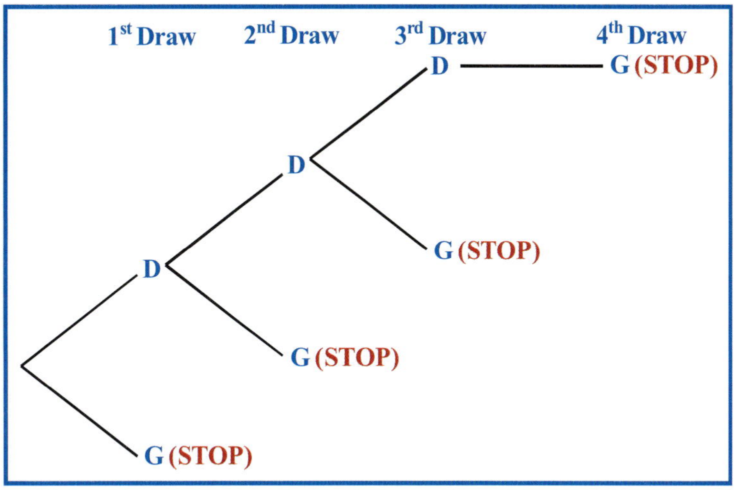

Figure 4.4

$\Rightarrow$ sample space S = {G, DG, DDG, DDDG}. Recall that the random variable X represents the number of draws required to select a good transistor. The numerical values of the random variable X along with the outcomes are given in the table 4.3.

Outcome	X
G	1
DG	2
DDG	3
DDDG	4

Table 4.3

Note: In most cases we can list only the values of the random variable without describing the sample space of the experiment. However, description of the sample space is useful in finding the probabilities corresponding to the values of the random variable.

4.3 Discrete Probability Distributions

> **Objectives**
>
> After completing this section you will be able to:
>
> **A.** Compute Discrete Probability Distribution.
>
> **B.** Discribe Properties of a Probability Distribution.
>
> **C.** Use Steps in constructing a Probability Distribution.
>
> **D.** Calculate probabilities using a Probability Distribution.

A. Discrete Probability Distribution

It is a two-column (row) table listing all distinct values the random variable takes along with the corresponding probabilities. A simple example to illustrate a discrete probability distribution is the experiment of rolling a fair die. Here the random variable X takes on values $\{1, 2, 3, 4, 5, 6\}$. Since every outcome in the experiment of rolling a fair die is equally likely, the corresponding probabilities will all equal $\frac{1}{6}$.

Sample Space:

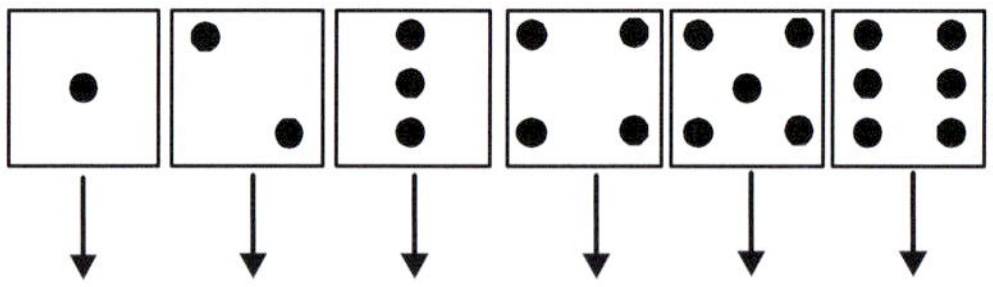

x	1	2	3	4	5	6
$P(X=x)$	$\frac{1}{6}$	$\frac{1}{6}$	$\frac{1}{6}$	$\frac{1}{6}$	$\frac{1}{6}$	$\frac{1}{6}$

Assuming X to be a discrete random variable, $P(X = x)$ means

P(the random variable X takes the value x).

Note that $P(X = x)$ can also be written as $p(x)$. Then, the function $p(x)$ is called the probability function of the random variable X and the set of all possible ordered pairs $(x, p(x))$, is called the *probability distribution* of the random variable X. In simple words, the listing of values of random variables along with their probabilities is called the probability distribution of a random variable and it may be expressed in any one of the following three ways,

i) Tabular form:

x	1	2	3	4	5	6
$P(X=x)$	$\frac{1}{6}$	$\frac{1}{6}$	$\frac{1}{6}$	$\frac{1}{6}$	$\frac{1}{6}$	$\frac{1}{6}$

Table 4.4

ii) Graphic form called the Probability Histogram: Note that the height of the bars in the histogram is equal to the corresponding probabilities. The values of the random variable are shown on the x axis and the corresponding probabilities are shown on the vertical axis.

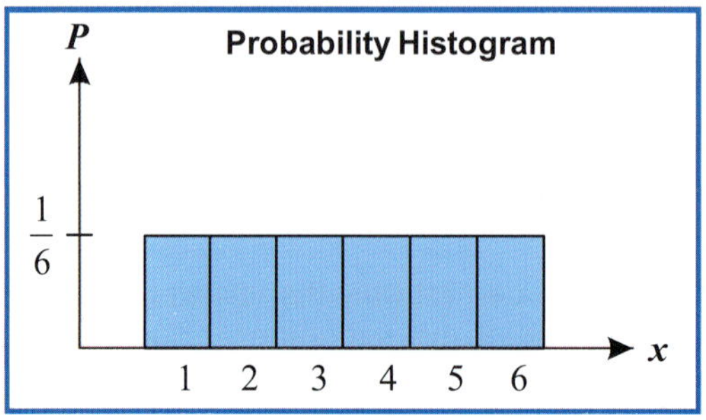

Figure 4.5

iii) Formula form: (will be discussed in more detail in Section 4.5)

$$P(X = x) = \frac{x}{15}, \qquad x = 1, 2, 3, 4, 5$$

B. Properties of a Probability Distribution

i) $0 \le p(x) \le 1$ Probability is a real number lying between zero and one. Zero being the probability corresponding to impossible event and one being the probability corresponding to a certain (or sure) event.

ii) $\sum p(x) = 1$ The sum of the probabilities in a probability distribution must be one.

Example 4: Does the following table represent a discrete probability distribution? Explain.

x	0	1	2
$p(x)$	$\frac{1}{5}$	$\frac{2}{5}$	$\frac{1}{5}$

Solution: In order to show that the above table describes a probability distribution, we must verify that

i) $0 \le p(x) \le 1$ and **ii)** $\sum p(x) = 1$.

All the values in the $p(x)$ row are between zero and one. Therefore property (i) is satisfied. The sum of the values in the $p(x)$ row $= \frac{1}{5} + \frac{2}{5} + \frac{1}{5} = \frac{4}{5} \neq 1$. Therefore condition (ii) is not satisfied. Hence the above table does not represent a probability distribution.

Example 5: Does the function $p(x) = \frac{x+1}{3}$ represent a discrete probability distribution for $x = 0, 1, 2$.

Solution: In order to check the properties (i) and (ii), we must first construct the probability distribution table.

x	0	1	2
$p(x)$	$p(0) = \frac{0+1}{3} = \frac{1}{3}$	$p(1) = \frac{1+1}{3} = \frac{2}{3}$	$p(2) = \frac{2+1}{3} = \frac{3}{3}$

Table 4.5

As we can see, all the probabilities in the $p(x)$ column are values between zero and one. Therefore property (i) is satisfied. In order to verify property (ii), we must show that

$\sum p(x) = 1$. But,

$$\sum p(x) = p(0) + p(1) + p(2) = \frac{1}{3} + \frac{2}{3} + \frac{3}{3} = 2 \neq 1$$

Hence, property (ii) is not satisfied. Therefore, the above formula does not represent a probability distribution.

C. Steps in constructing a Probability Distribution

1. Identify the random variable.
2. Construct the sample space if possible.
3. List all possible values of the random variable corresponding to each output.
4. List distinct values of the random variable along with the corresponding outcomes. Determine the probability of each value using the listed number of outcomes.
5. Make a 2-column table. Label the first column by X (the random variable), and the second column $P(X = x)$. Complete the first column by listing all the distinct numerical values that the random variable takes, and the 2nd column by listing all the corresponding probabilities.

Example 6: Construct a probability distribution for the number of girls in a family of 3 children.

Solution: ***Step 1*** Random Variable X: number of girls in a family of 3 children

Step 2 The sample space S = {BBB, BBG, BGB, GBB, GGB, GBG, BGG, GGG}

Step 3 The number of girls in a family of 3 children ranges between zero (zero girls in 3 children) to three (all 3 children are girls).

$$\begin{array}{lcccccccc} S: & \{BBB & BBG & BGB & GBB & GGB & GBG & BGG & GGG\} \\ & \downarrow & \downarrow & \downarrow & \downarrow & \downarrow & \downarrow & \downarrow & \downarrow \\ X: & 0 & 1 & 1 & 1 & 2 & 2 & 2 & 3 \end{array}$$

$$\Rightarrow X = \{0, 1, 2, 3\}$$

Step 4

X	Composition of Outcomes	Number of outcomes	Probability $P(X=x)$
0	BBB	1	$\frac{1}{8}$
1	GBB, BGB, BBG	3	$\frac{3}{8}$
2	GGB, GBG, BGG	3	$\frac{3}{8}$
3	GGG	1	$\frac{1}{8}$
Total		**8**	**1**

Table 4.6

Step 5 2-Rows Probability Distribution Table is presented below.

x	0	1	2	3
$P(X = x)$	$\frac{1}{8}$	$\frac{3}{8}$	$\frac{3}{8}$	$\frac{1}{8}$

Table 4.7

Example 7: Construct a probability distribution for the difference between the number of heads and the number of tails, when a coin is tossed 3 times. Draw the Probability Histogram.

Solution: ***Step* 1** Random Variable X: number of heads – number of tails

***Step* 2** The sample space S = {HHH, HHT, HTH, THH, HTT, THT, TTH, TTT}

***Step* 3**

S:	{HHH	HHT	HTH	THH	HTT	THT	TTH	TTT}
	↓	↓	↓	↓	↓	↓	↓	↓
X:	3	1	1	1	−1	−1	−1	−3

***Step* 4**

x	Composition of Outcomes	Number of outcomes	Probability $P(X=x)$
−3	TTT	1	$\frac{1}{8}$
−1	TTH, THT, HTT	3	$\frac{3}{8}$
1	HHT, HTH, THH	3	$\frac{3}{8}$
3	HHH	1	$\frac{1}{8}$
Total		**8**	**1**

Table 4.8

***Step* 5** 2-Column Probability Distribution Table and the Probability Histogram are presented below.

x	$P(X=x)$
−3	$\frac{1}{8}$
−1	$\frac{3}{8}$
1	$\frac{3}{8}$
3	$\frac{1}{8}$

Table 4.9

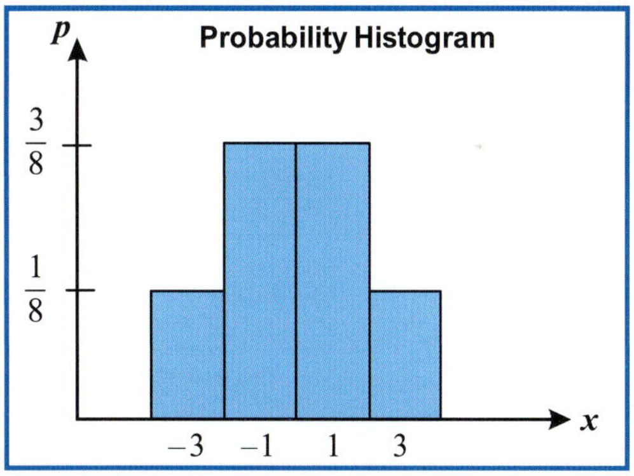

Figure 4.6

Example 8: Construct a probability distribution for a \$100 bet on a number in a game of Roulette. There are 38 members to bet on, and the net payoff for a bet on a number is 35 times. Draw the Probability Histogram.

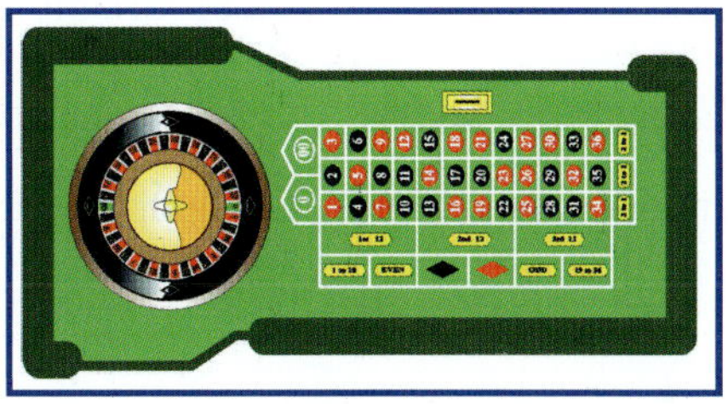

Figure 4.7

Solution: Suppose we bet on number 3 without loss of generality.

***Step* 1** Random Variable X: Winning, Losing

***Step* 2** The sample space $S = \{0, 00, 1, 2, \ldots, 36\} \rightarrow 38$ outcomes

***Step* 3** Recall note of example 2 Section 4.2.

$$\begin{array}{lcccccccc} S = & \{1 & 2 & 3 & \cdots & 35 & 36 & 0 & 00\} \\ & \downarrow & \downarrow & \downarrow & \downarrow & \downarrow & \downarrow & \downarrow & \downarrow \\ X: & -\$100 & -\$100 & \$3500 & -\$100 & -\$100 & -\$100 & -\$100 & -\$100 \end{array}$$

***Step* 4**

x	**Composition of Outcomes**	**Number of outcomes**	**Probability $P(X=x)$**
\$3500	3	1	$\frac{1}{38}$
$-$\$100	$1, 2, 4, 5, \ldots, 35, 36, 0, 00$	37	$\frac{37}{38}$
Total		**38**	**1**

Table 4.10

***Step* 5** 2-Column Probability Distribution Table and the Probability Histogram are presented below.

x	$P(X=x)$
\$3500	$\frac{1}{38}$
$-$\$100	$\frac{37}{38}$

Table 4.11

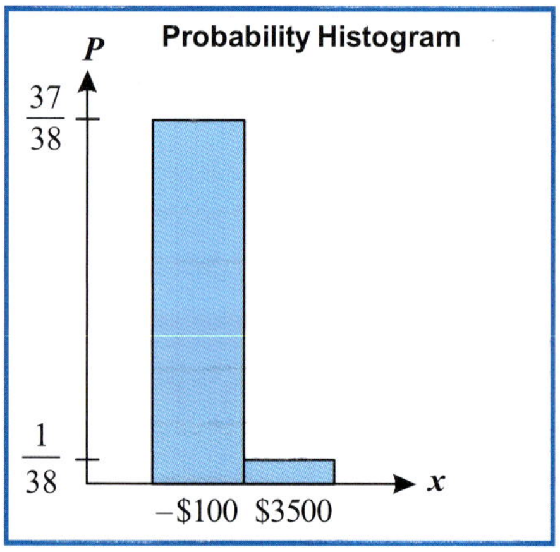

Figure 4.8

Example 9: A box contains 3 defective and 7 good transistors. A transistor is selected at random. If it is defective, it is discarded and a new transistor is tried. The process continues until a good transistor is selected. Let the random variable X represent the number of draws. Construct a discrete probability distribution of the random variable X.

Solution: ***Step* 1** Random Variable X: Number of draws required to select a good transistor

***Step* 2** From Example 3 (Section 4.2), the sample space $S = \{G, DG, DDG, DDDG\}$.

***Step* 3**

$$\begin{array}{lcccc} S = & \{G & DG & DDG & DDDG\} \\ & \downarrow & \downarrow & \downarrow & \downarrow \\ X: & 1 & 2 & 3 & 4 \end{array}$$

Step 4 The experiment ends in 1 draw if the transistor is a good one. The probability of drawing one good (G) transistor from 7 good and 3 defective ones is given by $P(x=1) = \dfrac{7}{7+3} = \dfrac{7}{10}$.

The experiment will end in 2 draws when the first transistor is defective and the second one is good. The probability of DG is $P(x=2) = P(\text{DG}) = P(\text{D}) \cdot P(\text{G} \mid \text{D}) = \dfrac{3}{10} \cdot \dfrac{7}{9} = \dfrac{7}{30}$.

The experiment will end in 3 draws when the first 2 transistors are defective and the third one is good. The probability of DDG is

$$P(x=3) = P(\text{DDG}) = P(\text{D}) \cdot P(\text{D}|\text{D}) \cdot P(\text{G}|\text{DD}) = \frac{3}{10} \cdot \frac{2}{9} \cdot \frac{7}{8} = \frac{7}{20}.$$

The experiment will end in 4 draws when the first 3 are defective and the fourth one is good. The probability of DDDG is:

$$P(x=3) = P(\text{DDDG}) = P(\text{D}) \cdot P(\text{D} \mid \text{D}) \cdot P(\text{D} \mid \text{DD}) \cdot P(\text{G} \mid \text{DDD}) = \frac{3}{10} \cdot \frac{2}{9} \cdot \frac{1}{8} \cdot \frac{7}{7} = \frac{1}{120}.$$

Step 5 Probability Distribution Table 4.12 is presented below.

x	1	2	3	4
$P(X=x)$	$\frac{7}{10}$	$\frac{7}{30}$	$\frac{7}{120}$	$\frac{1}{120}$

Table 4.12

Example 10: Two dice are rolled at random. Let the random variable X represent the sum of the dots on two dice. Construct the probability distribution of X.

Solution: **Step 1** Random Variable X: Sum of the dots on two dice

Step 2 The sample space:

$$\left\{\begin{matrix} (1,1) & (1,2) & (1,3) & (1,4) & (1,5) & (1,6) \\ (2,1) & (2,2) & (2,3) & (2,4) & (2,5) & (2,6) \\ (3,1) & (3,2) & (3,3) & (3,4) & (3,5) & (3,6) \\ (4,1) & (4,2) & (4,3) & (4,4) & (4,5) & (4,6) \\ (5,1) & (5,2) & (5,3) & (5,4) & (5,5) & (5,6) \\ (6,1) & (6,2) & (6,3) & (6,4) & (6,5) & (6,6) \end{matrix}\right\}$$

Step **3, 4**

Sum of the dots(x)	Outcomes	Total outcomes	Probability $P(X=x)$
2	(1,1)	1	$\frac{1}{36}$
3	(1,2), (2,1)	2	$\frac{2}{36}$
4	(1,3), (3,1), (2,2)	3	$\frac{3}{36}$
5	(1,4), (4,1), (2,3), (3,2)	4	$\frac{4}{36}$
6	(1,5), (5,1), (2,4), (4,2), (3,3)	5	$\frac{5}{36}$
7	(1,6), (6,1), (2,5), (5,2), (3,4), (4,3)	6	$\frac{6}{36}$
8	(2,6), (6,2), (3,5), (5,3), (4,4)	5	$\frac{5}{36}$
9	(3,6), (6,3), (4,5), (5,4)	4	$\frac{4}{36}$
10	(4,6), (6,4), (5,5)	3	$\frac{3}{36}$
11	(5,6), (6,5)	2	$\frac{2}{36}$
12	(6,6)	1	$\frac{1}{36}$
Total		**36**	**1**

Table 4.13

Step **5** Probability Distribution Table and Probability Histogram are presented below:

Sum of the dots (x)	2	3	4	5	6	7	8	9	10	11	12
Probability $P(X=x)$	$\frac{1}{36}$	$\frac{2}{36}$	$\frac{3}{36}$	$\frac{4}{36}$	$\frac{5}{36}$	$\frac{6}{36}$	$\frac{5}{36}$	$\frac{4}{36}$	$\frac{3}{36}$	$\frac{2}{36}$	$\frac{1}{36}$

Table 4.14

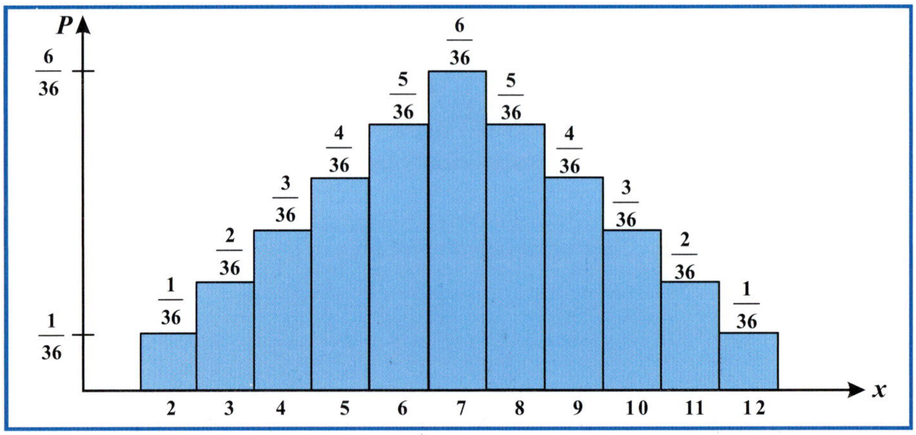

Figure 4.9

Example 11: Based on Alex's academic history, there is a 70% chance that he will complete a course for which he registered. Construct a probability distribution for the number of courses that Alex will complete, if he plans to register for 4 courses.

Solution: Notation X: A random variable, denotes the number of courses, Alex completed.

C: Completing a course.

$\bar{C}$: Not completing the course.

The set of values that the Random Variable can take on is $X = \{0, 1, 2, 3, 4\}$.

The sample space for this experiment is;

$$\left\{\begin{matrix} \bar{C}\bar{C}\bar{C}\bar{C}, C\bar{C}\bar{C}\bar{C}, \bar{C}C\bar{C}\bar{C}, \bar{C}\bar{C}C\bar{C}, \bar{C}\bar{C}\bar{C}C, \bar{C}\bar{C}CC, \bar{C}C\bar{C}C, \bar{C}CC\bar{C}, \\ C\bar{C}\bar{C}C, C\bar{C}C\bar{C}, CC\bar{C}\bar{C}, \bar{C}CCC, C\bar{C}CC, CC\bar{C}C, CCC\bar{C}, CCCC \end{matrix}\right\}$$

The Probability that Alex will complete "0 out of 4" courses.

$$P(X=0) = P(\bar{C})\cdot P(\bar{C})\cdot P(\bar{C})\cdot P(\bar{C}) = (0.3)^4 = 0.0081$$

The Probability that Alex will complete "1 out of 4" courses is

$P(X=1) = 4(0.7)(0.3)^3 = 0.0756$

The Probability that Alex will complete "2 out of 4" courses is

$P(X=2) = 6(0.7)^2(0.3)^2 = 0.2646$

The Probability that Alex will complete "3 out of 4" courses is

$P(X=3) = 4(0.7)^3(0.3)^1 = 0.4116$

The Probability that Alex will complete "4 out of 4" courses is

$P(X=4) = (0.7)^4 = 0.2401$

We just calculated the probabilities using counting techniques and the Multiplication Rule.

However, we can also calculate the probabilities using the Probability Tree Diagram.

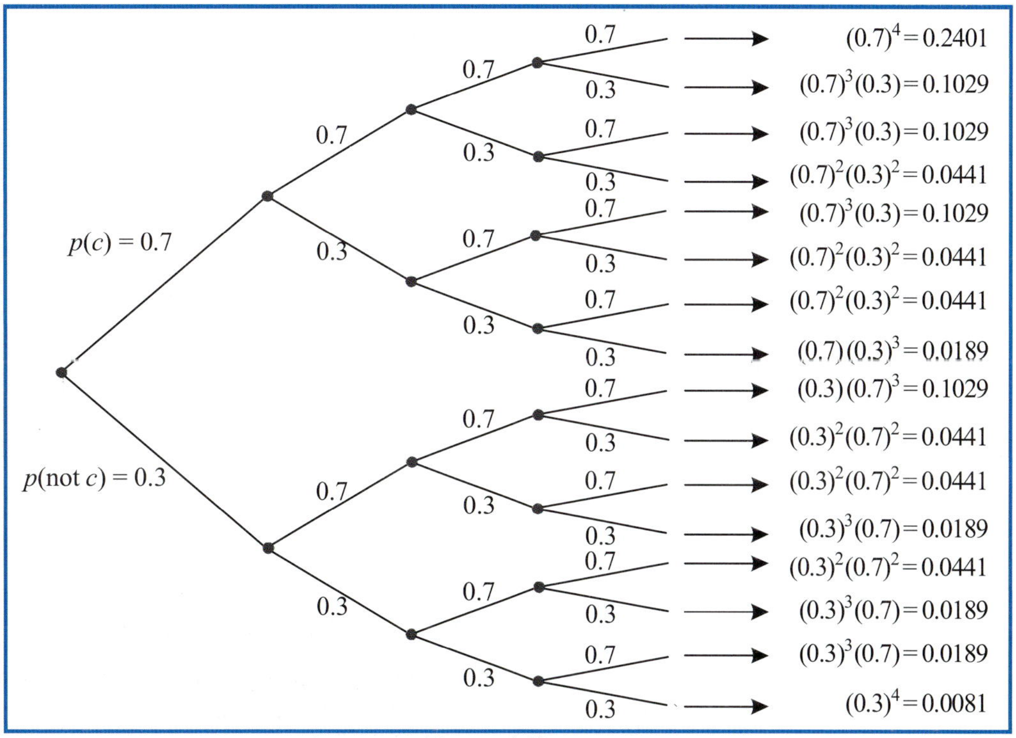

Figure 4.10

We are now ready to construct the Probability Distribution. X represents the number of courses Alex completes.

x	0	1	2	3	4
$P(X=x)$	0.0081	0.0756	0.2646	0.4116	0.2401

Table 4.15

D. Calculating Probabilities Using a Probability Distribution

Once a probability distribution is given or constructed, we can calculate probabilities of derived events by using the values in the probability column. The following example illustrates this concept.

Example 12: The given table is the probability distribution of the number of left handed (LH) students in a class of size 10, given that the probability that a person is left handed is 20%.

x	0	1	2	3	4	5	6	7	8	9	10
$P(X=x)$	0.1074	0.2684	0.3020	0.2013	0.0881	0.0264	0.0055	0.0008	0.0001	0	0

(a) What is the probability of having 3 LH students in a class?

(b) What is the probability of having at least 5 LH students in a class?

(c) What is the probability of having at most 4 LH students in a class?

(d) What is the probability of having more than 6 LH students in a class?

(e) What is the probability of having less than 4 LH students in a class?

Solution:

(a) $P(X=3) = 0.2013$

(b) $P(X \geq 5) = P(X=5) + P(X=6) + P(X=7) + P(X=8) + P(X=9) + P(X=10)$

$= 0.0264 + 0.0055 + 0.0008 + 0.0001 + 0 + 0$

$= 0.0328$

(c) $P(X \leq 4) = P(X=0) + P(X=1) + P(X=2) + P(X=3) + P(X=4)$

$= 0.1074 + 0.2684 + 0.3020 + 0.2013 + 0.0881$

$= 0.9672$

(d) $P(X > 6) = P(X=7) + P(X=8) + P(X=9) + P(X=10)$

$= 0.0008 + 0.0001 + 0.0000 + 0.0000$

$= 0.0009$

(e) $P(X < 4) = P(X=0) + P(X=1) + P(X=2) + P(X=3)$

$= 0.1074 + 0.2684 + 0.3020 + 0.2013$

$= 0.8791$

Try it yourself:

In exercises 1-3, construct the Discrete Probability Distribution, and draw the probability histogram.

1. Two dice are rolled. Let the random variable X be

(a) the difference between the larger and the smaller numbers of dots on the two dice.

(b) maximum of number of dots obtained on the two dice.

(c) minimum of number of dots on the two dice.

2. A Shipment of 8 Blu-Ray players contains 2 defective ones. Two players are selected at random. Let the random variable be the number of defective players in the sample.

3. Five percent of a shipment of 100 computer chips is defective. Two of the chips are randomly selected and tested. Let the random variable be the number of defective chips in the sample.

4. Does the function $p(x) = \frac{1}{x+1}$, where $x = -3, -2, 0$ and 1, form a proper probability distribution?

5. A manufacturer of metal pistons finds that on the average 12% of the pistons are rejected because they are either oversized or undersized. If random variable X represents the number of pistons rejected in a lot of 10 pistons, then by using the probability distribution of X (given below),

x	0	1	2	3	4	5	6	7	8	9	10
$P(X=x)$	0.2785	0.3798	0.2330	0.0847	0.0202	0.0034	0.0004	0	0	0	0

calculate the probability that a batch of 10 pistons will contain

(a) at least 2 rejects.

(b) at most 4 rejects.

(c) more than 5 rejects.

(d) exactly 6 rejects.

4.4 Mean, Variance and Standard Deviation

In Chapter 2, we calculated the mean, variance, and standard deviation of raw data as well as frequency tables. In this section we will describe the procedure of calculating the mean, variance and standard deviation of a discrete random variable using its probability distribution.

Objectives

After completing this section you will be able to compute:

A. Mean of a Random Variable.

B. Variance and Standard Deviation of a Random Variable.

Previously, the mean of raw data was calculated using the arithmetic mean $\bar{X} = \frac{\sum X_i}{n}$. However, the same formula can not be used to calculate the mean number of dots that show up when a die is rolled! One approach could be to roll the die many times, add the numbers that show up (1 through 6) and divide by the total number of rolls (Table 4.16). Displayed below are the results of such an experiment.

Number of Rolls	Experimental Mean
20	3.9500
40	3.8000
100	3.7500
500	3.5620
1000	3.5410
5000	3.4920

Table 4.16

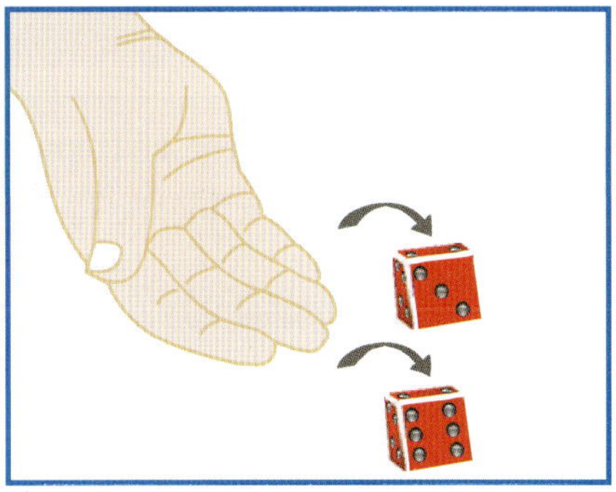

As we can see in Table 4.16, increasing the number of rolls produces values of the experimental mean that are successively closer to 3.5. In fact, if we define our random variable to be the number of dots on each roll, we can show that the theoretical mean of that random variable is in fact 3.5. It appears that in order to get this value by practice, it would require infinitely many rolls. Since this is an impossible task, the original formula for the mean cannot be used because we cannot divide by infinity. Hence a new approach is necessary to compute the mean of a random variable with a discrete probability distribution.

A. Mean of a Random Variable

The Mean of a random variable X with a discrete probability distribution $P(X)$, is the sum of the products of each value the random variable takes $x_1, x_2, \ldots, x_n$ and their corresponding probabilities $p(x_1), p(x_2), \ldots, p(x_n)$ respectively.

$$\text{Mean} = \mu = \sum_{i=1}^{n}\left[x_i\, p(x_i)\right]$$

$$= x_1 p(x_1) + x_2 p(x_2) + \ldots + x_n p(x_n) \qquad \text{or}$$

$$= x_1 p_1 + x_2 p_2 + \ldots + x_n p_n$$

The mean of a discrete probability distribution of a random variable X can be interpreted as the *Expected Value* of the random variable, denoted by $E(X)$:

$$\Rightarrow \quad E(X) = \mu = \sum\left[x \cdot p(x)\right]$$

The expected value of a probability distribution is the theoretical average of the random variable.

B. Variance and Standard Deviation of Random Variable

The *variance* of a probability distribution is calculated by using the formula;

$$\sigma^2 = \sum (x-\mu)^2\, p(x)$$

$$= \sum\left[x^2 p(x)\right] - \mu^2$$

$$= \left[x_1^2 p(x_1) + x_2^2 p(x_2) + \ldots + x_n^2 p(x_n)\right] - \mu^2$$

The standard deviation of a probability distribution is:

$$\sigma = \sqrt{\sigma^2} = \sqrt{\sum\left[x^2 p(x)\right] - \mu^2}$$

Remarks:

i) The variance and standard deviation can't be negative.

ii) The mean, variance, and standard deviation of a discrete probability distribution must be rounded to one more decimal place than the outcomes.

Example 13: Find the mean and standard deviation of the number of dots when a die is rolled.

Solution: First we define the random variable X as number of dots obtained when rolling a die.

The probability function of X is $p(x) = \dfrac{1}{6}$, $x = 1, 2, \ldots, 6$.

$$\mu = E(X) = \sum\left[x \cdot p(x)\right]$$

$$= x_1 p_1 + x_2 p_2 + x_3 p_3 + x_4 p_4 + x_5 p_5 + x_6 p_6$$

$$= (1)\left(\frac{1}{6}\right) + (2)\left(\frac{1}{6}\right) + (3)\left(\frac{1}{6}\right) + (4)\left(\frac{1}{6}\right) + (5)\left(\frac{1}{6}\right) + (6)\left(\frac{1}{6}\right) = \frac{21}{6} = 3.5$$

$$\sigma = \sqrt{\sum\left[x^2 \cdot p\right] - \mu^2}$$

$$= \sqrt{\left[1^2\left(\frac{1}{6}\right) + 2^2\left(\frac{1}{6}\right) + 3^2\left(\frac{1}{6}\right) + 4^2\left(\frac{1}{6}\right) + 5^2\left(\frac{1}{6}\right) + 6^2\left(\frac{1}{6}\right)\right] - (3.5)^2}$$

$$= \sqrt{15.167 - (3.5)^2} = 1.71$$

Example 14: Find the mean and variance of number of girls in a family with 3 children.

Solution: *Define the random variable X as: number of girls in a family with 3 children.*

From the example given in the discussion of Section 4.3C (Table 4.7), recall the probability distribution of X.

x	0	1	2	3
$P(x)$	$\frac{1}{8}$	$\frac{3}{8}$	$\frac{3}{8}$	$\frac{1}{8}$

Table 4.17

$$\mu = \sum[x \cdot p]$$

$$= x_1p_1 + x_2p_2 + x_3p_3 + x_4p_4$$

$$= (0)\left(\frac{1}{8}\right) + (1)\left(\frac{3}{8}\right) + (2)\left(\frac{3}{8}\right) + (3)\left(\frac{1}{8}\right) = 1.5$$

$$\sigma^2 = \sum\left[x^2 \cdot p\right] - \mu^2$$

$$= \left[0^2\left(\frac{1}{8}\right) + 1^2\left(\frac{3}{8}\right) + 2^2\left(\frac{3}{8}\right) + 3^2\left(\frac{1}{8}\right)\right] - (1.5)^2$$

$$= 3 - 2.25 = 0.75$$

Example 15: Find the expected value and standard deviation for the experiment of making a \$100 bet on a number in a game of Roulette.

Solution: *Define the random variable X as: dollar value of game.*

From Example 5 of Section 4.3, recall the probability distribution for a \$100 bet on a number.

x	$-\$100$	$\$3500$
$P(X=x)$	$\frac{37}{38}$	$\frac{1}{38}$

Table 4.18

$$E(X) = \sum[x \cdot p]$$

$$= x_1p_1 + x_2p_2$$

$$= (-100)\left(\frac{37}{38}\right) + (3500)\left(\frac{1}{38}\right) \approx -\$5.26$$

$$\sigma = \sqrt{\sum[x^2 \cdot p] - \mu^2}$$

$$= \sqrt{\left[(-100)^2\left(\frac{37}{38}\right) + (3500)^2\left(\frac{1}{38}\right)\right] - (-5.26)^2}$$

$$\approx \$576.26$$

Example 16: Find the expected value for the probability distribution given below.

x	1	2	3	4
$P(X=x)$	$\frac{7}{10}$	$\frac{7}{30}$	$\frac{7}{120}$	$\frac{1}{120}$

Solution:

$$\mu = E(X) = \sum[x \cdot p]$$

(Example 9, Section 4.3).

$$= x_1p_1 + x_2p_2 + x_3p_3 + x_4p_4$$

$$= (1)\left(\frac{7}{10}\right) + (2)\left(\frac{7}{30}\right) + (3)\left(\frac{7}{120}\right) + (4)\left(\frac{1}{120}\right) = \frac{165}{120}$$

$$= 1.375 \approx 1.4$$

Example 17: An Insurance company charges Lee \$1,200 a year for a \$1,000,000 life insurance policy. If the probability that the policy holder will survive one full year is 0.9993, then

(a) construct the probability distribution for this experiment and find the mean,

(b) draw the histogram for the distribution, and

(c) find the standard deviation.

Solutions: *Define the random variable X: dollar value of the insurance policy from Lee's perspective.*

(a)

x	p	$x \cdot p$	$x^2 \cdot p$
-1200	0.9993	-1199.16	1438992
998800	0.0007	699.16	698321008
Sum	**1**	**-500**	**699760000**

Table 4.19

$$\mu = \sum[x \cdot p] = (-1200)(0.9993) + (998800)(0.0007) = -500$$

(b)

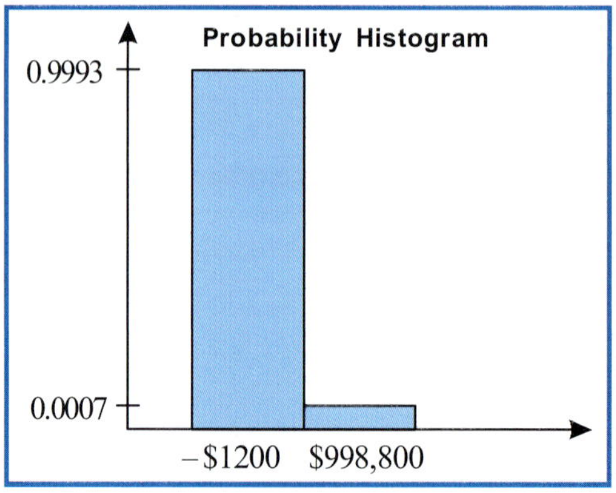

Figure 4.11

(c) $\sigma = \sqrt{\sum[x^2 \cdot p] - \mu^2} = \sqrt{699760000 - (-500)^2} \approx 26448.3$

Example 18: Kristine purchased a \$20 ticket from a private lottery. The rules of the game were as follows: She would win a \$1000 with a 1 in 3000 chance, \$500 with a 1 in a 1000 chance, \$300 with a 1 in 500 chance and \$20 with a 1 in 200 chance. Otherwise she would lose the original \$20 investment. Find the expected value and standard deviation of this game. Graph the distribution.

Solution: *We define the random variable* X = net winning amount in \$

The probability distribution of X is:

x	$P(X=x)$
$1000-20 = 980$	$\frac{1}{3000} = 0.0003333\ldots$
$500-20 = 480$	$\frac{1}{1000} = 0.001$
$300-20 = 280$	$\frac{1}{500} = 0.002$
$20-20 = 0$	$\frac{1}{200} = 0.005$
-20	$1-(0.00033+0.001+0.002+0.005) = 0.99167$

Table 4.20

Table for computing mean and Standard Deviation

x	p	$x \cdot p$	$x^2 \cdot p$
980	0.0003333	0.3267	320.1333
480	0.001	0.48	230.4
280	0.002	0.56	156.8
0	0.005	0	0
-20	0.99167	-19.8333	396.6667
Sums	1	-18.4666	1104
		Expected value(μ)	-18.4666
		Standard Deviation(σ)	27.6221

Table 4.21

$$E(X) = \sum[x \cdot p] = -18.5$$

$$\sigma = \sqrt{\sum[x^2 \cdot p] - \mu^2}$$

$$= \sqrt{1104 - (-18.47)^2}$$

$$= 27.6$$

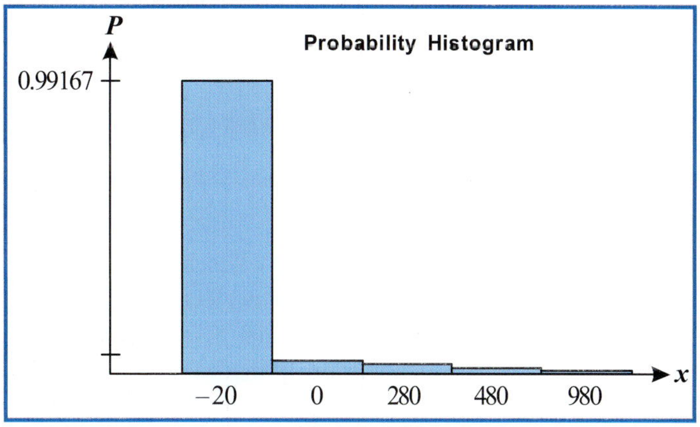

Figure 4.12

Example 19: Construct the probability Distribution of the number of tails obtained when four coins are tossed simultaneously. Also find the mean and the standard deviation of the distribution.

Solution: Let X represent the number of tails recorded in the experiment.

X: {0, 1, 2, 3, 4}.

$$\text{Sample space} = \begin{Bmatrix} \text{HHHH} & \text{HHHT} & \text{HHTH} & \text{HHTT} & \text{HTHH} & \text{HTHT} & \text{HTTH} & \text{HTTT} \\ \text{THHH} & \text{THHT} & \text{THTH} & \text{THTT} & \text{TTHH} & \text{TTHT} & \text{TTTH} & \text{TTTT} \end{Bmatrix}$$

Now, we can compute probabilities by counting the outputs corresponding to similar values of X.

x	0	1	2	3	4
p	0.0625	0.25	0.375	0.25	0.0625

Table 4.22

We construct the table given below to assist us with calculations of the mean and the variance.

x	p	$x \cdot p$	$x^2 \cdot p$
0	0.0625	0	0
1	0.25	0.25	0.25
2	0.375	0.75	1.5
3	0.25	0.75	2.25
4	0.0625	0.25	1
Sums	**1**	**2**	**5**

Table 4.23

$$\begin{aligned} E(X) &= \mu = \sum[x \cdot p] \\ &= x_1p_1 + x_2p_2 + x_3p_3 + x_4p_4 + x_5p_5 \\ &= 0 + 0.25 + 0.75 + 0.75 + 0.25 = 2 \end{aligned}$$

$$\begin{aligned} \sigma^2 &= \sum\left[x^2 \cdot p\right] - \mu^2 \\ &= (0 + 0.25 + 1.5 + 2.25 + 1) - 2^2 \\ &= 5 - 4 = 1 \end{aligned}$$

Try it yourself:

1. In the game of Roulette there are 18 red and 18 black numbers you can bet on. Suppose a \$1 bet is placed on a red number. If the number hits, you will win \$1 net. Otherwise, you lose the \$1 bet. What should you expect to win in one game? In 100 games?

2. The probabilities that a police officer will write up 0, 1, 2, 3 or 4 traffic citations on a Monday night are 0.01, 0.9, 0.04, 0.03 and 0.02, respectively. Let the random variable be the number of tickets. Find the expected value and the variance of the random variable X.

3. A wholesaler makes a profit of \$50 on each item that is received from the manufacturer in perfect condition. If it is received in less than perfect condition, then the wholesaler loses \$5 on each item. Let the random variable X be the total amount of profit made on each item and 0.7 be the probability of receiving an item in perfect condition. Find the expected value of the random variable X and the standard deviation.

4. An investor has \$1000 to invest. She has two options. In the first, she can invest in an account that pays 5% annually. In the other she can buy bonds. If she buys bonds, there is a 0.6 chance that she may lose \$300. Otherwise she wins \$700. Which option should she pick?

5. A random variable has a probability function presented in the table below.

x	-2	-2	0	1	2	3
$P(X=x)$	0.1	k	0.2	$2k$	0.3	k

Find the value of k and calculate the mean, variance and standard deviation.

4.5 Binomial Distribution

There are several different types of discrete probability distributions. Perhaps the most common is the *Binomial Distribution*. Binomial Distribution was discovered in 1700 by Swiss Mathematician James Bernoulli (1654-1705) and was first published in 1713, eight years after his death.

Before studying binomial distribution, we must discuss Bernoulli Trials and outline the conditions that must be satisfied in order for an experiment to be considered a binomial experiment.

Objectives

After completing this section you will be able to describe:

A. Bernoulli Trials (Processes).

B. Conditions for a Binomial Experiment.

C. Binomial Probability Distribution.

D. Mean and Standard Deviation of a Binomial Experiment.

A. Bernoulli Trials

An experiment that has either exactly two outcomes or two categories of outcomes denoted as "success" or "failure" in a series of fixed and independent trials is known as **Bernoulli Trials**.

If an experiment is repeated a number of times in similar conditions, then each trial (each repetition of the experiment) is known as Bernoulli Trial satisfying the following conditions:

i) Each trial results in only two mutually disjoint outcomes, called ***success*** (the occurrence of the desired event) and ***failure*** (otherwise).

ii) The trials are independent.

iii) The probability of success (p) is constant for each trial.

The process resulting from such an experiment is called the **Bernoulli Process**.

For example, if we toss a coin n times (which is fixed and finite) then the outcome of any trial is one of the mutually exclusive events, head (success) and tail (failure). Furthermore, all the trials are independent, since the result of any toss of a coin does not affect the result of other tosses. Moreover, the probability of success (heads) in any trial is $1/2$, which is constant for each trial.

B. Conditions for a Binomial Experiment

i) The experiment is performed repeatedly a finite and fixed number (say n) of times.

ii) The outcome of each trial may be classified into two mutually disjoint categories, called success (the occurrence of the event) and ***failure*** (the non-occurrence of the event).

iii) All the trials are independent. That is, the result of any trial is not affected by the preceding trials and doesn't affect the results of succeeding trials.

iv) The probability of success (the occurrence of the event) in any trial is p and is constant in each trial. The probability of failure (the non-occurrence of the event) is $q = 1 - p$.

Before we introduce and discuss the Binomial Probability Distributions, we must take a closer look at the requirements for a binomial experiment and what they really mean.

First and foremost, we must clarify the meaning of the word ***success*** in the context of binomial experiments. A success simply denotes the outcome of the event of interest and probability of success is the probability of that outcome. For example, a quality control manager may be interested in the proportion of defective computer chips coming out of the assembly line. In this example, a defective computer chip is considered a success in the context of a binomial experiment.

Next, for an experiment to qualify as a binomial experiment there must be a fixed number of trials known in advance. For example, tossing a coin 20 times is a binomial experiment since we know the number of trials in advance. In contrast, a quality control manager checking for defective computer chips from the production line is not a binomial experiment. Even though each chip he tests is either defective or non-defective. In this case, number of trials are not fixed in advance and therefore this experiment can't be considered as binomial experiment.

Binomial experiments also require independent trials. Suppose a researcher is conducting a survey in a supermarket. He asks two women to respond to his survey, where one is the employer and the other is an employee. Since the response given by the employer is likely to affect the response given by the employee, the responses (outcomes) are not independent.

Here are some examples of experiments that fulfill the requirements of a binomial experiment.

1. coin toss (H,T)	**2.** games of chance (win, lose)	**3.** pass, fail
4. late, on time	**5.** true, false	**6.** defective, non-defective

C. Binomial Probability Distribution

The outcomes of a binomial experiment together with their corresponding probabilities constitute a unique discrete probability distribution called the ***Binomial Probability Distribution***.

If the random variable X represents number of successes in fixed number (n) of Bernoulli trials, then it is called Binomial random variable and the formula for its distribution is

$$P(X = x) = C_x^n \, p^x q^{n-x}, \quad x = 0, 1, \ldots, n,$$

where

n = the number of trials

x = $0, 1, \ldots, n$.

p = the probability of success in a single trial and $q = 1 - p$.

$P(X = x)$ gives the probability of exactly x successes in n binomial trials.

Example 20: A fair die is rolled 10 times. Find the probability that a 6 will show up exactly 5 times.

Solution: *Define:*

Success = [getting a six on a single roll of a die.]

$$\Rightarrow \quad p = \frac{1}{6} \quad \text{and} \quad q = 1 - p = \frac{5}{6}$$

n = number of times die is rolled = 10.

Therefore,

$$P(X=x) = C_x^{10}\left(\frac{1}{6}\right)^x\left(\frac{5}{6}\right)^{10-x}, \quad x = 0, 1, \ldots, 10.$$

$$\text{Required probability} = P(X=5)$$

$$= C_5^{10}\left(\frac{1}{6}\right)^5\left(\frac{5}{6}\right)^{10-5}$$

$$= \frac{252 \times 3125}{6^{10}} = 0.013$$

D. Mean and Standard Deviation of a Binomial Experiment

The mean, variance and standard deviation of binomial distribution with parameter n and p are given by the formulas

mean	variance	standard deviation
$\mu = np$	$\sigma^2 = npq$	$\sigma = \sqrt{npq}$

The reader may also compute the above parameters by using the direct formulas.

$$\mu = \sum\left[xP(X=x)\right]$$

$$\sigma^2 = \sum\left[x^2 P(X=x)\right] - \mu^2$$

$$\sigma = \sqrt{\sum\left[x^2 P(X=x)\right] - \mu^2}$$

Note: $\begin{cases} q \le 1 \\ npq \le np \end{cases} \Rightarrow$ variance $\le$ mean

In Binomial distribution, variance is always less than the mean.

Example 21: A fair die is rolled 10 times.

(a) Find the mean and standard deviation of the experiment.

(b) Calculate the range for usual and unusual observation.

Solutions: $n = 10, p = \frac{1}{6}$ and $q = 1 - \frac{1}{6} = \frac{5}{6}$

(a) $\mu = np = (10)\left(\frac{1}{6}\right) = 1.667$ and $\sigma = \sqrt{(10)\left(\frac{1}{6}\right)\left(\frac{5}{6}\right)} = 1.179$

(b) Range of usual values: $\mu \pm 2\sigma$

$\Rightarrow \quad 1.667 \pm 2(1.179)$

→ min. usual value $= -0.690$

→ max. usual value $= 4.024$

Since X assumes only non negative integer values, the range for usual observations is $0 \le x \le 4$ and for unusual observations $x > 4$.

Example 22: There is a 25% chance that a US-Airways flight is late. In a group of 8 flights,

(a) find the probability that

i) 5 of them are late. **ii)** at least 3 of them are late.

iii) less than 4 of them are late. **iv)** at most 2 of them are late.

(b) find the mean and the standard deviation of this experiment.

(c) graph (Histogram) the distribution.

Solutions: The above experiment is of *Bernoulli type*. The number of trials is fixed ($n = 8$). There are two outcomes: *Success* (late flight) and *Failure* (on time flight). Assuming the arrivals of flights (late or on time) are independent of each other, let

X = number of late flights, $n = 8$, p = probability of success = P(of a late flight) = 0.25, and $q = 1 - 0.25 = 0.75$.

(a) X follows a Binomial Distribution with parameters $n = 8$ and $p = 0.25$. Hence,

$$P(X = x) = C_x^8\,(0.25)^x(0.75)^{8-x}\,,\quad x = 0, 1, \ldots, 8$$

i) $P(X = 5) = C_5^8\,(0.25)^5(0.75)^{8-5} = 0.0231$

ii)
$$\begin{aligned}P(X \geq 3) &= P(X = 3) + P(X = 4) + P(X = 5) + P(X = 6) + P(X = 7) + P(X = 8)\\ &= 1 - \{P(X = 0) + P(X = 1) + P(X = 2)\}\\ &= 1 - \left((0.75)^8 + C_1^8(0.25)(0.75)^7 + C_2^8(0.25)^2(0.75)^6\right) = 0.3215\end{aligned}$$

iii)
$$\begin{aligned}P(X < 4) &= P(X = 3) + P(X = 2) + P(X = 1) + P(X = 0)\\ &= C_3^8\,(0.25)^3(0.75)^{8-3} + C_2^8\,(0.25)^2(0.75)^{8-2}\\ &\quad + C_1^8\,(0.25)^1(0.75)^{8-1} + C_0^8\,(0.25)^0(0.75)^{8-0} = 0.8862\end{aligned}$$

iv)
$$\begin{aligned}P(X \leq 2) &= P(X = 0) + P(X = 1) + P(X = 2)\\ &= C_0^8\,(0.25)^0(0.75)^{8-0} + C_1^8\,(0.25)^1(0.75)^{8-1} + C_2^8\,(0.25)^2(0.75)^{8-2}\\ &= 0.6785\end{aligned}$$

(b) $\mu = np = 8 \times 0.25 = 2$, and $\sigma = \sqrt{npq} = \sqrt{8 \times 0.25 \times 0.75} = 1.225$

(c)

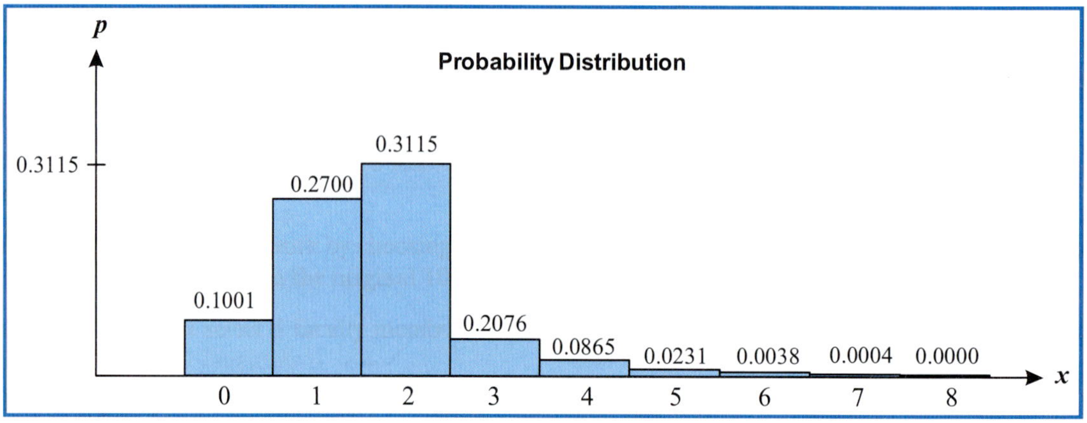

Figure 4.13

Example 23: There is 12% chance for a person to be left handed. In a randomly selected group of 10 people.

(a) Find the probability that:

i) exactly 4 are left-handed.

ii) at least 6 of them are left-handed.

iii) at most 3 of them are left-handed.

(b) Find the mean and standard deviation of number of left handed people in the group and calculate the range of usual number of left-handed people in a class of size 10.

Solutions: X = number of left handed people in a group of 10 person,

$n = 10, p = 0.12$ and $q = 1 - p = 0.88$.

x	$P(X=x) = C_x^n p^x (1-p)^{n-x}$
0	$C_0^{10}(0.12)^0(0.88)^{10-0} \approx 0.2785$
1	$C_1^{10}(0.12)^1(0.88)^{10-1} \approx 0.3798$
2	$C_2^{10}(0.12)^2(0.88)^{10-2} \approx 0.2330$
3	$C_3^{10}(0.12)^3(0.88)^{10-3} \approx 0.0847$
4	$C_4^{10}(0.12)^4(0.88)^{10-4} \approx 0.0202$
5	$C_5^{10}(0.12)^5(0.88)^{10-5} \approx 0.0033$
6	$C_6^{10}(0.12)^6(0.88)^{10-6} \approx 0.0004$
7	$C_7^{10}(0.12)^7(0.88)^{10-7} \approx 0.0000$
8	$C_8^{10}(0.12)^8(0.88)^{10-8} \approx 0.0000$
9	$C_9^{10}(0.12)^9(0.88)^{10-9} \approx 0.0000$
10	$C_{10}^{10}(0.12)^{10}(0.88)^{10-10} \approx 0.0000$

Table 4.24

Once the probability distribution is constructed, the above questions can be answered by simple addition of appropriate probabilities from table 4.24.

(a) **i)** $P(X=4) = 0.0202$

ii)
$$\begin{aligned} P(X \geq 6) &= P(X=6) + P(X=7) + \ldots + P(X=10) \\ &= 0.0004 + 0 + 0 + 0 + 0 \\ &= 0.0004 \end{aligned}$$

iii)
$$\begin{aligned} P(X \leq 3) &= P(X=0) + P(X=1) + P(X=2) + P(X=3) \\ &= 0.2785 + 0.3798 + 0.2330 + 0.0847 \\ &= 0.9760 \end{aligned}$$

(b) Mean $= \mu = np = 10(0.12) = 1.2$

Standard Deviation $= \sigma = \sqrt{npq} = \sqrt{(10)(0.12)(0.88)} = 1.03$.

The range of usual number of left-handed people in this experiment can be calculated using $\mu \pm 2\sigma \rightarrow 1.2 \pm 2(1.03) \rightarrow (-0.86, 3.26)$ or $0 \le x \le 3$.

Example 24: A study showed that 4% of American teenagers have tattoos. 30 teenagers are randomly selected. What is the probability that exactly 3 will have a tattoo?

Solution: $n = 30$, $p = 0.04$, and $q = 1 - p = 0.96$

$$P(X = 3) = C_3^{30}(0.04)^3(0.96)^{27} = 0.0863$$

Example 25: A HDTV is made from 100 components and in order for the HDTV to work properly all 100 components must work properly. Each component has a 0.005 probability of being defective. What is the probability that a HDTV will not work properly?

Solution: X = number of defective components, $n = 100$, $p = 0.005$, and $q = 1 - p = 0.995$.

In order for the TV to work properly, all 100 components must be defect free. Hence

$$\begin{aligned} P(\text{HDTV will not work properly}) &= P(\text{there is at least one defective component}) \\ &= 1 - P(\text{no defective component}) \\ &= 1 - P(X = 0) \\ &= 1 - C_0^{100}(0.005)^0(0.995)^{100} = 0.3942 \end{aligned}$$

Example 26: There are 10 green bottles sitting on a wall. The probability of a green bottle accidentally falling is 0.95. What is the probability that fewer than 8 of the green bottles will accidentally fall?

Solution: X = number of fallen green bottles, $n = 10$, $p = 0.95$, and $q = 1 - p = 0.05$.

x	0	1	2	3	4	5	6	7	8	9	10
$P(X=x)$	0.0000	0.0000	0.0000	0.0000	0.0000	0.0001	0.0010	0.0105	0.0746	0.3151	0.5987

Table 4.25

$$\begin{aligned} P(X < 8) &= (X = 7) + P(X = 6) + \ldots + P(X = 0) \\ &= 0.0105 + 0.0010 + 0.0001 + 0 + 0 + 0 + 0 + 0 = 0.0116 \end{aligned}$$

or
$$\begin{aligned} P(X < 8) &= 1 - P(X \ge 8) \\ &= 1 - [P(X = 8) + P(X = 9) + P(X = 10)] \\ &= 1 - [0.0746 + 0.3151 + 0.5987] = 0.0116 \end{aligned}$$

Example 27: The probability that a missile hits its target is 0.81. If 9 missiles are launched, what is the probability that less than 7 of them hit their targets?

Solution: X = number of missiles hitting their target, $n = 9$, $p = 0.81$, and $q = 1 - p = 0.19$.

$$\begin{aligned} P(X < 7) &= P(X = 6) + P(X = 5) + \ldots + P(X = 0) \\ &= C_6^9(0.81)^6(0.19)^3 + C_5^9(0.81)^5(0.19)^4 + \ldots + C_0^9(0.81)^0(0.19)^9 \\ &= 0.1627 + 0.0573 + 0.0134 + 0.0021 + 0.0002 + 0 + 0 = 0.2357 \end{aligned}$$

or $P(X < 7) = 1 - P(X \ge 7) = 1 - [P(X = 7) + P(X = 8) + P(X = 9)] = 0.2357$

Example 28: A fair coin is tossed four times. Find the mean, variance, and standard deviations of the number of heads obtained.

Solution: *We define the random variable X as: the number of heads obtained in four tosses of a fair coin.*

$$n = 4,\ p = \frac{1}{2},\ q = 1 - p = \frac{1}{2}$$

$$E(X) = \mu = np = 4 \times \frac{1}{2} = 2$$

$$\sigma^2 = npq = (4)\left(\frac{1}{2}\right)\left(\frac{1}{2}\right) = 1 \Rightarrow \sigma = \sqrt{npq} = \sqrt{1} = 1$$

Try it yourself:

1. Suppose each time you take a free throw shot, you have a 25% chance of making it. If you take 15 shots, what is the probability of

   (a) making exactly 5 shots?

   (b) making fewer than 3 shots?

2. A manufacturer claims that a new drug is 80% effective. If the drug is given to 12 patients, find the probability that at least 8 will experience relief.

3. A coin is tossed 16 times. Find the probability of getting 9 tails.

4. The probability that a patient recovers from a delicate brain surgery is 0.7. In seven randomly selected patients.

   (a) find the probability that

   i) more than 5 will survive.

   ii) less than 3 will survive

   iii) at most 4 will survive.

   iv) all will survive.

   v) none will survive.

   (b) find the mean and variance.

5. If a die is rolled 15 times, what is the probability of getting a 4 at least once? Construct the probability distribution for this experiment and draw the Histogram.

4.6: POISSON DISTRIBUTION

Objectives

After completing this section you will be able to describe:

- **A.** Probability Function of Poisson Distribution.
- **B.** Conditions for Poisson Experiment.
- **C.** Mean and Variance of a Poisson Distribution.
- **D.** Poisson Approximation to the Binomial Distribution.

Poisson Distribution is a discrete probability distribution with many practical applications. Poisson Distribution was derived in 1837 by a French mathematician Simeon Denis Poisson (1781-1840). It serves as a valuable mathematical model to describe queuing problems, such as:

i) The number of customers arriving at a cafeteria.

ii) The number of airplanes arriving at an airport.

iii) The number of cars arriving at a gas station.

iv) The number of users logging on to a particular website.

v) The number of customers arriving at a bank counter.

The Poisson Distribution is used for mutually independent events, occurring with a constant average rate μ per unit time or space, observed over a specified interval. In a Poisson Distribution, the random variable X is the number of occurrences of some specific event per unit (time, interval, length, volume, or space *etc.*).

A. Probability Function of Poisson Distribution

If the expected number of occurrences per unit time is μ, then the probability that there are exactly x occurrences is equal to

$$P(X=x) = p(x) = \frac{e^{-\mu}\mu^x}{x!}, \; x = 0, 1, 2, \ldots$$

Here e is the Euler's constant and is approximately equal to 2.7183.

Remarks:

i) Poisson distribution is a discrete probability distribution because random variable X assumes only non-negative integer values 0, 1, 2, 3, …

ii) In Poisson distribution, the random variable X refers to the number of occurrence of an event per unit while the parameter μ refers to the average or expected number of occurrencs per unit.

iii) A random variable X having a Poisson distribution with parameter μ is written as $X \sim$ Poisson (μ).

Note: The Binomial random variable X takes the values 0, 1, 2, …, n, that is, it has a finite domain, while a Poisson random variable takes the values 0, 1, 2,… that is, its domain is infinite.

If X has a Poisson distribution with parameter μ, then the probabilities that a random variable assumes value 0, 1, 2, … are respectively given by:

$$P(X=0) = e^{-\mu},\; P(X=1) = \frac{e^{-\mu}\mu^1}{1!},\; P(X=2) = \frac{e^{-\mu}\mu^2}{2!} \text{ and so on.}$$

B. Conditions for Poisson Experiment

As in the case of Binomial experiments, in Poisson experiment we also assume two outcomes:

$$\begin{Bmatrix} \text{success,} & \text{failure} \\ \text{(occurrence),} & \text{(non-occurrence)} \end{Bmatrix}$$

during the given unit of time.

The outcomes in non overlapping intervals are:

i) random in nature, **ii)** independent, and **iii)** uniformly distributed over the given unit interval.

C. Mean and Variance of a Poisson Distribution

mean	variance	standard deviation
μ	$\sigma^2 = \mu$	$\sigma = \sqrt{\mu}$

Example 29: Suppose on the average, Percy drinks 5 cups of coffee per hour, with an occasional deviation from his routine especially if he has a Statistics exam. Assuming that the number of cups follows a Poisson process, what is the probability that in a given hour, he will drink 8 cups of coffee?

Solution: Let X = number of coffee cups Percy drinks in an hour. Then X follows, a Poisson distribution with $\mu = 5$. Hence $P(X = x) = \dfrac{e^{-5}(5)^x}{x!}$.

$$\text{Required probability} = P(X = 8) = \frac{e^{-5} \cdot 5^8}{8!} \approx 0.0653$$

If we randomly pick an hour during the day, the number of cups of coffee he will drink is most likely 5. However, since there are 24 hours in a day, on an average day, there should be $(24)(0.0653) \approx 1.6$ hours during which he drinks 8 cups of coffee. Below, is the simulation of the probability distribution for this experiment. Extremely Rare outcomes have been omitted.

# of cups of coffee per hour	0	1	2	3	4	5	6	7
Probability	0.0067	0.0337	0.0842	0.1404	0.1755	0.1755	0.1462	0.1044
# of cups of coffee per hour	8	9	10	11	12	13	14	15
Probability	0.0653	0.0363	0.0181	0.0082	0.0034	0.0013	0.0005	0.0002

Table 4.26

Example 30: If the number of accidents on a particular freeway is modeled using a Poisson process and it is found that the average rate is 2.8 accidents per day, find the probability that on any given day, there are

(a) no accidents. **(b)** at least one accident. **(c)** exactly 3 accidents.

Solutions: Let X : represent the number of accidents per day on a given freeway. Then X is a Poisson random variable with parameter $\mu = 2.8$ and $P(X = x) = \dfrac{e^{-2.8}(2.8)^x}{x!}$, $x = 0, 1, 2, \ldots$

(a) $P(\text{no accident}) = P(X = 0)$

$$= \frac{e^{-2.8}(2.8)^0}{0!} = 0.0608$$

(b) $P(\text{at least 1 accident}) = P(X \geq 1)$

$$= 1 - P(X < 1)$$
$$= 1 - P(X = 0)$$
$$= 1 - \frac{e^{-2.8}(2.8)^0}{0!} = 0.9392$$

(In Poisson Distribution, we choose to calculate $P(X \geq 1)$ using the complement approach $1 - P(X < 1)$ instead of direct calculations $P(X \geq 1)$.)

(c) $P(\text{exactly 3 accidents}) = P(X = 3)$

$$= \frac{e^{-2.8}(2.8)^3}{3!} = 0.2225$$

Example 31: Customers arrive at a cafeteria during lunch hour according to a Poisson process at an average rate of 0.025 per second. Find the following probabilities.

(a) At most 3 customers will arrive in any given minute.

(b) At least 5 customers will arrive during an interval of 2 minutes.

(c) At most 10 customers will arrive during the interval of 5 minutes.

Solutions: To solve this problem, we must first convert the average number of customers per second (given in the example) to average per minute.

Conversions:

Average per second	Average per minute	Average per 2 minutes	Average per 5 minutes
$\mu = 0.025$	$\mu = 0.025 \times 60 = 1.5$	$\mu = 1.5 \times 2 = 3$	$\mu = 1.5 \times 5 = 7.5$

Table 4.27

Let X : denote the number of customers arriving per unit time.

(a) Unit: 1 minute - X has a Poisson distribution with parameter $\mu = 1.5$.

P(at most 3 customers will arrive in any given minute)

$$P(X \le 3) = P(X = 3) + P(X = 2) + P(X = 1) + P(X = 0)$$

$$= \frac{e^{-1.5}(1.5)^3}{3!} + \frac{e^{-1.5}(1.5)^2}{2!} + \frac{e^{-1.5}(1.5)^1}{1!} + \frac{e^{-1.5}(1.5)^0}{0!}$$

$$= 0.1255 + 0.2510 + 0.3347 + 0.2231 = 0.9343$$

(b) Unit: 2 minute - X has a Poisson distribution with parameter $\mu = 3$.

P(at least 5 customers will arrive during an interval of 2 minutes)

$$P(X \ge 5) = 1 - P(X \le 4)$$

$$= 1 - \left[\frac{e^{-3}(3)^4}{4!} + \frac{e^{-3}(3)^3}{3!} + \frac{e^{-3}(3)^2}{2!} + \frac{e^{-3}(3)^1}{1!} + \frac{e^{-3}(3)^0}{0!}\right]$$

$$= 1 - (0.1680 + 0.2240 + 0.2240 + 0.1494 + 0.0498)$$

$$= 1 - 0.8152 = 0.1848.$$

(c) Unit: 5 minute - X has a Poisson distribution with parameter $\mu = 7.5$.

P(at most 10 customers will arrive during an interval of 5 minutes)

$$P(X \le 10) = P(X = 10) + P(X = 9) + \ldots + P(X = 1) + P(X = 0)$$

$$= \frac{e^{-7.5}(7.5)^{10}}{10!} + \frac{e^{-7.5}(7.5)^9}{9!} + \ldots + \frac{e^{-7.5}(7.5)^1}{1!} + \frac{e^{-7.5}(7.5)^0}{0!}$$

$$= 0.0858 + 0.1144 + \ldots + 0.0041 + 0.0006 = 0.8622$$

Example 32: Vehicles arrive at an intersection according to a Poisson process at an average rate of 15 per minute. A traffic light cycle lasts about 40 seconds. Construct the Probability Distribution of the number of vehicles that arrive at an intersection per cycle.

Solution: To construct the Poisson Probability Distribution, we must first calculate the average number of vehicles that arrive at an intersection $\mu = 15 \cdot \frac{40}{60} = 10$. (Unit: 40 seconds)

x	$P(X=x)$	x	$P(X=x)$	x	$P(X=x)$
0	$\frac{e^{-10}(10)^0}{0!} \approx 0.0000$	9	$\frac{e^{-10}(10)^9}{9!} \approx 0.1251$	18	$\frac{e^{-10}(10)^{18}}{18!} \approx 0.0071$
1	$\frac{e^{-10}(10)^1}{1!} \approx 0.0005$	10	$\frac{e^{-10}(10)^{10}}{10!} \approx 0.1251$	19	$\frac{e^{-10}(10)^{19}}{19!} \approx 0.0037$
2	$\frac{e^{-10}(10)^2}{2!} \approx 0.0023$	11	$\frac{e^{-10}(10)^{11}}{11!} \approx 0.1137$	20	$\frac{e^{-10}(10)^{20}}{20!} \approx 0.0019$
3	$\frac{e^{-10}(10)^3}{3!} \approx 0.0076$	12	$\frac{e^{-10}(10)^{12}}{12!} \approx 0.0948$	21	$\frac{e^{-10}(10)^{21}}{21!} \approx 0.0009$
4	$\frac{e^{-10}(10)^4}{4!} \approx 0.0189$	13	$\frac{e^{-10}(10)^{13}}{13!} \approx 0.0729$	22	$\frac{e^{-10}(10)^{22}}{22!} \approx 0.0004$
5	$\frac{e^{-10}(10)^5}{5!} \approx 0.0378$	14	$\frac{e^{-10}(10)^{14}}{14!} \approx 0.0521$	23	$\frac{e^{-10}(10)^{23}}{23!} \approx 0.0002$
6	$\frac{e^{-10}(10)^6}{6!} \approx 0.0631$	15	$\frac{e^{-10}(10)^{15}}{15!} \approx 0.0347$	24	$\frac{e^{-10}(10)^{24}}{24!} \approx 0.0001$
7	$\frac{e^{-10}(10)^7}{7!} \approx 0.0901$	16	$\frac{e^{-10}(10)^{16}}{16!} \approx 0.0217$	25	$\frac{e^{-10}(10)^{25}}{25!} \approx 0.0000$
8	$\frac{e^{-10}(10)^8}{8!} \approx 0.1126$	17	$\frac{e^{-10}(10)^{17}}{17!} \approx 0.0218$	26	$\frac{e^{-10}(10)^{26}}{26!} \approx 0.0000$

Table 4.28

Note: Even though a Poisson Random Variable can take on infinitely many values, as you can see in the above example, after a certain value of the random variable, the corresponding probabilities become so small that for all practical purposes we could assume that their values are zero.

Example 33: The number of typographical errors in Statistics books have a Poisson Distribution with a mean of 2 typos per 150 pages. If Professor Badalian randomly selects 600 pages of his 3rd edition Statistics book,

(a) what would be the probability that there are 5 typos?

(b) what would be the probability that there are less than 4 typos?

Solutions: Let X = number of typos per 600 pages of a book. Then X follows a Poisson distribution with parameter $\mu = (2)(4) = 8$ typos per 600 pages.

(a) $P(X = 5) = \dfrac{e^{-8}(8)^5}{5!} \approx 0.0916$

(b) $P(X < 4) = P(0) + P(1) + P(2) + P(3)$

$$= \frac{e^{-8}(8)^0}{0!} + \frac{e^{-8}(8)^1}{1!} + \frac{e^{-8}(8)^2}{2!} + \frac{e^{-8}(8)^3}{3!} \approx 0.0424$$

Try it yourself:

1. The number of students that ace Professor Badalian's Statistics class each semester follow a Poisson Distribution with mean 16 per semester. Calculate the following probabilities.

   (a) $P(X = 0)$ **(b)** $P(X < 14)$ **(c)** $P(X > 12)$ **(d)** $P(X \leq 9)$

2. In a book of 520 pages, 390 typographical errors occur. Assuming Poisson distribution for the number of errors per page, find the probability that a random sample of 5 pages will contain no errors.

3. In a town 10 accidents took place in 50 days. Assuming that the number of accidents per day follows Poisson distribution, find the probability that there will be at least three per day.

4. At a checkout counter, customers arrive according to a Poisson process at an average rate of 1.5 per minute. Find the probabilities that

   (a) at most 4 will arrive in any given minute.

   (b) at least 3 will arrive during an interval of 2 minutes.

   (c) at most 15 will arrive during an interval of 6 minutes.

D. Poisson Approximation to the Binomial Distribution

Poisson Distribution is used to approximate the Binomial Distribution under the following conditions:

i) n, the number of trials (sample size) is large;

ii) p, the constant probability of success for each trial is small; and

iii) $np = \mu$, is constant.

Note: If in binomial distribution $n \geq 100$ and $np \leq 10$, then we can approximate binomial probabilities by Poisson probabilities with a fair degree of accuracy. However, in practice, the Poisson distribution gives reasonably good approximation when $n > 20$ and $p < 0.05$.

Example 34: In California daily 3 lottery, you pay \$1 and must select a set of three (3) numbers from a field of 0 to 9 (example: 1-3-5, 9-3-9, *etc.*,) marking one number per row on your playslip. To win at Daily 3, all of your numbers must match all of the winning numbers. Henry plays this game once every day. What is the probability that he will win once a month? (Assume 30 days in each month)

Solution: $n = 30$, $P(\text{winning}) = \dfrac{1}{10^3} = 0.001$, $\mu = np = (30)\left(\dfrac{1}{1000}\right) = 0.03$.

Since both conditions are satisfied ($n \geq 100$ and $np \leq 10$), we can use the Poisson Approximation to binomial probabilities.

$$P(X=1) = \frac{e^{-0.03} \cdot (0.03)^1}{1!} \approx 0.0291$$

Try it yourself:

1. An automatic machine makes paper clips from coils of wire. On the average, 1 in 400 paper clips are defective. If the paper clips are packed in boxes of 100.

   (a) What is the probability that any given box of clips will contain

   i) zero defective clips. **ii)** one or more defective clips.

   iii) less than two defective clips.

   (b) Find the mean and variance.

2. If 5% of the electric bulbs manufactured by a company are defective, use Poisson distribution as an approximation to the Binomial distribution to find the probability that in a sample of 100 bulbs

   (a) none is defective. **(b)** 5 bulbs will be defective.

4.7: Discrete Probability Distributions with Microsoft Excel®

In this section we will demonstrate the use of Microsoft Excel® in computing Discrete Probability Distributions, their Mean and Variances, and range of usual and unusual values.

Objectives

After completing this section you will be able to carryout the computations in Chapter 4 using Microsoft Excel®.

Illustration 1: Construct the Probability Distribution for the experiment of rolling a fair die and find the expected value (mean) and standard deviation of the distribution. What is the range of usual and unusual values for this experiment ($\mu \pm 2\sigma$)? Also, draw the probability histogram.

Solution: Solution of the above problem on Excel® worksheet is shown in the following sequence of tables.

	A	B	C	D
1	x	p	x*p	$x^{2*}p$
2	1	=1/6	=A2*B2	=A2^2*B2
3	2	=1/6		
4	3	=1/6		
5	4	=1/6		
6	5	=1/6		
7	6	=1/6		

Table 4.29 a

	A	B	C	D
1	x	p	x*p	$x^{2*}p$
2	1	0.166667	0.166667	0.166667
3	2	0.166667		
4	3	0.166667		
5	4	0.166667		
6	5	0.166667		
7	6	0.166667		

Table 4.29 b

	A	B	C	D	E	F	G
1	x	p	x*p	$x^{2*}p$		**Mean(μ)**	=C8
2	1	0.166667	0.166667	0.166667		**SD(σ)**	=SQRT(D8 – C8^2)
3	2	0.166667	0.333333	0.666667		$\mu - 2\sigma$	=G1 – 2*G2
4	3	0.166667	0.5	1.5		$\mu + 2\sigma$	=G1 + 2*G2
5	4	0.166667	0.666667	2.666667			
6	5	0.166667	0.833333	4.166667			
7	6	0.166667	1	6			
8		=SUM(B2:B7)	=SUM(C2:C7)	=SUM(D2:D7)	⟵	**Totals**	

Table 4.29 c

	A	B	C	D	E	F	G
1	x	p	x*p	$x^{2*}p$		**Mean(μ)**	3.5
2	1	0.166667	0.166667	0.166667		**SD(σ)**	1.70782513
3	2	0.166667	0.333333	0.666667		$\mu - 2\sigma$	0.08434974
4	3	0.166667	0.5	1.5		$\mu + 2\sigma$	6.91565026
5	4	0.166667	0.666667	2.666667			
6	5	0.166667	0.833333	4.166667			
7	6	0.166667	1	6			
8		**1**	**3.5**	**15.1667**	⟵	**Totals**	

Table 4.29 d

The Histogram for the above probability distribution is presented below.

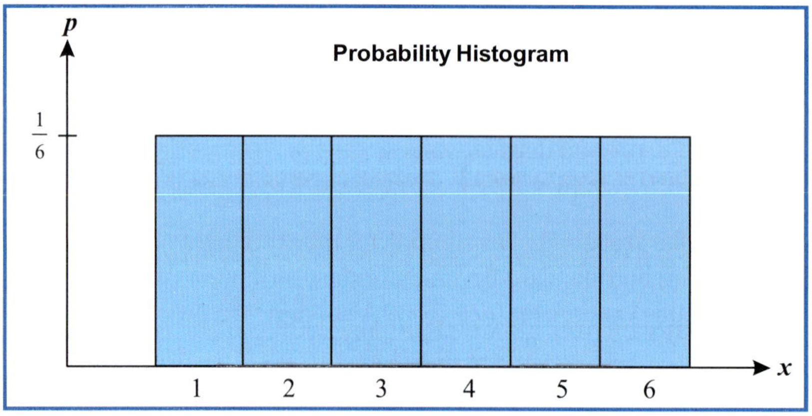

Figure 4.14

- **The Built in Shortcut Excel® Commands to Calculate Probabilities in a Binomial Experiment and Poisson Distribution**

(a) = BINOMDIST(number_s, trials, probability_s, cumulative), where

Number_s: is the number of successes in 'n' trials (x).

Trials: is the number of independent trials (n).

Probability_s: is the probability of success in each trial.

Cumulative: If cumulative is TRUE, then BINOMDIST calculates the probability that there are at most "x" successes; if FALSE, it calculates the probability that there are exactly "x" successes.

(b) = POISSON(x, mean, cumulative), where

x is number of times, an event of interest occurs.

mean is expected value of Poisson distribution.

Cumulative: If cumulative is TRUE, then POISSON calculates the probability that there are at most "x" occurrences; if FALSE, it calculates the probability that there are exactly "x" occurrences.

Illustration 2: Calculate $P(X = 5) = C_5^{10} \times \left(\frac{1}{6}\right)^5 \times \left(\frac{5}{6}\right)^{10-5}$ by writing a command in Excel®. Then use the built-in commands in Excel® to verify your answer.

Solution: =COMBIN(10,5)*(1/6)^5*(5/6)^(10 – 5) = 0.013

Verify: =BINOMDIST(5,10,1/6,FALSE) = 0.013

Illustration 3: There is a 12% chance for a person to be left handed. In a randomly selected group of 10 people from the population, find the probability that

(a) exactly 4 are left-handed

(b) at least 6 of them are left-handed

(c) at most 3 of them are left-handed.

(d) Find the mean and the standard deviation of this experiment and calculate the range of usual and unusual number of left-handed people in a class of size 10.

Solution: The following tables represent the computational commands and the output values of the binomial probability distribution. To compute the probabilities, we have used the combin(n, x) command, the power command (^), the multiplication command (*), and the drag down property in Excel®.

	A	B
1	X	P(X = x)
2	0	0.2785
3	1	0.3798
4	2	0.2330
5	3	0.0847
6	4	0.0202
7	5	0.0033
8	6	0.0004
9	7	0.0000
10	8	0.0000
11	9	0.0000
12	10	0.0000

Table 4.30 a

	A	B
1	X	P(X = x)
2	0	=COMBIN(10,F2)*(0.12^F2)*0.88^(10-F2)
3	1	=COMBIN(10,F3)*(0.12^F3)*0.88^(10-F3)
4	2	=COMBIN(10,F4)*(0.12^F4)*0.88^(10-F4)
5	3	=COMBIN(10,F5)*(0.12^F5)*0.88^(10-F5)
6	4	=COMBIN(10,F6)*(0.12^F6)*0.88^(10-F6)
7	5	=COMBIN(10,F7)*(0.12^F7)*0.88^(10-F7)
8	6	=COMBIN(10,F8)*(0.12^F8)*0.88^(10-F8)
9	7	=COMBIN(10,F9)*(0.12^F9)*0.88^(10-F9)
10	8	=COMBIN(10,F10)*(0.12^F10)*0.88^(10-F10)
11	9	=COMBIN(10,F11)*(0.12^F11)*0.88^(10-F11)
12	10	=COMBIN(10,F12)*(0.12^F12)*0.88^(10-F12)

Table 4.30 b

Once the probabilities are calculated, we can use them to answer the questions.

(a) $P(X = 4) = 0.0202 \approx 0.02$

(b) $P(X \geq 6) = P(X = 6) + P(X = 7) + \ldots + P(X = 10) = 0.0004 + 0 + 0 + 0 + 0 = 0.0004$

(c) $P(X \leq 3) = P(X = 0) + P(X = 1) + P(X = 2) + P(X = 3)$

$= 0.2785 + 0.3798 + 0.2330 + 0.0847 = 0.9761$

(d) Mean $= \mu = np = 10(0.12) = 1.2$

Standard Deviation $= \sigma = \sqrt{npq} = \sqrt{(10)(0.12)(0.88)} = 1.03$

The range of usual number of left-handed people in this experiment can be calculated using

$\mu \pm 2\sigma \rightarrow 1.2 \pm 2(1.03)$ or $(-0.86, 3.26)$

As mentioned earlier, we can also calculate the binomial probabilities using

BINOMDIST(number_s, trials,probability_s,cumulative) as well, instead of generating the entire distribution and adding probabilities (but that's boring and not as much fun).

Symbolic Question	Excel Command
$P(X = 4)$	= BINOMDIST(4,10,0.12,FALSE) = 0.0202
$P(X \geq 6) = 1 - P(X \leq 5)$	= 1 − BINOMDIST(5,10,0.12,TRUE) = 0.0004
$P(X \leq 3)$	= BINOMDIST(3,10,0.12,TRUE) = 0.9761

Table 4.31

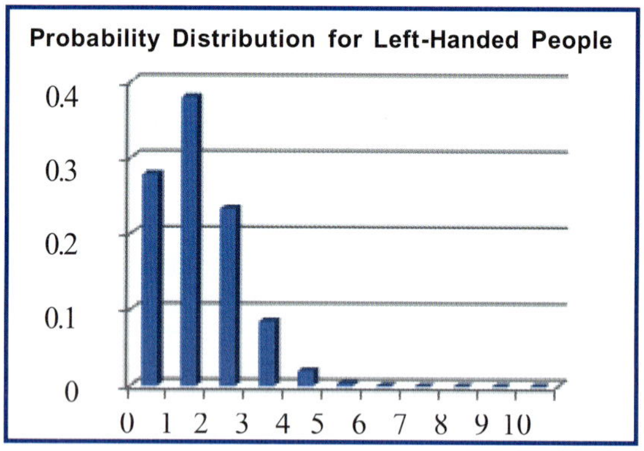

Figure 4.15

Same Answers!!

We can also create a bar chart for the above binomial experiment (See Figure 4.15).

Illustration 4: Vehicles arrive at an intersection at a rate of 15 per minute. A traffic light cycle lasts about 40 seconds. Construct the Poisson Probability Distribution of the number of vehicles that arrive at an intersection per cycle.

Solution: To construct the Poisson Probability Distribution, we must first calculate the average number of vehicles that arrive at an intersection. $\mu = 15 \cdot \frac{40}{60} = 10$

	A	B	C
1	x (# of cars arriving at the intersection)	Probability of exactly x arrivals	Probability of x or fewer arrivals
2	0	=POISSON(A2,10,FALSE)	=POISSON(A2,10,TRUE)
3	1		
4	2		
5	⋮		

Table 4.32 a

	A	B	C
1	x (# of cars arriving at the intersection)	Probability of exactly x arrivals	Probability of x or fewer arrivals
2	0	0.0000	0.0000
3	1		
4	2		
5	⋮		

Table 4.32 b

	A	B	C
1	x (# of cars arriving at the intersection)	Probability of exactly x arrivals	Probability of x or fewer arrivals
2	0	0.0000	0.0000
3	1	0.0005	0.0005
4	2	0.0023	0.0028
5	3	0.0076	0.0103
6	4	0.0189	0.0293
7	5	0.0378	0.0671
8	6	0.0631	0.1301
9	7	0.0901	0.2202
10	8	0.1126	0.3328
11	9	0.1251	0.4579
12	10	0.1251	0.5830
13	11	0.1137	0.6968
14	12	0.0948	0.7916
15	13	0.0729	0.8645
16	14	0.0521	0.9165
17	15	0.0347	0.9513
18	16	0.0217	0.9730
19	17	0.0128	0.9857
20	18	0.0071	0.9928
21	19	0.0037	0.9965
22	20	0.0019	0.9984
23	21	0.0009	0.9993
24	22	0.0004	0.9997
25	23	0.0002	0.9999
26	24	0.0001	1.0000
27	25	0.0000	1.0000
28	26	0.0000	1.0000
29			
30			

Table 4.32 c

Illustration 5: In California daily 3 lottery, you pay \$1, and must select a set of three (3) numbers from a field of 0 to 9 (e.g. : 1-3-5, 9-3-9, *etc.*) marking one number per row on your playslip. To win at Daily 3, ALL of your numbers must match ALL of the winning numbers. Henry plays this game once every day. What is the probability that he will win once a month? (Assume 30 days in each month)

Solution: Total options for Henry to select three numbers is $10 \times 10 \times 10 = 1000$. Since only one sequence of 3 numbers is displayed as a winning number, the probability of winning in a single day is $\frac{1}{1000}$ and average number of winning in a month (30 days) is $30 \times \frac{1}{1000} = 0.03$. The probability of winning once in a month will be computed by Poisson formula

$$P(X=1) = \frac{e^{-0.03} \times 0.03}{1!} = 0.0291$$

The Excel ® function for computing above probability is **=POISSON (1, 0.03, FALSE)**

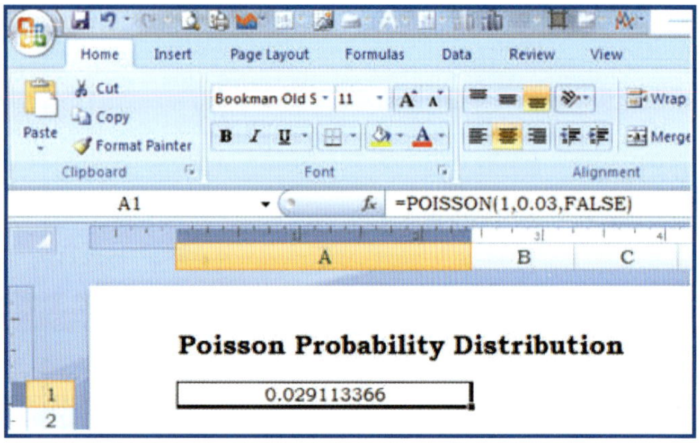

Figure 4.16

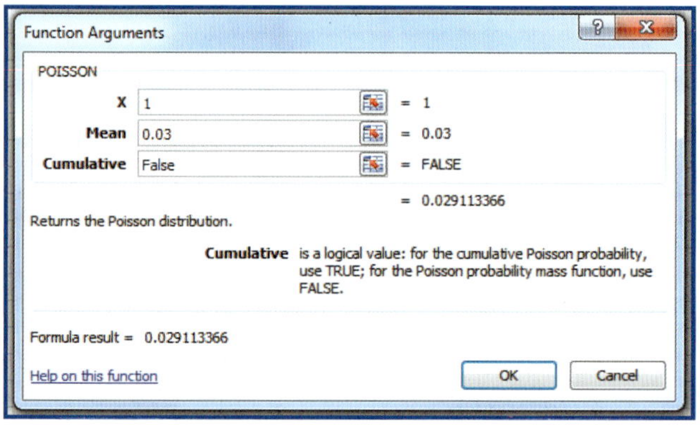

Figure 4.17

4.8: Extended Topics and Proofs in Discrete Probability Distributions (Optional)

Objectives

After completing this section you will be able to understand:

A. Properties of Expected Value.

B. Properties of Variance.

C. Binomial Distribution and its Properties.

D. Poisson Distribution and its Properties.

E. Other Discrete Probability Distribution.

A. Properties of Expected Value

i) The expected value of a constant is the constant itself $E(k) = k$.

ii) The expected value is a linear operator. This means that the expected value is closed under addition and scalar multiplication.

(a) $E(X+k) = E(X) + E(k) = E(X) + k$

(b) $E(kX) = kE(X)$

(c) $E(k_1X_1 + k_2X_2 + ... + k_nX_n)$

$= k_1E(X_1) + k_2E(X_2) + ... + k_nE(X_n)$

B. Properties of Variance

i) The variance of a constant is zero. $V(k) = 0$.

ii) $V(X+k) = V(X)$

iii) $V(kX) = k^2V(X)$

The following examples will demonstrate the above properties.

Example 35: Let the random variable X represent the IQ of 25,000 college students. The average IQ is 108, standard deviation is 16 and the variance is 256. Assume that it has been found that the measurement has been over-reported incorrectly by 6 points. Find the students real average IQ and real variance.

Solution: Set a new random variable as $Y = X - 6$. Using the properties of the Expected Value and the Variance, we have

$E(Y) = E(X-6) = E(X) - 6 = 108 - 6 = 102.$

$\text{Var}(Y) = \text{Var}(X-6) = \text{Var}(X) = 256$

Example 36: Given $E(X) = 24.6$ and $V(X) = 38.4$ find $E(10X)$ and Var $(10X)$.

Solution: $E(10X) = 10E(X) = (10)(24.6) = 246$

$V(10X) = 10^2V(X) = (100)(38.4) = 3840$

Example 37: Prove that if X is a random variable, then $\sigma^2(X) = E\left[(X-\mu)^2\right] = E\left(X^2\right) - \mu^2$

Proof:

$$\begin{aligned}\sigma^2(X) &= E\left[(X-\mu)^2\right]\\ &= E(X^2 - 2X\mu + \mu^2)\\ &= E(X^2) - 2\mu E(X) + E(\mu^2) \qquad (E(X) = \mu)\\ &= E(X^2) - 2\mu\cdot\mu + E(\mu^2)\\ &= E(X^2) - 2\mu^2 + \mu^2 = E(X^2) - \mu^2\end{aligned}$$

Example 38: Prove that If X and Y are independent, then $E(X\cdot Y) = E(X)\cdot E(Y)$

Proof: Assume X and Y are discrete random variables and $X = \{x_1, x_2, \ldots\}$ and $Y = \{y_1, y_2, \ldots\}$.

Now X and Y are independent random variables; therefore

$P(X = x_j, Y = y_k) = P(X = x_j)P(Y = y_k)$ where $j, k, \varepsilon\{1, 2, 3, \ldots\}$ and

$$\begin{aligned}E(XY) &= \sum\sum x_j y_k P\left(X = x_j, Y = y_k\right)\\ &= \sum\sum x_j y_k \cdot P\left(X = x_j\right)\cdot P\left(Y = y_k\right)\\ &= \sum\left[x_j \cdot P\left(X = x_j\right)\right]\cdot\sum\left[y_k \cdot P\left(Y = y_k\right)\right]\\ &= E(X)\cdot E(Y)\end{aligned}$$

C. Binomial Distribution and its Properties

The probability of getting x successes in n independent trials of a Bernoulli process is given by

$$P(X = x) = \binom{n}{x} p^x q^{n-x}, \quad x = 0, 1, 2, \ldots\ldots, n$$

i) By substituting all possible values of x into the above Binomial function we can obtain the probability distribution of X as given below:

X	0	1	2	...	x	...	n
P(X=x)	q^n	$c_1^n p^1 q^{n-1}$	$c_2^n p^2 q^{n-2}$	...	$c_x^n p^x q^{n-x}$	...	p^n

Table 4.33

Since the probabilities $P(0)$, $P(1)$, $P(2)$, … , $P(n)$ in the Table 4.33 are the successive terms of the binomial expansion $(q+p)^n$, the distribution is called ***binomial distribution***.

ii) Total probability is 1; that is,

$$\sum_{x=0}^{n} P(x) = P(0) + P(1) + P(2) + \ldots + P(n)$$

$$= q^n + \binom{n}{1} p^1 q^{n-1} + \binom{n}{2} p^2 q^{n-2} + \ldots + p^n$$

$$= (q+p)^n = 1, \qquad \text{since } p + q = 1$$

iii) n and p are known as the parameters of the binomial distribution. If we know n and p, we can compute all the probabilities of the binomial distribution.

iv) The notation $X \sim \mathrm{B}(n, p)$ is used to express a binomial random variable with parameters n and p.

- **Derivation of the Mean and Variance Formulas for a Binomial Distribution:**

Example 39: Show that the mean of a binomial distribution is np.

Proof: Need to show that $\mu = np$.

Recall that if $X \sim \mathrm{B}(n, p)$, then $P(X=x) = \binom{n}{x} p^x q^{n-x}, \quad x = 0, 1, 2, \ldots, n.$

We will start with the definition of an expected value.

$\mu = \mathrm{E}(X) = \sum_{x=0}^{n} [x \cdot P(X=x)]$, the first term of above number is equal to zero; therefore

$$\mu = \sum_{x=0}^{n}\left[x \cdot \binom{n}{x} p^x q^{n-x}\right] = \sum_{x=1}^{n}\left[x \cdot \frac{n!}{x!(n-x)!} p^x q^{n-x}\right]$$

$$= \sum_{x=1}^{n} \frac{x \cdot n(n-1)!}{x(x-1)!(n-x)!} \cdot p \cdot p^{x-1} \cdot q^{n-x}$$

$$= np \sum_{x=1}^{n} \frac{(n-1)!}{(x-1)!(n-x)!} \cdot p^{x-1} \cdot q^{n-x}$$

$$= np \sum_{x=1}^{n} \binom{x-1}{x-1} \cdot p^{x-1} \cdot q^{n-x}$$

$$= np\,(q+p)^{n-1} = np$$

Example 40: Show that the variance of a binomial probability distribution is npq.

$$\sigma^2 = \mathrm{V}(X) = npq$$

Proof: We know that $\sigma^2 = V(X) = E(X^2) - \mu^2$. Hence, in order to calculate σ^2 we need to find $E(X^2)$.

Notice that

$$E[X(X-1)] = E(X^2) - E(X)$$
$$= E(X^2) - \mu$$

Therefore, $E(X^2) = E[X(X-1)] + \mu$.

Using the definition of the expected value, we have

$$E[X(X-1)] = \sum_{x=0}^{n} x(x-1)\frac{n!}{x!(n-x)!}p^x q^{n-x}$$

The first and second terms of the above sum are equal to zero; therefore,

$$E[X(X-1)] = \sum_{x=2}^{n} x(x-1)\frac{n!}{x!(n-x)!}p^x q^{n-x}$$

$$= \sum_{x=2}^{n} x(x-1)\frac{n(n-1)(n-2)!}{x(x-1)(x-2)!(n-x)!}p^2 p^{x-2} q^{n-x}$$

Taking out factor $n(n-1)p^2$ from each term in the above sum

$$E[X(X-1)] = n(n-1)p^2 \sum_{x=2}^{n} \frac{(n-2)!}{(x-2)!(n-x)!}p^{x-2} q^{n-x}$$

$$= n(n-1)p^2 \sum_{x=2}^{n} \binom{n-2}{x-2} p^{x-2} q^{n-x}$$

$$= n(n-1)p^2 (q+p)^{n-2} = n(n-1)p^2$$

$$\Rightarrow E[X(X-1)] = n(n-1)p^2$$

Therefore

$$E[X^2] = E[X(X-1)] + \mu = n(n-1)p^2 + \mu = n(n-1)p^2 + np.$$

Hence $\sigma^2 = V(X) = E(X^2) - \mu^2$

$$= n(n-1)p^2 + np - \mu^2 = n(n-1)p^2 + np - n^2p^2$$
$$= n^2p^2 - np^2 + np - n^2p^2$$
$$= np(1-p) = npq$$

D. Poisson Distribution and its Properties

To derive the Poisson formula, let's start with writing the exponential function as its infinite power series expansion.

$$e^{\mu} = \sum_{x=0}^{\infty} \frac{\mu^x}{x!} = \frac{\mu^0}{1!} + \frac{\mu^1}{1!} + \frac{\mu^2}{2!} + \ldots$$

Now let's divide both sides by e^{μ}.

$$1 = \frac{\mu^0}{e^{\mu}0!} + \frac{\mu^1}{e^{\mu}1!} + \frac{\mu^2}{e^{\mu}2!} + \frac{\mu^3}{e^{\mu}3!} + \ldots$$

Since the sum of the terms on the right hand side equals 1, an essential condition of a probability distribution, each term on the right hand side is the probability of x occurrences with an average rate of occurrence equal to μ. Hence,

$$1 = \frac{\mu^0}{e^{\mu}0!} + \frac{\mu^1}{e^{\mu}1!} + \frac{\mu^2}{e^{\mu}2!} + \ldots = \sum_{x=0}^{\infty} \frac{e^{-\mu}\mu^x}{x!}$$

Where μ is the average rate of occurrence per unit time.

- **Derivation of the Mean and Variance of a Poisson Probability Distribution:**

Example 41: Given $P(X = x) = \dfrac{e^{-\mu}\mu^x}{x!}$, $x = 0, 1, 2, 3, \ldots$

Prove that $E(X) = \mu$.

Proof:

$$E(X) = \sum\left[x \cdot P(X = x)\right]$$

$$= \sum_{x=0}^{\infty}\left[x \cdot \frac{e^{-\mu}\mu^x}{x!}\right]$$; the first term in the expression is equal to zero.

$$= \sum_{x=1}^{\infty}\left[x \cdot \frac{e^{-\mu}\mu^x}{x(x-1)!}\right]$$

$$= \sum_{x=1}^{\infty}\left[\frac{e^{-\mu}\mu^x}{(x-1)!}\right],$$ Let $y = x - 1 \Rightarrow x = y + 1$

$$= \sum_{y=0}^{\infty}\frac{e^{-\mu}\mu^{y+1}}{y!} = \sum_{y=0}^{\infty}\frac{e^{-\mu}\mu^{y}\mu^{1}}{y!}$$

$$= \mu\sum_{y=0}^{\infty}\frac{e^{-\mu}\mu^{y}}{y!} = \mu \cdot 1 = \mu$$

Example 42: Show that the variance of a Poisson distribution is $\sigma^2 = \mu$.

Proof: As we know

$$\text{Var}(X) = E(X^2) - \mu^2, \text{ where}$$

$$E(X^2) = E(X(X-1) + X)$$

$$= E(X(X-1)) + E(X)$$

$$= \sum_{x=0}^{\infty} x(x-1)\frac{e^{-\mu}\mu^x}{x!} + \mu$$

The first two terms (corresponding to $x = 0$ and $x = 1$) are zero; therefore

$$E(X^2) = \sum_{x=2}^{\infty}\frac{x(x-1)e^{-\mu}\mu^2\mu^{x-2}}{x(x-1)(x-2)!} + \mu$$

$$= e^{-\mu}\mu^2 \sum_{x=2}^{\infty} \frac{\mu^{x-2}}{(x-2)!} + \mu$$

$$= e^{-\mu}\mu^2 \sum_{y=0}^{\infty} \frac{\mu^{y}}{y!} + \mu \text{, where } y = x - 2$$

$$= e^{-\mu}\mu^2 \cdot e^{\mu} + \mu = \mu^2 + \mu$$

Since $\sigma^2 = E(X^2) - \mu^2 = \mu^2 + \mu - \mu^2$, $\sigma^2 = \mu$

E. Other Discrete Probability Distribution

• Multinomial Distribution:

Multinomial Experiments are generalization of the Binomial Experiments with the following properties;

i) The experiment consists of repeated trials.

ii) Trials are independent.

iii) Each trial has k (discrete) number of possible outcomes.

iv) Probabilities corresponding to each particular outcome are constant.

Binomial Experiments are a special case of Multinomial Experiments. In Binomial Experiments there are only 2 outcomes, whereas in a Multinomial Experiments there are 2 or more outcomes.

The formula for the Multinomial Experiment

$$p(x_1, x_2, \dots, x_k) = \frac{n!}{x_1!x_2!\dots x_k!} \cdot p_1^{x_1} \cdot p_2^{x_2} \dots \cdot p_k^{x_k}$$

where $0 < p_i < 1$, $\sum_{i=1}^{k} p_i = 1$ and $x_i \in \{0, 1, 2, \dots, n\}$, $\sum_{i=1}^{k} x_i = n$.

If $X_1, X_2, \dots, X_k$ have a multinomial distribution with parameters n and $p_1, p_2, \dots, p_k$, then

i) the ***expected value*** of the random variable X_i is $E(X_i) = np_i$.

ii) the ***variance*** of the random variable is X_i is $V(X_i) = np_iq_i$, where $q_i = 1 - p_i$.

Example 43: Suppose a card is randomly drawn from an ordinary deck of playing cards with replacement. The experiment is repeated 6 times. What is the probability of drawing 2 clubs, 1 spade, 1 heart, and 2 diamonds?

Solution: In this experiment, $n = 6$, $x_1 = 2$, $x_2 = 1$, $x_3 = 1$, and $x_4 = 2$.

$$p_1 = p_2 = p_3 = p_4 = \frac{13}{52} = 0.25$$

$$p(2, 1, 1, 2) = \frac{6!}{(2!)(1!)(1!)(2!)}(0.25)^2(0.25)^1(0.25)^1(0.25)^2 = 0.043945 \approx 0.04$$

Example 44: There are 5 green, 4 red, 2 blue, and 3 yellow marbles in a box. If 5 marbles are randomly selected with replacement, what is the probability of getting 3 green, 1 red, and 1 blue marbles?

Solution: In this experiment, $n = 14$, $x_{\text{green}} = 3$, $x_{\text{red}} = 1$, $x_{\text{blue}} = 1$, and $x_{\text{yellow}} = 0$.

$$p_{\text{green}} = \frac{5}{14} \approx 0.36,\ p_{\text{red}} = \frac{4}{14} \approx 0.29,\ p_{\text{blue}} = \frac{2}{14} \approx 0.14, \text{ and } p_{\text{yellow}} = \frac{3}{14} \approx 0.21.$$

$$p(3 \text{ green, } 1 \text{ red, } 1 \text{ blue, } 0 \text{ yellow}) = \frac{5!}{(3!)(1!)(1!)(0!)}\left(\frac{5}{14}\right)^3\left(\frac{4}{14}\right)^1\left(\frac{2}{14}\right)^1\left(\frac{3}{14}\right)^0 \approx 0.04$$

- **Hyper Geometric Distribution:**

Suppose that a population contains N elements that possess one of two characteristics. Hence k of the elements might possess a certain characteristic like (defective, blue, ...) and $N - k$ of them a different characteristic. A sample of n elements is randomly selected from the population N and the random variable in question is denoted by X.

A random variable X is said to have a *Hyper Geometric* probability distribution with parameters (N, n, k) if

$$P(X = x) = \frac{\binom{k}{x}\binom{N-k}{n-x}}{\binom{N}{n}},\ x = 0, 1, 2, \ldots$$

If X is a random variable with a Hyper-Geometric probability distribution, then

$$E(X) = \mu = \frac{nk}{N} \text{ and } \sigma = \sqrt{n\left(\frac{k}{N}\right)\left(\frac{N-k}{N}\right)\left(\frac{N-n}{N-1}\right)}$$

Example 45: Six students were selected from a group of 15, of whom 8 were female and 7 were male.

(a) What is the probability that out of 6 selected students, there are only 2 females?

(b) Find the mean and variance of the distribution.

Solutions: **(a)** $N = 15$, $n = 6$, $k = 8$, and $x = 2$

$$P(X = 2) = \frac{\binom{8}{2}\binom{7}{4}}{\binom{15}{6}} = 0.1958$$

(b) $$E(X) = \mu = \frac{nk}{N} = \frac{(6)(8)}{15} = 3.2,\ \sigma^2 = (6)\left(\frac{8}{15}\right)\left(\frac{7}{15}\right)\left(\frac{9}{14}\right) = 0.96$$

Chapter 4: General Exercises

1. **In exercises a-c, draw Histogram of the following probability distributions.**

(a)

x	1	2	5	-10	8
$P(X=x)$	0.1	0.2	0.08	0.6	0.02

(b)

x	0	1	2	3	4
$P(X=x)$	0.08	0.1	0.1	0.1	0.62

(c)

x	20	30	40	50	60	-30
$P(X=x)$	0.17	0.17	0.17	0.17	0.17	0.15

2. An ice cream store has anywhere from 2 to 8 flavors of ice cream every day. The probability function for the number of flavors is given below:

Number of flavors	2	3	4	5	6	7	8
Corresponding probabilities	0.06	0.07	0.14	0.27	0.33	0.12	0.01

(a) Verify that this is a probability distribution. **(b)** Draw histogram.

3. The number of cartons of cookies that each Girl Scout sells has the probabilities given below:

Number of cartoons (x)	0	1	2	3	4
$P(X=x)$	0.10	0.05	0.28	0.34	0.23

(a) Is this a probability distribution?

(b) What is the average number of cartons a girl scout sells?

(c) What is the amount of variation? Find the range of usual values.

4. A florist buys 60 roses on Valentine's Day. The number of roses he sells has the following probability distribution.

Number of roses (x)	53	54	55	56	57	58	59	60
Probabilities $P(x)$	0.14	0.14	0.19	0.09	0.12	0.22	0.06	0.04

Find the expected value for the above probability distribution, and explain your answer.

5. The number of items left behind by students in a lecture hall after a large lecture has the following probability distribution.

Number of items left behind	0	1	2	3	4
Probability	0.35	0.12	0.10	?	0.13

Complete the above table and answer the following questions.

(a) How many forgotten items should the next class expect to find?

(b) What is the standard deviation for this function?

6. A certain driving school offers minimum six and maximum ten lessons. The number of lessons a person takes before he passes the driving test and their probabilities are given below:

Number of driving lessons	6	7	8	9	10	11	12
Probability	0.30	0.26	?	0.11	0.08	0.09	0.04

(a) What is the missing probability?

(b) How many lessons should the company be prepared to provide for an average person?

(c) What is the standard deviation of the data?

7. The following table shows the number of moving violations and the respective probabilities of an average person having this many violations in the last year.

Number of moving violations	0	1	2	3
Probability	0.20	0.45	0.24	?

(a) What percent of people had three moving violations in the past year?

(b) Find the mean and the standard deviation of the data.

8. The number of loads of laundry that a person does a week is given in the chart, along with the probabilities.

Number of loads of laundry	0	1	2	3	4	5
Probability that this many loads are done	0.03	0.14	0.44	0.2	0.11	0.08

(a) On average, how many loads of laundry does a person do?

(b) What is the variance of this distribution?

9. In a roll of film, there are 24 snapshots, but the store does not always print all of them. The following table gives the number of printed pictures and their probabilities.

Number of printed pictures	20	21	22	23	24
Probability	0.12	0.145	0.485	0.20	0.05

(a) If a person takes his film to this store, how many pictures should he expect to get?

(b) Find the variance and standard deviation of the number of pictures printed.

10. A study was conducted to find out the probability distribution of the number of households in the Los Angeles area watching the World Poker Tour on TV. After surveying 6 randomly selected households in LA and letting "x" represent the number of households, (ranging from 0 to 6) the following probabilities were observed.

x	0	1	2	3	4	5	6
$p(x)$	0.001	0.14	0.51	0.29	0.018	0.04	0.001

(a) Is this a valid probability distribution?

(b) Find the mean of the above probability distribution, the standard deviation, and the range of usual and unusual values.

(c) Use a bar chart to graph the distribution.

11. A random variable X has the following probability distribution.

x	0	1	2	3	4	5	6	7	8
$p(x)$	a	$3a$	$5a$	$7a$	$9a$	$11a$	$13a$	$15a$	$17a$

(a) Determine the value of a. (b) Find $P(x<3)$, $P(x \geq 3)$, and $P(0<x<5)$.

12. Let $p(x)$ be the probability function of a discrete random variable X which assumes the values x_1, x_2, x_3, and x_4, such that $2P(x_1) = 3P(x_2) = P(x_3) = 5P(x_4)$. Construct the probability distribution of X.

13. A Psychologist administered a personality inventory test for passive-aggressive traits to 160 test subjects. Individuals were rated on a score from 1 to 5, where 1 was extremely passive and 5 extremely aggressive. The results are shown below. Let the random variable X represent the scores of the test subjects. Construct a probability distribution for X and draw histogram.

Score (x)	1	2	3	4	5
Frequency (f)	27	36	44	33	20

14. Test the following function to determine if it is a probability function. $P(X=x)=\dfrac{x^2+5}{50}$, for $x=1, 2, 3$ or 4.

(a) Construct the probability distribution.

(b) Draw histogram.

(c) Find the variance for the distribution.

15. Let $P(X=x)=\dfrac{x}{15}$, with random variables $x=0, 1, 2, 3, 4$ and 5.

(a) Write and verify the probability distribution.

(b) Find the Expected Value (mean) and standard deviation of the distribution.

(c) Draw histrogram.

(d) Find $P(X=1$ or $X=2)$.

(e) Find the probability of at most 3: $P(X \le 3)$.

(f) What can you say about the areas of the rectangles (in the histogram), and the probabilities calculated in this exercise? Is there a relationship? Explain.

16. Assume that a Zoologist, randomly selects 14 newborn baby hippos and counts the number of girls selected, x. The probabilities corresponding to the 14 possible values of x are summarized in the given table.

x	$p(x)$	x	$p(x)$	x	$p(x)$
0	0.001	5	0.103	10	0.012
1	0.002	6	0.189	11	0.025
2	0.006	7	0.201	12	0.007
3	0.012	8	0.156	13	0.150
4	0.008	9	0.123	14	0.005

(a) Find the probability of selecting 2 or more baby hippo girls.

(b) Find the probability of selecting exactly 11 baby hippo girls.

(c) Find the probability of selecting at most 3 baby hippo girls.

(d) Find the probability of selecting at least 7 baby hippo girls.

17. Assume that a quality control manager randomly selects 11 light bulbs. The probabilities and the number of defective bulbs are summarized in the given table. Answer the question using the table. Round your answers to 4 decimal places.

x	0	1	2	3	4	5	6	7	8	9	10	11
$P(x)$	0.0014	0.0125	0.0513	0.1259	0.2060	0.2360	0.1931	0.1128	0.0462	0.0125	0.0021	0.0002

Find the probability of selecting

(a) at least 3 defective bulbs.

(b) at most 4 defective bulbs.

(c) less than 5 defective bulbs.

(d) more than 6 defective bulbs.

(e) 2 defective bulbs.

18. A woman wants to adopt four pets, either cats or dogs, or both, but she is not sure how many of each.

    (a) Construct the probability distribution for the number of cats she can adopt. (Use a tree diagram and assume equal probability for adopting a cat or a dog.)

    (b) What are the expected value and the standard deviation for the number of cats she adopts?

19. A man invests \$400 in a tech stock. The probability that the stock will rise tomorrow is 0.06. If it rises, his gain is \$8000, but if it falls, he loses his original investment (\$400). What is the expected gain in this transaction?

20. A man is playing basketball in his backyard. He has the same probability of making a shot as he does missing it.

    (a) Make a table showing the number of possible times he could make a basket in the next three throws and their probabilities. (Use a probability tree diagram).

    (b) How many baskets should he expect to make in the next three shots and what is the standard deviation?

21. A box contains 5 quarters and 6 nickels. Two coins are randomly selected without replacement. Let the random variable X be the total value of the coins selected. Construct the probability distribution of X and find the expected value and the variance of the distribution.

22. You play a coin game in which you win \$1.50 if the coin shows heads and you lose \$2 if it shows tails. Let the random variable be the amount you win in one game. Construct the discrete probability distribution for the random variable and calculate its expected value.

23. The Sky Ranch is a supplier of aircraft parts. Included in stock are eight altimeters that are correctly calibrated and two that are not. Three altimeters are randomly selected *without replacement*. Let the random variable X represent the numbers that are not correctly calibrated. Construct the probability distribution and find the mean and standard deviation for the random variable X.

24. Let the random variable X represent the number of boys in five children. Construct a table describing the probability distribution, and then find the mean and standard deviation.

25. In a dice game, you roll a die once for a \$5 wager. You win \$8.00 (net winning) for rolling a 3 or a 4, and lose the original wager otherwise. What is your expected gain in this game?

26. Construct the sample space for flipping 3 coins, list the outcomes, assign numerical values to the number of tails that can appear, and calculate the mean (expected value) and the standard deviation of the experiment.

27. Construct a probability distribution for the number of girls in a family of 4 children. Find the expected value, standard deviation, and graph the distribution.

28. Construct a probability distribution for the sum of the dots when 2 dice are rolled. Find the expected value and the standard deviation.

29. A coin is tossed four times. Find the mean, variance, and standard deviations of the number of heads that will be obtained.

30. Based on past results found, it is estimated that there is a 0.120 probability that a basketball post season contest will last a minimum of four games, a 0.253 probability that it will last five games, a 0.217 probability that it will last six games, and a 0.410 probability that it will last seven games. Construct the probability distribution and find the expected value and the variance for the numbers of post season games.

31. A local lottery prizes were listed along with the chances of *net* winning: \$8,000 (1 chance in 201,000), \$5000 (1 chance in 180,000), \$1500 (1 chance in 100,000), \$600 (1 chance in 60,000), \$200 (1 chance in 10,000), and a watch valued at \$64 (1 chance in 4000). Find the expected value if the cost of entering this lottery is \$5.

32. When you give a casino \$10 for a bet on the number 7 in roulette, you have a 1/38 probability of winning \$350 and a 37/38 probability of losing \$10. If you bet \$10 that the outcome is an odd number, the probability of winning \$10 is 18/38 and the probability of losing \$10 is 20/38.

(a) Construct the probability distribution.

(b) If you bet \$10 on the number 7, what is your expected value?

(c) If you bet \$10 that the outcome is an odd number, what is your expected value?

33. The probabilities that a batch of 4 transistors will contain 0, 1, 2, 3, and 4 defective components are 0.5521, 0.2970, 0.1307, 0.0191 and 0.0011, respectively. Find the variance for the probability distribution.

34. A contractor is considering a sale that promises a profit of \$26,000 with a probability of 0.7 or a loss of \$6000 with a probability of 0.3. What is the expected profit?

35. The following table lists the money that can be won together with the chances of winning each category in a game. (All winnings are net amounts)

\$4,500	1 in 15,000 chance
\$3,000	1 in 9,000 chance
\$900	1 in 5,000 chance
\$200	1 in 900 chance

Find the expected value of the amount won if you purchased one ticket for \$0.40.

36. An Insurance Company charges Lee \$2000 a year for a \$100,000 life insurance policy. The probability that she will live is 0.9884.

(a) Construct the probability distribution for this experiment and find the expected value for the distribution.

(b) Draw the histogram for the distribution. **(c)** Find the standard deviation.

(d) Find the range of usual and unusual values for this experiment.

37. Robert purchased a \$10 ticket from a private lottery. He would lose the original \$10 investment unless he follow the following rules.

Winning amount in (\$)	Chances
800	once in 2000
500	once in 1000
300	once in 300
10	once in 100

Find the mean and standard deviation of this game. Graph the distribution and label the axis as well as the graph.

38. A building contractor pays \$380 to bid on a contract. The probability of him getting the contract is 0.4, in which case he will make a net profit of \$15,000. On the other hand, if he does not get the contract, he loses the \$380. Find the contractor's expected net profit on the bid. Graph the distribution.

39. Find the expected value of winning when you make a \$100 bet on a number on the roulette table. Graph the distribution. (It is given that there are 38 numbers to bet on and the payoff is 36 times.)

40. Percy wants to start investing his money into mutual funds. He is looking into two companies that offer competitive plans. The following table summarizes the investment choices Percy has with each company. The first column in each table represents the net gain, and the second column is the probability column.

Company A		Company B	
Net Gain	Probability	Net Gain	Probability
1000	0.2	500	0.2
2000	0.1	3000	0.1
500	0.4	4000	0.1
−250	0.1	−10000	0.4
−1200	0.2	−400	0.2

Based on the above data, which company should Percy invest his money in?

41. Toss a coin 6 times and record the number of tails. Construct a probability distribution for the number of tails represented by X. Find the mean and the standard deviation of this experiment using the probability distribution.

42. The probability that a student will pass Prof. Badalian's Statistics class is 67%. For a class of size 46, graph the probability distribution.

(a) Find the probability that

i) half of the class will pass. **ii)** at least 20 will pass.

iii) at most 30 will pass. **iv)** draw the distribution.

(b) Calculate the mean and the standard deviation of probability distributions.

43. At a company that makes canned peaches, the probability that a can will have air in it is 0.01. If a person buys ten cans of peaches, what is the probability that two have air in them?

44. At a certain community college of 1200 students, 60% have a grade point average above 3.00. If 300 students are randomly selected, what is the probability that all 300 will have a grade point average above 3.00?

45. 9% of men cannot distinguish between the colors red and green. This is the type of color blindness that causes problems with traffic signals. If six men are randomly selected for a study of traffic signal perceptions, then compute the probability that exactly two of them cannot distinguish between red and green by using Binomial distribution formula.

46. A machine has 12 identical components which function independently. The probability that a component will fail is 0.02. The machine will stop working if more than three components fail. Find the probability that the machine works properly.

47. If 43% of all cars in a large community are foreign made, find the probability that out of 10 randomly selected cars 3 are foreign.

48. The probability that a person has blue eyes is 0.41. In a math class with 6 students, what is the probability that

(a) at most 3 have blue eyes? **(b)** exactly 6 have blue eyes? **(c)** at least 4 have blue eyes?

49. Agent Ray a notorious FBI Agent in charge of the Fringe Division has discovered that in a small town in the south 8% of all truck drivers experienced a 30 minutes blackout while driving through Hwy 51. To investigate this unexplained phenomena, he has randomly selected 10 drivers.

(a) Find the probability that in his random selection;

i) 3 or more drivers experienced blackouts. **ii)** at most 5 drivers experienced blackouts.

iii) none experienced blackouts.

(b) Find the mean and SD of this investigation. Find the range of usual number of blackouts in this city.

50. A survey showed that 40% of Los Angeles City College students own an iPhone. If ten students are randomly selected and surveyed, what is the probability that

(a) at most two will have an iPhone?

(b) at least 8 will have an iPhone?

(c) exactly one will have an iPhone?

51. If you toss a coin 14 times, what is the probability that you will get

(a) at most 2 heads?

(b) at least 8 heads?

(c) exactly 1 head?

52. City of Glendale mayor's office conducted a city-wide survey of adults and found that 20% smoke. A group of 9 adults were randomly selected. Find the probability that

(a) none smoke.

(b) at least 3 smoke.

(c) less than 4 smoke.

(d) at most 5 smoke.

53. A multiple choice test has 10 questions each with 4 possible answers, only one of which is correct.

(a) If Andy, who forgot to study for the test, guesses on all the questions, then compute the probability that he will answer

i) exactly 6 questions correctly.

ii) at least 4 questions correctly.

iii) at most 3 questions correctly.

iv) greater than 5 questions correctly.

(b) If the passing grade is 70%, what is the probability that Andy will pass the class by completely guessing at all the answers?

(c) Find the mean and SD of this experiment.

54. A recent study showed that 43% of all 16-years-old males in LA sag their pants. If a random sample of 8 16-year olds is selected, find the probability that

(a) exactly 3 sag their pants.

(b) more than 5 sag their pants.

(c) at most 4 sag their pants.

55. The probability that a flight will be delayed is 14%. In 15 randomly selected flights,

(a) find the probability that

i) six of them are delayed.

ii) at least 7 of them are delayed.

iii) at most 4 are delayed.

iv) more than 8 of them are delayed.

v) less than 4 of them are delayed.

(b) find the mean and variance of the number of delayed flights.

56. In 2001, according to the US Department of Labor, out of 135,073,000 employed people age 20 and over, 68,587,000 were men. If 800 employed people are selected at random, how many females should one expect to find? What is the standard deviation?

57. According to the US Census Bureau, 12.1% of the US population is below poverty level. If 500 families are surveyed, how many, on average, will be below poverty level. Find the variation.

58. In a bookstore, the probability that any given sale is a sale of a fiction is 0.68. If 18 randomly selected sales are checked, how many fiction sales should one expect to see? What is the standard deviation? Would it be unusual to see 16 fiction book sales out of 18?

59. According to a survey done at a community college, 62% of all students work full time. Find the mean number of community college students who work full time in a random sample of 16 students. Find the variance.

60. If $n = 34$ and $p = 0.4$, find the standard deviation, σ, for the binomial distribution. Round your answer to the nearest hundredth.

61. A company manufactures batteries in batches of 20 with a 1% defect rate. Find the mean and standard deviation of the number of defects per batch and find the range of usual values for the number of defects.

62. The probability of winning a certain lottery is $\frac{1}{184,528,125}$. For people who play the lottery a 1000 times, find the standard deviation of the number of wins.

63. The Providence Computer Supply Company knows that 16% of its computers will require warranty repairs within one month of shipment. In a typical month, 279 computers are shipped. If X is the random variable representing the number of computers requiring warranty repairs among the 279 sold in one month, find its mean and standard deviation.

64. In a college Ethics class, 40 percent of students are in favor of the Death Penalty and 60 percent are opposed. For groups of 260 Ethics college students, find

(a) the mean number of students who oppose the Death Penalty.

(b) the variance.

In exercises 65-69, use Poisson distribution and Microsoft Excel® for computational work.

65. The number of students (X) that passes Professor Badalian's Statistics class each semester follow a Poisson distribution with mean 6 per semester. Calculate the following probabilities.

(a) $P(X=0)$ **(b)** $P(X<14)$ **(c)** $P(X>12)$ **(d)** $P(X\leq 9)$

66. An airport traffic control tower gets an average of 24 arriving flights during the noon hour on Fridays. Statistics shows that the traffic control crew can handle up to 40 arriving flights during that hour. What is the probability that the traffic control tower will have to handle 46 arriving flights during the noon hour on a random Friday?

67. Suppose that the average number of claims for missing baggages per day is 6. Find the probability that in a given day, there will be

(a) No claims. **(b)** Exactly 6 claims. **(c)** At least 2 claims.

68. It is known from past experience that in a certain plant there are on the average 4 industrial accidents per month. Find the probability that in a given year there will be less than 4 accidents. Assume Poisson distribution.

69. During the rush hour on a toll road, the average number of cars going through the toll gate every minute is 4.5. Find the probability that during the rush hour period

(a) exactly 8 cars will go through during a 1 minute period.

(b) 5 or 6 cars will go through during a 1 minute period.

(c) at least 3 cars will go through during a 1 minute period.

In exercises 70-71, use Poisson approximation to Binomial distribution and Microsoft Excel® for computing required probabilities.

70. If 2% of flash drives manufactured by a company are defective, find the probability that in a sample of 75 flash drives exactly 4 are defective.

71. An experiment involves rolling a die 100 times and recording the number of 6's that occur. Find the probability of rolling

(a) no 6's, when a die is rolled 100 times. **(b)** twenty-five 6's.

(c) at least twenty 6's. **(d)** at most eighteen 6's.

NORMAL PROBABILITY DISTRIBUTION

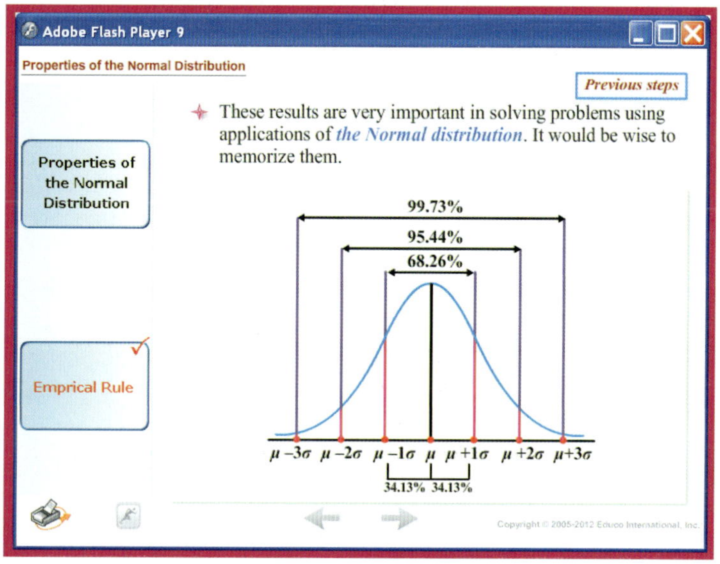

CHAPTER 5

NORMAL PROBABILITY DISTRIBUTION

Chapter Overview:

5.1: INTRODUCTION

In Chapter 4 we discussed discrete random variables, defined a discrete probability distribution as a rule which assigns a probability to every value of a discrete random variable, and presented the graph of a discrete probability distribution as a set of points that are not connected. We also introduced continuous random variables and defined them as variables that take on a continuous range of values (interval), hence making it impossible to assign a probability to each possible value since there are infinitely many values in any given interval. Therefore, to circumvent this, we instead assign probabilities to intervals of values that our continuous random variable takes. We will see that the best way to do this is to describe a continuous probability distribution by a curve, and assign the probabilities over intervals as the areas under the curve over those intervals. Finally, since these areas are not simple geometric shapes, we will employ tables that contain the appropriate areas and use these to answer any relevant probability questions.

5.2: Continuous Distribution

Objectives

After completing this section you will be able to:

A. Understand Continuous Random Variable.

B. Describe Probability Density Function of Continuous Random Variable.

C. Use Properties of Continuous Random Variable.

A. Example of a Continuous Random Variable

What is the probability that the body temperature of a person is less than 96? Body temperatures take on a continuous range of values. If we attempt to list all the possible values, then we can always find values between any two of them. Therefore, we cannot use a discrete random variable to denote body temperature. Instead, the body temperature of a person is modeled by a *continuous random variable*. Let's say that the curve to the right describes the probability distribution of this random variable. The probability that a person's body temperature is less than 96 is the area under the curve corresponding to the part of the x-axis where the values are less than 96.

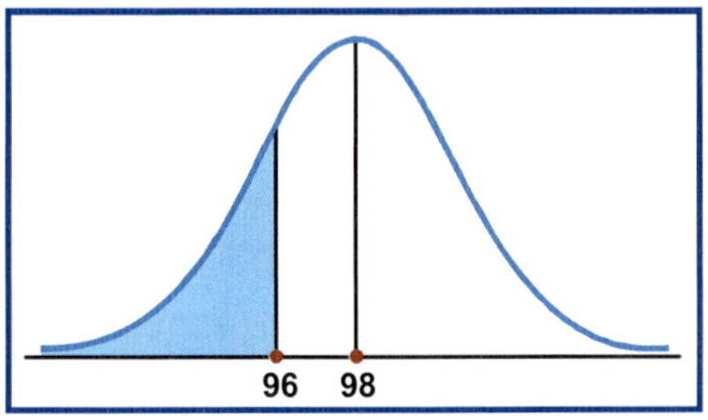

Figure 5.1

B. Probability Density Function of a Continuous Random Variable

Unlike a discrete probability distribution, a continuous probability distribution cannot be presented in a tabular form. It has either a formula form or a graphical form. These smooth curves representing continuous probability distributions are called *density curves* and the functions that define them are called *probability density functions* known as *pdfs*.

A discrete density polygon or histogram gets smoother and smoother as the sample size gets larger, and the class intervals become more numerous (increase in numbers) and narrower.

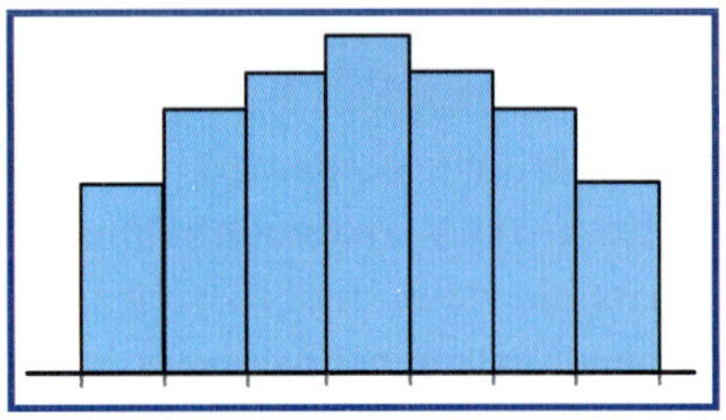

Figure 5.1 a

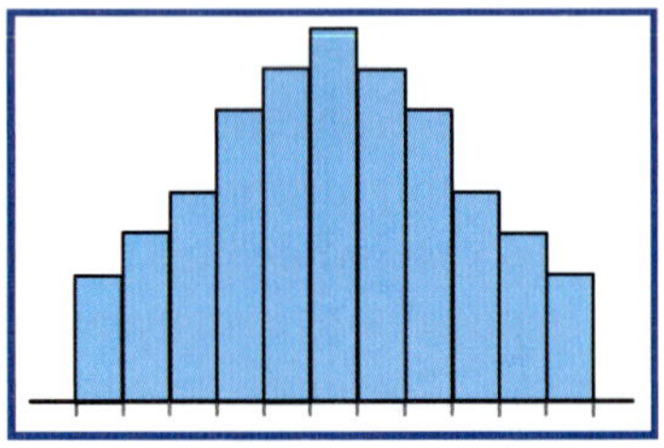

Figure 5.1 b

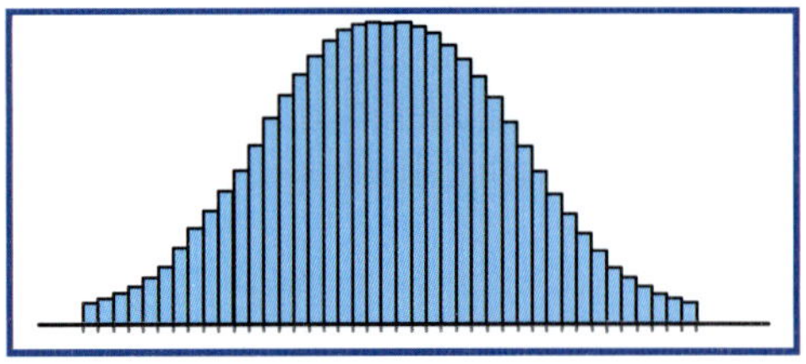

Figure 5.1 c

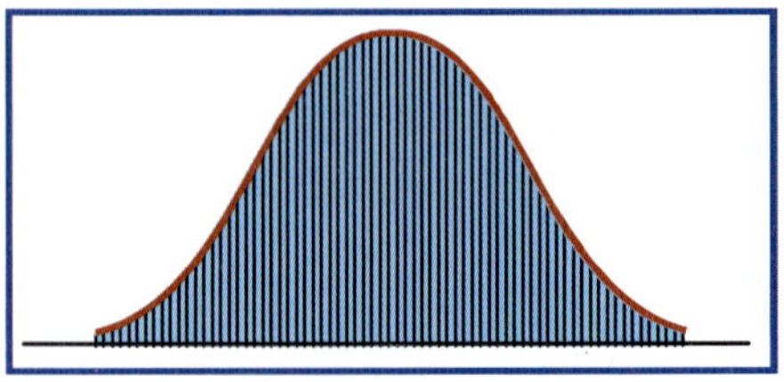

Figure 5.1 d

Ultimately, as the number of intervals goes to infinity (a very large number) and the width of each class tends to zero (a very small number), the *discrete density polygon* becomes a continuous smooth density curve.

• Properties of Probability Density Function

Let X be a continuous random variable taking values on the interval $[a, b]$. A function $f(x)$ is said to be the *probability density function* of the continuous random variable X if it satisfies the following properties.

i) $f(x) \geq 0$ for all x's in the interval $[a, b]$.

ii) For two distinct numbers c and d in the interval $[a, b]$,

$P(c < X < d) =$ [Area under the probability curve between $x = c$ and $x = d$]

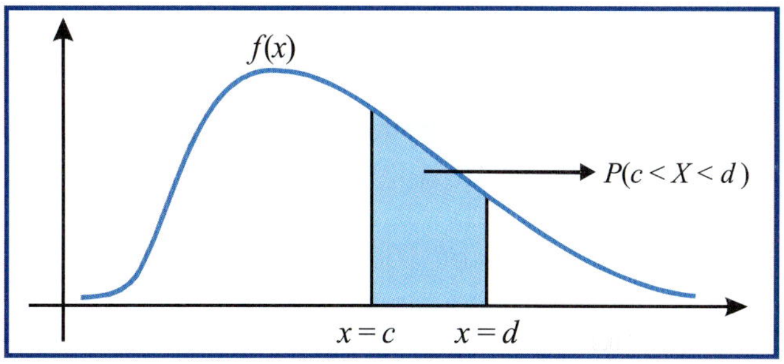

Figure 5.2

iii) The total area under the probability curve is 1.
That is, $P(a < X < b) = 1$

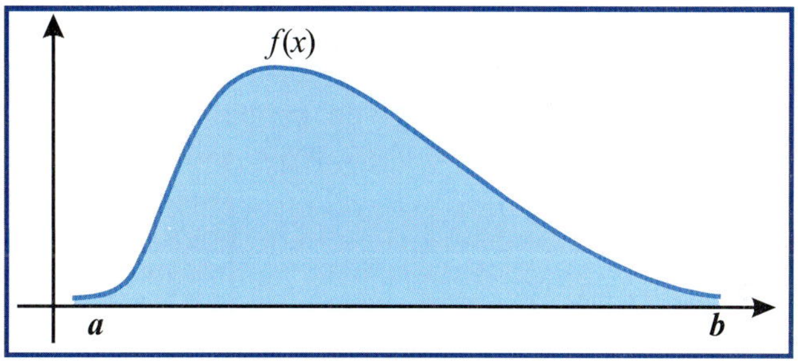

Figure 5.3

Note: Some probability curves are unbounded, meaning that they extend to infinity on both directions as they get closer and closer to the x-axis. Interestingly enough, if such functions are probability curves, then the total area under the curve is still 1.

$$P(-\infty < X < \infty) = 1$$

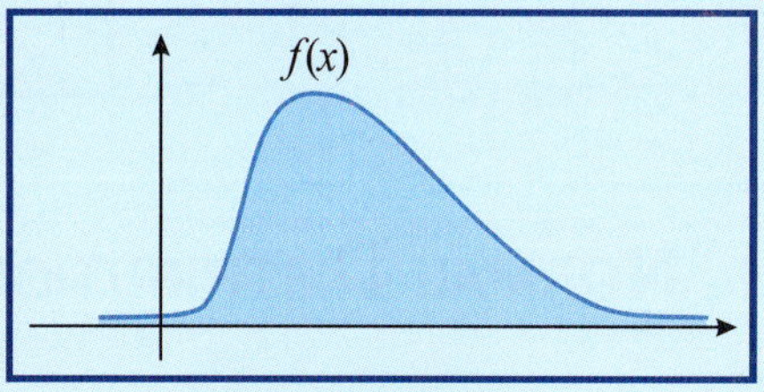

Figure 5.4

C. Properties of Continuous Random Variable

i) $P(X = c) =$ Area under the probability curve at $X = c$

$= 0$

because the area of the rectangular strip with width 0 is 0.

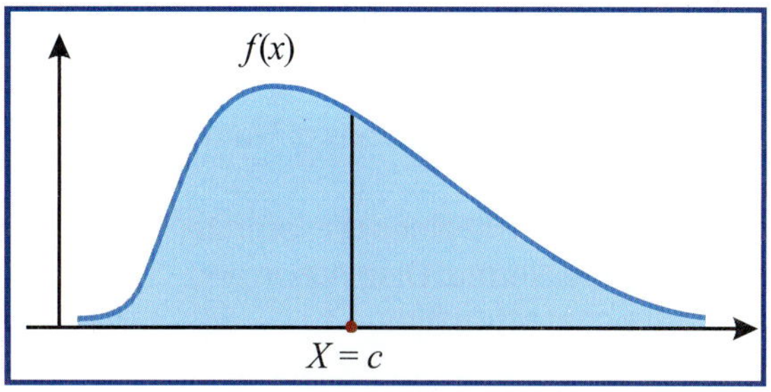

Figure 5.5

For a continuous random variable, the probability at a point is always zero, that is, $P(X = c) = 0$ for all c's. Hence in case of continuous random variable, we always talk of probabilities in an interval and not at a point (which is always zero).

ii) Since in a case of continuous random variable, the probability at a point is always zero, then it does not matter if one or both endpoints of the interval (c, d) are included or not.

Given that $P(X = c) = 0$ and $P(X = d) = 0$, then
$P(c \leq X \leq d) = P(c \leq X < d) = P(c < X \leq d) = P(c < X < d)$

iii) If $\boldsymbol{X}$ is a continuous random variable, then $P(X \leq 3) = P(X < 3)$.

However, if $\boldsymbol{X}$ is a discrete random variable taking positive integer values, then

$P(X \leq 3) = P(X = 0) + P(X = 1) + P(X = 2) + P(X = 3)$

and $P(X < 3) = P(X = 0) + P(X = 1) + P(X = 2)$.

Therefore, $P(X \leq 3) \neq P(X < 3)$.

iv) As stated in **i)**, if X is a continuous random variable, then $P(X = c) = 0$. However, in application, it is possible to be faced with the following scenario.

Let's consider the distribution of the heights of 100 athletes with a mean of 69 inches and standard deviation of 2.8 inches. If the continuous random variable X denotes the heights in inches, of the athletes, then by **i)** $P(X = 67.4) = 0$, a result which does not make much sense. Theoretically, it means that there is no athlete in a group of 100 athletes, whose height is exactly 67.4 inches, a result which sounds absurd. To bridge the gap between theoretical mathematics and real world application, we should be careful in interpreting such results and try to look into their practical implications. Since no measuring instrument can take measurements with exact magnitudes, from practical point of view $x = 67.4$, simply means that the height of an athlete is 67.4 measured to the nearest first decimal place. Practically, 67.4 inches can be thought of as the value lying in a small interval surrounding it, say, (67.395, 67.405), which has considerable non-zero area under the given probability curve. Hence, we may compute the required probability as: $P(X = 67.4) = P(67.395 < X < 67.405)$.

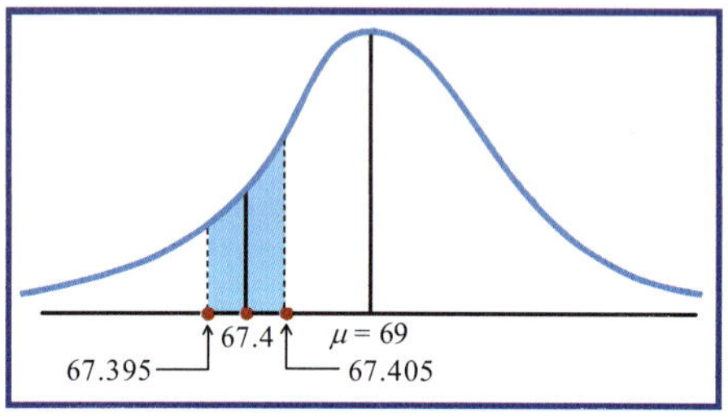

Figure 5.6

5.3: Normal Distribution

Objectives

After completing this section you will be able to:

A. Describe Applications of Normal Distribution.

B. Express The algebraic formula for the Normal Distribution and its characteristics.

C. Use Properties of the Normal Distribution.

A large number of statistical applications relating to the practices of economics, psychology, finance, physical and social sciences follow a continuous probability distribution called the *Normal Probability Distribution*.

In other words, when data is collected from a study, the histogram follows a bell-shaped curve. Therefore the probability questions that arise are answered based on the shape of this Normal Distribution.

The Normal Distribution was first discovered by De-Moivre (1667-1754) in 1733 while working on problems in the game of chance (gambling). Later Karl Friedrich Gauss (1777-1855) used this distribution to describe the errors of measurements in the calculation of orbits of then called heavenly bodies (planets). Today, the Normal Probability Distribution is the most important tool of statistical analysis in applied sciences.

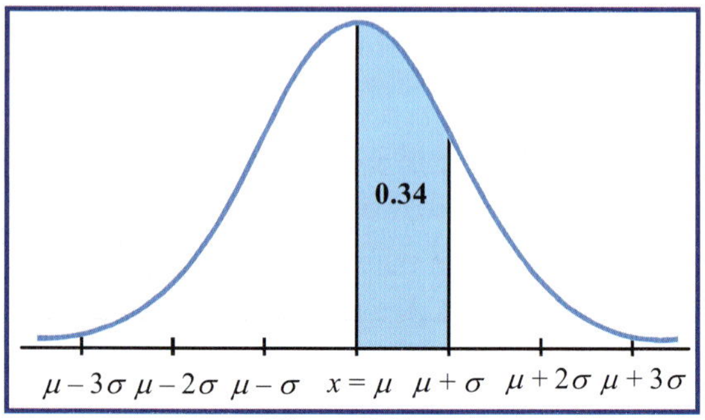

Figure 5.7

A. Applications of Normal Distribution

i) If X is a continuous random variable that follows a normal distribution with mean μ and standard deviation σ, then

$$P(\mu - 3\sigma < X < \mu + 3\sigma) = 0.9973$$

In other words, the probability that X lies within three standard deviations of the mean is 99.73%. Similarly the probability that X lies more than 3 standard deviations from the mean is 0.27%.

As we can see, the probability that X lies outside the 3 standard deviation region is negligible. Thus, in all probability applications of the normal distribution, we should expect a standard normal variable to lie between the limits $\pm 3\sigma$. This property of the normal distribution forms the basis of Large Sampling Theory. (The detailed study of the Sampling Theory will be presented in Chapter 6).

ii) Most Discrete Probability Distributions (*i.e.* Binomial Distribution, Poisson Distribution *etc.*) tend towards the normal distribution as the number of trials increases.

iii) The basis for decision theory and statistical estimation are based on the fundamental assumption that the parent population from which the samples are drawn follows the normal distribution (more detailed discussion in Chapters 6, 7, and 8).

iv) The normal distribution appears in the conclusion of the most important Theorem in Statistics, the Central Limit Theorem (Section 5.7).

v) Normal distribution is used in the Statistical Quality Control industry for setting the 3 σ control limits (known as upper and lower control limits) in the construction of Control Charts.

B. The Algebraic Formula for the Normal Distribution and its Characteristics

The Normal Distribution is described by the following probability density function (*pdf*).

$$f(x) = \frac{1}{\sigma\sqrt{2\pi}} e^{-\frac{(x-\mu)^2}{2\sigma^2}}, \quad -\infty < x < \infty,$$

where $\pi \approx 3.14$ and $e \approx 2.72$(Euler's constant)

Since the level of mathematics required to understand and use this formula is beyond the scope of the text, in practice, we use standard tables for calculating various probabilities under the normal curve created by the above *pdf*.

The Normal Distribution is completely determined by the mean μ and standard deviation σ of the population. This means that if we know the mean and standard deviation of the population under study, we can calculate any probability, using the Normal Distribution. The mean and standard deviation are known as the *parameters* of the normal distribution.

Notation: A random variable X that follows a Normal Distribution with mean μ and standard deviation σ is denoted by $X \sim N(\mu, \sigma)$.

C. Properties of the Normal Distribution

Calculating probabilities for a normally distributed Continuous Random Variable requires calculus, since the probability is the area bounded by the curve and the horizontal axis. Therefore, to resolve this obstacle, we need to use the geometric properties of the area under the normal curve to compute probabilities (areas) without the use of calculus.

i) The normal distribution is a continuous (smooth) symmetric bell-shaped curve.

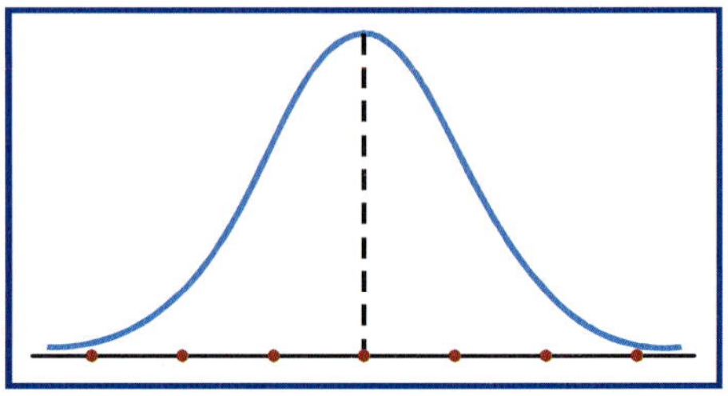

Figure 5.8

ii) The mean, median, and the mode of the normal distribution are equal and located in the center of the distribution.

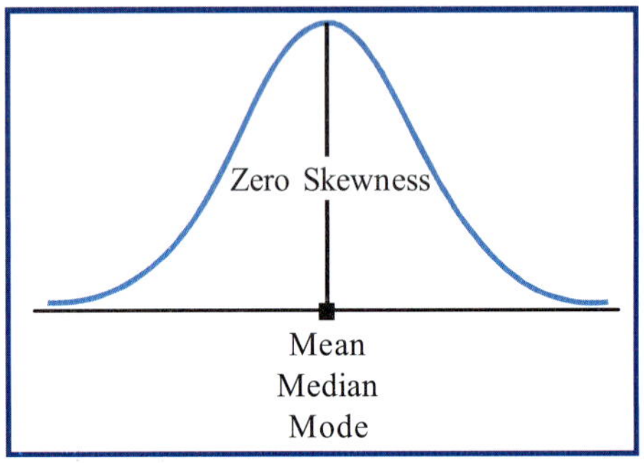

Figure 5.9

iii) The normal curve is symmetric about its mean. That is to say that its shape is the same on both sides of the center line.

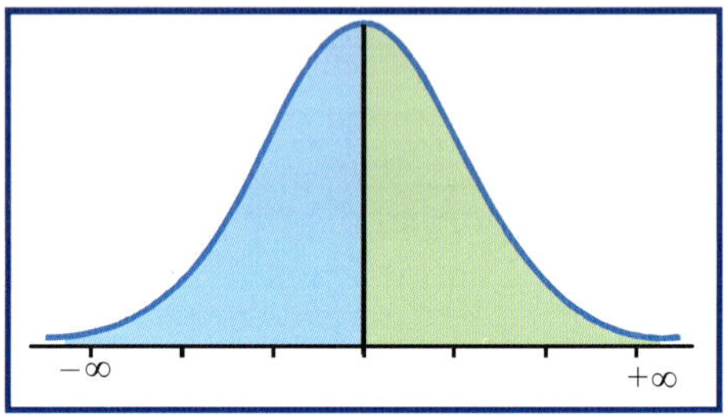

Figure 5.10

iv) The normal curve never meets the x-axis. That is to say that even though the curve gets closer and closer to the x axis as it moves further to the left and to the right, it never touches the axis.

Figure 5.11

v) The normal distribution has only one mode.

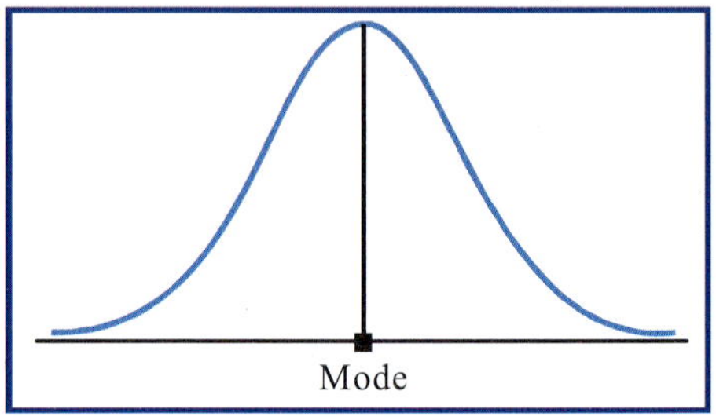

Figure 5.12

vi) The total area under the normal curve equals 1 (or 100%). Since the normal distribution is a symmetric distribution, the area of each half is 0.5 (50%).

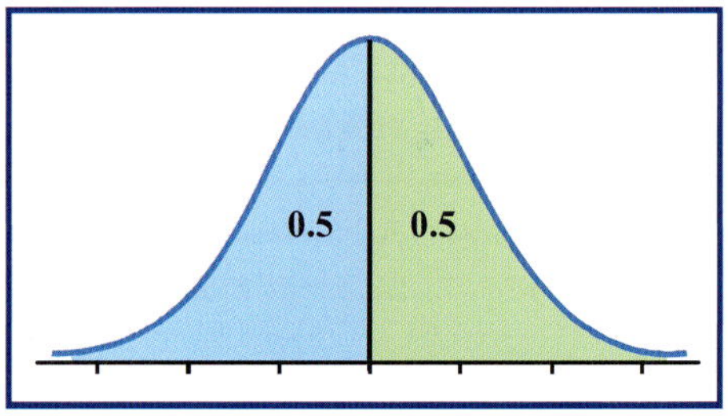

Figure 5.13

vii) Since the mode of a normal curve is at μ, the maximum value of the function that generates the normal curve occurs at $x = \mu$. It has been mathematically determined that this maximum value is inversely proportional to the standard deviation σ.

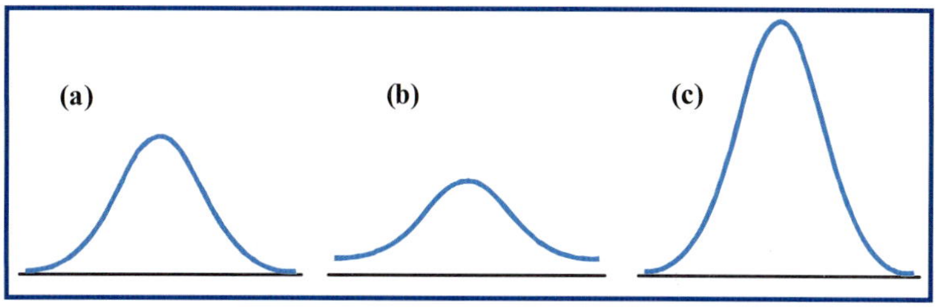

Figure 5.14 a, b, and c

Hence large values of σ decrease the maximum height of the normal curve and increase the spread of the curve along the x-axis. Similarly, smaller values of σ increase the maximum height of the normal curve and also result in the concentration of the probabilities about mean μ. The Figures 5.14 a, b, c give normal probability curves with moderate, large and small values of σ respectively.

viii) The area under the normal curve gives the probability of a value lying between two ordinates. For example, if the mean IQ of a population is 100 and the standard deviation is 15, then the probability that a randomly selected person from that population will have an IQ between 90 to 120 is the area under the normal curve between 90 to 120.

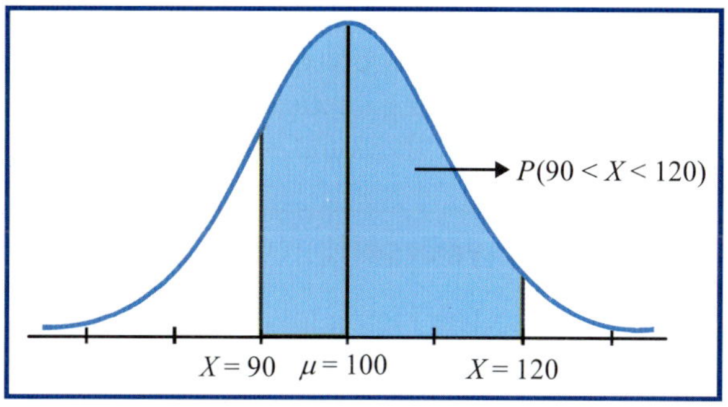

Figure 5.15

Example 1: Identify if the following figures represent a normal probability curve.

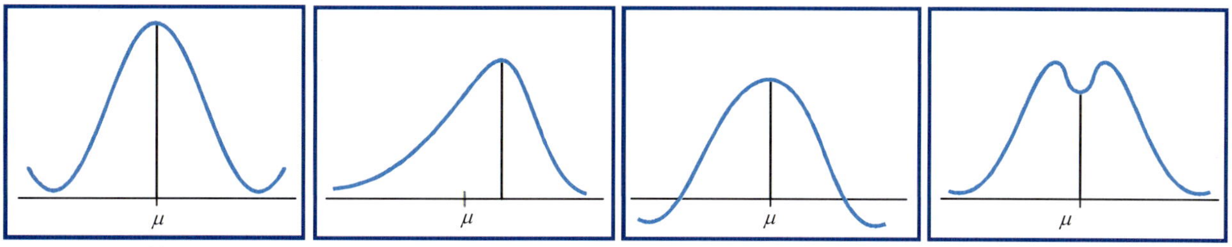

(a) (b) (c) (d)

Solutions:

(a) **No**, it is not a normal probability curve. A normal probability curve becomes closer and closer to the x-axis as increases on either side of μ. In this case, the tails of the curve are getting away from the x-axis.

(b) **No**, it is not a normal probability curve. The given curve is skewed to the left (negatively skewed) whereas the normal curve is symmetric and bell shaped.

(c) **No**, it is not a normal probability curve. A normal probability curve can never touch or cross the x-axis.

(d) **No**, it is not a normal probability curve. The normal curve is unimodal with the mode occurring at $x = \mu$. (Here the given curve is bimodal (two peaks)).

- ## Emprical Rule

The Emprical Rule is also known as 68-95-99.7 Rule.

The areas under the normal probability curve between the values of X at distances of 1σ, 2σ, and 3σ from the mean are given below:

i) The area between $X = \mu$ and $X = \mu \pm 1\sigma$ is: 0.6826

ii) The area between $X = \mu$ and $X = \mu \pm 2\sigma$ is: 0.9544

iii) The area between $X = \mu$ and $X = \mu \pm 3\sigma$ is: 0.9973

These results are very important in solving problems using applications of the normal distribution. It would be wise to memorize them.

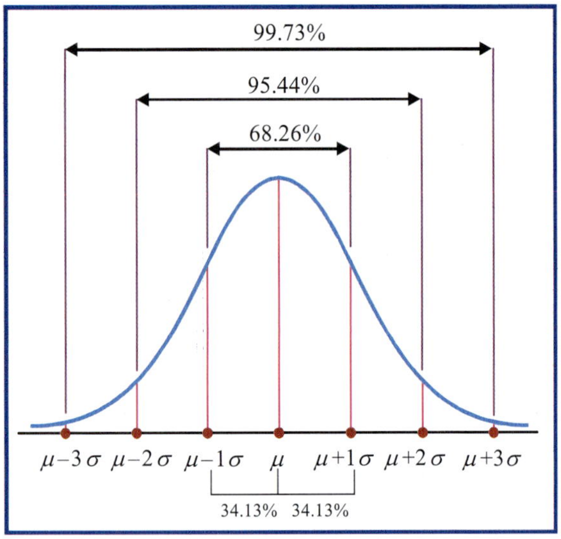

Figure 5.16

Try it yourself:

1. The Figure given below contains four different normal probability curves distinguished with different standard deviations. Arrange the standard deviations in ascending order. Make a conclusion about the spread of data these curves model based on your answer.

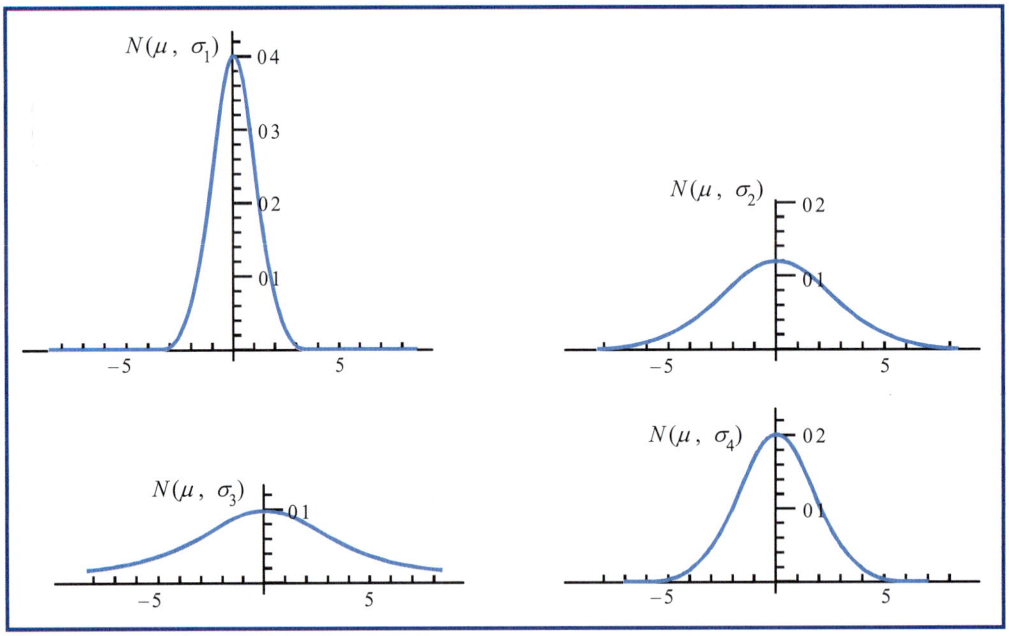

2. Find the area between the mean $X = \mu$ and $X = \mu + 1\sigma$.

3. Find the area between $X = \mu - 1\sigma$ and $X = \mu - 3\sigma$.

4. Find the area between $X = \mu - 1\sigma$ and $X = \mu + 2\sigma$.

5.4: The Standard Normal Distribution $N(0, 1)$

A. Z-Score (Standard Score)

For computing normal probabilities, we need first, standard scores. Standard scores are denoted by Z and are calculated by the formula: $Z = \dfrac{X - \mu}{\sigma}$

Standard scores (Z-values), give the number of standard deviations between the measurement X and the mean μ.

Definition:

A ***z-score*** *or* ***z-value*** *is the number of standard deviations, a given value from the mean.*

Objectives

After completing this section you will be able to:

A. Compute Z-Score (Standard Score).

B. Use Properties of Z-Scores.

C. Use Standard Normal Distribution.

D. Use Empirical Rule for Computing Normal Probabilities.

Example 2: Lee's score on the first statistics test is 78. Her class average and standard deviation are 68 and 4, respectively. Lee's friend, Erica who is in a different section of the same course, scored 82 on the first test. Erica's class average was 71 and standard deviation 6. Compare Lee's and Erica's performances relative to their sections.

Solution: A z-score measures how many standard deviations an observation (X) is above or below the mean. We compare Lee's and Erica's performances, by comparing their standard scores using $Z = \dfrac{X - \mu}{\sigma}$.

Lee's standard score $z_1 = \dfrac{78 - 68}{4} = 2.5$, Erica's standard score $z_2 = \dfrac{82 - 71}{6} = 1.83$

Since Lee's z-score is greater than Erica's and her test score is further away from the average than Erica's. Therefore Lee did better in the first test than Erica when comparing their scores with respective classes.

We will now demonstrate how standard scores can help us calculate probabilities given a normal distribution. The following examples show how to use standard scores in finding the areas (probabilities) under a normal curve.

B. Properties of Z-Scores

i) Z-values are independent of units of measurement. They are in standardized real numbers.

ii) The transformation of x scores to standard scores (z) enables us to compute probabilities in various applications of the normal distribution with different means and standard deviations.

iii) For any given value of z, the corresponding value of x is given by: $x = \mu + z\sigma$

For instance, suppose the average IQ of a population is 100 with a standard deviation of 15. A z-score of 2.3 corresponds to an IQ. $x = 2.3(15) + 100 = 134.5$. Thus $z = 2.3$ implies that the value of $x = 134.5$ is 2.3 standard deviations from the mean.

iv) A *positive* value of z means that value of x lies to the *right* of the mean and a *negative* value of z indicates that the value of x lies to the *left* of the mean

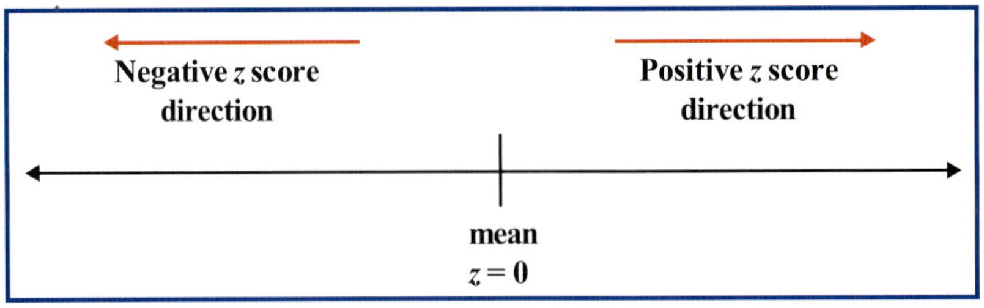

Figure 5.17

v) To find the standard score z given a x value, we use the formula $z = \dfrac{x-\mu}{\sigma}$.

When $x = \mu$, $z = \dfrac{\mu - \mu}{\sigma} = 0$

When $x > \mu$, $z = \dfrac{x-\mu}{\sigma} > 0$ (positive z-value)

When $x < \mu$, $z = \dfrac{x-\mu}{\sigma} < 0$ (negative z-value)

x	z
$x = \mu$	$z = 0$
$x > \mu$	$z > 0$
$x < \mu$	$z < 0$

Table 5.1

Common Z-values in application:

x	$z = \dfrac{x-\mu}{\sigma}$	x	$z = \dfrac{x-\mu}{\sigma}$
μ	0		
$\mu - \sigma$	-1	$\mu + \sigma$	1
$\mu - 2\sigma$	-2	$\mu + 2\sigma$	2
$\mu - 3\sigma$	-3	$\mu + 3\sigma$	3
$\mu - 1.65\sigma$	-1.65	$\mu + 1.65\sigma$	1.65
$\mu - 1.96\sigma$	-1.96	$\mu + 1.96\sigma$	1.96
$\mu - 2.33\sigma$	-2.33	$\mu + 2.33\sigma$	2.33
$\mu - 2.58\sigma$	-2.58	$\mu + 2.58\sigma$	2.58

Table 5.2

Interpretation of Z:

$$z = \frac{x-\mu}{\sigma} \Rightarrow x = \mu + z\sigma$$

Note: Usual z-values: $-3 \le z \le 3$

Unusual z-values: $z < -3$ or $z > 3$

Using the above conversion formulas, $z = 2.33$ implies that the corresponding value of x is 2.33 standard deviations to the right of the mean. Similarly, $z = -2.33$ means that the corresponding value of x is 2.33 standard deviations to the left of the mean.

Example 3: If a continuous random variable X has a normal distribution with mean $\mu = 100$ and standard deviation $\sigma = 15$, draw a rough sketch of

(a) The original nonstandard normal curve,

(b) The transformed (shifted) standard normal curve ($N(0, 1)$).

Solutions: **(a)** Figure 5.18 gives the graph of the original normal distribution with $\mu = 100$ and $\sigma = 15$. (Nonstandard)

(b) Figure 5.19 gives the graph of the standardized normal distribution after shifting the mean to zero and standard deviation to one.

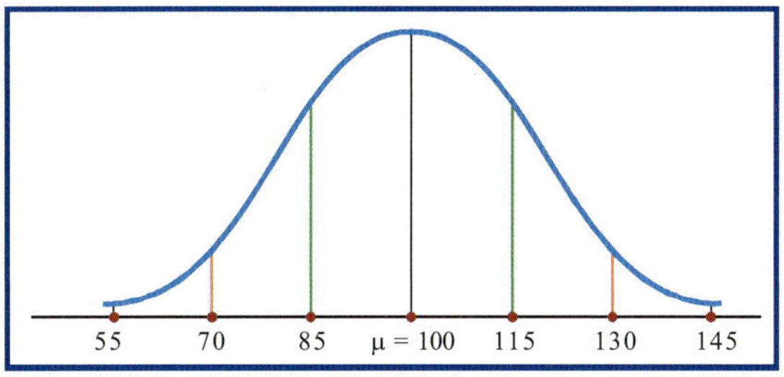

Figure 5.18

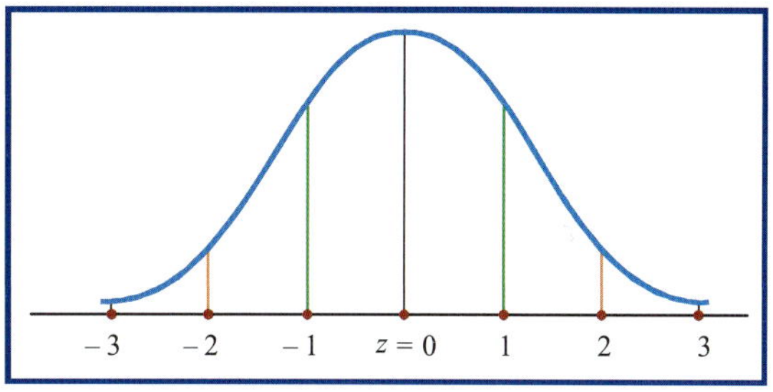

Figure 5.19

C. Standard Normal Distribution

If a continuous random variable X has a normal distribution, then the corresponding standard variable $Z = \dfrac{X - \mu}{\sigma}$, also has a normal distribution with $\mu = 0$ and $\sigma = 1$. The distribution is then known as the *Standard Normal Distribution.*

The notation for a Standard Normal Distribution is $N(0, 1)$. A random variable Z that follows a standard normal distribution is denoted by $Z \sim N(0, 1)$. In other words Z is normally distributed with mean 0 and standard deviation 1.

In the absence of Integral Calculus, if we want to construct a table to compute the areas, say $P(a < X < b)$, under any normal probability curve $N(\mu, \sigma)$, we would need to construct a table for each combination of mean and standard deviation… effectively an infinite number of tables. For obvious reasons we overcome this problem by standardizing the random variable. In practice, we standardize the variable X by using the conversion formula $\left(z = \dfrac{x - \mu}{\sigma}\right)$ and then construct the table for the areas under the standard normal curve. Hence, by rescaling the normal distribution axis, any normal distribution can be converted into standard normal distribution with mean 0 and standard deviation 1. Consequently we need only one table of areas (probabilities) under standard normal curve called the Standard Normal Distribution Table. Thus for any normal distribution with mean μ and standard deviation σ, this table can be used for obtaining the areas (probabilities) under the standard normal curve for almost any interval along the z-axis. Table 3 (at the end of the book) is known as the z-table, gives the areas under standard normal curve.

D. Empirical Rule for Computing Normal Probabilities

The Empirical Rule also known as 68-95-99.7% rule provides the procedure for computing normal probabilities associated with only whole number multiples of the standard deviation (σ), that is, within 1, 2, or 3 standard deviations of the mean. For normal probability curve with mean (μ) and standard deviation (σ), we have [Also See Figure 5.20].

$$P(\mu - \sigma < X < \mu + \sigma) = P(-1 < Z < 1)$$
$$= 0.6826 \approx 68\%$$
$$P(\mu - 2\sigma < X < \mu + 2\sigma) = P(-2 < Z < 2)$$
$$= 0.9544 \approx 95\%$$
$$P(\mu - 3\sigma < X < \mu + 3\sigma) = P(-3 < Z < 3)$$
$$= 0.9973 \approx 99.7\%$$

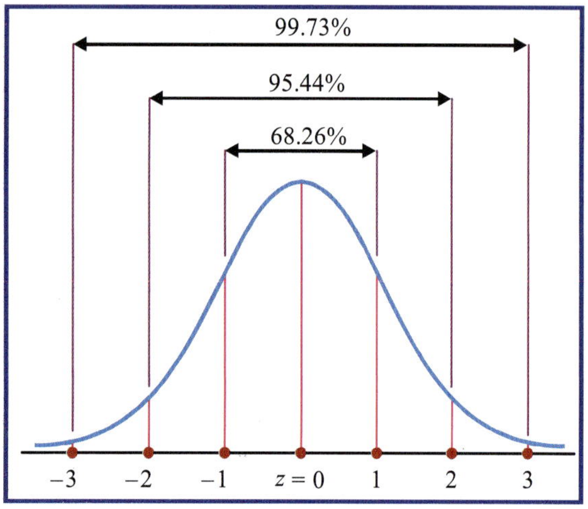

Figure 5.20

Example 4: Using the areas in the given Figure (the Empirical Rule), find the areas;

(a) Between $z = -2$ and $z = 1$

(b) To the right of $z = 1$

(c) To the right of $z = -1$

(d) To the left of $z = -2$

(e) To the left of $z = 1$

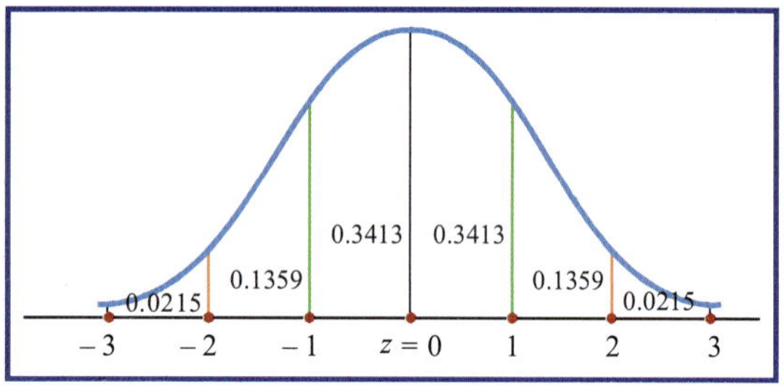

Solutions: **(a)** Area between $z = -2$ and $z = 1$ is the sum of the areas between -2 to -1, -1 to 0, and 0 to 1.

Area between $z = -2$ and $z = 1$

$$= 0.1359 + 0.3413 + 0.3413$$
$$= 0.8185$$

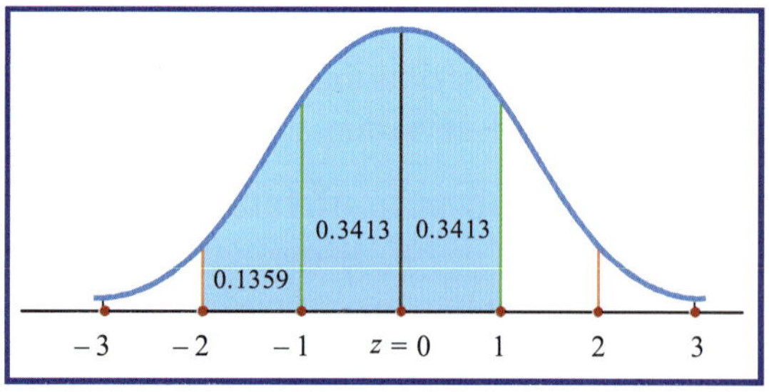

Figure 5.21

(b) (Area to the right of $z = 1$)

= Area to the right of 0 – the area between 0 and 1

$= 0.5 - 0.3413$

$= 0.1587$

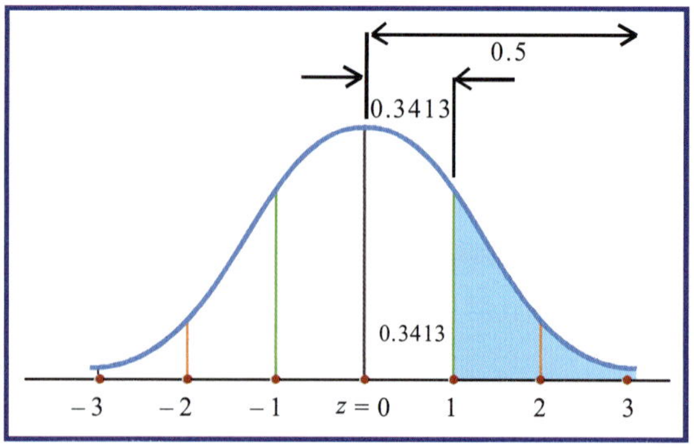

Figure 5.22

(c) (Area to the right of $z = -1$)

$=$ (Area between -1 and 0) + (area to the right of 0)

$=$ $0.3413 + 0.5$

$=$ 0.8413

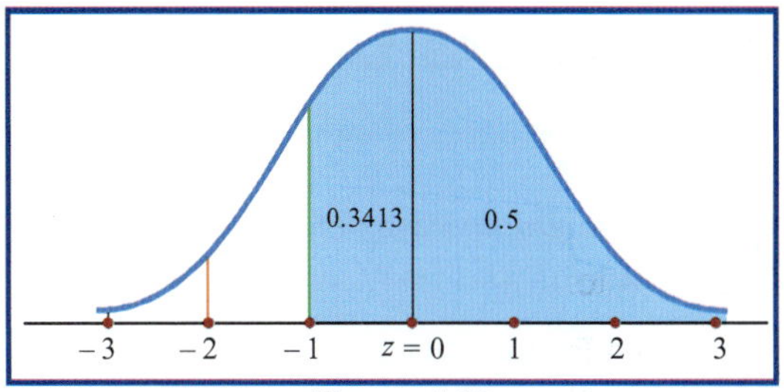

Figure 5.23

(d) (Area to the left of $z = -2$)

$=$ Area to the left of 0 – the area between -2 and 0

$=$ $0.5 - (0.3413 + 0.1359)$

$=$ 0.0228

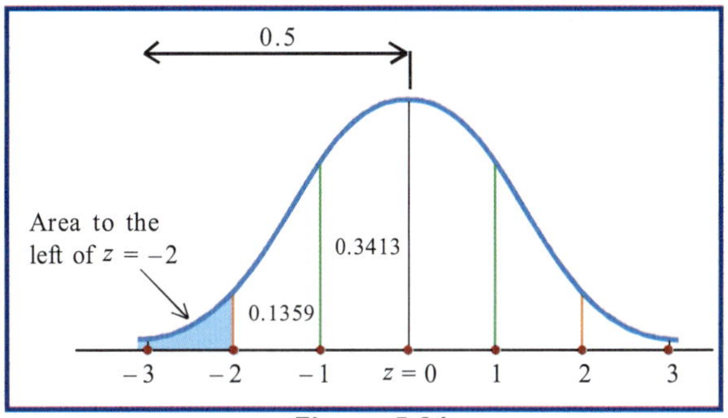

Figure 5.24

(e) (Area to the left of $z = 1$)

$=$ Area to the left of 0 + the area between 0 and 1

$=$ $0.5 + 0.3413$

$=$ 0.8413

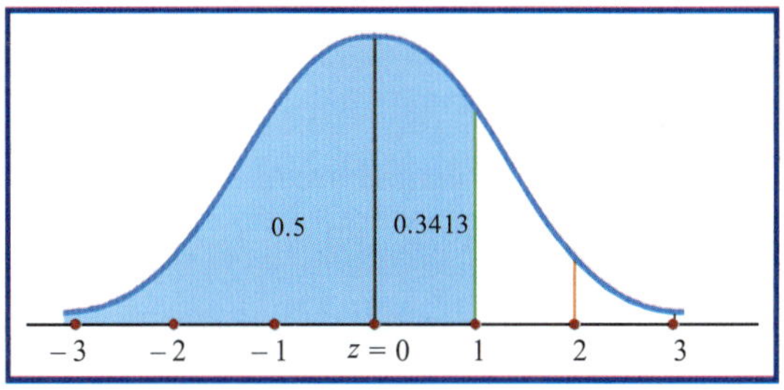

Figure 5.25

Note: The empirical rule is not helpful if we have to compute probabilities associated with some fractional multiples of σ such as:

$P(\mu - 1.8\sigma < X < \mu + 1.8\sigma) = P(-1.8 < Z < 1.8)$ or

$P(\mu - 2.34\sigma < X < \mu + 2.34\sigma) = P(-2.34 < Z < 2.34)$, and so on.

In such cases, we use the technique of computing areas under standard normal curve, discussed in the next section.

Try it yourself:

In exercises 1-6, use the areas in the given figure under the standard normal curve to approximate the required areas.

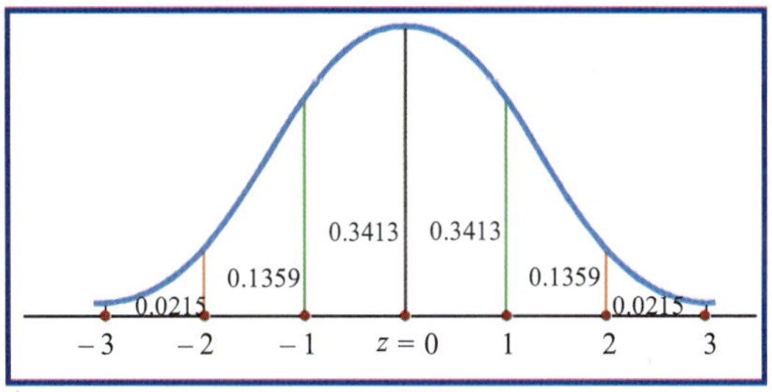

1. Between $z = -1$ to $z = 1$.
2. Between $z = 1$ and $z = 3$.
3. To the right of $z = 2$.
4. Between $z = -3$ to $z = 0$.
5. To the left of $z = -3$.
6. To the left of $z = -1$.
7. Describe the graphical characteristics of the normal curve.
8. Define Z-scores and discuss their use in normal distribution.
9. Describe the Empirical Rule for computing the probabilities under the normal probability curve.

5.5: Areas Under the Standard Normal Curve

Objectives

After completing this section you will be able to:

A. Use Table of Areas under the Standard Normal Curve.

B. Compute areas using the Standard Normal Curve.

C. Compute Z-Scores for a given Area Probability, Percentage, or Proportions.

A. Table of Areas under Standard Normal Curve

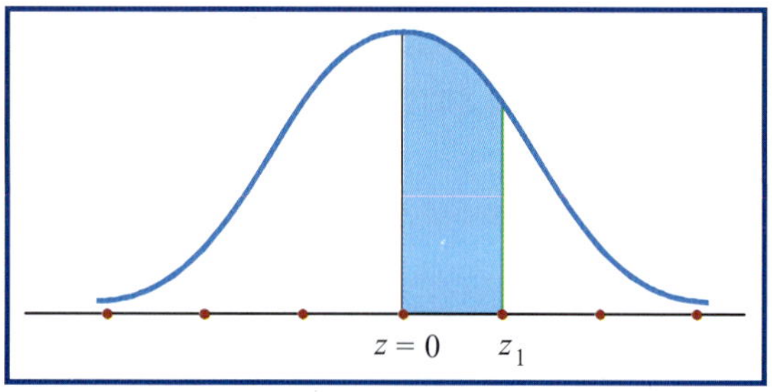

Figure 5.26

For $z_1 > 0$, $P(0 < Z < z_1)$ = Area under standard normal curve between $Z = 0$ and $Z = z_1$

= Shaded area in Figure 5.26

The areas (probabilities) such as the one in Figure 5.26, have been tabulated for different z-scores ranging from 0 to 3.9 using two decimal place accuracy. These areas (probabilities) are given in Table 3 at the end of the book.

In table 3, the z-values are listed in the left column and across the top row (second decimal place), while the shaded areas (Areas from $Z = 0$ to $Z = z_1$) in Figure 5.27 are given in the body of the table. The format of Table 3 (the z table) is given below:

Format of Table 3

Table of Areas $P(0 < Z < z)$

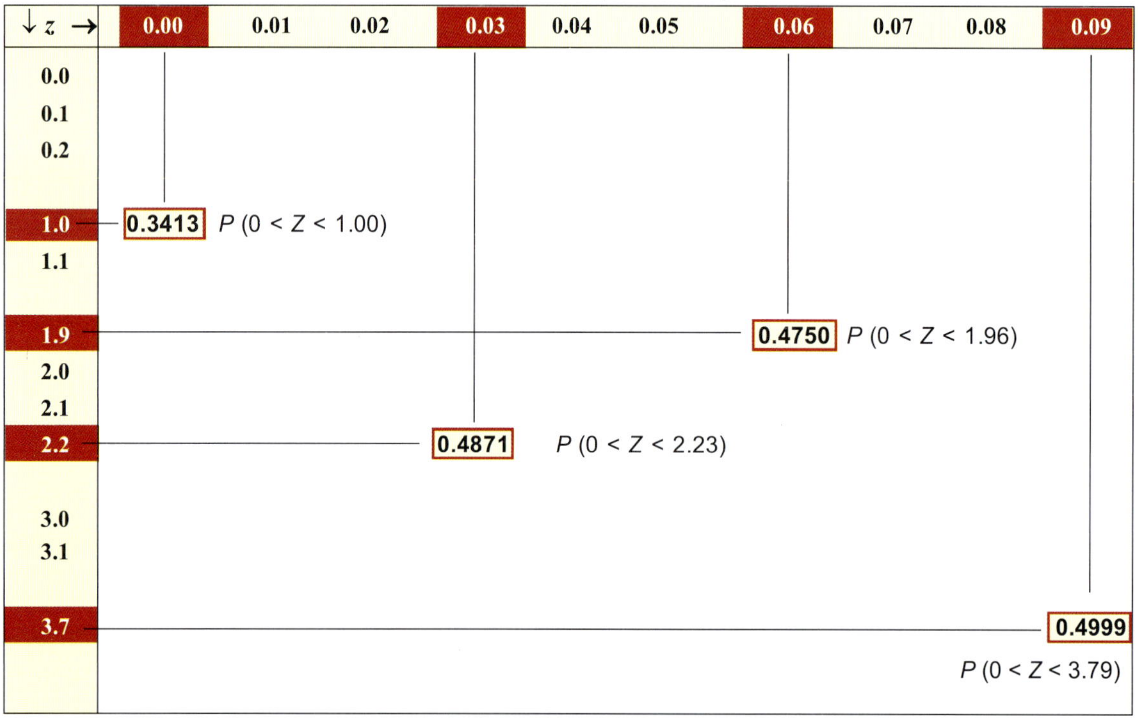

Figure 5.27

• How to Read the Standard Normal Table (z-table):

The z-table, gives the area between $z = 0$ (the mean of the standard normal variable) and any given value of z.

***Step* 1** Round off the z value to two decimal places.

***Step* 2** Locate the z value (with one decimal place) in the first column (z) of the table as shown in Figure 5.28.

***Step* 3** Move across the top row to the column that corresponds to the number in the second decimal place. The value that is in the intersection of both the row and the column is the desired area. For example, to find the area between $z = 0$ and $z = 1.26$, we look up the value that is in row labeled 1.2 and under the column labeled 0.06.

z	0.00	0.01	0.02	0.03	0.04	0.05	0.06	0.07	0.08	0.09
0.0										
0.1										
1.2							0.3962			

Figure 5.28

Therefore, the area from $Z = 0$ to $Z = 1.26$ is 0.3962 or 39.62%.

Example 5: Find the area (probability) between $Z = -2.587$ and $Z = 0$.

Solution: ***Step* 1** Rounding off the z value to two decimal places, we get $z = -2.59$.

***Step* 2** Since the normal curve is symmetrical about the mean, the area from $Z = -2.59$ to $Z = 0$ will be the same as the area between $Z = 0$ and $Z = 2.59$ (Figure 5.29). That is $P(-2.59 < Z < 0) = P(0 < Z < 2.59)$ by symmetry.

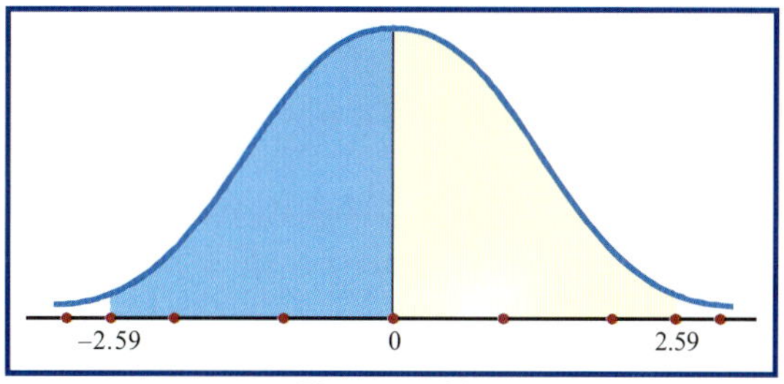

Figure 5.29

Thus, we locate the value 2.5 in the first (z) column.

***Step* 3** Move across the row labeled 2.5 and stop at the column labeled 0.09 (Figure 5.30).

z	0.00	0.01	0.09
2.5			0.4952

Figure 5.30

$P(-2.587 < Z < 0) = P(0 < Z < 2.59) = 0.4952.$

Therefore, the area (probability) between $Z = -2.587$ and $Z = 0$ is approximately 0.4952.

B. Computing Areas Using Standard Normal Curve

Suppose the area we are interested in is not between 0 and z_1. We can still use table 3, basic geometry and arithmetic to calculate the desired area (probability). We discuss below the various possibilities and the technique of calculating any given area (probability).

In terms of probabilities, the areas under the standard normal curve are given by:

$P(Z \geq a) = P(Z > a)$ = Area to the right of the vertical line at $z = a$.

$P(Z \leq a) = P(Z < a)$ = Area to the left of the vertical line at $z = a$.

$P(a \leq Z \leq b) = P(a < Z < b)$ = Area between the vertical lines at $z = a$ and $z = b$.

Remarks:

i) Since Z is a continuous random variable, it is immaterial whether one or both the end points in the corresponding interval are included or not.

ii) The values of Z to the left of the mean are negative and to the right of the mean are positive.

iii) Since the standard normal curve is symmetric about the mean, (Figure 5.31)

$P(Z > 0)$ = Area to the right of $(Z = 0) = 0.5$

$P(Z < 0)$ = Area to the left of $(Z = 0) = 0.5$

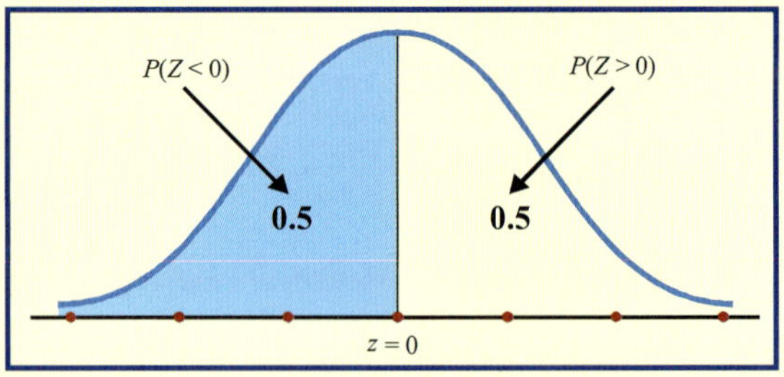

Figure 5.31

(a) Area to the right of a positive z-value, say z_1, that is $P(Z > z_1)$; $z_1 > 0$

$$\begin{aligned} P(Z > z_1) &= \text{Area to the right of } Z = z_1 \\ &= 0.5 - \text{Area between } (Z = 0 \text{ and } Z = z_1) \\ &= 0.5 - P(0 < Z < z_1) \end{aligned}$$

Note: You will read the $P(0 < Z < z_1)$ from Table 3.

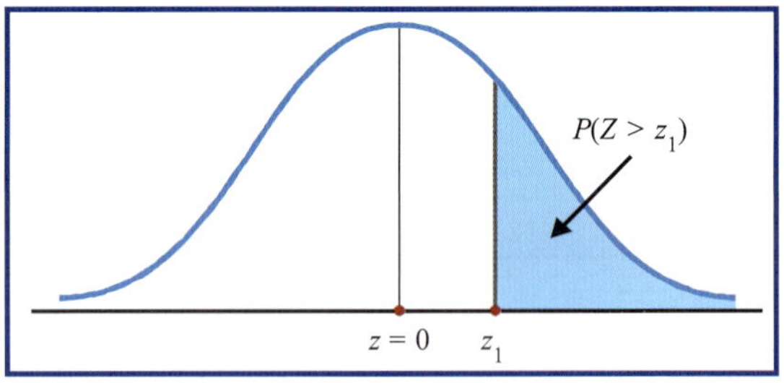

Figure 5.32

(b) Area to the left of a positive z-value, say z_1, that is $P(Z < z_1)$; $z_1 > 0$

$$\begin{aligned} P(Z < z_1) &= \text{Area to the left of } Z = z_1 \\ &= P(Z < 0) + P(0 < Z < z_1) \\ &= 0.5 + P(0 < Z < z_1) \end{aligned}$$

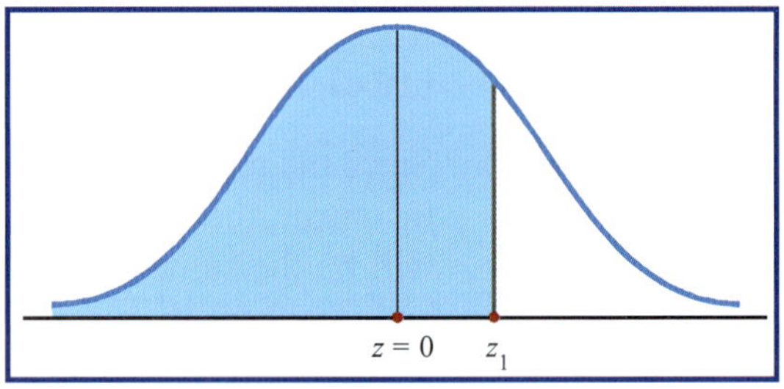

Figure 5.33

(c) Area between two positive z-values, that is, $P(z_1 < Z < z_2)$ in case $z_1 < z_2$

$$\begin{aligned} P(z_1 < Z < z_2) &= \text{Area between } Z = z_1 \text{ and } Z = z_2 \\ &= \text{Area between } Z = 0 \text{ and } z_2) \\ &\quad - (\text{Area between } Z = 0 \text{ and } z_1) \\ &= P(0 < Z < z_2) - P(0 < Z < z_1) \end{aligned}$$

If $z_2 < z_1$, interchange z_1, and z_2 to get the area.

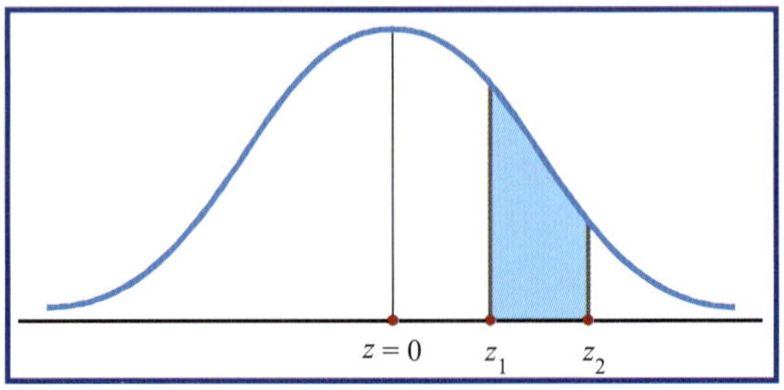

Figure 5.34

(d) Area between two negative z-values, that is, $P(-z_1 < Z < -z_2)$; $z_1 > 0$, $z_2 > 0$ and $z_1 > z_2$.

Because of the symmetric properties of the Normal Distribution

$$\begin{aligned} P(-z_1 < Z < -z_2) &= P(z_2 < Z < z_1) \\ &= (\text{Area between } Z = 0 \text{ and } z_1) \\ &\quad - (\text{Area between } Z = 0 \text{ and } z_2) \\ &= P(0 < Z < z_1) - P(0 < Z < z_2) \end{aligned}$$

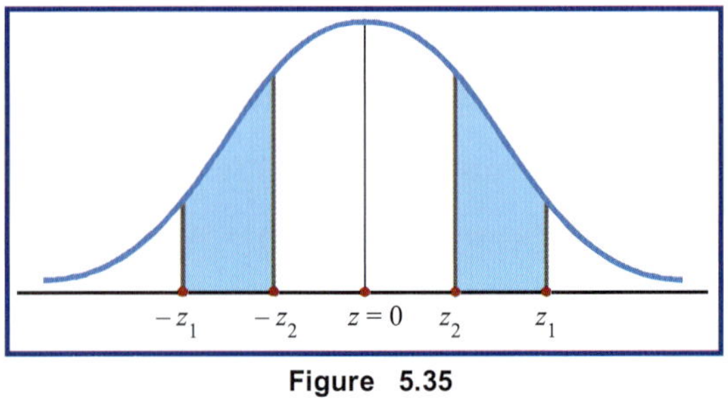

Figure 5.35

(e) Area to the right of a negative z-value, that is, $P(Z > -z_1)$; $z_1 > 0$

$$P(Z > -z_1) = \text{Area to the right of } Z = z_1$$
$$= P(-z_1 < Z < 0) + P(Z > 0)$$
$$= P(0 < Z < z_1) + 0.5$$

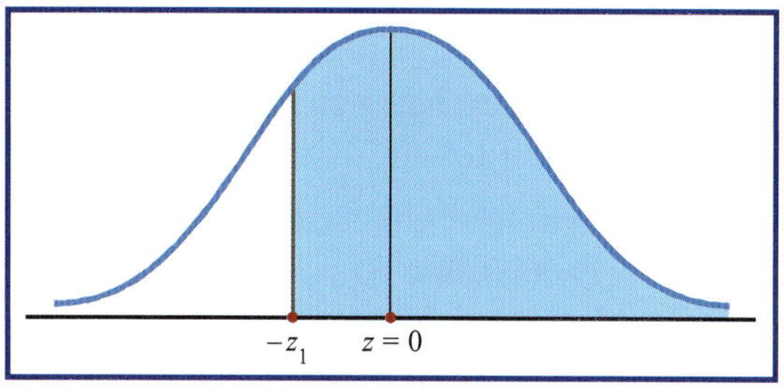

Figure 5.36

Note: $P(-z_1 < Z < 0) = P(0 < Z < z_1)$ (By Symmetry)

(f) Area to the left of a negative z-value, that is $P(Z < -z_1)$; $z_1 > 0$

$$P(Z < -z_1) = P(Z > z_1) \quad \text{(By Symmetry)}$$
$$= 0.5 - P(0 < Z < z_1)$$

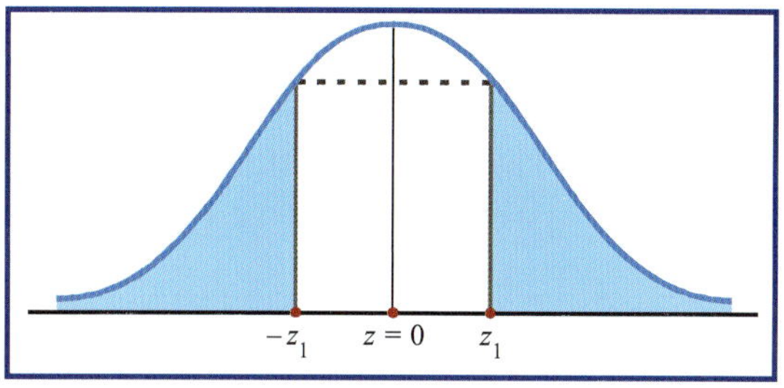

Figure 5.37

(g) Area between one negative z-value and one positive z-value, that is $P(-z_1 < Z < z_2)$; $z_1 > 0$, $z_2 > 0$.

$$P(-z_1 < Z < z_2) = P(-z_1 < Z < 0) + P(0 < Z < z_2)$$
$$= P(0 < Z < z_1) + P(0 < Z < z_2)$$

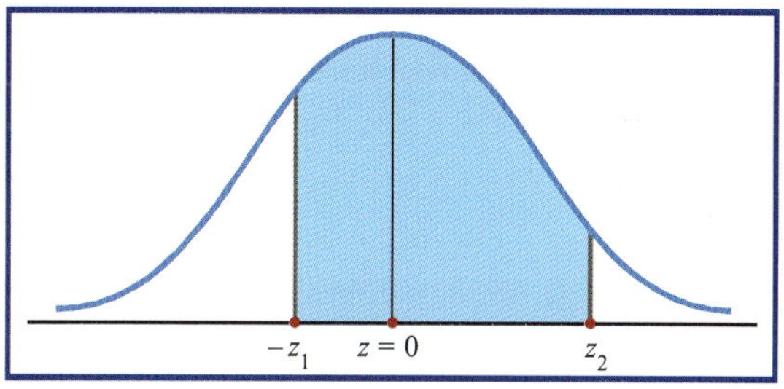

Figure 5.38

Note: If $z_1 = z_2$, then

$$P(-z_1 < Z < z_2) = P(-z_1 < Z < z_1)$$
$$= P(-z_1 < Z < 0) + P(0 < Z < z_1)$$
$$= P(0 < Z < z_1) + P(0 < Z < z_1)$$
$$= 2P(0 < Z < z_1)$$

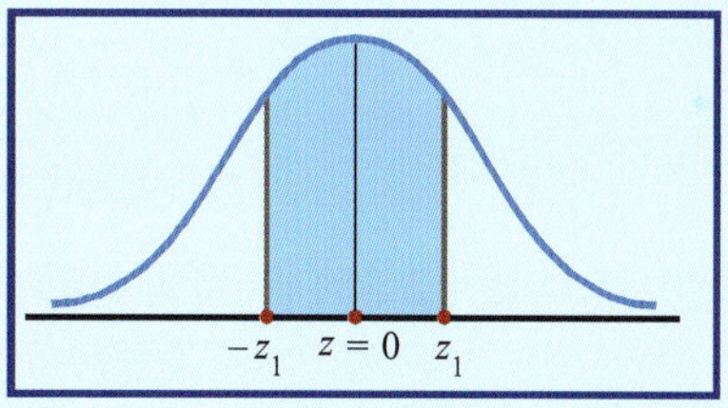

Figure 5.39

Hence, in all of the above possible cases, we have ultimately converted the desired areas to the form: $P(0 < Z < z_1)$; $z_1 > 0$ and these values can be obtained from Table 3, for any given value of z_1.

Example 6: Find the shaded areas under the following standard normal curves.

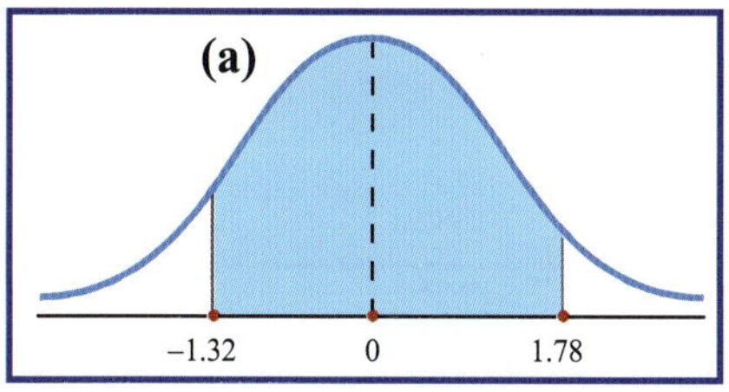

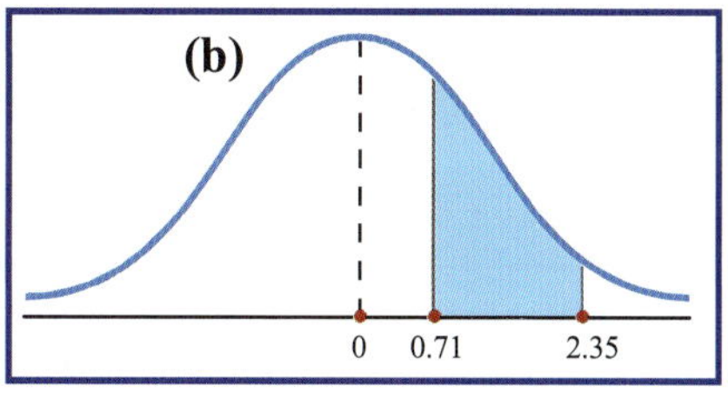

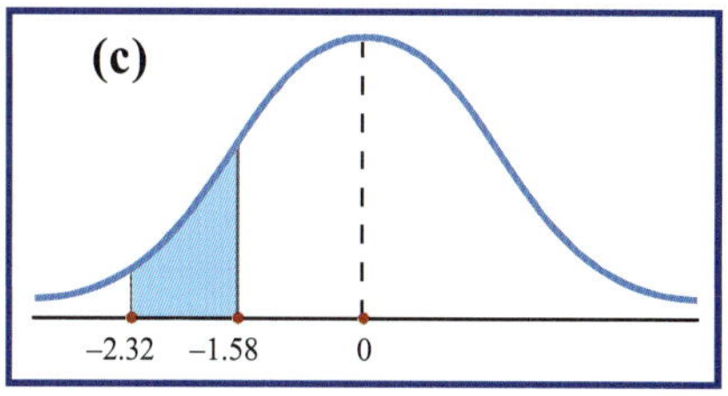

Solution: Recall that Table 3 gives areas from 0 to z. Therefore we express each area in terms of areas from 0 to z.

(a) The area between –1.32 and 1.78 is the **sum** of the areas from 0 to –1.32 (same as from 0 to 1.32) and 0 to 1.78.

Therefore, the shaded area is 0.4066 + 0.4625 = 0.8691 (From Table 3)

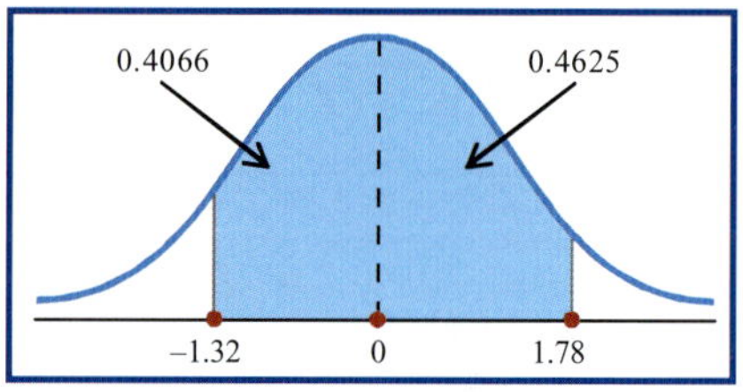

Figure 5.40

(b) The area between 0.71 and 2.35

= (The area from 0 to 2.35) – (The area from 0 to 0.71)

= 0.4906 – 0.2611 = 0.2295

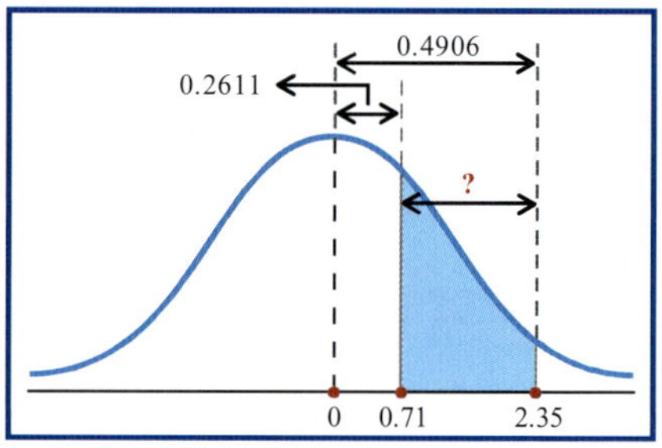

Figure 5.41

(c) The area between –2.32 and –1.58

= (The area between 0 and –2.32) – (The area between 0 and –1.58)

= (The area between 0 and 2.32) – (The area between 0 and 1.58)

= 0.4898 – 0.4429 = 0.0469

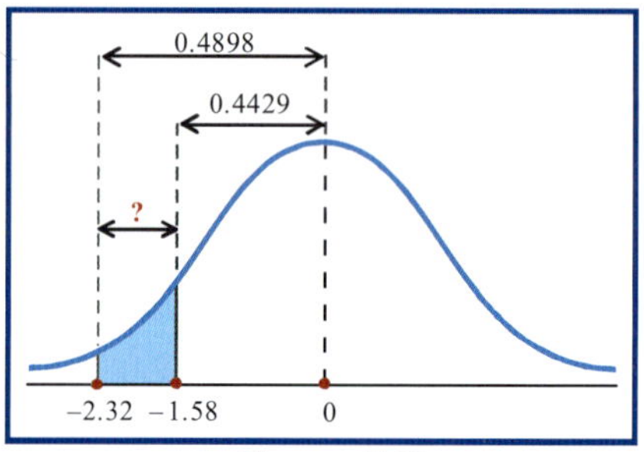

Figure 5.42

Remark: The above areas can be obtained more elegantly by expressing the areas in terms of probabilities.

i) $P(-1.32 < Z < 1.78) = P(-1.32 < Z < 0) + P(0 < Z < 1.78)$
$= P(0 < Z < 1.32) + P(0 < Z < 1.78)$
$= 0.4066 + 0.4625 = 0.8691$

ii) $P(0.71 < Z < 2.35) = P(0 < Z < 2.35) - P(0 < Z < 0.71)$
$= 0.4906 - 0.2611 = 0.2295$

iii) $P(-2.32 < Z < -1.58) = P(1.58 < Z < 2.32)$ (By Symmetry)
$= P(0 < Z < 2.32) - P(0 < Z < 1.58)$
$= 0.4898 - 0.4429 = 0.469$

Remark: To compute the areas under the standard normal probability curve, it is more convenient to express the desired areas in terms of probabilities, and use the symmetric property of the normal distribution. In the examples that follows, we will demonstrate this technique.

Example 7: Find the shaded areas under the following standard normal curves.

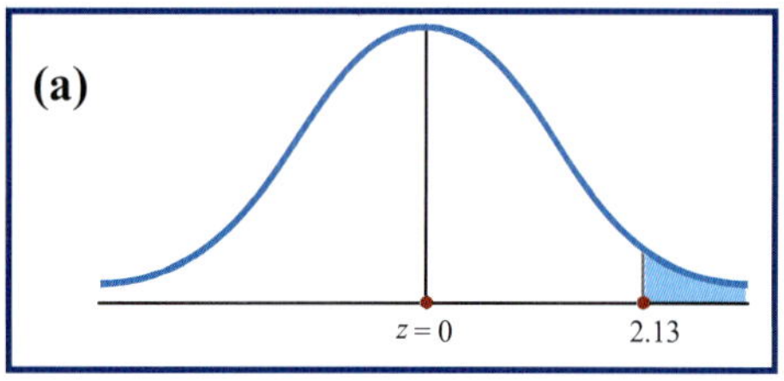

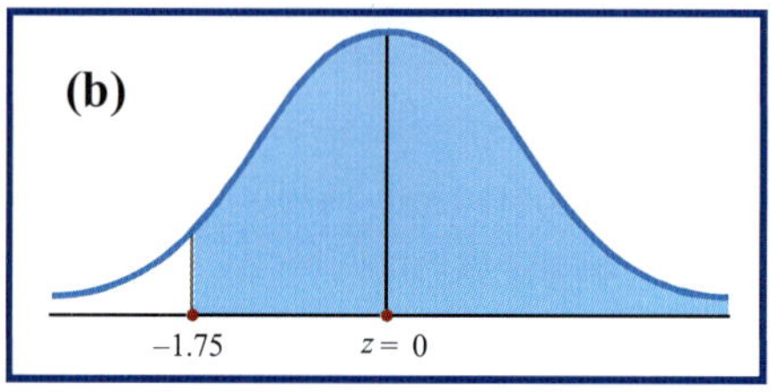

Solution: **(a)** The shaded area = $P(Z > 2.13)$

$= 0.5 - P(0 < Z < 2.13)$

$= 0.5 - 0.4834 = 0.0166$

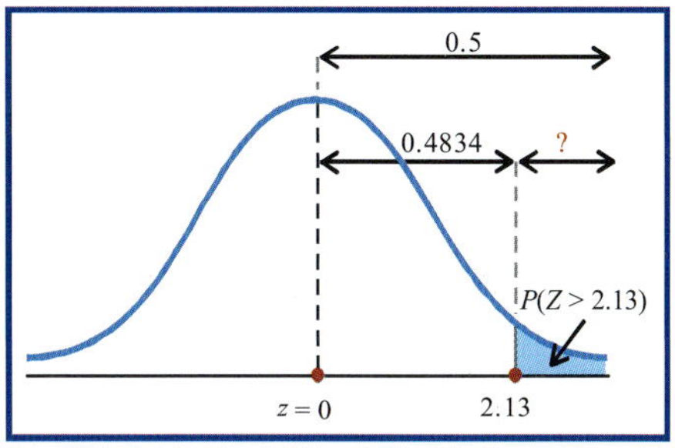

Figure 5.43

(b) The shaded area = $P(Z > -1.75)$

$= P(-1.75 < Z < 0) + 0.5$

$= P(0 < Z < 1.75) + 0.5$ (By Symmetry)

$= 0.4599 + 0.5 = 0.9599$

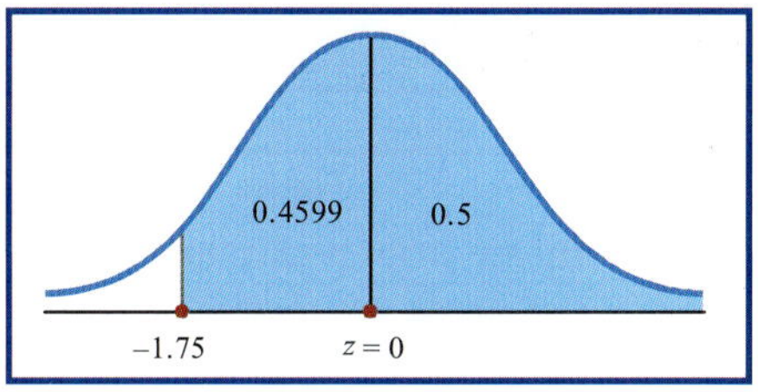

Figure 5.44

Example 8: Find the shaded areas under the following standard normal curves.

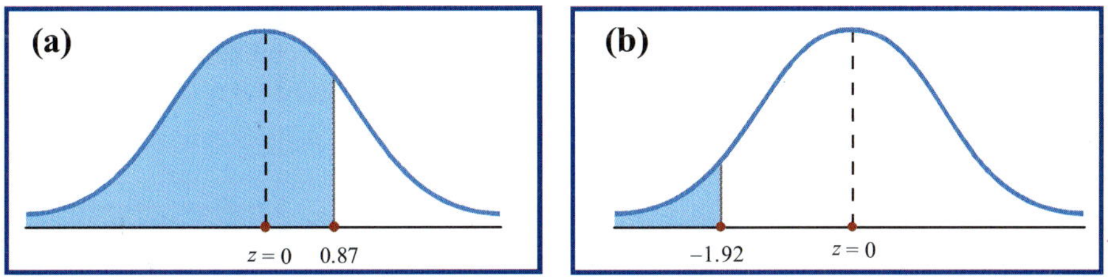

Solution: **(a)** The area to the left of 0.87 $= 0.5 + P(0 < Z < 0.87)$

$= 0.5 + 0.3078$

$= 0.8078$

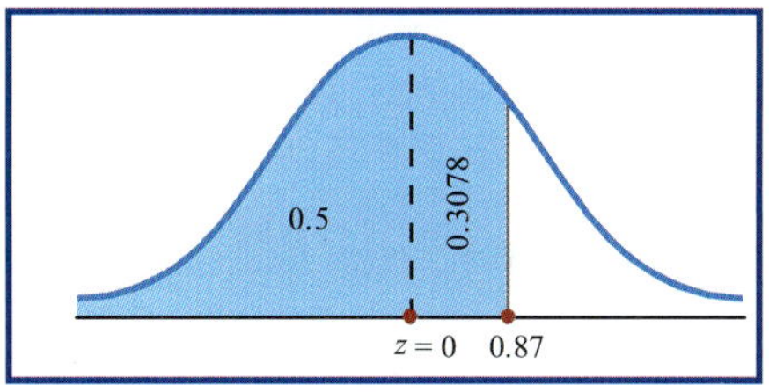

Figure 5.45

(b) The area to the left of -1.92

$= P(Z < -1.92)$

$= 0.5 - P(-1.92 < Z < 0)$

$= 0.5 - P(0 < Z < 1.92)$ (By Symmetry)

$= 0.5 - 0.4726 = 0.074$

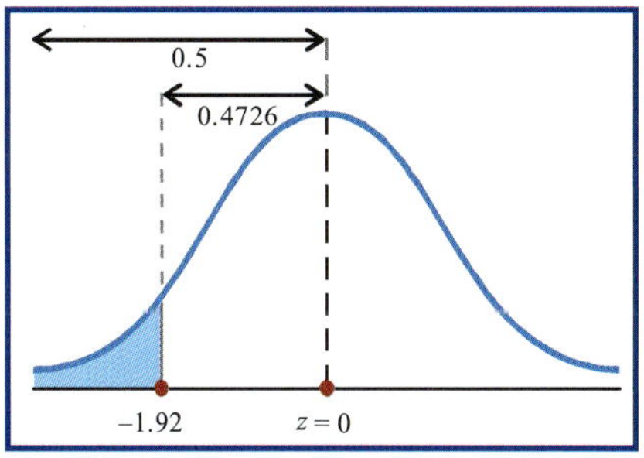

Figure 5.46

Example 9: For the standard normal random variable Z, compute the following probabilities.

(a) $P(Z < 1.38)$ **(b)** $P(Z < -2.34)$

(c) $P(1.18 < Z < 2.06)$

Solution: **(a)** $P(Z < 1.38) = 0.5 + P(0 < Z < 1.38)$

$= 0.5 + 0.4162 = 0.9162$

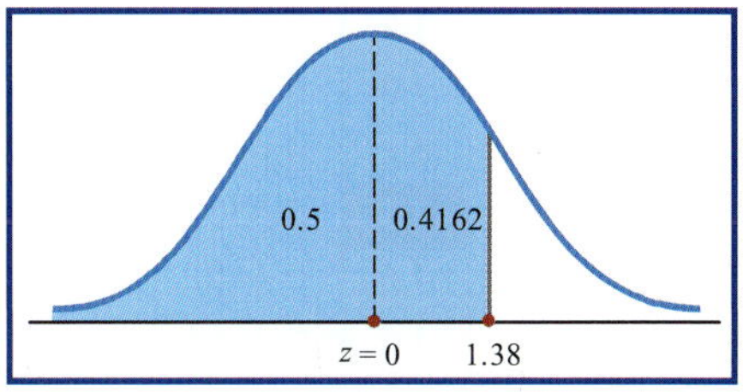

Figure 5.47

(b) $P(z < -2.34)$

$= 0.5 - P(-2.34 < Z < 0)$

$= 0.5 - P(0 < Z < 2.34)$ (By Symmetry)

$= 0.5 - 0.4904 = 0.0096$

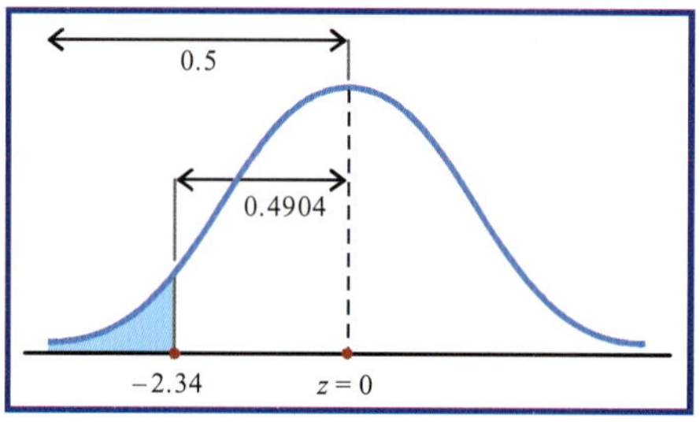

Figure 5.48

(c) $P(1.18 < Z < 2.06) = P(0 < Z < 2.06) - P(0 < Z < 1.18)$

$= 0.4803 - 0.3810 = 0.0993$

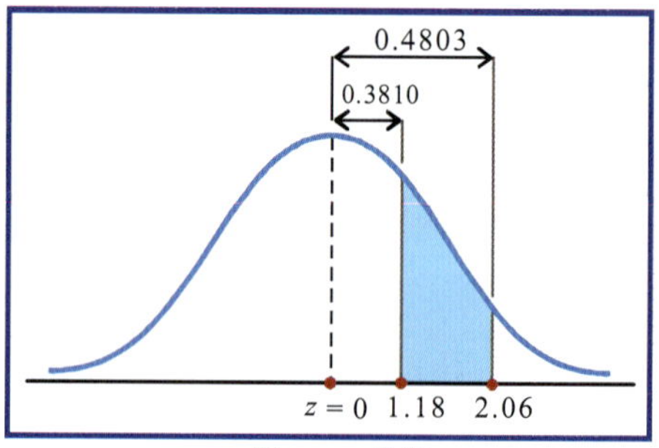

Figure 5.49

In the above examples, we have used the Standard Normal Distribution Table (z-table) to compute the areas (probabilities) under the standard normal curve. Now we show how to use Table 3 to find z-values for a given area (probability or percentage).

C. Computing z-Scores for a Given Area (Probability, Percentage, or Proportions)

Recall that the body (inside the first column and top row) of Table 3 contains probability or area values. These areas range from 0 to 0.4999. Now suppose the area from 0 to z is given to be 0.4406 as shown in Figure 5.50.

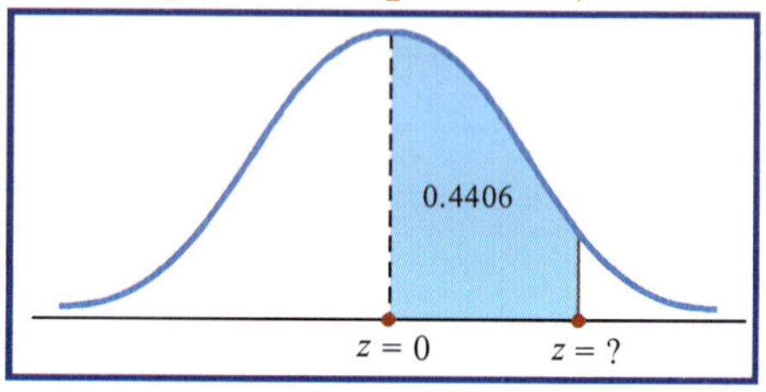

Figure 5.50

In order to find the z-value that correponds to the area 0.4406, we must first find 0.4406 in the body of the table, and then look for a z-value in the left hand margin of the row and the top margin of the column that corresponds to 0.4406. The z-value found in this manner is the z-score that corresponds to the area 0.4406. Hence, $z = 1.56$ as shown here.

z \ z	0.00	0.01	0.02	0.03	0.04	0.05	0.06	0.07	0.08	0.09
0.0	0.0000	0.0040	0.0080	0.0120	0.0160	0.0199	0.0239	0.0279	0.0319	0.0359
0.1	0.0398	0.0438	0.0478	0.0517	0.0557	0.0596	0.0636	0.0675	0.0714	0.0753
0.2	0.0793	0.0832	0.0871	0.0910	0.0948	0.0987	0.1026	0.1064	0.1103	0.1141
0.3	0.1179	0.1217	0.1255	0.1293	0.1331	0.1368	0.1406	0.1443	0.1480	0.1517
0.4	0.1554	0.1591	0.1628	0.1664	0.1700	0.1736	0.1772	0.1808	0.1844	0.1879
0.5	0.1915	0.1950	0.1985	0.2019	0.2054	0.2088	0.2123	0.2157	0.2190	0.2224
0.6	0.2257	0.2291	0.2324	0.2357	0.2389	0.2422	0.2454	0.2486	0.2517	0.2549
0.7	0.2580	0.2611	0.2642	0.2673	0.2704	0.2734	0.2764	0.2794	0.2823	0.2852
0.8	0.2881	0.2910	0.2939	0.2967	0.2995	0.3023	0.3051	0.3078	0.3106	0.3133
0.9	0.3159	0.3186	0.3212	0.3238	0.3264	0.3289	0.3315	0.3340	0.3365	0.3389
1.0	0.3413	0.3438	0.3461	0.3485	0.3508	0.3531	0.3554	0.3577	0.3599	0.3621
1.1	0.3643	0.3665	0.3686	0.3708	0.3729	0.3749	0.3770	0.3790	0.3810	0.3830
1.2	0.3849	0.3869	0.3888	0.3907	0.3925	0.3944	0.3962	0.3980	0.3997	0.4015
1.3	0.4032	0.4049	0.4066	0.4082	0.4099	0.4115	0.4131	0.4147	0.4162	0.4177
1.4	0.4192	0.4207	0.4222	0.4236	0.4251	0.4265	0.4279	0.4292	0.4306	0.4319
1.5	0.4332	0.4345	0.4357	0.4370	0.4382	0.4394	**0.4406**	0.4418	0.4429	0.4441
1.6	0.4452	0.4463	0.4474	0.4484	0.4495	0.4505	0.4515	0.4525	0.4535	0.4545
1.7	0.4554	0.4564	0.4573	0.4582	0.4591	0.4599	0.4608	0.4616	0.4625	0.4633

Data from Table 3

However, sometimes the given area is not exactly equal to one of the values (probabilities) found in the body of the table. For example, suppose the area between 0 and z is given to be 0.4684. We cannot locate this number inside the z-table. In such cases, we locate the number (probability) closest to the given value. In this case, 0.4686 is closest to 0.4684, and its corresponding z value is 1.86 as shown in the table given below.

What if the given area between 0 to z is equally close to two neighboring numbers in the body of the table? For example, if the area between 0 and z is given to be 0.4500, then the numbers 0.4495 and 0.4505 are equally close to 0.4500. In such cases, the z value that corresponds to the area (probability) 0.4500 is the ***average*** of the z values that correspond to 0.4505 and 0.4495.

$$z = \frac{1.64 + 1.65}{2} = 1.645$$

z \ z	0.00	0.01	0.02	0.03	0.04	0.05	0.06	0.07	0.08	0.09
0.0	0.0000	0.0040	0.0080	0.0120	0.0160	0.0199	0.0239	0.0279	0.0319	0.0359
0.1	0.0398	0.0438	0.0478	0.0517	0.0557	0.0596	0.0636	0.0675	0.0714	0.0753
0.2	0.0793	0.0832	0.0871	0.0910	0.0948	0.0987	0.1026	0.1064	0.1103	0.1141
0.3	0.1179	0.1217	0.1255	0.1293	0.1331	0.1368	0.1406	0.1443	0.1480	0.1517
0.4	0.1554	0.1591	0.1628	0.1664	0.1700	0.1736	0.1772	0.1808	0.1844	0.1879
0.5	0.1915	0.1950	0.1985	0.2019	0.2054	0.2088	0.2123	0.2157	0.2190	0.2224
0.6	0.2257	0.2291	0.2324	0.2357	0.2389	0.2422	0.2454	0.2486	0.2517	0.2549
0.7	0.2580	0.2611	0.2642	0.2673	0.2704	0.2734	0.2764	0.2794	0.2823	0.2852
0.8	0.2881	0.2910	0.2939	0.2967	0.2995	0.3023	0.3051	0.3078	0.3106	0.3133
0.9	0.3159	0.3186	0.3212	0.3238	0.3264	0.3289	0.3315	0.3340	0.3365	0.3389
1.0	0.3413	0.3438	0.3461	0.3485	0.3508	0.3531	0.3554	0.3577	0.3599	0.3621
1.1	0.3643	0.3665	0.3686	0.3708	0.3729	0.3749	0.3770	0.3790	0.3810	0.3830
1.2	0.3849	0.3869	0.3888	0.3907	0.3925	0.3944	0.3962	0.3980	0.3997	0.4015
1.3	0.4032	0.4049	0.4066	0.4082	0.4099	0.4115	0.4131	0.4147	0.4162	0.4177
1.4	0.4192	0.4207	0.4222	0.4236	0.4251	0.4265	0.4279	0.4292	0.4306	0.4319
1.5	0.4332	0.4345	0.4357	0.4370	0.4382	0.4394	0.4406	0.4418	0.4429	0.4441
1.6	0.4452	0.4463	0.4474	0.4484	**0.4495**	**0.4505**	0.4515	0.4525	0.4535	0.4545
1.7	0.4554	0.4564	0.4573	0.4582	0.4591	0.4599	0.4608	0.4616	0.4625	0.4633
1.8	0.4641	0.4649	0.4656	0.4664	0.4671	0.4678	**0.4686**	0.4693	0.4699	0.4706
1.9	0.4713	0.4719	0.4726	0.4732	0.4738	0.4744	0.4750	0.4756	0.4761	0.4767

Data from Table 3

As seen above, it is quite simple to find the z-value when the area (probability) between 0 and z is given. However, for areas other than 0 to z, the following steps must be used to find Z-values.

***Step* 1** Draw the normal curve and shade the required area.

***Step* 2** By simple manipulations, [see Section 5.3 (B)], from the given area compute the probability of the region from 0 to z.

***Step* 3** Locate in Table 3, the value (area, probability) found in ***Step* 2**, that is either exactly equal, or closest, to it.

***Step* 4** Read the z-value corresponding to the number located in ***Step* 3** from the left most column and top row.

Example 10: Find z, for the probability values given in the diagrams below:

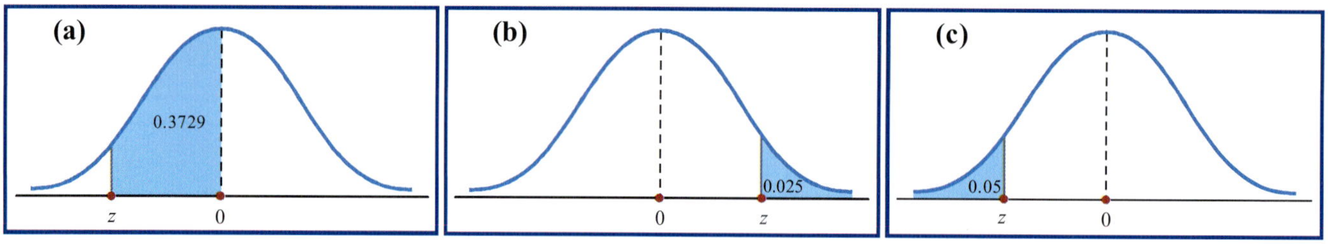

Solution: **(a)** ***Step 1*** The picture of the region and its area are already shown in Figure (a).

Step 2 The area between 0 and z, is 0.3729.

Step 3 The body of Table 3 has a number exactly equal to 0.3729.

z	0.00	0.01	0.02	0.03	0.04	0.05	0.06	0.07	0.08	0.09
0.0	0.0000	0.0040	0.0080	0.0120	0.0160	0.0199	0.0239	0.0279	0.0319	0.0359
0.1	0.0398	0.0438	0.0478	0.0517	0.0557	0.0596	0.0636	0.0675	0.0714	0.0753
0.2	0.0793	0.0832	0.0871	0.0910	0.0948	0.0987	0.1026	0.1064	0.1103	0.1141
0.3	0.1179	0.1217	0.1255	0.1293	0.1331	0.1368	0.1406	0.1443	0.1480	0.1517
0.4	0.1554	0.1591	0.1628	0.1664	0.1700	0.1736	0.1772	0.1808	0.1844	0.1879
0.5	0.1915	0.1950	0.1985	0.2019	0.2054	0.2088	0.2123	0.2157	0.2190	0.2224
0.6	0.2257	0.2291	0.2324	0.2357	0.2389	0.2422	0.2454	0.2486	0.2517	0.2549
0.7	0.2580	0.2611	0.2642	0.2673	0.2704	0.2734	0.2764	0.2794	0.2823	0.2852
0.8	0.2881	0.2910	0.2939	0.2967	0.2995	0.3023	0.3051	0.3078	0.3106	0.3133
0.9	0.3159	0.3186	0.3212	0.3238	0.3264	0.3289	0.3315	0.3340	0.3365	0.3389
1.0	0.3413	0.3438	0.3461	0.3485	0.3508	0.3531	0.3554	0.3577	0.3599	0.3621
1.1	0.3643	0.3665	0.3686	0.3708	**0.3729**	0.3749	0.3770	0.3790	0.3810	0.3830
1.2	0.3849	0.3869	0.3888	0.3907	0.3925	0.3944	0.3962	0.3980	0.3997	0.4015
1.3	0.4032	0.4049	0.4066	0.4082	0.4099	0.4115	0.4131	0.4147	0.4162	0.4177

Data from Table 3

Step 4 The z-value that corresponds to the area 0.3729 is 1.14. Since z in the Figure (a) is on the left side of 0, it must be negative. Thus, $z = -1.14$.

(b) ***Step 1*** The picture of the region and its area are already shown in Figure (b).

Step 2 The area between 0 and z equals $0.5 - 0.025 = 0.475$.

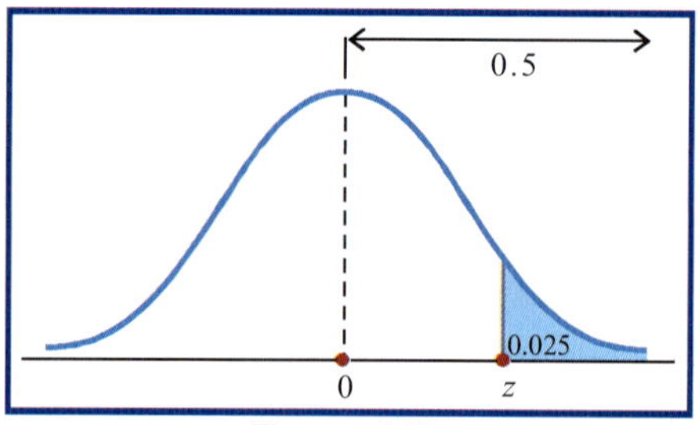

Figure 5.51

Step 3 The body of Table 3 has a number exactly equal to 0.4750.

Step 4 The z-value that corresponds to 0.4750 is 1.96. Thus $z = 1.96$.

z	0.00	0.01	0.06
1.9			**0.4750**

Data from Table 3

(c) ***Step* 1** The picture of the region and its area are already shown in Figure (c).

***Step* 2** The area between 0 and z equals $0.5 - 0.05 = 0.45$.

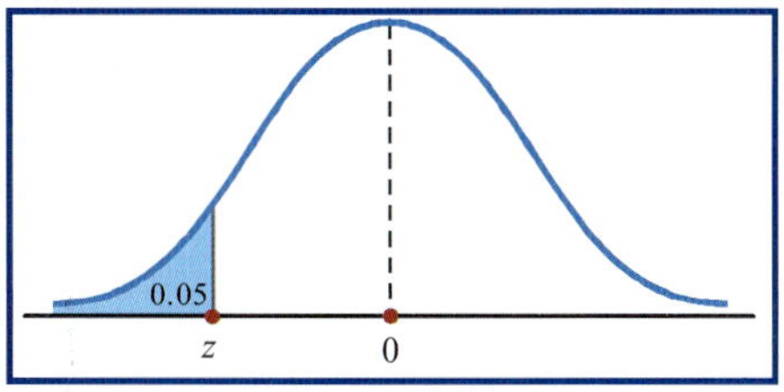

Figure 5.52

***Step* 3** The number 0.4500 is not in the body of Table 3. However, the numbers closest to 0.4500 are 0.4495 and 0.4505.

***Step* 4** The z-value that corresponds to 0.4500 is the average of the z-values corresponding to 0.4495 and 0.4505. Therefore, $z = \dfrac{1.64 + 1.65}{2} = 1.645$. Since z in Figure (c) is on the left side of 0, so it is negative $z = -1.645$.

Example 11: Find z, for the following areas under the standard normal curve.

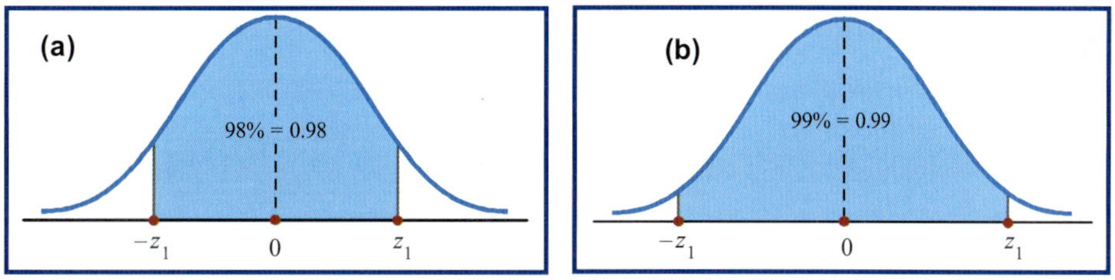

Solution: **(a)** ***Step* 1** The picture of the region is already shown in Figure (a).

***Step* 2** By symmetry of the normal curve, the area between 0 and $z_1 = \dfrac{0.98}{2} = 0.4900$.

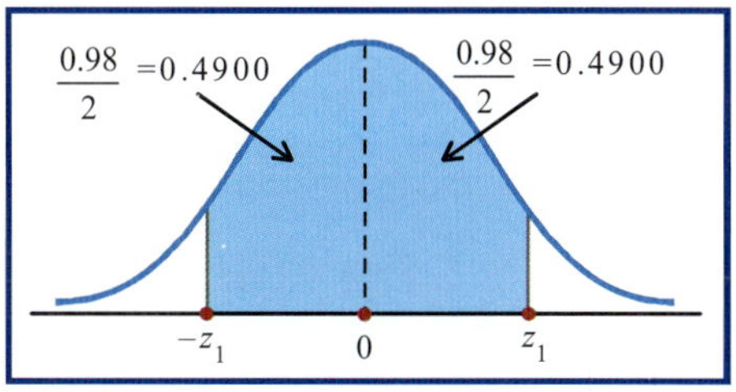

Figure 5.53

***Step* 3** The number 0.4900 is not in the body of Table 3. The number in the table closest to 0.4900 is 0.4901.

***Step* 4** The z-value that corresponds to 0.4901 is 2.33. Thus, -2.33 and 2.33 are the $-z_1$ and z_1 shown in the Figure 5.53. (Alternative notation: $z = \pm 2.33$)

(b) ***Step* 1** The Figure (b) is the required diagram.

***Step* 2** By symmetry of the normal curve, the area between 0 and $z_1 = \dfrac{0.99}{2} = 0.4950$.

***Step* 3** The number 0.4950 is not in the body of Table 3. The numbers in the table closest to 0.4950 are 0.4949 and 0.4951.

***Step* 4** The z-value that corresponds to 0.4950 is the average of z-values corresponding to the areas 0.4949 ($z = 2.57$) and 0.4951 ($z = 2.58$). Therefore, $z_1 = \dfrac{2.57 + 2.58}{2} = 2.575$.

Hence the required z-values are -2.575 and 2.575.

Example 12: Compute z for the percentiles given:

(a) P_{95} (95th percentile) **(b)** P_{40} (40th percentile)

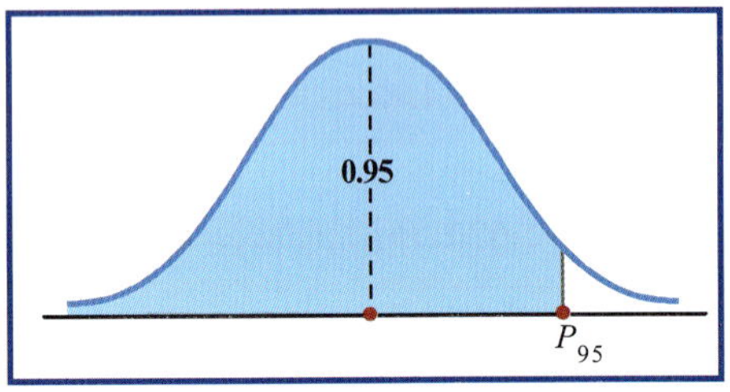

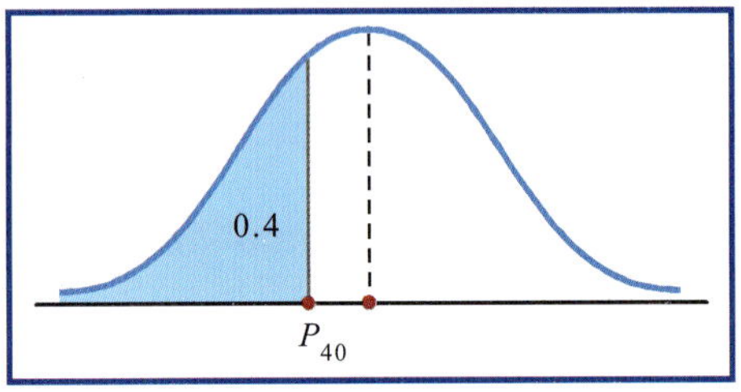

Solution: **(a)** Let the z-value for P_{95} be z_1 so that $P(Z < z_1) = 0.95$

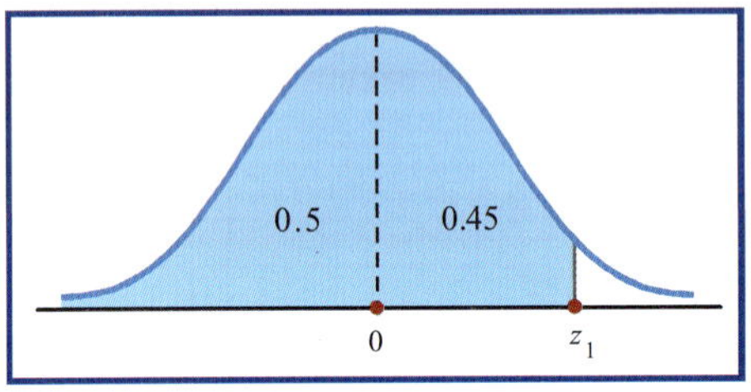

Figure 5.54

***Step* 1** z_1 has to be on the right of $z = 0$, otherwise the area to the left will always be less than 50% = 0.5 (Figure 5.54)

***Step* 2** $P(0 < Z < z_1) = 0.95 - 0.5 = 0.45$

***Step* 3** The number 0.4500 is in the midpoint of the numbers 0.4495 and 0.4505 in Table 3. The Z-values for these numbers are 1.64 and 1.65.

***Step* 4** Thus, to find the z-value for the area 0.45, we must calculate the midvalue of 1.69 and 1.65.

$$z = \frac{1.64 + 1.65}{2} = 1.645$$

Therefore, the z-value for P_{95} is 1.645.

(b) Let the z-value for P_{40} be z_2 so that $P(Z < z_2) = 0.40$

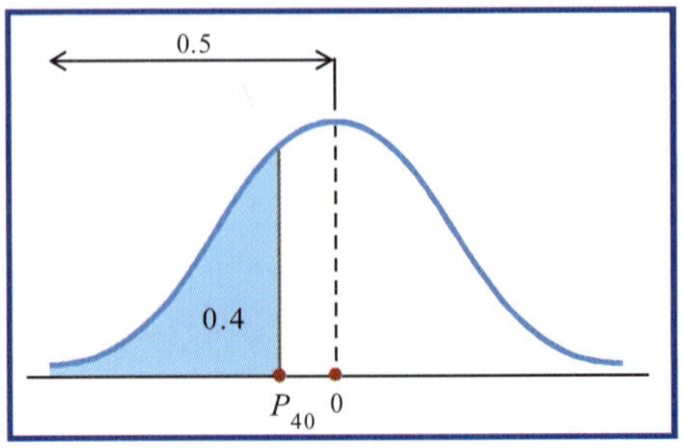

Figure 5.55

Step 1 From Figure 5.55, it is clear that z_2 is located to the left of $z = 0$ and must be negative.

Step 2 $P(z_2 < Z < 0) = 0.5 - 0.4 = 0.10$.

Step 3 The value in Table 3 that is closest to 0.10 is 0.0987.

Step 4 The z-value that corresponds to 0.0987 in Table 3 is 0.25. Thus, the z-value for P_{40} is -0.25 or $P_{40} = -0.25$.

Try it yourself:

In exercises 1-6, use Table 3 to compute the probability:

1. Between $z = 0$ and $z = 1.58$

2. Between $z = -2.73$ and $z = 1.58$

3. Between $z = 1.12$ and $z = 2.48$

4. Between $z = -2.33$ and $z = -1.02$

5. To the right of $z = -1.95$

6. To the left of $z = -1.66$

In exercises 7-12, use Table 3 to compute the probability (area) for the standard normal random variable Z.

7. $P(Z > 1.96)$

8. $P(Z < 1.14)$

9. $P(Z > -1.07)$

10. $P(Z < -2.16)$

11. $P(1.12 < Z < 2.64)$

12. $P(-1.92 < Z < -0.68)$

In exercises 13-19, find z-scores for the given probabilities represented by the following areas.

13. The area between 0 and z is 0.4495.

14. The area to the right of z is 0.1271.

15.

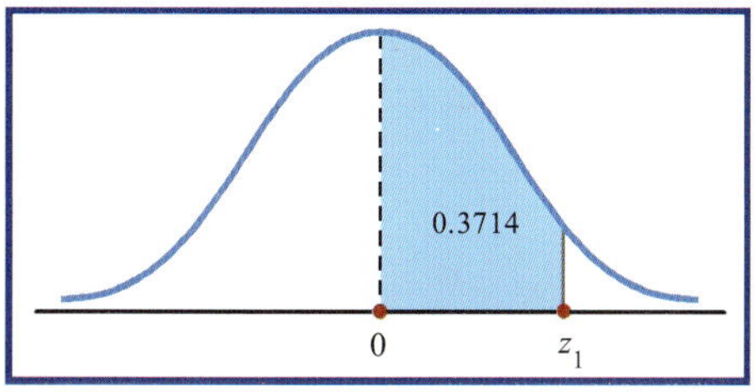

16.

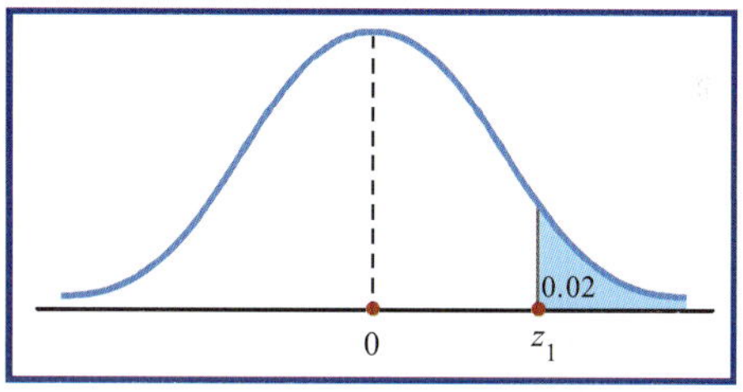

17.

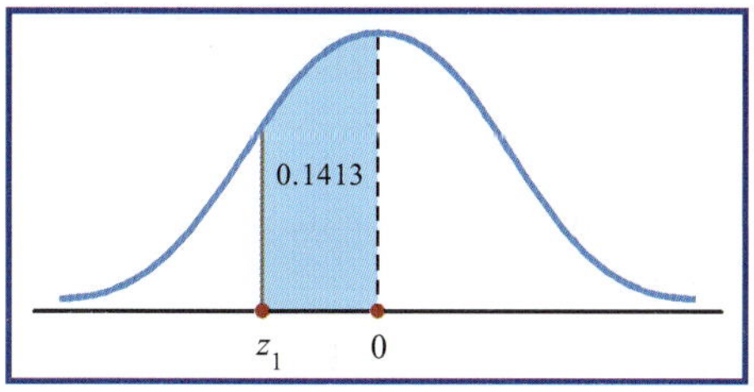

18.

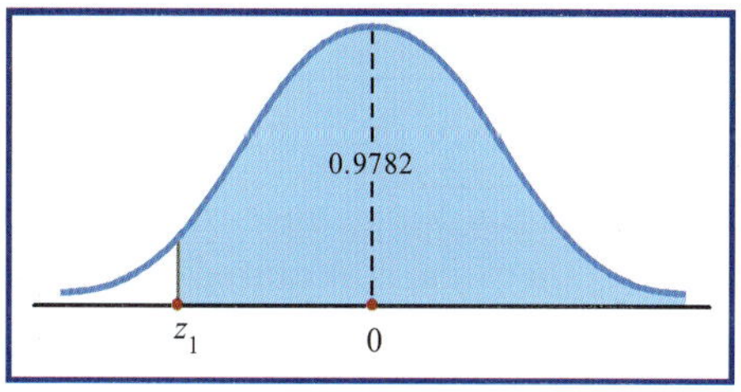

19. The middle 95% area.

In exercises 20-25, compute the z-scores for the given percentiles and quartiles.

20. P_{90} **21.** Q_3 **22.** P_{98} **23.** P_{35} **24.** P_3 **25.** Q_1

5.6: THE NON-STANDARD NORMAL DISTRIBUTION $N(\mu,\sigma)$

Objectives

After completing this section you will be able to:

A. Compute probabilities (areas) under a Normal Probability Curve with mean μ and standard deviation σ.

B. Compute x for a given Area (Probability, Percentage).

C. Use Application of the Normal Distribution.

A. Finding Probabilities (Areas) under a Normal Probability Curve with Mean μ and Standard Deviation σ.

The table of areas under normal probability curve is available in terms of standard normal variable Z (Table 3). Hence for applied problems with non-standard normal variable X, we must first transform the x-values to z-values using the transformation $z = \dfrac{x-\mu}{\sigma}$.

Suppose we want to compute the probabilities $P(a < X < b)$, $P(X > c)$ or $P(X < d)$. The areas (probabilities) in terms of X are equivalent to the areas corresponding to values of Z. Thus

i) $$P(a < X < b) = P\left(\frac{a-\mu}{\sigma} < \frac{X-\mu}{\sigma} < \frac{b-\mu}{\sigma}\right)$$

$$= P(z_1 < Z < z_2), \text{ where } z_1 = \frac{a-\mu}{\sigma} \text{ and } z_2 = \frac{b-\mu}{\sigma}$$

ii) $$P(X > c) = P\left(\frac{X-\mu}{\sigma} > \frac{c-\mu}{\sigma}\right)$$

$$= P\left(Z > \frac{c-\mu}{\sigma}\right) = P(Z > z_3), \text{ where } z_3 = \frac{c-\mu}{\sigma}$$

iii) $$P(X < d) = P\left(\frac{X-\mu}{\sigma} < \frac{d-\mu}{\sigma}\right)$$

$$= P\left(Z < \frac{d-\mu}{\sigma}\right) = P(Z < z_4), \text{ where } z_4 = \frac{d-\mu}{\sigma}$$

Remark: z_1, z_2, z_3 and z_4 may be positive or negative depending on the values of μ, σ, a, b, c, and d. The areas in terms of $\boldsymbol{Z}$ can now be obtained as discussed in Section (5.3B).

***Step* 1** Identify the values of mean (μ) and standard deviation (σ).

***Step* 2** Draw the picture of the normal curve, locate the value(s) of $\boldsymbol{X}$ and shade the region for which the area is required.

***Step* 3** Convert the x-values (non-standard normal distribution) to z-values (standard normal distribution) by the transformation $Z = \dfrac{X-\mu}{\sigma}$, and shade the region identical to that in Step 2, in terms of z-values.

***Step* 4** Compute the desired area (probability) under standard normal curve as outlined in Section 5.3. The area (probability) obtained in this step is the required area under $N(\mu, \sigma)$ distribution.

Process Summary:

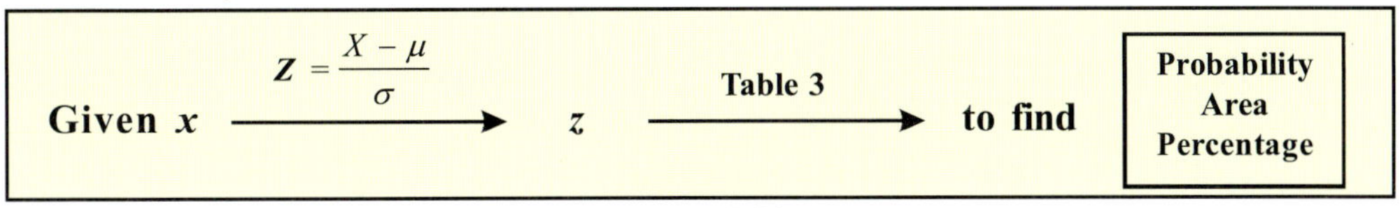

Figure 5.56

Example 13: A non-standard normal random variable (X) has mean $\mu = 50$ and standard deviation $\sigma = 5$. Find the probability of x-values that lie between 42 and 59.

Solution: ***Step 1*** We are given $X \sim N(\mu = 50, \sigma = 5)$ and we want to find $P(42 < X < 59)$.

Step 2 Picture is given in Figure 5.57.

Step 3

x	$z = \frac{x - \mu}{\sigma}$
42	$\frac{42 - 50}{5} = -1.6$
59	$\frac{59 - 50}{5} = 1.8$

Table 5.3

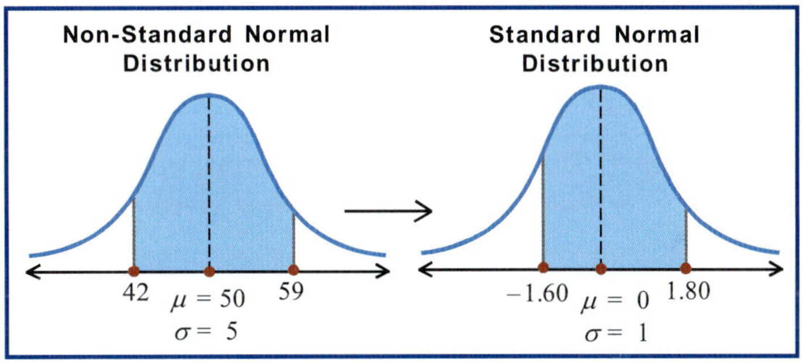

Figure 5.57

Step 4

$$\begin{aligned} P(42 < X < 59) &= P(-1.6 < Z < 1.8) \\ &= P(-1.6 < Z < 0) + P(0 < Z < 1.8) \\ &= P(0 < Z < 1.6) + P(0 < Z < 1.8) \quad \text{(By Symmetry)} \\ &= 0.4452 + 0.4641 = 0.9093 \quad \text{(From Table 3)} \end{aligned}$$

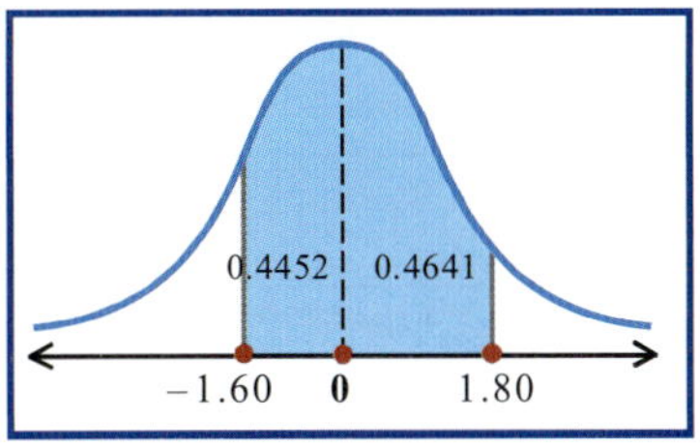

Figure 5.58

Thus, the percentage of x-values that fall between 42 and 59 is 90.93%.

Example 14: For a non-standard normal distribution with $\mu = 300$ and $\sigma = 60$, find the probability that X assumes values less than 211.

Solution: We are given $X \sim N(\mu = 300, \sigma = 60)$ and we want $P(X < 211)$.

When $X = 211$, $Z = \frac{X - \mu}{\sigma} = \frac{211 - 300}{60} = -1.48$

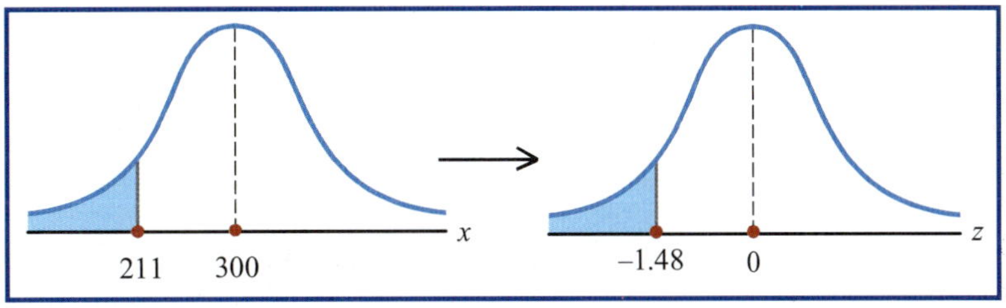

Figure 5.59

$$P(X < 211) = P(Z < -1.48)$$
$$= 0.5 - P(-1.48 < Z < 0)$$
$$= 0.5 - P(0 < Z < +1.48) \quad \text{(By Symmetry)}$$
$$= 0.5 - 0.4306 = 0.0694$$

Thus the probability that X-values fall below 211 is approximately 0.0694.

B. Finding x for a Given Area (Probability, Percentage)

In the previous section, we learned how to find the z-value corresponding to the area under the standard normal curve. In the following section, we will demonstrate how to find x-values (real application based value) when the area (percentage, probability) under the normal curve is given.

How to find x?

Step **1** Draw a picture of the normal curve, indicating μ, σ, x and the region whose area (probability, percentage) is given.

Step **2** Convert the picture in Step 1 to the standard normal curve, indicating $Z = \dfrac{X - \mu}{\sigma}$ for **X**.

Step **3** Use the method of Section 5.3 to find Z.

Step **4** Find x by solving the equation $z = \dfrac{x - \mu}{\sigma}$ for $x \Rightarrow x = \mu + z\sigma$.

Process Summary:

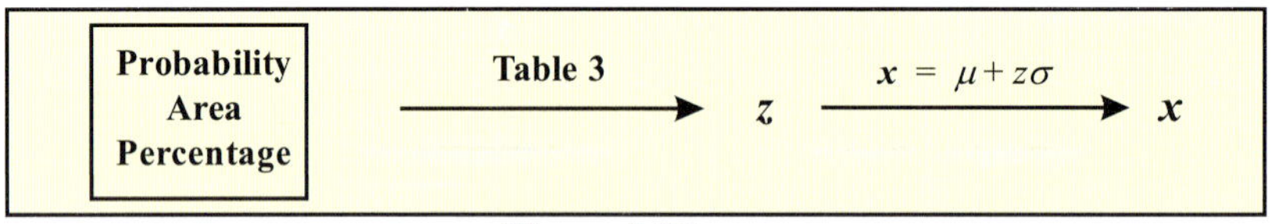

Figure 5.60

Example 15: Given a non-standard normal random variable (**X**) with the mean 100 and standard deviation 15, find **X** $= a$ when the area between 100 and a is 0.4484.

Solution: $X = a$ can be on the left or right side of the mean $\mu = 100$.

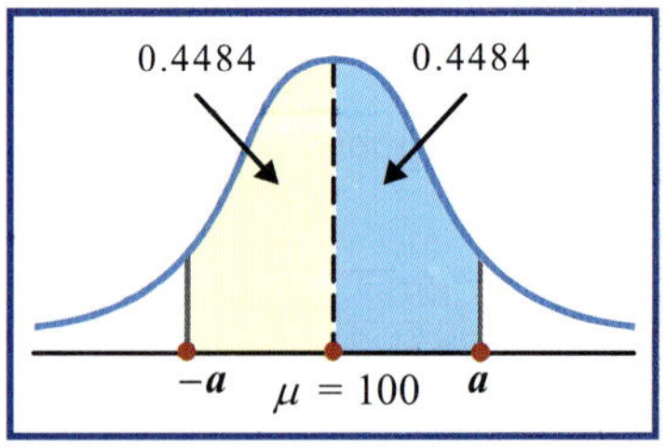

Figure 5.61

Case **(i):** $X = a > \mu$

Since $X = a$ lies to the right of μ, the corresponding z-value will also be positive (as shown in Figure 5.62a)

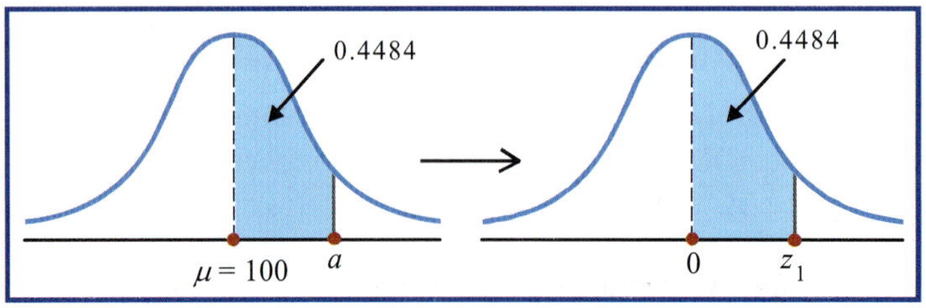

Figure 5.62 a **Figure 5.62 b**

From the problem and Figure 5.62 a, b, it is clear that $P(100 < X < a) = 0.4484$.

Transforming $\boldsymbol{X}$ to $\boldsymbol{Z}$, we have

$$P(100 < X < a) = P\left(\frac{100-100}{15} < Z < \frac{a-100}{15}\right)$$

$$= P(0 < Z < z_1), \text{ where } \frac{a-100}{15} = z_1 \qquad (5.18)$$

$$= 0.4484 \Rightarrow z_1 = 1.63 \quad \text{(From Table 3)}$$

Substituting $z_1 = 1.63$ in (5.18), we have $\frac{a-100}{15} = 1.63 \Rightarrow a = 124.45$

***Case* (ii):** $X = a < \mu$

Since $X = a$ lies to the left of μ, the corresponding z-value will be negative (as shown in Figure 5.63 a)

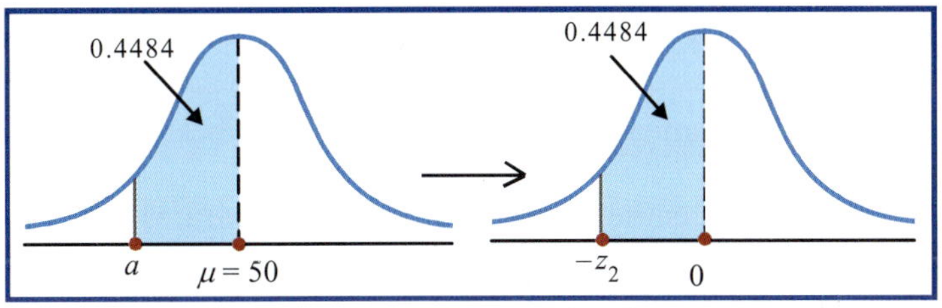

Figure 5.63 a **Figure 5.63 b**

From the problem and Figure 5.63 a, it is clear that $P(a < X < 100) = 0.4484$.

Transforming $\boldsymbol{X}$ to $\boldsymbol{Z}$, we have

$$P(a < X < 100) = P\left(\frac{a-100}{15} < Z < \frac{100-100}{15}\right)$$

$$= P(-z_2 < Z < 0), \text{ where } \frac{a-100}{15} = -z_2 \; (z_2 > 0) \qquad (5.19)$$

$$= P(0 < Z < z_2) \quad \text{(By symmetry)}$$

$$= 0.4484 \Rightarrow z_2 = 1.63 \quad \text{(From Table 3)}$$

Substituting $z_2 = 1.63$ in (5.19), we have $\frac{a-100}{15} = -1.63 \Rightarrow a = 75.55$

Thus, there are two x-values (75.55 and 124.45) such that the area between 0 and x is 0.4484.

Example 16: Given $N(100, 15)$, find the value of $\boldsymbol{X}$ that has 85% of the area bounded by the normal curve to its left.

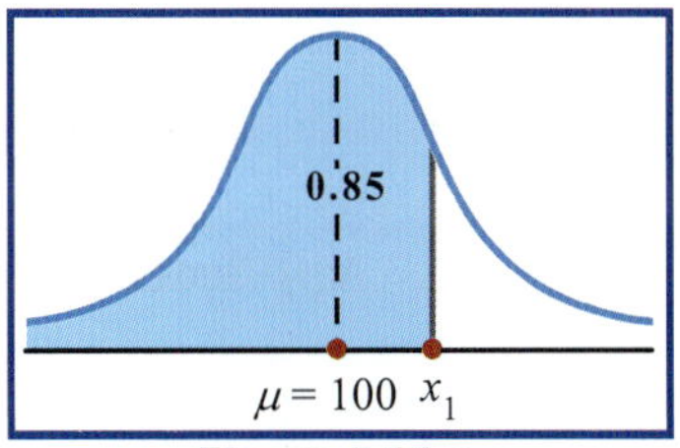

Figure 5.64 a

Solution: ***Step* 1** Let the value of $\boldsymbol{X}$ be x_1 so that $P(X < x_1) = 0.85$.

Since the area to the left of x_1 is more than 0.5, x_1 is located to the right of mean $\mu = 100$ as shown in Figure 5.64 (a). Hence, the corresponding value of z will also be positive.

***Step* 2** Converting Figure 5.64 (a) to the standard normal curve $N(0, 1)$, we get Figure 5.64 (b).

***Step* 3** When $X = x_1$, $\Rightarrow$ $Z = \frac{x_1 - \mu}{\sigma} = \frac{x_1 - 100}{15} = z_1$

From Figures 5.64 (a) and (b),

$N(100, 15)$	$N(0, 1)$
$P(X < x_1) = 0.85$	$P(Z < z_1) = 0.85$

Table 5.4

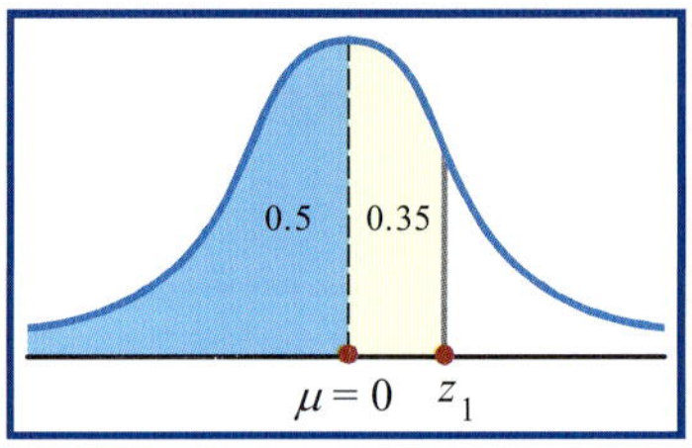

Figure 5.64 b

***Step* 4** $P(Z < z_1) = P(Z < 0) + P(0 < Z < z_1) = 0.85.$

Since $P(Z < 0) = 0.5$, then $P(0 < Z < z_1) = 0.85 - 0.5 = 0.35.$

$P(0 < Z < z_1) = 0.35 \Rightarrow z_1 = 1.04$ (From Table 3)

Since $\dfrac{x_1 - 100}{15} = z_1 \Rightarrow x_1 = (1.04)(15) + 100 = 115.60.$

This means that, given a nonstandard normal distribution with $\mu = 100$ and $\sigma = 15$, 85% of values lies to the left of 115.60.

Example 17: For a non-standard normal random variable $\boldsymbol{X}$, with $\mu = 400$ and $\sigma = 80$, find P_{70}.

Solution: P_{70}(70th percentile) is the X-value that separates the bottom 70% of the distribution (area under the bell shape curve) from the top 30%.

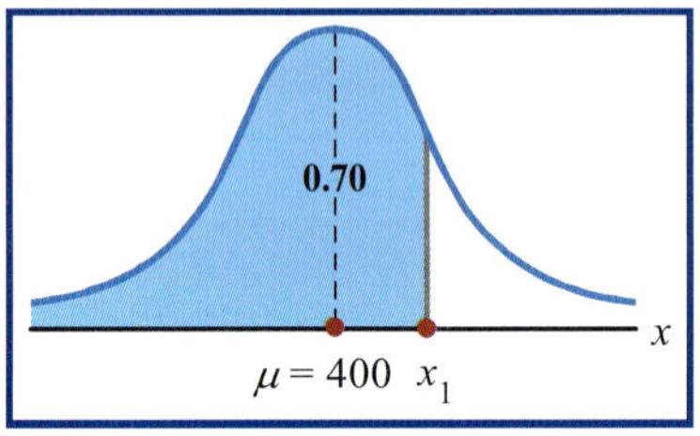

Figure 5.65 a

Let $P_{70} = z_1$.

***Step* 1** From Figure 5.65 a, we get $P(X < x_1) = 0.70$. It is clear that x_1 is located to the right of the mean $\mu = 400$. Hence the corresponding Z value will be positive. (Figure 5.65 b)

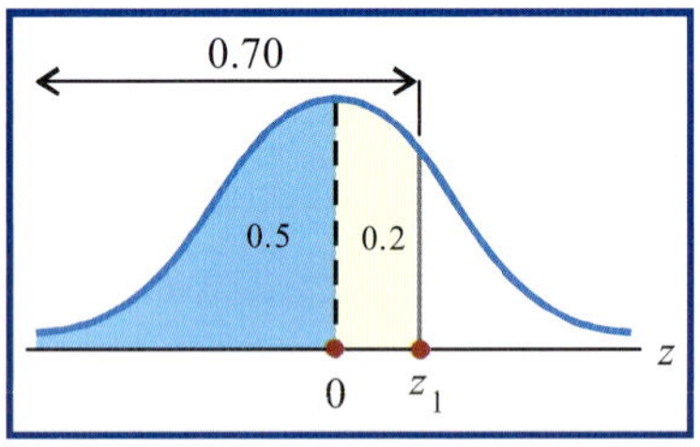

Figure 5.65 b

***Step* 2** Converting the non-standard normal distribution (Figure 5.65 a) to a standard normal distribution (Figure 5.65 b),

we get $Z = \dfrac{X - \mu}{\sigma} = \dfrac{x_1 - 400}{80} = z_1.$

Given $P(Z < z_1) = 0.70 \Rightarrow P(0 < Z < z_1) = 0.20$ (Figure 5.65 b)

***Step* 3** The number (area) in table 3, closest to 0.20 is 0.1985. The z value that corresponds to the area 0.1985 is 0.52.

***Step* 4** Since $z_1 = 0.52$ from Step 3, we can find x_1 using the formula $x_1 = z_1\sigma + \mu$.

$x_1 = (0.52)(80) + 400 = 441.6$. Thus, $P_{70} = x_1 = 441.6$. This means that 70% of data in the distribution in this example lies below 441.95.

C. Application of the Normal Distribution

The normal distribution is used to solve problems that involve normal random variables as well as to approximate solutions from other random variables, such as the Binomial and Poisson distributions. Calculation of probabilities is much easier if the distribution is normal compared to the Binomial or Poisson Distributions. The normal distribution also serves as a fundamental assumption in more advanced theories in Statistics such as the theory of Estimation and Decision theory. Here are some examples of the application of the normal distribution.

Example 18: A manufacturer of halogen light bulbs claims that the lives of the light bulbs are normally distributed with an average life of 900 hours and standard deviation of 50 hours. Find the probability that a randomly selected light bulb will last more than 820 hours.

Solution: Let the random variable (**X**) be the life (number of hours) of a light bulb. We are given that $X \sim N(900, 50)$. The required probability that a randomly selected bulb will last for more than 820 hours is given by $P(X > 820)$.

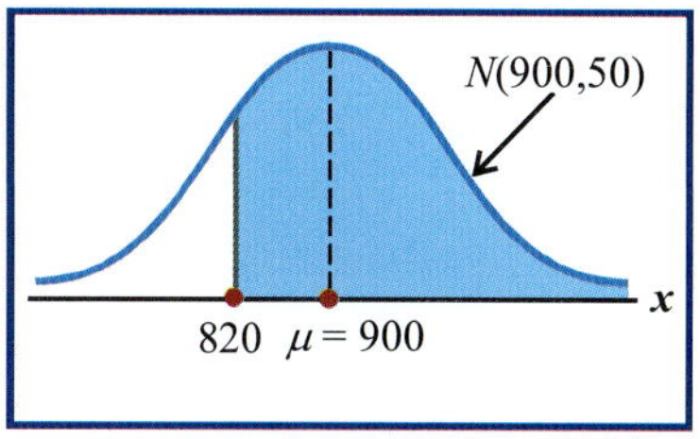

Figure 5.66 a

We start with transforming the nonstandard normal random variable (**X**) to a standard normal random variable (**Z**).

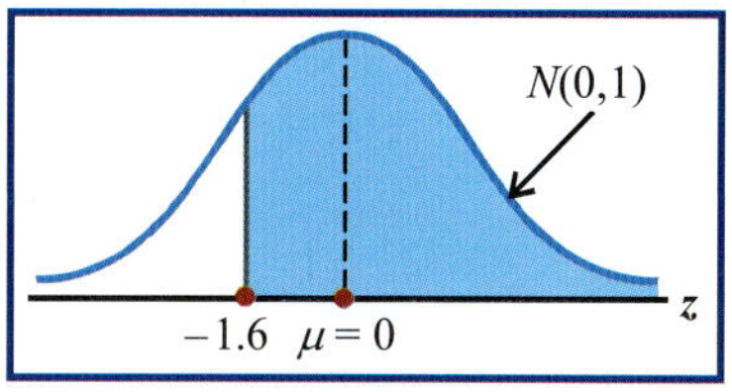

Figure 5.66 b

$$
\begin{aligned}
P(X > 820) &= P\left(\frac{X - 900}{50} > \frac{820 - 900}{50}\right) \\
&= P(Z > -1.6) \\
&= P(-1.6 < Z < 0) + P(Z > 0) \\
&= P(0 < Z < 1.6) + 0.5 \qquad \text{(By Symmetry)} \\
&= 0.4452 + 0.5 \\
&= 0.9452
\end{aligned}
$$

Thus, the probability that a light bulb will survive for more than 820 hours is 0.9452.

Example 19: The mean amount of soft drink in plastic bottles is a nonstandard normal random variable (X) with $\mu = 16$ ounces and standard deviation $\sigma = 0.6$ ounces. What percentage of bottles contain less than 15 ounces?

Solution: Let **X** denote the amount (in ounces) of the soft drink in a plastic bottle. We are given that $X \sim N(\mu, \sigma)$ where $\mu = 16$ ounces and $\sigma = 0.6$ ounces. The percentage of bottles that contain less than 15 ounces is given by $P(X < 15)$.

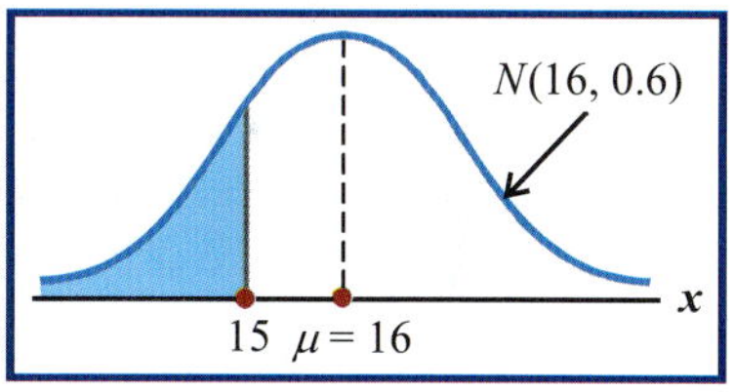

Figure 5.67 a

We start with transforming the nonstandard random variable (**X**) to a standard random variable (**Z**).

When $X = 15$, $Z = \dfrac{X - \mu}{\sigma} = \dfrac{15 - 16}{0.6} = -1.67.$

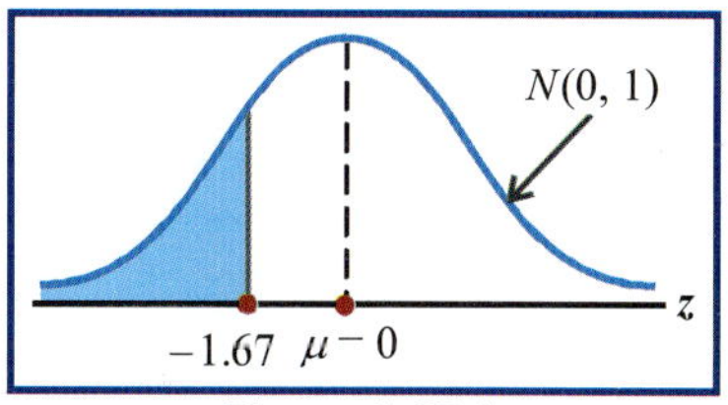

Figure 5.67 b

Therefore

$$
\begin{aligned}
P(X < 15) &= (Z < -1.67) \\
&= 0.5 - P(-1.67 < Z < 0) \qquad \text{[Figure 5.67 b]} \\
&= 0.5 - P(0 < Z < 1.67) \qquad \text{By Symmetry} \\
&= 0.5 - 0.4525 = 0.0475 \qquad \text{From Table 3}
\end{aligned}
$$

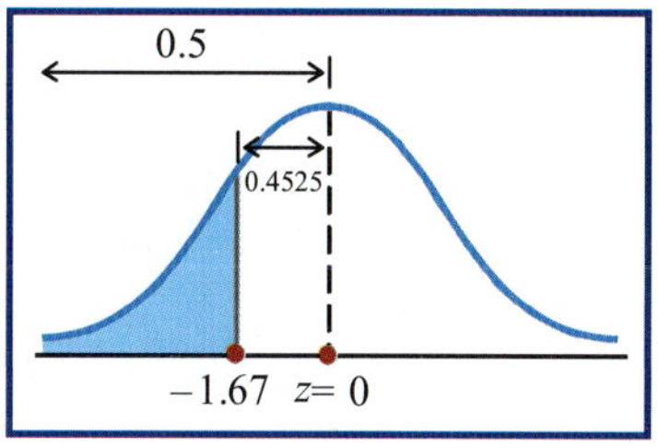

Figure 5.67 c

Thus, the percentage of bottles containing less than 15 ounces of soft drink is 0.0475 = 4.75%.

Example 20: Given an average IQ to be 100 with a standard deviation of 15, find the probability that if a person is randomly selected, his or her IQ will be greater than 134.

Solution: Let X denote the IQ of a person. We are given that $X \sim N(\mu, \sigma)$ where $\mu = 100$ and $\sigma = 15$. The probability (percentage) that a randomly selected person will have an IQ greater than 134 is given by $P(X > 134)$.

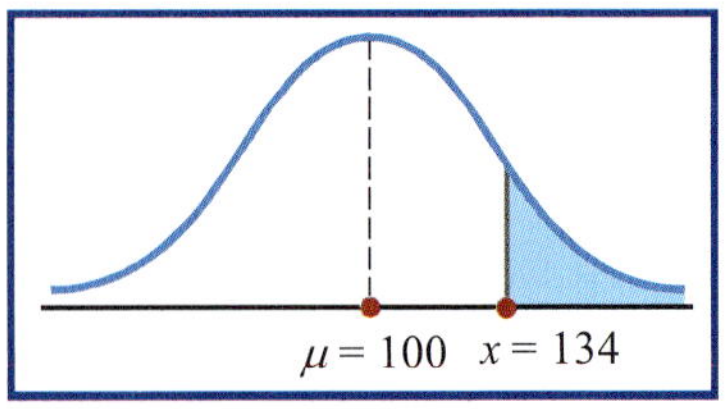

Figure 5.68 a

$$\text{When } X = 134, \quad Z = \frac{X - \mu}{\sigma}$$

$$= \frac{134 - 100}{15}$$

$$= 2.2666... \approx 2.27$$

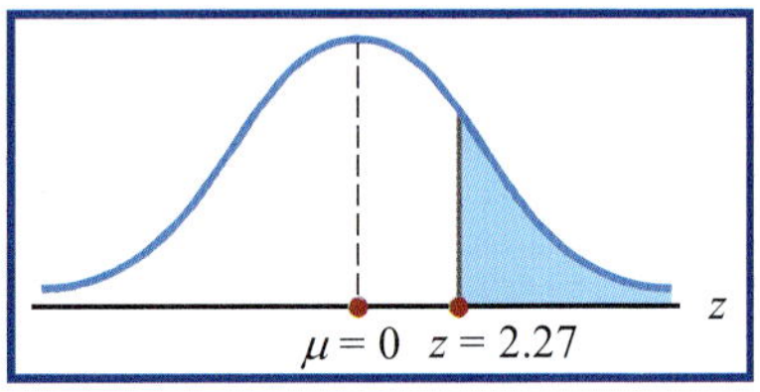

Figure 5.68 b

$$P(X > 134) = (Z > 2.27)$$

$$= 0.5 - P(0 < Z < 2.27)$$

$$= 0.5 - 0.4884 = 0.0116$$

Thus, the probability that a randomly selected person's IQ will be greater than 134 is 0.0116 .

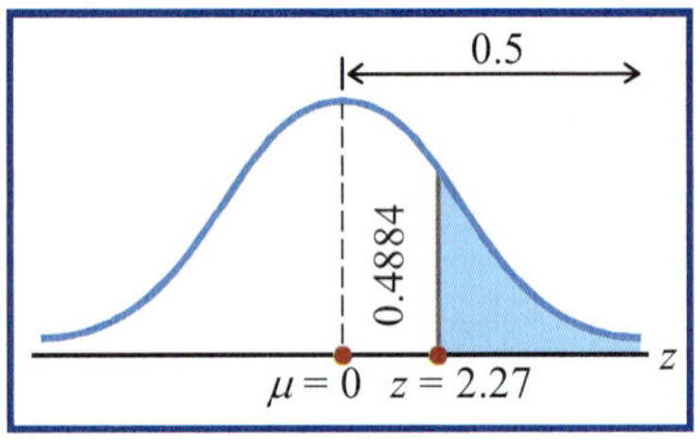

Figure 5.68 c

Example 21: IQ examination scores of 400 members of a club are normally distributed with a mean of 160 and standard deviation 13. How many of the members of the club have IQ scores between 150 and 180.

Solution: Let X denote the IQ scores of the members of the club. We are given $X \sim N(160, 13)$. The probability (p) that a member of the club has an IQ score between 150 and 180 is given by:

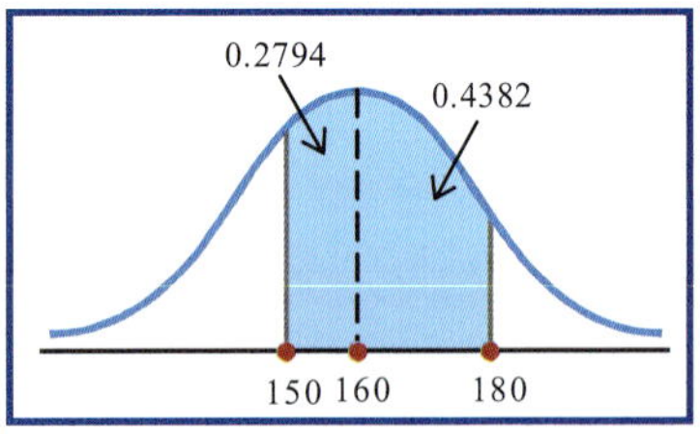

Figure 5.69 a

$p = P(150 < X < 180)$ (See Figure 5.77)

$= P\left(\frac{150 - 160}{13} < Z < \frac{180 - 160}{13}\right)$ (Transformation from nonstandard normal RV)

$= P(-0.77 < Z < 1.54)$

$= P(-0.77 < Z < 0) + P(0 < Z < 1.54)$

$= P(0 < Z < 0.77) + P(0 < Z < 1.54)$ (By Symmetry)

$= 0.2794 + 0.4382 = 0.7176$ (From Table 3)

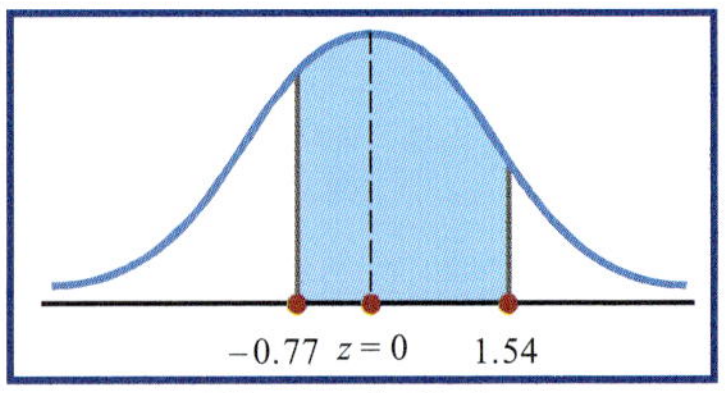

Figure 5.69 b

Thus, the probability that a club member IQ is between 150 and 180 is 0.7176.

The number of members whose IQ scores lie between 150 and 180 are:

$$400 \times P(150 < X < 180)$$

$$= 400(0.7176) \approx 287$$

Therefore, about 287 members of this 400-member club have IQ scores between 150 and 180.

Example 22: Suppose the verbal SAT scores are normally distributed with a mean of 520 and a standard deviation of 70. If Andy's score is at the 90th percentile, what is his score?

Solution: Let $\boldsymbol{X}$ denote Andy's score in the verbal portion of the SAT. Then $X \sim N(520, 70)$.

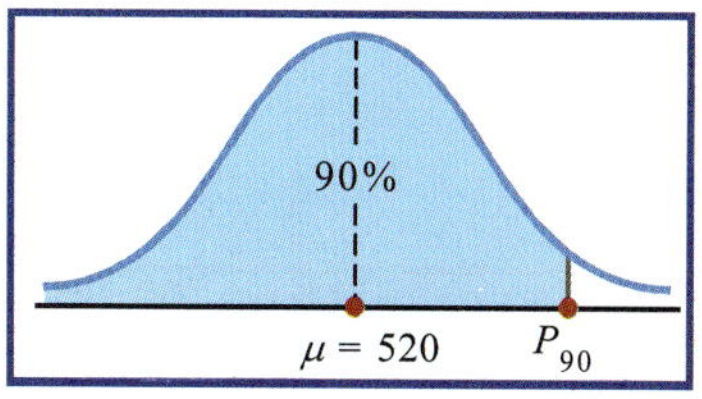

Figure 5.70 a

Recall that $P_{90} = X$ value that separates the bottom 90% of the distribution from the top 10%.

Let $P_{90} = x_1$. Therefore $P(X < x_1) = 0.90$.

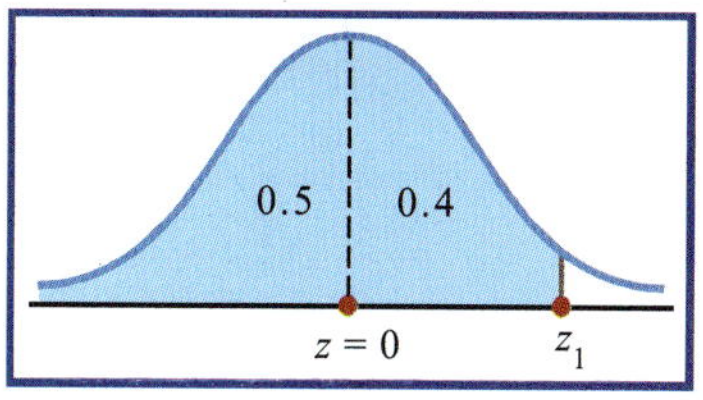

Figure 5.70 b

Since x_1 is located to the right of $\mu = 520$, the corresponding value of $\boldsymbol{Z}$ will be positive.

When $X = x_1$, $Z = \dfrac{x_1 - \mu}{\sigma} = \dfrac{x_1 - 520}{70} = z_1$

From Figure 5.70 b, we get;

$P(Z < z_1) = 0.90$ or $P(0 < Z < z_1) = 0.40$. In Table 3, the number closest to 0.4000 is 0.3997.

z \ z	0.00	0.01	0.02	0.03	0.04	0.05	0.06	0.07	0.08	0.09
0.0	0.0000	0.0040	0.0080	0.0120	0.0160	0.0199	0.0239	0.0279	0.0319	0.0359
0.1	0.0398	0.0438	0.0478	0.0517	0.0557	0.0596	0.0636	0.0675	0.0714	0.0753
0.2	0.0793	0.0832	0.0871	0.0910	0.0948	0.0987	0.1026	0.1064	0.1103	0.1141
0.3	0.1179	0.1217	0.1255	0.1293	0.1331	0.1368	0.1406	0.1443	0.1480	0.1517
0.4	0.1554	0.1591	0.1628	0.1664	0.1700	0.1736	0.1772	0.1808	0.1844	0.1879
0.5	0.1915	0.1950	0.1985	0.2019	0.2054	0.2088	0.2123	0.2157	0.2190	0.2224
0.6	0.2257	0.2291	0.2324	0.2357	0.2389	0.2422	0.2454	0.2486	0.2517	0.2549
0.7	0.2580	0.2611	0.2642	0.2673	0.2704	0.2734	0.2764	0.2794	0.2823	0.2852
0.8	0.2881	0.2910	0.2939	0.2967	0.2995	0.3023	0.3051	0.3078	0.3106	0.3133
0.9	0.3159	0.3186	0.3212	0.3238	0.3264	0.3289	0.3315	0.3340	0.3365	0.3389
1.0	0.3413	0.3438	0.3461	0.3485	0.3508	0.3531	0.3554	0.3577	0.3599	0.3621
1.1	0.3643	0.3665	0.3686	0.3708	0.3729	0.3749	0.3770	0.3790	0.3810	0.3830
1.2	0.3849	0.3869	0.3888	0.3907	0.3925	0.3944	0.3962	0.3980	**0.3997**	0.4015
1.3	0.4032	0.4049	0.4066	0.4082	0.4099	0.4115	0.4131	0.4147	0.4162	0.4177
1.4	0.4192	0.4207	0.4222	0.4236	0.4251	0.4265	0.4279	0.4292	0.4306	0.4319

Data from Table 3

The z-value that corresponds to this area is 1.28. Substituting in the formula for z and solving for x_1, we get: $\dfrac{x_1 - 520}{70} = 1.28 \quad \Rightarrow \quad x_1 = (1.28)(70) + 520 = 609.6$

Thus, Andy must have scored about 610 on the test.

Note: Another interpretation is that 90% of scores in the above experiment are less than 610 and 10% of scores are above 610.

Example 23: GRE (Graduate Record Exam) scores are normally distributed with a mean of 1000 and a standard deviation of 285 [$N(1000, 285)$]. What GRE score will put you in the top 1% of all graduate applications?

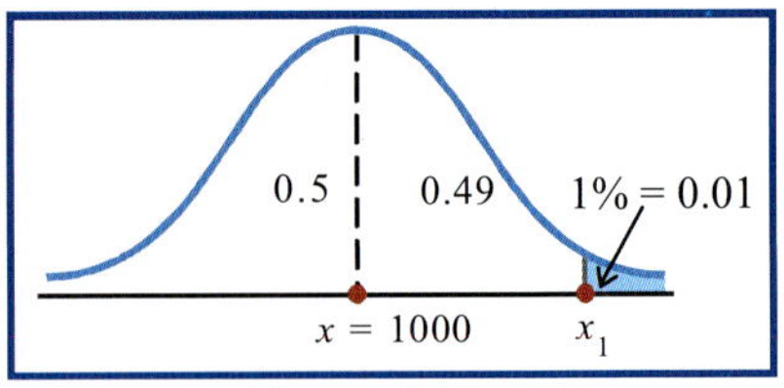

Figure 5.71 a

Solution: Let X denote the GRE score of a graduate applicant. Therefore $X \sim N(1000, 285)$.

The problem is to find the value of $X = x_1$, such that 1% of the area under the normal curve lies to its right. That is, $P(X > x_1) = 0.01$. Since x_1 lies to the right of the mean, the corresponding value of $\boldsymbol{Z}$ will be positive.

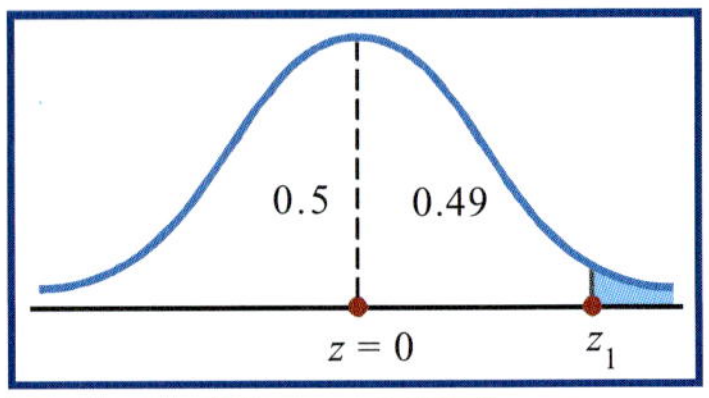

Figure 5.71 b

When $X = x_1$, $Z = \dfrac{X - \mu}{\sigma} = \dfrac{x_1 - 1000}{285} = z_1$.

From Figure 5.71, it is clear that

$$P(0 < Z < z_1) = 0.5 - 0.01 = 0.49.$$

The number in Table 3 that is closest to 0.49 is 0.4901 which corresponds to $z_1 = 2.33$.

Substituting $z_1 = 2.33$ into $\dfrac{x_1 - 1000}{285} = z_1$, we get $x_1 = 2.33 \times 285 + 1000 = 1664.05 \approx 1664$.

Hence, any GRE score above 1664 will put you in the top 1% of all graduate applicants.

Example 24: The scores on a short quiz in a Statistics course were 0, 1, 2, 3, …, 10, points, depending on the number of correctly answered questions out of 10. The mean score was 6.7 with a standard deviation of 1.2. Assuming the scores are normally distributed, determine

(a) the percentage of students scoring 6 points.

(b) the maximum score of the lowest 10% of the class.

Solution: Let the variable X denote the score on the quiz. We are given that $X \sim N(\mu, \sigma)$, where $\mu = 6.7$ and $\sigma = 1.2$.

(a) Since X is given to be normally distributed, that is, X is a continuous random variable, its probability at a particular point $X = 6$ is zero [$P(X = 6) = 0$].

Hence, theoretically the probability (percentage, area) of the students scoring exactly 6 points is zero. However, this does not make much sense from an application point of view. In practice, we may regard the score of 6 as ranging between 5.5 and 6.5 and compute the probability $P(5.5 < X < 6.5) = P(X = 6)$

Therefore

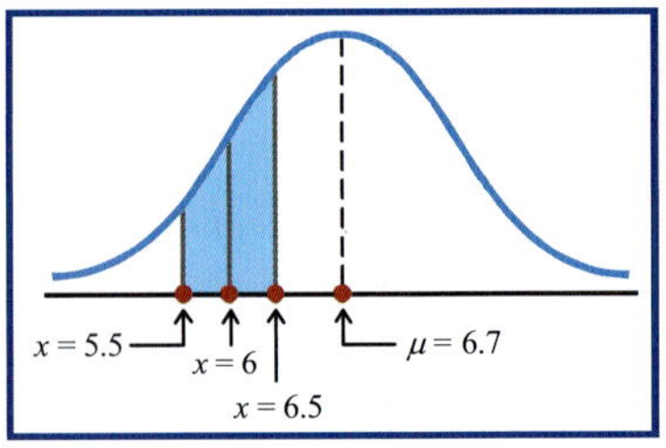

Figure 5.72

$$
\begin{aligned}
P(X = 6) &= P(5.5 < X < 6.5) \\
&= P\left(\frac{5.5 - 6.7}{1.2} < Z < \frac{6.5 - 6.7}{1.2}\right) \\
&= P(-1 < Z < -0.17) \\
&= P(-1 < Z < 0) - P(-0.17 < Z < 0) \\
&= P(0 < Z < 1) - P(0 < Z < 0.17) \qquad \text{(By Symmetry)} \\
&= 0.3413 - 0.0675 = 0.2738
\end{aligned}
$$

Hence, we may regard that the percentage of students scoring 6 points is approximately 27%.

(b) Let x_1 be the maximum score of the lowest 10% of the class as shown in Figure 5.73.

We want to find x_1 so that $P(X < x_1) = 0.10$ as shown in Figure 5.73.

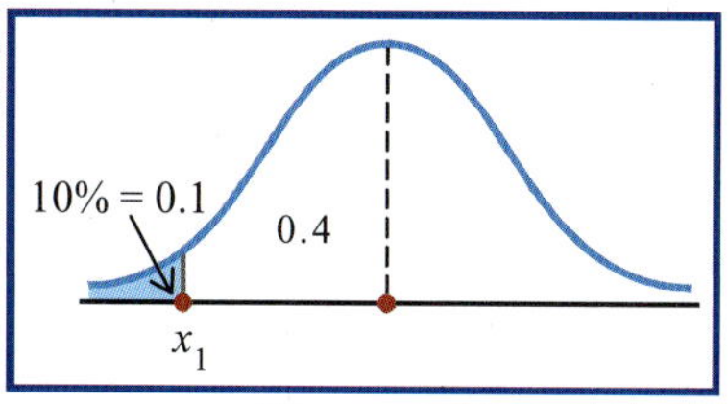

Figure 5.73

When $X = x_1$, $Z = \dfrac{x_1 - \mu}{\sigma} = \dfrac{x_1 - 6.7}{1.2} = -z_1, (z_1 > 0)$

$$P(X < x_1) = 0.10$$

$$\Rightarrow \quad P(Z < -z_1) = 0.10$$

$$\Rightarrow \quad P(-z_1 < Z < 0) = 0.5 - 0.1 = 0.4$$

$$\Rightarrow \quad P(0 < Z < z_1) = 0.4 \qquad \text{(By Symmetry)}$$

$$\Rightarrow \quad z_1 = 1.28 \qquad \text{(Using Table 3)}$$

Substituting $z_1 = 1.28$ into $\dfrac{x_1 - 6.7}{1.2} = -z_1$ we get $x_1 = 5.164$. Hence, the maximum score of the lowest 10% of the class is 5.

Try it yourself:

1. A non-standard normal random variable (X) has mean $\mu = 100$ and standard deviation $\sigma = 15$ *i.e.*, $X \sim N(100, 15)$. Find the probability of x-values that lie

   (a) between 95 and 128.
   (b) between 110 and 132.
   (c) below 112.
   (d) above 138.
   (e) above 85.
   (f) below 78.

2. Given $N(75, 10)$, find $X = a$, from the given figure:
   Discuss both possible cases: $X = a > \mu$ and $X = a < \mu$.

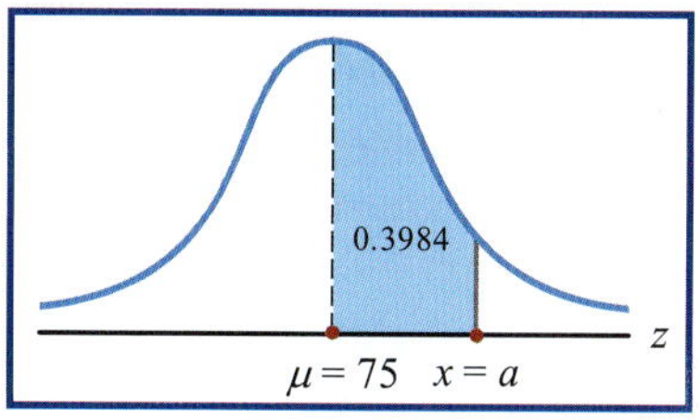

3. Given $N(75, 10)$, find the value of X in the given diagram.

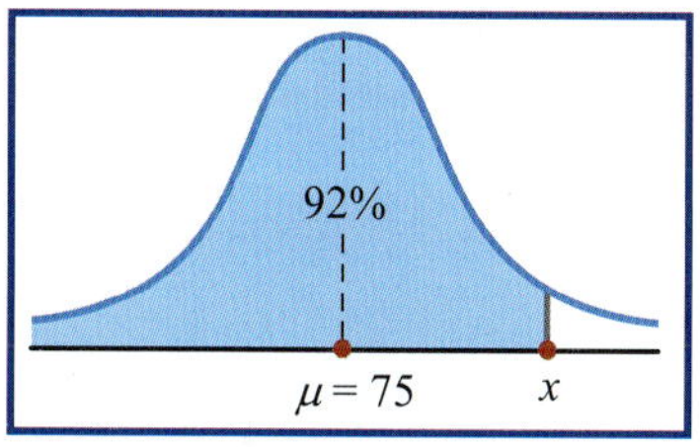

4. For a normal non-standard random variable X, with $\mu = 950$ and $\sigma = 210$, find

   (a) P_{95}, **(b)** P_{99}, **(c)** P_3, and **(d)** P_{40}.

5. Given $N(73, 20)$, find the value of X that has 95% area to its left.

6. Given $N(5.6, 1.2)$, find the values that separate the top and bottom 1% from the rest of the distribution.

7. The scores on a placement exam are normally distributed with a mean of 24 and standard deviation of 5. Find the probability of one's score to be

   (a) more than 30.
   (b) between 30 and 32.
   (c) more than 32.

8. The family incomes in a community are normally distributed with a mean of \$36,000 and standard deviation of \$5000. Find the probability that randomly selected family's income in the community is

(a) under \$42,000 **(b)** over \$30,000.

9. IQs of students at a private college are normally distributed with a mean of 130 and a standard deviation of 10.

(a) Find the percentage of students with IQs

i) below 125. **ii)** above 145.

iii) between 125 and 145. **iv)** exactly 145.

(b) Find the IQ that will put a student in the top 5% of all the students.

10. Given $N(100, 15)$, find

(a) P_{18}, **(b)** P_{93}, **(c)** P_{38} and **(d)** Q_3.

5.7: Normal Approximation to the Binomial Distribution

Objectives

After completing this section you will be able to:

A. Check Conditions for Normal Approximation to the Binomial Distribution.

B. Use Continuity Correction used for Approximation.

C. Approximate Binomial Distribution using a Normal Distribution.

In this section, we will discuss the process and conditions used to approximate the Binomial Distribution by the Normal Distribution.

A. Conditions for Normal Approximation to the Binomial Distribution

Given the following conditions **i)** and **ii)**, Normal Distribution is the limiting case of the Binomial Probability Distribution:

i) The number of trials n should be large. That is, $n \to \infty$.

ii) Neither p nor q could be very small.

From a practical point of view, when the sample size in a Binomial Experiment is large, using the discrete Binomial probability formula $P(X = x) = C_x^n\, p^x q^{n-x}$ to calculate probabilities becomes tedious and time consuming. For example if $n = 150$, calculating $P(X > 80)$ would mean that we have to find a value for each of the following probabilities using the Binomial formula.

$$P(X = 81) + P(X = 82) + P(X = 83) + \ldots + P(X = 150)$$

As one can see, these are a lot of probabilities to compute, and doing so would be quite impractical.

However, in such cases where $np > 5$, $nq > 5$ and n is sufficiently large, we can approximate the Binomial probabilities using the Normal Distribution. In this case, the mean and the standard deviation can be approximated using $\mu = np$ and $\sigma = \sqrt{npq}$.

The following Binomial Experiments and their graphs will demonstrate how a Binomial Distribution approaches the Normal Distribution as n becomes larger.

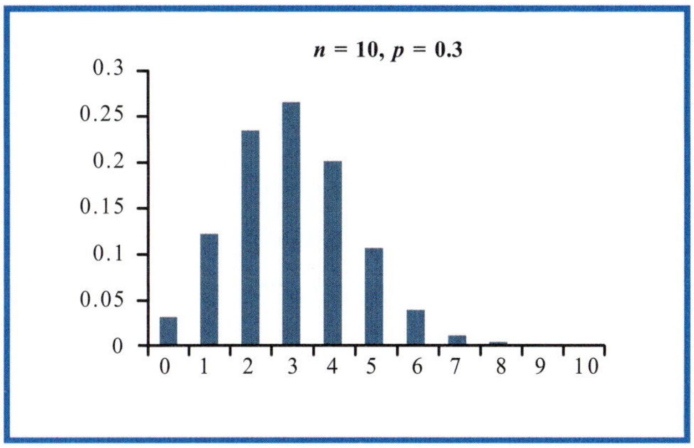

Figure 5.74 a

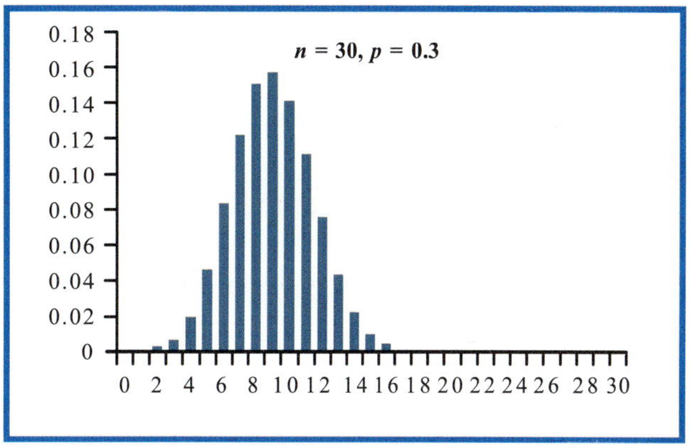

Figure 5.74 b

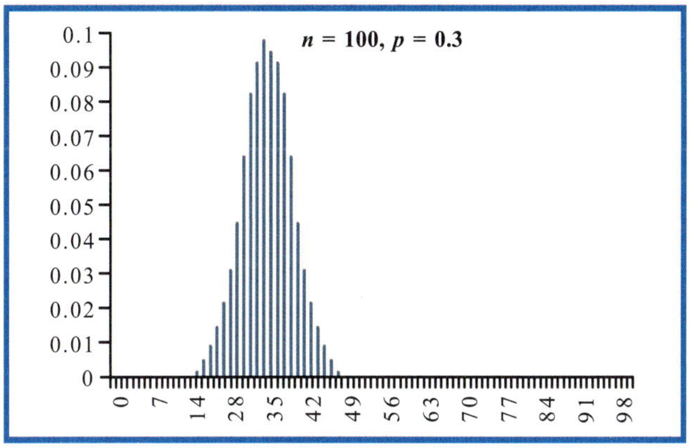

Figure 5.74 c

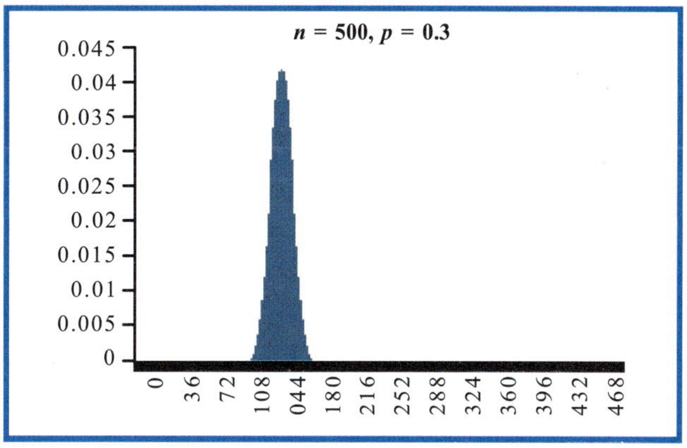

Figure 5.74 d

Notice that as the number of trials in a Binomial experiment increases, the Binomial Distribution behaves and looks more like a Normal Distribution (Symmetric Distribution).

B. Continuity Correction Used for Approximation

The task of approximating a Binomial Distribution using a Normal Distribution is carried out by applying a concept called the Continuity Correction. The Continuity Correction (0.5 units on each side of X) is a correction that enables a discrete probability distribution to be approximated by a continuous probability distribution.

The continuity correction signifies that for any discrete value of X, the neighborhood of X ($X - 0.5$ to $X + 0.5$) is used to approximate a Binomial probability by using Normal Distribution. For example, if we are interested in finding the probability that $X = 4$ (where X is a discrete random variable), then the correction is $P(3.5 < X < 4.5)$, here X is a continuous random variable.

In general, while approximating a discrete random variable with a continuous random variable, (given $np > 5$, and $nq > 5$), we have

$$P(X = a) \approx P(a - 0.5 < X < a + 0.5) \qquad (5.1)$$

$$P(X \leq a) \approx P(X < a + 0.5) \qquad (5.2)$$

$$P(X \geq a) \approx P(X > a - 0.5) \qquad (5.3)$$

$$P(X < a) \approx P(X < a - 0.5) \quad (5.4)$$

$$P(X > a) \approx P(X > a + 0.5) \quad (5.5)$$

$$P(a \le X \le b) \approx P(a - 0.5 < X < b + 0.5) \quad (5.6)$$

The approximations in (5.1) to (5.6) are called the ***Continuity Corrections*** because we are using a continuous probability distribution to calculate the probability of a discrete random variable.

C. Procedure to Approximate a Binomial Distribution Using a Normal Distribution

***Step* 1** Check to see if the Normal Approximation can be used- *i.e.*, $np > 5$, and $nq > 5$.

***Step* 2** Find the mean $\mu = np$ and the standard deviation $\sigma = \sqrt{npq}$.

***Step* 3** Express the question in probability notation. For example: $P(X \le a)$.

***Step* 4** Rewrite the problem using the continuity correction. For example: $P(X < a + 0.5)$.

***Step* 5** Transform the non-standard random variable $X \sim N(\mu,\sigma)$ to standard random variable $Z \sim N(0,1)$.

***Step* 6** Draw the Normal curve and shade the appropriate area (probability). Use the *z*-table and properties of the Normal Distribution to calculate the desired probability (shaded area).

Example 25: Edutech International is a company that publishes textbooks. From past experience, it appears that on an average, 12 books are rejected out of every 100 books. What is the probability that in a box of 50 books,

(a) 8 will be rejected. **(b)** At most 10 will be rejected.

Solutions: Here $n = 50$, $p = \dfrac{12}{100} = 0.12$ and $q = 1 - p = 0.88$

$\Rightarrow np = 50(0.12) = 6 > 5$ and $nq = 50(0.88) = 44 > 5$

Define *X*: Number of books rejected in the lot of 50 books. Then $X \sim B\,(50, 0.12)$. Since np and nq both are greater than 5, we can approximate the Binomial distribution with a Normal distribution having parameters

$\mu = np = 50(0.012) = 6$, and $\sigma = \sqrt{npq} = \sqrt{(50)(0.12)(0.88)} \approx 2.30$.

(a) Therefore; $P(X = 8) = P(7.5 < X < 8.5\ )$

$$= P\left(\frac{7.5-6}{2.30} < Z < \frac{8.5-6}{2.30}\right)$$

$$= P(0.65 < Z < 1.09)$$

$$= 0.3621 - 0.2422$$

$$= 0.1199$$

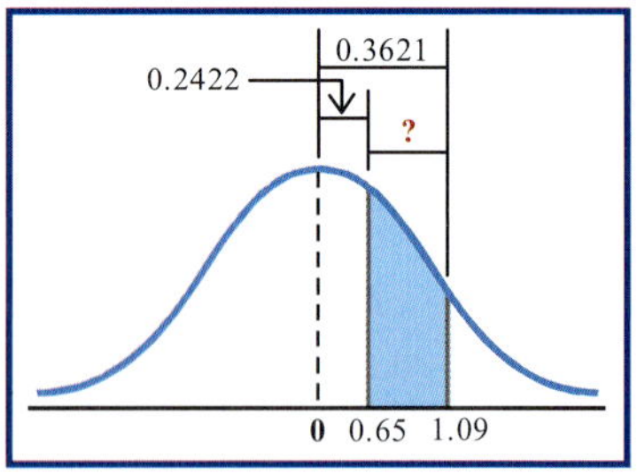

Figure 5.75

(b) $P(X \le 10) = P(X < 10.5)$ (Using continuity correction (5.2))

$$= P\left(\frac{X-6}{2.3} < \frac{10.5-6}{2.3}\right)$$

$$= P(Z < 1.96)$$

$$= 0.5 + 0.4750$$

$$= 0.9750$$

Example 26: A coin is tossed 100 times. Find the probability that there are:

(a) between 40 to 60 tails (inclusive),

(b) at least 55 tails,

(c) at most 40 tails,

(d) less than 53 tails,

(e) exactly 60 tails.

Solutions: Tossing a coin is a binomial experiment. The number of trials is 100 and the probability of success (getting tails) is 0.5

Here $n = 100$, $p = 0.5$ and $q = 1 - p = 1 - 0.5 = 0.5$

$\Rightarrow \quad np = 100 \times (0.5) = 50 > 5$ and $nq = 100 \times (0.5) = 50 > 5$

Therefore, we can use the Normal Distribution to approximate binomial probabilities.

In this case $\mu = np = (100)(0.5) = 50$, and $\sigma = \sqrt{npq} = \sqrt{(100)(0.5)(0.5)} = 5$.

(a) $P(40 \le X \le 60) = P(39.5 < X < 60.5)$ (Using continuity correction (5.1))

When $X = 39.5$, $Z = \dfrac{39.5 - 50}{5} = -2.1$

When $X = 60.5$, $Z = \dfrac{60.5 - 50}{5} = 2.1$

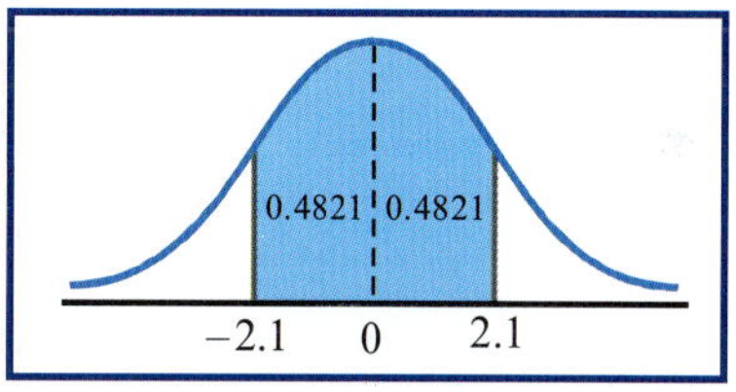

Figure 5.76

$$\begin{aligned} P(40 \le X \le 60) &= P(39.5 < X < 60.5) \\ &= P(-2.1 < Z < 2.1) \\ &= 2P(0 < Z < 2.1) \quad \text{(By Symmetry)} \\ &= 2(0.4821) = 0.9642 \end{aligned}$$

Thus, the probability of getting tails between 40 to 60 trials in 100 tosses of a fair coin is approximately $0.9642 \approx 96\%$.

(b)

$$\begin{aligned} P(X \ge 55) &= P(X > 54.5) \quad \text{(Using continuity correction, 5.3)} \\ &= P\left(Z > \frac{54.5 - 50}{5}\right) \\ &= P(Z > 0.9) \\ &= P(Z > 0) - P(0 < Z < 0.9) \\ &= 0.5 - 0.3159 = 0.1841 \end{aligned}$$

Thus, the probability of getting at least 55 tails in 100 tosses of a fair coin is approximately 0.1841.

(c)

$$\begin{aligned} P(X \le 40) &= P(X < 40.5) \quad \text{(Using continuity correction, 5.2)} \\ &= P\left(\frac{X - 50}{5} < \frac{40.5 - 50}{5}\right) \\ &= P(Z < -1.9) \\ &= P(Z > 1.9) \quad \text{(By Symmetry)} \end{aligned}$$

$$= P(Z > 0) - P(0 < Z < 1.9)$$
$$= 0.5 - 0.4713 = 0.0287$$

Thus, the probability of getting at most 40 tails in 100 tosses of a fair coin is approximately 0.0287.

(d) $P(X < 53) = P(X < 52.5)$ (Using continuity correction, 5.4)

$$= P\left(\frac{X - 50}{5} < \frac{52.5 - 50}{5}\right)$$
$$= P(Z < 0.5)$$
$$= P(Z < 0) + P(0 < Z < 0.5)$$
$$= 0.5 + 0.1915 = 0.6915$$

Thus, the probability of getting less than 53 tails in 100 tosses of a fair coin is approximately 0.6915.

(e) $P(X = 60) = P(59.5 < X < 60.5)$ (Using continuity correction, 5.1)

$$= P\left(\frac{59.5 - 50}{5} < Z < \frac{60.5 - 50}{5}\right)$$
$$= P(1.9 < Z < 2.1)$$
$$= P(Z < 2.1) - P(Z < 1.9)$$
$$= 0.4821 - 0.4713$$
$$= 0.0108$$

Try it yourself:

1. Two percent of flash drives manufactured in a certain plant are defective. Estimate the probability that of 5000 randomly selected flash drives, exactly 125 are defective.
2. Estimate the probability of obtaining at least 60 fives in 300 tosses of a fair die.
3. A multiple choice test consists of 50 questions. There are 5 possible answers for each question of which one is correct. If all answers are random guesses, estimate the probability of getting at least 20% correct.
4. If 68% of households in a town watch the CNN channel, find the probability that in a random sample of 200 households in the town
   (a) between 120 and 130 inclusive watch CNN.
   (b) at least 150 watch CNN. **(c)** less than 125 watch CNN.
5. The probability that a person is left-handed is 12%. In a class of 100 students, find the probability that
   (a) at most 15 are left-handed. **(b)** at least 8 are left-handed.
   (c) 18 are left-handed. **(d)** less than 20 are left-handed.
6. A coin is tossed 40 times. A person randomly guessing on all the outcomes predicts correctly on 16 tosses. What is the probability of being correct more than 16 times?

5.8: Normal Probability Distributions with Microsoft Excel®

Objectives

After completing this section you will be able to:

A. Use Standard Excel® Functions Available for Normal Distribution.

B. Generate Random Numbers.

A. Standard Excel® Functions Available for Normal Distribution

- **NORMSDIST(z):**

Returns the cumulative probability of standard normal variate. It calculates a probability $P(Z \le z)$, where Z is standard normal variate.

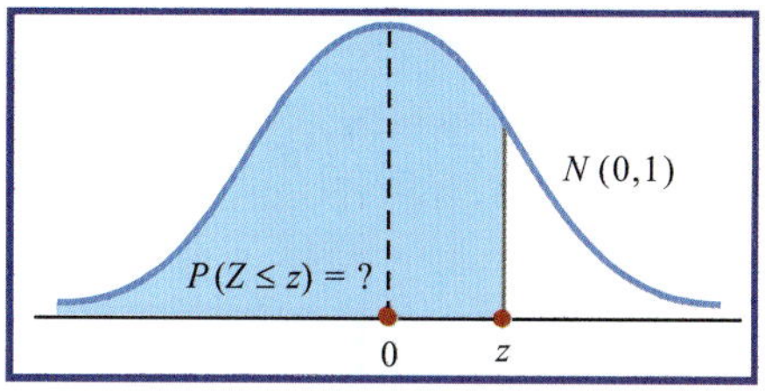

Figure 5.77

- **NORMSINV(p):**

Returns the inverse of the standard normal cumulative distribution. In other words it returns z, when $P(Z \le z) = p$.

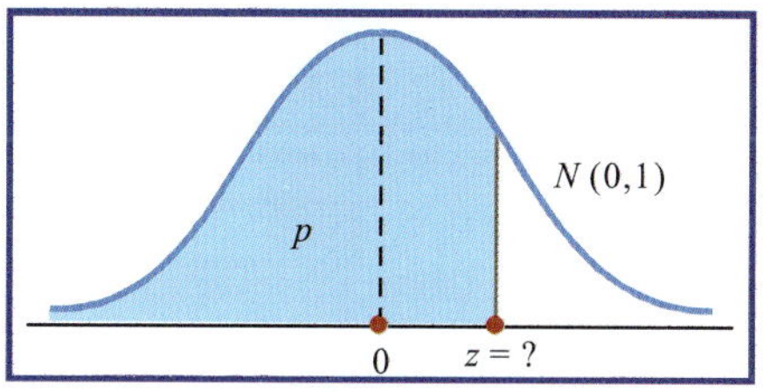

Figure 5.78

- **NORMDIST(x, mean, standard_dev, cumulative):**

Return cumulative probability of normal distribution with specific x, mean and standard deviation(sd). In other words it returns $P(X \le x)$, where $X \sim N(\mu, \sigma)$.

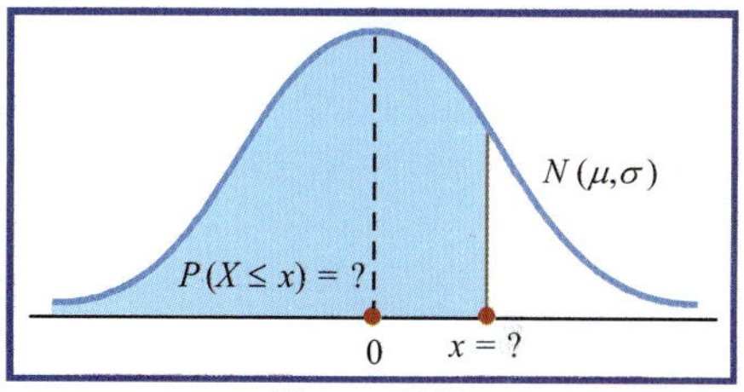

Figure 5.79

- **NORMINV(probability, mean, standard_dev):**

Returns the inverse of the normal cumulative probability(p) for the specified mean and SD. In other words if $X \sim N(\mu, \sigma)$ and $P(X \le x) = p$, then it returns x for specified values of p, μ and σ.

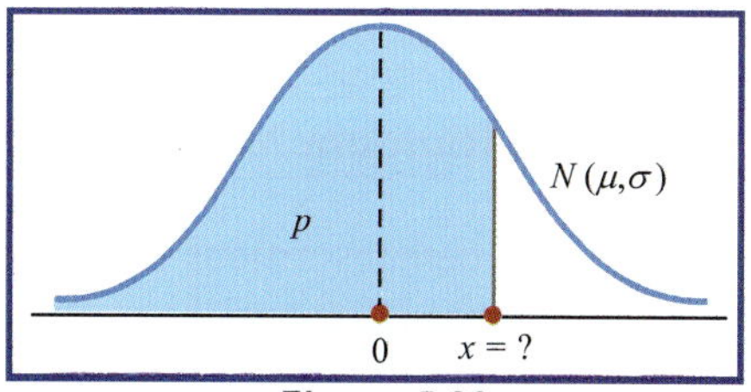

Figure 5.80

- **STANDARDIZE(x, mean, standard_dev):**

Returns a 'z' value from a given 'x' value. Normalizes a 'x' value. In other words, it computes

$Z = \dfrac{x - \mu}{\sigma}$ for specified values of x, mean(μ) and standard deviation(σ).

Illustration 1: Find $P(Z < 1.38)$.

Solution: The Excel® command to calculate $P(Z < 1.38)$ is **=NORMSDIST(1.38)**.

This returns a value of 0.91621 corresponding to the shaded area. $P(Z < 1.38) = 0.91621$

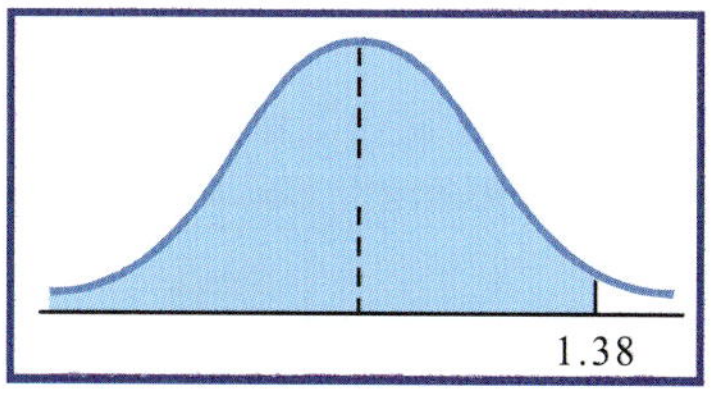

Figure 5.81

Illustration 2: Find a z-score separating the bottom 95% from the top 5% (also known as P_{95}).

Solution: = **NORMSINV(0.95)**

This will return a z value = 1.64485

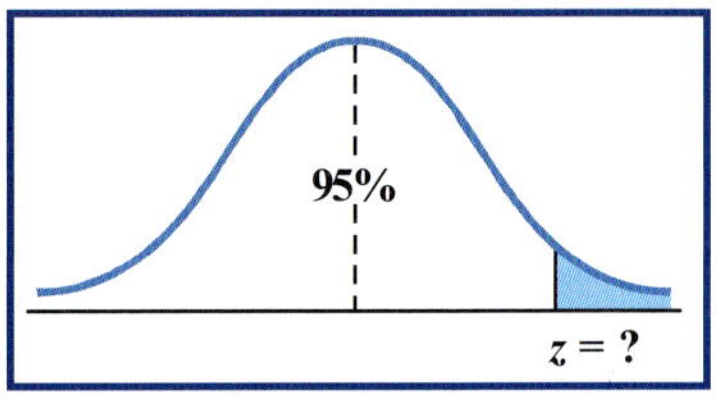

Figure 5.82

Illustration 3: Use the **NORMDIST** command to solve the following illustration.

Given an average IQ to be 100 with a standard deviation of 15, find the probability that, if you select a person at random, his or her IQ will be greater than 128.

Solution: In this example, our aim is to find, $P(X > 128)$, where $X \sim N(100, 15)$.

The Excel command for computing this probability is

=(1 – NORMDIST(128,100,15,TRUE)). This function returns value 0.0309.

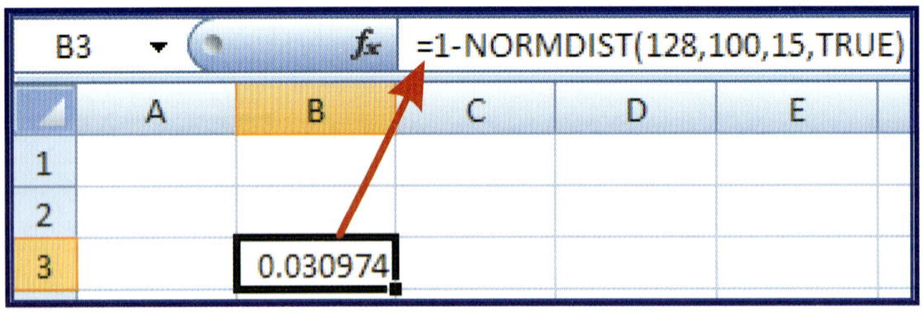

Figure 5.83 a

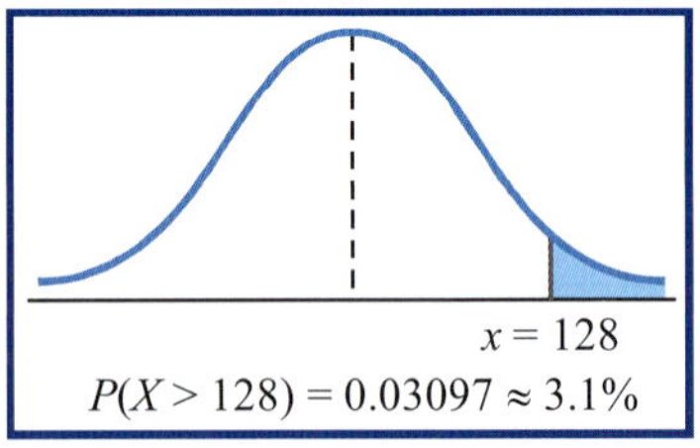

Figure 5.83 b

Illustration 4: Give an average IQ of 100 and a standard deviation 15, find the minimum IQ score that will put you in the top 5% of the population.

Solution: We have to find x such that $P(X > x) = 0.05$

$\Rightarrow\ 1 - P(X \leq x) = 0.05$

$\Rightarrow\ P(X \leq x) = 0.95$ and Excel command for computing x is:

=NORMDIST(0.95,100,15). This function returns value $x = 124.67$

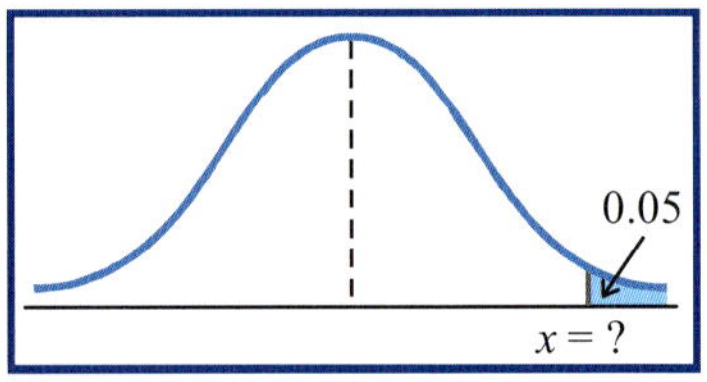

Figure 5.84

B. Random Number Generation

The demonstrations given below explain the procedure for generating random numbers from the Normal Distribution.

Illustration 5: Generate a sequence of normally distributed 20 random scores with mean of 73 and standard deviation 9.

Solution: The Excel® formula for computing a single random score from $N(73, 9)$ is: =NORMINV(rand( ),73,9). By copying this formula 20 times, we have a set of 20 random numbers.

Sample Number	1	2	. . .	19	20
Final Score	=NORMINV(RAND(),73,9)	=NORMINV(RAND(),73,9)	. . .	=NORMINV(RAND(),73,9)	=NORMINV(RAND(),73,9)

Sample Number	1	2	3	4	5	6	7	8	9	10	11	12	13	14	15	16	17	18	19	20
Final Score	71	79.9	73.1	90.6	84.8	77.3	62.8	82.2	69.7	77.8	87.5	76	71.8	59.2	57.4	70.5	75.6	65.5	85.3	85

Figure 5.85

Illustration 6: *Generating a Random Sample of Observations from a Normally Distributed Population*

Generate 20 random values for a Normally Distributed population with a mean of 100 and a standard deviation of 15.

Solution: Go to **the Data tab**. Select **Data Analysis** from the Analysis Group.

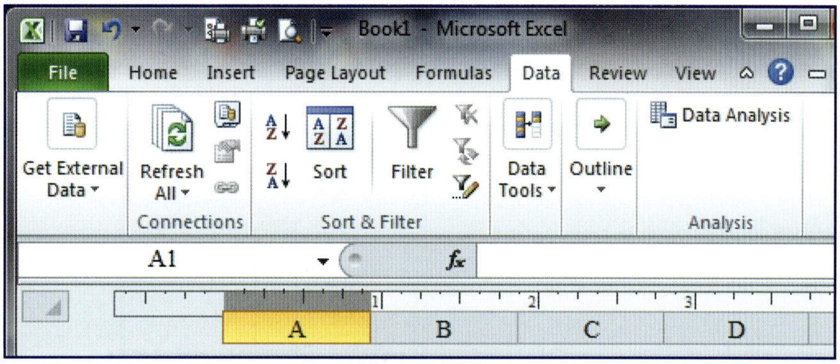

Figure 5.86

Select **Random Number Generation**.

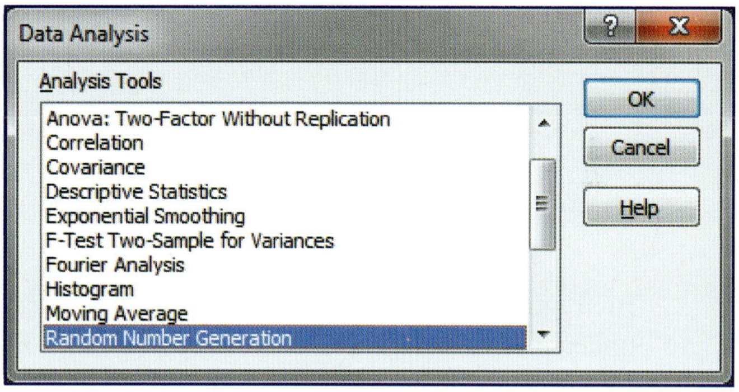

Figure 5.87

Fill in the appropriate boxes. You do not need to type anything for Number of Variables and Random Seed.

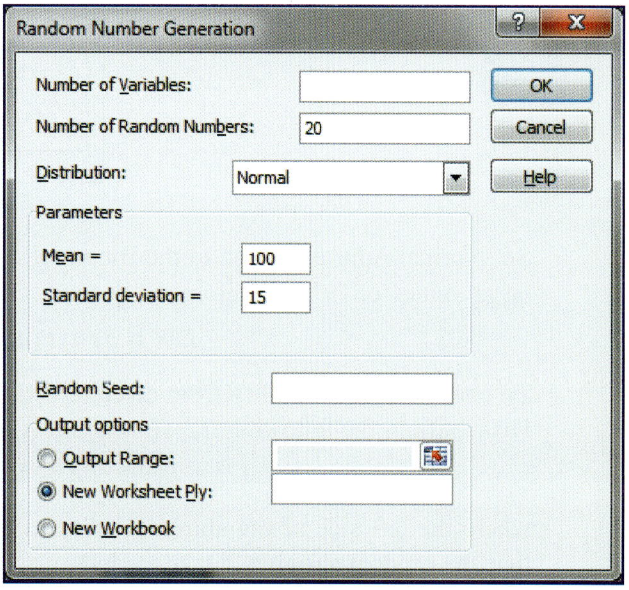

Figure 5.88

The output after using this procedure is:

106.34	97.34	101.34	98.67
76.74	89.90	116.82	103.28
112.57	92.55	80.95	123.88
107.11	96.59	107.61	77.04
121.72	71.49	93.63	151.41

The user will get different set of random numbers at every new trial.

5.9: Extended Topics and Proofs in Continuous Probability Distributions (Optional)

Objectives

After completing this section you will be able to:

A. Understand Transformation from a Discrete Random Variable to a Continuous Random Variable.

B. Compute Probabilities (Areas) under Simple Curves.

A. Transformation from a Discrete Random Variable to a Continuous Random Variable

The following discussion outlines the mathematical procedure that establishes this transformation.

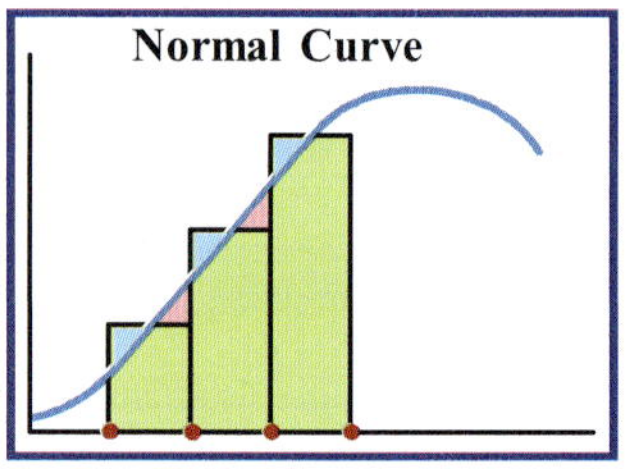

Figure 5.89 a

Zooming in on the above rectangles, we can construct Figure 5.89(b).

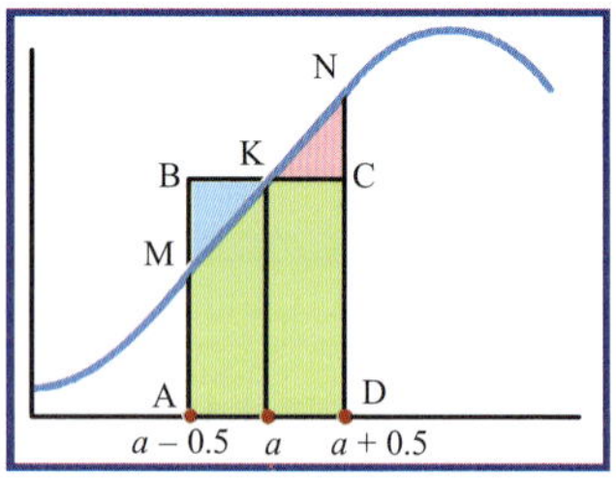

Figure 5.89 b

In Figure 5.89(b) the area of the Rectangle (ABCD) is the binomial probability for $X = a$. The area under the normal curve, between $X = a - 0.5$ and $X = a + 0.5$ is the area of the trapezoidal (AMND).

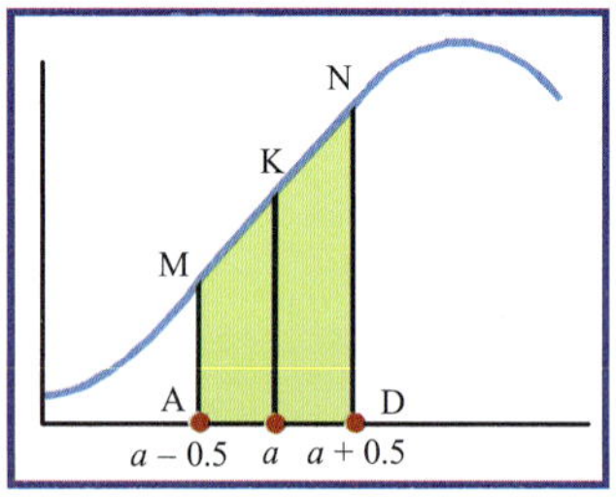

Figure 5.89 c

So, Normal Approximation to the Binomial Distribution also means: Area (ABCD) ≈ Area (AMND) for large values of "n". Therefore,

$$P(X = a) \approx P(a - 0.5 \le X \le a + 0.5)$$

To convert the probability from discrete nodes into an area to be calculated using a Normal Probability Distribution, the following transformation is used

$$P(10 \le X \le 12) = P(9.5 \le X \le 12.5),$$

where the left side of the above equality means the probability that the random variable X takes values $x = 10$, $x = 11$, $x = 12$, and the right side represents the area under the normal curve between the x-axis and the ordinates 9.5 and 12.5.

The adjacent Figure 5.90 will simplify this concept by the use of Geometry.

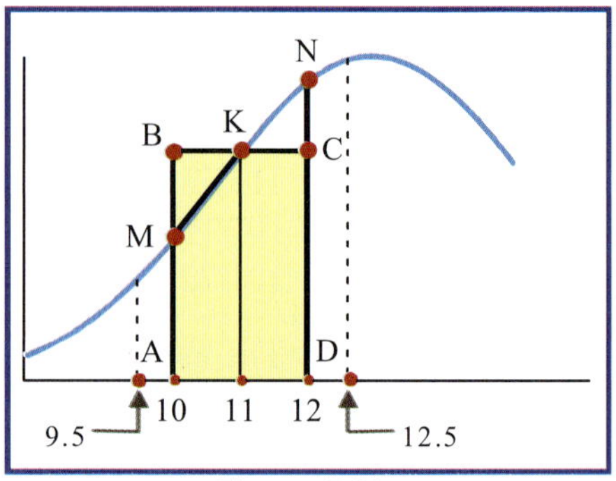

Figure 5.90

B. Computation of Probabilities (Areas) under Simple Curves

In most cases, the functions modeling continuous probability distributions are difficult to handle without the knowledge of calculus. Therefore, we give below a few examples of very simple continuous probability distributions (density functions).

Example 27: A probability density function is defined as follows: $f(x)=\frac{1}{5}$ for $x = 2$ to $x = b$. Find b.

Solution: Changing the formula form to graphical form (graphing the probability density function), we get the density curve (Figure 5.91).

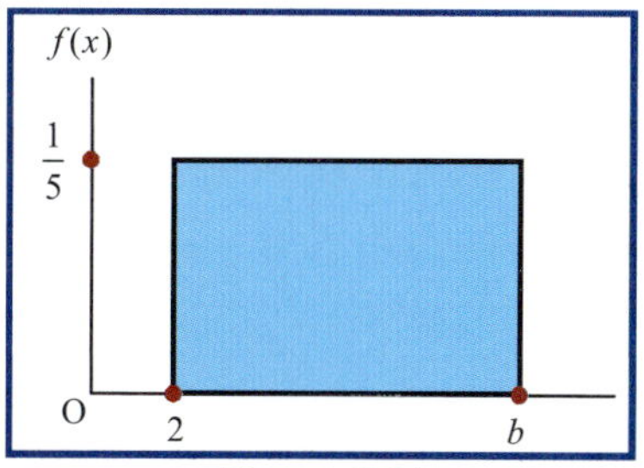

Figure 5.91

Since the total area under the density curve must be 1, Area of the rectangle = 1. Recall the area of a rectangle: Area = length × width.

Therefore, $\frac{1}{5}(b-2)=1 \Rightarrow b=7$. Thus, for $f(x)$ to be a density function (probability distribution),

Uniform Distribution:

The graph of the *probability density function* (probability distribution) in Figure 5.91 is a rectangle. Such probability density functions are known as ***Uniform Distribution***. In general, the *pdf* of a continuous random variable which has a *uniform distribution* on the interval (a, b) is given by

$$f(x)=\frac{1}{b-a}, \text{ for } x = a \text{ to } x = b$$

A uniformly distributed continuous random variable $\boldsymbol{X}$ is denoted by $X \sim U(a, b)$. The graph of uniform distribution on (a, b) is given in Figure 5.92.

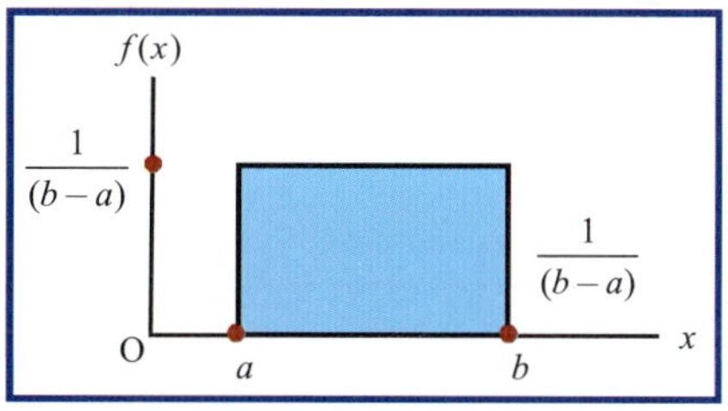

Figure 5.92

Example 28: For a density function $f(x) = \frac{2(x+2)}{5}$ for $x = 0$ to $x = 1$,

(a) Show that $P(0 < X < 1) = 1$

(b) Find $P\left(\frac{1}{3} < X < \frac{1}{2}\right)$

(c) Find $P\left(X > \frac{1}{4}\right)$

Solutions: $f(x) = \dfrac{2(x+2)}{5} = \dfrac{2}{5}x + \dfrac{4}{5}$, which is of the form $f(x) = y = mx + b$ and hence is the equation of a straight line.

We start with graphing the line.

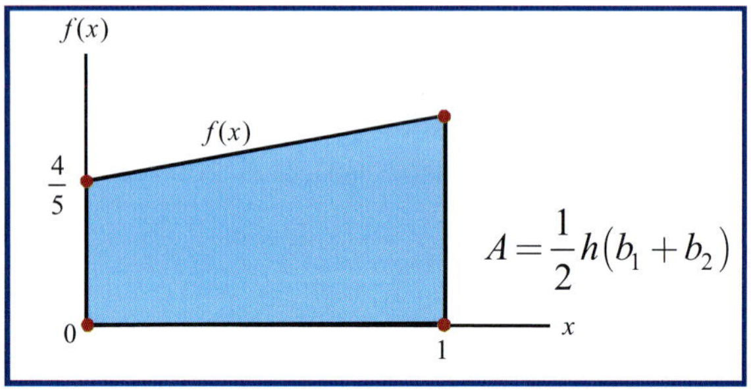

Figure 5.93

(a) $P(0 < X < 1)$ = Area under the density curve between $x = 0$ and $x = 1$

= Area of the shaded trapezoid

$$= \frac{1}{2}(h)[f(0) + f(1)] = \frac{1}{2}(1)\left(\frac{4}{5} + \frac{6}{5}\right) = 1$$

(b) $P\left(\frac{1}{3} < X < \frac{1}{2}\right)$ = Area under the density curve between $x = \frac{1}{3}$ and $x = \frac{1}{2}$

= Area of the shaded trapezoid

$$= \frac{1}{2}(h)\left[f\left(\frac{1}{3}\right) + f\left(\frac{1}{2}\right)\right] = \frac{1}{2}\left(\frac{1}{2} - \frac{1}{3}\right)\left(\frac{14}{15} + 1\right) = 0.16$$

Note: $f\left(\frac{1}{3}\right) = \dfrac{2\left(\frac{1}{3} + 2\right)}{5} = \dfrac{14}{15}$ and $f\left(\frac{1}{2}\right) = \dfrac{2\left(\frac{1}{2} + 2\right)}{5} = 1$

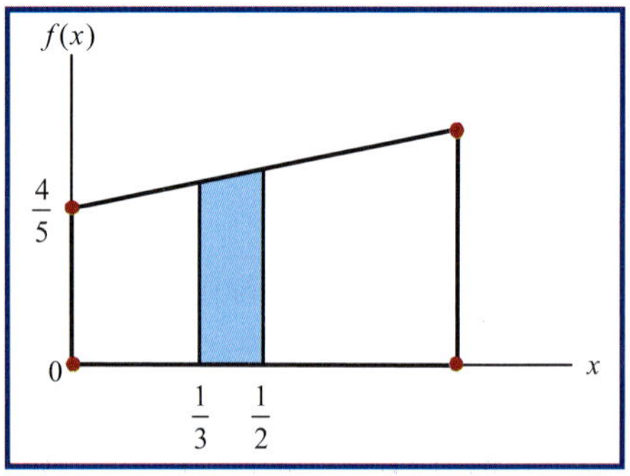

Figure 5.94

(c) $P\left(X > \frac{1}{4}\right)$

= Area under the density curve between

$x = \frac{1}{4}$ and $x = 1$

$$= \frac{1}{2}(h)\left[f\left(\frac{1}{4}\right) + f(1)\right] = \frac{1}{2}\left(1 - \frac{1}{4}\right)\left(\frac{9}{10} + \frac{6}{5}\right) = 0.79$$

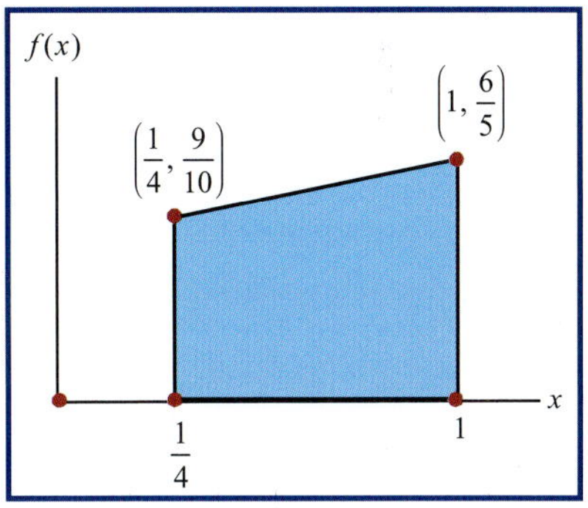

Figure 5.95

Note: $f\left(\frac{1}{4}\right) = \frac{2\left(\frac{1}{4}+2\right)}{5} = \frac{9}{10}$ and $f(1) = \frac{2(1+2)}{5} = \frac{6}{5}$

Example 29: The amount of soft drink dispensed by a machine into a cup is a continuous random variable X. The probability distribution of X is defined as follows:

$$f(x) = \frac{1}{4} \text{ for } x = 8 \text{ to } x = 12 \text{ ounces.}$$

Find the probability that the machine will fill a cup with

(a) more than 9.1 ounces but less than 11.5 ounces.

(b) at least 10.5 ounces.

(c) at most 11.2 ounces.

Solutions: The density curve is shown in Figure 5.96.

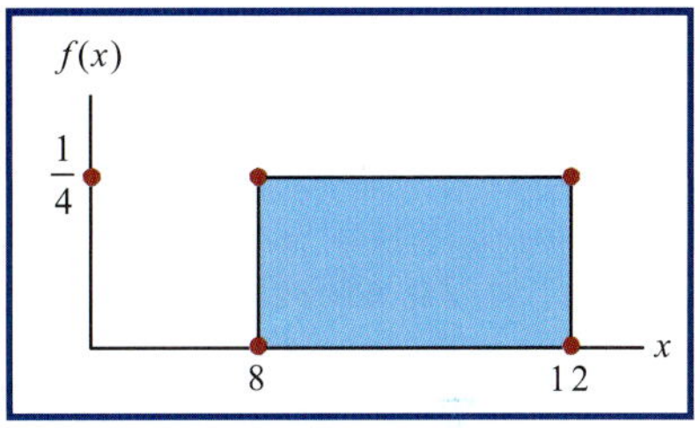

Figure 5.96

(a) $P(9.1 < X < 11.5)$

= Area under the density curve between $x = 9.1$ and $x = 11.5$ (Figure 5.97)

= Area of the shaded rectangle

$= \frac{1}{4}(11.5 - 9.1) = 0.6$

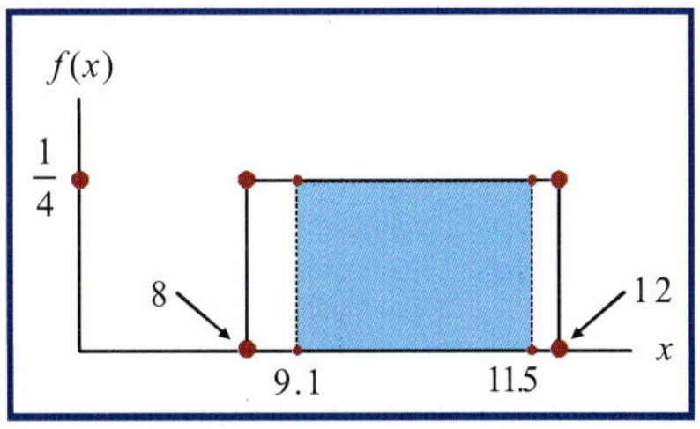

Figure 5.97

(b) $P(X > 10.5)$

= Area under the density curve between $x = 10.5$ and $x = 12$ (Figure 5.98)

= Area of the shaded rectangle

$= \frac{1}{4}(12 - 10.5) = 0.375$

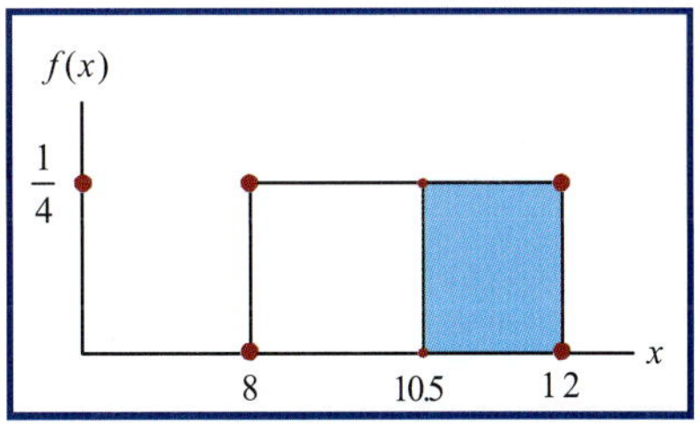

Figure 5.98

(c) $P(X < 11.2) =$ Area under the density curve between $x = 8$ and $x = 11.2$ Figure 5.99

$= $ Area of the shaded rectangle

$= \frac{1}{4}(11.2 - 8) = 0.8$

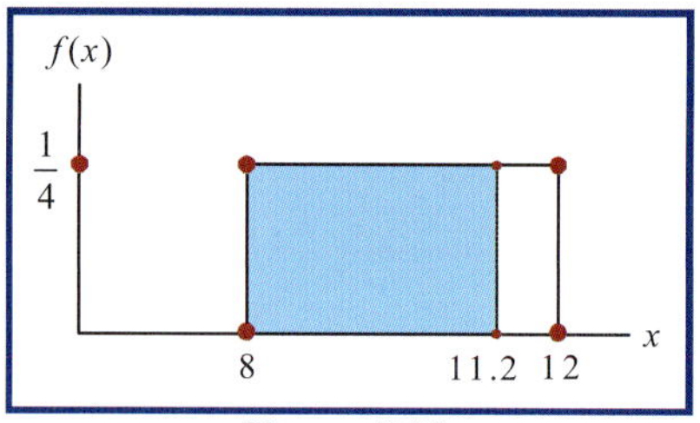

Figure 5.99

Try it yourself:

In exercises 1-4, find the unknown part of the probability density function.

1. $f(x) = \frac{1}{3}$ for $x = 0$ to $x = b$

2. $f(x) = \frac{2}{3}$ for $x = a$ to $x = 5$

3. $f(x) = \frac{a}{4}$ for $x = 0$ to $x = 2$

4. $f(x) = a$ for $x = 2$ to $x = 9$

In exercises 5-8, use the probability density function $f(x) = \frac{1}{5}$ for $x = 1$ to $x = 6$.

5. Show that $P(1 < X < 6) = 1$

6. Find $P(3 < X < 5)$

7. Find $P(X > 3.5)$

8. Find $P(X < 4)$

In exercises 9-12, use the probability density function $f(x) = \frac{x}{2}$ for $x = 0$ to $x = 2$.

9. Show that $P(0 < X < 2) = 1$

10. Find $P\left(\frac{1}{3} < X < \frac{3}{2}\right)$

11. Find $P\left(X > \frac{3}{2}\right)$

12. Find $P(X > 0.7)$

In exercises 13-16, use the probability density function $f(x) = \frac{2}{3}(x+1)$ for $x = 0$ to $x = 1$.

13. Show that $P(0 < X < 1) = 1$

14. Find $P\left(\frac{1}{3} < X < \frac{1}{2}\right)$

15. Find $P\left(X > \frac{1}{4}\right)$

16. Find $P\left(X < \frac{1}{5}\right)$

17. The amount of coffee dispensed by a machine into a cup is a continuous random variable X. The probability density function of x is defined as $f(x)=\frac{1}{3}$ for x = 7 to x = 10 ounces. Find the probability that the machine will fill a cup with

(a) at least 8.5 ounces. **(b)** at most 9 ounces.

(c) between 7.5 and 9.2 ounces.

18. The duration (X) of baby's nap (in minutes) is a continuous random variable with the probability density curve shown in the given figure.

Find the probability that the baby sleeps for

(a) more than 30 minutes.

(b) less than 45 minutes.

(c) between 25 and 45 minutes.

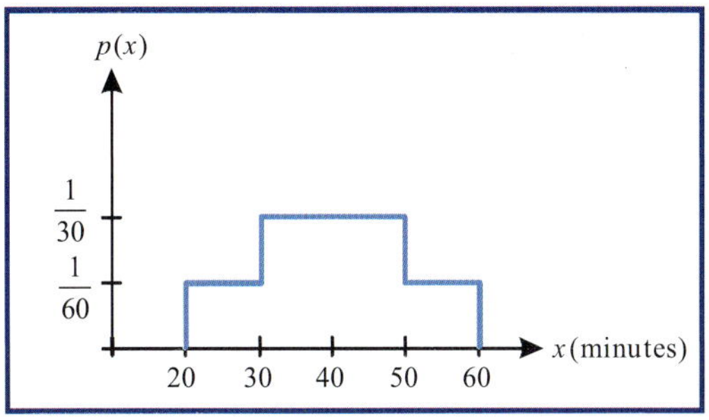

Other Special Continuous *probability density functions*:

Gamma Distribution:

$$X \sim G(\theta,\kappa),\ p(x)=\frac{1}{\theta^{\kappa}\Gamma(\kappa)}x^{\kappa-1}e^{-x/\theta},\ \ 0<x<\infty,\ (\theta,\kappa)>0.$$

The Gamma Function satisfies the following properties:

i) $\Gamma(\kappa)=(\kappa-1)\Gamma(\kappa-1)\quad \kappa>1$

ii) $\Gamma(n)=(n-1)!\qquad n=1, 2, \ldots$

iii) $\Gamma\left(\frac{1}{2}\right)=\sqrt{\pi}$

Exponential Distribution:

$$X \sim EXP(\theta),\ \ p(x)=\frac{1}{\theta}e^{-x/\theta},\ \ x>0,\ \ \theta>0.$$

Weibull Distribution:

$$X \sim WEI(\theta,\beta),\ p(x) = \frac{\beta x^{\beta-1}}{\theta^{\beta}}e^{-\left(\frac{x}{\theta}\right)^{\beta}},\ x>0,\ \kappa>0,\ \theta>0\ \text{and}\ \beta>0.$$

Pareto Distribution:

$$X \sim PAR(\theta,\kappa),\ \ p(x)=\frac{\kappa}{\theta}\left(1+\frac{x}{\theta}\right)^{-(\kappa+1)},\ x>0,\ \kappa>0\ \text{and}\ \theta>0.$$

Chapter 5: General Exercises

1. Find the area of the shaded regions for the following standard normal curve.

(a)

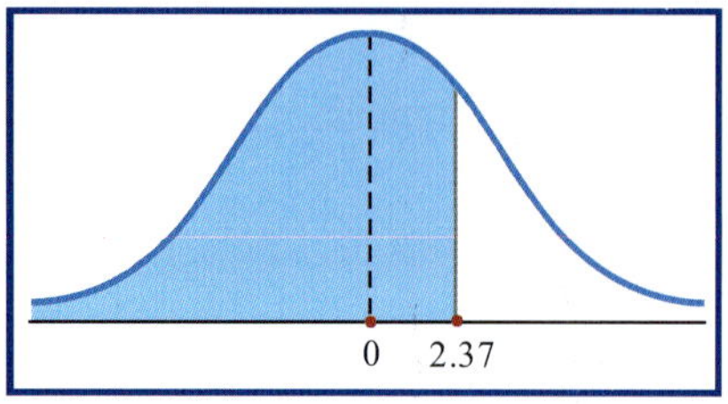

(b)

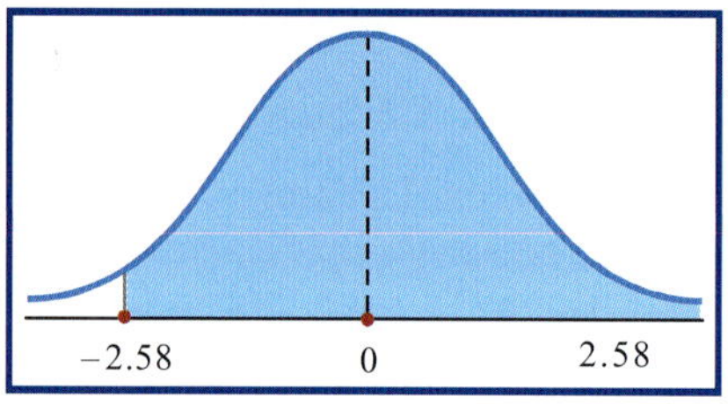

2. Given $N(0, 1)$, find the indicated Z scores. **Recall:** $N(0, 1)$ is the mathematical Notation for a standard normal distribution with mean $\mu = 0$, and standard deviation $\sigma = 1$.

(a) The area of the shaded region is 0.9761.

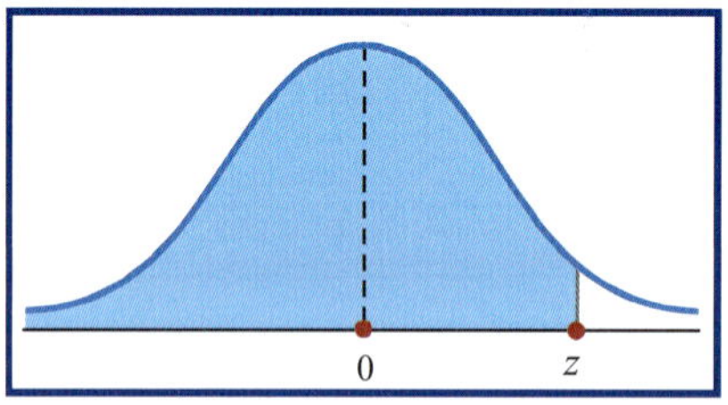

(b) The area of the shaded region is 0.1492.

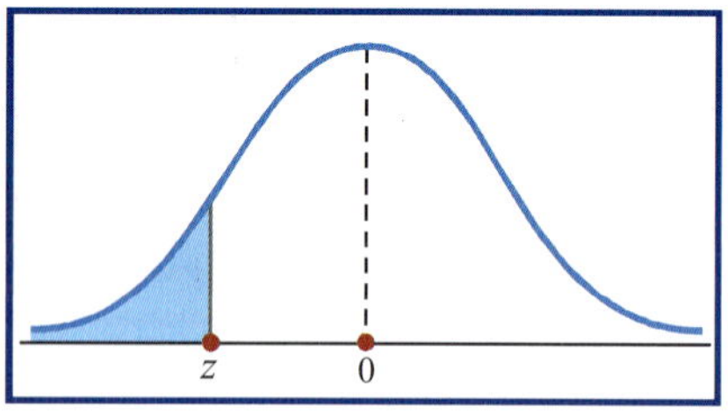

3. If z follows a standard normal distribution, find c such that

(a) $P(-c < Z < c) = 0.8164$ **(b)** $P(-c < Z < c) = 0.9900$ **(c)** $P(Z < c) = 0.7734$

(d) $P(Z > c) = 0.2177$ **(e)** $P(Z < c) = 0.1823$ **(f)** $P(Z > c) = 0.8830$

4. Given a standard normal distribution, compute the z-score that has

(a) 62.93% of the distribution's area to its left.

(b) 28.43% of the distribution's area to its left.

(c) 90% of the distribution's area in between $-z$ and z.

5. If $Z \sim N(0, 1)$, find the probability that Z

(a) is between −2.36 and −1.05 **(b)** is less than −1.79

(c) is greater than 0.37 **(d)** is between −1.50 and 2.37

(e) is between −2.16 and −1.23 **(f)** is less than 2.08

6. Find the following probabilities by shading the appropriate region under the standard normal curve.

(a) $P(Z < 0.28)$ **(b)** $P(Z > -1.74)$ **(c)** $P(-1.15 < Z < 2.04)$

(d) $P(Z < -2.06)$ **(e)** $P(Z > -1.84)$ **(f)** $P(Z < 1.33)$

(g) $P(Z > 2.49)$ **(h)** $P(1.18 < Z < 2.35)$ **(i)** $P(-1.77 < Z < 2.21)$

(j) $P(-2.21 < Z < -1.77)$ **(k)** $P(Z < -2.5)$ **(l)** $P(Z < -1.89)$

(m) $P(Z > 1.9)$ **(n)** $P(Z > -2.25)$ **(o)** $P(-0.06 < Z < 1.45)$

(p) $P(-1.32 < Z < 2.74)$ **(q)** $P(1.86 < Z < 2.23)$ **(r)** $P(-2.82 < Z < -1.06)$

7. Readings on a Voltmeter are normally distributed with a mean of 0 volts and standard deviation of 1 volt. Find the voltage readings for the following percentages. **Note:** Voltage readings are represented as Z-scores, since we have $N(0, 1)$.

(a) Q_3 **(b)** P_{64} **(c)** P_{46} **(d)** P_{93} **(e)** P_{12}

(f) Q_1 (g) P_{40} (h) P_{20} (i) P_{83} (j) P_2

(k) Find the reading that separates the bottom 2.5% from the rest.

(l) If 2% of the voltages are rejected because they are too high but all other voltages are acceptable, find the voltage that separates the rejected voltages from the rest.

8. An engineer wants to examine voltmeters that give readings in the bottom 10%. Find the reading (z score) that separates the bottom 10% from the top 90%. Assume $N(0, 1)$.

9. Given $N(23.4, 6.2)$, find the probability that X is less than 23.4.

10. For $N(120, 18)$, find $P(110 < X < 140)$.

11. Mr. Vartan is 72 inches tall in a region where the average height of adult males is 67 inches with a standard deviation of 2 inches. Mr. Mardir is 74 inches tall in a region where the average height of adult males is 69 inches with a standard deviation of 2.5 inches. Compare the heights of Mr. Vartan and Mr. Mardir relative to their regions.

12. Scores on a Statistics test are distributed as $N(0, 1)$. Find the minimum score that will put a student in the top 5%.

13. Assume that women's weights are normally distributed with $\mu = 143$ *lbs* and a standard deviation $\sigma = 29$ *lbs*. Find the indicated values.

(a) Q_3 (b) P_{30} (c) P_{97} (d) P_{15} (e) P_{12}

14. In a population, the IQ scores are normally distributed with $\mu = 100$ and $\sigma = 15$. Use the emprical distribution to estimate the probability that a randomly selected person has an IQ between 70 and 115.

15. In a Non-Standard Normal Distribution, find σ when $\mu = 100$ and 2.68% of the area lies to the right of $X = 105$.

16. Scores on a test are normally distributed with a mean 70 and a standard deviation 11.5. Find P_{81}, which separates the bottom 81% from the top 19%.

17. Suppose that replacement times for washing machines are normally distributed with a mean of 9.5 years and a standard deviation of 1.8 years. Find the replacement time that separates the top 18% from the bottom 82%.

18. The volumes of Hemi engines are normally distributed with a mean of 380 cubic inches and a standard deviation of 20 cubic inches. What is the probability that the volume of a randomly selected Hemi engine from the production line will be less than 370 cubic inches?

19. Suppose that replacement times for timing belts in cars are normally distributed with a mean of 5.3 years and a standard deviation of 2.5 years. Find the replacement time for the top 12% of all timing belts.

20. IQ scores are normally distributed with mean of 100 and a standard deviation of 15.

(a) Find the probability that a randomly selected IQ score will be less than 129.

(b) Find the probability that a randomly selected IQ score will be less than 93.

(c) What IQ score will put you in the top 1%?

(d) What percentage of IQ scores is considered usual?

21. GRE (Graduate Record Exam) scores are normally distributed with mean of 1000 and a standard deviation of 285.

(a) Find the probability that a randomly selected GRE score will be greater than 800.

(b) Find the probability that a randomly selected GRE score will be less than 1100.

(c) Find the probability that a randomly selected GRE score will be less than 700.

(b) What GRE score will put you in the top 40%?

22. The ages of a group of university faculty are given in the following table:

23	45	67	37	69	44	36	55
29	50	60	47	58	42	54	77
43	51	53	40	45	58	55	64
35	24	36	47	53	44	38	39

(a) Calculate the 35th percentile, P_{35}.

(b) Calculate the third quartile.

(c) What percentile corresponds to $x = 47$?

23. Scores on the biology portion of the Medical College Admissions Test are normally distributed with a mean of 9.1 and a standard deviation of 2.1.

(a) What percentage of scores is above 8.5, the minimum score required for admission to the Providence School of Medicine?

(b) What percentage of scores is between 7.5 and 9.0?

(c) What is the value of Q_3, the score separating the bottom 75% from the top 25%?

(d) What is the value of Q_1?

24. The average density of a certain stone is 1.7 g/ml, with the standard deviation of 0.06 g/ml. An archeologist wants to reject all samples that are more than 1.86 g/ml and less than 1.59 g/ml. What percent of his samples will he reject, given that the densities are normally distributed?

25. The cars on a used car lot are rated with an average score of 187 points and a standard deviation of 5 points. Suppose, a customer wants a car with a minimum rating of 177 but cannot afford cars rated over 200. What percent of cars on this lot is unsuitable for this customer? Assume the ratings are normally distributed.

26. A florist buys long-stem roses with an average length of 24 inches and a standard deviation of 3.6 inches. He sends back roses that are shorter than 19.4 inches or longer than 29 inches. What percent of roses go back to the grower, assuming that the lengths are normally distributed?

27. A certain modeling agency only hires women who have an average height of 5'8" with a standard deviation of 2". The agency will automatically reject any applicant shorter than 5'5" or taller than 6'0". What percent of women will go home disappointed, assuming that the heights are normally distributed? **Hint:** Change all heights in inches.

28. Suppose a study has found that it takes women an average of 43 minutes to get ready in the morning, with a standard deviation of 12 minutes. If 70 women are selected at random, how many of them will take longer than 45 minutes to get ready?

29. In order to enroll in Parliament magnet school, a minimum score of 600 points is required on the entrance exam. The average score for incoming students is 580 with a standard deviation of 70.2. If 100 students apply for admission, what percent will be rejected? Assume that the scores are normally distributed.

30. An ESL program accepts students who have scored between 45 and 55 points on their last ESL placement test. The mean score on the placement exam is 48 points with a standard deviation of 5 points, and the scores are normally distributed. What percentage of the students will not be accepted into this program?

In exercise 31-34, use Microsoft Excel® to find the mean and standard deviation of the following raw data sets.

31. The following table gives the average monthly pound prices for Red Delicious Apples for the past eleven years, as given by the Department of Agriculture.

Month/ Year	Jan	Feb	Mar	Apr	May	Jun	Jul	Aug	Sep	Oct	Nov	Dec
1992	0.876	0.886	0.899	0.913	0.925	0.962	0.991	1.015	0.933	0.765	0.753	0.764
1993	0.810	0.817	0.802	0.802	0.815	0.835	0.854	0.904	0.939	0.850	0.798	0.778
1994	0.789	0.810	0.804	0.803	0.806	0.822	0.848	0.881	0.874	0.745	0.735	0.722
1995	0.765	0.789	0.793	0.784	0.813	0.833	0.864	0.901	0.923	0.863	0.853	0.834
1996	0.877	0.877	0.894	0.915	0.921	0.954	0.976	0.998	1.006	0.949	0.907	0.886
1997	0.907	0.912	0.914	0.895	0.912	0.914	0.918	0.935	0.933	0.881	0.864	0.897
1998	0.922	0.960	0.962	0.949	0.974	0.955	1.000	0.990	0.971	0.902	0.878	0.854
1999	0.860	0.870	0.852	0.870	0.881	0.893	0.905	0.921	0.972	0.919	0.902	0.918
2000	0.952	0.974	0.960	0.957	0.927	0.918	0.940	0.928	0.922	0.899	0.833	0.816
2001	0.808	0.830	0.845	0.834	0.848	0.890	0.892	0.898	0.874	0.894	0.915	0.893
2002	0.877	0.892	0.915	0.910	0.921	0.938	0.968	0.977	1.011	1.001	0.980	0.985

Find $P(X < 0.912)$.

32. Given the following raw data,

2029	12	264	2314	323	358	13126	275
1698	4290	406	1500	2597	180	1723	2134
609	243	1523	2187	2865	153	1061	3952
63	169	7176	1565	828	4492	3104	147

(a) find $P(X < 2187)$

(b) find $P(X > 828)$

(c) find $P(180 < X < 3104)$

(d) find $P(X > 2865)$

33. The following table gives the number of homicides in the state of New York from 1990 to 1995, as presented by the US Department of Justice, Bureau of Justice Statistics.

1969	1919	1820	2092	2228	2166
2013	1958	1786	1683	1907	2016
2244	2246	2605	2571	2397	2420
2016	1550	1353	1093	924	903

What is the probability that the number of homicides will be

(a) between 1200 and 1800?

(b) less than 1500?

(c) greater than 1000?

34. Given the following raw data set,

1010	980	770	1340	1410	570
1020	790	970	1230	910	950
1390	1470	1140	1060	940	1290

find the following probabilities.

(a) $P(600 < X < 1000)$ **(b)** $P(X < 700)$ **(c)** $P(X > 1100)$

(d) $P(X > 1200)$ **(e)** $P(1350 < X < 1510)$ **(f)** $P(X < 900)$

(g) $P(1140 < X < 1400)$

35. An employer wants to promote the top 10% of his staff based on how many hours a week they work. Assume that the number of work hours is normally distributed with an average of 38 hours per week and a standard deviation of 5 hours. What is the minimum number of hours an employee must work to be eligible for a promotion?

36. In order to be accepted into a graduate program, your G.P.A. must be in the top 4% of all applicants. If G.P.A.'s are normally distributed with a mean $\mu = 3.4$ and a standard deviation $\sigma = 0.2$, what is the minimum required GPA for admission?

37. The combined math and verbal scores for students taking the SAT-I test are normally distributed with a mean of 998 and a standard deviation of 202. If a college includes a minimum score of 900 among its requirements, what percentage of students does not satisfy that requirement?

38. IQ scores are normally distributed with a mean of 108 and a standard deviation of 18. Mensa is an organization for people with high IQs, and eligibility requires an IQ above 132.5.

(a) If an applicant is randomly selected, find the probability that he or she meets the Mensa requirement.

(b) In a typical region of 100,000 people, how many would be eligible for Mensa?

39. Scores for the combined math and verbal portions of the SAT-I test are normally distributed with a mean of 1017 and a standard deviation of 207. If the College of Narnia wants to consider only those applicants in the top 42%, what is the minimum required score for admission?

40. In LA country, weights of discarded paper by households each week are normally distributed with a mean of 9.4 lbs. and a standard deviation of 4.2 lbs.

(a) For a randomly selected household, find the probability that the weight of the discarded paper is less than 15 lbs.

(b) How many pounds of discarded paper will put a household in the top 1%?

41. According to data from the College Entrance Examination Board, scores on the SAT-I test have a mean of 1017, and Q_1 is 880. The scores have a distribution that is approximately normal. Find the standard deviation, then use it to calculate P_{98}.

42. A survey indicates that people use their computers an average of 2.4 years before upgrading to a new one. The standard deviation is 0.5 years. If a computer owner is selected at random, find the probability that he or she will upgrade in less than 2 years. Assume that the population is normally distributed.

43. A survey indicates that for each trip to the supermarket, a shopper spends an average of 45 minutes with a standard deviation of 12 minutes shopping. The length of time spent in the store is normally distributed. Find the probability that a randomly selected shopper will be in the store

(a) between 24 and 54 minutes. **(b)** more than 39 minutes. **(c)** less than 50 minutes.

44. Scores for a civil service exam are normally distributed, with a mean of 75 and a standard deviation of 6.5. To be eligible for civil service employment, you must score in the top 5%. What is the lowest score you can earn and still be eligible for employment?

45. In a recent year, the ACT scores for high school students were normally distributed, with a mean of 24.3 and a standard deviation of 4.2. A student who took the ACT is randomly selected. Find the probability that the student's ACT score is

(a) less than 20. **(b)** between 20 and 29. **(c)** more than 29.

46. Suppose the scores of students on a test are normally distributed with mean score of 70 points and a standard deviation of 10 points. If the professor decides to give A's to the top 10 percent of the students and F's to the bottom 53 percent, find the cut off scores to get A's and F's.

47. The mean income of a worker in a professional field is $23,000 with a standard deviation of $1,400. Assuming that the distribution of income is normal, what is the probability that a randomly selected worker will earn less than $22,500?

48. The real estate test scores in Maryland are normally distributed with a mean of 60 and standard deviation of 8. Find the probability of scoring

(a) more than 70. **(b)** between 55 and 65. **(c)** more than 65.

49. The annual salaries of teachers in Georgia are normally distributed with a mean of $35,000 and a standard deviation of $5000. Find the probability that a teacher's annual income in Georgia is

(a) under $42000. **(b)** over $32000.

50. The dollar amount of dairy products consumed per day by teenagers in the US is normally distributed with a mean of $3.50 and a standard deviation of $1.50. Find the probability that a teenager selected at random consumes

(a) over $4.25 in dairy products per day.

(b) less than $5.25 in dairy products per day.

(c) between $2.80 and $4.25 in dairy products per day.

51. The heights of new born babies are normally distributed with a mean of 20 inches and a standard deviation of 2 inches. Find the probability that the height of a randomly selected new born baby is

(a) between 19 and 22 inches. **(b)** more than 18 inches. **(c)** less than 19 inches.

52. A quality control manager claims that the diameters of bolts manufactured must be between 2.08 and 2.10 inches, in order to meet company standards. If the diameter of bolts manufactured are normally distributed with $\mu = 2.09$ inches and $\sigma = 0.005$ inches, what percent of bolts does not meet the company standards?

53. A new brand of flashlight battery is guaranteed to last for at least one year. If the lifetime of these batteries is normally distributed with a mean of 1.6 years and a standard deviation of 0.5 years, what proportion of batteries

(a) will not meet the guarantee? **(b)** will meet the guarantee?

54. The diameters of grapefruits at a farm are normally distributed with a mean of 5 inches and standard deviation of 2 inches. Approximately what percent of grapefruits have diameter

(a) between 3 and 9 inches. **(b)** between 7 and 11 inches.

(c) more than 3 inches. **(d)** more than 7 inches.

(e) less than 7 inches. **(f)** less than 3 inches.

55. The life expectancy of a popular brand of transistor is normally distributed with a mean of 8 years and a standard deviation of 6 months. What percentage of the transistors will last

(a) more than 6.5 years? **(b)** at least 9.5 years?

(c) more than 6.5 but less than 9.5 years?

In exercises 56-67, compute probabilities by using the Normal Approximation to the Binomial Distribution.

56. 3.4 percent of compressor wheels manufactured by Power Automotive are defective.

(a) Estimate the probability that out of 20,000 randomly selected compressor wheels exactly 700 are defective.

(b) Estimate the probability that out of 20,000 randomly selected compressor wheels at most 690 are defective.

57. Find the probability that in 150 rolls of a fair die we will obtain at least 30 sixes.

58. Find the probability that in 200 rolls of a fair die we will obtain at least 40 fives.

59. Among the members of a bowling league, 12% are widowed. If 150 bowling league members are selected at random, find the probability that 10 or more will be widowed.

60. Two percent of hair dryers produced in a certain plant are defective. Estimate the probability that in 10,000 randomly selected hair dryers exactly 225 are defective.

61. Twenty-nine percent of Americans believe that commercial travel to the moon will occur in their lifetime. What is the probability that, among 200 randomly selected Americans, at least 50 believe that commercial moon travel will occur in their lifetime?

62. A machine manufactures a very large number of transistors. Suppose that 10% of the transistors produced by the machine are defective. What is the probability that a random sample of 400 chips will contain 43 defective transistors?

63. The probability that a patient recovers from a delicate surgery is 0.95. Out of 100 randomly selected patients undergoing this surgery, find the probability that

(a) at least 90 patients will recover.

(b) at most 93 patients will recover.

(c) between 92 and 97 (inclusive) patients will recover.

(d) fewer than 94 patients will recover.

(e) exactly 96 patients will recover.

64. If 42% of the households in a city are cable subscribers, find the probability that in a random sample of 500 households in the city

(a) between 200 and 220 are cable subscribers. (Inclusive).

(b) at least 190 are cable subscribers.

(c) exactly 195 are cable subscribers.

65. The probability that a person is left handed is 10%. In a class of 65 students, find the probability that at least 9 students are left handed.

66. It is known that 40% of the homes in a community have DVRs in the master bedroom. If 200 homes are selected at random, find the probability that

(a) at least 65 have DVRs in the master bedroom.

(b) at most 90 have DVRs in the master bedroom.

67. It is reported that 60% of the housewives like to shop at discount stores. Out of 100 randomly selected housewives, find the probability that

(a) exactly 65 shop at discount stores.

(b) more than 55 shop at discount stores.

(c) fewer than 52 shop at discount stores.

6

THE THEORY OF ESTIMATION AND CONFIDENCE INTERVALS

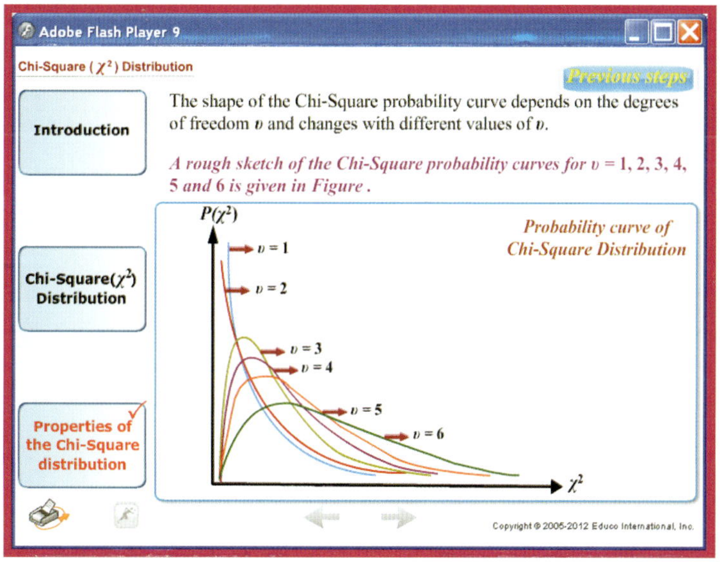

CHAPTER 6

THE THEORY OF ESTIMATION AND CONFIDENCE INTERVALS

Chapter Overview:

1. *Introduction*
2. *Sampling Distribution of the Mean*
3. *Central Limit Theorem and its Applications*
4. *Sampling Distribution of the Proportion and its Properties*
5. *Estimation of the Population Mean μ (σ Known)*
6. *Estimation of the Population Mean μ (σ Unknown)*
7. *Estimation of the Population Proportion p (Large Sample)*
8. *Estimation of the Population Variance and Standard Deviation (σ^2/σ)*
9. *Theory of Estimation with Microsoft Excel®*
10. *Extended Topics and Proofs in the Theory of Estimation (Optional)*

6.1: INTRODUCTION

The main objective in Statistics is to allow a user to make inferences about characteristics of a population (numerical descriptive values called parameters) based on information obtained in a sample.

Traditionally, the study of Inferential Statistics has been divided into two parts:

i) Point Estimation, where we study the values of population parameters such as μ, σ, and p (known as Target Parameters) using corresponding sample statistics (known as Point Estimators) $\bar{x}$, s , and $\hat{p}$.

Recall: $\overline{X}$: sample mean, S : sample standard deviation, and $\hat{p}$: sample proportion or percent.

ii) Hypothesis Testing (also known as Decision Theory), where we must accept or reject assertions about population parameters based on information obtained from a sample.

In this chapter we will focus on the Theory of Estimation. Hypothesis Testing will be presented in Chapter 7.

Suppose we want to estimate the average amount of sugar μ in a can of soda. We can estimate the amount of sugar in two distinct forms. First, we could use a single numerical value, which is close to the unknown population parameter μ, for instance 37 grams. This type of estimate is called a ***point estimate***, because a single value is used to estimate μ.

Definition: *A point estimate is the value of a sample statistic that is used to estimate a population (target) parameter.*

Each sample selected from a population yields a different sample statistics. For instance, the value assigned to a population mean, μ based on a point estimate, depends on choice of sample observations. Thus, the value assigned to μ usually differs from the true value of the population mean.

On the other hand, we may say that μ will fall between two values, for instance 35 and 41 grams. In this form of estimation, two estimated values are used to construct an interval (35, 41) that is intended to capture the parameter of interest (target parameter), in this case the mean amount of sugar in a soda can. Hence, the estimate is called an ***interval estimate*** or a ***confidence interval***. The upper and lower endpoints of a confidence interval are called the **upper** and **lower confidence limits** respectively. In interval estimation, an interval is constructed around the point estimate, and it is understood that this interval is likely to contain the corresponding population parameter. The likelihood (accuracy) of a confidence interval is called a ***confidence level*** or ***confidence coefficient***. Simply put, the confidence level (confidence coefficient) identifies the percentage of time, (in repeated sampling) the constructed interval will contain the target population parameter. If the confidence level is high, (Say 98% or 99%.), then we can be confident that there is a strong chance (probability) the confidence interval will contain the target population parameter. Suppose LCL and UCL denote the lower and upper confidence limits respectively for a target parameter. Then the probability that the confidence interval contains the target parameter is $(1 - \alpha)$, known as the confidence level or confidence coefficient.

$$P(\text{LCL} < \text{Target Parameter} < \text{UCL}) = 1 - \alpha \tag{6.1}$$

The resulting random interval defined by (LCL,UCL) is known as a confidence interval.

Here is a list of some popular confidence levels:

Confidence level	$1-\alpha$	α
90%	0.9	0.1
95%	0.95	0.05
98%	0.98	0.02
99%	0.99	0.01

Remark: **The need for Interval Estimation given point estimates**

Calculating point estimates for population parameters gives us the best single value estimate of that parameter. However, we have no indication of how good our point estimate is and no way of measuring its accuracy. Because of this deficiency, statisticians have developed another type of estimation called Interval Estimation (Confidence Intervals) which is based on a confidence level (degree of accuracy).

The Mathematical Theory of Estimation requires the in-depth study and understanding of Sampling Distributions and the Central Limit Theorem. However, the in-depth study of these topics requires advanced mathematics and is beyond the scope of this text. Therefore, in this section we will briefly mention some basic results and treat these concepts in more detail in Section 6.7. Before discussing the concepts of sampling distributions, we must first define a few terms from sampling theory.

- **Population:** The entire group of individuals or items under study in any statistical investigation is known as the population. For example, if an inquiry is intended to determine the average per capita income of people in a particular city, the population will consist of all the employed people in that city.

- **Sample:** A part of the population, selected to investigate its properties, is called a sample. The number of units in a sample is known as the ***sample size***. **Sampling** is a tool which enables us to draw conclusions about the characteristics of a population based on the information obtained from corresponding samples.

  Simple Random Sampling (S.R.S.) Simple random sampling is the process in which "a sample is so drawn that each and every unit in the population has an equal and independent chance of being included in the sample." If the selection process is carried out with replacement, then it is known as simple random sampling with replacement (**srswr**) and if it is carried out without replacement, it is called simple random sampling without replacement (**srswor**).

- **Parameters:** The characteristics like the mean μ, standard deviation σ, variance σ^2, and population proportion p are known as ***population parameters***.

- **Statistics:** Numerical characteristics of the sample mean $\bar{x}$, standard deviation s, variance s^2, and sample proportion $\hat{p}$ are known as ***sample parameters*** or ***statistics***. They are also defined as functions of sample observations.

	Mean	**Standard Deviation**	**Variance**	**Proportion**
Parameters	μ	σ	σ^2	p
Statistics	$\bar{x}$	s	s^2	$\hat{p}$

Table 6.1

Remark: The point estimates (sample statistics) mentioned above can be calculated from a given sample by using the following formulas.

Sample mean	**Sample Standard Deviation**	**Sample Variance**	**Sample Proportion**
$\bar{x} = \dfrac{\sum x}{n}$	$s = \sqrt{\dfrac{n\sum x^2 - \left(\sum x\right)^2}{n(n-1)}}$	$s^2 = \dfrac{n\sum x^2 - \left(\sum x\right)^2}{n(n-1)}$	$\hat{p} = \dfrac{x}{n}$

Remark: Generally, the population parameters are unknown and their estimates provided by the appropriate sample statistics are used. The sample statistics are functions of the sample observations and vary from sample to sample. For example, the statistics $\bar{x}$ can be written as a function of the sample observations $x_1, x_2, \ldots, x_n$ as: $\bar{x} = \bar{x}(x_1, x_2, \ldots, x_n)$.

Note: We use notations such as $\bar{x}$, s, $\hat{p}$, or s^2 when discussing sample statistics of single samples, and $\bar{X}$, S, $\hat{P}$, or S^2 when discussing sampling distributions, where the sample statistics behaves as a ***random variables***.

6.2: Sampling Distribution of the Mean

Objectives

After completing this section you will be able to understand:

A. Sampling Distribution.

B. Standard Error.

C. Sampling Distribution of the Mean.

D. Properties of the Sampling Distribution of the Mean.

If we draw a simple random sample of size n from a given finite population of size N, then the total number of possible samples is

$C_n^N = \dfrac{N!}{n!(N-n)!} = k$. For each of these k samples we can compute a statistics such as $\bar{x}$, s, s^2, or p.

	Statistics			
Sample Number	$\bar{X}$	S	S^2	$\hat{p}$
1	$\bar{x}_1$	s_1	s_1^2	$\hat{p}_1$
2	$\bar{x}_2$	s_2	s_2^2	$\hat{p}_2$
3	$\bar{x}_3$	s_3	s_3^2	$\hat{p}_3$
4	$\bar{x}_4$	s_4	s_4^2	$\hat{p}_4$
⋮	⋮	⋮	⋮	⋮
k	$\bar{x}_k$	s_k	s_k^2	$\hat{p}_k$

Table 6.2

A. Sampling Distribution

For each of these k samples, we can compute a statistics, in particular, the mean $\bar{x}$, the standard deviation s etc. as given in Table 6.2. The set of the values of the statistics so obtained, one for each sample, constitutes the sampling distribution of that particular statistics. For example, the values $\bar{x}_1, \bar{x}_2, \ldots, \bar{x}_k$ determine the sampling distribution of the statistics $\bar{X}$.

Note: The sampling distribution of a statistic is also called the probability distribution of the corresponding statistic.

B. Standard Error

The standard deviation of the sampling distribution of a statistic is known as its *Standard Error* (SE). The standard error of the sampling distribution of statistic $\bar{X}$ is given by

$$\text{SE}(\bar{X}) = \sqrt{\text{var}(\bar{X})} = \sigma_{\bar{X}}.$$

The standard errors of the sampling distributions of some of the well-known statistics, where n is the sample size, σ is the population standard deviation, p is the population proportion and $q = 1 - p$, are given in Table 6.3.

Statistics	Standard Error
$\bar{X}$	$\dfrac{\sigma}{\sqrt{n}}$
$\hat{P}$	$\sqrt{\dfrac{pq}{n}}$
S	$\dfrac{\sigma}{\sqrt{2n}}$ (for large sample)
S^2	$\sigma^2\sqrt{\dfrac{2}{n}}$ (for large sample)

Table 6.3

> **Note:** The above formulas for the standard error are obtained in random sampling from an infinite population or from a very large population so that the sample size *n* is relatively small when compared to the population size *N* and consequently *n/N* is negligible. Otherwise, the above error formulas will need to be adjusted to compensate for finite population correction factor (See section 6.10.).

C. Sampling Distribution of the Mean

Let's consider a population of 5 members {2, 3, 6, 8 and 11}.

Select all possible samples of size 2 which can be drawn from this population with replacement. Since the samples are drawn with replacement, there will be in all 5 × 5 = 25 samples and a particular digit may occur more than once. Compute the means $\bar{x}_1, \bar{x}_2, ..., \bar{x}_{25}$ for each of the 25 samples. Since the samples are drawn with simple random sampling method, each of the 25 samples has an equal chance (1/25) of being selected. Thus, $\bar{X}$ is a random variable which can take the values $\bar{x}_1, \bar{x}_2, ..., \bar{x}_{25}$ each with probability 1/25. The probability distribution of the sample means is the sampling distribution of the mean $\bar{X}$ and is given in Table 6.4.

Sample observation	Mean	Probability	Sample observation	Mean	Probability
(2, 2)	2.0	$\frac{1}{25}$	(6, 6)	6.0	$\frac{1}{25}$
(2, 3), (3, 2)	2.5	$\frac{2}{25}$	(2, 11), (11, 2)	6.5	$\frac{2}{25}$
(3, 3)	3.0	$\frac{1}{25}$	(3, 11), (11, 3) (6, 8), (8, 6)	7.0	$\frac{4}{25}$
(2, 6), (6, 2)	4.0	$\frac{2}{25}$	(8, 8)	8.0	$\frac{1}{25}$
(3, 6), (6, 3)	4.5	$\frac{2}{25}$	(6, 11), (11, 6)	8.5	$\frac{2}{25}$
(2, 8), (8, 2)	5.0	$\frac{2}{25}$	(8, 11), (11, 8)	9.5	$\frac{2}{25}$
(3, 8), (8, 3)	5.5	$\frac{2}{25}$	(11, 11)	11.0	$\frac{2}{25}$

Table 6.4

> **Note:** The probability distribution of sample means is called the **sampling distribution of the mean.**

D. Properties of the Sampling Distributiocn of the Mean

Sampling Distribution of the Mean has three important properties.

- **Property 1:** Sample mean $\bar{X}$ is an unbiased estimate of the population mean μ, that is $E(\bar{X}) = \mu$, for any sample size.

Note: A statistic is said to be unbiased if its expected value equals to the corresponding population parameter.

- **Property 2:** If we draw simple random samples of size n
  i) from an infinite or very large population or
  ii) from a finite population with replacement (srswr) which amounts to sampling from an infinite population, then

$$\operatorname{var}\left(\bar{X}\right)=\sigma_{\bar{X}}^{2}=\frac{\sigma^{2}}{n} \quad \Rightarrow \quad \operatorname{SD}\left(\bar{X}\right)=\operatorname{SE}\left(\bar{X}\right)=\sigma_{\bar{X}}=\frac{\sigma}{\sqrt{n}}.$$

Note: var $\left(\bar{X}\right)$ is inversely proportional to the sample size n or standard error of estimate SE $\left(\bar{X}\right)$ is inversely proportional to the square root of the sample size. Hence, the larger the sample size, the smaller is the SE $\left(\bar{X}\right)$; and consequently, the more efficient is $\bar{X}$ as an estimate of μ.

- **Property 3:** If a simple random sample of size n is drawn from a normally distributed population, then the sampling distribution of the mean $\bar{X}$ is also normal with mean $\mu_{\bar{X}}$ and standard deviation $\sigma_{\bar{X}}$ given by

$$\mu_{\bar{X}}=\mu \text{ and } \sigma_{\bar{X}}=\frac{\sigma}{\sqrt{n}} \tag{6.2}$$

independent of the size of the sample, that is, $\bar{X}\sim N\left(\mu,\frac{\sigma}{\sqrt{n}}\right)$, for all n.

Example 1: Given the following population, verify the sampling distribution of the mean by listing all samples of size 2 with replacement (to preserve independence), and show that $\mu_{\bar{X}}=\mu$.

The original population

3	5	8	9	12	15

Solution: The mean of the original population $\mu=\dfrac{3+5+8+9+12+15}{6}=\dfrac{52}{6}=8.67$

All samples of size two selected from the original population with replacement and their corresponding means are listed below.

All samples of size 2	Sample means
(3, 3), (3, 5), (3, 8), (3, 9), (3, 12), (3, 15), (5, 3), (5, 5), (5, 8), (5, 9), (5, 12), (5, 15), (8, 3), (8, 5), (8, 8), (8, 9), (8, 12), (8, 15), (9, 3), (9, 5), (9, 8), (9, 9), (9, 12), (9, 15), (12, 3), (12, 5), (12, 8), (12, 9), (12, 12), (12, 15), (15, 3), (15, 5), (15, 8), (15, 9), (15, 12), (15, 15)	(3), (4), (5.5), (6), (7.5), (9), (4), (5), (6.5), (7), (8.5), (10), (5.5), (6.5), (8), (8.5), (10), (11.5), (6), (7), (8.5), (9), (10.5), (12), (7.5), (8.5), (10), (10.5), (12), (13.5), (9), (10), (11.5), (12), (13.5), (15)
Total of all sample means	**312**

Table 6.5

The mean of all the sample means is $\mu_{\bar{X}}=\dfrac{312}{36}=8.67$

Therefore $\mu_{\bar{X}}=\mu$. This verifies that the sample mean is an unbiased (best) estimate of the population mean.

Example 2: A population consists of five numbers (2, 3, 6, 8, 11). Consider all possible samples of size two which can be drawn with replacement from this population.

(a) Find the mean and standard deviation of the sampling distribution.

(b) Compute the mean and standard deviation of the population.

(c) Verify that:

i) sample mean is an unbiased (best) estimate of the population mean $\mu_{\bar{X}} = \mu$.

ii) standard deviation of the sampling distribution $\sigma_{\bar{X}} = \frac{\sigma}{\sqrt{n}}$.

Solutions: Table 6.6 lists all possible random samples of size two taken from the population with replacement, the mean of each sample, and information necessary to compute the standard deviation of the sampling distribution.

Sample Number	Sample values (X)	Sample Mean $(\bar{X})$	$\bar{X} - E(\bar{X})$	$[\bar{X} - E(\bar{X})]^2$
1	(2,2)	2.0	2.0 – 6 = –4.0	16.00
2	(2,3)	2.5	–3.5	12.25
3	(2,6)	4.0	–2.0	4.00
4	(2,8)	5.0	–1.0	1.00
5	(2,11)	6.5	0.5	0.25
6	(3,2)	2.5	–3.5	12.25
7	(3,3)	3.0	–3.0	9.00
8	(3,6)	4.5	–1.5	2.25
9	(3,8)	5.5	–0.5	0.25
10	(3,11)	7.0	1.0	1.00
11	(6,2)	4.0	–2.0	4.00
12	(6,3)	4.5	–1.5	2.25
13	(6,6)	6.0	0	0
14	(6,8)	7.0	1.0	1.00
15	(6,11)	8.5	2.5	6.25
16	(8,2)	5.0	–1.0	1.00
17	(8,3)	5.5	–0.5	0.25
18	(8,6)	7.0	1.0	1.00
19	(8,8)	8.0	2.0	4.00
20	(8,11)	9.5	3.5	12.25
21	(11,2)	6.5	0.5	0.25
22	(11,3)	7.0	1.0	1.00
23	(11,6)	8.5	2.5	6.25
24	(11,8)	9.5	3.5	12.25
25	(11,11)	11.0	5.0	25.00
Total		**150.0**		**135**

Table 6.6

(a) The mean of the sampling distribution of the mean is

$$E(\bar{X}) = \frac{\sum \bar{X}}{25} = \frac{150}{25} = 6$$

The variance of the sampling distribution of the mean is

$$\text{var}(\bar{X}) = \frac{1}{25}\sum\left[\bar{X} - E(\bar{X})\right]^2 = \frac{135}{25} = 5.40 \quad \Rightarrow \quad \text{SD}(\bar{X}) = \sqrt{\text{var}(\bar{X})} = \sqrt{5.40} = 2.32$$

(b) Given the population (2, 3, 6, 8, 11), we have

X	2	3	6	8	11	$\sum X = 30$
X^2	4	9	36	64	121	$\sum X^2 = 234$

Table 6.7

$$\mu = \frac{\sum X}{N} = \frac{30}{5} = 6 \text{ (Population Mean) and}$$

$$\sigma = \sqrt{\frac{N\sum X^2 - \left(\sum X\right)^2}{N^2}} = \sqrt{\frac{5(234) - 30^2}{5^2}} \approx 3.286 \text{ (Standard Deviation)}$$

(c) **i)** We can see that $E(\bar{X}) = \mu_{\bar{X}} = 6 = \mu$. Hence, sample mean is unbaised estimate of the population mean.

ii) $\text{SD}(\bar{X}) = \sigma_{\bar{X}} = \frac{\sigma}{\sqrt{n}} \approx \frac{3.286}{\sqrt{2}} \approx 2.32$; therefore, we have verified that $\sigma_{\bar{X}} = \frac{\sigma}{\sqrt{n}}$.

Remark: Sampling Distribution without replacement: If we have to draw samples without replacement, then we shall exclude those samples where the values are duplicated namely the samples (2, 2), (3, 3), (6, 6), (8, 8), and (11, 11). In this case the variance of the sampling distribution of $\bar{X}$ is

$$\text{SD}(\bar{X}) = \sigma_{\bar{X}} = \sqrt{\frac{\sigma^2}{n}\left(\frac{N-n}{N-1}\right)}. \qquad (6.3)$$

Example 3: A population of 400 has a mean of 50 and standard deviation of 5. If samples of size 30 are randomly selected without replacement, what is the mean and standard deviation of the distribution of all sample means?

Solution: The mean of the sampling distribution of the means is $\mu_{\bar{X}} = \mu = 50$.

The standard deviation of the sampling distribution of the means is

$$\sigma_{\bar{X}} = \frac{5}{\sqrt{30}}\sqrt{\frac{400-30}{400-1}} \approx 0.8791$$

6.3: Central Limit Theorem and its Applications

A. Central Limit Theorem

Objectives

After completing this section you will be able to:

A. Interpret and apply the Central Limit Theorem (CLT).

B. Normalize (Standardize) the random variable $\bar{X}$.

Sampling Distribution of Means (Large Samples)

Property 3 of the sampling distribution of means states that if $X_1, X_2, \ldots, X_n$ is a random sample from a normal population with mean μ and variance σ^2, then the sample mean $\bar{X}$ is also normally distributed with mean μ and standard deviation $\sigma/\sqrt{n}$, that is

$$\bar{X} \sim N\left(\mu, \frac{\sigma}{\sqrt{n}}\right). \qquad (6.4)$$

This result is true even if the population from which the samples are drawn is not normal, provided the sample size is sufficiently large ($n > 30$). This result is known as **Central Limit Theorem (CLT)**.

Note: The CLT applies to infinite populations, also to finite populations when n is sufficiently large. We cannot say exactly how large n must be so that the CLT can be applied; however in most applications $n = 30$ is usually regarded as sufficiently large.

Assume all samples of size n are selected from any population with mean μ and standard deviation σ.

1. If $n \geq 30$, then the sampling distribution of the Means can be approximated by a normal distribution with mean μ and standard deviation $\sigma/\sqrt{n}$. Formally speaking, for sufficiently large n, $\bar{X} \sim N\left(\mu_{\bar{X}} = \mu, \sigma_{\bar{X}} = \frac{\sigma}{\sqrt{n}}\right)$.
2. If $n < 30$ and the original population has a normal distribution, then the sampling distribution of the means also follows normal distribution with mean μ and standard deviation $\sigma/\sqrt{n}$.
3. If $n < 30$ and the original population is not normally distributed, then we can not say anything about the distribution of the sample means.

B. Normalization (Standardization) of $\bar{X}$

The Central Limit Theorem states that for large samples the sampling distribution of the means can be approximated with a normal distribution. Formally speaking, if $\bar{X}$ is the mean of a sufficiently large random sample of size n from a large population with mean μ and standard deviation σ, then $Z_{\bar{X}} = \frac{\bar{X} - \mu_{\bar{X}}}{\sigma_{\bar{X}}} = \frac{\bar{X} - \mu}{\sigma/\sqrt{n}}$ has approximately a standard normal distribution.

$$Z_{\bar{X}} = \frac{\bar{X} - \mu_{\bar{X}}}{\sigma_{\bar{X}}} = \frac{\bar{X} - \mu}{\sigma/\sqrt{n}} \sim N(0,1)$$

Remark: It is important to remember that, when working with an individual value x from a normally distributed population, we use the standardization formula $z = \frac{x - \mu}{\sigma}$; however, when working with the mean of a sample from a population, then we must use the standardization formula $Z_{\bar{X}} = \frac{\bar{X} - \mu}{\sigma/\sqrt{n}}$.

Example 4: It is known that the average national IQ (Intelligent Quotient) is about 100 with standard deviation 15. Assuming that the population of all IQ's is normally distributed, find

(a) the probability that if a person is randomly selected from the population, his or her IQ will be greater than 118.

(b) the probability that 20 randomly selected people will have an average IQ that is greater than 118.

Solutions: **(a)** $P(X > 118) = ?$

$$P(X > 118) = P\left(Z > \frac{118-100}{15}\right)$$

$$= P(Z > 1.2) = 0.1151$$

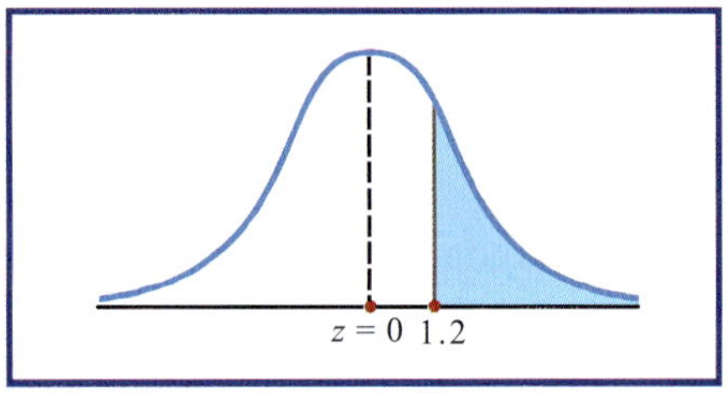

Figure 6.1

The probability that a randomly selected person from the population will have an IQ greater than 118 is about 12%.

(b) $P(\bar{X} > 118) = ?$

$$P(\bar{X} > 118) = P\left(Z > \frac{118-100}{15/\sqrt{20}}\right)$$

$$= P(Z > 5.37) \approx 0.000$$

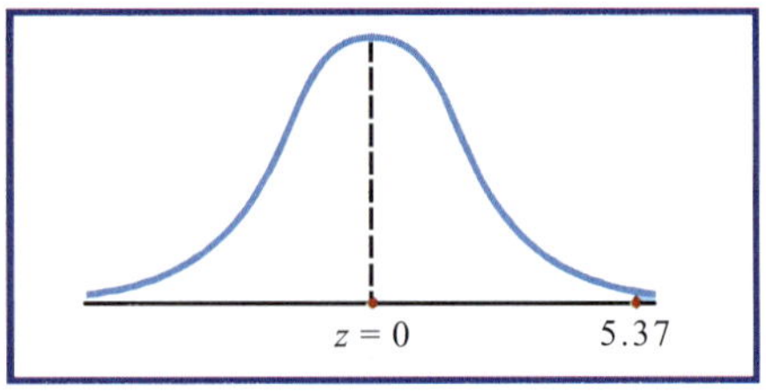

Figure 6.2

The probability that 20 randomly selected people from the population will have an average IQ greater than 118 is approximately 0%.

Example 5: A random sample of size 35 is taken from a large population with $\mu = 98$ and $\sigma = 15$. Find the probability that the sample mean will be

(a) greater than 102. **(b)** less than 96. **(c)** between 96.5 and 99.8.

Solutions: Since the sample size $n = 35$ is greater than 30, by the Central Limit Theorem, the sampling distribution of the mean ($\bar{X}$) is approximately normal with $\mu_{\bar{X}} = \mu = 98$ and

$$\sigma_{\bar{X}} = \frac{\sigma}{\sqrt{n}} = \frac{15}{\sqrt{35}} \approx 2.54.$$

(a) $$P(\bar{X} > 102) = P\left(Z > \frac{102-98}{2.54}\right) = P(Z > 1.57) = 0.5 - 04418 = 0.0582 \approx 6\%$$

(b) $$P(\bar{X} < 96) = P\left(Z < \frac{96-98}{2.54}\right) = P(Z < -0.79)$$

$$= P(Z > 0.79) \quad \text{(Using symmetry)}$$

$$= 0.5 - P(0 < Z < 0.79) = 0.5 - 0.2852 = 0.2148 \approx 21\%$$

(c) $P(96.5<\bar{X}<99.8) = P\left(\frac{96.5-98}{2.54}<Z<\frac{99.8-98}{2.54}\right)$

$= P(-0.59 < Z < 0.71) = P(0 < Z < 0.59) + P(0 < Z < 0.71)$

$= 0.2224 + 0.2611 = 0.4835$

Example 6: There is a large supply of light bulbs with an average life of 900 hours and a standard deviation of 60 hours.

(a) A light bulb is drawn at random. Find the probability that it will last for more than 920 hours. (Assume normal distribution).

(b) A random sample of 36 light bulbs is drawn. Find the probability that the sample average is more than 920 hours.

Solutions: **(a)** We have $X \sim N(900, 60)$.

$$P(X>920) = P\left(Z>\frac{920-900}{60}\right)$$

$$= P(Z > 0.33)$$

$$= 0.5 - P(0 < Z < 0.33)$$

$$= 0.5 - 0.1293 = 0.3707$$

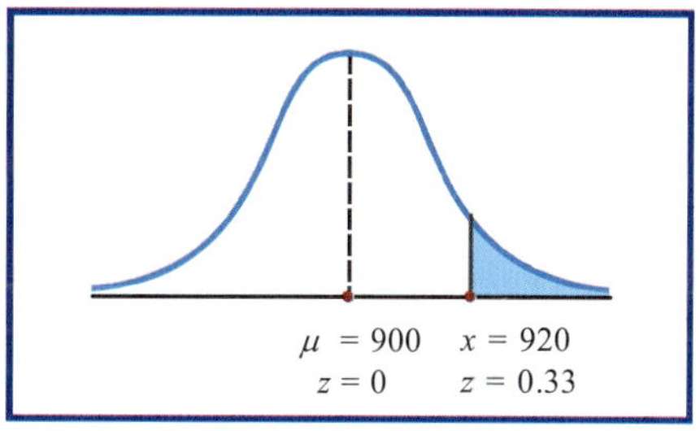

Figure 6.3

Thus, the probability that a light bulb drawn at random will last more than 920 hours is approximately 37%.

(b) The sample size $n>30$, hence by the CLT, $\bar{X} \sim N(\mu_{\bar{X}},\sigma_{\bar{X}})$ where $\mu_{\bar{X}} = \mu = 900$ hours, and $\sigma_{\bar{X}} = \frac{\sigma}{\sqrt{n}} = \frac{60}{\sqrt{36}} = 10$ hours.

$$P(\bar{X}>920) = P\left(Z>\frac{920-900}{10}\right)$$

$$= P(Z > 2.00)$$

$$= 0.5 - P(0 < Z < 2.00) = 0.5 - 0.4772 = 0.0228$$

Thus, the probability that the average life of a sample of 36 light bulbs is more than 920 hours is approximately 2%.

Example 7: Find the 95th percentile for the sampling distribution of the means when samples of size 64 are selected at random from a large population with $\mu = 120$ and $\sigma = 16$.

Solution: Since $n > 30$, by the CLT $\bar{X} \sim N\left(120,\frac{16}{\sqrt{64}}\right)$

$P_{95} = 95^{th}$ percentile of the sampling distribution of $\bar{X} \Rightarrow P(\bar{X} < P_{95}) = 0.95$.

When $\bar{X} = P_{95}$, $Z_{\bar{X}} = \frac{\bar{X}-\mu}{\sigma/\sqrt{n}} = \frac{P_{95}-120}{16/\sqrt{64}} = z_1$ (6.5)

$P(Z < z_1) = 0.95 \Rightarrow P(0 < Z < z_1) = 0.45$

$z_1 = \dfrac{1.64 + 1.65}{2} = 1.645$ (From z-Table)

Substituting $z_1 = 1.645$ into (6.5), we get

$\dfrac{P_{95} - 120}{16/\sqrt{64}} = 1.645 \Rightarrow P_{95} = 120 + 2(1.645) = 123.29.$

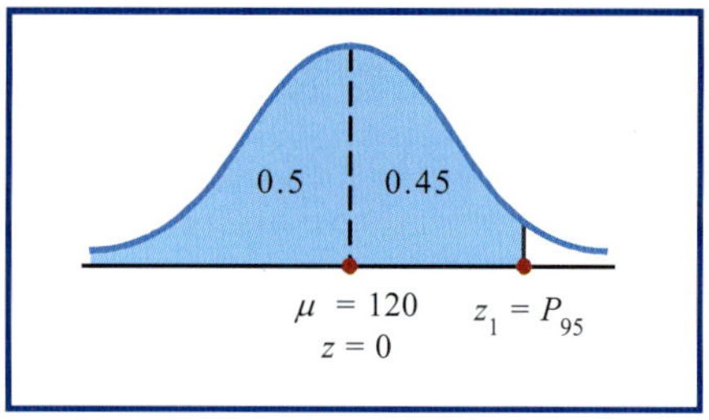

Figure 6.4

Thus, the 95th percentile of the sampling distribution of means of all samples of size 64 is approximately 123.29.

Try it yourself:

1. Given the population {1, 3, 5, 7, 9} and sample size $n = 2$ (samples are drawn with replacement)

(a) Find

(i) the mean μ and standard deviation σ of the population.

(ii) the sampling distribution of the sample mean $\bar{X}$.

(iii) the mean $\mu_{\bar{X}}$ and standard deviation $\sigma_{\bar{X}}$ of the sampling distribution of means.

(b) Show that $\mu_{\bar{X}} = \mu$ and $\sigma_{\bar{X}} = \dfrac{\sigma}{\sqrt{n}}$.

In exercises 2 and 3, find the mean $\mu_{\bar{X}}$ and the standard error $\sigma_{\bar{X}}$ of the distribution of all sample means.

2. A population of 100 has a mean of 20 and standard deviation of 5. All random samples of size 25 are selected.

3. All random samples of size 20 are selected from a large population with mean $\mu = 26.5$ and standard deviation $\sigma = 5.28$.

4. Given that $\mu = 500$, $\sigma = 30$, $N = 200$, and $n = 25$, find $P(\bar{X} > 520)$.

5. IQ examination scores of 400 members of a club are normally distributed with a mean of 160 and standard deviation of 13. If a random sample of 36 club members is selected, find the probability that the mean IQ of the sample is less than 150.

6. In a community college, the Statistics test scores of a large population are normally distributed with a mean of 74 and a standard deviation of 6. Find the probability that the mean score of a random sample of size 8 will lie between 55 and 71.

7. The IQ's of students at a college have a mean of 106 and a standard deviation of 18. Find Q_1 and Q_3 of the sampling distribution of the means of all samples of size 80.

Note: The original population of IQ's need not be normally distributed since the sample size is well over 30.

6.4: Sampling Distribution of the Proportion and its Properties

A. Sampling Distribution of the Proportion

Objectives

After completing this section you will be able to understand and apply:

A. Sampling Distribution of the Proportion.

B. Properties of the Sampling Distribution of $\hat{p}$.

Suppose we have a binomial experiment so that the following conditions are satisfied:

i) The number of trials n is fixed.

ii) All the trials are independent.

iii) Each trial results in 2 mutually exclusive outcomes, success and failure.

iv) The probability of success p is a constant for each trial, $q = 1 - p$ is the probability of failure (complement of probability of success).

Let X be the observed number of successes in a simple random sample from the population with proportion of successes p. We introduce the notation $\hat{p}$, read as **p-hat**, for the sample proportion.

Notation:

p = Population proportion (percentage)

$\hat{p}$ (Observed proportion of successes in the sample of size n) $= \dfrac{x}{n}$

$\hat{q} = 1 - \hat{p} =$ Proportion of failures in a sample of size n

Suppose we randomly select independent samples of size n from a large population and compute the sample proportion $\hat{p}$ of successes for each of these samples. The set of values of these sample proportions gives the sampling distribution of the proportion $\hat{p}$.

We present below, few important properties of the sampling distribution of sample proportion $\hat{p}$ when sampling is from an infinite population or it is a simple random sampling with replacement.

B. Properties of the Sampling Distribution of $\hat{p}$

- **Property 1:** The mean $\mu_{\hat{p}}$ of the sampling distribution of proportions $\hat{p}$ is given by $\mu_{\hat{p}} = p$, where p is the proportion of successes in the population.

- **Property 2:** The standard deviation $\sigma_{\hat{p}}$ of the sampling distribution of proportion $\hat{p}$ is given by

$$\sigma_{\hat{p}} = \sqrt{\frac{p(1-p)}{n}} = \sqrt{\frac{pq}{n}} \qquad (6.6)$$

- **Property 3:** If all large samples of size n are taken from a large population, the distribution of the sample proportions $\hat{p}$ is approximately normal with mean $\mu_{\hat{p}}$ and standard deviation $\sigma_{\hat{p}}$. That is, for large n

$$\hat{p} \quad N\left(\mu_{\hat{p}}, \sigma_{\hat{p}}\right) = N\left(p, \sqrt{\frac{pq}{n}}\right) \tag{6.7}$$

Example 8: A population consists of 6 numbers (3, 5, 8, 9, 12, 15). Consider all possible samples of size 2, which can be drawn with replacement from this population.

(a) Compute the proportion (probability) of getting an even number from the population (3, 5, 8, 9, 12, 15).

(b) List all possible samples of size 2 which can be drawn with replacement from this population, and calculate the probability of getting an even number for each sample.

(c) Show that $\hat{p}$ is an unbiased estimate of the population proportion- *i.e.*, $E(\hat{p}) = p$.

(d) Compute var $(\hat{p})$.

Solutions: **(a)** $p = P(\text{even number}) = \frac{2}{6} \approx 0.333$

(b)

Sample	P**(even number)** $= \hat{p}$	**Sample**	P**(even number)** $= \hat{p}$	**Sample**	P**(even number)** $= \hat{p}$
(3, 3)	0	(8, 3)	0.5	(12, 3)	0.5
(3, 5)	0	(8, 5)	0.5	(12, 5)	0.5
(3, 8)	0.5	(8, 8)	1	(12, 8)	1
(3, 9)	0	(8, 9)	0.5	(12, 9)	0.5
(3, 12)	0.5	(8, 12)	1	(12, 12)	1
(3, 15)	0	(8, 15)	0.5	(12, 15)	0.5
(5, 3)	0	(9, 3)	0	(15, 3)	0
(5, 5)	0	(9, 5)	0	(15, 5)	0
(5, 8)	0.5	(9, 8)	0.5	(15, 8)	0.5
(5, 9)	0	(9, 9)	0	(15, 9)	0
(5, 12)	0.5	(9, 12)	0.5	(15, 12)	0.5
(5, 15)	0	(9, 15)	0	(15, 15)	0

Table 6.8

(c) $E(\hat{p}) = \mu_{\hat{p}} = \frac{\sum \hat{p}}{36} = \frac{12}{36} = 0.333 \Rightarrow E(\hat{p}) = p$

(d) $\text{var}(\hat{p}) = \frac{n\sum \hat{p}^2 - \left(\sum \hat{p}\right)^2}{n^2}$, where $\sum \hat{p} = 12$, $\sum \hat{p}^2 = 8$ and $n = 36$

$$= \frac{36 \times 8 - (12)^2}{36^2} = 0.111$$

We can also calculate $E(\hat{p})$ and the variance by constructing the sampling distribution of proportions.

$$E(\hat{p}) = \sum XP = 0.333$$

$$\text{var}(\hat{p}) = E(\hat{p}^2) - E^2(\hat{p})$$

$$= \sum XP^2 - \left(\sum XP\right)^2$$

$$= 0.222 - 0.111$$

$$= 0.111$$

X	f	$P = \frac{f}{n}$	$X \cdot P$	$X^2 \cdot P$
0	16	0.4444	0	0
0.5	16	0.4444	0.2222	0.1111
1	4	0.1111	0.1111	0.1111
Totals	$n = 36$		**0.3333**	**0.2222**

Table 6.9

Example 9: A random sample of size 400 is drawn from a population where the proportion (probability) of success $p = 0.30$.

(a) What is the mean and standard deviation of the sampling distribution of proportion $\hat{p}$?

(b) What is the probability that the proportion of successes in the sample will be greater than 0.27?

Solutions: **(a)** The mean of sampling distribution of $\hat{p}$ is given by $\mu_{\hat{p}} = p = 0.30$.

The standard deviation of sampling distribution of $\hat{p}$ is

$$\sigma_{\hat{p}} = \sqrt{\frac{p(1-p)}{n}} = \sqrt{\frac{(0.30)(0.70)}{400}} \approx 0.023.$$

(b) Since the sample size $n = 400 > 30$, then $Z = \frac{\hat{p} - \mu_{\hat{p}}}{\sigma_{\hat{p}}} = \frac{\hat{p} - 0.30}{0.023} \sim N(0, 1)$.

Therefore,

$$P(\hat{p} > 0.27) = P\left(Z > \frac{0.27 - 0.30}{0.023}\right)$$

$$= P(Z > -1.3)$$

$$= P(-1.3 < Z < 0) + 0.5 \quad \text{Figure 6.5}$$

$$= P(0 < Z < 1.3) + 0.5 \quad \text{By Symmetry}$$

$$= 0.4032 + 0.5 \quad \text{(From } z\text{-Table)}$$

$$= 0.9032$$

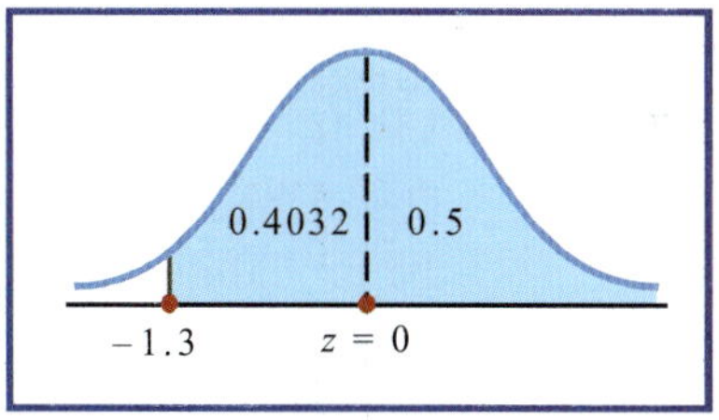

Figure 6.5

Thus, the probability that the proportion of successes in the sample will be greater than 0.27 is 0.9032.

Example 10: If a random sample of 400 people is selected from the adult population of a big city, what is the probability that less than 50% of the sample favor the new property tax rates, if the true proportion favoring the new tax rates for the entire population is 0.48?

Solution: We are given $n = 400$, $p = 0.48 \Rightarrow q = 1 - p = 0.52$.

Since n is large, $\hat{p} \sim N(\mu_{\hat{p}}, \sigma_{\hat{p}})$ where $\mu_{\hat{p}} = p = 0.52$, and $\sigma_{\hat{p}} = \sqrt{\frac{pq}{n}} = \sqrt{\frac{(0.48)(0.52)}{400}} = 0.025$.

$$
\begin{aligned}
P(\hat{p} < 0.5) &= P\left(Z < \frac{0.5 - 0.48}{0.025}\right) \\
&= P(Z < 0.8) \\
&= 0.5 + P(0 < Z < 0.8) \\
&= 0.5 + 0.2881 = 0.7881
\end{aligned}
$$

Thus, the probability that less than 50% of the sample favor the new property tax rates in the city is 0.7881.

Try it yourself:

1. Given p = 0.34 and n = 500, compute the standard deviation of the sampling distribution of $\hat{p}$.

2. A random sample of 500 is drawn from a population with the proportion of successes $p = 0.28$. Find the probability that the proportion $\hat{p}$ of successes in the sample is between 0.24 and 0.32.

3. A cable company believes that 30% of the households in a large community subscribe to TV cable services. A random sample of 500 households is obtained and the sample proportion is computed. Find the probability that the sample proportion is

   (a) greater than 0.33. **(b)** less than 0.34. **(c)** between 0.28 and 0.40.

4. An electronic goods retailer receives a shipment of 200 portable Blu-ray players. The manufacturer claims that only 2% of the Blu-ray players are defective. Find the probability that the shipment contains 14 or more defective Blu-ray players.

Remark: Sampling Distribution states that when it is possible to list **all** samples of size "***n***" from a population, the mean and standard deviation of the population can be determined by using the sample means ($\bar{x}$'s), the sample size, and the standard deviation of all the sample means $\sigma_{\bar{x}}$.

$$\mu = \mu_{\bar{x}} \text{ and } \sigma = \sigma_{\bar{x}} \cdot \sqrt{n}$$

However, in application and research, it is usually impossible to list all the samples of size ***n*** from a population. Hence, statistical procedures were developed to carry out the task of estimating population parameters (target parameters) using information from a single randomly selected sample. The *Central Limit Theorem* and the *Theory of Estimation* are two such procedures. We have already discussed the Central Limit Theorem, so the rest of this chapter will be dedicated to presenting the Theory of Estimation.

6.5: Estimating the Population Mean μ (σ Known)

In this section, we will present the requirements and procedures for using sample data to find a point estimate and Confidence Interval Estimate of the population mean μ, when the population standard deviation σ is known. We will also present a method for finding the sample size required to estimate the mean of a population. Requirements and conditions to use a *normal distribution* for constructing a confidence interval for μ are:

Objectives

After completing this section you will be able to:

- **A.** Obtain $100(1-\alpha)\%$ Confidence Interval for μ, when σ is known.
- **B.** Interpret Confidence Interval.
- **C.** Obtain Sample Size for Estimating the Population Mean.

Case i) If the population standard deviation σ is known, the sample size is large ($n \geq 30$), and we have no information about the distribution of the parent population (the population from which the sample is selected), then we can use a normal distribution to construct a **confidence interval** for μ, since due to the Central Limit Theorem, the Sampling Distribution of $\overline{X}$ is approximately normal with a mean μ and standard deviation $\sigma/\sqrt{n}$.

Note: In this case, although the population is not necessarily normally distributed, the fact that the sample size is sufficiently large, it compensates for the assumption that distribution of $\overline{X}$ is approximately normal.

Case ii) If the population standard deviation σ is known, the sample size is small ($n < 30$), and the parent population is normally distributed, then we can use a normal distribution to construct a **confidence interval** for μ.

Note: The sampling distribution of $\overline{X}$ is normal with a mean μ and standard deviation $\sigma/\sqrt{n}$.

Remark: If the population standard deviation σ is unknown, the sample size ($n < 30$), and the distribution of the parent population is unknown, then we must use a nonparametric method to construct a confidence interval for μ. Non-parametric Statistics will be discussed in Chapter 11.

The following diagram summarizes the above conditions and requirements.

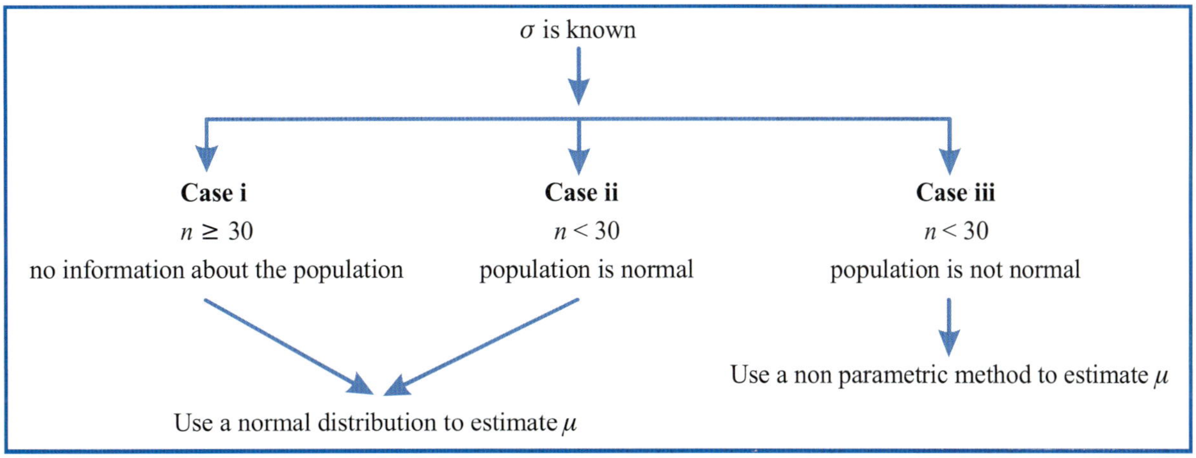

Figure 6.6

A. Construction of $100(1 - \alpha)\%$ Confidence Interval for μ when σ Known

1. Let $\overline{X}$ be the point estimate (statistic) that is normally distributed with mean μ and standard deviation $\dfrac{\sigma}{\sqrt{n}}$.

2. Let E be the Margin of Error. $E = z_{\alpha/2} \cdot \dfrac{\sigma}{\sqrt{n}}$

3. Let $1 - \alpha$ be the probability that the confidence interval contains the target parameter μ. Then, the $100(1 - \alpha)\%$ ***confidence interval*** for the mean of a population is

$$\overline{x} - z_{\alpha/2}\frac{\sigma}{\sqrt{n}} < \mu < \overline{x} + z_{\alpha/2}\frac{\sigma}{\sqrt{n}} \quad \text{or} \quad \overline{x} - E < \mu < \overline{x} + E$$

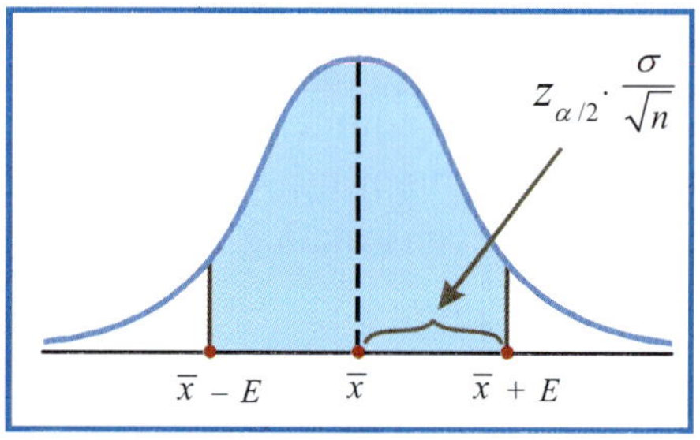

Figure 6.7

Derivation of the Confidence Interval Formula for the Mean μ

We want $P\left(\left|\overline{X} - \mu\right| < E\right) = 1 - \alpha$ as shown below.

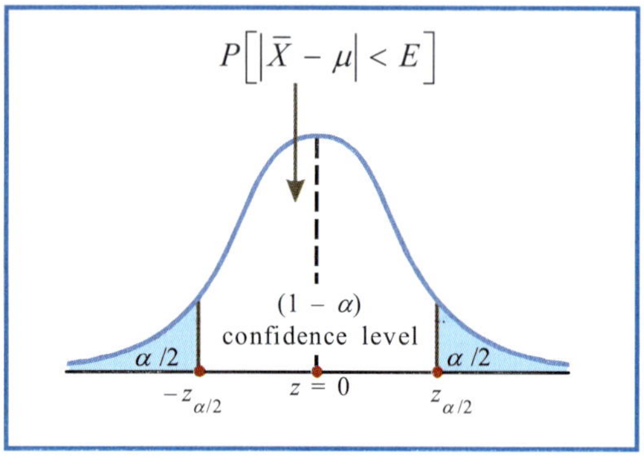

Figure 6.8

From Figure (6.9 a), we can see that

$$P\left(-z_{\alpha/2} < Z < z_{\alpha/2}\right) = 1 - \alpha \qquad (6.8)$$

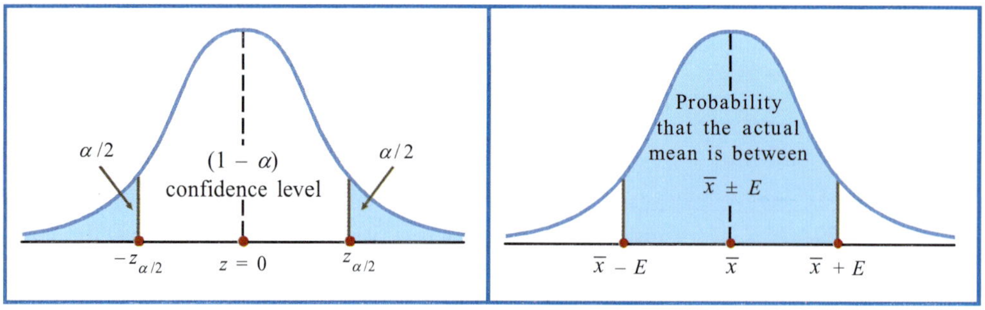

Figure 6.9 a

Figure 6.9 b

Recall that

$$Z = \frac{\overline{X} - \mu}{\sigma / \sqrt{n}} \qquad (6.9)$$

Substituting (6.9) into (6.8), we have

$$P\left(-z_{\alpha/2} < \frac{\overline{X} - \mu}{\sigma/\sqrt{n}} < z_{\alpha/2}\right) = 1 - \alpha \tag{6.10}$$

Solving the inequality in (6.10) for μ results in:

$$\begin{aligned} P\left(-z_{\alpha/2} < \frac{\overline{X} - \mu}{\sigma/\sqrt{n}} < z_{\alpha/2}\right) &= P\left(-z_{\alpha/2}\frac{\sigma}{\sqrt{n}} < \overline{X} - \mu < z_{\alpha/2}\frac{\sigma}{\sqrt{n}}\right) \\ &= P\left(-\overline{X} - z_{\alpha/2}\frac{\sigma}{\sqrt{n}} < -\mu < -\overline{X} + z_{\alpha/2}\frac{\sigma}{\sqrt{n}}\right) \\ &= P\left(\overline{X} + z_{\alpha/2}\frac{\sigma}{\sqrt{n}} > \mu > \overline{X} - z_{\alpha/2}\frac{\sigma}{\sqrt{n}}\right) \\ &= P\left(\overline{X} - z_{\alpha/2}\frac{\sigma}{\sqrt{n}} < \mu < \overline{X} + z_{\alpha/2}\frac{\sigma}{\sqrt{n}}\right) \end{aligned}$$

Therefore

$$P\left(\overline{X} - z_{\alpha/2}\frac{\sigma}{\sqrt{n}} < \mu < \overline{X} + z_{\alpha/2}\frac{\sigma}{\sqrt{n}}\right) = 1 - \alpha \tag{6.11}$$

B. Interpretation of the Confidence Interval

If we repeatedly take independent random samples of the same size from a normally distributed population, approximately $100(1 - \alpha)\%$ of times, the population mean will fall within the upper and lower limits of the **confidence interval**. For instance, assume that a 95% confidence interval for the mean IQ of a population is between 98 and 107 ($98 < \mu < 107$). This means that in repeated sampling approximately 95% of the time (95 out of every 100 attempts) the population mean μ will fall between 98 and 107. Mathematically speaking, $P(98 < \mu < 107) = 0.95$

In this example the ***confidence level*** $100(1 - \alpha)\% = 95\% \Rightarrow 1 - \alpha = 0.95 \Rightarrow \alpha = 0.05$

Note: Since the most common confidence level in industry and research is 95%, we will use a 95% confidence level as our default probability unless otherwise stated.

Round-Off Rule for Confidence Intervals Used to Estimate μ

1. When using raw data to construct a confidence interval for μ, round the confidence interval limits to one more decimal place than is used for the original raw data.
2. When raw data is not given and summary statistics are used instead, round the confidence interval limits to the same number of decimal places used for the sample mean $\overline{x}$.

Note: Summary Statistics include $\overline{x}$, s, and n.

Steps for Constructing $100(1-\alpha)\%$ Confidence Interval for μ (σ Known)

$$\bar{x} - z_{\alpha/2}\frac{\sigma}{\sqrt{n}} < \mu < \bar{x} + z_{\alpha/2}\frac{\sigma}{\sqrt{n}}$$

***Step* 1** From the given confidence level (confidence coefficient) $100(1-\alpha)\%$, we must first find

$$\frac{1-\alpha}{2} = \frac{CL}{2}.$$

Once the area $\frac{1-\alpha}{2} = \frac{CL}{2}$ is calculated, we can then (using the z-Table) find the corresponding z value, (labelled as $z_{\alpha/2}$) so that $P(0 < Z < z_{\alpha/2}) = \frac{1-\alpha}{2}$.

> ***Remark:*** The notation $z_{\alpha/2}$ means that the area to the right of the z value is $\frac{\alpha}{2}$.
>
> 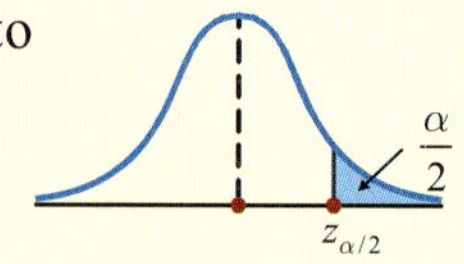
>

Below are some common confidence levels with their corresponding critical values $z_{\alpha/2}$.

Confidence Coefficient $100(1-\alpha)\%$	90%	95%	98%	99%
Decimal form of the Confidence Level (CL)	0.90	0.95	0.98	0.99
$\frac{1-\alpha}{2} = \frac{CL}{2}$	$\frac{0.9}{2} = 0.4500$	$\frac{0.95}{2} = 0.4750$	$\frac{0.98}{2} = 0.4900$	$\frac{0.99}{2} = 0.4950$
Critical value $z_{\alpha/2}$	1.645	1.96	2.33	2.575

Table 6.10

***Step* 2** Compute the Margin of Error (E) using $z_{\alpha/2}$, σ, and n.

$$E = z_{\alpha/2} \cdot \sigma_{\bar{x}} = z_{\alpha/2}\frac{\sigma}{\sqrt{n}}$$

***Step* 3** $100(1-\alpha)\%$ confidence interval for μ is

$$\bar{x} - E < \mu < \bar{x} + E \quad \text{or} \quad \bar{x} - z_{\alpha/2}\frac{\sigma}{\sqrt{n}} < \mu < \bar{x} + z_{\alpha/2}\frac{\sigma}{\sqrt{n}}.$$

Example 11: Given a 95% confidence level, compute $z_{\alpha/2}$.

Solution: Given confidence level $= 1 - \alpha = 0.95$, $P(0 < Z < z_{\sigma/2}) = \frac{1-\alpha}{2} = 0.475$.

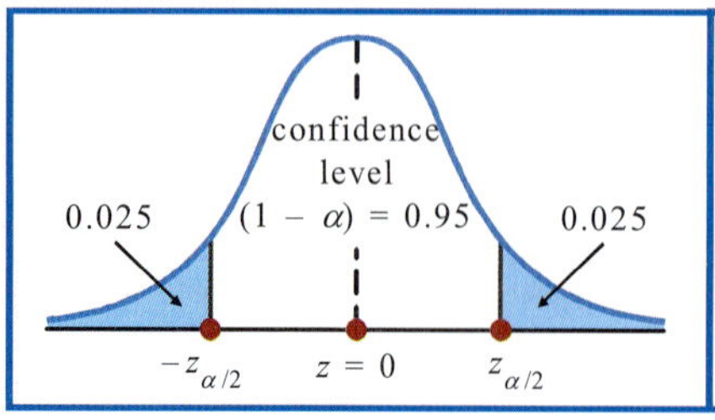

Figure 6.10 a

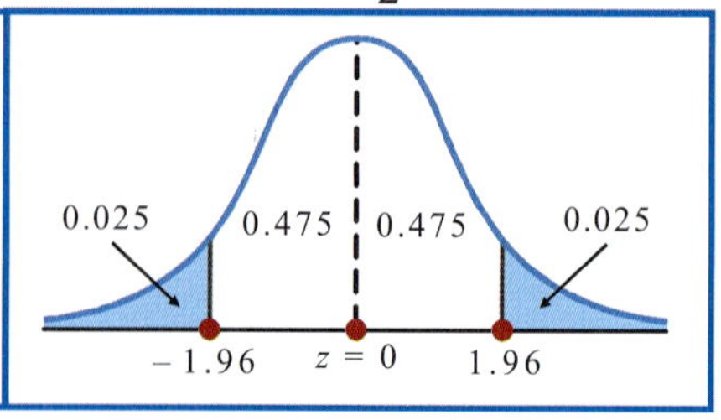

Figure 6.10 b

Using the z-table, we can easily read the value of $z_{\alpha/2}$ to be 1.96.

Z	0	0.01	0.02	0.03	0.04	0.05	0.06	0.07	0.08	0.09
0	0.0000	0.0040	0.0080	0.0120	0.0160	0.0199	0.0239	0.0279	0.0319	0.0359
0.1	0.0398	0.0438	0.0478	0.0517	0.0557	0.0596	0.0636	0.0675	0.0714	0.0753
0.2	0.0793	0.0832	0.0871	0.0910	0.0948	0.0987	0.1026	0.1064	0.1103	0.1141
0.3	0.1179	0.1217	0.1255	0.1293	0.1331	0.1368	0.1406	0.1443	0.1480	0.1517
0.4	0.1554	0.1591	0.1628	0.1664	0.1700	0.1736	0.1772	0.1808	0.1844	0.1879
0.5	0.1915	0.1950	0.1985	0.2019	0.2054	0.2088	0.2123	0.2157	0.2190	0.2224
0.6	0.2257	0.2291	0.2324	0.2357	0.2389	0.2422	0.2454	0.2486	0.2517	0.2549
0.7	0.2580	0.2611	0.2642	0.2673	0.2704	0.2734	0.2764	0.2794	0.2823	0.2852
0.8	0.2881	0.2910	0.2939	0.2967	0.2995	0.3023	0.3051	0.3078	0.3106	0.3133
0.9	0.3159	0.3186	0.3212	0.3238	0.3264	0.3289	0.3315	0.3340	0.3365	0.3389
1	0.3413	0.3438	0.3461	0.3485	0.3508	0.3531	0.3554	0.3577	0.3599	0.3621
1.1	0.3643	0.3665	0.3686	0.3708	0.3729	0.3749	0.3770	0.3790	0.3810	0.3830
1.2	0.3849	0.3869	0.3888	0.3907	0.3925	0.3944	0.3962	0.3980	0.3997	0.4015
1.3	0.4032	0.4049	0.4066	0.4082	0.4099	0.4115	0.4131	0.4147	0.4162	0.4177
1.4	0.4192	0.4207	0.4222	0.4236	0.4251	0.4265	0.4279	0.4292	0.4306	0.4319
1.5	0.4332	0.4345	0.4357	0.4370	0.4382	0.4394	0.4406	0.4418	0.4429	0.4441
1.6	0.4452	0.4463	0.4474	0.4484	0.4495	0.4505	0.4515	0.4525	0.4535	0.4545
1.7	0.4554	0.4564	0.4573	0.4582	0.4591	0.4599	0.4608	0.4616	0.4625	0.4633
1.8	0.4641	0.4649	0.4656	0.4664	0.4671	0.4678	0.4686	0.4693	0.4699	0.4706
1.9	0.4713	0.4719	0.4726	0.4732	0.4738	0.4744	**0.4750**	0.4756	0.4761	0.4767
2	0.4772	0.4778	0.4783	0.4788	0.4793	0.4798	0.4803	0.4808	0.4812	0.4817
2.1	0.4821	0.4826	0.4830	0.4834	0.4838	0.4842	0.4846	0.4850	0.4854	0.4857
2.2	0.4861	0.4864	0.4868	0.4871	0.4875	0.4878	0.4881	0.4884	0.4887	0.4890

Table 6.11

Example 12: A group of 100 randomly selected college students have an average GPA of 3.15. It is known that the standard deviation of the GPA of all college students is 0.3.

(a) What is the point estimate of the mean GPA of all college students?

(b) Use a 99% confidence level to construct the Interval Estimate of the true mean GPA of all college students.

Solutions: In this example, σ is known and the sample size $n \geq 30$. Hence, we can use the normal distribution.

The given summary statistics are $\overline{x} = 3.15$, $\sigma = 0.3$, and $n = 100$.

(a) The point estimate of the mean GPA of all college students is 3.15; that is,

Point Estimate of $\mu = \overline{x} = 3.15$

(b) The confidence level $= 1 - \alpha = 0.99$

***Step* 1** Find the $z_{\alpha/2}$ value corresponding to a 99% confidence level.

$$\frac{1-\alpha}{2} = \frac{0.99}{2} = 0.495$$

$$P(0 < Z < z_{\alpha/2}) = \frac{1-\alpha}{2} = \frac{\text{CL}}{2} = \frac{0.99}{2} = 0.495$$

Now using z-table, the $z_{\alpha/2}$ value that correspond to the area 0.495 is 2.575.

Therefore, $z_{\alpha/2} = 2.575$.

***Step* 2** Compute the Margin of Error.

$$E = z_{\alpha/2} \cdot \sigma_{\overline{x}} = z_{\alpha/2} \cdot \frac{\sigma}{\sqrt{n}} = 2.575 \cdot \frac{0.3}{\sqrt{100}} = 0.07725$$

***Step* 3** Compute the 99% confidence interval for μ.

$$\overline{x} - E < \mu < \overline{x} + E$$

$$3.15 - 0.07725 < \mu < 3.15 + 0.07725$$

$$3.07 < \mu < 3.23$$

Conclusion: If we take all possible samples of 100 such college students and construct a 99% confidence interval for μ about each sample mean $\overline{x}$, we can expect that 99% of these intervals will actually include μ and 1% will not.

Conclusion Summary: Thus, we are 99% confident that the mean GPA of all college students is between 3.07 and 3.23.

Example 13: A publishing company has just published a Technology Based Statistics textbook. The company's marketing director wants to know the average price of all such textbooks before he makes a recommendation on the price at which to sell this textbook. The company's research department has collected information about the sales price of 10 randomly selected comparable textbooks. The prices (in $) are as follows

85	110	96	94	120	93	93	100	106	112

It is known that the standard deviation of the prices of all such textbooks is $11.46, and the population of such textbook prices is normally distributed.

(a) What is the point estimate of the mean price of all such technology based statistics textbooks?

(b) Construct a 95% confidence interval for the mean price of all such textbooks.

Solutions: Since the parent population is normally distributed, we can use the normal distribution to carry out the required procedure (even though $n < 30$). We have $n = 10$ and $\sigma = 11.46$. The standard deviation of $\overline{X}$ is

$$\sigma_{\overline{x}} = \text{Error of estimate} = \frac{\sigma}{\sqrt{n}} = \frac{11.46}{\sqrt{10}} = 3.623970199 \approx 3.6$$

(a) From the given data, the point estimate of the mean price of all such textbooks is

$$\text{Point Estimate of } \mu = \overline{x} = \frac{\sum x}{n} = \frac{1009}{10} = 100.9$$

(b) The confidence level is 95% or 0.95.

***Step* 1** Find the $z_{\alpha/2}$ value corresponding to a 95% confidence level.

We divide the confidence level in half, and use the z-table to find $z_{\alpha/2}$.

$$P\left(0 < Z < z_{\alpha/2}\right) = \frac{1-\alpha}{2} = \frac{\text{CL}}{2} = \frac{0.95}{2} = 0.475$$

$$\Rightarrow \quad z_{\alpha/2} = 1.96$$

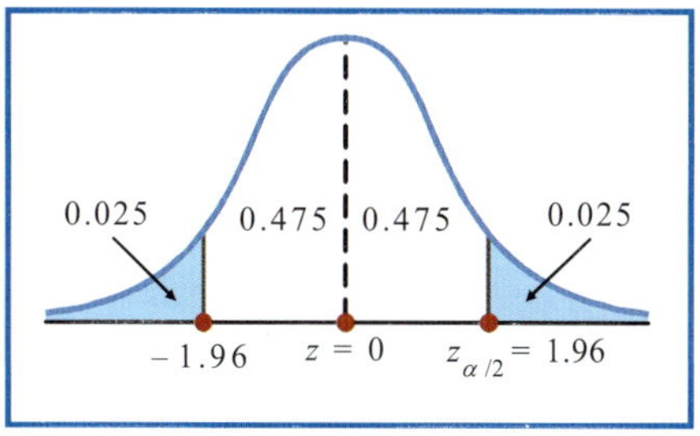

Figure 6.11

***Step* 2** Compute the Margin of Error.

$$E = z_{\alpha/2} \cdot \sigma_{\overline{x}} = z_{\alpha/2} \cdot \frac{\sigma}{\sqrt{n}} = 1.96 \cdot \frac{11.46}{\sqrt{10}} = 7.103$$

Note: To prevent round off errors, it is recommended to carry one extra decimal place than required during the calculations. Round off rules must only be applied to the final answer.

***Step* 3** Compute the 95% confidence interval for μ, where $\overline{X}$ = 100.9 and E = 7.103.

$$\overline{x} - E < \mu < \overline{x} + E$$

$$100.9 - 7.103 < \mu < 100.9 + 7.103$$

$$\$93.8 < \mu < \$108.0$$

Conclusion: If we take all possible samples of 10 Technology Based Statistics textbooks and construct a 95% confidence interval for μ around each sample mean $\overline{x}$, we can expect that 95% of these intervals will actually include μ and 5% will not.

Conclusion Summary: Thus, we are 95% confident that the mean price of all Technology Based Statistics textbooks is between \$93.8 and \$108.0.

Notational Discrepancy: In the Theory of Estimation, the notation for the critical value $z_{\alpha/2}$ can be misleading. Based on the *z*-Table provided in this text, we read the *z*-values (critical values) using the area $\frac{1-\alpha}{2}$, but use the area $\frac{\alpha}{2}$ in writing them!

For example, the *z*-value for a $100(1-\alpha)\% = 95\%$ confidence level is found using the $\frac{1-\alpha}{2} = 0.475$ area but its notation is based on $\frac{\alpha}{2} = 0.025$ area in the tail region of the distribution.

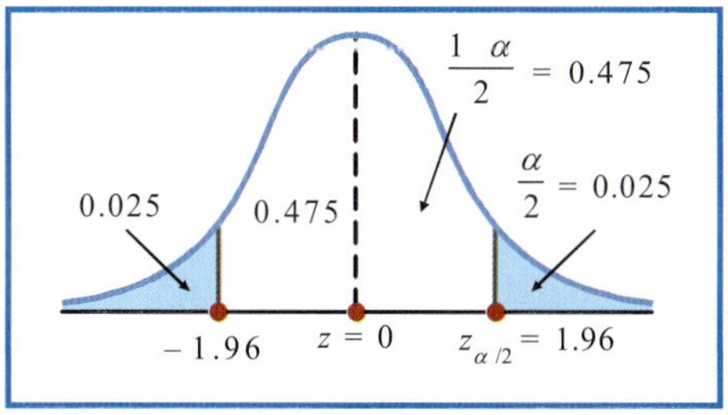

Figure 6.12

Z	0	0.01	0.02	…	0.06	0.07	0.08	0.09
0	0.0000	0.0040	0.0080		0.0239	0.0279	0.0319	0.0359
0.1	0.0398	0.0438	0.0478	…	0.0636	0.0675	0.0714	0.0753
0.2	0.0793	0.0832	0.0871		0.1026	0.1064	0.1103	0.1141
0.3	0.1179	0.1217	0.1255		0.1406	0.1443	0.1480	0.1517
⋮		⋮				⋮		
1.8	0.4641	0.4649	0.4656		0.4686	0.4693	0.4699	0.4706
1.9	0.4713	0.4719	0.4726		**0.4750**	0.4756	0.4761	0.4767
2	0.4772	0.4778	0.4783	…	0.4803	0.4808	0.4812	0.4817
2.1	0.4821	0.4826	0.4830		0.4846	0.4850	0.4854	0.4857
2.2	0.4861	0.4864	0.4868		0.4881	0.4884	0.4887	0.4890

Table 6.12

When finding *z*-scores, you should think $z_{(1-\alpha)/2}$ but write $z_{\alpha/2}$.

Example 14: In a random sample of size 25 selected from a normally distributed population, the average cholesterol level of a group of adults is 39.63. If the standard deviation σ of the population is known to be 5.28, construct a 94% confidence interval for the mean cholesterol level μ of the population.

Solution: We have $n = 25$, $\bar{x} = 39.63$, and $\sigma = 5.28$.

(a) Point Estimate of $\mu = \bar{x} = 39.63$.

(b) Construct the 94% confidence interval for μ.

Step 1 Find $z_{\alpha/2}$ for a 94% confidence level.

$$\frac{1-\alpha}{2} = \frac{\text{CL}}{2} = \frac{0.94}{2} = 0.47 \Rightarrow P(0 < Z < z_{\alpha/2}) = 0.47 \Rightarrow z_{\alpha/2} = 1.88$$

Step 2 Compute the margin of error $E = z_{\alpha/2} \cdot \sigma_{\bar{x}} = z_{\alpha/2} \frac{\sigma}{\sqrt{n}} = 1.88\left(\frac{5.28}{\sqrt{25}}\right) = 1.985.$

Step 3 The 94% confidence interval for μ is

$$\bar{x} - E < \mu < \bar{x} + E \Rightarrow 39.63 - 1.985 < \mu < 39.63 + 1.985 \Rightarrow 37.64 < \mu < 41.62.$$

Conclusion: If we take all possible samples of 25 such adults, and construct a 94% confidence interval for μ around each sample mean, we can expect 94% of these intervals to include μ and 6% not.

Conclusion Summary: We are 94% confident that the mean cholesterol level of all adults in the city is between 37.64 and 41.62.

C. Sample Size for Estimating the Population Mean

In statistical research, especially in Sampling Theory and Design of Experiments, determination of sample size has been a major obstacle. In fact, "How many measurements should be included in the sample?" is the most common question in statistical research.

It has been experimentally and theoretically verified that the minimum sample size depends on (a) desired accuracy of the estimate to be (confidence level) and (b) the margin of error E.

To derive a formula for the sample size, we can start with the margin of error and solve for n.

$$E = z_{\alpha/2}\frac{\sigma}{\sqrt{n}} \Rightarrow \sqrt{n} = \frac{z_{\alpha/2}\sigma}{E} \Rightarrow n = \left(\frac{z_{\alpha/2}\sigma}{E}\right)^2 \quad (6.12)$$

As you can see, the minimum sample size necessary in a research study depends on the critical value $z_{\alpha/2}$, the population standard deviation σ, and the margin of error E.

Remark: If the population standard deviation σ and sample standard deviation s are not known, by using the range rule of thumb, an estimate of σ can be obtained as given here.

$$\hat{\sigma} \simeq \frac{\text{Range}}{4} \quad (6.13)$$

Example 15: A manufacturing firm wants to estimate the average amount of sales for its products in a month. If the standard deviation of the population is known to be \$10, find the necessary sample size if the maximum error is not to exceed \$3 with a 99% confidence level.

Solution: We have $\sigma = 10$, $E = 3$, and $1 - \alpha = 0.99 \Rightarrow z_{\alpha/2} = 2.575$

Substituting the above information in the sample size formula (6.12) we have

$$n = \left(\frac{z_{\alpha/2}\,\sigma}{E}\right)^2 = \left(\frac{2.575 \times 10}{3}\right)^2 = 73.67 \approx 74$$

Example 16: Professor Badalian wants to estimate the mean final exam grade for all his statistics classes. If standard deviation of final grades from a previous study is 8.3, then how many such sample grades must he use for this experiment to be 95% confident that the sample mean is within 2 points of the true student population mean?

Solution: $\sigma = 8.3$, $E = 2$, and the $z_{\alpha/2}$ value for a 95% confidence level is 1.96.

Using the formula $n = \left(\frac{z_{\alpha/2}\,\sigma}{E}\right)^2$, we have $n = \left(\frac{(1.96)(8.3)}{3}\right)^2 = 66.16 \approx 67.$

Example 17: A random sample of hotdogs of a particular brand has an average fat content of 3 grams. How large a sample should be taken to be 90% confident that the sample mean is within 0.18 grams of the true mean? (Assume that the fat content ranges from 1.75 grams to 4.15 grams.)

Solution: We are given $\bar{x} = 3$, $E = 0.18$, and the confidence coefficient $(1 - \alpha) = 0.90$. Therefore the corresponding $z_{\alpha/2} = 1.645$. Since σ is not known, we estimate it from the given range by using the range rule of thumb.

$$\hat{\sigma} \approx \frac{\text{Range}}{4} = \frac{4.15 - 1.75}{4} = 0.6 \text{ grams}$$

Substituting all the given and calculated values in to the sample size formula gives us

$$n = \left(\frac{z_{\alpha/2}\,\sigma}{E}\right)^2 = \left(\frac{1.645 \times 0.6}{0.18}\right)^2 = 30.06 \approx 31$$

Example 18: A researcher wishes to estimate the mean of a population by using a sufficiently large sample. There is a 0.95 probability (95% confidence level), that the sample mean will not differ from the true mean by more than 25% of the standard deviation. How large a sample should be taken?

Solution: Given the error "E" to be 25% of the standard deviation, we have

$(0.25)(\sigma) = \frac{\sigma}{4}$. $(1 - \alpha) = 0.95 \Rightarrow z_{\alpha/2} = 1.96$; and

given a 95% confidence level $1 - \alpha = 0.95$, $z_{\alpha/2} = 1.96$.

Hence $n = \left(\frac{z_{\alpha/2}\,\sigma}{E}\right)^2 = \left(\frac{1.96 \times \sigma}{\sigma / 4}\right)^2 = (1.96 \times 4)^2 \approx 61.466 \approx 62$

Try it yourself:

1. Construct a 95% confidence interval for μ given

   (a) $\bar{x} = 50.73$, $\sigma = 8.24$, and $n = 75$.

   (b) $\bar{x} = 120$, $\sigma = 18$, and $n = 10$. Assume that the parent population is normally distributed.

   (c) $\bar{x} = 16.8$, $s = 4.1$, and $n = 16$. Assume that the parent population is normally distributed.

2. A random sample of 64 credit sales at a gas station showed an average of \$38.56. The standard deviation of credit sales is known to be $\sigma = \$4$.

   (a) Find the standard error of the mean $\sigma_{\bar{x}}$.

   (b) Find the margin of error $E = z_{\alpha/2} \cdot \sigma_{\bar{x}}$ with 98% confidence.

   (c) Construct a 98% confidence interval for the average amount of all credit sales.

3. The record of randomly selected 25 employees in a large office reveal that the average number of sick days used during the past year is 8 with a standard deviation of 2. Assuming that the population is normally distributed, construct a 90% confidence interval for the average number of sick days used by the employees during the past year.

4. The weights (in pounds) of 10 randomly selected new born babies at a local hospital are:

6.5	5.8	7.2	7.0	8.2	6.2	6.7	7.5	8.5	7.8

   Assuming that the standard deviation $\sigma = 0.93$ and that the population is normally distributed, construct a 90% confidence interval for the average weight of new born baby.

5. How large a sample is needed if we wish to be 98% confident that our sample mean $\bar{x}$ will be within 2 units of the true mean μ?

6. A doctor estimates that the cholesterol levels of middle age men range from 180 to 320 milligram per deciliter (mg/dL). How large a sample must be taken in order to be 95% confident that the error of his estimate of the mean cholesterol level will not exceed 12 mg/dL?

7. The average outstanding balance of loans issued by a bank varies from month to month. From past experience it is known that the amounts are normally distributed with $\sigma = \$5000$. The bank wishes to estimate the average by drawing a random sample with a 0.95 probability that the mean of the sample will deviate from the true mean by no more than \$600. What should be the minimum sample size?

6.6: Estimation of the Population Mean μ (σ Unknown)

Objectives

After completing this section you will be able to:

A. Understand Student *t*-Distribution.

B. Compute, Confidence Interval by using the *t*-distribution.

In Section 6.5 we constructed the confidence interval for μ assuming that the population standard deviation σ was known. In this section, we will develop a corresponding method for estimating the population mean μ when σ is unknown. Once again, there are three cases when constructing the confidence interval for μ when σ is unknown.

Case i) If σ is unknown, $n < 30$, and the population from which the sample is selected is normally distributed, then we can use the student *t*-distribution to construct a confidence interval for μ.

Note: The *t*-distribution will be explained next.

Case ii) If σ is unknown, and $n \geq 30$, then, independent of the distribution of the original population, we can use the student *t*-distribution to construct a confidence interval for μ.

Case iii) If σ is unknown, $n < 30$, and the population is not normally distributed, then we must use a nonparametric method to estimate μ.

Note: Nonparametric methods will be discussed in Chapter 11.

The following diagram summarizes the above cases.

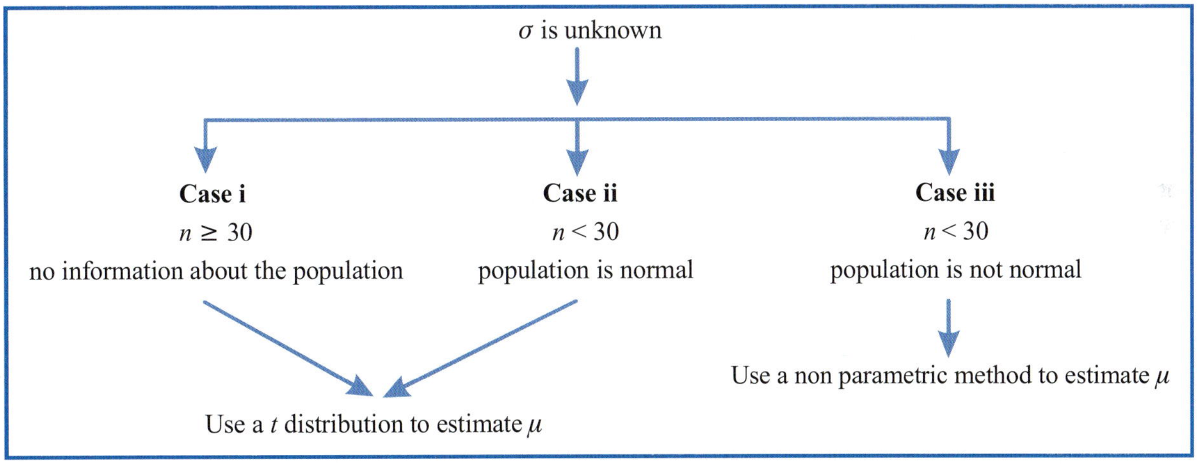

Figure 6.13

Before we begin our discussion, we must briefly introduce the student *t*-distribution. (For more detailed treatment of the *t*-distribution see section 6.10.)

A. The student *t*-distribution

Prior to 1908, when constructing a confidence interval for μ with σ unknown, it was customary to replace σ by s for small samples and then apply the normal distribution even for small samples. In 1908, William S. Gosset, an Irish brewery employee, who used to write under the pen name ***student,*** proved that the sampling distribution of the statistics, $\dfrac{\bar{X}-\mu}{s/\sqrt{n}}$ for small samples, was not normal and introduced a new distribution called the ***student t-distribution***.

• Properties of the *t*-distribution:

1. The t-distribution is a symmetric bell shaped curve about the mean.

2. The curve never touches the horizontal axis.

3. The total area under the t-distribution curve is 100% or 1.

4. The t-distribution approaches the normal distribution as the sample size increases, or as the degree of freedom (df) $\to \infty$.

5. The mean of the t-distribution is zero and its standard deviation is always greater than 1.

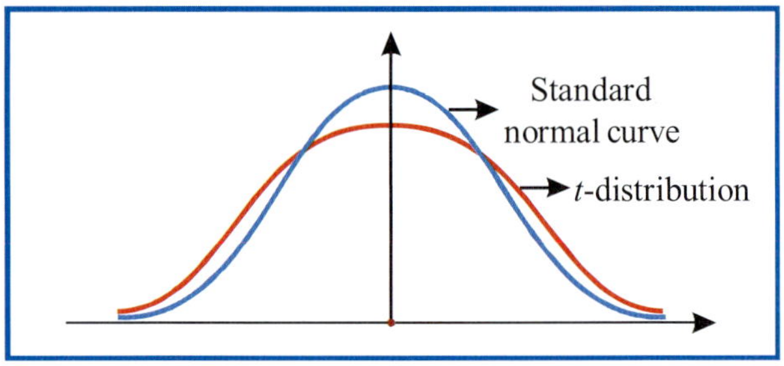

Figure 6.14

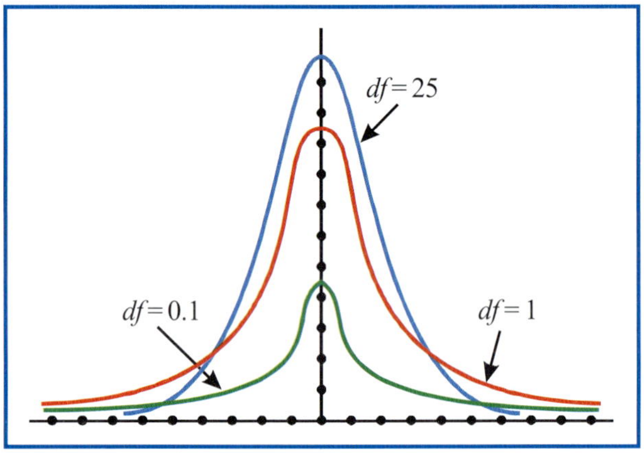

Figure 6.15

As we can see, the t-distribution is quite similar to the normal distribution. However, the t-distribution curve generally has a lower height and a wider spread (larger standard deviation) than the standard normal curve. The shape of a t-distribution curve and reading probabilities associated with it depend on the number of degrees of freedom (df). (See table)

Degree of freedom (df)	$\mu_{t\text{-dist}}$	$\sigma_{t\text{-dist}}$
υ (nu) $= n - 1$	0	$\sqrt{\dfrac{df}{df-2}} = \sqrt{\dfrac{\upsilon}{\upsilon-2}}$

(6.14)

Remark: Degrees of freedom for the t-distribution is defined as $n - 1$. The number of df is the number of observations that can be chosen freely. Suppose we know that the mean of 5 values is 30.

$$\frac{\sum_{i=1}^{5} x_i}{5} = \frac{x_1 + x_2 + x_3 + x_4 + x_5}{5} = 30 \Rightarrow \sum_{i=1}^{5} x_i = 150$$

Now, how many values out of 5 can we choose freely so that the sum of these values is 150? The answer is 4. We can freely choose 4 out of 5 values. For instance, let $x_1 = 25$, $x_2 = 23$, $x_3 = 35$, and $x_4 = 18$. In order for $x_1 + x_2 + x_3 + x_4 + x_5$ to add up to 150, x_5 **must be** 49. As you can see, once we have chosen 4 values freely, the 5th value is automatically determined. Consequently the number of df for this example is $df = n - 1 = 5 - 1 = 4$.

Note:	Table 4 at the end of the book lists the t-values (critical values of t) for different values of $\frac{\alpha}{2} = \frac{1\text{-}CLT}{2}$ (area in the right tail of the t-distribution curve) and degrees of freedom. (Figure 6.16)

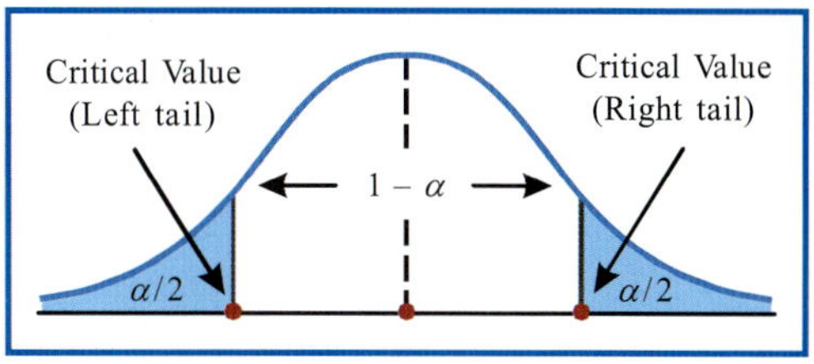

Figure 6.16

• Reading the *t*-distribution Table:

In Table 4, the top row indicates the area $\alpha/2$ in the right tail of the t-distribution curve as shaded in Figure 6.17. The first column shows the degrees of freedom ($n - 1 = \upsilon$). The body of the table (enclosed by the top row and the first column) gives the value

$$t = t_{n-1,\,\alpha/2} = t_{\upsilon,\,\alpha/2},\ \text{where } P(t > t_{\upsilon,\,\alpha/2}) = \frac{\alpha}{2}$$

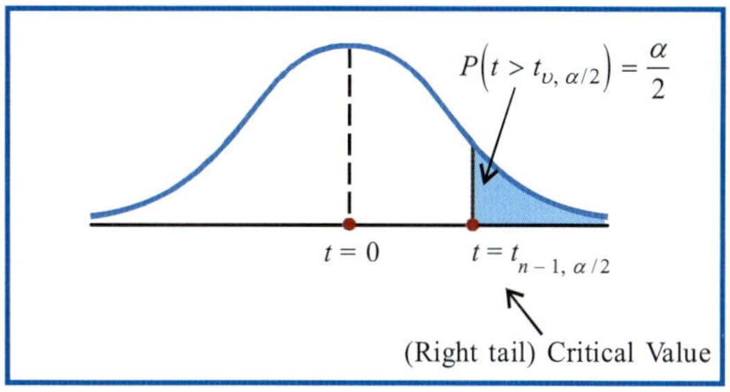

Figure 6.17

$\alpha/2$ values

Area in the right tail under the t-distribution curve

degrees of freedom (υ)

df	0.1	0.05	0.025	0.01	0.005	0.001
1	3.0777	6.3138	12.7062	31.8205	63.6567	318.3088
2	1.8856	2.9200	4.3027	6.9646	9.9248	22.3271
3	1.6377	2.3534	3.1824	4.5407	5.8409	10.2145
4	1.5332	2.1318	2.7764	3.7469	4.6041	7.1732
5	1.4759	2.0150	2.5706	3.3649	4.0321	5.8934
6	1.4398	1.9432	2.4469	3.1427	3.7074	5.2076
7	1.4149	1.8946	2.3646	2.9980	3.4995	4.7853
8	1.3968	1.8595	2.3060	2.8965	3.3554	4.5008
9	1.3830	1.8331	2.2622	2.8214	3.2498	4.2968
10	1.3722	1.8125	2.2281	2.7638	3.1693	4.1437

Table 6.13

Example 19: Find the t-value for 12 degrees of freedom ($\upsilon = 12$) given a 95% confidence level.

Solution: To read the t-table, we need the degrees of freedom and the $\alpha/2$ value (area in the right tail of the t-distribution curve). In this example, the degrees of freedom is given. Thus we only need to find the value of $\alpha/2$. Recall that $\frac{\alpha}{2} = \frac{1-CL}{2} = \frac{1-0.95}{2} = 0.025$. We can now use Table 4 to read the t-value corresponding to 12 degrees of freedom and a 0.025 significance level. We trace the degrees of freedom (12) in the first column of the t-table and the area of 0.025 in the top row. The number or entry in the intersection of the row with 12 *df* and the column with 0.025 heading (significance level), which is 2.1788, gives the corresponding t-value.

Area in the right tail under the *t*-distribution curve

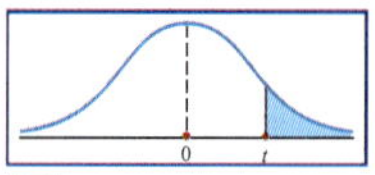

df \ α/2	0.1	0.05	0.025	0.01	0.005	0.001
1	3.0777	6.3138	12.7062	31.8205	63.6567	318.3088
2	1.8856	2.9200	4.3027	6.9646	9.9248	22.3271
3	1.6377	2.3534	3.1824	4.5407	5.8409	10.2145
4	1.5332	2.1318	2.7764	3.7469	4.6041	7.1732
5	1.4759	2.0150	2.5706	3.3649	4.0321	5.8934
6	1.4398	1.9432	2.4469	3.1427	3.7074	5.2076
7	1.4149	1.8946	2.3646	2.9980	3.4995	4.7853
8	1.3968	1.8595	2.3060	2.8965	3.3554	4.5008
9	1.3830	1.8331	2.2622	2.8214	3.2498	4.2968
10	1.3722	1.8125	2.2281	2.7638	3.1693	4.1437
11	1.3634	1.7959	2.2010	2.7181	3.1058	4.0247
12	1.3562	1.7823	**2.1788**	2.6810	3.0545	3.9296
13	1.3502	1.7709	2.1604	2.6503	3.0123	3.8520
14	1.3450	1.7613	2.1448	2.6245	2.9768	3.7874
15	1.3406	1.7531	2.1314	2.6025	2.9467	3.7328

Table 6.14

Therefore, the *t*-value with 12 degrees of freedom and a 0.025 significance level (95% confidence level) is 2.1778.

$$t_{n-1,\,\alpha/2} = t_{\upsilon,\,\alpha/2} = t_{12,\,0.025} = 2.1778$$

Remark: Because of the symmetric nature of the *t*-distribution curve, the *t*-value for 12 degrees of freedom and 0.025 area in the left tail is −2.1788.

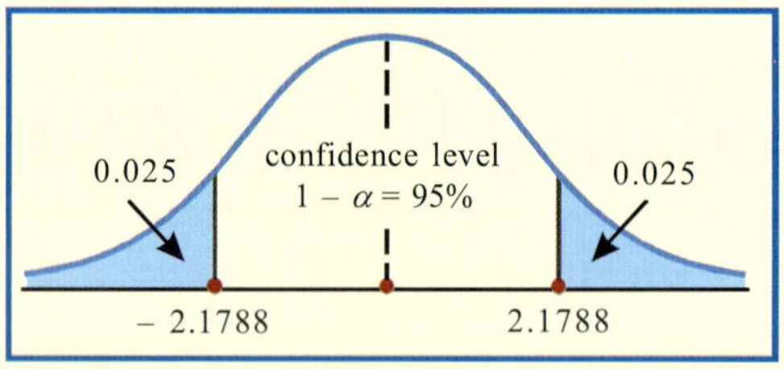

Figure 6.18

Example 20: Find the $t_{n-1,\,\alpha/2}$ if the confidence level is 95% and the sample size is 20.

Solution: The value of $\frac{\alpha}{2} = \frac{1-\text{CL}}{2} = \frac{1-0.95}{2} = \frac{0.05}{2} = 0.025.$

The degrees of freedom is $n - 1 \to 20 - 1 = 19.$

Use the area 0.025 in the right tail of the *t*-distribution curve with 19 degrees of freedom to read the corresponding critical value (t) from the student *t*-distribution.

$t_{19,\,0.025} = 2.093.$

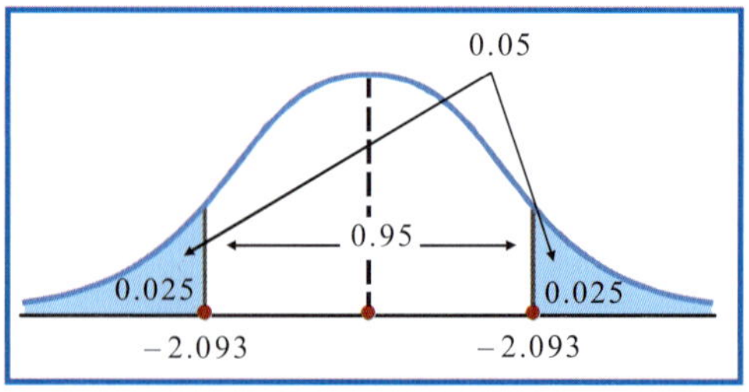

Figure 6.19

Area in the right tail under the t-distribution curve

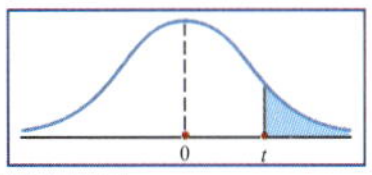

$\alpha/2$

df	0.10	0.05	0.025	0.01	0.005	0.001
1	3.0777	6.3138	12.7062	31.8205	63.6567	318.3088
2	1.8856	2.9200	4.3027	6.9646	9.9248	22.3271
	⋮				⋮	
17	1.3334	1.7396	2.1098	2.5669	2.8982	3.6458
18	1.3304	1.7341	2.1009	2.5524	2.8784	3.6105
19	1.3277	1.7291	**2.0930**	2.5395	2.8609	3.5794
20	1.3253	1.7247	2.0860	2.5280	2.8453	3.5518

Table 6.15

B. Confidence Interval for μ Using the t-distribution ($n < 30$ and σ Unknown)

Recall the conditions specified under cases (i) and (ii) in the beginning of this section (Figure 6.13).

When these conditions are satisfied, we can use the t-distribution to construct a **confidence interval** for the population mean μ.

Note: When the population standard deviation σ is unknown, we can replace it by the sample standard deviation s, which is its best estimate. Therefore, the standard deviation of the sampling distribution for $\bar{X}$ when σ is unknown will be $s_{\bar{X}} = \dfrac{s}{\sqrt{n}}$. ($s_{\bar{X}}$ is the point estimate of $\sigma_{\bar{X}}$)

Definition: *The* $100(1-\alpha)\%$ *confidence interval for* μ *using the t-distribution is*

$\bar{x} \pm t_{\upsilon,\,\alpha/2}\, s_{\bar{x}}$, *where* $s_{\bar{x}} = \dfrac{s}{\sqrt{n}}$, *with the margin of error* $E = t_{\upsilon,\,\alpha/2}\, s_{\bar{x}} = t_{\upsilon,\,\alpha/2} \dfrac{s}{\sqrt{n}}$.

Confidence interval formula for μ, σ unknown

$$\bar{x} - E < \mu < \bar{x} - E$$

$$\bar{x} - t_{n-1,\,\alpha/2}\, s_{\bar{x}} < \mu < \bar{x} - t_{n-1,\,\alpha/2}\, s_{\bar{x}}$$

$$\bar{x} - t_{n-1,\,\alpha/2} \frac{s}{\sqrt{n}} < \mu < \bar{x} - t_{n-1,\,\alpha/2} \frac{s}{\sqrt{n}} \qquad (6.15)$$

Example 21: Dr. Tiago wants to estimate the mean GPA of all incoming freshmen. The mean GPA of a random sample of 25 freshmen is 2.58 with a standard deviation of 0.38. Assuming that the GPA of all incoming freshmen is normally distributed, construct a 99% confidence interval for the population mean μ.

Solution: Since σ is unknown, $n < 30$ and the population is normally distributed, we must use the t-distribution to construct a confidence interval for μ. The summary of the information is

$$n = 25,\ \bar{x} = 2.58,\ s = 0.38,\ \text{CL} = 99\% \Rightarrow 1 - \alpha = 0.99 \Rightarrow \frac{\alpha}{2} = \frac{1-\text{CL}}{2} = \frac{1-0.99}{2} = 0.005.$$

In order to construct a 99% confidence interval for μ, we must find the t-value corresponding to $\alpha/2 = 0.005$ and $\upsilon = n - 1 = 25 - 1 = 24$ degrees of freedom.

Therefore, $t_{\upsilon,\,\alpha/2} = t_{24,\,0.005} = 2.797$.

The margin of error

$$E = t_{n-1,\,\alpha/2}\frac{s}{\sqrt{n}} = 2.797\left(\frac{0.38}{\sqrt{25}}\right) \approx 0.213.$$

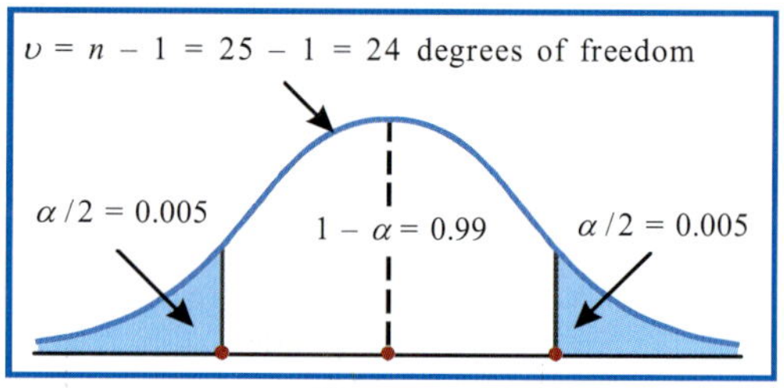

Figure 6.20

Therefore, the 99% confidence interval for μ is

$$\bar{x} - E < \mu < \bar{x} + E \Rightarrow 2.58 - 0.213 < \mu < 2.58 + 0.213 \Rightarrow 2.37 < \mu < 2.79.$$

Hence, Dr. Tiago can be 99% confident that the mean GPA of all incoming freshmen is between 2.37 and 2.79.

Example 22: Fifty-eight randomly selected adults who buy New Age music CD's were asked the dollar amount they usually spend on buying CD's per year. The sample produced a mean of \$760 and a standard deviation of \$120. Construct a 95% confidence interval for the true mean cost of buying New Age CD's per year.

Solution: Since σ is unknown, $n \geq 30$ and there is no information about the distribution of the population, we must use the t-distribution to construct a confidence interval for μ. Here is the summary of the information.

$n = 58$, $\bar{x} = 760$, $s = 120$ and CL = 95%.

Since $1 - \alpha = 0.95$, then $\frac{\alpha}{2} = \frac{1 - \text{CL}}{2} = \frac{1 - 0.95}{2} = 0.025$.

Area in the right tail under the t-distribution curve

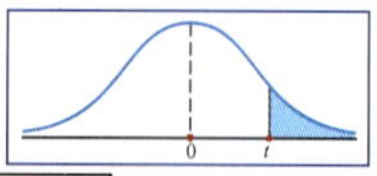

df	0.1	0.05	0.025	0.01	0.005	0.001
42	1.3020	1.6820	2.0181	2.4185	2.6981	3.2960
43	1.3016	1.6811	2.0167	2.4163	2.6951	3.2909
44	1.3011	1.6802	2.0154	2.4141	2.6923	3.2861
45	1.3006	1.6794	2.0141	2.4121	2.6896	3.2815
46	1.3002	1.6787	2.0129	2.4102	2.6870	3.2771
47	1.2998	1.6779	2.0117	2.4083	2.6846	3.2729
48	1.2994	1.6772	2.0106	2.4066	2.6822	3.2689
49	1.2991	1.6766	2.0096	2.4049	2.6800	3.2651
50	1.2987	1.6759	2.0086	2.4033	2.6778	3.2614
51	1.2984	1.6753	2.0076	2.4017	2.6757	3.2579
52	1.2980	1.6747	2.0066	2.4002	2.6737	3.2545
53	1.2977	1.6741	2.0057	2.3988	2.6718	3.2513
54	1.2974	1.6736	2.0049	2.3974	2.6700	3.2481
55	1.2971	1.6730	2.0040	2.3961	2.6682	3.2451
56	1.2969	1.6725	2.0032	2.3948	2.6665	3.2423
57	1.2966	1.6720	**2.0025**	2.3936	2.6649	3.2395
58	1.2963	1.6716	2.0017	2.3924	2.6633	3.2368

Table 6.16

In order to construct a 95% confidence interval for μ, we must find the t value corresponding to $\alpha/2 = 0.025$ and $\upsilon = n - 1 = 58 - 1 = 57$ degrees of freedom. Using the t-table, we have 6.18, $t_{\upsilon,\, \alpha/2} = t_{57,\, 0.025} = 2.0025$.

The margin of error $E = t_{n-1,\, \alpha/2} \dfrac{s}{\sqrt{n}} = 2.0025 \left(\dfrac{120}{\sqrt{58}}\right) \approx 31.6.$

Therefore, the 95% confidence interval for μ is:

$$\bar{x} - E < \mu < \bar{x} + E \;\Rightarrow\; 760 - 31.6 < \mu < 760 + 31.6$$

$$\$728 < \mu < \$792$$

Thus, there is a 95% chance that the mean annual expenditure on new age CD's by all adults in the population is between \$728 and \$792.

Example 23: A random sample of 12 packaged chickens is selected from a chicken processing plant. The amount of fat content (in ounces) each chicken package contains is listed here

5.2	6.1	5.1	5.7	5.9	6.7	5.0	6.5	6.2	7.0	6.4	5.3

Assuming that the population is normally distributed, construct a 90% confidence interval for the average fat content in each chicken package produced in this processing plant.

Solution: We have $n < 30$, σ is unknown, and the parent population is normally distributed. Therefore, we must use a t-distribution to construct the confidence interval for μ.

$\bar{x}$ and s can be calculated using the sample raw data given in the problem:

$$\bar{X} = \frac{\sum x}{n} = 5.925, \text{ and } s = \sqrt{\frac{n\sum x^2 - \left(\sum x\right)^2}{n(n-1)}} \;\; 0.669$$

We also know $n = 12$ which implies that, $df = n - 1 = 12 - 1 = 11$,

and $\dfrac{\alpha}{2} = \dfrac{1 - \text{CL}}{2} = \dfrac{1 - 0.9}{2} = 0.05.$

Therefore, $t_{n-1,\, \alpha/2} = t_{11,\, 0.05} = 1.7959$

Area in the right tail under the t-distribution curve

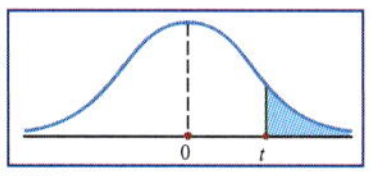

df	0.100	0.05	0.025	0.01	0.005	0.001
1	3.0777	6.3138	12.7062	31.8205	63.6567	318.3088
2	1.8856	2.9200	4.3027	6.9646	9.9248	22.3271
3	1.6377	2.3534	3.1824	4.5407	5.8409	10.2145
4	1.5332	2.1318	2.7764	3.7469	4.6041	7.1732
5	1.4759	2.0150	2.5706	3.3649	4.0321	5.8934
6	1.4398	1.9432	2.4469	3.1427	3.7074	5.2076
7	1.4149	1.8946	2.3646	2.9980	3.4995	4.7853
8	1.3968	1.8595	2.3060	2.8965	3.3554	4.5008
9	1.3830	1.8331	2.2622	2.8214	3.2498	4.2968
10	1.3722	1.8125	2.2281	2.7638	3.1693	4.1437
11	1.3634	**1.7959**	2.2010	2.7181	3.1058	4.0247
12	1.3562	1.7823	2.1788	2.6810	3.0545	3.9296
13	1.3502	1.7709	2.1604	2.6503	3.0123	3.8520

Table 6.17

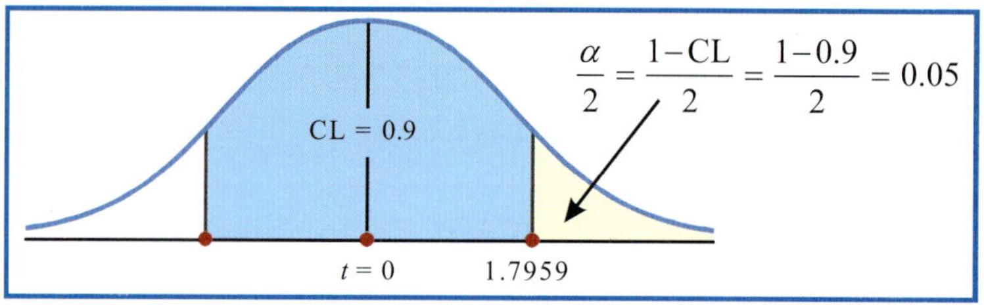

Figure 6.21

We now have all the information required to construct a 90% confidence interval for μ.

$$\text{The margin of error } E = t_{n-1,\,\alpha/2} \cdot \frac{s}{\sqrt{n}}$$

$$= 1.7959\left(\frac{0.669}{\sqrt{12}}\right) = 0.3468$$

Therefore, the 90% confidence interval for μ is $(\bar{x} - E,\ \bar{x} + E)$.

$$\bar{x} - E < \mu < \bar{x} + E$$

$$5.925 - 0.3468 < \mu < 5.925 + 0.3169$$

$$5.578 < \mu < 6.272$$

Hence, we can be 90% confident that the mean amount of fat content in all chicken packages is between 5.578 and 6.272 ounces.

Example 24: A random sample of 8 different airlines showed an average of 12 delayed flights per week with a standard deviation of 3. Construct a 95% confidence interval for the true average number of delayed flights per week for all airlines. Assume that the population is normally distributed.

Solution: We have $\bar{x} = 12$, $s = 3$, $n = 8$ and $\frac{\alpha}{2} = \frac{1-CL}{2} = \frac{1-1.095}{2} = 0.025$, which implies that $t_{7,\,0.025} = 2.3646$.

Therefore, the 95% confidence interval for μ is $\bar{x} - E < \mu < \bar{x} + E$, where

$$E = t_{7,\,0.025}\frac{s}{\sqrt{n}} = 2.3646\frac{3}{\sqrt{8}}$$

Ultimately we have

$$12 - 2.3646\frac{3}{\sqrt{8}} < \mu < 12 + 2.3646\frac{3}{\sqrt{8}} \Rightarrow 9 < \mu < 15$$

Therefore, we are 95% confident that the average number of delayed flights per week for all airlines is between 9 to 15.

Try it yourself:

In exercises 1-5, find the *t*-value with degrees of freedom $\upsilon = 10$ and the given area.

1. The area to the right of t is 0.005.
2. The area left of t is 0.01.

3. The area between $-t$ and t is 0.92.
4. The area to the right of t is 0.1.
5. The area between $-t$ and t is 98%.
6. A random sample of size 16 is drawn from a normally distributed population with σ unknown. The sample mean and standard deviation are 60 and 5, respectively. Construct a 99% confidence interval for the population mean μ.
7. The average and standard deviation of daily commuting distances of 64 randomly selected employees in a large law office are 18 miles and 5 miles, respectively. Construct a 98% confidence interval for the mean daily commuting distance in this office. Assume that the population is normally distributed.
8. A random sample of 70 UPS packages has an average weight of 1.85 lb with a standard deviation of 0.2 lb, respectively.

   (a) Construct a 99% confidence interval for μ.

   (b) Construct a 90% confidence interval for μ.

6.7: Estimation of the Population Proportion *p* (Large Sample)

Objectives

After completing this section you will be able to obtain:

A. $100(1-\alpha)\%$ Confidence Interval for Population Proportion *p*.

B. Sample Size for estimating the Population Proportion.

In this section, we will estimate the population proportion or percentage (percentage is obtained by multiplying the proportion by 100) and construct its interval estimate (confidence interval). Recall that the population proportion is denoted by p and the sample proportion is denoted by $\hat{p}$. The sample proportion $\hat{p}$ is a sample statistics, and it possesses a sampling distribution. From the sampling distribution section earlier, we know that

1. the sampling distribution of the sample proportion, $\hat{p}$, is approximately normal.
2. the mean, $\mu_{\hat{p}}$, of the sampling distribution of $\hat{p}$ is equal to population proportion p; that is $E\left(\hat{p}\right) = p$.

   Hence the sample proportion of successes $\hat{p}$ is the best point estimate of the population proportion p; that is Point Estimate of $p = \hat{p} = \dfrac{x}{n}$, where x is the observed number of successes in a sample of size n.

3. The standard deviation, $\sigma_{\hat{p}}$, of the sampling distribution of the sample proportion, $\hat{p}$, is

$$\sigma_{\hat{p}} = \sqrt{\frac{pq}{n}} \text{ where } q = 1 - p$$

Note: In the case of a proportion, a sample is considered to be large if *np* and *nq* are both greater than 5.

When estimating the value of a population proportion, we naturally do not know the values of p and q. As a result, we cannot compute $\sigma_{\hat{p}}$. Therefore, in the estimation of the population proportion, we use $s_{\hat{p}}$ as an estimate of $\sigma_{\hat{p}}$.

$$\sigma_{\hat{p}} \approx s_{\hat{p}} = \sqrt{\frac{\hat{p}\hat{q}}{n}} \qquad (6.16)$$

Hence, the margin of error for estimating the population proportion is

$$E = z_{\alpha/2}\,\sigma_{\hat{p}} \approx z_{\alpha/2}\,s_{\hat{p}} = z_{\alpha/2}\sqrt{\frac{\hat{p}\hat{q}}{n}}\,. \qquad (6.17)$$

Since the point estimate from a random sample is seldom equal to the population proportion, we need to construct an interval estimate (confidence interval) within which the true population proportion p lies.

A. 100(1 – α)% confidence interval for population proportion p

The $100(1 - \alpha)\%$ confidence limits for the population proportion, p, are given by

$\hat{p} \pm E$ where E = margin of error $= z_{\alpha/2}\sqrt{\dfrac{\hat{p}\hat{q}}{n}}$

Hence, the $100(1 - \alpha)\%$ confidence interval for population proportion, p, is given by:

$$\hat{p} - E < p < \hat{p} + E$$

$$\hat{p} - z_{\alpha/2}\sqrt{\frac{\hat{p}\hat{q}}{n}} < p < \hat{p} + z_{\alpha/2}\sqrt{\frac{\hat{p}\hat{q}}{n}}$$

Round off Rule:	It is common practice to round off the confidence interval limits for the true population proportion "p" to three decimal places.

Steps in constructing 100(1 – α)% confidence interval for p

***Step* 1** Compute the sample proportion $\hat{p} = \dfrac{x}{n}$, where x is the observed number of successes and n is the sample size. If $\hat{p}$ is given, then skip **Step 1**.

***Step* 2** For the given confidence level $(1 - \alpha)$, find $z_{\alpha/2}$ so that $P(0 < Z < z_{\alpha/2}) = \dfrac{1-\alpha}{2} = \dfrac{\text{CL}}{2}$.

***Step* 3** Compute $\hat{\sigma}_{\hat{p}} = s_{\hat{p}} = \sqrt{\dfrac{\hat{p}\hat{q}}{n}} = \sqrt{\dfrac{\hat{p}(1-\hat{p})}{n}}$.

***Step* 4** Compute the margin of Error, $E = z_{\alpha/2}\,\sigma_{\hat{p}} \approx z_{\alpha/2}\,s_{\hat{p}} = z_{\alpha/2}\sqrt{\dfrac{\hat{p}\hat{q}}{n}}$.

***Step* 5** Construct the required confidence interval for p : $\hat{p} - E < p < \hat{p} + E$.

Example 25: In a random sample of 150 customers, 80 ordered chicken sandwiches for lunch. Construct a 95% confidence interval for the true proportion of customers who buy chicken sandwiches for lunch.

Solutions: We have $n = 150$ and $x = 80$.

Step 1 Compute the sample proportion $\hat{p}$. $\hat{p} = \frac{x}{n} = \frac{80}{150} \approx 0.533$

The point estimate of population proportion p is $\hat{p} = 0.533$.

Step 2 Compute $z_{\alpha/2}$: Corresponding to a 95% confidence level, $z_{\alpha/2} = 1.96$.

Step 3 Compute $\hat{\sigma}_{\hat{p}}$: $\hat{\sigma}_{\hat{p}} = s_{\hat{p}} = \sqrt{\frac{\hat{p}\hat{q}}{n}} = \sqrt{\frac{(0.533)(1-0.533)}{150}} \approx 0.0407$.

Step 4 Compute Margin of Error, E:

$$E = z_{\alpha/2}\,\sigma_{\hat{p}} \approx z_{\alpha/2}\,s_{\hat{p}} = z_{\alpha/2}\sqrt{\frac{\hat{p}\hat{q}}{n}} = 1.96\sqrt{\frac{(0.533)(0.467)}{150}} \approx 0.0798$$

Step 5 95% confidence interval for p:

$$\hat{p} - E < p < \hat{p} + E$$

$$0.533 - 0.0798 < p < 0.533 + 0.0798$$

$$0.453 < p < 0.613$$

Thus, we can be 95% confident that the proportion of all customers who buy chicken sandwiches for lunch is between 45.3% to 61.3%.

Example 26: Find a 99% confidence interval for the true percentage of DUI arrests during the weekends, if in a random sample of 15 weekends, $\hat{p} = 48\%$.

Solution: We have $n = 15$ and $\hat{p} = 0.48$.

Step 1 We can skip this step, since $\hat{p}$ is given: $\hat{p} = 48\%$.

Step 2 The critical value for a 99% confidence level is: $z_{\alpha/2} = 2.575$.

Step 3 Compute $\hat{\sigma}_{\hat{p}}$: $\hat{\sigma}_{\hat{p}} = s_{\hat{p}} = \sqrt{\frac{\hat{p}\hat{q}}{n}} = \sqrt{\frac{(0.48)(1-0.48)}{15}} \approx 0.1290$

Step 4 Compute margin of error, E: $E = z_{\alpha/2}\,s_{\hat{p}} = (2.575)(0.1290) \approx 0.3322$

Step 5 99% confidence interval for p:

$$\hat{p} - E < p < \hat{p} + E$$

$$0.48 - 0.3322 < p < 0.48 + 0.3322$$

$$0.148 < p < 0.812$$

Thus, we can be 99% confident that the proportion of all DUI arrests during the weekends is between 14.8% to 81.2%.

Note: The reason for the unusual large range for the population proportion in this example is the size of the sample. Even though $np > 5$ in this example, the sample size is still not large enough to produce a more accurate confidence interval.

Example 27: In a CD manufacturing plant, it is found that the defect rate in a sample of 35 randomly selected CD's is 29%. Construct a 90% confidence interval for the true defect rate of the entire production line population (p).

Solution: ***Step 1*** We can skip this step, since $\hat{p}$ is given: $\hat{p}$ = 29%.

Step 2 The critical value for a 90% confidence level is: $z_{\alpha/2} = 1.645$

Step 3 Compute $\hat{\sigma}_{\hat{p}}$: $\hat{\sigma}_{\hat{p}} = s_{\hat{p}} = \sqrt{\frac{\hat{p}\hat{q}}{n}} = \sqrt{\frac{(0.29)(0.71)}{35}} \approx 0.0767$

Step 4 Compute Margin of Error(E): $E = z_{\alpha/2}\, s_{\hat{p}} = (1.645)(0.0767) \approx 0.1262$

Step 5 99% confidence interval for p:

$$0.29 - 0.1262 < p < 0.29 + 0.1262$$

$$\hat{p} - E < p < \hat{p} + E$$

$$0.164 < p < 0.416$$

Thus, we can be 90% confident that the proportion of the defect rate for all CDs manufactured is between 16.4% to 41.6%.

Table 6.18 shows the relationship between the sample size and confidence limits for the above example.

Sample Size n	Lower Confidence Limit	Upper Confidence Limit
10	0.054 = 5.4%	0.526 = 52.6%
50	0.184 = 18.4%	0.396 = 39.6%
100	0.215 = 21.5%	0.365 = 36.5%
500	0.257 = 25.7%	0.323 = 32.3%
2000	0.273 = 27.3%	0.307 = 30.7%
5000	0.279 = 27.9%	0.301 = 30.1%

Table 6.18

Notice that as the sample size gets larger, the confidence limits get closer to each other, thus a more accurate measure of the interval estimate of the population proportion is created. Therefore, we can decrease the width of the confidence interval by increasing the sample size.

B. Sample Size for estimating the population proportion

Similar to estimating the sample size for a study about the mean of a population, we can also determine the sample size for estimating the population proportion. We can derive the sample size formulas for estimating the population proportion using the margin of error in estimating p.

$$\text{Margin of Error} = E = z_{\alpha/2}\sqrt{\frac{\hat{p}\hat{q}}{n}} \Rightarrow E = \frac{z_{\alpha/2}\sqrt{\hat{p}\hat{q}}}{\sqrt{n}} \Rightarrow \sqrt{n} = \frac{z_{\alpha/2}\sqrt{\hat{p}\hat{q}}}{E} \Rightarrow n = \frac{z_{\alpha/2}^2\hat{p}\hat{q}}{E^2}$$

Sample size formula for estimating the population proportion ($\hat{p}$ known)

$$n = \frac{z_{\alpha/2}^2 \hat{p}\hat{q}}{E^2} \tag{6.18}$$

When no estimate of p is available ($\hat{p}$ is unknown), under certain conditions, the sample size n can be found by letting $\hat{p} = \hat{q} = 0.5$ (The proof is in section 6.10).

Sample size formula (6.18) for estimating the population proportion ($\hat{p}$ unknown)

$$n = \frac{z_{\alpha/2}^2 (0.5)(0.5)}{E^2} = \frac{z_{\alpha/2}^2 (0.25)}{E^2} \tag{6.19}$$

Round off rule for Sample Size: Always round the sample size to the next whole number.

Example 28: Kristine (A forensic scientist) wants to estimate the percentage of unsolved crimes in her division. How many samples must she include in her study in order to be 99% confident that the error is no more than 2%?

Solution: The confidence level $1 - \alpha = 0.99 \Rightarrow z_{\alpha/2} = 2.575$. Since no estimate of p is available, we let $\hat{p} = \hat{q} = 0.5$, and $E = 0.02$. Substituting the information into (6.19), we get

$$n = \frac{z_{\alpha/2}^2 (0.25)}{E^2} \Rightarrow n = \frac{(2.575)^2 (0.25)}{0.02^2} = 4144.14 \approx 4145$$

Thus, if Kristine takes a sample of 4145 cases, there is a 99% chance that the estimate of p will be within 2% of the true population proportion p.

Example 29: In a random sample of 500 teenagers, 150 have an i-Phone. How large a sample is required, if we want to be 95% confident that our estimate of p is within 3% of the true population proportion?

Solution: The confidence level $1 - \alpha = 0.95 \Rightarrow z_{\alpha/2} = 1.96$.

$\hat{p} = \frac{x}{n} = \frac{150}{500} = 0.3 \Rightarrow \hat{q} = 1 - \hat{p} = 1 - 0.3 = 0.7$, and the margin of error $E = 0.03$.

Substituting all the above data into $n = \frac{z_{\alpha/2}^2 \hat{p}\hat{q}}{E^2}$, we have

$$n = \frac{z_{\alpha/2}^2 \hat{p}\hat{q}}{E^2} = \frac{(1.96)^2 (0.3)(0.7)}{0.03^2} = 896.37 \approx 897.$$

Thus, if you take a sample of 897 teenagers, there is 95% chance that the estimate of p will be within 3% of the true population proportion.

Example 30: Suppose a sample of 300 auto parts produced by a machine showed a 3% rate of defect. How large a sample should the quality control manager select in order to be 90% confident that the sample proportion is within 2% of the true population proportion?

Solution: The confidence level $1 - \alpha = 0.90 \Rightarrow z_{\alpha/2} = 1.645$. From the preliminary sample, $\hat{p} = 0.03 \Rightarrow \hat{q} = 0.97$, and the margin of error $E = 0.02$. Substituting all the information into $n = \dfrac{z_{\alpha/2}^2 \hat{p}\hat{q}}{E^2}$, we have

$$n = \frac{z_{\alpha/2}^2 \hat{p}\hat{q}}{E^2} = \frac{(1.645)^2(0.03)(0.97)}{0.02^2} = 196.8633 \approx 197.$$

Thus, if the quality control manager takes a sample of 197 items, there is a 90% chance that the estimate of p will be within 2% of the population proportion.

Try it yourself:

In exercises 1 and 2, construct a 98% confidence interval for the true proportion *p*.

1. $\hat{p} = 0.28$, $n = 160$ **2.** $x = 80$, $n = 490$

In exercises 3 and 4, $x = 96$, $n = 300$.

3. Construct a 99% confidence interval for p.

4. Construct a 88% confidence interval for p.

5. Construct a 90% confidence interval for the percentage of all students who have laptops.

6. In a random sample of 150 customers at a large department store, 75% used credit cards for their purchases. Construct a 92% confidence interval for the true proportion of all credit card purchases.

7. In a random sample of 200 patients of an Internist, 140 are referrals from other local family practice physicians. How large a sample is needed if we wish to be 97% confident that our sample proportion is within 4% of the true proportion of the referrals?

8. A consumer agency wants to estimate the proportion of all drivers who wear seatbelts while driving. How large should the sample size be so that the 96% confidence interval for the population proportion has a margin of error of 1%?

6.8: Estimation of the Population Variance and Standard Deviation (σ^2/σ)

Objectives

After completing this section you will be able to understand:

- **A.** Chi-Square (χ^2) Distribution.
- **B.** Critical Values $(\chi_R^2$ and $\chi_L^2)$ of Chi-Square Distribution.
- **C.** Interval Estimate of σ^2 for large samples.
- **D.** Confidence Interval for σ and σ^2 for small samples.

In order to construct a confidence interval for population standard deviation σ and variance σ^2 we must use a Chi-Square (χ^2) Distribution. Therefore, we will start this section with a brief introduction to the Chi-Square Distribution.

A. Chi-Square (χ^2) Distribution

Similar to *t*-distribution, the Chi-Square Distribution is characterized by its degrees of freedom denoted by υ (the Greek letter υ is pronounced as 'Nu'). Chi-Square distribution has $\upsilon = n - 1$ degrees of freedom. A random variable having a Chi-Square distribution with υ degrees of freedom is expressed as $\chi^2 \sim \chi_\upsilon^2 = \chi_{n-1}^2$. A Chi-Square distribution curve starts at the origin and lies entirely to the right of the vertical axis.

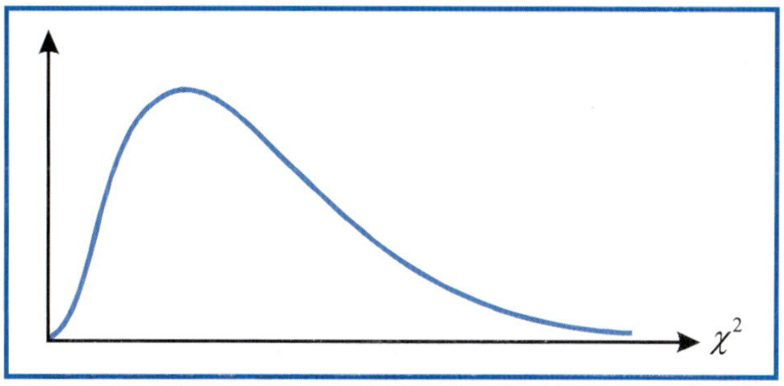

Figure 6.22

Definition: *The χ^2 distribution has only one parameter, called the degrees of freedom $\upsilon = n - 1$. The shape of the Chi-Square distribution curve is skewed to the right for small df and becomes symmetric as the df increases. The Chi-Square distribution lies entirely to the right of the vertical axis, hence a random variable that follows a χ^2 distribution takes on non-negative values.*

- **Properties of the Chi-Square Distribution:**

1. Since $p(\chi^2) \geq 0$, no portion of the probability curve lies below the χ^2 axis.
2. For Chi-Square distribution with υ degrees of freedom,
$$\text{Mean} = \upsilon,\ \text{variance} = 2\upsilon,\ \text{and Mode} = \upsilon - 2 \qquad (6.20)$$
3. For Chi-Square distribution with υ degrees of freedom,
$$\text{Skewness} = \frac{\text{Mean} - \text{Mode}}{\text{Standard deviation}} = \frac{\upsilon-(\upsilon-2)}{\sqrt{2\upsilon}} = \sqrt{\frac{2}{\upsilon}} \qquad (6.21)$$
4. For $\upsilon \geq 1$, Skewness > 0. Hence, the Chi-Square distribution is positively skewed for $\upsilon \geq 1$. Skewness tends to zero for large values of υ.
5. For large degrees of freedom, the Chi-Square distribution tends to be a symmetric distribution.
6. The shape of the Chi-Square probability curve depends on the degrees of freedom ν and changes with different values of υ. A rough sketch of the Chi-Square probability curves for $\upsilon = 1, 2, 3, 4, 5$ and 6 is given in Figure 6.23.

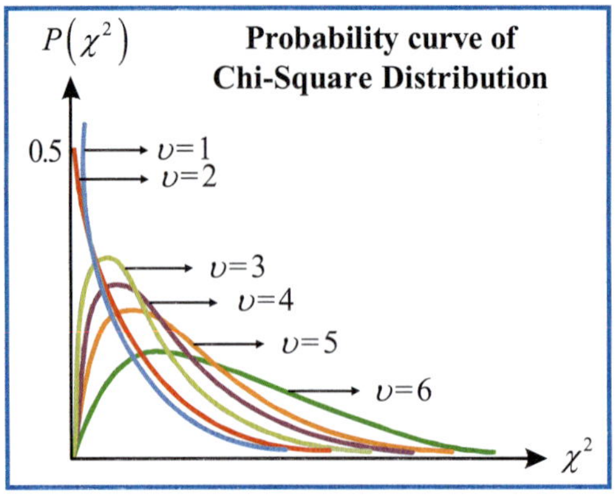

Figure 6.23

B. Critical Values $\left(\chi_R^2 \text{ and } \chi_L^2\right)$ of the Chi-Square Distribution

Let's define χ_R^2 as the right tail critical value of Chi-Square distribution where $P\left(\chi^2 > \chi_R^2\right) = \frac{\alpha}{2}$ and χ_L^2 as the left tail critical value of Chi-Square distribution where $P\left(\chi^2 < \chi_L^2\right) = \frac{\alpha}{2}$ or $P\left(\chi^2 > \chi_L^2\right) = 1 - \frac{\alpha}{2}$. Therefore $P\left(\chi_L^2 < \chi^2 < \chi_R^2\right) = 1 - \alpha$.

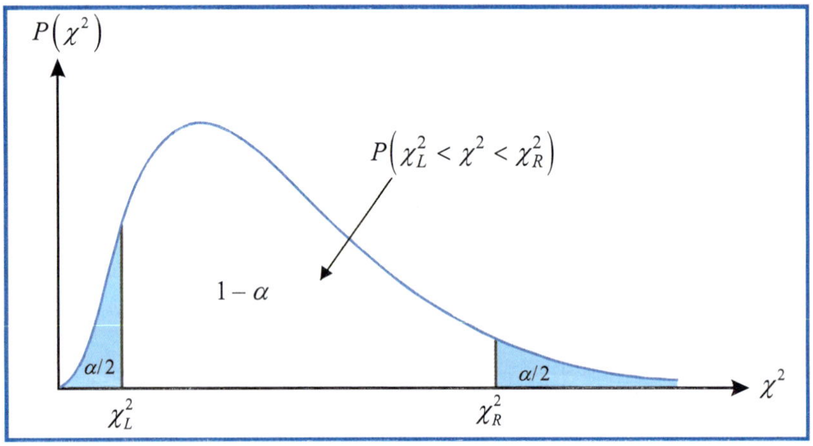

Figure 6.24

Chi-Square values for a given probability (area in the right tail of the Chi-Square distribution) and υ degrees of freedom have been tabulated in **Table 7** at the end of the book.

In Interval Estimation, we always work with a two sided distribution as shown in the Figure 6.24. Therefore, we must find both critical values χ_L^2 and χ_R^2. To find the critical values for the χ^2 distribution, we must calculate $\frac{\alpha}{2} = \frac{1-\text{CL}}{2}$ (area in the right tail of the Chi-Square distribution) and the degrees of freedom $\upsilon = n - 1$. Once we know the area in the right tail of the Chi-Square distribution and the degrees of freedom, we can find the value of χ_R^2 from **Table 7**.

To find the value for χ_L^2, we must calculate $1 - \frac{\alpha}{2} = 1 - \frac{1-\text{CL}}{2} = \frac{1+\text{CL}}{2}$ and the degrees of freedom $\upsilon = n - 1$.

Notation:	$\chi_R^2 = \chi_{n-1}^2\left(\frac{\alpha}{2}\right) = \chi_{n-1}^2\left(\frac{1-\text{CL}}{2}\right)$, $\chi_L^2 = \chi_{n-1}^2\left(1-\frac{\alpha}{2}\right) = \chi_{n-1}^2\left(\frac{1+\text{CL}}{2}\right)$

Example 31: Find the critical values of a Chi-Square distribution when the confidence level is 99% and $n = 11$.

Solution: $\alpha = 99\%$ and $n = 11$

To find χ_R^2:

In order to find χ_R^2 value from the χ^2 table, we must find the area to the right of the χ_R^2 value $\alpha/2$, and the degree of freedom $n - 1$.

Area to the right of χ_R^2 for a two tail calculation

$$= \frac{\alpha}{2} = \frac{1-\text{CL}}{2} = \frac{1-0.99}{2} = 0.005.$$

Degree of freedom $= \upsilon = n - 1 = 11 - 1 = 10$.

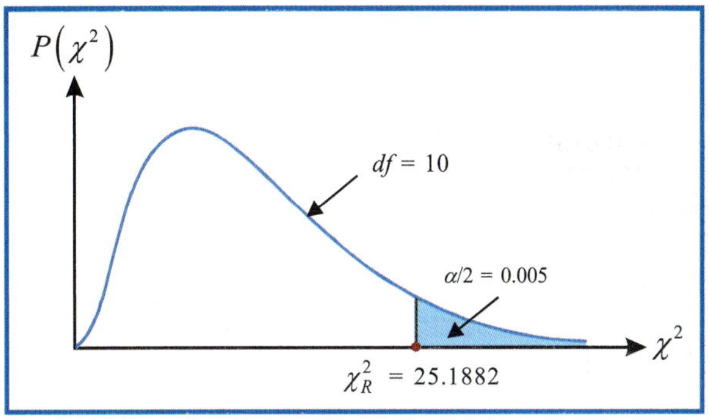

Figure 6.25

Now to find χ_R^2, we locate the $df = 10$, in the left outmost column of the χ^2 table, and the area in the right tail of the χ^2 curve = 0.005 in the top row. The intersection of the row for 10 df and the column for 0.005 area is the value for $\chi_R^2 = 25.1882$.

To compute the χ_L^2 value, we use the same df and an area of $1 - \frac{\alpha}{2}$; which is the area to the right of χ_L^2 under χ^2 curve.

Degree of freedom $= \upsilon = n - 1 = 10$, and the area to the right of $\chi_L^2 = 1 - \frac{\alpha}{2} = 1 - 0.005 = 0.995$.

Therefore, χ_L^2 is the value in the χ^2 table corresponding to the intersection of the row with $df = 10$ and the column with an area of 0.995 at $\chi_R^2 = 2.156$.

***Alternative Process* for Computing the Chi-Square Right (χ_R^2):**

We must find the area to the right of the χ_R^2, and the degrees of freedom.

i) Area to the right of χ_R^2

$$= \frac{\alpha}{2} = \frac{1-\text{CL}}{2} = \frac{1-0.99}{2} = 0.005.$$

ii) Degrees of freedom $\upsilon = n - 1 = 11 - 1 = 10$.

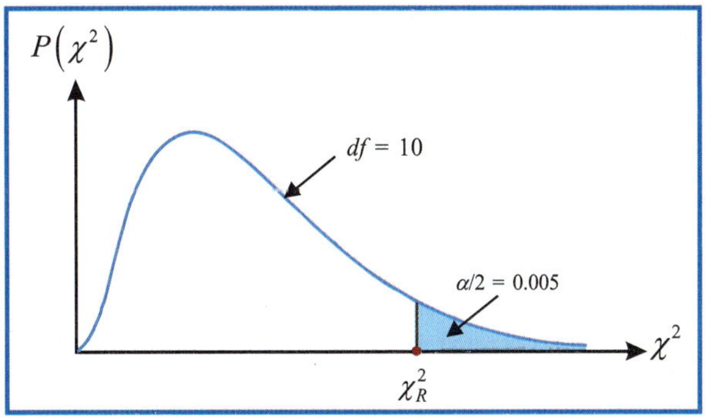

Figure 6.26

Therefore, to read the χ_R^2, we locate the degrees of freedom $\upsilon = 10$ in the first column of χ^2 table and the area in the right tail of the χ^2 curve $\frac{\alpha}{2} = 0.005$ in the first row of the χ^2 table.

The intersection of these two values ($\upsilon = 10$, $\frac{\alpha}{2} = 0.005$) in the body of table 6.19 is 25.1882.

(Format of Chi-Square distribution Table)

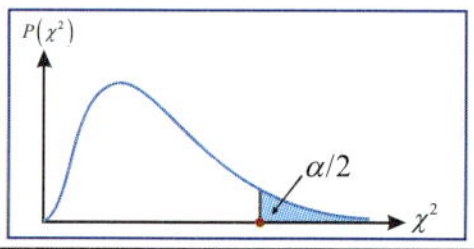

df	0.995	0.99	0.975	0.95	0.9	0.75	0.5	0.25	0.1	0.05	0.025	0.01	0.005
1	0	0.0002	0.0010	0.0039	0.0158	0.1015	0.4549	1.3233	2.7055	3.8415	5.0239	6.6349	7.8794
2	0.0100	0.0201	0.0506	0.1026	0.2107	0.5754	1.3863	2.7726	4.6052	5.9915	7.3778	9.2103	10.5966
3	0.0717	0.1148	0.2158	0.3518	0.5844	1.2125	2.3660	4.1083	6.2514	7.8147	9.3484	11.3449	12.8382
4	0.2070	0.2971	0.4844	0.7107	1.0636	1.9226	3.3567	5.3853	7.7794	9.4877	11.1433	13.2767	14.8603
5	0.4117	0.5543	0.8312	1.1455	1.6103	2.6746	4.3515	6.6257	9.2364	11.0705	12.8325	15.0863	16.7496
6	0.6757	0.8721	1.2373	1.6354	2.2041	3.4546	5.3481	7.8408	10.6446	12.5916	14.4494	16.8119	18.5476
7	0.9893	1.2390	1.6899	2.1673	2.8331	4.2549	6.3458	9.0371	12.0170	14.0671	16.0128	18.4753	20.2777
8	1.3444	1.6465	2.1797	2.7326	3.4895	5.0706	7.3441	10.2189	13.3616	15.5073	17.5345	20.0902	21.9550
9	1.7349	2.0879	2.7004	3.3251	4.1682	5.8988	8.3428	11.3888	14.6837	16.9190	19.0228	21.666	23.5894
10	2.1559	2.5582	3.247	3.9403	4.8652	6.7372	9.3418	12.5489	15.9872	18.3070	20.4832	23.2093	**25.1882**
11	2.6032	3.0535	3.8157	4.5748	5.5778	7.5841	10.3410	13.7007	17.2750	19.6751	21.9200	24.7250	26.7568
12	3.0738	3.5706	4.4038	5.226	6.3038	8.4384	11.3403	14.8454	18.5493	21.0261	23.3367	26.2170	28.2995

Table 6.19

For computing Chi-Square Left $\left(\chi_L^2\right)$ value, we use the same degrees of freedom, and the complement of the area calculated for χ_R^2

$$1-\frac{\alpha}{2}=1-\left(\frac{1-\text{CL}}{2}\right)=\frac{1+\text{CL}}{2}=\frac{1+0.99}{2}$$
$$=0.995$$

Therefore, χ_L^2 can be found at the intersection of 10 degrees of freedom row and 0.995 column from the χ^2 Table 6.20- i.e., $\chi_L^2=2.1560$.

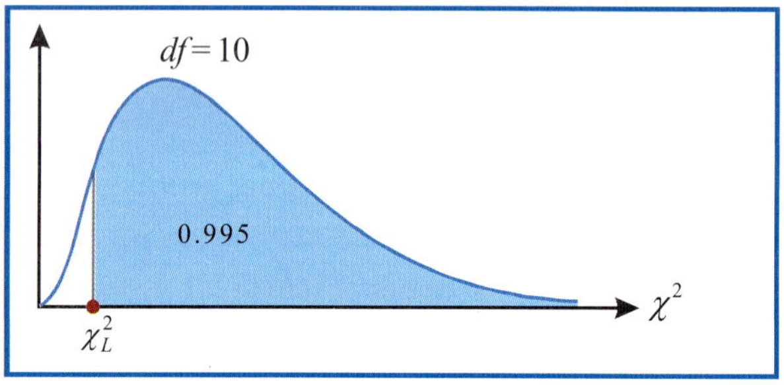

Figure 6.27

Area in the right tail of the Chi-Square distribution

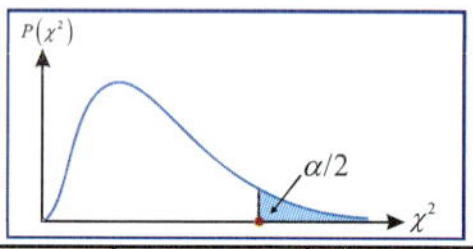

df	0.995	0.99	0.975	0.95	0.9	0.75	0.5	0.25	0.1	0.05	0.025	0.01	0.005
1	0	0.0002	0.0010	0.0039	0.0158	0.1015	0.4549	1.3233	2.7055	3.8415	5.0239	6.6349	7.8794
2	0.0100	0.0201	0.0506	0.1026	0.2107	0.5754	1.3863	2.7726	4.6052	5.9915	7.3778	9.2103	10.5966
3	0.0717	0.1148	0.2158	0.3518	0.5844	1.2125	2.3660	4.1083	6.2514	7.8147	9.3484	11.3449	12.8382
4	0.2070	0.2971	0.4844	0.7107	1.0636	1.9226	3.3567	5.3853	7.7794	9.4877	11.1433	13.2767	14.8603
5	0.4118	0.5543	0.8312	1.1455	1.6103	2.6746	4.3515	6.6257	9.2364	11.0705	12.8325	15.0863	16.7496
6	0.6757	0.8721	1.2373	1.6354	2.2041	3.4546	5.3481	7.8408	10.6446	12.5916	14.4494	16.8119	18.5476
7	0.9893	1.2390	1.6899	2.1673	2.8331	4.2549	6.3458	9.0371	12.0170	14.0671	16.0128	18.4753	20.2777
8	1.3444	1.6465	2.1797	2.7326	3.4895	5.0706	7.3441	10.2189	13.3616	15.5073	17.5345	20.0902	21.9549
9	1.7349	2.0879	2.7004	3.3251	4.1682	5.8988	8.3428	11.3888	14.6837	16.9190	19.0228	21.6660	23.5894
10	**2.1560**	2.5582	3.2470	3.9400	4.8652	6.7372	9.3418	12.5490	15.9870	18.3070	20.4830	23.2090	25.1880
11	2.6030	3.0535	3.8157	4.5748	5.5778	7.5841	10.3410	13.7010	17.2750	19.6750	21.9200	24.7250	26.7570

Table 6.20

C. Interval Estimate of σ^2 for Large Samples

In a random sample of size n from an infinite (very large) population with variance σ^2, the variance of the sampling distribution of s^2.

$$\text{var}\left(s^2\right) = \sigma_{s^2}^2 = \frac{2\sigma^4}{n} \quad \Rightarrow \quad \sigma_{s^2} = \sigma^2\sqrt{\frac{2}{n}} \tag{6.22}$$

If σ^2 is unknown, for large samples, we can use $\hat{\sigma}^2 \approx s^2 \quad \Rightarrow \quad \hat{\sigma}_{s^2} = s^2 \cdot \sqrt{\frac{2}{n}}$ and

$$\text{Margin of error } E = z_{\alpha/2} \cdot \hat{\sigma}_{s2} = z_{\alpha/2}\, s^2 \cdot \sqrt{\frac{2}{n}} \tag{6.23}$$

Hence, $100(1 - \alpha)\%$ confidence interval for the population variance σ^2 is given by:

$$s^2 - E < \sigma^2 < s^2 + E$$

$$s^2 - z_{\alpha/2} s^2 \sqrt{\frac{2}{n}} < \sigma^2 < s^2 + z_{\alpha/2} s^2 \sqrt{\frac{2}{n}}$$

$$s^2\left(1 - z_{\alpha/2}\sqrt{\frac{2}{n}}\right) < \sigma^2 < s^2\left(1 + z_{\alpha/2}\sqrt{\frac{2}{n}}\right)$$

Therefore, the $100(1 - \alpha)\%$ confidence interval for the population standard deviation σ is given by:

$$s\sqrt{1 - z_{\alpha/2}\sqrt{\frac{2}{n}}} < \sigma < s\sqrt{1 + z_{\alpha/2}\sqrt{\frac{2}{n}}} \tag{6.24}$$

D. Confidence Interval for σ and σ^2 for Small Samples based on the Chi-Square Distribution

To obtain an interval estimate of σ or σ^2 for small samples, the following assumptions must be made.

(a) The sample is a simple random sample,

(b) The parent population from which the sample is drawn has a normal distribution. It is known that for a random sample of size n from a normal $N(\mu, \sigma)$ population, the statistics is

$$\chi^2 = \frac{(n-1)s^2}{\sigma^2} \quad \chi_{n-1}^2 .$$

Unlike the normal distribution or the t-distribution, the Chi-Square distribution is not symmetric. Accordingly, the confidence interval for σ^2 cannot be obtained by using the formula $\sigma^2 \pm E$; therefore, we need to compute the lower and upper confidence limits separately.

To obtain the $100(1 - \alpha)\%$ confidence interval for σ^2, we start with

$$P\left(\chi_L^2 < \chi^2 < \chi_R^2\right) = 1 - \alpha \tag{6.25}$$

Substituting $\chi^2 = \dfrac{(n-1)s^2}{\sigma^2}$ into (6.25), we get

$$P\left(\chi_L^2 < \frac{(n-1)s^2}{\sigma^2} < \chi_R^2\right) = 1 - \alpha \quad \Rightarrow \quad P\left(\frac{(n-1)s^2}{\chi_R^2} < \sigma^2 < \frac{(n-1)s^2}{\chi_L^2}\right) = 1 - \alpha.$$

Therefore, $100(1-\alpha)\%$ confidence interval for σ^2 is given by:

$$\frac{(n-1)s^2}{\chi_R^2} < \sigma^2 < \frac{(n-1)s^2}{\chi_L^2} \tag{6.26}$$

Similarly, taking the square root of (6.26), $100(1-\alpha)\%$ confidence interval for population standard deviation σ is given by:

$$\sqrt{\frac{(n-1)s^2}{\chi_R^2}} < \sigma < \sqrt{\frac{(n-1)s^2}{\chi_L^2}} \tag{6.27}$$

Where $\chi_R^2 = \chi_{n-1}^2\left(\frac{\alpha}{2}\right) = \chi_{n-1}^2\left(\frac{1-\text{CL}}{2}\right)$ and $\chi_L^2 = \chi_{n-1}^2\left(1-\frac{\alpha}{2}\right) = \chi_{n-1}^2\left(\frac{1+\text{CL}}{2}\right)$

Example 32: From a random sample of 150 employees in a large corporation, the standard deviation of the monthly salaries was found to be \$580. Construct a 99% confidence interval for the standard deviation of the monthly salaries of all the employees in the company.

Solution: Since the sample was taken from a large population, we must use the formula

$$s\sqrt{1 - z_{\alpha/2}\sqrt{\frac{2}{n}}} < \sigma < s\sqrt{1 + z_{\alpha/2}\sqrt{\frac{2}{n}}} \tag{6.28}$$

to construct a 99% confidence interval for σ.

We have $n = 150$, $s = 580$, and $z_{\alpha/2} = 2.575$.

Substituting all the data into (6.28), we get

$$580\sqrt{1-(2.575)\sqrt{\frac{2}{150}}} < \sigma < 580\sqrt{1+(2.575)\sqrt{\frac{2}{150}}} = 486.19 < \sigma < 660.62$$

Hence, with a 99% confidence, we can conclude that the standard deviation of the monthly salaries of all employees in this company is between \$486.19 and \$660.62.

Example 33: Given a random sample of size 20 from a normally distributed population, construct a 95% confidence interval for σ if the sample variance $s^2 = 28$.

Solution: We are given $n = 20 \Rightarrow$ degrees of freedom $\upsilon = n - 1 = 20 - 1 = 19$.
Sample variance $s^2 = 28$.

For a 95% confidence level with 19 degrees of freedom, we have

$$\chi_R^2 = \chi_{20-1}^2\left(\frac{1-0.95}{2}\right) = \chi_{19}^2\ (0.025) = 32.852 \text{ and } \chi_R^2 = \chi_{20-1}^2\left(\frac{1+0.95}{2}\right) = \chi_{19}^2\ (0.975) = 8.907.$$

Therefore, the 95% confidence interval for population standard deviation σ is given by:

$$\sqrt{\frac{(n-1)s^2}{\chi_R^2}} < \sigma < \sqrt{\frac{(n-1)s^2}{\chi_L^2}}$$

$$\Rightarrow \sqrt{\frac{(20-1)\times 28}{32.852}} < \sigma < \sqrt{\frac{(20-1)\times 28}{8.907}}$$

$$4.02 < \sigma < 7.73$$

Hence, with 95% confidence, we can conclude that the population standard deviation lies between 4.02 and 7.73.

Example 34: The following are the ages of 22 Mathematics faculty in a large community college.

56	62	61	54	52	32	25	31	48	45	53
52	42	35	37	29	52	54	43	75	36	50

Assuming the population is normally distributed, find

(a) the point estimate of σ^2 and **(b)** a 98% confidence interval for σ^2.

Solutions: From the given data, $\sum x = 1024$ and $\sum x^2 = 50862$

(a) The point estimate:

$$\sigma^2 = s^2 = \frac{n\sum x^2 - \left(\sum x\right)^2}{n(n-1)} = \frac{22(50862) - 1024^2}{22(21)} = 152.3549784 \approx 152.355.$$

(b) Given a 98% confidence interval, with $\upsilon = 21$, the critical values are as follows:

$$\chi_R^2 = \chi_{22-1}^2\left(\frac{1-0.98}{2}\right) = \chi_{21}^2(0.01) = 38.932 \text{ and } \chi_L^2 = \chi_{22-1}^2\left(\frac{1+0.98}{2}\right) = \chi_{21}^2(0.99) = 8.897$$

Therefore, the 98% confidence interval for population variance σ^2 is given by:

$$\frac{(n-1)s^2}{\chi_R^2} < \sigma^2 < \frac{(n-1)s^2}{\chi_L^2}$$

$$\Rightarrow \quad \frac{(22-1)(152.355)}{38.932} < \sigma^2 < \frac{(22-1)(152.355)}{8.897}$$

$$82.18 < \sigma^2 < 359.61$$

Hence, with 98% confidence, we can say that the population variance lies between 82.18 and 359.61.

Try it yourself:

1. In a random sample of size 42 from a normally distributed population, the sample variance is 2.14. Use this information to find point estimate of σ^2 and construct a

   (a) 90% confidence interval for σ^2. **(b)** 98% confidence interval for σ^2.

2. In a random sample of size 200 from a large population, the sample standard deviation is 45.56. Construct a 95% confidence interval for σ^2.

3. A random sample of 10 purchases at a store revealed the following amounts (in \$).

5.00	3.5	1.25	2.30	3.00	4.20	5.30	1.50	1.75	2.45

   Assuming the parent population is normally distributed, find a 95% confidence interval for σ.

4. Here are the weights (in lbs.) of 15 randomly selected students.

80	76	90	106	94
86	110	96	112	114
108	115	99	100	107

   Obtain the point estimate of the population variance and construct a 99% confidence interval for σ^2. (Assume the parent population is normally distributed.)

6.9: Theory of Estimation with Microsoft Excel®

Objectives

After completing this section you will be able to carry out the computations in Chapter 6 using Microsoft Excel®.

The use of spreadsheets and technology in recent years in Statistics has significantly reduced the tedium of unnecessary calculations and has allowed researchers to focus more on the results and what they mean. This is evermore true in the Theory of Estimation. The following demonstrations will show how to use the power of technology, more specifically spreadsheets, to produce results and interpret them.

We will start with listing the commands in Excel® that will enable you to calculate critical values such as t, z and χ^2.

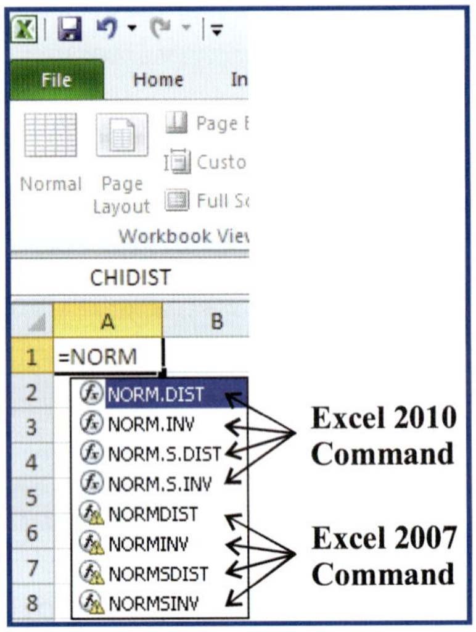

Figure 6.28

General Note: The commands with a dot are from Excel 2010 and without a dot are from Excel 2007 and earlier.

A. Normal Distribution

Standard Excel® Functions:

- **NORM.DIST(x, mean, standard_dev, cumulative)**

This function will return probability (area) of normal distribution for different values of x, mean, standard deviation and cumulative probabilities. Here the cumulative option assumes two values TRUE and FALSE. When, we use the cumulative option = TRUE, it will compute $P(X \leq x)$ and for cumulative option = FALSE, it will compute probability at $X = x$ by using continuity correction. That is, $P(x - 0.5 < X < x + 0.5)$.

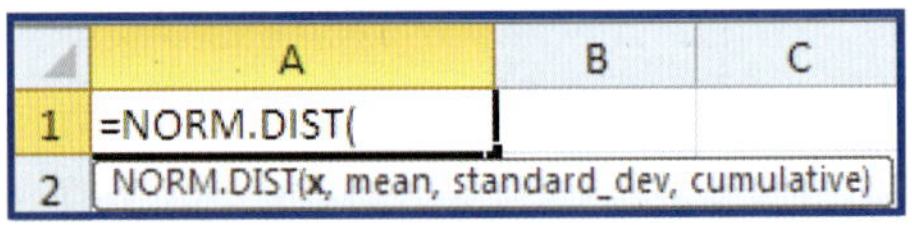

Figure 6.29

Illustration 1: X follows a normal distribution with mean 100 and standard deviation 15. Compute $P(X > 143)$.

Solution:

$$P(X > 143) = 1 - P(X \leq 143)$$
$$= 1 - \text{NORM.DIST}(143, 100, 15, \text{TRUE})$$
$$= 0.002074098 \approx 0.0021$$

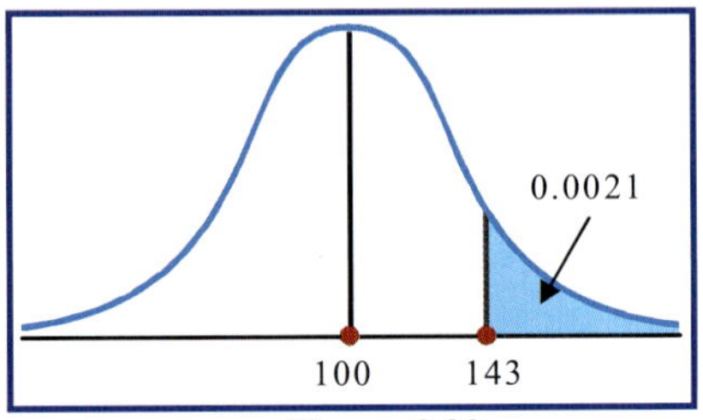

Figure 6.30

Illustration 2: Given $N(100, 15)$, find $P(X = 143)$.

Solution:

$$P(X = 143) = P(142.5 < X < 143.5)$$
$$= \text{NORM.DIST}(143, 100, 15, \text{FALSE})$$
$$= 0.000436869 \approx 0.0004$$

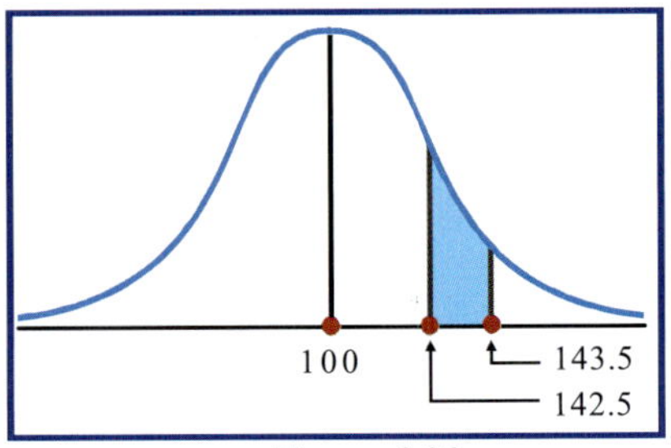

Figure 6.31

- **NORM.INV(probability, mean, standard deviation)**

This function returns the value of x such that $P(X < x) = p$, where X follows a normal distribution.

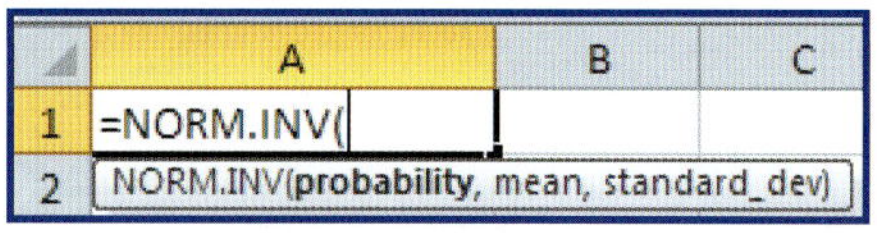

Figure 6.32 a

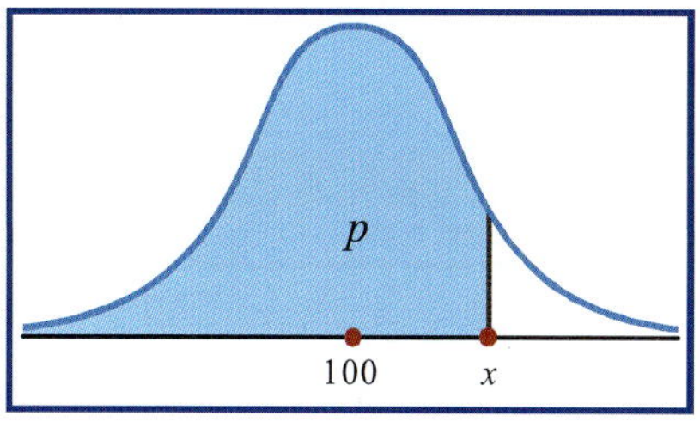

Figure 6.32 b

Illustration 3: Given a $N(100, 15)$, compute P_{95}.

Solution: P_{95} = NORM.INV(0.95, 100,15) = 124.673

- **NORM.S.DIST(*z*, cumulative)**

This function is similar to NORM.DIST( ). The only difference is that it returns probability (area) under standard normal curve.

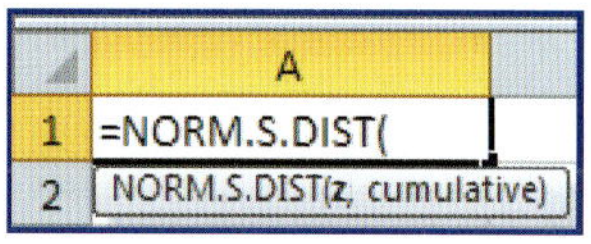

Figure 6.33

Illustration 4: Given a standard normal distribution, compute $P(Z < 1.96)$.

Solution: $P(Z < 1.96)$ = NORM.S.DIST(1.96,TRUE) $\Rightarrow$ 0.9750

$\Rightarrow P(Z < 1.96) = 0.9750$

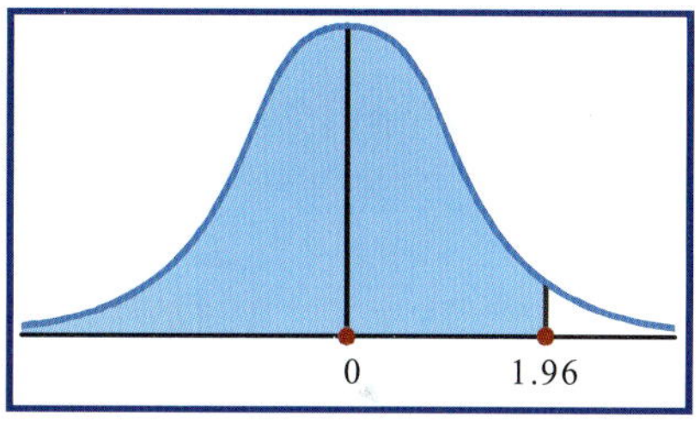

Figure 6.34

B. Student *t*-Distribution

Standard Excel® Function related to *t*-distribution:

- **T.DIST(*x*, deg_freedom, cumulative)**

This function will return probability (area) of the t-distribution for different values of x, degrees of freedom and cumulative option. Here the cumulative option assumes two values TRUE and FALSE. When we use the cumulative option = TRUE, then it computes $P(T \leq x)$; and when cumulative option = FALSE, it computes probability at $T = x$ by using continuity correction. That is, $P(x - 0.5 < T < x + 0.5)$.

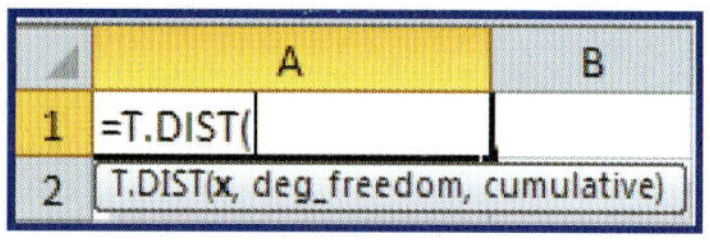

Figure 6.35

Illustration 5: T follows a t-distribution with degrees of freedom 12. Compute $P(T < 2.1788)$.

Solution:

$$P(T < 2.1788) = \text{T.DIST}(2.1788,12,\text{TRUE}) = 0.974999426 \approx 0.9750$$

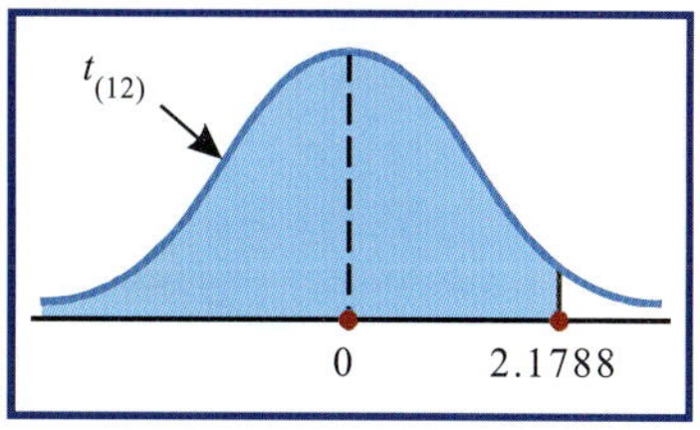

Figure 6.36

- **T.DIST.2T(x, deg_freedom)**

This function returns the area (probability) to the left of $x = -t$ and to the right of $x = +t$ as shown below (area of the shaded region) where $x = |t|$.

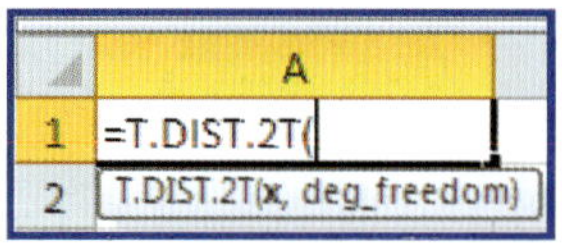

Figure 6.37 a

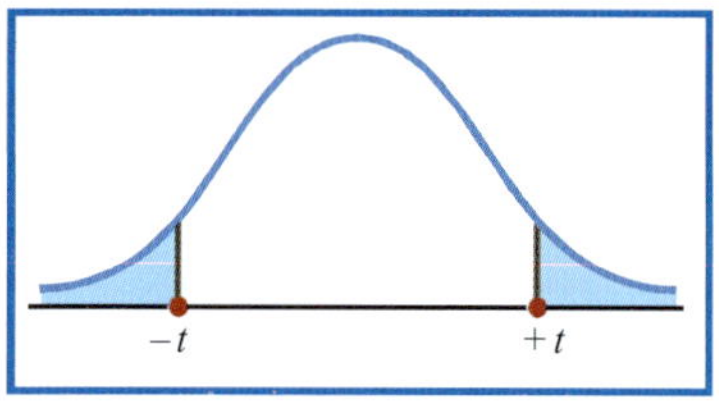

Figure 6.37 b

Illustration 6: Compute $P(-2.1788 < T < 2.1788)$, where the random variable T follows a t-distribution with 12 degrees of freedom.

Solution:

$$\begin{aligned} P(-2.1788 < T < 2.1788) &= 1 - P(T < -2.1788 \text{ or } T > 2.1788) \\ &= 1 - \text{T.DIST.2T}(2.1788,12) \\ &= 1 - 0.050001149 \\ &\approx 0.9499 \end{aligned}$$

- **T.DIST.RT(x, deg_freedom)**

This function returns the area (probability) to the right of a given $t = x$ value.

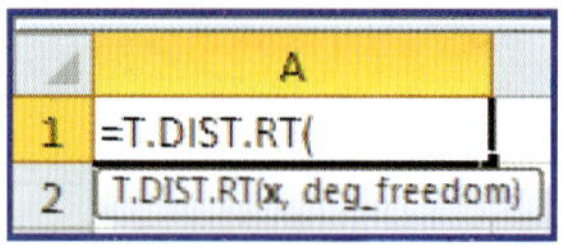

Figure 6.38 a

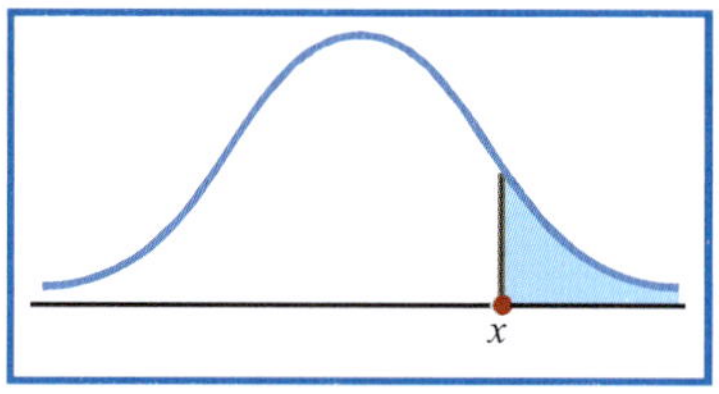

Figure 6.38 b

Illustration 7: Compute $P(T > 2.1788)$, where T follows a t-distribution with 12 degrees of freedom.

Solution:

$$\begin{aligned} P(T > 2.1788) &= \text{T.DIST.RT}(2.1788,12) \\ &= 0.02500074 \approx 0.025 \end{aligned}$$

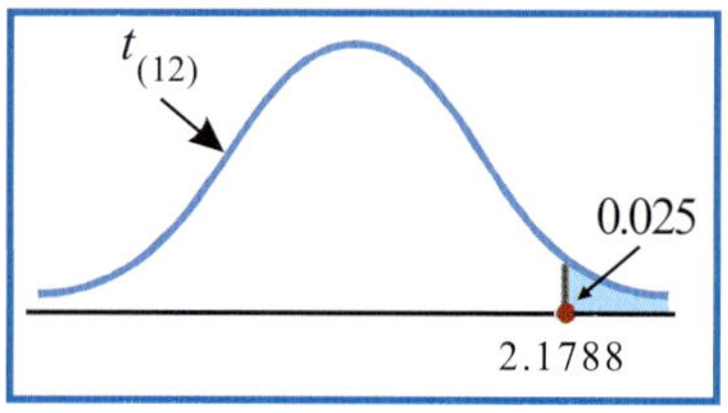

Figure 6.39

- **T.INV(probability, deg_freedom)**

This function returns the t-value for a given probability (area) to the left of $t = x$ and a given degrees of freedom.

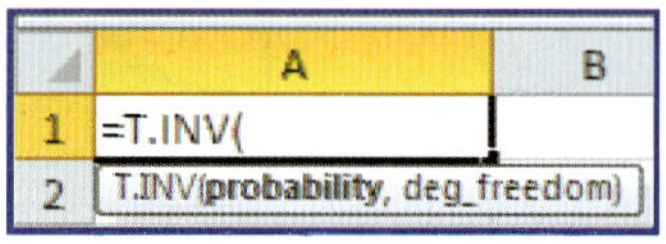

Figure 6.40 a

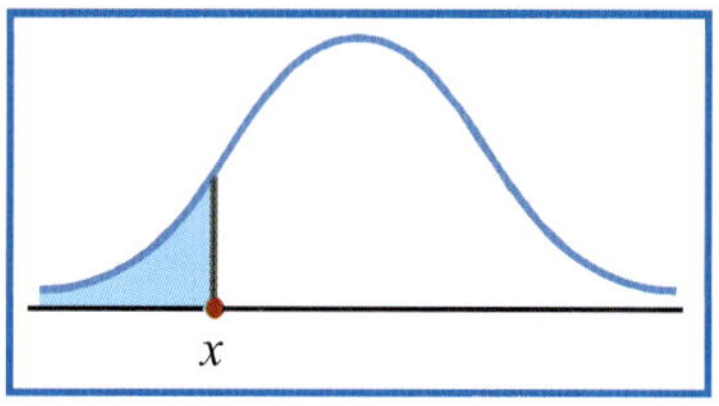

Figure 6.40 b

Illustration 8: $P(T < t) = 0.975$. Find t.

$$= \text{T.INV}(0.975,12) = 2.17881283$$

$$\Rightarrow t = 2.1788$$

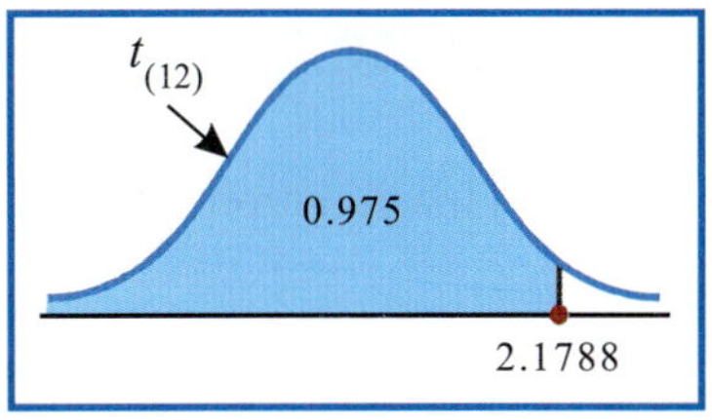

Figure 6.41

- **T.INV.2T(probability, deg_freedom)**

This function returns $|t|$ given the area (probability) of the shaded region in both tails of the t-distribution.

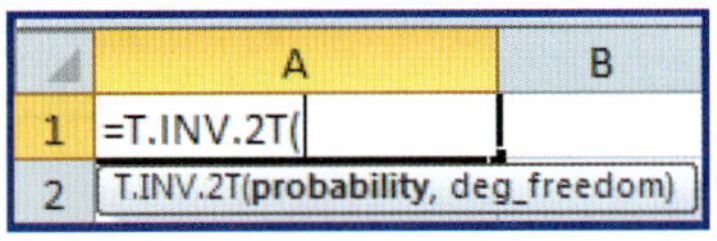

Figure 6.42 a

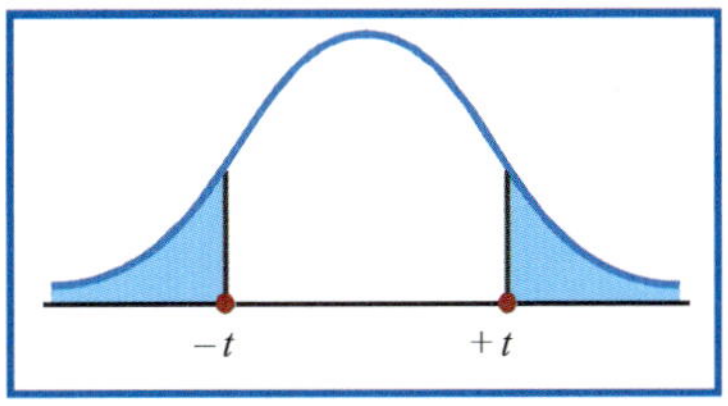

Figure 6.42 b

Illustration 9: Find the critical value for a 95% confidence level, given a t-distribution with 12 degrees of freedom.

Solution: The critical value for a 95% confidence level is:

$$= \text{T.INV.2T}(0.05,12) = 2.17881283 = 2.1788$$

$$\Rightarrow t = 2.1788$$

C. Chi-Square Distribution

Standard Excel® Functions:

- **CHISQ.DIST(probability, deg_freedom, cumulative)**

This function will return the probability (area) of the Chi-Square distribution for different values of x, degree of freedom and cumulative option. Here cumulative option assumes two values TRUE and FALSE. When we use the cumulative option = TRUE,

Figure 6.43

it computes $P\left(\chi^2 < x\right)$; and when the cumulative option = FALSE, it computes the probability at $\chi^2 = x$ by using continuity correction. That is, $P\left(x - 0.5 < \chi^2 < x + 0.5\right)$.

Illustration 10: Find $P\left(\chi^2 < 23.337\right)$, where χ^2 represents a Chi-Square random variable with 12 degrees of freedom. That is $\chi^2 \sim \chi^2_{(12)}$.

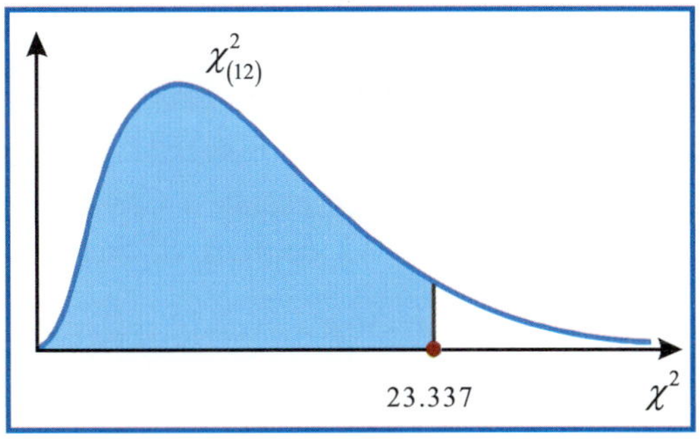

Figure 6.44

Solution: $P\left(\chi^2 < 23.337\right) = \text{CHISQ.DIST}(23.337,12,\text{TRUE})$

$$= 0.975003$$

- **CHISQ.DIST.RT(probability, deg_freedom)**

This function returns the probability (area) to the right of a given $\chi^2 = x$ value.

Figure 6.45 a

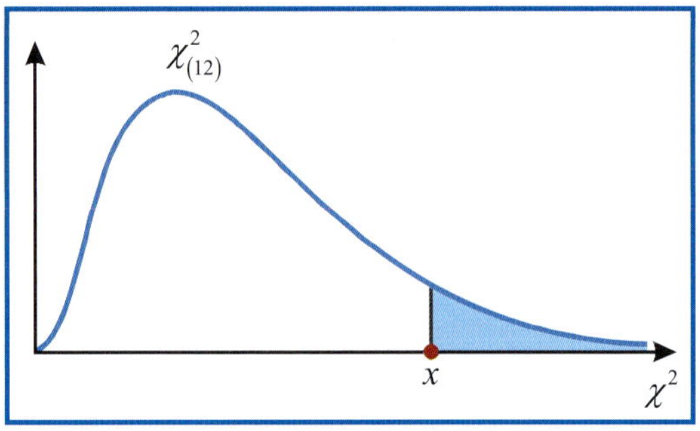

Figure 6.45 b

Illustration 11: Find $P(\chi^2 > 23.337)$ where χ^2 represents a Chi-Square random variable with 12 degrees of freedom.

Solution:

$$P(\chi^2 > 23.337) = \text{CHISQ.DIST.RT}(23.337,12)$$
$$= 0.024997409$$

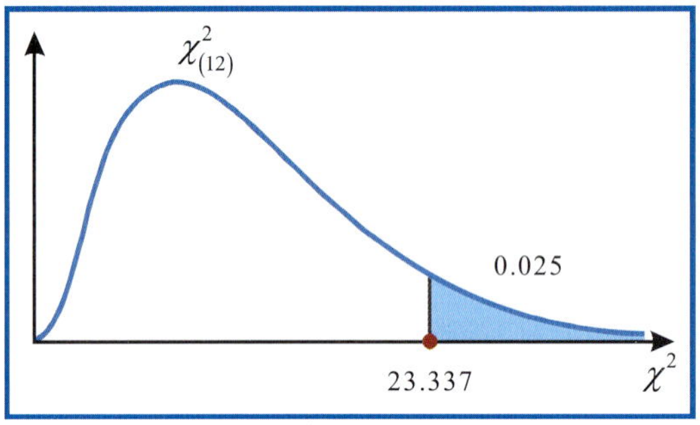

Figure 6.46

- **CHISQ.INV(probability, deg_freedom)**

This function returns the χ^2 value for a given area (probability) in the left tail of Chi-Square distribution. That is, it returns

χ^2_L value when probability = α and returns χ^2_R when probability = $1 - \alpha$.

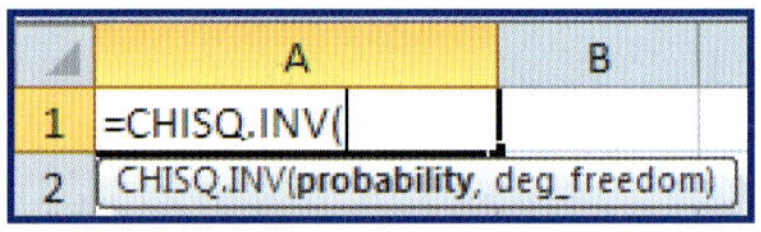

Figure 6.47

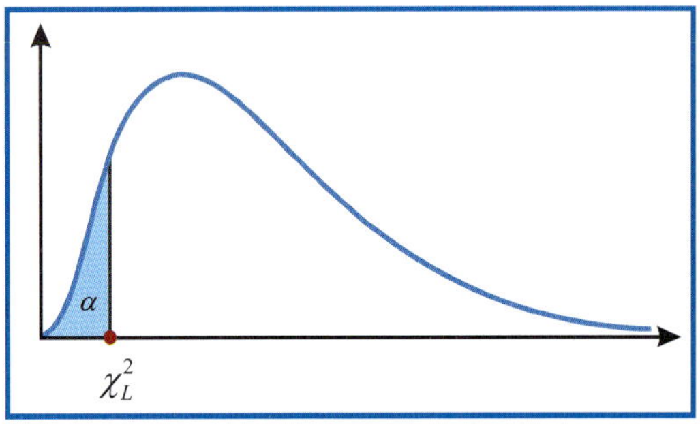

Figure 6.48 a

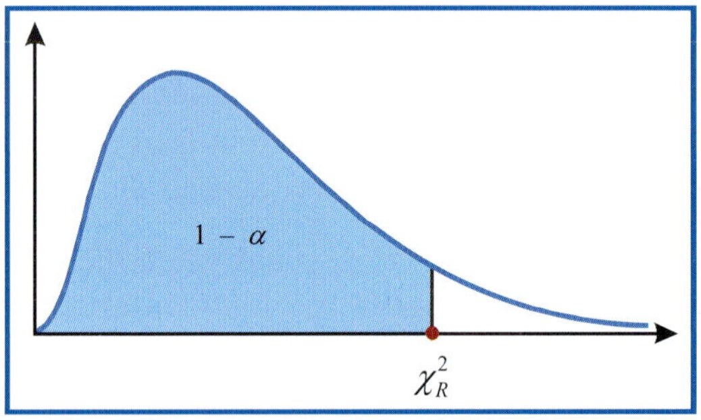

Figure 6.48 b

Illustration 12: Find χ^2_L given $P(\chi^2 < \chi^2_L) = 0.025$, where χ^2 represents a Chi-Square random variable with 12 degrees of freedom.

Solution:

$$\chi^2_L = \text{CHISQ.INV}(0.025,12)$$
$$= 4.403788507 \approx 4.4038$$

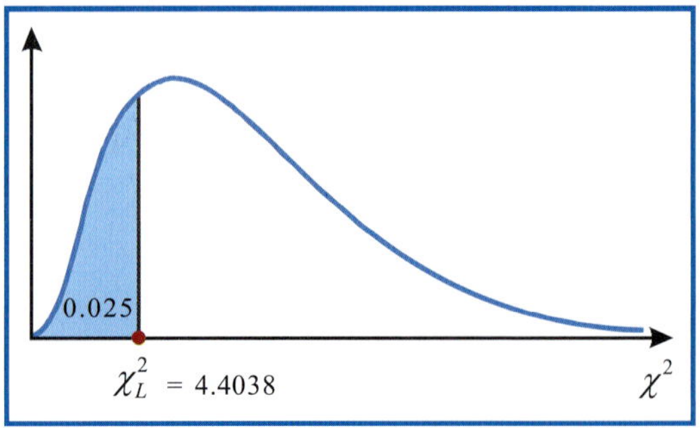

Figure 6.49

Illustration 13: Find χ_R^2 given $P\left(\chi^2 > \chi_R^2\right) = 0.025$, where χ^2 represents a Chi-Square random variable with 12 degrees of freedom.

Solution:

$$P\left(\chi^2 > \chi_R^2\right) = 0.025$$

$$\Rightarrow \quad P\left(\chi^2 < \chi_R^2\right) = 1 - 0.025 = 0.975$$

$$\Rightarrow \quad \chi_R^2 = \text{CHISQ.INV}(0.975,12)$$

$$= 23.33666416 \approx 23.3367$$

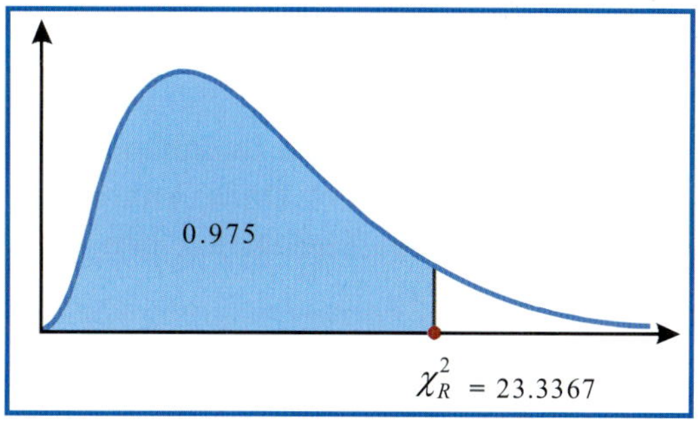

Figure 6.50

Confidence Interval Calculations Using Microsoft Excel®:

Illustration 14: The yearly income in ($) of 10 randomly selected Medical Doctors are as follows:

85000	140000	190000	100000	360000	75000	210000	170000	90000	50000

Construct a 95% confidence interval for the mean salary of Medical Doctors. Assume that the parent population is normally distributed.

Solution: The formula for obtaining a 95% confidence interval for computing the mean salary of medical doctors is

$$\overline{x} - t_{n-1,\,\alpha/2}\frac{s}{\sqrt{n}} < \mu < \overline{x} + t_{n-1,\,\alpha/2}\frac{s}{\sqrt{n}}\ .$$

We start with calculating the sample mean $\overline{x}$ and the sample standard deviation s for the sample data in the problem using the following commands.

To find the sample mean: = AVERAGE( )

To find the sample standard deviation: = STDEV( )

Next, we find the t-value (critical value) with degrees of freedom $\upsilon = n - 1$ using the t-table.

Now we must calculate the Margin of Error $E = t_{n-1,\,\alpha/2}\dfrac{s}{\sqrt{n}}$.

Once the Margin of Error is calculated, we can construct the LCL and UCL as $\overline{x} - E$ and $\overline{x} + E$ respectively. The display below shows the Excel® commands used to find the LCL and UCL.

	A	B	C	D
1	85000		x bar	=AVERAGE(A1:A10)
2	140000		s	=STDEV(A1:A10)
3	190000		n	=COUNT(A1:A10
4	100000		$dof\ (n-1)$	=D3 – 1
5	360000		t value for 95% CL ($\alpha = 0.05$)	=T.INV.2T(0.05,9)
6	75000		Margin of Error (E)	=D5*D2/SQRT(D3)
7	210000		LCL (x bar – E)	=D1 – D6
8	170000		UCL (x bar + E)	=D1 + D6
9	90000			
10	50000			

Table 6.21 a

	A	B	C	D
1	85000		x bar	147000
2	140000		s	91687.87633
3	190000		n	10
4	100000		$dof\ (n-1)$	9
5	360000		t value for 95% CL ($\alpha = 0.05$)	2.262157158
6	75000		Margin of Error (E)	65589.5554
7	210000		LCL (x bar – E)	81410.4446
8	170000		UCL (x bar + E)	212589.5554
9	90000			
10	50000			

Table 6.21 b

Therefore \$81,410 < μ < \$212590.

Illustration 15: Carry out the calculations for the previous demonstration using the Descriptive Statistics package under the Data Analysis group.

Solution: Input the data in column A. Select **the Data Tab.** Select **Data Analysis** from **Analysis Group.** The Dialog box (Figure 6.51) will appear.

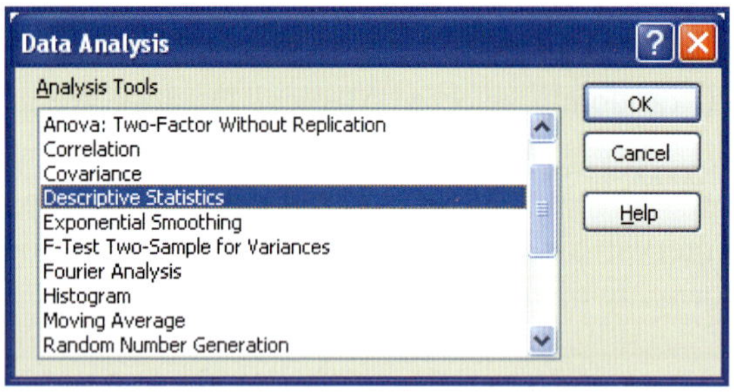

Figure 6.51

Select **Descriptive Statistics.** Press **OK.** The dialog box (Figure 6.52) should appear.

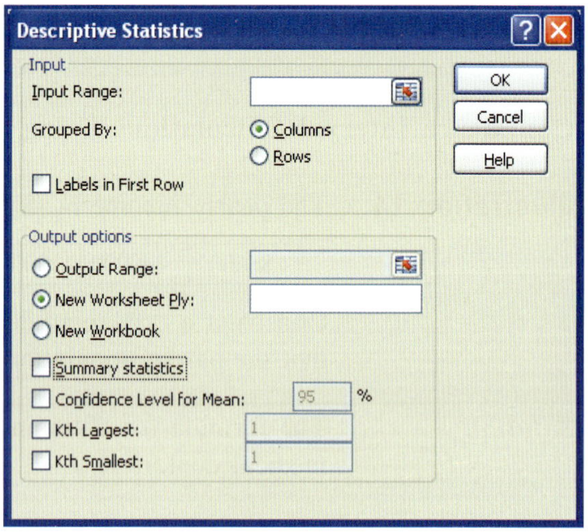

Figure 6.52

For the input range, select the inputted data in column A. Select the column option under **Grouped By:** since the data is in a column. Check the box for **Labels in First row**. Select **New Workbook**. Check the **Summary Statistics** box. Check the **Confidence Level** for **Mean box** and input the required Confidence Level.

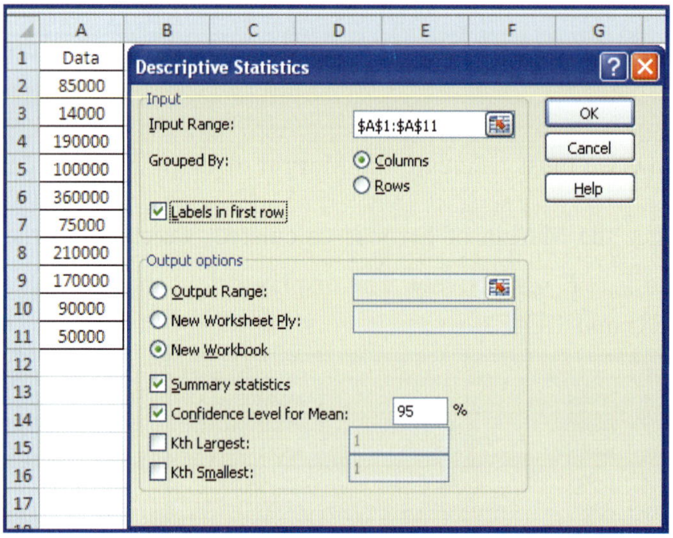

Figure 6.53

Press OK.

The following dialog box (Table 6.22) will appear.

	A	B
1	*Data*	
2		
3	Mean	147000
4	Standard Error	28994.2523
5	Median	120000
6	Mode	#N/A
7	Standard Deviation	91687.87633
8	Sample Variance	8406666667
9	Kurtosis	2.527208958
10	Skewness	1.476067885
11	Range	310000
12	Minimum	50000
13	Maximum	360000
14	Sum	1470000
15	Count	10
16	Confidence Level(95.0%)	65589.5554

Table 6.22

Recall the confidence interval formula for μ: $\bar{x} - E < \mu < \bar{x} - E$.

In Table 6.22, $\bar{x}$ is denoted by "Mean" = 147000 and the margin of error

$\left(E = t_{n-1,(\alpha/2)}\frac{s}{\sqrt{n}}\right)$ is denoted by "Confidence Level (95%)" = 65589.55553.

Using the values of $\bar{x}$ and E given in Table 6.22, the 95% confidence interval for μ is

$$147000 - 65589.55553 < \mu < 147000 + 65589.55553$$

$$81410.44 < \mu < 212589.56$$

Illustration 16: In a CD manufacturing plant, it is found that the defect rate in a sample of 35 randomly selected CD's is 29%. Construct a 95% confidence interval estimate of the true defect rate p for the entire production line.

Solution: The formula for computing 95% confidence interval of the true defect rate p for the entire production line is

$$\hat{p} - z_{\alpha/2}\sqrt{\frac{\hat{p}\hat{q}}{n}} < p < \hat{p} + z_{\alpha/2}\sqrt{\frac{\hat{p}\hat{q}}{n}}.$$

We have $\hat{p} = 0.29$, $\hat{q} = 1 - \hat{p} = 0.71$, $n = 35$, and $z_{\alpha/2} = 1.96$

	A	B	C	D
1	*p* hat	0.29	LCL	=B1 – B4*SQRT(B1*B2/B3)
2	*q* hat	=1 – B1	UCL	=B1 + B4*SQRT(B1*B2/B3)
3	*n*	35		
4	*z*-value	1.96		

Table 6.23 a

	A	B	C	D
1	*p* hat	0.29	LCL	0.13966842
2	*q* hat	0.71	UCL	0.44033158
3	*n*	35		
4	*z*-value	1.96		

Table 6.23 b

$$0.140 < p < 0.440$$

Illustration 17: The following collected data represents 12 randomly selected coolant level readings.

2.5	1.8	3.5	4.1	2.2	0.9	1.6	2.3	2.7	1.7	0.8	2.8

Assuming the population is normally distributed, construct a 95% confidence interval for σ.

Solution:

$$\sqrt{\frac{(n-1)s^2}{\chi_R^2}} < \sigma < \sqrt{\frac{(n-1)s^2}{\chi_L^2}}$$

	A	B	C	D
1	2.5		n	=COUNT(A1:A12)
2	1.8		$n-1$	=D1 – 1
3	3.5		s^2	=VAR.S(A1:A12)
4	4.1		$(1-\alpha)$ = 95%	α = 0.05
5	2.2		χ_R^2	=CHISQ.INV(0.975,11)
6	0.9		χ_L^2	=CHISQ.INV(0.025,11)
7	1.6			
8	2.3		LCL	=SQRT(D2*D3/D5)
9	2.7		UCL	=SQRT(D2*D3/D6)
10	1.7			
11	0.8			
12	2.8			

Table 6.24 a

	A	B	C	D
1	2.5		n	12
2	1.8		$n-1$	11
3	3.5		s^2	0.946287879
4	4.1		$(1-\alpha)$ = 95%	α = 0.05
5	2.2		χ_R^2	21.92004926
6	0.9		χ_L^2	3.815748252
7	1.6			
8	2.3		LCL	0.689107883
9	2.7		UCL	1.65165039
10	1.7			
11	0.8			
12	2.8			

Table 6.24 b

95% confidence interval for σ is $0.69 < \sigma < 1.65$.

Illustration 18: Professor Badalian wants to estimate the mean Final Exam grade for all his statistics classes. How many sample final scores must he use for this experiment to be 95% confident that the sample mean is within 2 points of the true student population mean. $\sigma = 8.3$ is given from a previous study.

Solution: $\sigma = 8.3$, $E = 2$, and the z-value for a 95% confidence level is 1.96

Using the formula $n = \left(\frac{z_{\alpha/2}\sigma}{E}\right)^2$, we have $n = \left(\frac{(1.96)(8.3)}{2}\right)^2 = 66.16 \approx 67$

Using Excel®;

	A	B	C	D	E	F
1	σ	8.3				
2	E	2		n	=(B3*B1/B2)^2	=CEILING(E2,1)
3	$z(\alpha/2)$	1.96				

Table 6.25 a

	A	B	C	D	E	F
1	σ	8.3				
2	E	2		n	66.162	67
3	$z(\alpha/2)$	1.96				

Table 6.25 b

The minimum sample size necessary to conduct this study is 67.

6.10: Extended Topics and Proofs in the Theory of Estimation (Optional)

A. Sampling Distributions of Mean

Theorem 6.1: Let $X_1, X_2, \ldots, X_n$ be a random sample of size n from a normally distributed population with mean μ and standard deviation σ. Then prove

that $\bar{X} = \dfrac{\sum_{i=1}^{n} X_i}{n}$ is normally distributed with

mean $\mu_{\bar{X}} = \mu$ and standard deviation

$$\sigma_{\bar{X}} = \frac{\sigma}{\sqrt{n}}$$

Proof :

i) $X_1, X_2, \ldots, X_n$ is a random sample from a normally distributed population with mean μ and standard deviation σ.

ii) are independent and normally distributed with $E(X_i) = \mu$ and $\sigma(X_i) = \sigma$. Therefore,

$$\bar{X} = \frac{\sum_{i=1}^{n} X_i}{n} = \frac{1}{n}(X_1) + \frac{1}{n}(X_2) + \cdots + \frac{1}{n}(X_n)$$

Since $\frac{1}{n}$ is a constant, we can conclude that $\bar{X}$ is a linear combination of $X_1, X_2, \ldots, X_n$.

Hence,
$$E(\bar{X}) = E\left[\frac{1}{n}(X_1) + \frac{1}{n}(X_2) + \ldots + \frac{1}{n}(X_n)\right]$$

$$= \frac{1}{n}(\mu) + \ldots + \frac{1}{n}(\mu) = \frac{1}{n}(n\mu) = \mu$$

and
$$\sigma(\bar{X}) = \sqrt{\operatorname{var}(\bar{X})}$$

$$= \sqrt{\operatorname{var}\left[\frac{1}{n}X_1 + \ldots + \frac{1}{n}X_n\right]}$$

$$= \sqrt{\frac{1}{n^2}\operatorname{var}(X_1) + \ldots + \frac{1}{n^2}\operatorname{var}(X_n)}$$

$$= \sqrt{\frac{1}{n^2}\left[\sigma^2 + \ldots + \sigma^2\right]} = \sqrt{\frac{1}{n^2}n\sigma^2} = \frac{\sigma}{\sqrt{n}}$$

Objectives

After completing this section you will be able to obtain:

- **A.** Sampling Distributions of Mean.
- **B.** Properties of Distribution of Proportion.
- **C.** Unbaised Estimator of Population Variation.
- **D.** Bias of a Point Estimator and Error in Estimation.
- **E.** Confidence Interval for μ, p.
- **F.** Sample Size Derivation from the Margin of Error.
- **G.** Estimation of Population Standard Deviation σ or Variance σ^2.
- **H.** Testing for Normality.

Note: $\bar{X}$ is a statistic that is a function of the random variables observed in a sample. Hence, the statistic itself is a random variable. The probability distribution of a statistic (random variable) is called the sampling distribution of that statistic.

- **Standardization of $\bar{X}$**

Notation:

i) $V(X_i) = \sigma^2$

ii) $V(\bar{X}_i) = \dfrac{\sigma^2}{n} = \sigma_{\bar{X}}^2$

The Z value for $\bar{X}$: $\bar{X} \sim N(\mu_{\bar{X}}, \sigma_{\bar{X}}^2) \Rightarrow Z_{\bar{X}} = \dfrac{\bar{X} - \mu_{\bar{X}}}{\sigma_{\bar{X}}} = \dfrac{\bar{X} - \mu}{\sigma / \sqrt{n}} \sim N(0,1)$

Example 35: An ice cream machine can be adjusted so that it dispenses an average of μ ounces per cone. From a previous study, it has been verified that the amount of ice cream dispensed by the machine is normally distributed with $\sigma = 1.0$ ounce. A sample of 12 ice creams is randomly selected on a given day. Find the probability that the sample mean will be within 0.2 ounces of the true mean μ.

Solution: We know that all 12 randomly selected ice cream are normally distributed with mean μ and $\sigma = 1$. Based on the sampling distribution of the means $\mu_{\bar{X}} = \mu$ and $\sigma_{\bar{X}} = \dfrac{\sigma}{\sqrt{n}} = \dfrac{1}{\sqrt{12}} = \dfrac{1}{2\sqrt{3}}$.

$$P(|\bar{X} - \mu| \le 0.2) = P[-0.2 \le \bar{X} - \mu \le 0.2]$$

$$= P\left[-\frac{0.2}{\sigma/\sqrt{n}} \le \frac{\bar{X} - \mu}{\sigma/\sqrt{n}} \le \frac{0.2}{\sigma/\sqrt{n}}\right]$$

$$= P\left[-\frac{0.2}{\frac{1}{2\sqrt{3}}} \le Z_{\bar{X}} \le \frac{0.2}{\frac{1}{2\sqrt{3}}}\right], \text{ where } Z_{\bar{X}} = \frac{\bar{X} - \mu}{\sigma/\sqrt{n}}$$

$$= P[-0.69 \le Z_{\bar{X}} \le 0.69]$$

$$= 2P[0 \le Z_{\bar{X}} \le 0.69] \quad \text{(Using Symmetry)}$$

$$= 2(0.2549) = 0.5298$$

B. Properties of Distribution of Proportion

Example 36: Prove that $E(\hat{p}) = p$.

Solution: $E(X) = np$ and $\text{var}(X) = npq$. We know that $\hat{p}$ = sample proportion of successes = $\dfrac{x}{n}$.

Therefore, $E(\hat{p}) = E\left(\dfrac{X}{n}\right) = \dfrac{1}{n} E(X) = \dfrac{1}{n} \cdot np = p$

Example 37: Prove that $\text{var}(\hat{p}) = \dfrac{pq}{n}$.

Solution: $$\text{var}(\hat{p}) = \text{var}\left(\frac{X}{n}\right) = \frac{1}{n^2}\text{var}(X) = \frac{1}{n^2}\cdot npq = \frac{pq}{n}$$

> ***Remark:*** In simple random sampling from a finite population of size N, (or in srswor), the variance of the sampling distribution of proportion $\hat{p}$ is given by:
>
> $$\text{var}(\hat{p}) = \frac{pq}{n}\left(\frac{N-n}{N-1}\right) \Rightarrow \text{S.E.}(\hat{p}) = \sigma_{\hat{p}} = \sqrt{\frac{pq}{n}\left(\frac{N-n}{N-1}\right)}$$

In practice, the population proportion p and consequently $q = 1 - p$ are unknown. Hence, an estimate of $\text{var}(\hat{p})$ is obtained by replacing p with its unbiased estimate $\hat{p}$. Thus,

$$\hat{\sigma}^2_{\hat{p}} = \frac{\hat{p}\hat{q}}{n} \Rightarrow \sigma_{\hat{p}} = \sqrt{\frac{\hat{p}\hat{q}}{n}}$$

C. Unbaised Estimator of Population Variation

Theorem 6.2: $S^2 = \dfrac{\sum_{i=1}^{n}(X_i - \bar{X})^2}{n-1}$ is an unbiased estimator for σ^2.

Proof: In order to show that S^2 is an unbiased estimator of σ^2, we must show that $E(S^2) = \sigma^2$.

We begin with $\text{Var}(X_i) = E(X_i - E(X_i))^2 = E(X_i - \mu)^2 = \sigma^2$.

Now,
$$\text{Var}(\bar{X}) = \text{Var}\left(\frac{1}{n}\sum X_i\right)$$
$$= \frac{1}{n^2}\sum \text{Var}(X_i) \quad \text{(Using independent property of observations)}$$
$$= \frac{1}{n^2}\sum \sigma^2 = \frac{n\sigma^2}{n^2} = \frac{\sigma^2}{n}$$

Also,
$$E(\bar{X}^2) = \text{Var}(\bar{X}) + E^2(\bar{X}) = \frac{\sigma^2}{n} + \mu^2.$$

Therefore,
$$E(S^2) = E\left(\frac{1}{n-1}\sum(X_i - \bar{X})^2\right)$$
$$= \frac{1}{n-1}E\left(\sum X_i^2 - n\bar{X}^2\right) = \frac{1}{n-1}\left(\sum E(X_i^2) - nE(\bar{X}^2)\right)$$

$$= \frac{1}{n-1}\left(\sum\left(\text{Var}(X_i) + E^2(X_i)\right) - n\left(\frac{\sigma^2}{n} + \mu^2\right)\right)$$

$$= \frac{1}{n-1}\left(\sum\left(\sigma^2 + \mu^2\right) - n\left(\frac{\sigma^2}{n} + \mu^2\right)\right)$$

$$= \frac{1}{n-1}\left(n\left(\sigma^2 + \mu^2\right) - \sigma^2 - n\mu^2\right) = \frac{1}{n-1}(n-1)\sigma^2 = \sigma^2$$

D. Bias of a Point Estimator and Error in Estimation

Definition: *Let $\hat{T}$ be a point estimator for a parameter T. Then $\hat{T}$ is an unbiased estimator if $E(\hat{T}) = T$. Otherwise $\hat{T}$ is said to be biased.*

Definition: *The bias of a point estimator $\hat{T}$ is given by $B(\hat{T}) = E(\hat{T}) - T$.*

- **Error of Estimation $|\hat{T} - T|$:**

The distance between an estimator such as $\bar{X}$ (or in general $\hat{T}$) and its target parameter μ (or in general T) is represented as $|\bar{X} - \mu|$ (or in general $|\hat{T} - T|$).

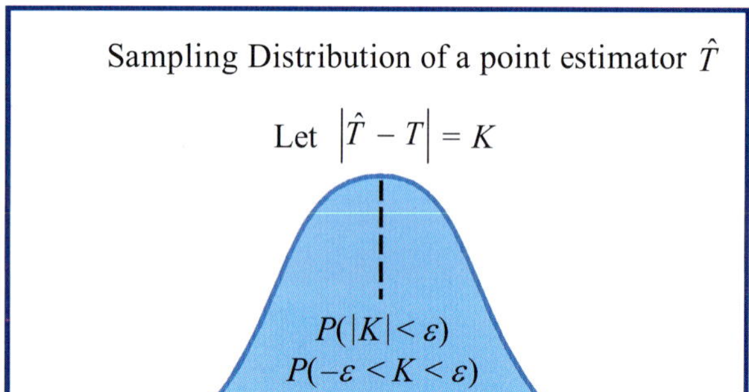

Figure 6.54 a

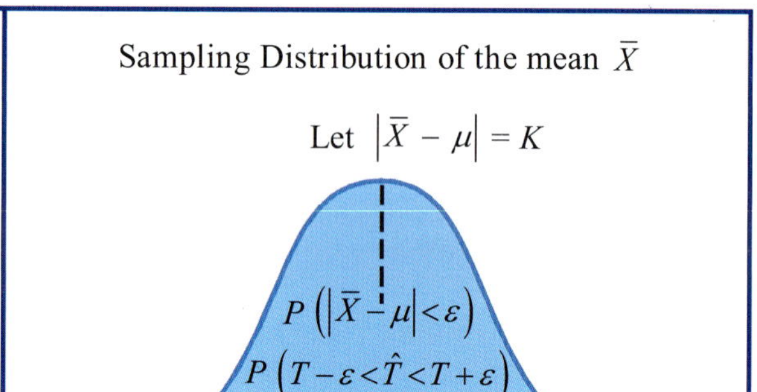

Figure 6.54 b

E. Confidence Interval for μ, p and Sample Size for given Margin of Error

- **Formal Definition of a Confidence Interval**

Suppose that $\hat{T}_L$ and $\hat{T}_U$ are the random lower and upper confidence limits, respectively, for a parameter T. Then, $P(\hat{T}_L < T < \hat{T}_U) = 1 - \alpha$.

The probability $1 - \alpha$ is called the ***confidence level*** or ***confidence coefficient***. The resulting random interval defined by $\left(\hat{T}_L, \hat{T}_U\right)$ is called a ***two-sided Confidence Interval***.

- **Estimating the Population Mean μ, s Known, Large Sample**

Remark: The confidence interval for μ

$$\bar{x} - z_{\alpha/2}\frac{\sigma}{\sqrt{n}} < \mu < \bar{x} + z_{\alpha/2}\frac{\sigma}{\sqrt{n}}$$

can only be used for large samples in sampling from an infinite population or in case of simple random sampling without replacement (SRSWR). In case of SRSWR from a finite population of size N, the confidence interval level for μ, for large samples, is given by:

$$\bar{x} - z_{\alpha/2}\frac{\sigma}{\sqrt{n}}\sqrt{\frac{N-n}{N-1}} < \mu < \bar{x} + z_{\alpha/2}\frac{\sigma}{\sqrt{n}}\sqrt{\frac{N-n}{N-1}}$$

where the margin of error $E = z_{\alpha/2}\frac{\sigma}{\sqrt{n}}\sqrt{\frac{N-n}{N-1}}$. The factor $\sqrt{\frac{N-n}{N-1}}$ is called the **finite population correction factor**.

Remark: If σ^2 of large samples is not known, its estimate provided by the sample variance s^2 is used to obtain the confidence limits for μ.

$$\bar{x} - z_{\alpha/2}\frac{s}{\sqrt{n}} < \mu < \bar{x} + z_{\alpha/2}\frac{s}{\sqrt{n}}$$

F. Sample Size Derivation from the Margin of Error

In case of simple random sampling without replacement from a finite population of size N, the sample size n for estimating μ with confidence level $(1 - \alpha)$ and margin of error E is given by:

$$E = z_{\alpha/2}\frac{\sigma}{\sqrt{n}}\sqrt{\frac{N-n}{N-1}}$$

$$E^2 = \frac{z_{\alpha/2}^2\sigma^2}{n}\left(\frac{N-n}{N-1}\right)$$

$$n(N-1)E^2 = z_{\alpha/2}^2\,\sigma^2(N-n)$$

$$nNE^2 - nE^2 = z_{\alpha/2}^2\,\sigma^2 N - z_{\alpha/2}^2\,\sigma^2 n$$

$$nNE^2 - nE^2 + z_{\alpha/2}^2\,\sigma^2 n = z_{\alpha/2}^2\,\sigma^2 N$$

$$n\left(NE^2 - E^2 + z_{\alpha/2}^2\sigma^2\right) = z_{\alpha/2}^2\,\sigma^2 N$$

$$n = \frac{z_{\alpha/2}^2\sigma^2 N}{NE^2 - E^2 + z_{\alpha/2}^2\sigma^2}$$

• Estimation of Population Proportion *p* and Sample Size

> ***Remark***: In SRSWR from a finite population of size N, $(1 - \alpha)$ confidence limits for p, are:
>
> $$\hat{p} - z_{\alpha/2}\sqrt{\frac{\hat{p}\hat{q}}{n}\left(\frac{N-n}{N-1}\right)} < p < \hat{p} + z_{\alpha/2}\sqrt{\frac{\hat{p}\hat{q}}{n}\left(\frac{N-n}{N-1}\right)}$$

Example 38: A factory is producing 50,000 pairs of shoes daily. From a sample of 500 pairs, 2% were found to be of sub-standard quality. Estimate the number of pairs that can be reasonably expected to be spoiled in the daily production and construct its 95% confidence interval.

Solution: We are given:

n	N	$\hat{p}$	$\hat{q}$
500	50,000	0.02	0.98

Table 6.26

The expected number of pairs of spoiled shoes in a day's production of 50,000 pairs of shoes is given by:

$$N\hat{p} = 50{,}000 \times 0.02 = 1000$$

The 95% confidence limits for p are:

$$\hat{p} \pm E = \hat{p} \pm 1.96\sqrt{\frac{\hat{p}\hat{q}}{n}\left(\frac{N-n}{N-1}\right)}$$

$$= 0.02 \pm 1.96\sqrt{\frac{(0.02)(0.98)}{500}\left(\frac{50000-500}{50000-1}\right)}$$

$$= 0.02 \pm 0.0122 = (0.0078,\ 0.0322)$$

$$\Rightarrow\ 0.0078 < p < 0.0322 \quad \Rightarrow \quad 0.78\% < p < 3.22\%$$

• Sample Size for Estimating Population Proportion *p*

We know that $P\left[|\hat{p} - p| \le E\right] = 1 - \alpha$. For large samples, $\hat{p} \sim N\left(\mu_{\hat{p}} = p,\ \sigma_{\hat{p}} = \sqrt{\frac{pq}{n}}\right)$.

For $(1 - \alpha)$ confidence coefficient, we have $P\left[\left|\frac{\hat{p} - p}{\sqrt{\frac{pq}{n}}}\right| \le z_{\alpha/2}\right] = 1 - \alpha$.

Therefore,

$$E = z_{\alpha/2}\sqrt{\frac{pq}{n}} \approx z_{\alpha/2}\sqrt{\frac{\hat{p}\hat{q}}{n}}. \qquad (6.29)$$

In simple random sampling without replacement from a finite population of size N, we have:

$$E = z_{\alpha/2}\sqrt{\frac{\hat{p}\hat{q}}{n}\left(\frac{N-n}{N-1}\right)} \qquad (6.30)$$

Solving (6.29) for n, we get the required sample size.

$$n = \frac{z_{\alpha/2}^2 (\hat{p})(\hat{q})}{E^2}$$

Solving (6.30) for n, we get the required sample size for a simple random sample without replacement from a finite population of size

$$E = z_{\alpha/2}\sqrt{\frac{\hat{p}\hat{q}}{n}\left(\frac{N-n}{N-1}\right)} \Rightarrow E^2 = \frac{z_{\alpha/2}^2 \hat{p}\hat{q}(N-n)}{n(N-1)} \Rightarrow E^2 n(N-1) = z_{\alpha/2}^2 \hat{p}\hat{q}\,(N-n)$$

$$\Rightarrow n = \frac{z_{\alpha/2}^2 \hat{p}\hat{q}N}{E^2(N-1) + z_{\alpha/2}^2 \hat{p}\hat{q}}$$

Remark: When no estimate of p is available, we take its estimate as $\hat{p} = 0.5$.

Proof: We know that $\sigma_{\hat{p}} = \sqrt{\frac{pq}{n}}$ and $p + q = 1$.

$\sigma_{\hat{p}}$ is maximum when pq is maximum. $pq = p(1 - p)$ is maximum if and only if $p = q = 0.5$.

Note: The reason for the above conclusion requires the knowledge of differential calculus. Let $f = p(1 - p)$. We can maximize f by taking its derivative and solving for p when $f' = 0$.

$$\frac{d}{dp}\left(p^2 - p\right) = 2p - 1$$

Once we solve $2p - 1 = 0$, we see that $p = 0.5$.

G. Estimation of Population Standard Deviation σ or Variance σ^2

Definition: *Let $X_1, X_2, \ldots, X_n$ be defined as in Theorem 6.1. Then* $Z_1 = \frac{X_1 - \mu}{\sigma}$, $Z_2 = \frac{X_2 - \mu}{\sigma}, \ldots, Z_n = \frac{X_n - \mu}{\sigma}$ *are independent, standard, normal, random variables; and* $\sum_{i=1}^{n} Z_i^2 = \sum_{i=1}^{n}\left(\frac{X_i - \mu}{\sigma}\right)^2$ *has a Chi-Square$\left(\chi^2\right)$ Distribution with n degrees of freedom.*

Suppose we want to estimate the population variance σ^2 based on a random sample $X_1, X_2, \ldots, X_n$ from a normally distributed population. The sample variance S^2 is a good estimator of σ^2 where

$$S^2 = \frac{1}{n-1}\sum_{i=1}^{n}(X_i - \bar{X})^2$$

Theorem 6.3: Let $X_1, X_2, \ldots, X_n$ be a random sample from a normally distributed population with mean μ and variance σ^2. Then random variable $\chi^2 = \frac{(n-1)S^2}{\sigma^2} = \frac{\sum_{i=1}^{n}\left(X_i - \bar{X}\right)^2}{\sigma^2}$ has a Chi-Square Distribution with $n - 1$ degrees of freedom.

Using this information prove that if $X_1, X_2, \ldots, X_n$ is a random sample from normal distribution with mean μ and variance σ^2, then $100(1 - \alpha)\%$ confidence interval for σ^2 is

$$\left(\frac{(n-1)S^2}{\chi^2_{n-1,\,\alpha/2}}, \frac{(n-1)S^2}{\chi^2_{n-1,\,1-\alpha/2}}\right) \quad \text{or} \quad \left(\frac{(n-1)S^2}{\chi^2_R}, \frac{(n-1)S^2}{\chi^2_L}\right).$$

Proof: Assume that we have a random sample $X_1, X_2, \ldots, X_n$ from a normally distributed population with mean μ and standard deviation σ, where μ and σ both are unknown, then choose two numbers χ^2_L and χ^2_R such that

$$P\left[\chi^2_L \le \frac{(n-1)S^2}{\sigma^2} \le \chi^2_R\right] = 1 - \alpha.$$

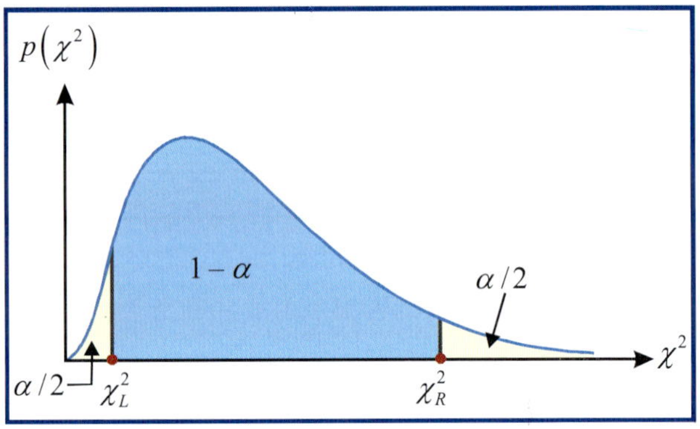

Figure 6.55

As shown in the Figure 6.52, $\chi^2_L = \chi^2_{n-1,1-\alpha/2}$ and $\chi^2_R = \chi^2_{n-1,\,\alpha/2}$, we get:

$$P\left[\chi^2_{n-1,1-\alpha/2} \le \frac{(n-1)S^2}{\sigma^2} \le \chi^2_{n-1,\,\alpha/2}\right] = 1 - \alpha$$

$$\Rightarrow \quad P\left[\frac{1}{\chi^2_{n-1,1-\alpha/2}} \ge \frac{\sigma^2}{(n-1)S^2} \ge \frac{1}{\chi^2_{n-1,\,\alpha/2}}\right] = 1 - \alpha$$

$$\Rightarrow \quad P\left[\frac{(n-1)S^2}{\chi^2_{n-1,1-\alpha/2}} \ge \sigma^2 \ge \frac{(n-1)S^2}{\chi^2_{n-1,\,\alpha/2}}\right] = 1 - \alpha$$

or

$$P\left[\frac{(n-1)S^2}{\chi^2_{n-1,\,\alpha/2}} \le \sigma^2 \le \frac{(n-1)S^2}{\chi^2_{n-1,1-\alpha/2}}\right] = 1 - \alpha$$

H. Testing for Normality

Majority of the procedures or methods in Inferential Statistics require that the sample is selected from a normally distributed population. In this section, we outline the procedure for determining whether the normality requirement is satisfied.

- **Procedure**

1. Visual inspection of the distribution of the population: (Histogram)

Construct a histogram for that the given data set. If the histogram is approximately bell shaped, then we can conclude that the given data set appears to come from a normally distributed population. If not, then we conclude that the sample does not come from a normally distributed population.

2. Outliers

Find the outliers by using $Q_1 - 2 \times IQR$ and $Q_3 + 2 \times IQR$. Reject normality, if there is at least one outlier.

3. Normal Quantile (Probability) Plot

Normal Quantile Plot is a more advanced and accurate display of a scatter plot of points (x, y) where each x value is from the original population, and each y value is the corresponding z score that is a quantile value found from the standard normal distribution. Once the scatter plot is constructed, the interpretation is quite similar to correlation analysis.

Illustration 1: The following data set represents final scores for 55 randomly selected statistics students. Determine if the given data set comes from a normally distributed population.

0	164	110	73	195	150	90	70	122	122	71
56	178	71	57	118	90	92	120	65	65	150
145	186	65	50	75	75	154	65	122	122	51
150	40	74	150	100	0	22	93	105	105	140
130	10	0	80	98	40	58	0	100	4	30

Solution: ***Step* 1**

Construct the Histogram

0	30	65	90	120	150	186
0	40	65	90	122	150	195
0	40	65	92	122	150	
0	50	65	93	122	150	
4	51	70	98	122	154	
10	56	71	100	130	164	
22	57	71	100	140	178	
	58	73	105	145		
		74	105			
		75	110			
		75	118			
		80				

Classes	Frequency
0-30	7
30-60	8
60-90	12
90-120	11
120-150	8
150-180	7
180-210	2

Table 6.27

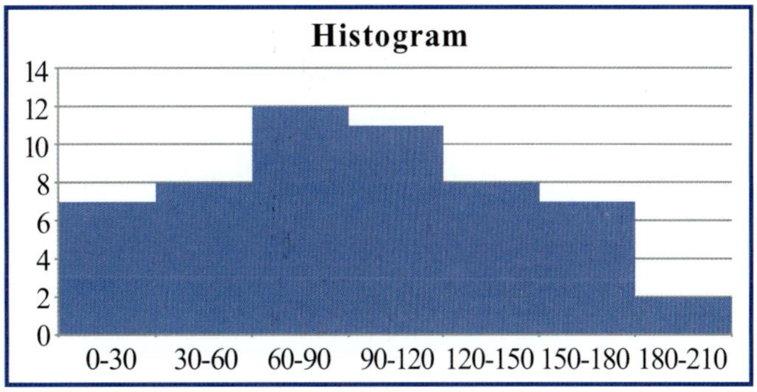

Figure 6.56

The histogram has an approximately bell shaped distribution. Therefore, the given data set appears to come from a normally distributed population.

Step 2 **Outlier Analysis**

$Q_1 = 57.5$, and $Q_3 = 122$. $IQR = Q_3 - Q_1 = 64.5$.

$Q_1 - 2 \times IQR = -71.5$, and $Q_3 + 2 \times IQR = 251$. This implies that any data value less than -71.5 and greater than 251 will be considered an outlier. Since the range of the given data set is between 0 to 95, we conclude that there are no outliers. Once again, we have verified that the given data set appears to come from a normally distributed population.

Step 3 **Construct the Normal Quantile (Probability) Plot**

i) Organize the data in an ascending order in Excel®.

ii) Find the cumulative areas to the left of the corresponding data values.

> **Note:** The cumulative areas are $\frac{1}{2n}, \frac{3}{2n}, \frac{5}{2n}, \ldots$, where n is the sample size.

$$\frac{1}{2(55)}, \frac{3}{2(55)}, \frac{5}{2(55)}, \ldots$$

iii) Find the z scores corresponding to the cumulative left areas.

$$-2.36, -1.92, -1.69, \ldots$$

We can now construct the Normal Probability Plot (Scatter Plot) using the population values and their corresponding scores.

X	Areas	Z	X	Areas	Z	X	Areas	Z
0	0.0091	−2.36	65	0.3364	−0.42	110	0.6636	0.42
0	0.0273	−1.92	70	0.3545	−0.37	118	0.6818	0.47
0	0.0455	−1.69	71	0.3727	−0.32	120	0.7000	0.52
0	0.0636	−1.52	71	0.3909	−0.28	122	0.7182	0.58
4	0.0818	−1.39	73	0.4091	−0.23	122	0.7364	0.63
10	0.1000	−1.28	74	0.4273	−0.18	122	0.7545	0.69
22	0.1182	−1.18	75	0.4455	−0.14	122	0.7727	0.75
30	0.1364	−1.10	75	0.4636	−0.09	130	0.7909	0.81
40	0.1545	−1.02	80	0.4818	−0.05	140	0.8091	0.87
40	0.1727	−0.94	90	0.5000	0.00	145	0.8273	0.94
50	0.1909	−0.87	90	0.5182	0.05	150	0.8455	1.02
51	0.2091	−0.81	92	0.5364	0.09	150	0.8636	1.10
56	0.2273	−0.75	93	0.5545	0.14	150	0.8818	1.18
57	0.2455	−0.69	98	0.5727	0.18	150	0.9000	1.28
58	0.2636	−0.63	100	0.5909	0.23	154	0.9182	1.39
65	0.2818	−0.58	100	0.6091	0.28	164	0.9364	1.52
65	0.3000	−0.52	105	0.6273	0.32	178	0.9545	1.69
65	0.3182	−0.47	105	0.6455	0.37	186	0.9727	1.92
						195	0.9909	2.36

Table 6.28

Excel® Commands Display

	A	B	C	D
1		*X*	**Areas**	***Z***
2	1	0	=A2/110	=NORM.S.INV(C2)
3	3	0	=A3/110	=NORM.S.INV(C3)
4	5	0	=A4/110	=NORM.S.INV(C4)
5	7	0	=A5/110	=NORM.S.INV(C5)
6	9	4	=A6/110	=NORM.S.INV(C6)
7	11	10	=A7/110	=NORM.S.INV(C7)
8	13	22	=A8/110	=NORM.S.INV(C8)
9	15	30	=A9/110	=NORM.S.INV(C9)
10	17	40	=A10/110	=NORM.S.INV(C10)
11	19	40	=A11/110	=NORM.S.INV(C11)
12	21	50	=A12/110	=NORM.S.INV(C12)
13	23	51	=A13/110	=NORM.S.INV(C13)

Figure 6.57

We can now construct the Normal Quantile Plot (scatter plot) for the bivariate data (X, Z).

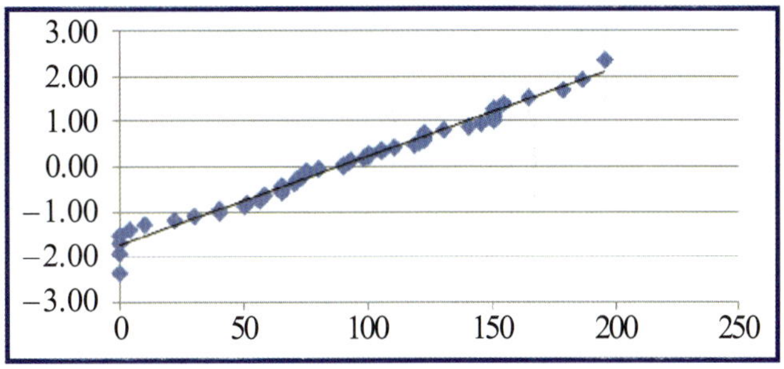

Figure 6.58

Since the points seem to lie reasonably close to a line, we can conclude that the given data set appears to come from a normally distributed population.

Chapter 6: General Exercises

Sampling Distribution and the Central Limit Theorem

1. A population consists of 5 numbers (2, 3, 6, 12, and 15). Consider all possible samples of size 2 which can be drawn with replacement from this population.

(a) Construct the sample space.

(b) Find the mean and standard deviation of the population.

(c) Verify that: Sample mean is an unbiased estimate of the population mean.

(d) Find the standard deviation of the sampling distribution of the means $\sigma_{\bar{X}} = \sigma / \sqrt{n}$.

2. Given the population $\{2, 4, 6, 8\}$ and a sample size $n = 2$ (**Note:** Samples are drawn with replacement), find

(a) the mean μ and standard deviation σ of the population.

(b) the sampling distribution of $\overline{X}$.

(c) $\mu_{\overline{X}}$ and show that $\mu_{\overline{X}} = \mu$.

(d) $\sigma_{\overline{X}}$ and show that $\sigma_{\overline{X}} = \sigma / \sqrt{n}$.

3. In selecting a sample of size 3 in simple random sampling without replacement (srswor) from the population $\{1, 5, 8, 12, 15 \text{ and } 19\}$,

(a) show that the sample mean is an unbiased estimate of the population mean by enumerating all possible samples.

(b) calculate the standard error of the sample means and verify that it agrees with the formula for S.E. $(\overline{X})$ in srswor.

(c) verify that the value obtained in (b) is less than the S.E. $(\overline{X})$ obtained from sampling the same population with replacement.

4. Find the mean $\mu_{\overline{X}}$ and the standard error $\sigma_{\overline{X}}$ of the distribution of all sample means when all random samples of size 49 are selected from a large population with mean $\mu = 1000$ and standard deviation $\sigma = 100$.

5. The average size of a two-bedroom apartment in Los Angeles is about 860 square feet with a standard deviation of 60 square feet. Let $\overline{X}$ be the mean size of a two bedroom apartment in LA for a random sample of 25 apartments. Find the mean and standard deviation of the sampling distribution of $\overline{X}$.

6. In a recent Graduate Record Exam (GRE), the mean score of all participants was 1028. Assume that the distribution of GRE scores of all examinees is normal with a mean of 1028 and a standard deviation of 218. Let $\overline{X}$ be the mean GRE score of a random sample of certain examinees. Calculate the mean and standard deviation of $\overline{X}$ when the sample size is

(a) 18. **(b)** 80. **(c)** 2000. **(d)** 10,000.

7. Among all college students who took General Education (GE) classes during the school year, the distribution of time spent studying per week is approximately normal with a mean of 18 hours and a standard deviation of 4.2 hours. Let $\overline{X}$ be the average time spent studying per week for 20 randomly selected college students who took GE classes during the school year. Calculate the mean and standard deviation of the sampling distribution of $\overline{X}$.

8. Given that $\mu = 400$, $\sigma = 50$, $N = 800$, and $n = 25$, find $P(\overline{X} > 385)$.

9. The amount of soft drink in plastic bottles is normally distributed with mean 16 ounces and standard deviation of 0.6 ounces.

(a) If a plastic bottle is selected at random from a large population, find the probability that it contains more than 15.8 ounces of soft drink.

(b) If a random sample of 25 is selected from a large population, find the probability that the mean amount of soft drink is more than 15.8 ounces.

10. Assume that the weights of all packages of a certain brand of chocolate are normally distributed with a mean of 28 ounces and a standard deviation of 0.2 ounce. Find the probability that mean weight, $(\overline{X})$, of a random sample of 24 packages of this brand of chocolate will be between 27.2 and 27.6 ounces.

11. The real estate examination scores of a large population of participants are normally distributed with a mean of 60 and standard deviation of 8. Find the probability that the mean score of a random sample of 49 will lie between 58 and 62.

12. The gas bills for all households in a city have a normal distribution with a mean of \$36 and a standard deviation of \$8. Find the probability that the mean amount of gas bills for a random sample of 85 households selected from this city will be

(a) between \$31 and \$34. **(b)** more than \$38. **(c)** within \$4 of the population mean.

13. The IQ's of all students enrolled at a large university are approximately normally distributed with a mean of 104 and a standard deviation of 18. Find the probability that the mean IQ of a random sample of 24 students selected from this university is

(a) 120 or higher. **(b)** 101 or lower.

(c) 100 to 135. **(d)** 120 to 138.

14. The IQ's of students at a large university have a mean of 130 with standard deviation of 10. Find Q_1 and Q_3 of the sampling distribution of the means of all samples of size 35.

15. The heights of adult females in the United States have a mean of 66 inches with standard deviation of 2 inches. Find the 90th percentile of the sampling distribution of the means of all samples of size 36.

16. Find the mean $(\mu_{\bar{X}})$ and standard deviation $(\sigma_{\bar{X}})$ of the distribution of all sample means when a population of 500 has a mean of 50 and standard deviation of 5. All random samples of size 39 are selected.

17. A random sample of 500 is drawn from a population with the proportion of successes $\hat{p}$ = 0.28. Find the probability that the sample proportion $\hat{p}$ of successes is

(a) greater than 0.34. **(b)** less than 0.31. **(c)** between 0.31 and 0.39.

18. A local politician believes that 23% of all government contracts go to minority owned businesses. A random sample of 200 contracts is reviewed and the proportion of contracts that were given to minority owned businesses was computed. Find the probability that the sample proportion is

(a) more than 0.2. **(b)** less than 25%. **(c)** between 0.22 and 0.26.

19. A natural gas distributer in a large city claims that 55% of the households use natural gas for heating. Find the probability that between 50% to 60% of a random sample of 400 households in the city use natural gas for heating.

20. A survey of all private-and public college districts in Narnia showed that 84% of them offer retirement plans to their employees. Let $\hat{p}$ be the sample proportion in a random sample of 48 such districts that offer retirement plans to their employees. Find the probability that the value of $\hat{p}$ will be

(a) between 0.73 and 0.80. **(b)** greater than 0.92. **(c)** less than 0.70.

Confidence Interval Estimation for μ, σ Known

21. A random sample of 100 final exams showed an average of 73.5 with a standard deviation of 5.8.

(a) Find the standard error of the mean $\sigma_{\bar{X}}$.

(b) Find the margin of error $E = z_{\alpha/2}\sigma_{\bar{X}}$ using a 99% confidence level.

(c) Construct a 99% confidence interval for the average amount of all exam scores.

22. Construct a 95% confidence interval for μ given the following information:

$\overline{x} = 60$, $\sigma = 9.3$, and $n = 45$.

$\overline{x} = 15.5$, $\sigma = 3.35$, and $n = 83$.

$\overline{x} = 135$, $\sigma = 15.24$, and $n = 25$. Assume that the parent population is normally distributed.

$\overline{x} = 12.3$, $\sigma = 1.6$, and $n = 15$. Assume that the parent population is normally distributed.

23. 85 randomly selected college students had a mean IQ of 102. The standard deviation of the IQ's of all college students is known to be 18. Construct an 88% confidence interval for the average IQ of all college students.

24. The weights of 100 randomly selected 7-year-old girls are normally distributed with a mean of 49.7 pounds. The population standard deviation is known to be 8.1 pounds. Find a 95% confidence interval for the population mean of the weights of 7-year-old girls.

25. Suppose, the average time a person can hold his breath under water is 1.2 minutes. Assuming $\sigma = 40$ seconds and 400 people were tested, find a 95% confidence interval for the population mean μ.

26. The following are the ages of 22 mathematics faculty in a large community college:

56	63	60	52	54	32	25	31	48	45	24
68	42	38	41	29	52	54	49	36	31	69

Assuming $\sigma = 14$, construct a 96% confidence interval for μ. (Population is normally distributed).

27. Sixty vehicles are randomly selected from a junk yard. The sample has a mean weight of 3009 lbs. Construct a 94% confidence interval for the true mean weight, μ, of all vehicles in the junk yard, when $\sigma = 403$ lbs.

28. A certain cereal brand claims that it has 60 grams of raisins in each box. Suppose 40 randomly selected boxes were tested, and it turned out to be that the average weight of raisins in each box was 55.6, with a standard deviation of 4.3. Using a 98% confidence level, construct a confidence interval for the true mean weight of raisins in each box.

29. The following table gives average hourly earnings (in dollars) of production workers, as given by the U.S. Department of Labor.

Year	Jan	Feb	Mar	Apr	May	Jun	Jul	Aug
2002	10.48	10.51	10.52	10.53	10.53	10.51	10.51	10.55
2003	10.76	10.76	10.79	10.79	10.82	10.76	10.76	10.78
2004	11.06	11.06	11.04	11.07	11.09	11.03	11.05	11.04
2005	11.35	11.36	11.34	11.38	11.35	11.33	11.39	11.39
2006	11.7	11.68	11.71	11.74	11.72	11.75	11.73	11.76

Using a 95% confidence level, construct a confidence interval estimate of the population mean amount of money a production worker earns. Assume $\sigma = 0.65$.

30. When 23 randomly selected athletes were chosen, it was found that they averaged 15 hours of workout per week. The population standard deviation is 2 hours. Assuming that the population is normally distributed, construct a 90% confidence interval estimate of the average number of workout hours for all athletes.

31. The finance manager in Santiago BMW dealership randomly selected 80 loan applications and found an average FICO score of 683. Assuming that the population standard deviation is known to be 38, construct a 99% confidence interval estimate of the population mean.

32. Twenty randomly selected college students were surveyed and it was found that they speak, on the average, 2.9 languages. The parent population is normally distributed with a standard deviation of 0.6. Construct a 92% confidence interval for the true mean number of languages a college student speaks.

Sample Size for Means

33. How many college students must be randomly selected to estimate the mean GPA of college students in the United States. We want 98% confidence that the sample mean is within 0.3 points of the population mean, and the population standard deviation is 0.6.

34. How many pedestrians must be randomly selected to estimate the mean crossing time of the pedestrians in Los Angeles? We want 99% confidence that the sample mean is within 3 seconds of the population mean. The population standard deviation is known to be 4 seconds.

35. **CNN** and **USA Today** asked people if they thought human cloning should be legal. If the standard deviation of the sample is 70, and the surveyors want 95% confidence that the sample mean is within 4.4 units of the population mean, how many people should be surveyed?

36. A researcher wants to estimate the effectiveness of a new headache medicine. How many users must she randomly select to estimate the mean number of hours it takes for the drug to take effect? She wants 95% confidence that the sample mean is within 0.5 hours of the population mean, and the population standard deviation is 1.12 hours.

37. In measuring patients' reaction times to a specific drug, a Psychologist, estimates that the population standard deviation is 48 seconds. How large a sample must she collect in order to be 90% confident that the error of her estimate of the mean reaction time will not exceed 12 seconds?

38. A researcher wishes to estimate the average expenditure for repairing highways in his city. He wants to be 95% confident that the sample mean is within $500 of the population mean. How large of a sample size is necessary if the population standard deviation is known to be $750.

Confidence Interval Estimation for μ, σ Unknown

39. $n = 20$, $\bar{x} = 32.8$, and $s = 4.0$. Construct a 90% confidence interval for μ. (Assume the population is normally distributed.)

40. $n = 10$, $\bar{x} = 65$, and $s = 6.2$. Construct a 95% confidence interval for μ. (Assume the population is normally distributed.)

41. A data set includes 81 body temperatures for which $\bar{x} = 98.18°F$ and $s = 0.64°F$.

    (a) What is the best point estimate (unbiased estimator) of the mean body temperature of the population?

    (b) Construct a 95% confidence interval estimate of the mean body temperature of the population.

42. A data set includes 70 GPA's for which $\bar{x} = 3.25$ and $s = 0.24$.

    (a) What is the best point estimate (unbiased estimator) of the mean GPA of the population?

    (b) Construct a 90% confidence interval estimate of the mean GPA of the population.

43. A sample of 15 cats selected from a normally distributed population showed an average weight of 8.7 pounds with a sample standard deviation of 1.4 pounds. Construct a 99% confidence interval estimate of the population mean.

44. Twelve randomly selected college freshmen were asked about the average time they spend on Facebook every day. The following table lists their responses in hours.

3.8	4.6	1.8	1.5	2.7	3.5	5.9	4	3.2	2.8	3.7	4.4

Construct a 95% confidence interval estimate of the average time a college freshman spends on Facebook. Assume the population is normally distributed.

45. The following is a table of unemployment rates in California for the month of February from 10 randomly selected years, as given by the US Department of Labor.

Unemployment Rates	9.0	9.6	9.3	7.8	7.6	6.6	6.1	5.6	5.0	4.7

Using this table, construct a 99% confidence interval estimate of the true mean unemployment rate in California in the month of February. Assume the population is normally distributed.

46. A nutritionist wants to estimate the mean number of servings of vegetables a person consumes per day. The following are result of surveys collected from 15 randomly selected participants.

1.2	3.5	6.9	0.6	3.4	5.5	3.8	6.3	1.9	2.0	5.4	3.6	5.2	1.3	2.6

Construct a 99% confidence interval for the true mean number of servings of vegetables a person consumes per day. Assume the population is normally distributed.

47. The following are times (in hours) it takes for a sample of a certain species of butterflies to lose their wings.

10	12	10.6	9.8	11.6	10.4	13	8.75	9.8	10.7	11.4

Find a 99% confidence interval for the true mean number of hours it takes for these types of butterflies to lose their wings. Assume the population is normally distributed.

48. Most couples marry in Las Vegas, Nevada, on New Year's Day and Valentine's Day. Samples were collected from 8 randomly selected years. It was found that the average number of couples getting married in Las Vegas on New Year's Day was 278 with a standard deviation of 40; and on Valentine's Day, the mean was 304 with a standard deviation of 48. Construct 90% confidence interval estimates of the true mean number of couples getting married on those two days. Assume the population is normally distributed.

49. At a community college, students have the option of registering for classes online or over the LAN Line. The average amount of time it takes to register over the LAN Line is 12.7 minutes, with a standard deviation of 3.4 minutes, as taken from a sample of 29 randomly selected students. For 26 randomly selected online users, the mean time was 9.6 minutes, with a standard deviation of 1.8 minutes. Using a 99% confidence level, construct the confidence intervals of the true mean time for both types of registration. Assume the population is normally distributed.

50. At an auto paint shop, the average time it takes to paint a car is 4.3 hours with a standard deviation of 1 hour, as observed from a sample of 12 randomly selected cars. Construct a 98% confidence interval estimate of the average number of hours it takes to paint a car in this shop. Assume the population is normally distributed.

51. A random sample of 14 items is drawn from a normally distributed population with σ unknown. The sample mean and standard deviation are 4.6 and 0.7, respectively. Construct a 95% confidence interval estimate of the population mean μ.

52. A high school principal randomly selected six students to take an aptitude test. Their scores were as follows.

76.5	85.2	77.9	83.6	71.9	88.6

Assuming the population is normally distributed, find a 90% confidence interval estimate of the mean score for all students taking the aptitude test.

53. The following table gives the number of hours in a week a teacher spends correcting papers, as reported by 24 randomly selected teachers.

12.2	14.7	17.1	9.8	16.8	21.1	17.9	16.3	12.6	13.7	13.8	12.8
13.5	11.9	6.9	20.8	17.4	18.5	14.3	12.7	16.9	22.7	14.4	15.9

Construct a 98% confidence interval for the true mean number of hours school teachers spend correcting papers. Assume that the parent population is normally distributed.

Confidence Interval Estimation for Population Proportion

54. $n = 86$ and $\hat{p} = 0.36$. Construct a 95% confidence interval for p.

55. $n = 120$ and $x = 36$. Construct a 90% confidence interval for p.

56. In a sample of 400 college students, it was found that 368 of them drank at least one soft drink every day.

(a) Estimate the percentage (proportion) of college students that consume at least one soft drink a day.

(b) Construct the 95% confidence interval estimate of the proportion of all college students who drink at least one soft drink every day.

57. **CNN/*USA Today*/Gallup Poll** asked 10,000 adults in January 2008 about how worried they were that they or someone in their families would become a victim of terrorism. They obtained the following results.

Very	Some-what	Not Too much	Not At All	No Opinion
25%	30%	15%	25%	5%

Construct a 95% confidence interval for the true proportion of people who are

(a) very worried. **(b)** not at all worried.

58. According to the National Science Foundation, out of the total 10,981,600 employed U.S. scientists and engineers, 459,300 of them have attained their highest degrees in Mathematical Sciences. Construct a 99% confidence interval for the true proportion of U.S. scientists and engineers who have an advanced degree in Mathematics.

59. In 8000 randomly selected traffic violations in a county, 2350 were for running a stop sign. Find a 99% confidence interval for the true proportion of all tickets written for running a stop sign in that county.

60. When 345 adults were asked which flavor of ice cream they liked best, 107 said chocolate, 98 said vanilla, 59 said strawberry, and the rest named another flavor. Construct 90% confidence intervals for the true proportions of people favoring

(a) vanilla.

(b) chocolate.

61. Of 367 randomly selected Merchant Marines, 30 indicated that they plan to work on oil tankers. Find a 95% confidence interval for the true proportion of all Merchant Marines who plan to work on oil tankers.

62. Of 234 students from Los Angeles University, 12.82% commute using the local subway system. Construct a 90% confidence interval for the true percentage of all Los Angeles University students who use the local subway system to commute.

63. The mayor of a large city wants to emphasize the usefulness of public transportation. He claims that 95% of people who use it are never late to work. A survey of 700 randomly selected public transportation users found that only 80% of them were never late to work. Using a 99% confidence level, construct a confidence interval for the true proportion of public transportation users who are never late to work.

64. 70 randomly selected families were asked how often they go grocery shopping. The following table represents the results.

Number of times they shop for groceries	% of families who shop that often
Daily	7
Once a week	45
Twice a week	33
Three times a week	9
Every two weeks	5
Less often than every two weeks	1

Construct a 95% confidence interval for a true proportion of families who do grocery shopping

(a) twice a week. **(b)** daily. **(c)** less often than every two weeks.

65. In a poll of 90 randomly selected married men, 68% said that they were happily married, 25% said they were not, and the rest had no opinion. Construct a 99% confidence interval for the true proportion of married men who are happy.

66. In an NBC/*Wall Street Journal* poll conducted in January 2003, 500 randomly selected adults were asked if they favor or oppose the affirmative action policy in admissions to public universities. Here are the results.

Opinion	%
Favor	26
Oppose	65
Not sure	9

Using a 95% confidence level, construct a confidence interval for the true proportion of adults

(a) favoring affirmative action. **(b)** opposing affirmative action.

67. 180 randomly selected high school students were asked to indicate the number of hours they spend studying outside the classroom per week. The number of hours are given in the table provided.

Number of hours	Number of students
0-9	29
10-19	45
20-29	56
30-39	41
40+	9

Construct a 94% confidence interval for the true proportion of high school students who

(a) study 0-9 hours a week outside the classroom.

(b) study 20-29 hours a week outside the classroom.

68. One hundred randomly selected teachers were polled and 67 of them said they enjoyed their job. Find a 98% confidence interval for the true proportion of teachers who enjoy their job.

69. In a random sample of 200 homebuyers, 120 preferred Variable Interest Rate (VIR) instead of Fixed Interest Rate (FIR) for their mortgage. Construct a 98% confidence interval for the true proportion of home buyers who prefer VIR.

70. In a random sample of 500 items from a large consignment, 20 items were found to be defective. Construct a 99% confidence interval for the true proportion of defective items in the consignment.

Sample Size for Estimating the Population Proportion

71. You want to poll people to see what proportion of them approves of the current president. Suppose you want to be 99% confident that your sample proportion is within 3% of the true proportion. How many people should you poll for your sample?

72. Use the given data to find the minimum sample size required to estimate the population proportion.

(a) Margin of error: 0.085; confidence level: 98%; from a prior study, $\hat{p}$ is estimated by 0.34.

(b) Margin of error: 0.024; confidence level: 90%; from a prior study, $\hat{p}$ is estimated by 0.82.

(c) $E = 0.04$, and $1 - \alpha = 0.95$. $\hat{p}$ is unknown.

(d) $E = 0.013$, and $1 - \alpha = 0.90$. $\hat{p}$ is unknown.

73. A nutritionist is interested in determining how many people over the age of 21 carry on with their day without eating lunch. If he wishes to be 95% confident that his estimate will contain the actual population proportion, how large of a sample will be necessary if he wishes to have an error no more than 2.5%? A previous study found that 18% of people over the age of 21 skipped lunch.

74. The marketing department for a computer company wants to estimate, with 99% confidence, the percentage of people who own an iPad. Find the minimum sample size necessary if $E = 0.043$. A previous study showed that 48% of people interviewed had an iPad.

75. In a random sample of 100 customers at a large department store, 88 used credit cards for their purchases. How large a sample is needed if we wish to be 92% confident that our sample proportion is within 1% of the population proportion of the credit card purchases?

76. In a random sample of 120 Asthma sufferers, 85 prefer a particular brand of pain killer. Find the minimum sample size necessary to ensure that the error will not exceed 2%.

Confidence Interval for Estimating Population Standard Deviation or Variation

77. $n = 40$ and $s = 5.8$. Construct a 95% confidence interval for σ. Assume that the original population is large.

78. $n = 20$ and $s = 18.5$. Construct a 99% confidence interval for σ. Assume that the original population is normally distributed.

79. Given a random sample of size 85, taken from a large population, construct the 95% confidence interval estimate of the population variance σ^2. Assume $s^2 = 28$.

80. In a random sample of size 110 from a large population, the sample standard deviation $s = 0.62$. Construct a 99% confidence interval estimate for

(a) σ.

(b) σ^2.

In exercises 81-92, assume that the parent population is normally distributed.

81. Construct a 95% confidence interval estimate of standard deviation of the number of pages a calculus textbook has if a sample of 6 randomly selected calculus books has a standard deviation of 17 pages.

82. The prices for a certain model car varies depending on the features and packages included. Suppose a randomly selected sample of 28 cars had a standard deviation of \$4,320. Construct a 98% confidence interval for the true variance of the prices of different model cars.

83. Construct a 90% confidence interval for the true standard deviation of weights of blocks of wood if a sample of 15 randomly selected blocks had a standard deviation of 0.35 lbs.

84. The time it took for twenty kindergarteners to eat their lunches had a standard deviation of 8.7 minutes. Using a 95% confidence level, construct an interval that reflects the true standard deviation for the time it takes a kindergartener to eat.

85. At a supermarket, customers complain that the Express-Lane takes longer than the regular lanes. The variation of time spent in the Express-Lane for twelve randomly selected customers was 2.37 min. For 7 randomly selected customers in the regular lane, the variation was 2.98 minutes. Construct the 98% confidence interval for true standard deviations for both types of lanes.

86. In a video game store, the standard deviation of the number of video games sold last month calculated from taking data for 7 randomly selected days was 9.5 units. Construct a 90% confidence interval for true standard deviations of video games sold in general.

87. A tattoo artist develops a survey to test attitudes of people in the medical field toward getting a tattoo. 16 randomly selected medical professionals are given the survey. Their mean score is 76.7 with a standard deviation of 8.4.

(a) Construct a 99% confidence interval for the standard deviation of the population.

(b) Construct a 90% confidence interval for σ^2 of the population.

88. The amount of money (in dollars) 10 randomly selected customers spent daily on breakfast at Café Barcelona are given below.

13.49	15.69	11.35	19.85	17.35	13.29	12.29	11.35	15.69	14.36

Construct a 99% confidence interval estimate for σ^2.

89. The following represents the times (in hours) for the battery life of a sample of 10 randomly selected cellular phones when they were not in use.

12.0	15.5	9.8	20.6	13.5	12.9	17.9	19.6	15.7	21.6

Construct a 95% confidence interval for the standard deviation σ of the population.

90. The chart below gives the highest percentages of discounts advertised in stores at a mall during Christmas time.

25	75	50	40	15
25	50	75	45	25
10	65	35	75	20

Construct a 95% confidence interval estimate of the variation of discount rates one can expect at this mall during Christmas.

91. A football coach randomly selected ten players and recorded the time each player took to perform a certain drill. The times (in minutes) were as follows:

9	5	7	12	14	14	5	11	6	9

Construct an 88% confidence interval estimate of the population variation σ^2.

92. The variation of tire pressure in a sample of 20 BMW tires was found to be 33.64 psi. Given this information, construct a 99% confidence interval estimate of σ.

7

HYPOTHESIS TESTING (ONE POPULATION)

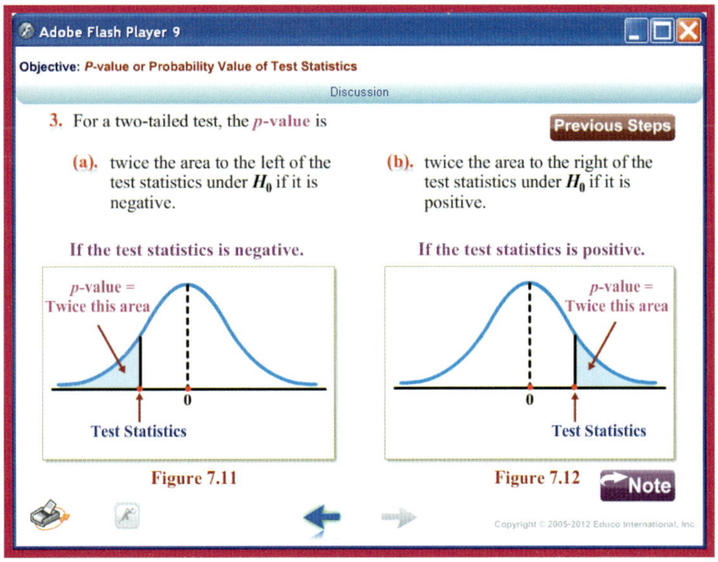

CHAPTER

7

HYPOTHESIS TESTING (ONE POPULATION)

Chapter Overview:

7.1: INTRODUCTION

We make decisions on a daily basis based on claims made by others, people or companies. Some of these decisions are quite serious; therefore testing their validity is of great importance.

For instance, a pharmaceutical company has just released a drug to treat headaches. In a TV ad it claims that this new drug has an average reaction time of 20 minutes. A sample of 100 patients suffering from headache are given this drug, and it is found that the average reaction time in these patients is 22 minutes with a standard deviation of 3 minutes. Is there enough evidence to reject the claim that this new drug has an average reaction time of 20 minutes?

Here are some other claims Hypothesis Testing can address:

i) "**Majority**" of Americans are coffee drinkers.

ii) The "**variation**" of rainfall in the Amazon is about 138.3 units.

iii) The "**average**" GPA in a community college is more than 2.8.

The bold words indicate the population parameters the claim is addressing. In these examples the population parameters are proportion, variance and mean respectively.

- ### Hypothesis Testing:

The theory of Hypothesis Testing was pioneered by J. Neyman and E.S. Pearson. It employs statistical techniques to arrive at decisions in certain situations with an element of uncertainty, using a sample of fixed size. The focus of this chapter is to develop the procedure for testing claims (hypotheses) made about the population mean μ, population proportion p, population standard deviation σ, and variance σ^2.

7.2: Fundamentals of Hypothesis Testing

Objectives

After completing this section you will be able to understand:

- **A.** Statistical Hypothesis.
- **B.** Type I and Type II Errors in Testing of Hypothesis.
- **C.** Test Statistics used for Testing of Hypothesis.
- **D.** Rejection Region.
- **E.** Significance Level.
- **F.** One Tail and Two Tail Test.
- **G.** Critical Values.
- **H.** *P*-value or Probability Value of Test Statistics.
- **I.** Procedure for Hypothesis Testing.
- **J.** Interpretation.

A. Statistical Hypothesis

A statistical hypothesis is a claim or statement which must be verified using statistical techniques.

- **Hypothesis Testing (Test of Significant Difference)**

Hypothesis testing is the statistical procedure for testing a hypothesis (claim) about the population parameter(s) using sample information.

Null Hypothesis

The Null Hypothesis, denoted by H_0(H-null), according to Fisher, "is the hypothesis which is tested for possible rejection under the assumption that it is true." To understand this definition better, try comparing it to the premise in a criminal trial by jury.

"You are assumed innocent until proven guilty."

"H_0 is always assumed true, until proven false."

Note: The researcher in the theory of hypothesis plays the role of the prosecutor.

The null hypothesis is a statement that the value of a population parameter such as the mean μ, proportion p, and standard deviation σ are equal to some specific values. Here are some examples:

$$H_0 : \mu = 86 \qquad H_0 : p = 0.42 \qquad H_0 : \sigma = 18$$

Formal Definition:

Null hypothesis is a statement, claim, or conjecture about a population parameter that is to be tested.

Alternative Hypothesis (Research Hypothesis)

Any hypothesis which contradicts the null hypothesis (a hypothesis that is true only when the null hypothesis is false) is called an alternative or research hypothesis. The alternative hypothesis is usually denoted by H_1, H_A, or H_l. In this text, we will use H_1 notation to represent the alternative hypothesis. The symbolic form of the alternative hypothesis must use one of the following symbols: $<$ or $>$ or $\neq$. Here are some examples.

$$H_1 : \mu \neq 92 \qquad H_1 : p < 0.28 \qquad H_1 : \sigma > 15$$

Important Notes:

(a) Traditionally the symbols $\leq$ and $\geq$ were used in the null hypothesis H_0, but recently, most research papers and professional journals only use the ***equal*** symbol for H_0. We will conduct the test of hypothesis by assuming that the population mean, proportion, standard deviation, or variance is ***equal*** to some specified value so that we can work with a unique distribution having a distinct value.

(b) Based on the nature of continuous distributions, we can never theoretically support a claim that a parameter is equal to some specified value. For example, if you believe that majority of Americans drink coffee, state your claim as $p > 0.5$. In this case, we will assume that $p = 0.5$, in the hope that it will be rejected so that the claim $p > 0.5$ can be supported.

(c) In practice, the original claim could become the null hypothesis, the alternative hypothesis, or it might not correspond to either one. To identify the null and the alternative hypotheses:

i) Start with identifying the original claim or hypothesis and express it symbolically using the symbols <, >, ≠, ≤, ≥, or =.

ii) Construct the symbolic form of the opposing statement that must be true if the original claim is false, known as the ***counter claim***.

iii) Of the two symbolic statements created (original claim and the counter claim), classify the statement that contains the symbols <, >, or ≠ as the alternative hypothesis, and let the null hypothesis represent the symbolic expression that the parameter equals the value that is being tested. For example, if the original claim is that "the mean G.P.A. of college students is at least 2.8 ($\mu \geq 2.8$)," the symbolic form of the counter claim will be $\mu < 2.8$. Note that if the original claim is false, then $\mu < 2.8$. In this example, the alternative hypothesis is $\mu < 2.8$, but the null hypothesis is $\mu = 2.8$.

(d) It is a common mistake to state the claim in terms of sample statistics. For example, if the claim states that the average IQ is less than 130, the notation $\bar{x} < 130$ would be incorrect. The correct notation is $\mu < 130$. All claims must be set up in terms of population parameters not sample statistics.

Example 1: Given the following original claims, determine

(a) the counter claim.

(b) the null and alternative hypothesis symbolically.

i)	$\mu \geq 78$	**ii)**	$\mu < 54$	**iii)**	$\mu \leq 12.6$	**iv)**	$\mu = 5$
v)	$\mu > 3.2$	**vi)**	$\mu \neq 15$	**vii)**	$p \geq 0.28$	**viii)**	$p < 0.78$
ix)	$p \leq 0.12$	**x)**	$p = 0.50$	**xi)**	$p > 0.32$	**xii)**	$p \neq 0.15$

Solution:

(a) The following table lists the corresponding ***counter claims***:

i)	$\mu < 78$	**ii)**	$\mu \geq 54$	**iii)**	$\mu > 12.6$	**iv)**	$\mu \neq 5$
v)	$\mu \leq 3.2$	**vi)**	$\mu = 15$	**vii)**	$p < 0.28$	**viii)**	$p \geq 0.78$
ix)	$p > 0.12$	**x)**	$p \neq 0.50$	**xi)**	$p \leq 0.32$	**xii)**	$p = 0.15$

Table 7.1

(b) The following table lists the Null and Alternative hypotheses.

i)	$H_0: \mu = 78(\text{oc})$ $H_1: \mu < 78(\text{cc})$	**ii)**	$H_0: \mu = 54(\text{cc})$ $H_1: \mu < 54(\text{oc})$	**iii)**	$H_0: \mu = 12.6(\text{oc})$ $H_1: \mu > 12.6(\text{cc})$	**iv)**	$H_0: \mu = 5(\text{oc})$ $H_1: \mu \neq 5(\text{cc})$
v)	$H_0: \mu = 3.2(\text{cc})$ $H_1: \mu > 3.2(\text{oc})$	**vi)**	$H_0: \mu = 15(\text{cc})$ $H_1: \mu \neq 15(\text{oc})$	**vii)**	$H_0: p = 0.28(\text{oc})$ $H_1: p < 0.28(\text{cc})$	**viii)**	$H_0: p = 0.78(\text{cc})$ $H_1: p < 0.78(\text{oc})$
ix)	$H_0: p = 0.12(\text{oc})$ $H_1: p > 0.12(\text{cc})$	**x)**	$H_0: p = 0.50(\text{oc})$ $H_1: p \neq 0.50(\text{cc})$	**xi)**	$H_0: p = 0.32(\text{cc})$ $H_1: p > 0.32(\text{oc})$	**xii)**	$H_0: p = 0.15(\text{cc})$ $H_1: p \neq 0.15(\text{oc})$

Table 7.2

B. Type-I and Type-II Errors in Testing a Hypothesis

In testing of a hypothesis, same as trial by a jury, we arrive at a decision to reject (guilty) or fail to reject (not guilty) a null hypothesis H_0. Since this decision is based on a sample from the population, there is an element of risk -- the risk of arriving at a wrong decision. In any hypothesis testing procedure, the four possible mutually disjoint and exhaustive decisions may be expressed in the following dichotomous table.

	Decision from Sample	
	Reject H_0	**Fail to Reject H_0**
H_0 is true	wrong decision Type I error	correct decision
H_0 is false	correct decision	wrong decision Type II error

Table 7.3

We make a type I error by rejecting a true null hypothesis and type II error by failing to reject a false null hypothesis. (Further Discussion on this topic can be found in Section 7.8: Extended Topics and Proofs in Hypothesis Testing)

C. Test Statistics used for Testing of Hypothesis

To test the null hypothesis H_0, we compute a statistic from sample data under the assumption that H_0 is true. This sample statistic is called the ***Test Statistics*** and is the standardized value which is used in making a decision about the null hypothesis. The sample values like the mean $\bar{x}$ and proportion $\hat{p}$ are converted to scores like the standard normal z-score or the t-score, under H_0. These values are used to test the significant difference between the observed sample statistics and the corresponding parameter value.

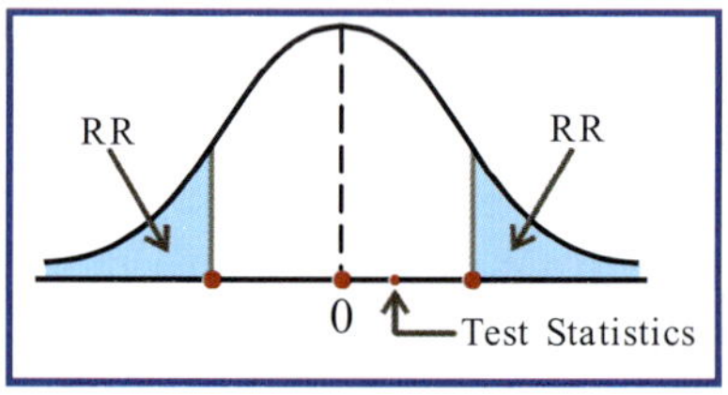

Figure 7.1

We will use the following test statistics in this chapter.

Test statistics for mean	**Test statistics for proportion**	**Test statistics for standard deviation or variance**
$Z = \dfrac{\bar{X}-\mu}{\dfrac{\sigma}{\sqrt{n}}}$ or $T = \dfrac{\bar{X}-\mu}{\dfrac{s}{\sqrt{n}}}$	$Z = \dfrac{\hat{p}-p}{\sqrt{\dfrac{pq}{n}}}$	$\chi^2 = \dfrac{(n-1)s^2}{\sigma^2}$

Table 7.4

Example 2: A survey of 20 randomly selected statistics final exam grades from a normally distributed population showed a standard deviation of 6 points. Find the value of the test statistics for the claim that the standard deviation of all statistics final exam scores is less than 8 points.

Solution: The given claim $\sigma < 8$ results in the following null and alternative hypotheses: $H_0 : \sigma = 8$ and $H_1 : \sigma < 8$. Since we conduct the test of hypothesis under the assumption that the null hypothesis is true ($\sigma = 8$), we get the following test statistics.

$$\chi^2 = \frac{(n-1)s^2}{\sigma^2} = \frac{(20-1)(6^2)}{8^2} = 10.69$$

D. Rejection Region

The statistics which lead to the rejection of H_0 give us a region known as the Rejection or Critical Region (RR, CR) while those which lead to the non-rejection of H_0 give us a region known as the Acceptance Region. Thus, if the test statistics falls in the Rejection Region, H_0 is *rejected*; and if it falls in the Acceptance Region, we *fail to reject* H_0 (which may be taken to be true).

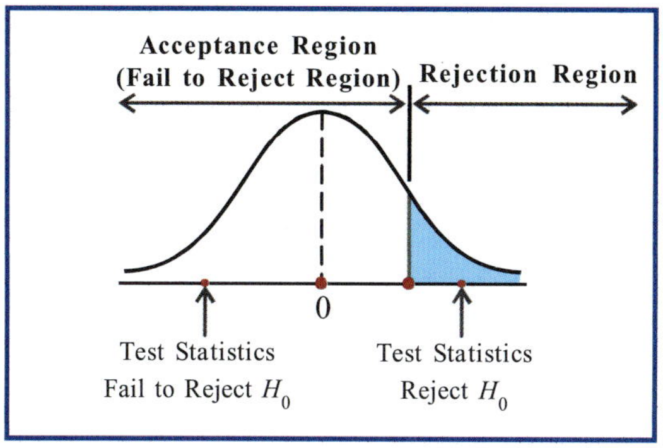

Figure 7.2

E. Significance Level

Formal Definitions:

i) *The maximum size of the type* I *error, which we are prepared to risk, is known as the level of significance or significance level. It is usually denoted by* α *and is given by*

$$P(\text{Rejecting } H_0 \text{ when } H_0 \text{ is true}) = \alpha.$$

ii) *The significance level* α *is the probability that the test statistics will fall in the rejection region.*

In practice, the commonly used significance level is 5% ($\alpha = 0.05$). Adopting a 5% level of significance means that if the hypothesis tests were performed using 100 different samples of the same size, 5 of them would lead to rejecting a correct null hypothesis. In other words, we would commit a type I error 5% of the time. Another interpretation of a significance level is its relationship to a confidence level. This means that if the significance level is 5%, we are 95% confident that our decisions to reject H_0 is correct.

Informal Definition:

The significance level α *is the area of the rejection region. Therefore, in a one tail test, the area of the rejection region is* α; *and for a two tail test,* α *is divided equally between the two tails that constitute the rejection region. Hence, the area of the rejection region in each tail is* $\alpha/2$.

Note: In research, the ***level of significance*** α is always fixed in advance before collecting the sample information.

Example 3: Construct the rejection region for the following claims, and find the corresponding critical values. Assume that the region is a part of the standard normal curve.

(a) $\mu < 53$, $\alpha = 0.05$.

(b) $\mu = 3.6$, $\alpha = 0.01$.

(c) $p \leq 0.28$, $\alpha = 0.10$.

Solution: **(a)** Since the original claim $\mu < 53$ is classified under H_1, the rejection region will be left-tailed. For $\alpha = 0.05$, we have $P(Z - z_\alpha) = 0.05 \Rightarrow P(-z_\alpha < Z < 0) = 0.45 \Rightarrow z_\alpha = -1.645$.

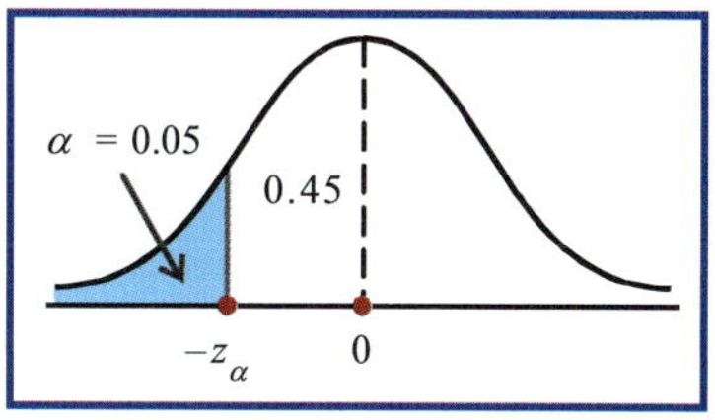

Figure 7.3

(b) Since the original claim $\mu = 3.6$ is classified under H_0, the rejection region based on $H_1 : \mu \neq 3.6$ will be two-tailed. For $\alpha = 0.01$ we have

$$P(Z > z_{\alpha/2}) = 0.005 \Rightarrow P(0 < Z < z_{\alpha/2}) = 0.495 \Rightarrow z_{\alpha/2} = 2.575$$

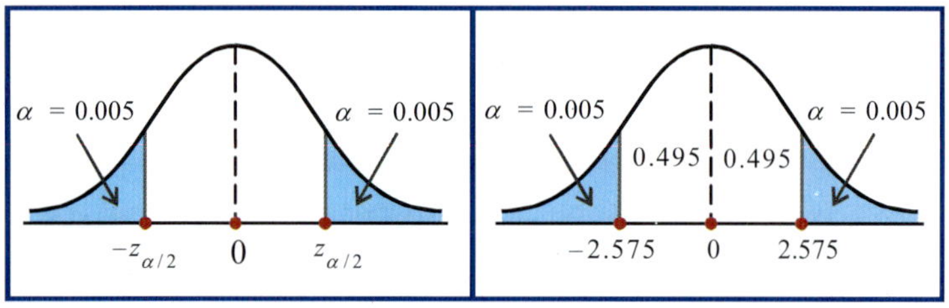

Figure 7.4

(c) Since the original claim $p \leq 0.28$ is classified under the null hypothesis as $H_0 : p = 0.28$, the rejection region based on $H_1 : p > 0.28$ will be right-tailed. For $\alpha = 0.10$ we have

$$P(Z > z_{\alpha}) = 0.1 \Rightarrow P(0 < Z < z_{\alpha}) = 0.4 \Rightarrow z_{\alpha} = 1.28$$

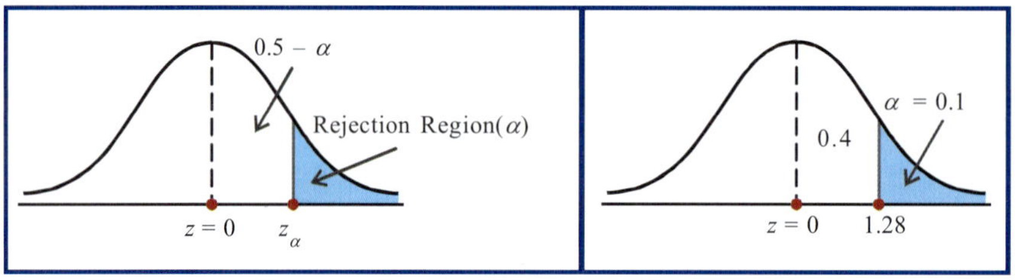

Figure 7.5

Decision: The standard hypothesis testing procedure requires that we always test the null hypothesis H_0. Therefore our initial decision must be one of the following:

(a) Reject the null hypothesis. **(b)** Fail to reject the null hypothesis.

F. One-Tailed and Two-Tailed Tests

A test of any statistical hypothesis where the alternative hypothesis is **one-tailed** (stated as greater or less than a parameter) is called a **one-tailed test.**

- ### Right-Tailed Test

To test a population parameter i.e., the mean of a population- where the original claim is $\mu \leq \mu_0$(μ_0 is an arbitrary value), the alternative hypothesis $\mu > \mu_0$ represents a right-tailed test. Note that, in a right-tailed test, the inequality symbol ">" in the alternative hypothesis points to the right and the rejection region is in the extreme right region (tail) under the distribution curve.

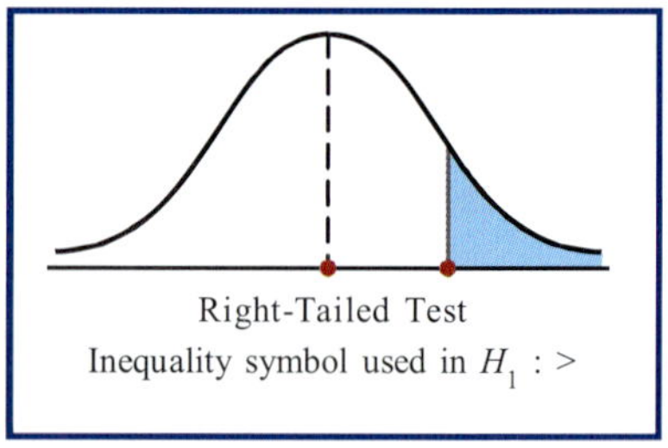

Figure 7.6

Original Claim	H_0	→	H_1
$\mu \leq \mu_0$ (Recall: μ_0 is some arbitrary constant)	$\mu = \mu_0$	against the alternative hypothesis	$\mu > \mu_0$

Table 7.5

- ### Left-Tailed Test

To test a population parameter i.e., the proportion of a population- where the original claim is $p \geq p_0$, the alternative hypothesis $p < p_0$ represents a left-tailed test. Note that, in a left-tailed test, the inequality symbol "$<$" in the alternative hypothesis points to the left and the rejection region is in the extreme left region (tail) under the distribution curve.

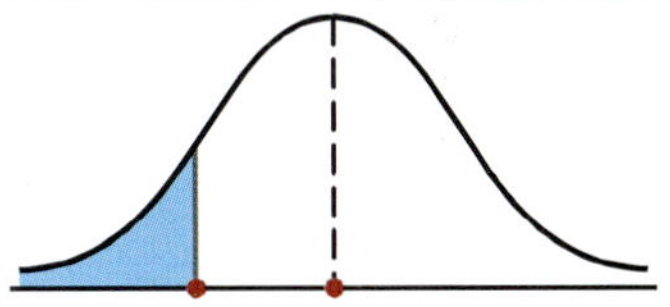

Left-Tailed Test
Inequality symbol used in H_1 : $<$

Figure 7.7

Original Claim	H_0	$\rightarrow$	H_1
$p \geq p_0$	$p = p_0$	against the alternative hypothesis	$p < p_0$

Table 7.6

Let θ represent a general population parameter and θ_0 an arbitrary constant. Then, in a right-tailed test ($H_1 : \theta > \theta_0$), the critical region lies entirely in the right-tail of the corresponding sampling distribution; while for the left-tailed test ($H_1 : \theta < \theta_0$), the critical region lies entirely in the left-tail of the sampling distribution of test statistics.

- ### Two-Tailed Test

A test of hypothesis where H_1 contains the $\neq$ symbol is known as a **two-tailed test**. In such cases, the rejection region lies in the two extreme regions under the probability curve. Furthermore, these two regions are symmetric in the sense that they have the same area.

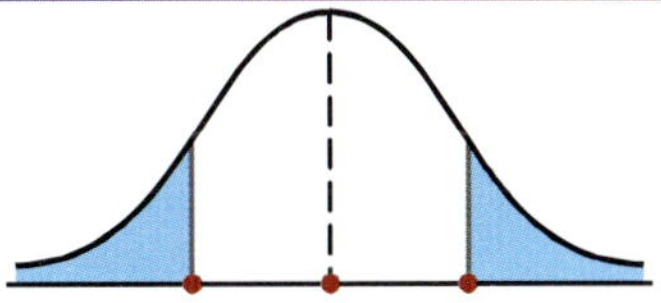

Two-Tailed Test
Inequality symbol used in H_1 : $\neq$

Figure 7.8

Example 4: Determine the rejection regions for the claims in Example 1.

Solution: Since the rejection region is determined by the direction of the inequality symbol in the alternative hypothesis, we have:

(i)	$H_0 : \mu = 78(\text{oc})$ $H_1 : \mu < 78(\text{cc})$ Left-tailed	**(ii)**	$H_0 : \mu = 54(\text{cc})$ $H_1 : \mu < 54(\text{oc})$ Left-tailed	**(iii)**	$H_0 : \mu = 12.6(\text{oc})$ $H_1 : \mu > 12.6(\text{cc})$ Right-tailed	**(iv)**	$H_0 : \mu = 5(\text{oc})$ $H_1 : \mu \neq 5(\text{cc})$ Two-tailed
(v)	$H_0 : \mu = 3.2(\text{cc})$ $H_1 : \mu > 3.2(\text{oc})$ Right-tailed	**(vi)**	$H_0 : \mu = 15(\text{cc})$ $H_1 : \mu \neq 15(\text{oc})$ Two-tailed	**(vii)**	$H_0 : p = 0.28(\text{oc})$ $H_1 : p < 0.28(\text{cc})$ Left-tailed	**(viii)**	$H_0 : p = 0.78(\text{cc})$ $H_1 : p < 0.78(\text{oc})$ Left-tailed
(ix)	$H_0 : p = 0.12(\text{oc})$ $H_1 : p > 0.12(\text{cc})$ Right-tailed	**(x)**	$H_0 : p = 0.50(\text{oc})$ $H_1 : p \neq 0.50(\text{cc})$ Two-tailed	**(xi)**	$H_0 : p = 0.32(\text{cc})$ $H_1 : p > 0.32(\text{oc})$ Right-tailed	**(xii)**	$H_0 : p = 0.15(\text{cc})$ $H_1 : p \neq 0.15(\text{oc})$ Two-tailed

Table 7.7

G. Critical Values

The value of the test statistic which separates the rejection (critical) region from values which lead to the acceptance region of H_0 is called the **critical value** (CV). In other words the critical value is the value that separates the rejection region from the acceptance region and is found using an appropriate distribution table. The critical value depends on

(a) the significance level. **(b)** the nature of the null hypothesis. **(c)** the sampling distribution.

The following table summarizes four tests of hypotheses with their corresponding distributions and test statistics for testing all levels of tests (left-tailed, right-tailed and two-tailed tests) with significance level (α).

	Purpose	Distribution used	Test Statistics
(i)	Testing μ, σ known	Standard Normal Distribution N(0,1)	$Z = \dfrac{\bar{X} - \mu}{\frac{\sigma}{\sqrt{n}}}$
(ii)	Testing μ, σ unknown	t - Distribution with $df = n - 1$	$T = \dfrac{\bar{X} - \mu}{\frac{s}{\sqrt{n}}}$
(iii)	Testing p	Standard Normal Distribution N(0,1)	$Z = \dfrac{\hat{p} - p}{\sqrt{\frac{pq}{n}}}$
(iv)	Testing σ (or σ^2)	Chi-Square Distribution with $df = n - 1$	$\chi^2 = \dfrac{(n-1)s^2}{\sigma^2}$

Table 7.8

Note: We will refer to the test statistics as the computed value of the test statistics and the critical value as the tabulated value of the test statistics.

H. *P*-value or Probability Value of Test Statistics

The probability that the value of test statistics is at least as extreme as its computed value on the basis of the sample data under H_0, is called its ***p*-value**.

1. For a left-tailed test, the ***p*-value** is the area to the left of the test statistics under H_0.

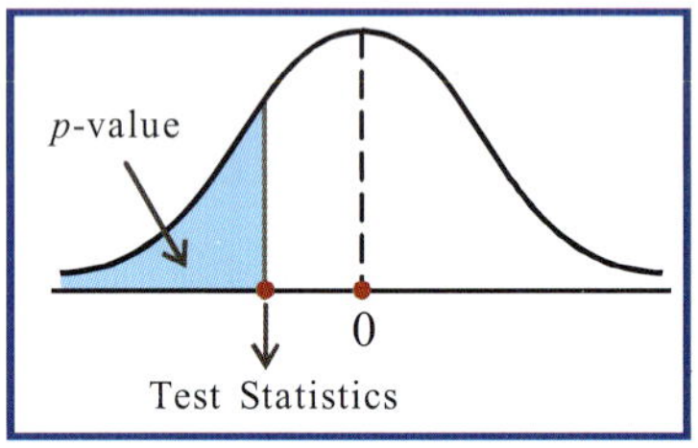

Figure 7.9

2. For a right-tailed test, the ***p*-value** is the area to the right of the test statistics under H_0.

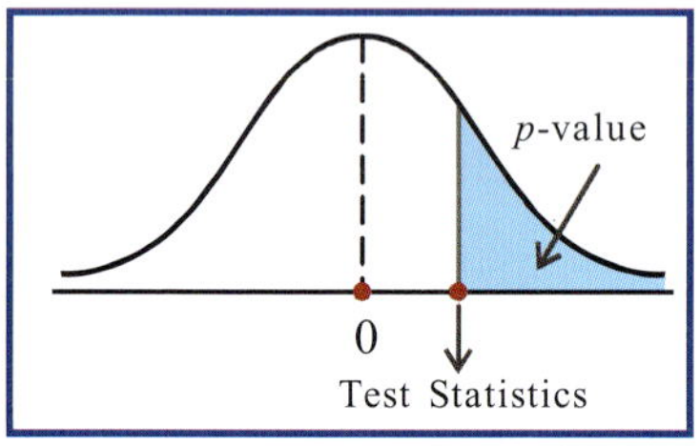

Figure 7.10

3. For a two-tailed test, the ***p*-value** is

   (a) twice the area to the left of the test statistics under H_0 if it is negative.

If the test statistics is negative.

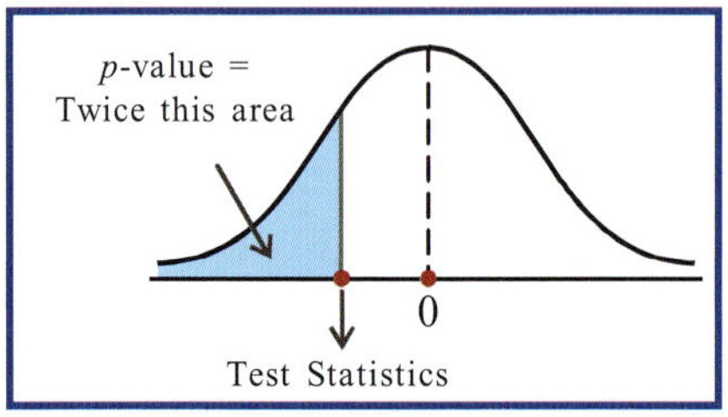

Figure 7.11

   (b) twice the area to the right of the test statistics under H_0 if it is positive.

If the test statistics is positive.

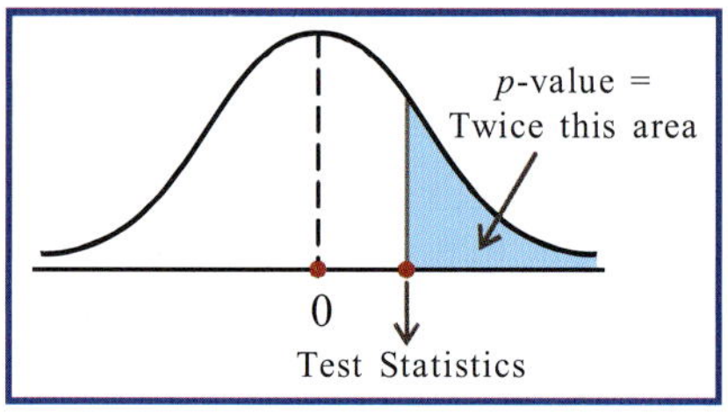

Figure 7.12

Note: The p-value for a two-tailed test is twice the area on either tail (left or right) of the computed value of test statistics under H_0.

I. Procedure for Hypothesis Testing

We can use any of the following two methods for testing of a statistical hypothesis.

1. Classical Method

2. P-value Method

We summarize below the various steps in testing of a hypothesis in each of these methods.

***Step* 0** Write down the sample statistics such as n, $\bar{x}$, s, $\hat{p}$. Note that if instead of sample statistics raw data is provided, you must use formulas from descriptive statistics to find $\bar{x}$, s or $\hat{p}$.

Recall: $\bar{x} = \dfrac{\sum x}{n}$, $s = \sqrt{\dfrac{n\sum x^2 - \left(\sum x\right)^2}{n(n-1)}}$ and $\hat{p} = \dfrac{x}{n}$

Moreover, record the original claim and construct the counter claim.

Classical Method:

***Step* 1** Set up the significance level, **null** and **alternative hypotheses**. Also declare whether the test is Left-Tailed, Right-Tailed or Two-Tailed.

***Step* 2** Calculate the appropriate **Test Statistics** under H_0.

***Step* 3** Construct the **Rejection Region** and compute the corresponding **Critical Value(s)** from the appropriate table.

***Step* 4** State the **Decision**, and write the conclusion in simple words.

P-value Method:

***Step* 1** Set up the significance level, **null** and **alternative hypotheses**. Also declare whether the test is Left-Tailed, Right-Tailed or Two-Tailed.

***Step* 2** Calculate the appropriate **Test Statistics** under H_0.

***Step* 3** Find the ***p*-value** of the computed test statistics.

***Step* 4** **Decision:** If ***p*-value** $< \alpha$, reject H_0 at α significance level and if ***p*-value** $> \alpha$, fail to reject H_0 at α significance level. Write the conclusion in simple words.

J. Interpretation

If the computed value of the test statistic lies in the rejection region, we must **Reject** H_0 at level of significance α. If the computed value of the test statistic lies outside the rejection region, we **Fail to Reject** H_0 at an "α" significance level.

Note: Whenever we reject the null hypothesis, we have a certain degree of confidence (which depends on the level of significance used) in our decision. Thus, at 'α' significance level, the degree of confidence in our decision to reject H_0 is $(1 - \alpha)$. This is also called the confidence coefficient.

Conclusion: Once a decision is made, we must then write a conclusion in simple language. The following flow chart summarizes all possible conclusions **based on our decision**.

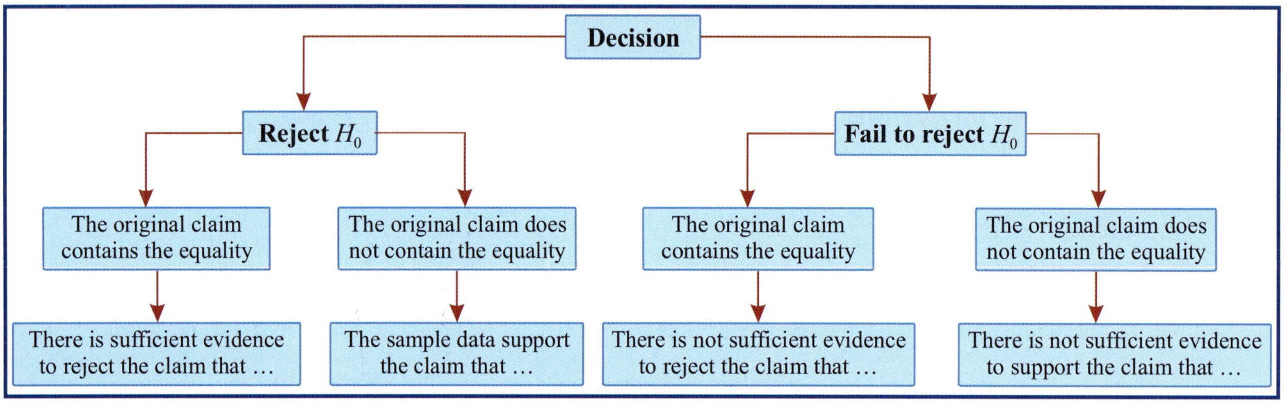

Figure 7.13

Note: Some textbooks use the term "accept the null hypothesis" instead of "fail to reject the null hypothesis". In this text, we will use the term "FTR H_0." However, if you are faced with using the term "accept the null hypothesis," keep in mind that by accepting the null hypothesis you are not proving the validity of null hypothesis; you are basically asserting that the sample evidence is not statistically strong enough to reject the null hypothesis. To understand the difference between accepting or failing to reject H_0, think of the final decision in a trial by jury (the verdict). When the jury does not find enough evidence to render a guilty verdict (convicting a suspect), it returns a not guilty verdict. This does not mean that the suspect is innocent; it just means that there was not enough evidence to convict the suspect. Therefore, when we fail to reject the null hypothesis, instead of saying there is sufficient evidence to accept H_0, we must say that the evidence is not strong enough to reject the null hypothesis (the final verdict can never be "innocent". It can only be "not guilty" or "guilty.")

Testing Claims about Population Parameters using Confidence Intervals

In Chapter 6, we discussed how to construct confidence intervals for population mean, proportion, variance, and standard deviation. We can also use confidence intervals to test claims about the values of the population parameters. The confidence interval for any parameter is expected to contain the true value of the parameter with certain degree of confidence depending on the level of significance.

Suppose we want to test a claim about a parameter at "α" level of significance. There are two cases:

(a) For any two-tailed test, we compute the ***confidence interval*** for the parameter, using the confidence coefficient $1 - \alpha$.

(b) For one-tailed (right-tailed or left-tailed) test, we compute the ***confidence interval*** for the parameter, using the confidence coefficient $1 - 2\alpha$.

Thus, for example, at α significance level, the confidence limits for the population mean μ, using the z-test, are:

Two-tailed Test	One-tailed Test
$\bar{x} \pm z_{\alpha/2} \dfrac{\sigma}{\sqrt{n}}$	$\bar{x} \pm z_{\alpha} \dfrac{\sigma}{\sqrt{n}}$

Table 7.9

The claim about the population parameter is rejected at α significance level if its value lies outside the corresponding confidence interval. On the other hand, we fail to reject the claim about the parameter if its value lies within the confidence interval.

7.3: Testing a Claim About a Mean μ (σ Known)

In this section we discuss the hypothesis testing method used to test claims about a population mean μ, assuming the population standard deviation σ is known. The process of hypothesis testing (test of significant difference) will help us determine whether the difference between the two means (sample and true mean) is due to chance variation and the claim is true or whether the difference is real and the claim is false.

Objectives

After completing this section you will be able to define:

A. Testing claim about the population mean μ, assuming σ is known.

B. Critical values, when significance level is α.

A. Testing Claims about the Population Mean μ assuming σ is Known

Notation:

Sample Size	Sample Mean	Population Mean (Mean of all Sample Means)	Population Standard Deviation
n	$\bar{x}$	$\mu_{\bar{x}} = \mu$	σ

Table 7.10

- **Requirements for this test**

1. The population is normally distributed or the sample is large ($n > 30$).

2. n is the size of a **simple** random sample.

3. σ is known.

Test Statistics: $Z = \dfrac{\bar{X} - \mu_{\bar{x}}}{\sigma_{\bar{x}}} = \dfrac{\bar{X} - \mu}{\sigma / \sqrt{n}} \sim N(0, 1)$ for large samples.

B. Critical Values based on α Significance Level

Critical values are determined using the standard normal distribution.

1. For a Right-Tailed Test: If $P(Z > z_\alpha) = \alpha$, then we must use probability = $(0.5 - \alpha)$ to determine the critical value z_α Figure 7.14.

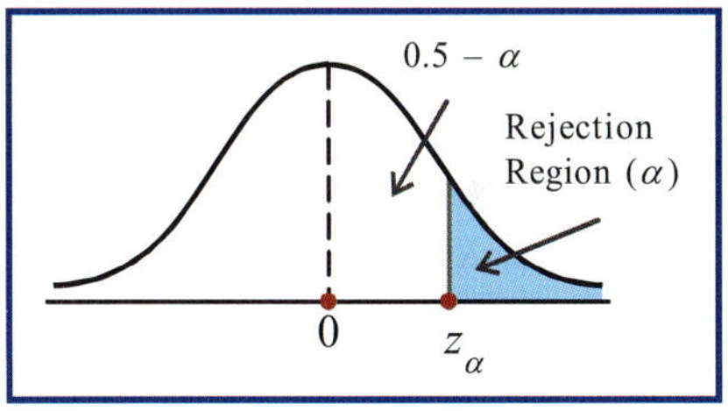

Figure 7.14

2. For a Left-Tailed Test: If $P(Z < -z_\alpha) = \alpha$, then we must use probability = $(0.5 - \alpha)$ to determine the critical value z_α Figure 7.15.

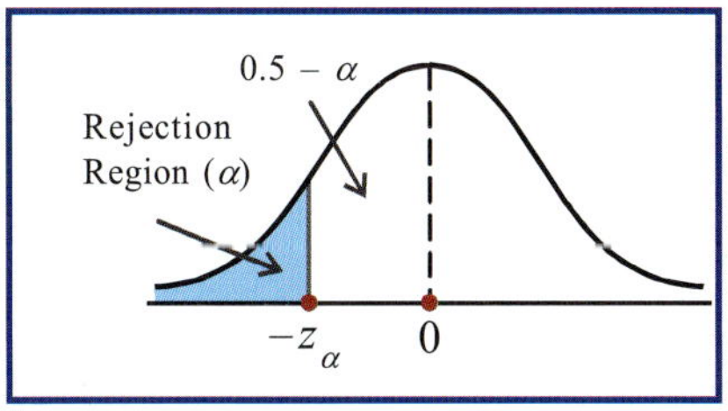

Figure 7.15

3. For a Two-Tailed Test: If $P(|Z| > z_{\alpha/2}) = \alpha$

$$\Rightarrow \begin{cases} P(Z > z_{\alpha/2}) = \alpha / 2 \\ P(Z < -z_{\alpha/2}) = \alpha / 2 \end{cases}, \text{ then we must use}$$

probability = $(0.5 - \alpha/2)$ to determine the critical value $z_{\alpha/2}$ Figure 7.16.

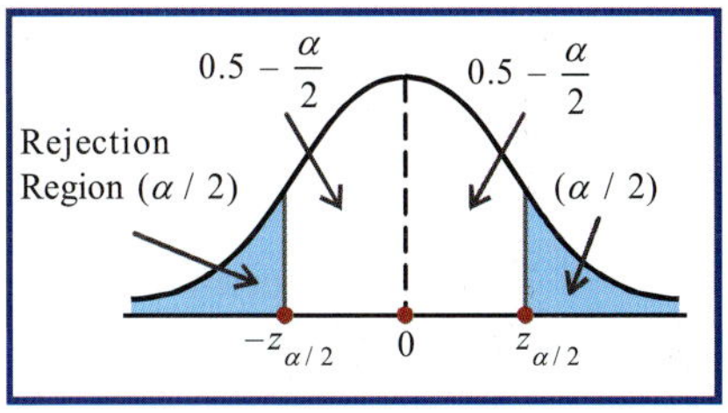

Figure 7.16

Summary for the z-test: σ known

Sample Size (n)	Population Distribution	Test Statistics and its sampling distribution
$n > 30$	Any	$Z = \frac{\bar{X} - \mu}{\sigma / \sqrt{n}} \sim N(0, 1)$; σ known (if σ is unknown, take $\hat{\sigma} = s$)
$n \leq 30$	Normal	$Z = \frac{\bar{X} - \mu}{\sigma / \sqrt{n}} \sim N(0, 1)$; σ known
$n \leq 30$	Unknown	Use non-parametric tests (Chapter 12)

Table 7.11

Example 5: Zion Memorial claims that the mean recovery time for emergency patients is 2.26 days. To test this claim, a researcher found that among 50 randomly selected patients from the emergency room, the mean recovery time was 2.12 days.

(a) Assuming that the standard deviation of all recovery times is 1.3 days, use a 0.01 significance level to test the claim

i) using the classical method. **ii)** using the P-value method.

(b) Construct a 99% confidence interval to verify the results obtained from the test of hypothesis in part (a).

Solution: **Step 0** We are given $\sigma = 1.3$, $\bar{x} = 2.12$, $n = 50$, and $\alpha = 0.01$.

The original claim: $\mu = 2.26$

The counter claim: $\mu \neq 2.26$

(a) i) Classical Method: Since n is large and σ is given, we use the normal test based on the statistics $\bar{X}$.

Step 1 Set up the Null and Alternative Hypotheses.

H_0 : $\mu = 2.26$ (oc)

H_1 : $\mu \neq 2.26$ (cc) (Two-Tail Test)

Step 2 The value of test statistics based on the sample information provided, and the hypothesized value of the population mean μ is

$$z = \frac{\bar{x} - \mu}{\sigma / \sqrt{n}} = \frac{2.12 - 2.26}{1.3 / \sqrt{50}} = -0.76.$$

Step 3 Since H_1 : $\mu \neq 2.26$, the test is two-tailed; hence the rejection region will be in both tails of the normal distribution curve. The critical values ($\pm z_{\alpha/2}$) for a 0.01 significance level are ± 2.575. (Figure 7.17)

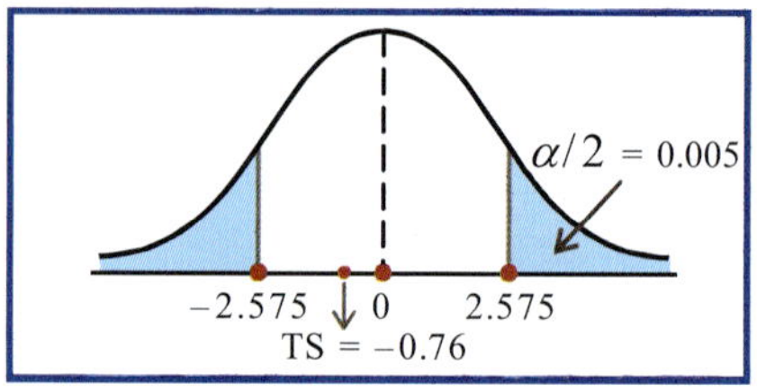

Figure 7.17

Recall : $\alpha = 0.01 \Rightarrow \frac{\alpha}{2} = 0.005$, therefore $z_{\alpha/2} = z_{0.005} = 2.575$

Step 4 **Decision:** Since the calculated value of test statistics ($z = -0.76$) lies outside of the rejection region, we fail to reject the null hypothesis at a 0.01 significance level.

Conclusion: There is sufficient evidence to support the claim that the mean recovery time for emergency patients is 2.26 days.

ii) *P*-value Method:

***Step* 1** Set up the Null and Alternative Hypotheses.

$H_0: \mu = 2.26$ (oc)

$H_1: \mu \neq 2.26$ (cc) (Two-Tail Test)

***Step* 2** The value of test statistics based on the sample information provided, and the hypothesized value of the population mean μ is

$$z = \frac{\bar{x} - \mu}{\sigma / \sqrt{n}} = \frac{2.12 - 2.26}{1.3/\sqrt{50}} = -0.76$$

***Step* 3** Since the test is two-tailed, the region showing the extreme values is twice the area to the left of the test statistics.

$$\begin{aligned} p\text{-value} &= \text{Area of the shaded region} \\ &= 2 \times P(Z < -0.76) \\ &= 2(0.2236) = 0.4472 \end{aligned}$$

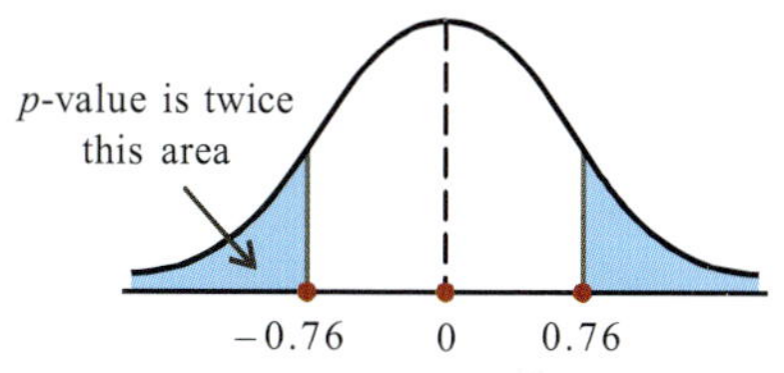

Figure 7.18

***Step* 4** **Decision:** Since the p-value is greater than $\alpha = 0.01$, we fail to reject the null hypothesis H_0 at 1% level of significance. *Conclusion:* The sample data provides enough evidence to support the claim that the mean recovery time for emergency patients is 2.26 days.

In short:

Hypotheses	**Test Statistics**
$H_0: \mu = 2.26$ $H_1: \mu \neq 2.26$	$z = \frac{\bar{x} - \mu}{\sigma / \sqrt{n}} = -0.76$
Rejection Region and Critical Value	**Conclusion**
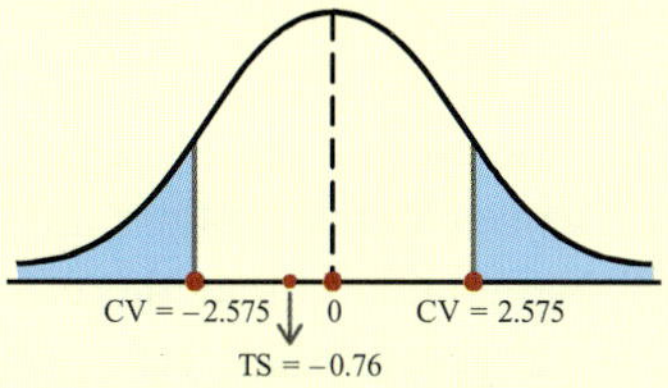	We fail to reject H_0. The mean recovery time for emergency patients is 2.26 days.

(b) 99% Confidence interval of the mean recovery time for emergency patients

Using the sample statistics provided in the example and a 99% confidence level, we have

$$\bar{x} - z_{\alpha/2}\frac{\sigma}{\sqrt{n}} < \mu < \bar{x} + z_{\alpha/2}\frac{\sigma}{\sqrt{n}}$$

$$2.12 - 2.575\frac{1.3}{\sqrt{50}} < \mu < 2.12 + 2.575\frac{1.3}{\sqrt{50}}$$

$$1.65 < \mu < 2.59$$

The confidence interval tells us that there is 0.99 probability that the population mean lies in the interval 1.65 to 2.59, which agrees with the results obtained in the test of hypothesis.

Example 6: Professor Badalian claims that the mean IQ of college students is greater than 110. To verify his claim, 150 college students are randomly selected and it is found that their mean IQ is 111.8.

(a) Assuming that the population standard deviation $\sigma = 7.2$, use a 0.05 significance level to test Prof. Badalian's claim.

i) using the classical method, **ii)** using the *P*-value method.

(b) Construct a 90% confidence interval to verify the results obtained from the test of hypothesis in part (a).

Solution: ***Step 0*** We are given: $\sigma = 7.2$, $\overline{x} = 111.8$, $n = 150$, and $\alpha = 0.05$.

The original claim: $\mu > 110$

The counter claim: $\mu \leq 110$

(a) i) Classical Method: Since n is large and σ is given, we use the z-test based on the statistics $\overline{X}$.

Step 1 Set up the Null and Alternative Hypotheses

Recall: The Null Hypothesis is always equal to some specific value to allow the researcher to work with a single distribution with a distinct value:

$H_0 : \mu = 110$ (cc)

$H_1 : \mu > 110$ (oc) (Right-Tail Test)

Step 2 The value of test statistics based on the sample information provided, the hypothesized value of the population mean $\mu = 110$ is

$$z = \frac{\overline{x} - \mu}{\sigma / \sqrt{n}} = \frac{111.8 - 110}{7.2 / \sqrt{150}} = 3.06$$

Step 3 Since $H_1 : \mu > 110$, the test is right-tailed, hence the rejection region will be in the right-tail of the normal distribution. The critical value (z_α) for the right-tail test at 5% level of significance is $z_\alpha = 1.645$.

Recall : $\alpha = 0.05$, therefore $z_{\alpha/2} = z_{0.05} = 1.645$

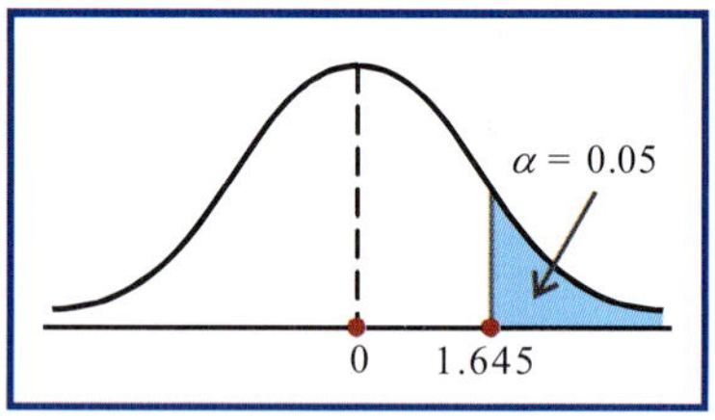

Figure 7.19

Step 4 **Decision:** Since the calculated value of test statistics ($z = 3.06$) lies inside the rejection region, we reject the null hypothesis at a 0.05 significance level. *Conclusion:* The data provides sufficient evidence against the Null Hypothesis (or in favor of the alternative hypothesis). We conclude that there is sufficient evidence to support the claim that the average IQ is greater than 110.

ii) *P*-value Method:

Step 1 Set up the Null and Alternative Hypotheses

Recall: The Null Hypothesis is always equal to some specific value to allow the researcher to work with a single distribution with a distinct value:

$H_0 : \mu = 110$ (cc)

$H_1 : \mu > 110$ (oc) (Right-Tail Test)

Step 2 The value of test statistics based on the sample information provided, the hypothesized value of the population mean $\mu = 110$ is

$$z = \frac{\overline{x} - \mu}{\sigma / \sqrt{n}} = \frac{111.8 - 110}{7.2 / \sqrt{150}} = 3.06$$

Step 3 Since this is a right-tailed test, the region showing the extreme values is the area to the right of the test statistics.

$$\begin{aligned} p\text{-value} &= \text{Area of the shaded region} \\ &= P(Z > 3.06) \\ &= 0.5 - 0.4989 = 0.0091 \end{aligned}$$

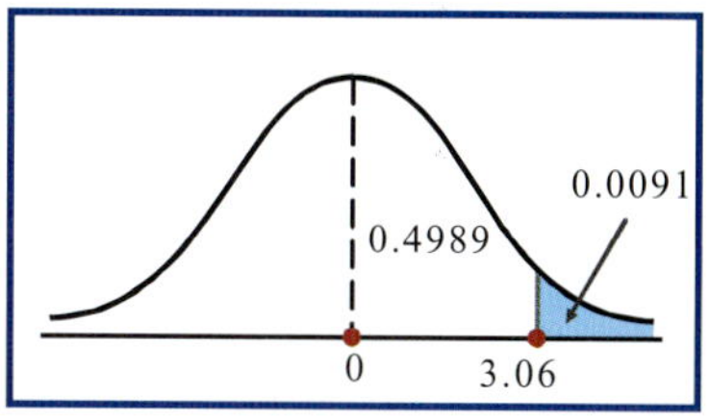

Figure 7.20

Step 4 **Decision:** Since the p-value is less than α, we reject the null hypothesis at 5% level of significance. *Conclusion:* There is sufficient evidence to support the claim that the average IQ is greater than 110.

In short:

Hypotheses	**Test Statistics**
$H_0: \mu = 110$ $H_1: \mu > 110$	$z = \dfrac{\bar{x} - \mu}{\sigma / \sqrt{n}} = 3.06$
Rejection Region and Critical Value	**Conclusion**
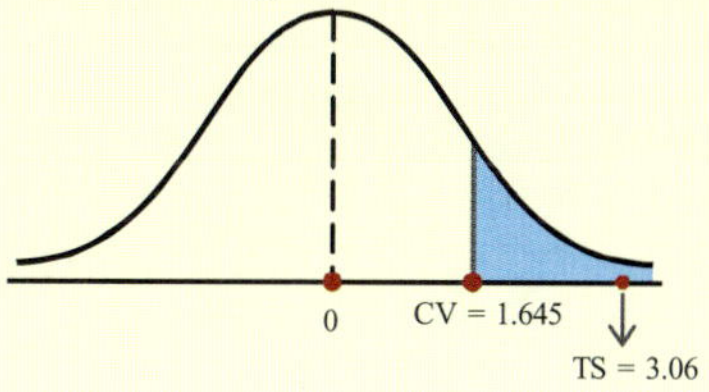	Reject H_0. The average IQ is greater than 110.

(b) 90% Confidence interval of the mean recovery time for emergency patients

Using the sample statistics provided in the example and a 90% confidence level, we have

$$\bar{x} - z_{\alpha/2}\frac{\sigma}{\sqrt{n}} < \mu < \bar{x} + z_{\alpha/2}\frac{\sigma}{\sqrt{n}}$$

$$111.8 - 1.645\frac{7.2}{\sqrt{150}} < \mu < 111.8 + 1.645\frac{7.2}{\sqrt{150}}$$

$$110.83 < \mu < 112.77$$

The confidence interval tells us that there is 0.90 probability that the population mean lies in the interval 110.83 to 112.77, which agrees with the results from the test of hypothesis. The average IQ doesn't seem to be significantly greater than 110.

Example 7: An insurance agent has claimed that the average age of the policy holders who insure through him is less than the average age of all other policy holders in the company, which is 30.5 years. A random sample of policy holders who have been insured through him gave the following age distribution.

Age group	16-20	21-25	26-30	31-35	36-40
No. of persons	12	22	20	30	16

(a) Calculate the sample mean and standard deviation of this distribution.

(b) Test the agent's claim at $\alpha = 0.01$. (Use both classical and the P-value methods.)

(c) Construct a 98% confidence interval to verify the results obtained from the test hypothesis in part (b).

Solution: **(a)** Recall the formulas for the mean and standard deviation for grouped frequency from Chapter 2.

$$\bar{x} = \frac{\sum f \cdot x}{\sum f},$$

and

$$s = \sqrt{\frac{n\left(\sum f \cdot x^2\right) - \left(\sum f \cdot x\right)^2}{n(n-1)}}.$$

Age Group	Frequency	Midpoint(x)	$f \cdot x$	$f \cdot x^2$
16-20	12	18	216	3888
21-25	22	23	506	11638
26-30	20	28	560	15680
31-35	30	33	990	32670
36-40	16	38	608	23104
	$n = \sum f = 100$		$\sum(f \cdot x) = 2880$	$\sum(f \cdot x^2) = 86980$

Table 7.12

Step 0 Sample Mean $= \bar{x} = \dfrac{\sum f \cdot x}{\sum f} = \dfrac{2880}{100} = 28.80$ years

Standard Deviation $= s = \sqrt{\dfrac{n\left(\sum f \cdot x^2\right) - \left(\sum f \cdot x\right)^2}{n(n-1)}} = \sqrt{\dfrac{100(86980) - 2880^2}{(100)(99)}}$

$= 6.384957 \approx 6.38$

(b) The original claim: $\mu < 30.5$

The counter claim: $\mu \geq 30.5$

Classical Method:

Step 1 Set up the Null and Alternative Hypotheses

$H_0: \mu = 30.5$ (cc)

$H_1: \mu < 30.5$ (oc) (Left-Tail Test)

Step 2 Since the sample is large, we can replace σ with $\hat{\sigma} = s = 6.38$ years.

The value of test statistics based on the calculated statistics and the hypothesized value of the population mean μ is

$$z = \frac{\bar{x} - \mu}{\sigma / \sqrt{n}} = \frac{28.8 - 30.5}{6.38/\sqrt{100}} = -2.66$$

Step 3 Since $H_1: \mu < 30.5$, the test is left-tailed; hence the rejection region will be in the left-tail of the normal distribution. The critical value (z_α) for the left-tail test at 1% level of significance is $z_\alpha = -2.33$.

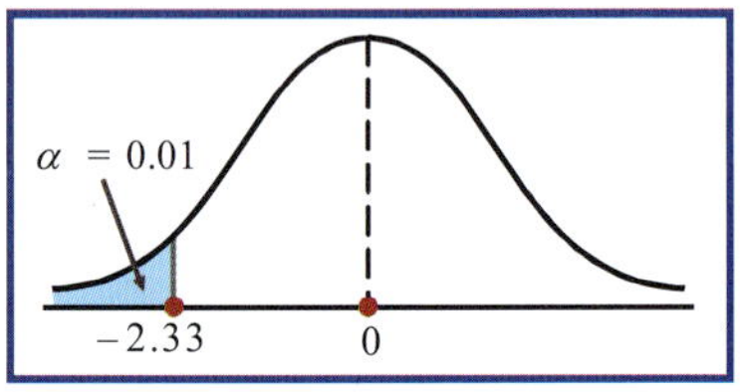

Figure 7.21

Recall: $\alpha = 0.01$, therefore $z_\alpha = z_{0.01} = 2.33$

Step 4 **Decision:** Since the calculated value of test statistics ($z = -2.66$) lies inside the rejection region, we reject the null hypothesis at a 0.01significance level.

Conclusion: The sample data provides sufficient information against H_0 (or in favor of H_1). Hence, we can conclude that there is sufficient evidence to support the agent's claim that the average age of the policy holders who insure through him is less than the average age of all other policy holders in the company, which is 30.5 years.

P-value Method:

***Step* 1** Set up the Null and Alternative Hypotheses

$H_0: \mu = 30.5$ (cc)

$H_1: \mu < 30.5$ (oc) (Left-Tail Test)

***Step* 2** Since the sample is large, we can replace σ with $\hat{\sigma} = s = 6.38$ years.

The value of test statistics based on the calculated statistics and the hypothesized value of the population mean μ is

$$z = \frac{\bar{x} - \mu}{\sigma / \sqrt{n}} = \frac{28.8 - 30.5}{6.38/\sqrt{100}} = -2.66$$

***Step* 3** **The *p*-value:** Since this is a left-tailed test; the region showing the extreme values is the area to the left of the test statistics.

$$\begin{aligned} p\text{-value} &= \text{Area of the shaded region} \\ &= P(Z < -2.66) \\ &= 0.5 - 0.4961 = 0.0039 \end{aligned}$$

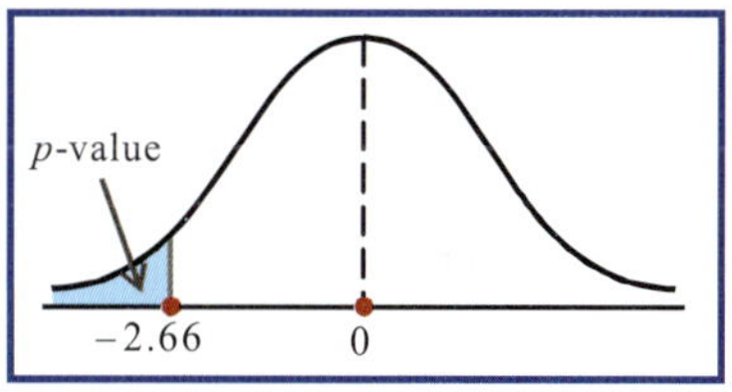

Figure 7.22

***Step* 4** **Decision:** Since p-value $< \alpha = 0.01$, we reject H_0 at 1% level of significance.

Conclusion: There is sufficient evidence to support the agent's claim that the average age of the policy holders who insure through him is less than the average age of all other policy holders in the company, which is 30.5 years.

In short:

Hypotheses	**Test Statistics**
$H_0: \mu = 30.5$ $H_1: \mu < 30.5$	$z = \frac{\bar{x} - \mu}{\sigma / \sqrt{n}} = -2.66$
Rejection Region and Critical Value	**Conclusion**
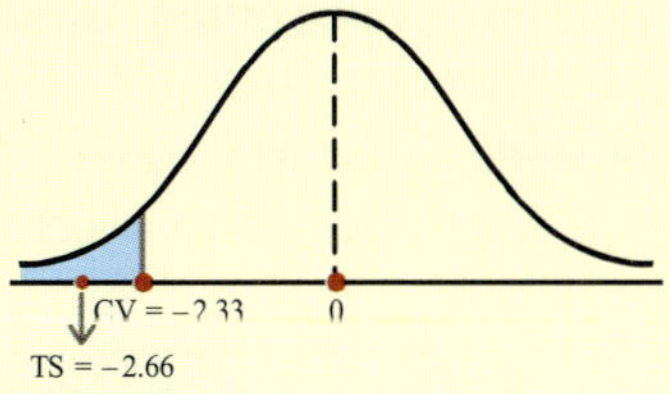	Reject H_0. The average age of the policy holders who insure through him is less than the average age of all other policy holders in the company, which is 30.5 years.

(c) 98% Confidence interval of the mean age of the policy holders who insure through the agent

The 98% confidence interval for the true average age of the policy holders through the agent is

$$\bar{x} - z_{\alpha/2}\frac{\sigma}{\sqrt{n}} < \mu < \bar{x} + z_{\alpha/2}\frac{\sigma}{\sqrt{n}}$$

$$28.8 - 2.33\frac{6.38}{\sqrt{100}} < \mu < 28.8 + 2.33\frac{6.38}{\sqrt{100}}$$

$$27.31 < \mu < 30.29$$

The constructed 98% confidence interval validates the conclusion obtained in the test of hypothesis in part (b). The average age of a policy holder with this agent does seem to be less than 30.5.

Example 8: The mean lifetime of a sample of 100 electric light bulbs manufactured by a company is found to be 1570 hours. The population standard deviation is known to be 120 hours. If μ is the mean lifetime of all the bulbs manufactured by the company,

(a) test the claim that the mean lifetime of a light bulb is 1600 hours. Use the classical method and a 0.05 significance level.

(b) construct a 95% confidence interval to verify the results obtained from the test hypothesis in part (a).

Solution: ***Step* 0** We are given $\sigma = 120$, $\bar{x} = 1570$, $n = 100$, and $\alpha = 0.05$.

The original claim: $\mu = 1600$

The counter claim: $\mu \neq 1600$

(a) **Classical Method:** Since n is large and σ is given, we use the normal test based on the statistics $\bar{X}$.

***Step* 1** Set up the Null and Alternative Hypotheses

H_0 : $\mu = 1600$ (oc)

H_1 : $\mu \neq 1600$ (cc) (Two-Tail Test)

***Step* 2** The value of test statistics based on the sample information provided and the hypothesized value of the population mean μ is

$$z = \frac{\bar{x} - \mu}{\sigma / \sqrt{n}} = \frac{1570 - 1600}{120/\sqrt{100}} = -2.5$$

***Step* 3** Since H_1 : $\mu \neq 1600$, the test is two-tailed; hence the rejection region is in both tails of the standard normal distribution (curve) with an area of $\alpha/2 = 0.025$ in each tail. The critical values ($\pm z_{\alpha/2}$) for a two-tailed test at 5% level of significance are $z_{\alpha/2} = \pm 1.96$.

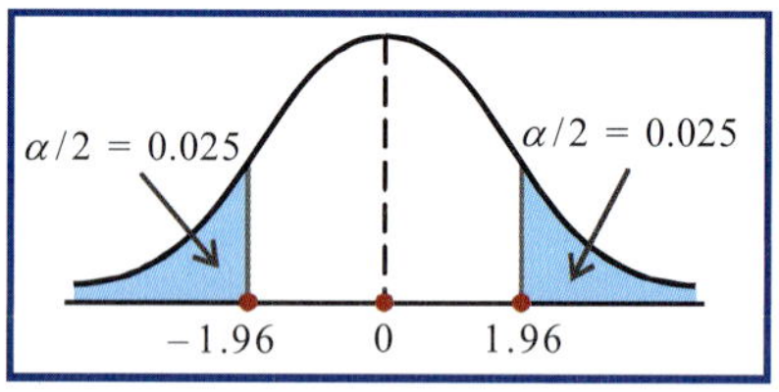

Figure 7.23

Recall : $\alpha = 0.05$, therefore $z_{\alpha/2} = z_{0.025} = 1.96$

***Step* 4** **Decision:** Since the calculated value of test statistics ($z = -2.5$) lies inside the rejection region, we reject the null hypothesis at a 0.05 significance level. *Conclusion:* The data provides sufficient evidence against the Null Hypothesis in favor of the alternative hypothesis. There is sufficient evidence to reject the claim that the mean lifetime of a light bulb is 1600 hours.

In short:

Hypotheses	**Test Statistics**
H_0: $\mu = 1600$ H_1: $\mu \neq 1600$	$z = \frac{\bar{x} - \mu}{\sigma / \sqrt{n}} = -2.5$
Rejection Region and Critical Value 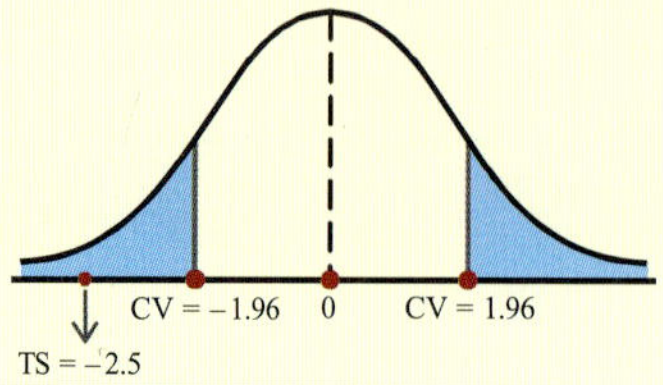	**Conclusion** Reject H_0. The mean lifetime of a light bulb is not equal to 1600 hours.

(b) 95% Confidence interval for the mean lifetime of all electric light bulbs

$$\bar{x} - z_{\alpha/2}\frac{\sigma}{\sqrt{n}} < \mu < \bar{x} + z_{\alpha/2}\frac{\sigma}{\sqrt{n}}$$

$$1570 - 1.96\frac{120}{\sqrt{100}} < \mu < 1570 + 1.96\frac{120}{\sqrt{100}}$$

$$1546.5 < \mu < 1593.5$$

There is a 0.95 probability that the mean lifetime of all light bulbs lies between 1546.5 and 1593.5 hours, which agrees with the results from the test of hypothesis. The mean lifetime of a light bulb does not seem to be 1600 hours.

Try it yourself:

Test the following original claims using the given sample information. Use both the traditional and the *p*-value methods. Construct an appropriate confidence interval to verify the results obtained in the test of hypothesis.

1. $\mu > 30$ (oc), $\bar{x} = 33.2$, $\sigma = 2.8$ and $n = 50$. Use a 5% significance level.

2. $\mu < 218$ (oc), $\bar{x} = 215$, $\sigma = 23$ and $n = 80$. Use a 1% significance level.

3. $\mu = 15$ (oc), $\bar{x} = 13.6$, $\sigma = 1.8$ and $n = 46$. Use a 1% significance level.

4. $\mu \geq 62.8$ (oc), $\bar{x} = 56.8$, $\sigma = 18$ and $n = 300$. Use a 2% significance level.

5. $\mu \neq 12$ (oc), $\bar{x} = 14.2$, $\sigma = 2.8$ and $n = 20$. Use a 5% significance level. Assume that the parent population is normally distributed.

6. Fifteen cartons of sugar, each nominally containing 1 pound, were randomly selected from a large batch of cartons. The weights of sugar (in Pounds) they contained were:

1.02	1.05	1.08	1.03	1.00	1.06	1.08	1.01	1.04	1.07	1.00	1.02	1.09	1.01	0.99

   Does this support the hypothesis, at 5%, that the mean weight for the whole batch is over 1.00 pound? Use $\sigma = 0.035$.

7. It is claimed that on the average a patient suffering from leukemia survives 1080 days from the time of diagnosis. A researcher's random sample of 22 such patients indicated an average survival time as 767 days with a standard deviation $\sigma = 180$ days. Test the claim at $\alpha = 0.01$. Assume the population is normally distributed.

8. A pharmaceutical company claims that the mean time for a drug to take effect is 24 minutes. In a sample of 400 trials, the mean time is 26 minutes with $\sigma = 4$ minutes. Test the company's claim at 10% level of significance.

7.4: Testing a Claim About a Mean μ (σ Unknown)

Objectives

After completing this section you will be able to understand:

A. Testing a claim about the population mean (μ), assuming (σ) is unknown.

B. Assumptions for Student's *t*-test.

In this section we will discuss the methods in testing a claim about a population mean μ, given that the population standard deviation σ is unknown. In research, it is virtually impossible to know the population standard deviation. Hence, the methods of this section are more realistic and are used more often in statistical studies. The test of hypothesis in this section is known as the *t*-test because we use the Student's *t*-Distribution as outlined in Chapter 6.

A. Testing a claim about the Population Mean (μ), assuming (σ) is Unknown

Notation:

Sample Size	Sample Mean	Population Mean (mean of all sample means)
n	$\overline{x}$	$\mu_{\overline{x}} = \mu$

Table 7.13

If n is small, ($n \leq 30$) then the sampling distribution of the statistic $\dfrac{\overline{X} - \mu}{s/\sqrt{n}}$ is far from normality. Thus, the normal test cannot be applied if $n \leq 30$, even if the parent population is normally distributed. Therefore, in the test of hypothesis for small samples, we will use the student's *t*-distribution.

Student's *t*-Distribution:

If $X_1, X_2, \ldots, X_n$ is random sample of size n from a normal population $N(\mu, \sigma)$, then student's *t*-statistic is defined as $t = \dfrac{\overline{X} - \mu}{s/\sqrt{n}}$ and it follows *t*-distribution with $n - 1$ degrees of freedom.

B. Assumptions for Student's *t*-test

1. The sample size is small, $n \leq 30$.
2. The sample is a simple random sample.
3. The parent population from which the sample is drawn is normally distributed.
4. The population standard deviation σ is unknown.

Note: The test for single mean μ with σ unknown is quite similar to that for σ known with the following differences:

i) **Test Statistics:** For a *t*-test, the test statistics under H_0 is $t = \dfrac{\overline{X} - \mu}{s/\sqrt{n}} \sim t_{n-1}$

instead of $Z = \dfrac{\overline{X} - \mu}{\sigma/\sqrt{n}} \sim N(0, 1)$, in the case of a *z*-test.

ii) **Critical Value(s):** The critical value(s) in this section will be determined from the *t*-distribution table instead of a normal distribution table.

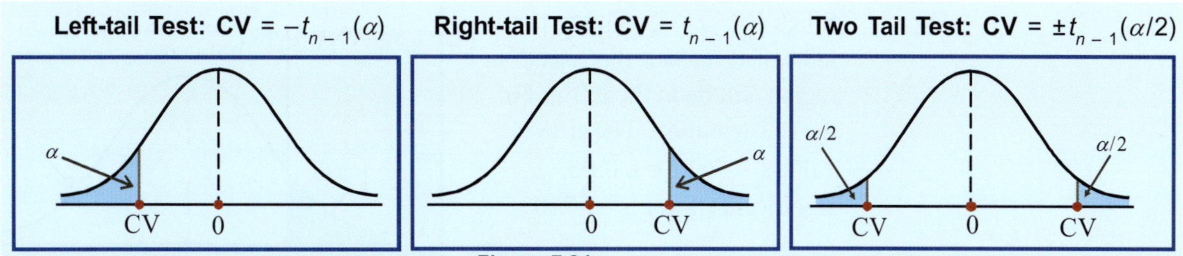

Figure 7.24

Summary for the t-test: σ unknown

Sample Size (n)	Population Distribution	Test Statistics and its sampling distribution
$n \le 30$	Normal	$t = \dfrac{\bar{X} - \mu}{s/\sqrt{n}} \sim t_{n-1}$; σ unknown
$n > 30$	Any	$t = \dfrac{\bar{X} - \mu}{s/\sqrt{n}} \sim t_{n-1}$; σ unknown
$n \le 30$	Unknown	Use non-parametric tests (Appendix 3)

Table 7.14

Remark: Using the t-table, the p-value may only be estimated. Therefore, to find the exact p-value using the t-table, we must use a computer or an advanced calculator. In this text, we will use Microsoft Excel to find p-values that cannot be exactly determined using the t-table.

Example 9: A medical student claims that the average HDL-Cholesterol level among the adult population of Smallvile is at least 39.63 milligram per deciliter (mg/dL). In order to test his claim, he randomly selects 25 adults and determines that their mean cholesterol level is 37.87 with a standard deviation of 8. Assume a 0.05 significance level and a normally distributed population.

(a) Test the claim

i) using the classical method.

ii) using the P-value method.

(b) Construct a 90% confidence interval to verify the results obtained from the test of hypothesis in part (a).

Solution: **(a)** ***Step* 0** We are given $s = 8$, $\bar{x} = 37.87$, $n = 25$ and $\alpha = 0.05$.

The original claim: $\mu \ge 39.63$

The counter claim: $\mu < 39.63$

Since n is small, σ is unknown, and the parent population is normally distributed, we use Student's t-statistics to test H_0.

i) Classical Method:

***Step* 1** Set up the Null and Alternative Hypotheses

$H_0: \mu = 39.63$ (oc)

$H_1: \mu < 39.63$ (cc) (Left-Tail Test)

***Step* 2** The value of test statistics based on the sample information provided and the hypothesized value of the population mean μ is

$$t = \frac{\bar{x} - \mu}{s/\sqrt{n}} = \frac{37.87 - 39.63}{8/\sqrt{25}} = -1.1$$

***Step* 3** Since $H_1: \mu < 39.63$, the test is left-tailed. Hence the rejection region will be in the left-tail of the t-distribution. The critical value $t_{n-1}(\alpha)$ for a 0.05 significance level is -1.7109.

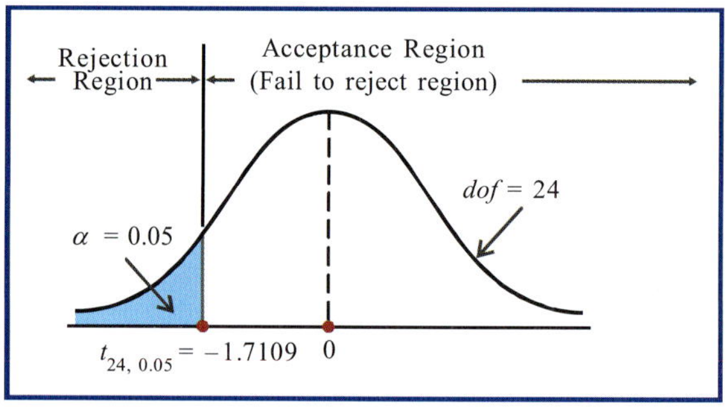

Figure 7.25

Recall : $\alpha = 0.05$ and $v = 24$;
therefore $t_{n-1}(\alpha) = t_{24}(0.05) = 1.7109$.

***Step* 4** **Decision:** Since the calculated value of test statistics ($t = -1.1$) lies outside of the rejection region, we fail to reject the null hypothesis at a 0.05 significance level.
Conclusion: The sample data does not provide sufficient evidence to reject H_0. There is sufficient evidence to support the claim that the mean HDL-Cholesterol level among the adult population of the city of Smallvile is at least 39.63.

ii) ***P*-value Method:**

***Step* 1** Set up the Null and Alternative Hypotheses

$H_0: \mu = 39.63$ (oc)

$H_1: \mu < 39.63$ (cc) (Left-Tail Test)

***Step* 2** The value of test statistics based on the sample information provided and the hypothesized value of the population mean μ is

$$t = \frac{\overline{x} - \mu}{s/\sqrt{n}} = \frac{37.87 - 39.63}{8/\sqrt{25}} = -1.1$$

Step* 3** **The *p*-value:** Since this is a left-tailed test, the region showing the extreme values is the area to the left of the test statistics under the ***t-curve.

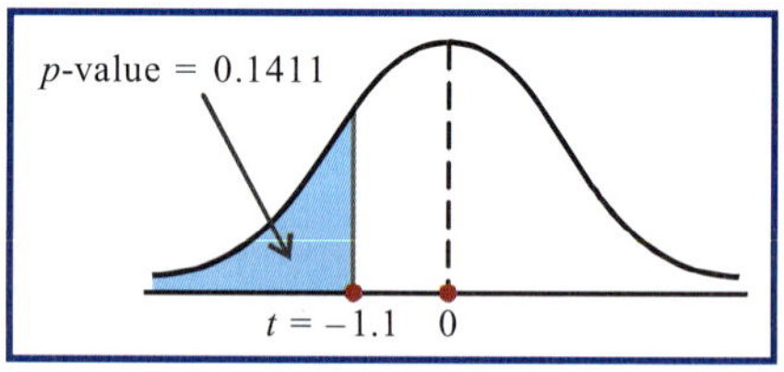

Figure 7.26

***p*-value** $= P(t < -1.1 \text{ for } v = 24)$
$= P(t > 1.1 \text{ for } v = 24)$ By symmetry
$= 0.1411$ Using the Excel Command: T.DIST.RT(1.1,24)

Note: To find the *p*-value if Microsoft Excel® is not available.

The closest *t*-value to $t = 1.1$ on the *t*-table with 24 degrees of freedom is $t = 1.318$.

From the *t*-table with 24 degrees of freedom, we have

$P(t < -1.318) = P(t > 1.318) = 0.10$

and $P(t < -1.1) = P(t > 1.1) > P(t > 1.318) = 0.10 \Rightarrow$ *p*-value > 0.10.

Step* 4** **Decision:** Since the ***p-value $> \alpha = 0.05$, we fail to reject H_0.
Conclusion: There is sufficient evidence to support the claim that the mean HDL-Cholesterol level among the adult population of the city of Smallvile is at least 39.63.

In short:

Hypotheses	Test Statistics
$H_0: \mu = 39.63$ $H_1: \mu < 39.63$	$t = \dfrac{\overline{x} - \mu}{s/\sqrt{n}} = -1.1$
Rejection Region and Critical Value	**Conclusion**
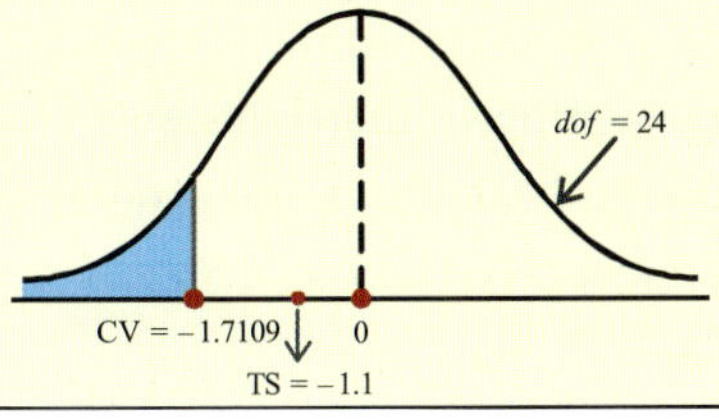	Fail to reject H_0. The mean HDL-Cholesterol level among the adult population of the city of Smallvile is at least 39.63.

(b) **90% confidence interval for the mean HDL-Cholesterol level among the adult population of Smallvile**

$$\overline{x} - t_{n-1,\,\alpha/2} \frac{s}{\sqrt{n}} < \mu < \overline{x} + t_{n-1,\,\alpha/2} \frac{s}{\sqrt{n}}$$

$$37.87 - 1.7109 \frac{8}{\sqrt{25}} < \mu < 37.87 + 1.7109 \frac{8}{\sqrt{25}}$$

$$35.13 < \mu < 40.61$$

There is a 0.90 probability that the mean HDL-Cholesterol level lies between 35.13 and 40.61 mg/dL, which agrees with the results obtained in the test of hypothesis.

Example 10: Scientists working on an experiment with a new drug claimed that, when the drug is added to the feed of chicks in an attempt to promote growth, an average increase of at most 4 ounces in one week is observed. To test this observation, the drug is added to the feed of 8 randomly selected chicks, and the increase in their weights (in ounces) in one week are recorded.

3.92	**3.76**	**3.90**	**3.95**	**3.85**	**3.75**	**3.97**	**3.82**

(a) Test the scientist's claim using a 0.01 significance level. Assume the population is normally distributed. Use the classical method.

(b) Construct a 98% confidence interval to verify the results obtained from the test of significant difference (Test of Hypothesis).

Solution: **(a)** **Classical Method:**

***Step* 0** Given the raw data set, we must first find the sample mean and standard deviation. Using the formulas for the mean and standard deviation of raw data, we have $\overline{x} = 3.865$, $s = 0.083666$ and $n = 8$.

The original claim: $\mu \leq 4$

The counter claim: $\mu > 4$

Since n is small, σ is unknown, and the parent population is normally distributed, we use Student's ***t***-statistics to test H_0.

***Step* 1** Set up the Null and Alternative Hypotheses

$H_0: \mu = 4$ (oc)

$H_1: \mu > 4$ (cc) (Right-Tail Test)

***Step* 2** The value of test statistics based on sample statistics obtained from the given raw data and the hypothesized value of the population mean μ is

$$t = \frac{\bar{x} - \mu}{s/\sqrt{n}} = \frac{3.865 - 4}{0.083666/\sqrt{8}} = -4.564$$

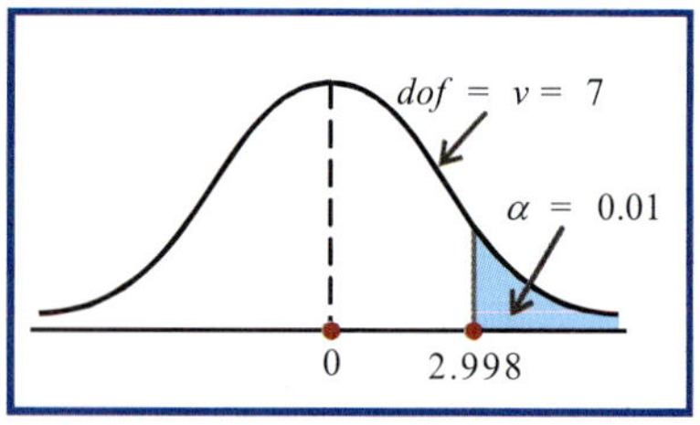

Figure 7.27

***Step* 3** Since $H_1 : \mu > 4$, the test is right-tailed. Hence the rejection region will be in the right-tail of the t-curve. The critical value t_{n-1} for a 0.01 significance level and $df = 7$ will be 2.998.

Recall : $\alpha = 0.01$ and $v = 7$; therefore $t_{n-1}(\alpha) = t_7(0.01) = 2.998$.

***Step* 4** **Decision:** Since the calculated value of test statistics ($t = -4.564$) lies outside of the rejection region, we fail to reject the null hypothesis at a 0.01 significance level.
Conclusion: The sample data does not provide sufficient evidence to reject H_0. There is sufficient evidence to support the claim that when the drug is added to the feed of chicks in an attempt to promote growth, an average increase of at most 4 ounces in one week is observed.

In short:

Hypotheses	**Test Statistics**
$H_0: \mu = 4$ $H_1: \mu > 4$	$t = \frac{\bar{x} - \mu}{s/\sqrt{n}} = -4.564$
Rejection Region and Critical Value 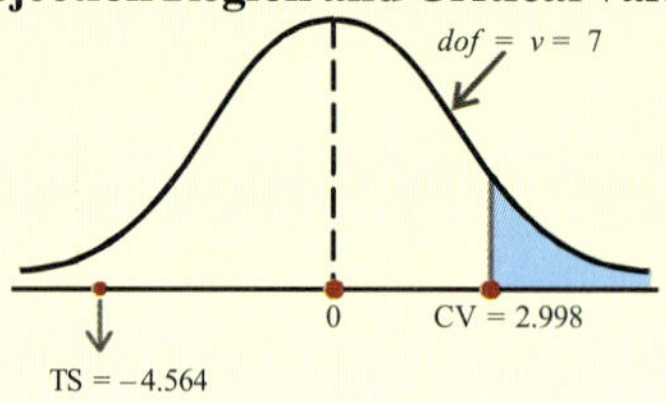	**Conclusion** Fail to reject H_0. The drug, when added to the feed of chicks, increases their growth at most 4 ounces in one week.

(b) 98% confidence interval for the mean growth

$$\bar{x} - t_{n-1(\alpha/2)} \frac{s}{\sqrt{n}} < \mu < \bar{x} + t_{n-1(\alpha/2)} \frac{s}{\sqrt{n}}$$

$$3.865 - 2.998 \frac{0.084}{\sqrt{8}} < \mu < 3.865 + 2.998 \frac{0.084}{\sqrt{8}}$$

$$3.78 < \mu < 3.95$$

There is a 0.98 probability that the average increase lies between 3.78 and 3.95 ounces, which agrees with the results obtained in the test of hypothesis.

Example 11: The following table summarizes the number of goals scored by 14 randomly selected soccer players in *La Liga* (Spanish premier soccer league) and their corresponding teams.

Player	Goals	Team
Diego Forlán	8	Atlético Madrid
Giuseppe Rossi	18	Villarreal CF
Luis Fabiano	10	Sevilla FC
David Villa	18	FC Barcelona
Manu	9	Getafe CF
Julio Baptista	9	Málaga CF
KarimBenzema	15	Real Madrid
Pedro	13	FC Barcelona
MesutÖzil	6	Real Madrid
Diego Castro	9	Sporting Gijón
Lionel Messi	31	FC Barcelona
Cristiano Ronaldo	40	Real Madrid
Kaká	7	Real Madrid
Andrés Iniesta	8	FC Barcelona

You are trying to settle a bet with a friend about the population mean number of goals scored in La Liga.

(a) Test whether the population mean number of goals is less than 15. Use $\alpha = 0.1$, and assume the parent population has a normal distribution

i) using the classical method, **ii)** using the P-value method.

(b) Construct an 80% confidence interval to verify the results obtained from the test of hypothesis in part (a).

Solution: **(a)** **Recall:** For ungrouped data $\overline{x} = \frac{\sum x}{n}$, and $s = \sqrt{\frac{n\left(\sum x^2\right) - \left(\sum x\right)^2}{n(n-1)}}$

***Step* 0** We are given $n = 14$ and $\alpha = 0.1$. Since s is not given, we use the raw data provided to calculate the sample standard deviation.

x	8	18	10	18	9	9	15	13	6	9	31	40	7	8	$\sum x = 201$
x^2	64	324	100	324	81	81	225	169	36	81	961	1600	49	64	$\sum x^2 = 4159$

Table 7.15

$$\overline{x} = \frac{\sum x}{n} = \frac{201}{14} = 14.4, \text{ and}$$

$$s = \sqrt{\frac{n\left(\sum x^2\right) - \left(\sum x\right)^2}{n(n-1)}} = \sqrt{\frac{14(4159) - (201)^2}{14(13)}} = 9.9$$

The original claim: $\mu < 15$

The counter claim: $\mu \geq 15$

Since n is small, σ is unknown, and the parent population is normally distributed, we use Student's $\boldsymbol{t}$-statistics to test H_0.

i) **Classical Method:**

***Step* 1** Set up the Null and Alternative Hypotheses

$H_0 : \mu = 15$ (cc)

$H_1 : \mu < 15$ (oc) (Left-Tail Test)

***Step* 2** The value of test statistics based on the sample information provided and the hypothesized value of the population mean μ is

$$t = \frac{\bar{x} - \mu}{s/\sqrt{n}} = \frac{14.4 - 15}{9.9/\sqrt{14}} = -0.2$$

***Step* 3** Since $H_1 : \mu < 15$, the test is left-tailed. Hence, the rejection region will be in the left-tail of the t-distribution curve. The critical value $t_{n-1}(\alpha)$ for a 0.1 significance level and $df = 13$ will be –1.3502.

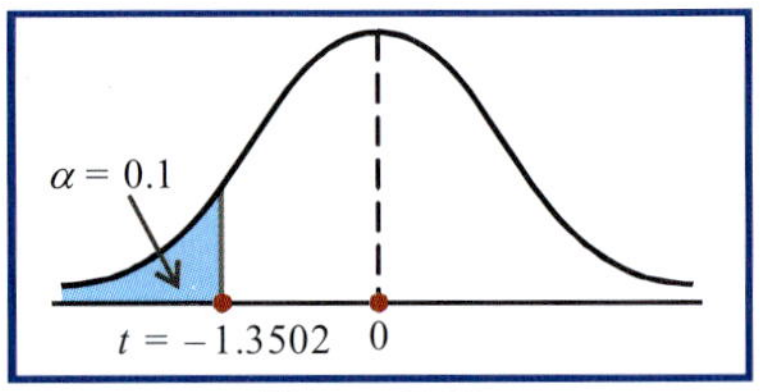

Figure 7.28

Recall : $\alpha = 0.1$ and $v = 13$; therefore $t_{n-1}(\alpha) = t_{13}(0.1) = 1.3502$.

***Step* 4** **Decision:** Since the calculated value of test statistics ($t = -0.2$) lies outside of the rejection region, we fail to reject the null hypothesis at a 0.1 significance level.

Conclusion: The sample data does not provide sufficient evidence to reject H_0. There is insufficient evidence to conclude that the population mean number of goals per player in La Liga is less than 15.

ii) ***P*-value Method:**

***Step* 1** Set up the Null and Alternative Hypotheses

$H_0 : \mu = 15$ (cc)

$H_1 : \mu < 15$ (oc) (Left-Tail Test)

***Step* 2** The value of test statistics based on the sample information provided and the hypothesized value of the population mean μ is

$$t = \frac{\bar{x} - \mu}{s/\sqrt{n}} = \frac{14.4 - 15}{9.9/\sqrt{14}} = -0.2$$

***Step* 3** Since the test is left-tailed, the region showing the extreme values is the area to the left of the test statistics.

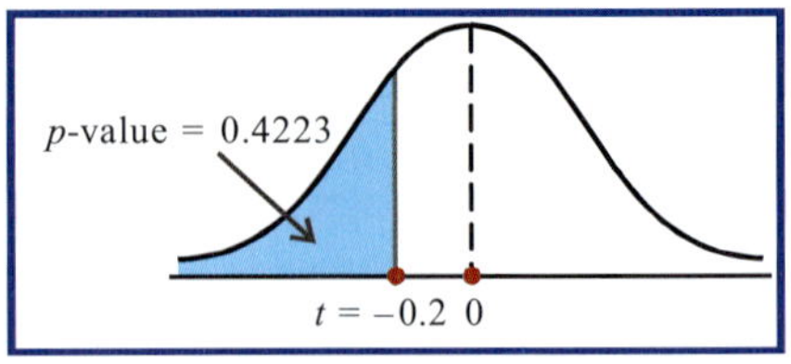

Figure 7.29

p-value $= P(t < t_{data})$

$= P(t < -0.2)$

$= P(t > 0.2)$ By symmetry

$= 0.4223$ Using the Excel Command: T.DIST.RT(0.2,13)

Step* 4** **Decision:** Since ***p-value = 0.4223 is greater than $\alpha = 0.1$, we fail to reject H_0.

Conclusion: There is insufficient evidence to support the claim that the average number of goals per player in La Liga is less than 15.

In short:

Hypotheses	**Test Statistics**
$H_0: \mu = 15$ $H_1: \mu < 15$	$t = \dfrac{\overline{x} - \mu}{s/\sqrt{n}} = -0.2$
Rejection Region and Critical value	**Conclusion**
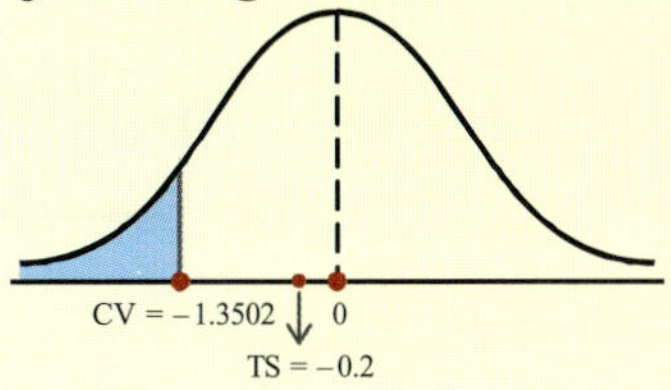	Fail to reject H_0. The average number of goals per player in La Liga is not less than 15.

(b) 80% confidence interval for the mean number of goals scored by a footballer in La Liga

critical values for a 80% confidence level are $t_{n-1(\alpha/2)} = t_{13}(0.1) = \pm 1.3502$.

$$\overline{x} - t_{n-1(\alpha/2)} \frac{s}{\sqrt{n}} < \mu < \overline{x} + t_{n-1(\alpha/2)} \frac{s}{\sqrt{n}}$$

$$14.4 - 1.3502 \frac{9.9}{\sqrt{14}} < \mu < 14.4 + 1.3502 \frac{9.9}{\sqrt{14}}$$

$$10.83 < \mu < 17.97$$

There is a 0.80 probability that the average number of goals lies between 10.83 and 17.97, which agrees with the results obtained in the test of hypothesis.

Example 12: To check the claim that the average height of 18-year-old males is 69 inches, a random sample of sixty, 18-year-old males is selected. The average and standard deviation of heights came out to be 67.8 and 6 inches respectively.

(a) Test the hypothesis at $\alpha = 0.05$.

i) Use classical method.

ii) Use the *P*-value method.

(b) Construct a 95% confidence interval to verify the results obtained from the test of significant difference (Test of Hypothesis).

Solution:

***Step* 0** We are given $s = 6$, $\overline{x} = 67.8$, $n = 60$ and $\alpha = 0.05$.

The original claim: $\mu = 69$

The counter claim: $\mu \neq 69$

(a) Since σ is unknown and n is large, we use Student's ***t***-statistics to test H_0, even though no information is provided about the distribution of the original population.

i) Classical Method:

***Step* 1** Set up the Null and Alternative Hypotheses.

$H_0 : \mu = 69$ (oc)

$H_1 : \mu \neq 69$ (cc) (Two-Tail Test)

***Step* 2** The value of test statistics based on the sample information provided and the hypothesized value of the population mean μ is

$$t = \frac{\overline{x} - \mu}{s/\sqrt{n}} = \frac{67.8 - 69}{6/\sqrt{60}} = -1.55$$

***Step* 3** Since $H_1 : \mu \neq 69$, the test is two-tailed. Hence, the rejection region will be in the both tails of the t-curve. The critical values for a 0.05 significance level are $t_{n-1, \alpha/2} = t_{59}(0.025) = \pm 2.001$.

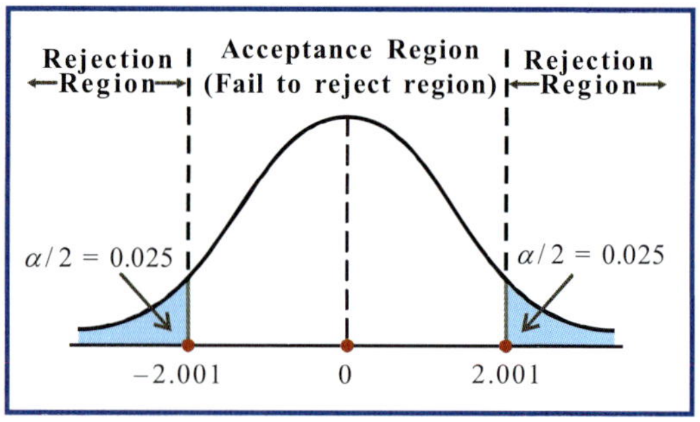

Figure 7.30

Recall : $\alpha = 0.05 \Rightarrow \alpha/2 = 0.025$ and
for $v = 59$, $t_{n-1, \alpha/2} = t_{59}(0.025) = 2.001$

***Step* 4** **Decision:** Since the calculated value of test statistics ($t = -1.55$) lies outside of the rejection region, we fail to reject the null hypothesis at a 0.05 significance level.
Conclusion: The sample data does not provide sufficient evidence to reject H_0. There is sufficient evidence to conclude that the average height of 18-year-old male is 69 inches.

ii) ***P*-value Method:**

***Step* 1** Set up the Null and Alternative Hypotheses.

$H_0 : \mu = 69$ (oc)

$H_1 : \mu \neq 69$ (cc) (Two-Tail Test)

***Step* 2** The value of test statistics based on the sample information provided and the hypothesized value of the population mean μ is

$$t = \frac{\bar{x} - \mu}{s/\sqrt{n}} = \frac{67.8 - 69}{6/\sqrt{60}} = -1.55$$

***Step* 3** **The *p*-value:**

$H_1 : \mu \neq 69$ inches represents a two-tailed rejection region.
Therefore, the ***p***-value of the test statistics is given by:

$= P(|t| > 1.55)$

$= P(t > 1.55 \text{ or } t < -1.55)$ Definition of an absolute value

$= 2 \cdot P(t > 1.55)$ By symmetry

$= 0.1265$ Using the Excel command: T.DIST.2T(1.55,59)

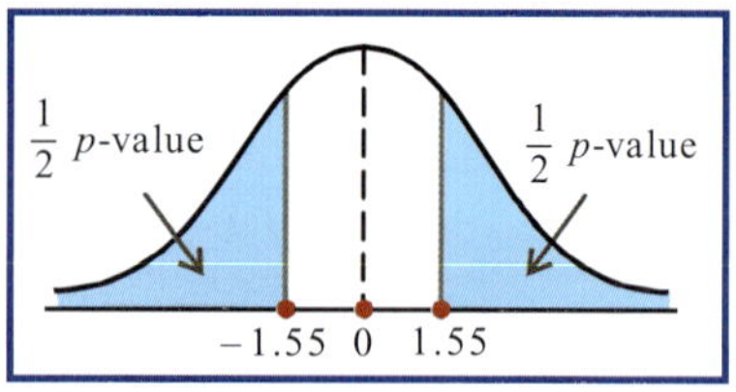

Figure 7.31

Step* 4** **Decision:** Since ***p-value = 0.1265 is greater than $\alpha = 0.05$, we fail to reject H_0.
Conclusion: There is sufficient evidence to support the claim that the average height of 18-year-old males is 69 inches.

In short:

Hypotheses	**Test Statistics**
$H_0: \mu = 69$ $H_1: \mu \neq 69$	$t = \frac{\bar{x} - \mu}{s/\sqrt{n}} = -1.55$
Rejection Region and Critical Value 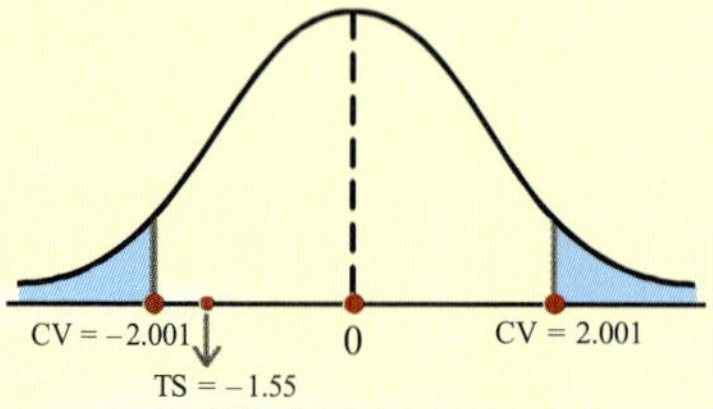	**Conclusion** Fail to reject H_0. The average height of 18-year-old males is 69 inches.

(b) 95% confidence interval for the average height of 18-year-old male

Critical values for a 95% confidence level are $t_{n-1(\alpha/2)} = t_{59}(0.025) = \pm 2.001$.

$$\overline{x} - t_{n-1\,,\,\alpha/2}\frac{s}{\sqrt{n}} < \mu < \overline{x} + t_{n-1\,,\,\alpha/2}\frac{s}{\sqrt{n}}$$

$$67.8 - 2.001\frac{6}{\sqrt{60}} < \mu < 67.8 + 2.001\frac{6}{\sqrt{60}}$$

$$66.25 < \mu < 69.35$$

There is a 0.95 probability that the average height of 18-year-old males lies between 66.25 and 69.35 inches, which agrees with the results obtained in the test of hypothesis.

Try it yourself:

In exercises 1 to 4, test the indicated original claims given the sample information.

1. $\mu \leq 8.6$ (oc), $\overline{x} = 8.2$, $s = 0.3$ and $n = 50$. Use a 5% significance level.

2. $\mu = 185$ (oc), $\overline{x} = 170$, $s = 10.5$ and $n = 8$. Use a 1% significance level, and assume that the parent population is normally distributed.

3. $\mu = 15$ (oc), $\overline{x} = 13.6$, $s = 1.8$ and $n = 20$. Use a 10% significance level, and assume that the parent population is normally distributed.

4. $\mu < 62.8$ (oc), $\overline{x} = 56.8$, $s = 6.4$, $n = 300$. Use a 5% significance level.

5. A manufacturer claims that the mean breaking strength of the springs it makes is 575 lbs. A distributer tests a random sample of 36 springs and finds the mean braking strength to be 571 lbs. with a standard deviation of 10 lbs. Should the distributer be concerned about possible complaints of weaker springs? Assume a 0.01 significance level. Use both the Classical and ***P***-value methods to test the manufacturers claim.

6. The scores in the quantitative SAT of students in Narnia school district are normally distributed with a mean of 625. A random sample of 20 students who took the SAT after completing a new coaching program yielded a mean score of 632 with a standard deviation 15. Test at 5% significance level, that new coaching program is successful.

7.5: TESTING A CLAIM ABOUT A PROPORTION p

Objectives

After completing this section you will be able to apply:

A. Test of Significant Difference for a Single Proportion.

B. Assumptions for testing a Single Proportion.

A political candidate may be interested in knowing whether there has been an increase in the proportion (percentage) of voters favoring him or her in an upcoming election. A cereal manufacturer might like to know whether or not there has been a drop in the proportion of consumers who prefer its brand. A hospital might like to test a manufacturer's claim that its drug cures 80% of the users. These examples are only a few situations where we need to test a **claim** or a **hypothesis** regarding a population **proportion** or **percentage**. The procedure of testing a hypothesis about p for a large population or a normally distributed population is similar to the one we used for testing a population mean μ. In this section we will learn how to perform the Z-test, for a hypothesis about the population proportion p.

A. Test of Significant Difference for a Single Proportion

Let us consider a simple random sample of size n from a large population with proportion of successes p. If x is the observed number of successes in the sample, then the sample proportion, $\hat{p}$, is given by

$\hat{p} = \dfrac{x}{n}$. Recall that the sampling distribution of proportion $\hat{p}$ is $N\left(\mu_{\hat{p}}, \sigma_{\hat{p}}\right)$, where n is large ($n > 30$)

and $\sigma_{\hat{p}} = \sqrt{\dfrac{pq}{n}}$. That is, $Z = \dfrac{\hat{p}-p}{\sqrt{\dfrac{pq}{n}}} \sim N(0, 1)$ for large n.

Remark: Recall that if $X \sim B(n, p)$, then for large n, the binomial distribution can be approximated by normal distribution.

$$X \sim N(\mu = np, \sigma = \sqrt{npq}) \text{ or } Z = \frac{X-\mu}{\sigma} = \frac{X-np}{\sqrt{npq}} \sim N(0, 1), \text{ if } n \text{ is large.}$$

B. Assumptions for Testing a Single Proportion

1. The sample is a simple random sample.
2. The conditions of the *Binomial Experiment* or *Binomial Probability Model* are satisfied.
3. Sample is large ($n > 30$) or the conditions of the binomial approximation using the normal distribution ($np > 5$ and $nq > 5$) are satisfied.

Remark: All examples and exercises in this textbook that involve testing claims about the population proportion satisfy the criteria above. Therefore, using the normal approximation to the binomial distribution is always applicable.

Notation:

Sample Size $= n$

Sample proportion: $\hat{p} = \dfrac{x}{n}$, where x is the number of favorable outcomes.

Population Proportion: determined by the null hypothesis $= p$ and $q = 1 - p$.

For instance, if 18 out of 46 students are left-handed, we have $x = 18$ and $n = 46$. Note that n is always greater than x in order for the probability to be between 0 and 1. Then $\hat{p} = \frac{x}{n} = \frac{18}{46} = 0.39$.

- **Test Statistics for Testing a Claim about a Population Proportion**

$$Z = \frac{\hat{p} - p}{\sqrt{\frac{pq}{n}}} \sim N(0,\ 1), \text{ where } p \text{ is the value assigned to the null hypothesis.}$$

The rest of the testing procedure is identical to testing of the single mean, σ known, discussed in Section 7.3 and we can use either the classical method or the *P*-value method.

Example 13: To check whether the popularity of a political candidate is down from 65%, a random sample of 500 voters is surveyed. 310 out of 500 favor the candidate.

(a) Test the hypothesis about the proportion of voters who favor this candidate at $\alpha = 0.05$.

i) Use the classical method. **ii)** Use the *P*-value method.

(b) Construct a 90% confidence interval to verify the results obtained from part (a).

Solution:

***Step* 0** We are given $x = 310$, $n = 500$ and $\alpha = 0.05$.
The original claim: $p < 0.65$
The counter claim: $p \geq 0.65$

(a) i) Classical Method:

***Step* 1** Set up the Null and Alternative Hypotheses.

$H_0:\ p = 0.65$ (cc)

$H_1:\ p < 0.65$ (oc) (Left-Tail Test)

***Step* 2** Under H_0, the test statistics is $z = \frac{\hat{p} - p}{\sqrt{\frac{pq}{n}}} \sim N(0,\ 1)$ since n is large.

$\hat{p}$ = Sample proportion of voters who favor the candidate $= \frac{x}{n} = \frac{310}{500} = 0.62$

$p = 0.65$ (Under H_0) $\Rightarrow q = 1 - p = 1 - 0.65 = 0.35$

Therefore, Test Statistics $= z = \frac{0.62 - 0.65}{\sqrt{\frac{(0.65)(0.35)}{500}}} = -1.41$

***Step* 3** Since in H_1, $p < 0.65$, the rejection region is in the left-tail of the standard normal curve with $\alpha = 0.05$ (area of the rejection region in the left-tail). The critical value $(-z_\alpha)$ for a left-tailed test at 5 % level of significance is $z_\alpha = z_{0.05} = -1.645$.

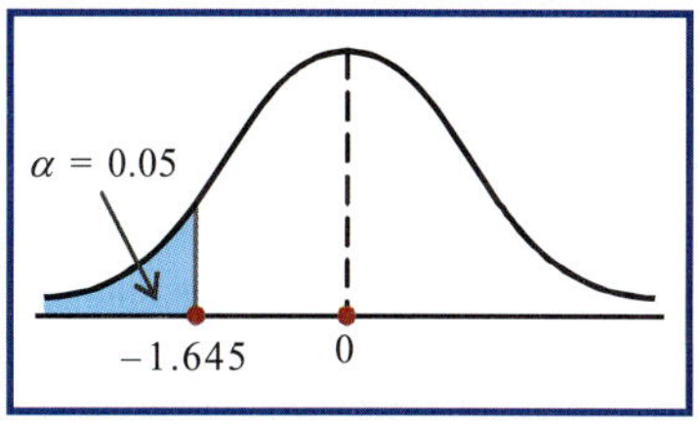

Figure 7.32

***Step* 4** **Decision:** Since the test statistics ($z = -1.41$) lies outside the rejection region, we fail to reject the null hypothesis at a 0.05 significance level. *Conclusion:* There is not sufficient evidence to support the claim that the proportion of voters who favor this candidate is below 65%.

ii) ***P*-value Method:**

***Step* 1** Set up the Null and Alternative Hypotheses.

$H_0 : p = 0.65$ (cc)

$H_1 : p < 0.65$ (oc) (Left-Tail Test)

***Step* 2** Under H_0, the test statistics is $z = \dfrac{\hat{p} - p}{\sqrt{\dfrac{pq}{n}}} \sim N(0, 1)$ since n is large.

$\hat{p}$ = Sample proportion of voters who favor the candidate $= \dfrac{x}{n} = \dfrac{310}{500} = 0.62.$

$p = 0.65$ (Under H_0) $\Rightarrow q = 1 - p = 1 - 0.65 = 0.35.$

Therefore, Test Statistics $= z = \dfrac{0.62 - 0.65}{\sqrt{\dfrac{(0.65)(0.35)}{500}}} = -1.41$

***Step* 3** Based on the alternative hypothesis $H_1 : p < 0.65$, the test is left-tailed; therefore, the region showing the extreme values is the area to the left of the test statistics.

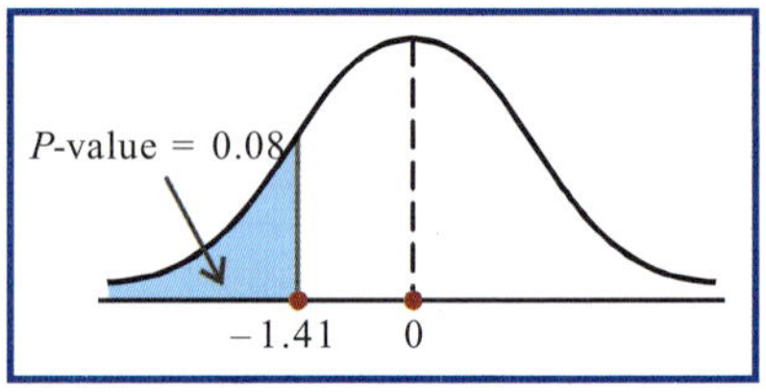

Figure 7.33

P-value = Area of the shaded region

$= P(Z < -1.41) = 0.5 - 0.4207 = 0.0793 \approx 0.08.$

***Step* 4** **Decision:** Since the p-value is greater than $\alpha = 0.05$, we fail to reject the null hypothesis H_0 at 5% level of significance. *Conclusion:* The sample data does not provide sufficient evidence to reject H_0. Hence we conclude that proportion of voters who favor this candidate is below 65%.

In short:

Hypotheses	**Test Statistics**
$H_0 : p = 0.65$ $H_1 : p < 0.65$	$z = \dfrac{\hat{p} - p}{\sqrt{\dfrac{pq}{n}}} = -1.41$
Rejection Region and Critical Value 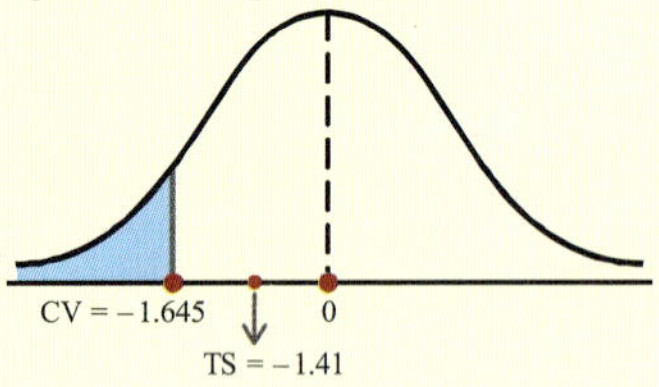 	**Conclusion** Fail to reject H_0. The proportion of voters who favor this candidate is not below 65%.

(b) To construct a 90% confidence interval for population proportion p, we use the formula

$$\hat{p} - z_{\alpha/2}\sqrt{\frac{\hat{p}\hat{q}}{n}} < p < \hat{p} + z_{\alpha/2}\sqrt{\frac{\hat{p}\hat{q}}{n}}$$

From part (a), we know that $\hat{p} = 0.62$; hence $\hat{q} = 1 - \hat{p} = 1 - 0.62 = 0.38$, $n = 500$ and $z_{\alpha/2} = 1.645$.

$$0.62 - 1.645\sqrt{\frac{(0.62)(0.38)}{500}} < p < 0.62 + 1.645\sqrt{\frac{(0.62)(0.38)}{500}}$$

$$0.584 < p < 0.656$$

The calculated confidence interval verifies the result from the test of hypothesis. The proportion of the voters is not significantly down from 65%.

Example 14: Among 850 randomly selected Facebook users, it was found that 136 of them use Facebook for marketing their business.

(a) Use a 0.01 significance level to test the claim that among Facebook users, more than 15% use it for marketing their business

i) using the classical method. **ii)** using the *P*-value method.

(b) Construct a 98% confidence interval to verify the results obtained from part (a).

Solution:

Step 0 We are given $x = 136$, $n = 850$ and $\alpha = 0.01$.

The original claim: $p > 0.15$

The counter claim: $p \leq 0.15$

(a) i) Classical Method:

Step 1 Set up the Null and Alternative Hypotheses.

$H_0: p = 0.15$ (cc)

$H_1: p > 0.15$ (oc) (Right-Tail Test)

Step 2 $\hat{p}$ = Sample proportion of voters who favor the candidate = $\frac{x}{n} = \frac{136}{850} = 0.16$

$p = 0.15$ (Under H_0) $\Rightarrow q = 1 - p = 1 - 0.15 = 0.85$

Therefore, Test Statistics $= z = \dfrac{0.16 - 0.15}{\sqrt{\dfrac{(0.15)(0.85)}{850}}} = 0.82$

Step 3 Since in H_1, $p > 0.15$, the test is right-tailed; hence the rejection region will be in the right-tail of the standard normal curve with $\alpha = 0.01$ (area of the rejection region in the right-tail). The critical value (z_α) for a right-tailed test at 1% level of significance is $z_\alpha = z_{0.01} = 2.33$.

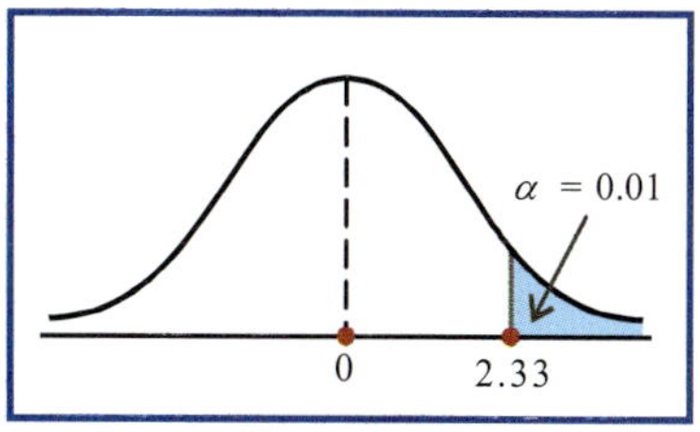

Figure 7.34

Step 4 **Decision:** Since the calculated value of test statistics ($z = 0.82$) lies outside the rejection region, we fail to reject the null hypothesis at a 0.01 significance level.
Conclusion: The data does not provide sufficient evidence against the null hypothesis in favor of the alternative hypothesis. There is not sufficient evidence to support the claim that among Facebook users, more than 15% use it for marketing their business.

ii) *P*-value Method:

Step 1 Set up the Null and Alternative Hypotheses.

$H_0: p = 0.15$ (cc)

$H_1: p > 0.15$ (oc) (Right-Tail Test)

Step 2 $\hat{p}$ = Sample proportion of voters who favor the candidate $= \dfrac{x}{n} = \dfrac{136}{850} = 0.16$

$p = 0.15$ (Under H_0) $\Rightarrow q = 1 - p = 1 - 0.15 = 0.85$

Therefore, Test Statistics $= z = \dfrac{0.16 - 0.15}{\sqrt{\dfrac{(0.15)(0.85)}{850}}} = 0.82$

Step 3 Since this is a right-tailed test, the region showing the extreme values is the area to the right of the test statistics.

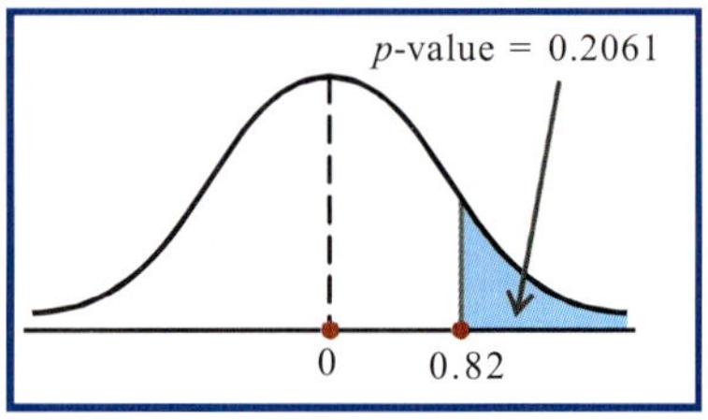

Figure 7.35

P-value = area of the shaded region

$= P(Z > 0.82) = 0.5 - 0.2939 = 0.2061 \approx 0.21.$

Step 4 **Decision:** Since the p-value is greater than $\alpha = 0.01$, we fail to reject the null hypothesis H_0 at 1% level of significance. *Conclusion:* There is not sufficient evidence to support the claim that among Facebook users, more than 15% use it for marketing their business.

In short:

Hypotheses	**Test Statistics**
$H_0: p = 0.15$ $H_1: p > 0.15$	$z = \dfrac{\hat{p} - p}{\sqrt{\dfrac{pq}{n}}} = 0.82$
Rejection Region and Critical Value 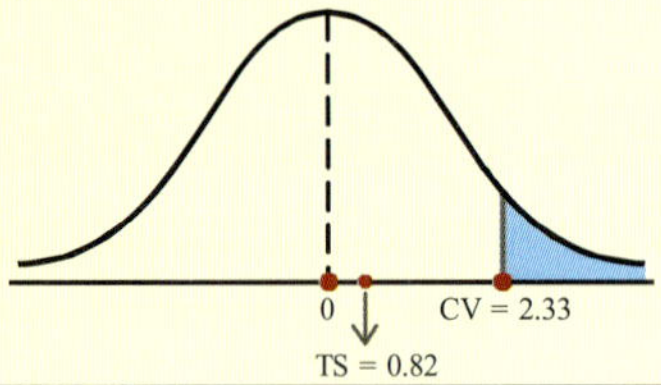 	**Conclusion** Fail to reject H_0. Among Facebook users, not more than 15% use it for marketing their business.

(b) To construct a 98% confidence interval for population proportion p, we use the formula

$$\hat{p} - z_{\alpha/2}\sqrt{\frac{\hat{p}\hat{q}}{n}} < p < \hat{p} + z_{\alpha/2}\sqrt{\frac{\hat{p}\hat{q}}{n}}$$

From part (a), we know that $\hat{p} = 0.16$; hence $\hat{q} = 1 - \hat{p} = 1 - 0.16 = 0.84$, $n = 850$ and $z_{\alpha/2} = 2.33$.

$$0.16 - 2.33\sqrt{\frac{(0.16)(0.84)}{850}} < p < 0.16 + 2.33\sqrt{\frac{(0.16)(0.84)}{850}}$$

$$0.131 < p < 0.189$$

The calculated confidence interval does not support the claim that among Facebook users more than 15% use it for marketing their business.

Example 15: An independent research company claims that 4% of heavy coffee drinkers suffer from coronary heart disease. In a random sample of 752 heavy coffee drinkers 5% suffer from coronary heart disease.

(a) Test the claim at a 5% significance level, using

i) the classical method **ii)** the P-value method.

(b) Construct a 95% confidence interval to verify the results obtained from part (a). Explain.

Solution: **Step 0** We are given $\hat{p} = 0.05$, $n = 752$ and $\alpha = 0.05$.

The original claim: $p = 0.04$

The counter claim: $p \neq 0.04$

(a) i) Classical Method:

Step 1 Set up the Null and Alternative Hypotheses.

$H_0: p = 0.04$ (oc)

$H_1: p \neq 0.04$ (cc) (Two-Tail Test)

Step 2 $\hat{p} = 0.05$. $p = 0.04$ (Under H_0) $\Rightarrow q = 1 - p = 0.96$

$$\text{Therefore, Test Statistics} = z = \frac{0.05 - 0.04}{\sqrt{\frac{(0.04)(0.96)}{752}}} = 1.40$$

Step 3 Since $H_1: p \neq 0.04$, the test is two-tailed; hence the rejection region will be in both tails of the standard normal curve with $\alpha = 0.05$ (area of the rejection regions in both tails). The critical values for $(\pm z_{\alpha/2})$ for two tailed test at 5% level of significance are $z_{\alpha/2} = z_{0.025} = \pm 1.96$.

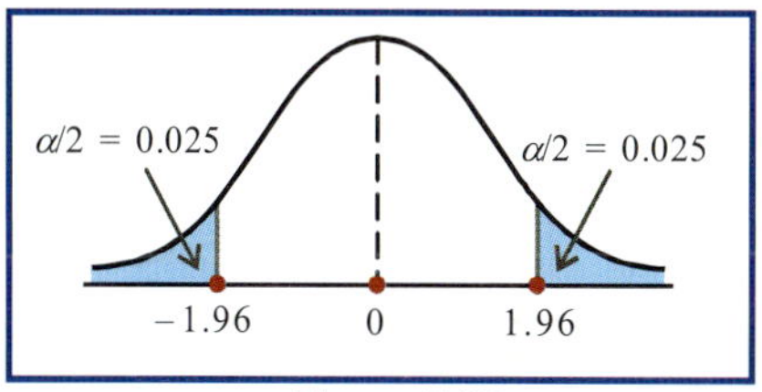

Figure 7.36

Step 4 **Decision:** Since the calculated value of test statistics ($z = 1.40$) lies outside the rejection region, we fail to reject the null hypothesis at a 0.05 significance level. *Conclusion:* There is sufficient evidence to support the claim that 4% of heavy coffee drinkers suffer from coronary heart disease.

ii) *P*-value Method:

Step 1 Set up the Null and Alternative Hypotheses.

$H_0: p = 0.04$ (oc)

$H_1: p \neq 0.04$ (cc) (Two-Tail Test)

Step 2 $\hat{p} = 0.05$. $p = 0.04$ (Under H_0) $\Rightarrow q = 1 - p = 0.96$

$$\text{Therefore, Test Statistics} = z = \frac{0.05 - 0.04}{\sqrt{\frac{(0.04)(0.96)}{752}}} = 1.40$$

Step 3 Since the test is two-tailed, the region showing the extreme values is twice the area to the right of the test statistics.

P-value $= 2 \times$ area of the shaded region

$= 2 \times P(Z > 1.40) = 2 \times 0.0808 = 0.1616.$

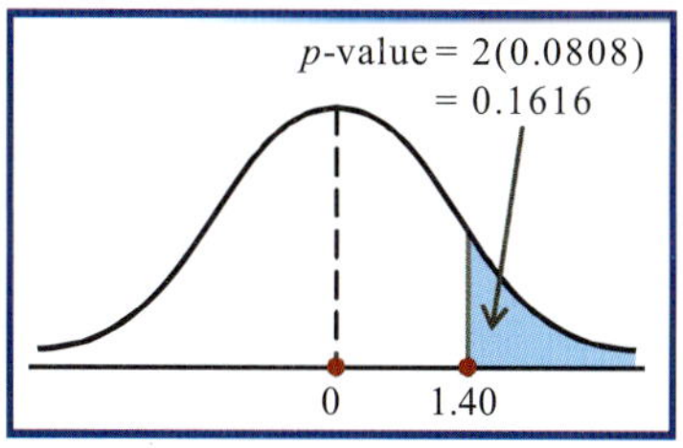

Figure 7.37

Step 4 **Decision:** Since the p-value is greater than $\alpha = 0.05$, we fail to reject the null hypothesis H_0 at 5% level of significance. *Conclusion:* Data provides sufficient evidence to support the claim that 4% of heavy coffee drinkers suffer from coronary heart disease.

In short:

Hypotheses	**Test Statistics**
$H_0: p = 0.04$ $H_1: p \neq 0.04$	$z = \dfrac{\hat{p} - p}{\sqrt{\dfrac{pq}{n}}} = 1.40$
Rejection Region and Critical Value 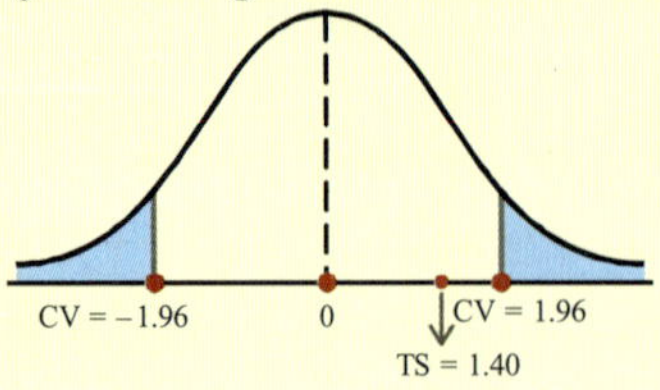	**Conclusion** Fail to reject H_0. 4% of heavy coffee drinkers suffer from coronary heart disease.

(b) To construct a 95% confidence interval for population proportion p, we use the formula

$$\hat{p} - z_{\alpha/2}\sqrt{\frac{\hat{p}\,\hat{q}}{n}} < p < \hat{p} + z_{\alpha/2}\sqrt{\frac{\hat{p}\,\hat{q}}{n}}$$

From part (a), we know that $\hat{p} = 0.05$; hence $\hat{q} = 1 - \hat{p} = 0.95$. $n = 752$ and $z_{\alpha/2} = 1.96$.

$$0.05 - 1.96\sqrt{\frac{(0.04)(0.96)}{752}} < p < 0.05 + 1.96\sqrt{\frac{(0.04)(0.96)}{752}}$$

$$0.036 < p < 0.064$$

The calculated confidence interval supports the claim that 4% of heavy coffee drinkers suffer from coronary heart disease.

Try it yourself:

In exercises 1 to 8, test the indicated original claims given the sample information.

1. $p = 0.43$ (oc), $\hat{p} = 0.4$ and $n = 40$. Use a 10% significance level.
2. $p < 0.36$ (oc), $\hat{p} = 0.29$ and $n = 22$. Use a 5% significance level.
3. $p \geq 0.80$ (oc), $\hat{p} = 0.91$ and $n = 50$. Use a 1% significance level.
4. $p \neq 0.4$ (oc), $\hat{p} = 0.53$ and $n = 100$. Use a 8% significance level.
5. $p = 0.64$ (oc), $x = 18$ and $n = 28$. Use a 1% significance level.
6. $p \leq 0.7$ (oc), $x = 20$ and $n = 34$. Use a 5% significance level.
7. $p \geq 0.33$ (oc), $x = 65$ and $n = 200$. Use a 5% significance level.
8. $p \neq 0.88$ (oc), $x = 35$ and $n = 48$. Use a 2% significance level.
9. An employment agency claims that 80% of all female applicants who have children prefer part time positions. In a random sample of 160 such applicants, 110 preferred part-time positions. Test the agency's hypothesis at 1% significance level. Carry out the test of hypothesis using both the classical and the p-value methods. Assume normal approximation is justified.
10. It is claimed that banks employ more females than males. In a random sample of 200 bank employees, 55% are female. Use a 0.05 significance level to test the agency's hypothesis.

7.6: Testing a Claim About Variance and Standard Deviation (σ^2/σ)

A. Chi-Square Test for Testing $H_0 : \sigma^2 = \sigma_0^2$ or $\sigma = \sigma_0$

> **Objectives**
>
> After completing this section you will be able to use:
>
> **A.** Chi-Square Test for testing $H_0 : \sigma^2 = \sigma_0^2$ or $H_0 : \sigma = \sigma_0$.
>
> **B.** Assumptions for the Chi-Square Test.
>
> **C.** Rejection Region and Critical Values for a test based on the Chi-Square.

Suppose we want to test if the given normally distributed population has a specific variance σ^2 or standard deviation σ. If $X_1, X_2, \ldots, X_n$ is a simple random sample of size n from the given population, then under the Null Hypothesis $H_0 : \sigma^2 = \sigma_0^2$ or $H_0 : \sigma = \sigma_0$.

(Note: σ_0 is some arbitrary constant.)

The statistics $$\chi^2 = \frac{(n-1)\,s^2}{\sigma_0^2} = \frac{\sum_{i=1}^{n}\left(X_i - \overline{X}\right)^2}{\sigma_0^2}$$

follows a *Chi-Square Distribution* with $n - 1$ degrees of freedom, where $s^2 = \frac{1}{n-1}\sum_{i=1}^{n}\left(X_i - \overline{X}\right)^2$,

is the *sample variance*.

B. Assumptions for the Chi-Square Test

1. The sample is a simple random sample.
2. The population from which the sample is drawn has a normal distribution.

- **Test Statistics:**

Under $H_0 : \sigma^2 = \sigma_0^2$ or $H_0 : \sigma = \sigma_0$, the *test statistics* is $\chi^2 = \frac{(n-1)\,s^2}{\sigma_0^2} \sim \chi_{n-1}^2$

(Chi-square distribution with $n - 1$ *df*)

C. Rejection Region and Critical Values for the Chi-Square Distribution

The rejection region and the corresponding critical values based on the alternative hypothesis is outlined in the following table.

Alternative Hypothesis (H_1)	Rejection Region	Critical Values (χ^2)
$H_1 : \sigma > \sigma_0$	Right-Tailed $\left[\chi^2 > \chi_{n-1}^2(\alpha)\right]$	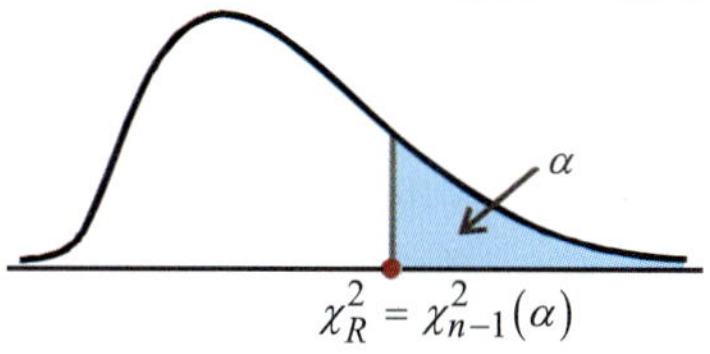
$H_1 : \sigma < \sigma_0$	Left-Tailed $\left[\chi^2 < \chi_{n-1}^2(1-\alpha)\right]$	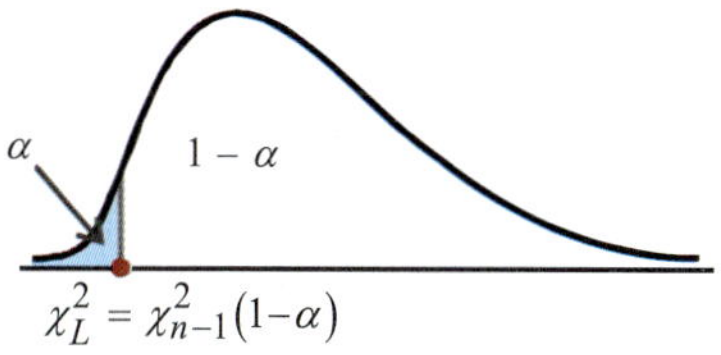

Table Cont.

$H_1: \sigma \neq \sigma_0$	Two-Tailed $\left[\chi^2 > \chi^2_{n-1,\,\alpha/2} \text{ or } \chi^2 < \chi^2_{n-1(1-\alpha/2)}\right]$	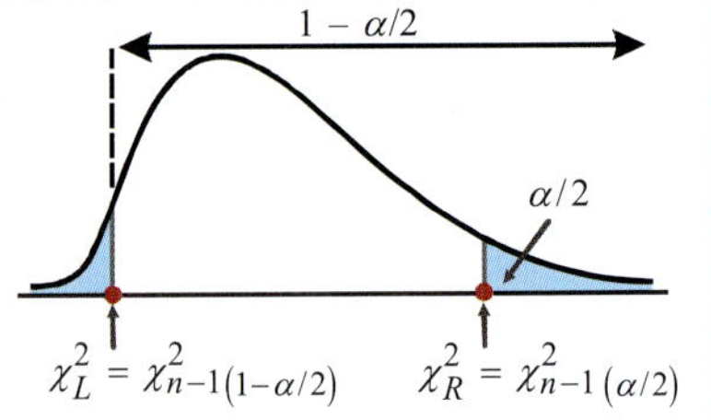

Table 7.16

Notation	**Recall:** Chi-Square values are always based on the area to their right under the χ^2 curve.
$\chi^2_{n-1}\ (\alpha)$ Degrees of Freedom = df (points to $n-1$) Area to the right of the Chi-Square value (points to α)	Left-Tailed Test: $\chi^2_L = \chi^2_{n-1\ (1-\alpha)}$ Right-Tailed Test: $\chi^2_R = \chi^2_{n-1}\ (\alpha)$ Two-Tailed Test: $\chi^2_L = \chi^2_{n-1\ (1-\alpha/2)}$ $\chi^2_R = \chi^2_{n-1\ (\alpha/2)}$

- **Formally Stated**

One Tailed Test: Let $\chi^2_v\ (\alpha)$ be the value of Chi-Square for v degrees of freedom such that the area under the χ^2 curve to its right is α. That is $P(\chi^2 > \chi^2_v\ (\alpha)) = \alpha$. $\chi^2_v\ (\alpha)$ is known as the upper right-tail critical value of the Chi-Square Distribution for v degrees of freedom and an α significance level. Similarly, $\chi^2_v\ (1-\alpha) = \chi^2_L = P\ (\chi^2 > \chi^2_L\ (1-\alpha)) = 1-\alpha$.

Two Tailed Test: Let $\chi^2_v\ (\alpha/2)$ be the value of Chi-Square for v degrees of freedom such that the area under the χ^2 curve to its right is $(\alpha/2)$. That is $P(\chi^2 > \chi^2_v\ (\alpha/2)) = \alpha/2$. $\chi^2_v\ (\alpha/2)$ is known as the upper right-tail critical value of the Chi-Square Distribution for v degrees of freedom and an α significance level. Similarly, $\chi^2_v\ (1-\alpha/2) = \chi^2_L = P\ (\chi^2 > \chi^2_L\ (1-\alpha/2)) = 1-\alpha/2$.

- **Conclusion**

Right-tailed Test H_1 : $\sigma^2 > \sigma^2_0$ or $\sigma > \sigma_0$: Reject H_0 at an α significance level if calculated χ^2 (test statistics) > tabulated $\chi^2_{n-1}\ (\alpha)$ (critical value).

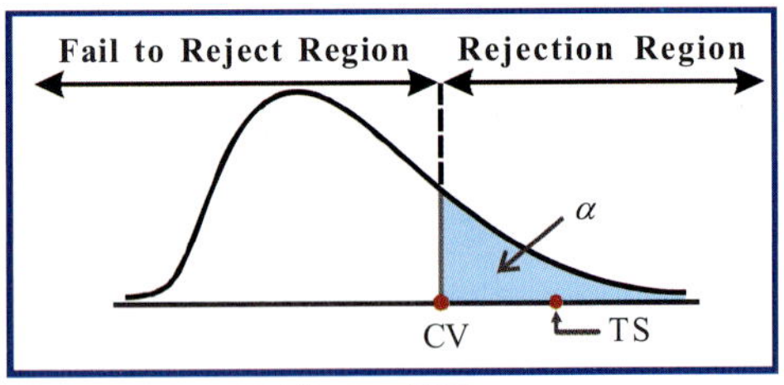

Figure 7.38

Recall: Calculated value = Test Statistics
Tabulated value = Critical Value

Left-tailed Test H_1 : $\sigma^2 < \sigma^2_0$ or $\sigma < \sigma_0$: Reject H_0 at an α significance level if calculated χ^2 < tabulated $\chi^2_{n-1\ (1-\alpha)}$.

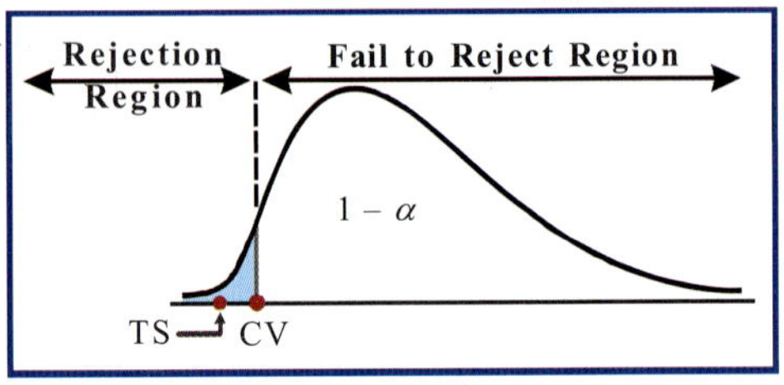

Figure 7.39

Two Tailed Test $H_1: \sigma^2 \neq \sigma_0^2$ or $\sigma \neq \sigma_0$:

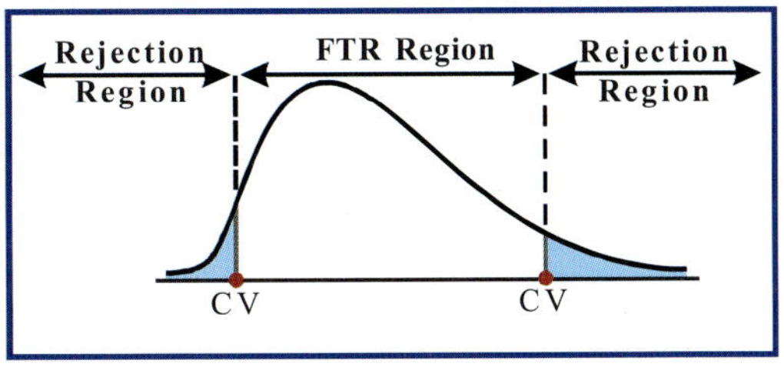

Figure 7.40 a

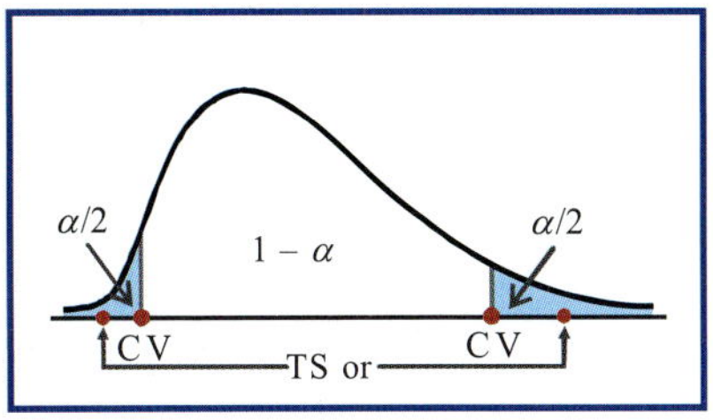

Figure 7.40 b

Reject H_0 at an α significance level if calculated $\chi^2 <$ tabulated $\chi^2_{n-1\ (1-\alpha/2)}$ or calculated $\chi^2 >$ tabulated $\chi^2_{n-1\ (\alpha/2)}$.

> **Note:** We will only use the traditional hypothesis testing method in this section. The *p*-value method for Chi-Square test can only be done using a statistical software. Without technology, we can only use an approximation of the ***p***-value to determine the validity of the original claim.

Example 16: A random sample of 15 values drawn from a normally distributed population has a standard deviation $s = 6.4$.

(a) Test the claim that the population standard deviation is at most 5. Use $\alpha = 0.10$.

(b) Verify the conclusion from part (a) by constructing a confidence interval for σ using an appropriate confidence level.

Solution:

***Step* 0** We are given $n = 15$, $s = 6.4$ and $\alpha = 0.10$.

The original claim is $\sigma \leq 5$; hence, the counter claim is $\sigma > 5$.

> **Note:** For testing a claim about σ, we use the Chi-Square test.

(a) ***Step* 1** Set up the Null and Alternative Hypotheses.

$H_0: \sigma = 5$ (oc)

$H_1: \sigma > 5$ (cc) (Right-Tail Test)

***Step* 2** Under H_0, $\sigma = 5$ and the test statistics is $\dfrac{(n-1)s^2}{\sigma^2}$.

$$\text{Test statistics} = \chi^2 = \frac{(n-1)s^2}{\sigma^2} = \frac{(15-1)(6.4)^2}{5^2} = 22.94$$

***Step* 3** Since in $H_1: \sigma > 5$, the rejection region will be in the right-tail of the Chi-Square curve with $\alpha = 0.1$ (area of the rejection region in the right-tail), and 14 degrees of freedom.

The critical value ($\chi^2_{n-1}(\alpha) = \chi^2_R$) for a right-tailed test at 10% level of significance is $\chi^2_{n-1}(\alpha) = \chi^2_{14}(0.1) = 21.064$.

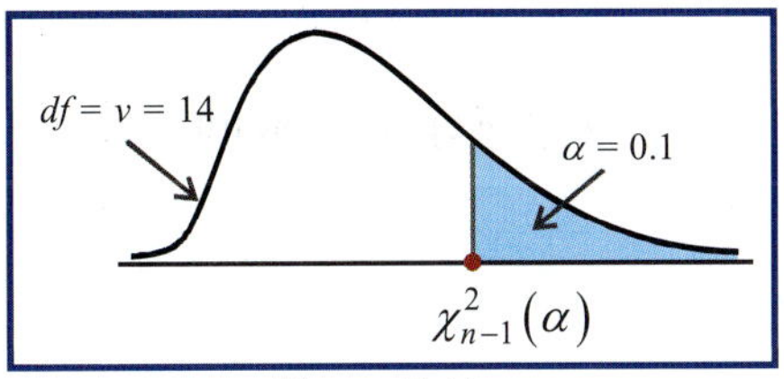

Figure 7.41 a

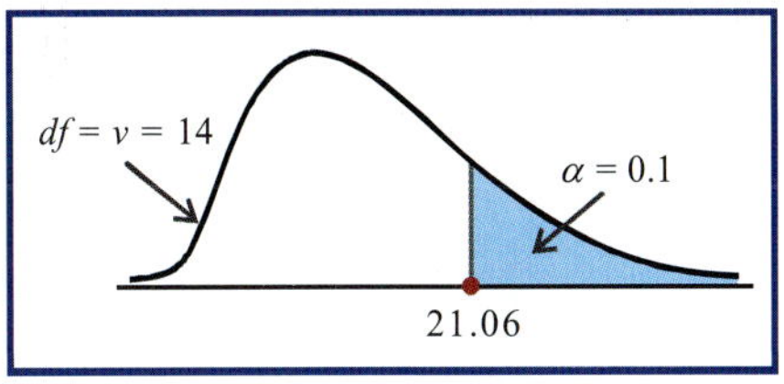

Figure 7.41 b

Step **4** **Decision:** Since the calculated χ^2 value (test statistics) lies inside the rejection region, we reject the null hypothesis at a 0.1 significance level. *Conclusion:* There is sufficient evidence against the claim that the standard deviation is at most 5.

In short:

Hypotheses	**Test Statistics**
$H_0: \sigma = 5$ $H_1: \sigma > 5$	$\chi^2 = \dfrac{(n-1)s^2}{\sigma^2} = 22.94$
Rejection Region and Critical Value 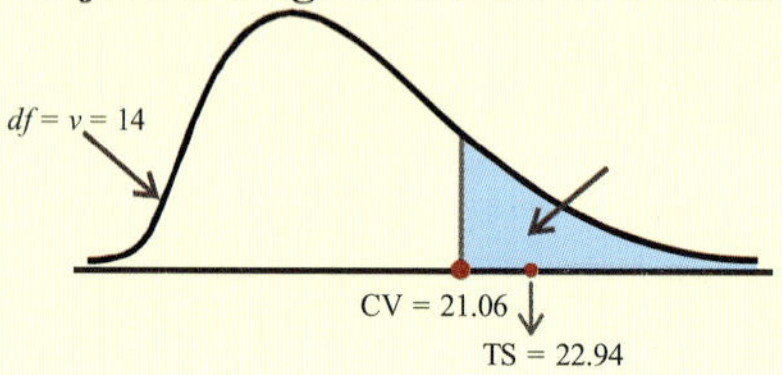	**Conclusion** Reject H_0. The standard deviation is at least 5.

(b) To verify the conclusion from part (a), we must construct the $1 - 2\alpha = 0.8 = 80\%$ confidence interval for σ.

Confidence Interval formula for σ	**Critical Values**
$\sqrt{\dfrac{(n-1)s^2}{\chi_R^2}} < \sigma < \sqrt{\dfrac{(n-1)s^2}{\chi_L^2}}$	$\chi_R^2 = \chi_{n-1}^2(\alpha) = \chi_{14}^2(0.1) = 21.064$ $\chi_L^2 = \chi_{n-1\ (1-\alpha)}^2 = \chi_{14}^2(0.9) = 7.790$

$$\sqrt{\frac{(15-1)(6.4)^2}{21.064}} < \sigma < \sqrt{\frac{(15-1)(6.4)^2}{7.790}} \Rightarrow 5.22 < \sigma < 8.58$$

Since $\sigma = 5$ is not in the estimated interval above, we must reject $H_0: \sigma = 5$. Our results from the calculated confidence interval above, agrees with the results obtained in the hypothesis testing conclusion.

Example 17: The average gasoline prices from 10 randomly selected cities across the United States are listed below.

3.455	3.509	3.376	3.483	4.106	3.915	3.586	3.858	3.107	3.936

(a) Test the claim that the standard deviation of gas prices across the country is \$0.3. Assume normality, and use a 0.05 significance level.

(b) Construct a 95% confidence interval for σ and verify the conclusion in part (a).

Solution: *Step* **0** We are given $n = 10$ and $\alpha = 0.05$. Since s^2 is not given, we use the raw data provided to calculate the sample variance.

x	3.455	3.509	3.376	3.483	4.106	3.915	3.586	3.858	3.107	3.936	$\sum x = 36.331$
x^2	11.937	12.313	11.398	12.131	16.859	15.327	12.859	14.884	9.653	15.492	$\sum x^2 = 132.854$

Table 7.17

$$s^2 = \frac{n\sum x^2 - \left(\sum x\right)^2}{n(n-1)} = \frac{(10)(132.854) - (36.331)^2}{(10)(9)} = 0.096$$

The original claim is $\sigma = 0.3 \Rightarrow$ counter claim is $\sigma \neq 0.3$.

(a) ***Step 1*** Set up the Null and Alternative Hypotheses.

$H_0 : \sigma = 0.3$ (oc)

$H_1 : \sigma \neq 0.3$ (cc) (Two-Tail Test)

Step 2 Under $H_0 : \sigma = 0.3$ and the test statistics is $\dfrac{(n-1)s^2}{\sigma^2}$.

$$\text{Test statistics} = \chi^2 = \frac{(n-1)s^2}{\sigma^2} = \frac{(10-1)(0.096)}{(0.3)^2} = 9.6$$

Step 3 Since in $H_1 : \sigma \neq 0.3$, the rejection region will be in both tails of the Chi-Square curve with $\alpha = 0.05$ and 9 degrees of freedom. The critical values for a two tailed test at 5% level of significance are

$$\chi^2_R = \chi^2_{n-1\ (\alpha/2)} = \chi^2_9\ (0.025) = 19.023$$

and $\chi^2_L = \chi^2_{n-1\ (1-\alpha/2)} = \chi^2_9\ (0.975) = 2.700$

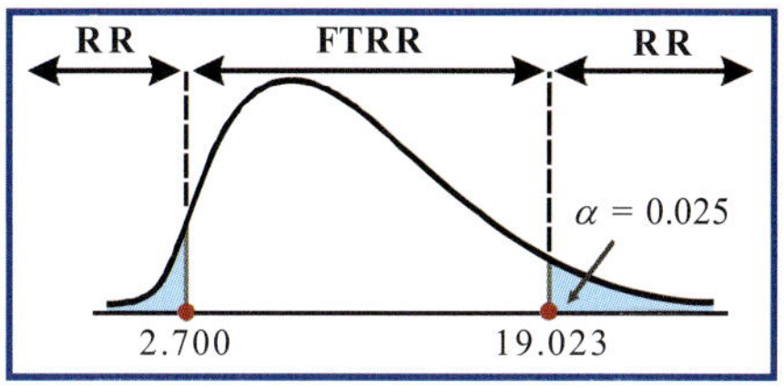

Figure 7.42

Step 4 **Decision:** Since the calculated χ^2 value (test statistics) lies outside the rejection region, we fail to reject the null hypothesis at a 0.05 significance level. *Conclusion:* There is sufficient evidence to support the claim that the standard deviation of gas prices across the country is 0.3.

In short:

Hypotheses	**Test Statistics**
$H_0 : \sigma = 0.3$ $H_1 : \sigma \neq 0.3$	$\chi^2 = \dfrac{(n-1)s^2}{\sigma^2} = 9.6$
Rejection Region and Critical Value	**Conclusion**
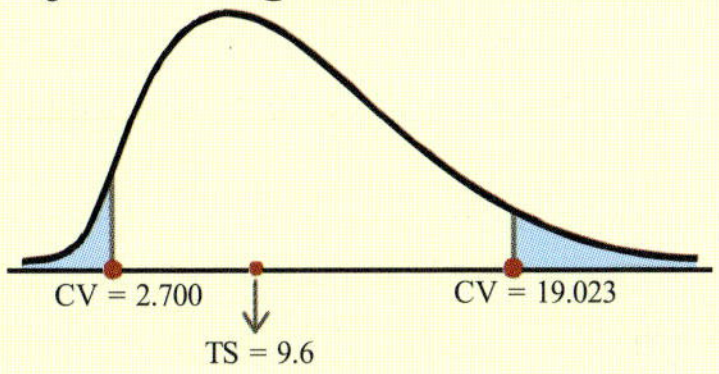	Reject H_0. The standard deviation of gas prices across the country is 0.3.

(b) To verify the conclusion from part (a), we must construct the $1 - \alpha = 0.95 = 95\%$ confidence interval for σ.

Confidence Interval formula for σ	**Critical Values**
$\sqrt{\dfrac{(n-1)s^2}{\chi^2_R}} < \sigma < \sqrt{\dfrac{(n-1)s^2}{\chi^2_L}}$	$\chi^2_R = \chi^2_{n-1\ (\alpha/2)} = \chi^2_9\ (0.025) = 19.023$ $\chi^2_L = \chi^2_{n-1\ (1-\alpha/2)} = \chi^2_9\ (0.975) = 2.700$

$$\sqrt{\frac{(10-1)(0.096)}{19.023}} < \sigma < \sqrt{\frac{(10-1)(0.096)}{2.700}} \Rightarrow 0.213 < \sigma < 0.566$$

Since $\sigma = 0.3$ lies in the estimated interval above we fail to reject $H_0 : \sigma = 0.3$. Our results from the calculated confidence interval above agrees with the results obtained in the hypothesis testing conclusion.

Example 18: A random sample of size 52 has a standard deviation of 8.6.

(a) Is this compatible with hypothesis that the sample is drawn from a normally distributed population with variance $\sigma^2 < 64$.

(b) Verify your results with constructing a 90% confidence interval for σ^2.

Solution:

***Step* 0** We are given: $n = 52$, $s = 8.6$ and $\alpha = 0.05$

The original claim: $\sigma^2 < 64$

The counter claim: $\sigma^2 \geq 64$

(a) ***Step* 1** Set up the Null and Alternative Hypotheses.

H_0 : $\sigma^2 = 64$ (cc)

H_1 : $\sigma^2 < 64$ (oc) (Left-Tail Test)

***Step* 2** Under H_0 : $\sigma^2 = 64$ and the test statistics is $\dfrac{(n-1)s^2}{\sigma^2}$

$$\text{Test statistics} = \chi^2 = \frac{(n-1)s^2}{\sigma^2} = \frac{(52-1)(8.6)^2}{64} = 58.94$$

***Step* 3** Since in H_1 : $\sigma^2 < 64$, the rejection region will be in the left-tail of the Chi-Square curve with $\alpha = 0.05$ and 51 *df*. The critical value $\chi^2_{n-1}(1-\alpha) = \chi^2_L$ for a left-tailed test at 5% level of significance is

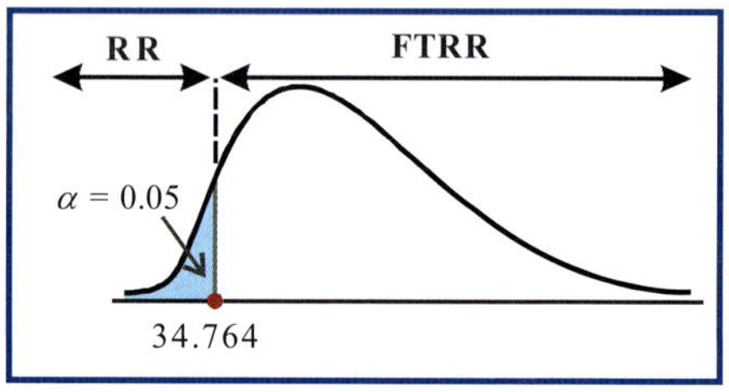

Figure 7.43

$$\chi^2_{n-1}(1-\alpha) = \chi^2_{51}(0.95)$$
$$= 35.5998$$

Using the Excel® command: CHISQ.INV(0.95,51)

Note: If Microsoft Excel® is not available and the desired *df* does not exist in the given Chi-Square table, then to find critical value, round the *df* to the nearest whole number on the table for approximation.

$$\chi^2_L = \chi^2_{n\text{-}1\,(1-\alpha)} = \chi^2_{51}(0.95) \approx \chi^2_{50}(0.95) = 34.764.$$

***Step* 4** **Decision:** Since the test statistics lies outside the rejection region, we fail to reject the null hypothesis at a 0.05 significance level. *Conclusion:* There is not sufficient evidence to support the claim that the variance is less than 64.

In short:

Hypotheses	**Test Statistics**
H_0: $\sigma = 64$ H_1: $\sigma < 64$	$\chi^2 = \dfrac{(n-1)s^2}{\sigma^2} = 58.94$
Rejection Region and Critical Value 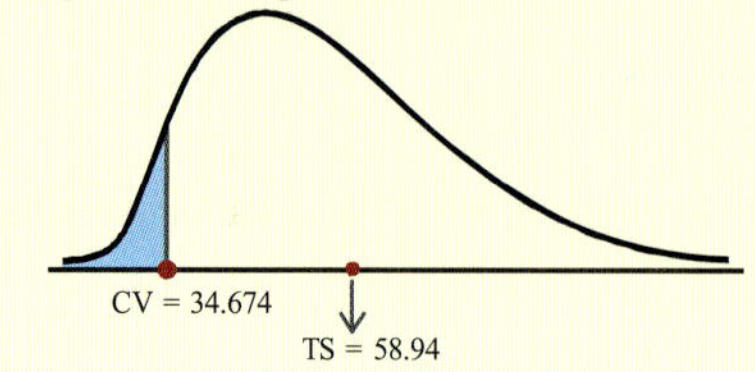	**Conclusion** Reject H_0. The variance is not less than 64.

(b) To verify the conclusion from part (a), we must construct the $1 - 2\alpha = 0.90 = 90\%$ confidence interval for σ^2.

Confidence Interval formula for σ^2	Critical Values
$\frac{(n-1)s^2}{\chi_R^2} < \sigma^2 < \frac{(n-1)s^2}{\chi_L^2}$	$\chi_R^2 = \chi_{n-1}^2(\alpha) = \chi_{51}^2(0.05) \approx \chi_{50}^2(0.05) = 67.505$ $\chi_L^2 = \chi_{n-1}^2(1-\alpha) = \chi_{51}^2(0.95) \approx \chi_{50}^2(0.95) = 34.764$
$\frac{(52-1)(8.6)^2}{67.505} < \sigma^2 < \frac{(52-1)(8.6)^2}{34.764} \Rightarrow 55.88 < \sigma^2 < 108.50$ Based on the confidence interval above, the variance does not seem to be less than 64. Hence, the results verify the conclusion from the test of hypothesis.	

Try it yourself:

In the following exercises, assume that the parent population is normally distributed.

1. $\sigma = 5$ (oc), $s = 4.2$ and $n = 81$. Use a 1% significance level to test the claim.
2. $\sigma < 0.46$ (oc), $s = 0.30$ and $n = 20$. Use a 5% significance level to test the claim.
3. $\sigma^2 \geq 324$ (oc), $s = 16.8$ and $n = 45$. Use a 5% significance level to test the claim.
4. $\sigma > 0.88$ (oc), $s = 0.92$ and $n = 50$. Use $\alpha = 0.01$ to test the claim.
5. An owner of a firm agrees to purchase the products of a factory if the produced items have standard deviations less than 0.707 inches in their length. A random sample of size 18 from the factory's lot gave a standard deviation of 0.174 inches. Does the sample data confirm to specifications?
6. In a random sample of 25 boxes of cereal, the variance is found to be 289 grams. Does the sample data provide evidence that the variance has changed from the previously known value of 300 grams? Use a 0.01 level of significance.
7. The standard deviation of the coolant level in a radiator of a car is believed to stay at about 1.5 inches. A random sample of the coolant level of 40 radiators showed a standard deviation of 2 inches. At $\alpha = 0.05$, is there sufficient evidence to support the belief?
8. A random sample of size 10 from a normally distributed population gave the following values.

65	72	68	74	77	61	63	69	73	71

Test the claim that the population standard deviation is greater than 5.7.

7.7: Hypothesis Testing with Microsoft Excel®

Objectives

After completing this section you will be able to carryout the computations in Chapter 7 using Microsoft Excel®.

Illustration 1: Zion Memorial claims that the mean recovery time for emergency patients is 2.26 days. To test this claim, a researcher found that among 50 randomly selected patients from the emergency room, the mean recovery time was 2.12 days with a standard deviation of 1.3 days. Calculate the Test Statistics using $\alpha = 0.01$.

Solution: The Test Statistics is 0.76 as shown in this figure.

Traditional Calculation for Test Statistics	Excel Command and output
$\frac{\bar{x}-\mu}{s/\sqrt{n}} = \frac{2.12-2.26}{1.3/\sqrt{50}} = -0.76$	=(B3-E2)/(B4/SQRT(B5))

	A	B	C	D	E
1					
2	Sample Data			μ	2.26
3	x bar	2.12		TS	–0.7615
4	σ	1.3			
5	n	50			

Illustration 2: Professor Badalian claims that the average score of all his students is at most 63. To validate his claim, he randomly selected 20 scores from all his classes.

88	75	43	94	21	67	55	63	74	86
64	70	78	84	46	30	72	81	69	66

Calculate the Test Statistics using

(a) manual Excel commands, and **(b)** the Descriptive Statistics package.

Solution: In order to calculate the test statistics, we must first calculate the sample statistics such as the $\bar{x}$, s, and n. We know that $\mu = 63$.

(a) Calculating the Test Statistics using Excel commands:

	A	B	C	D	E	F	G	H	I
1	88	75	43	94	21			Commands	Output
2	64	70	78	84	46		Mean	=AVERAGE(A1:E4)	66.3
3	67	55	63	74	86		SD	=STDEV(A1:E4)	19.12589867
4	30	72	81	69	66		n	=COUNT(A1:E4)	20
5							μ	63	63
6									
7							TS	=(I2–I5)/(I3/SQRT(I4))	0.771626416

Figure 7.44

(b) Calculating the Test Statistics using Descriptive Statistics package under the Data Analysis group:

	A	B	C	D	E
1	88	75	43	94	21
2	64	70	78	84	46
3	67	55	63	74	86
4	30	72	81	69	66

Figure 7.45

Select the Data tab, and click on the Data Analysis icon under the Analysis group. The following dialog box must appear.

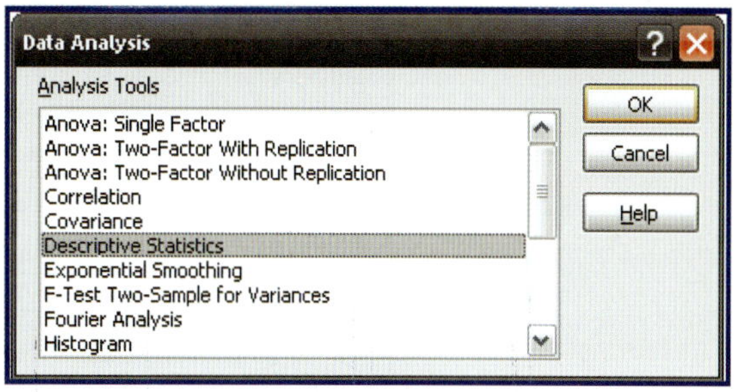

Figure 7.46

Organize the data in a row or column.

	A	B	C	D	E	F	G	H	I	J	K	L	M	N	O	P	Q	R	S	T
1	88	64	67	30	75	70	55	72	43	78	63	81	94	84	74	69	21	46	86	66

Figure 7.47

Click on Descriptive Statistics and press OK. Fill out all the required information as shown in Figure 7.48.

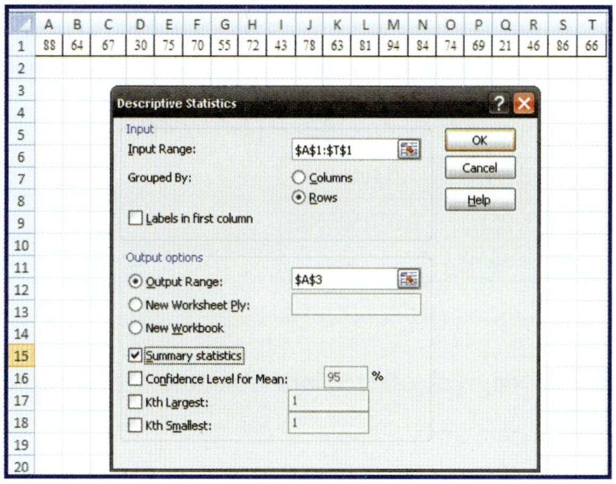

Figure 7.48

Press OK. Use the statistics provided by the descriptive statistics package to calculate the test statistics.

	A	B	C	D	E
1					
2	*Row 1*			μ	63
3				$\bar{x}$	66.3
4	Mean	66.3		s	19.125899
5	Standard Error	4.276681		n	20
6	Median	69.5			
7	Mode	#N/A		**TS**	**0.771626**
8	Standard Deviation	19.125899			
9	Sample Variance	365.8			
10	Kurtosis	0.5441175			
11	Skewness	-0.9402335			
12	Range	73			
13	Minimum	21			
14	Maximum	94			
15	Sum	1326			
16	Count	20			

Figure 7.49

Illustration 3: The Geographic National Agency claims that the standard deviation of rainfall in the Amazon is 23 inches. To validate this claim, the GNA has presented the amount of rainfall for 8 randomly selected seasons. Find the test Statistics.

40	58	46	37	60	70	55	64

Solution: **Test Statistics** $= \dfrac{(n-1)s^2}{\sigma^2} = \dfrac{(8-1)(11.696)^2}{23^2} = 1.81$

Here is the Excel® worksheet that illustrates the calculation of the test statistics.

	A	B	C	D
1	40		s	=STDEV(A1:A8)
2	58		n	8
3	46		$n-1$	=D2 – 1
4	37		s^2	=D1^2
5	60		σ^2	=23^2
6	70		TS	=D3*D4/D5
7	55			
8	64			

	A	B	C	D
1	40		s	11.6955425
2	58		n	8
3	46		$n-1$	7
4	37		s^2	136.7857143
5	60		σ^2	529
6	70		TS	1.810018904
7	55			
8	64			

Figure 7.50

- ***P*-value Calculations using Microsoft Excel®**

Illustration 4: Find the ***p***-value for $\boldsymbol{t}$ = 1.235 with 15 degrees of freedom. Assume a right-tailed test.

Solution: ***p***-value = $P(t > 1.235)$.

Excel command	Output
=T.DIST.RT(1.235,15) T.DIST.RT(x, deg_freedom)	0.117913224

Therefore, $P(t > 1.235) = 0.1179$.

Illustration 5: Find the ***p***-value for $\boldsymbol{t}$ = –2.463 with 25 degrees of freedom. Assume a left-tailed test.

Solution: ***p***-value = $P(t < -2.463)$. Since the t distribution is symmetric, $P(t < -2.463) = P(t > 2.463)$.

Excel command	Output
=T.DIST.RT(2.463, 25) T.DIST.RT(x, deg_freedom)	0.010506605

Therefore, $P(t < -2.463) = 0.0105$.

Illustration 6: Find the ***p***-value for χ^2 = 70.56 with 45 degrees of freedom. Assume a right-tailed test.

Solution: ***p***-value = $P(\chi^2 > 70.56)$.

Excel command	Output
=CHISQ.DIST.RT(70.56,45) CHISQ.DIST.RT(x, deg_freedom)	0.008803293

Therefore, $P(\chi^2 > 70.56) = 0.0088$.

Illustration 7: Find the ***p***-value for χ^2 = 12.40 with 24 degrees of freedom. Assume a left-tailed test.

Solution: ***p***-value = $P(\chi^2 < 12.40) = 1 - P(\chi^2 > 12.40)$.

Excel command	Output
=1 – CHISQ.DIST.RT(12.4,24) CHISQ.DIST.RT(x, deg_freedom)	0.024984781

Therefore, $P(\chi^2 < 12.40) = 0.0250$.

7.8: Extended Topics and Proofs in Hypothesis Testing: One Population (Optional)

A. Null and Alternative Hypothesis

Objectives

After completing this section, you will be able to understand:

A. Null and Alternative Hypothesis.

B. Type I and Type II Errors in Testing a Hypothesis.

C. Power of a Test.

- **Notation:**

To discuss concepts of hypothesis testing regarding all parameters, we will use θ to represent a particular parameter such as μ, p or σ and θ_0 to represent a particular value of θ (constant).

Remark: Because the null hypothesis must contain the equality sign, we sometimes express the null hypothesis in symbolic form as: $H_0 : \theta \geq \theta_0$ or $H_0 : \theta \leq \theta_0$. In each case, we conduct the test by assuming that $\theta = \theta_0$ is true. This is because the sampling distribution of the test statistics under H_0 must have a single distribution with specific value of the parameter θ. In this text, we have used the notation $H_0 : \theta = \theta_0$ for the $H_0 : \theta \leq \theta_0$ or $H_0 : \theta \geq \theta_0$ cases.

Reasoning: To test $H_0 : \theta \leq \theta_0$ against $H_1 : \theta > \theta_0$, the statement $H_0 : \theta = \theta_0$ against $H_1 : \theta > \theta_0$ is used instead, because if the data provides sufficient evidence to reject $H_0 : \theta = \theta_0$, (that is, accept $H_1 : \theta > \theta_0$), it obviously provides sufficient evidence to reject $\theta < \theta_0$. Similarly, to test $H_0 : \theta \geq \theta_0$ against $H_1 : \theta < \theta_0$, it will be sufficient to test $H_0 : \theta = \theta_0$ against $H_1 : \theta < \theta_0$.

All possible cases of the Null and Alternative Hypothesis for testing μ:

Original Claim	Counter Claim	Null Hypothesis (H_0)	Alternative Hypothesis (H_1)
$\mu < 5$	$\mu \geq 5$	$H_0 : \mu = 5$	$H_1 : \mu < 5$
$\mu > 5$	$\mu \leq 5$	$H_0 : \mu = 5$	$H_1 : \mu > 5$
$\mu \leq 5$	$\mu > 5$	$H_0 : \mu = 5$	$H_1 : \mu > 5$
$\mu \geq 5$	$\mu < 5$	$H_0 : \mu = 5$	$H_1 : \mu < 5$
$\mu = 5$	$\mu \neq 5$	$H_0 : \mu = 5$	$H_1 : \mu \neq 5$
$\mu \neq 5$	$\mu = 5$	$H_0 : \mu = 5$	$H_1 : \mu \neq 5$

Table 7.18

B. Type I and Type II Errors in Testing a Hypothesis

Recall: Type I and Type II Errors

We make a **type I** error by rejecting a true null hypothesis.

We make a **type II** error by not rejecting a false null hypothesis.

- **Notation:**

P [Type I Error] $= \alpha$	P [Type II Error] $= \beta$

An ideal test procedure would be one which is so planned as to safeguard against both these errors. But practically, in any given problem, it may not be possible to minimize both of these errors simultaneously; that is to attempt to decrease α results in an increase in β and *vice-versa*. In practice, in most of the decision making problems in business and social sciences, it is more risky to accept a wrong hypothesis than to reject a correct one because the consequences of type II error are likely to be more serious than the consequences of type I error. Since both errors cannot be reduced simultaneously for a given sample, a compromise is made by minimizing one serious error after fixing up the less serious error. Thus we fix α, the size of type I error, and then try to obtain a criterion which minimizes β, the size of type II error. We have

$$P\,[\text{Accept } H_0 \text{ when } H_0 \text{ is false}] + P\,[\text{Reject } H_0 \text{ when } H_0 \text{ is false}] = 1$$

$$\begin{aligned} P\,[\text{Reject } H_0 \text{ when } H_0 \text{ is false}] &= 1 - P\,[\text{Accept } H_0 \text{ when } H_0 \text{ is false}] \\ &= 1 - \beta \end{aligned}$$

Obviously, when H_0 is true, it ought not to be rejected. Hence minimizing β amounts to maximizing $(1 - \beta)$, which is called the *power of the test.*

C. Power of a Test

Definition:

Power of a test of hypothesis, denoted by $(1 - \beta)$, is the probability that the test will lead to rejection of H_0 when in fact H_0 is false.

The usual practice in hypothesis testing is to fix α, the size of type I error, and then try to obtain a criterion which minimizes β, the size of type II error, or maximizes $(1 - \beta)$, the power of the test. For any fixed α, an increase in the sample size (n) will result in a decrease in the value of β. Depending upon the seriousness of the consequences of type I error, we should select the maximum tolerable value of α. For more serious consequences, for example in medical research and industry, we should take smaller values of α and large sample size.

Example 19: For a candidate's political poll, 15 randomly selected voters were surveyed. We want to test $H_0 : p = 0.5$ against the alternative hypothesis $H_1 : p < 0.5$. Assume the test statistics is $\boldsymbol{T}$, the number of surveyed voters favoring the candidate. Calculate α if we select $\{T \le 3\}$ as the rejection region.

Solution: $\alpha = P\,[\text{Type I Error}] = P\,(\text{Rejecting } H_0 \text{ when } H_0 \text{ is true})$

$= P$ (value of the test statistics lies in the rejection region when H_0 is true)

$= P\,(T \le 3 \text{ when } p = 0.5)$

Assuming that $\boldsymbol{T}$ is a binomial random variable with $n = 15$ and $p = 0.5$, we have

$$\alpha = \sum_{T=0}^{3} \binom{15}{T} (0.5)^T (0.5)^{15-T}$$

$$= \binom{15}{0} (0.5)^0 (0.5)^{15} + \binom{15}{1} (0.5)^1 (0.5)^{14}$$

$$+ \binom{15}{2} (0.5)^2 (0.5)^{13} + \binom{15}{3} (0.5)^3 (0.5)^{12} = 0.018$$

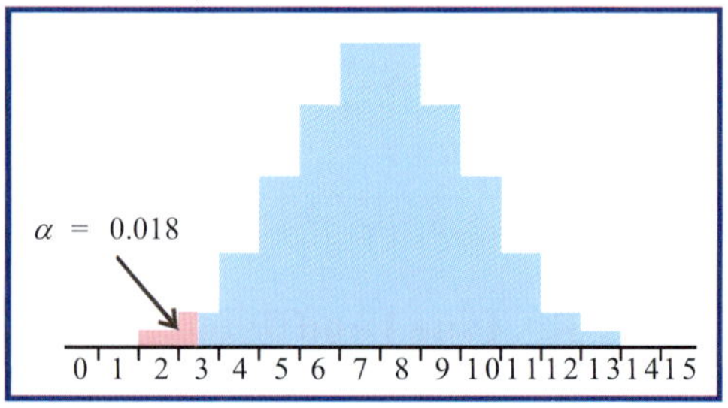

Figure 7.51

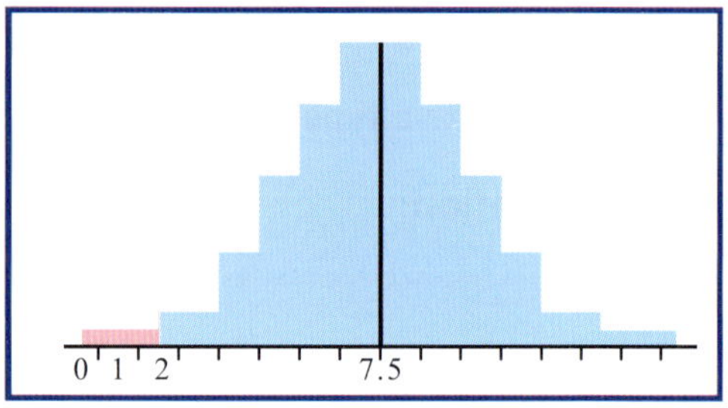

Figure 7.52

Similarly, if we decide to use a rejection region $\{T \le 2\}$, we will face a very small risk $\alpha = 0.004$ of concluding that the candidate will lose if, in fact, he is a winner.

Example 20: Testing $H_0 : p = 0.5$ against the alternative hypothesis $H_1 : p = 0.25$, calculate the probability of type-II error and power of test for Example 1.

Solution:

$$\beta = P(\text{Type II error}) = P\,(\text{Accepting } H_0 \text{ when } H_1 \text{ is true})$$

$$= P(\text{value of test statistics lies outside the rejection region when } H_1 \text{ is true})$$

$$= P(T > 3, \text{ when } p = 0.25) \text{ and}$$

$$1 - \beta = P(T \leq 3, \text{ when } p = 0.25)$$

$$= \sum_{T=0}^{3} \binom{15}{T} (0.25)^T (0.75)^{15-T} = 0.4613$$

Power of test = $1 - \beta = 0.4613$

There is a 46.13% chance that the test of hypothesis will lead us to conclude that the candidate is a winner, if $p = 0.25$.

Critical (Significant) Values:

1. The significant values of $|Z|$, namely, 1.96 (for $\alpha = 0.05$) and 2.575 (for $\alpha = 0.01$) for two-tailed test and 1.645 for ($\alpha = 0.05$) and 2.33 for ($\alpha = 0.01$) for right-tailed test should be committed to memory. The decision about applying a two-tailed test or a single-tailed test will be determined by the alternative hypothesis.

2. If $Z \sim N(0, 1)$, from the normal probability table, we get

$$P(|Z| \leq 3) = P(-3 \leq Z \leq 3) = 2P(0 \leq Z \leq 3) = 2\,(0.4987) = 0.9974$$

$$\Rightarrow P(|Z| > 3) = 1 - P(|Z| \leq 3) = 0.0026$$

Hence, in all probability, we should expect a standard normal variable to lie within the limits ± 3. The value of $|Z| = 3$ is regarded as critical or significant value of Z at all levels of significance.

χ^2 Test:

1. The test statistics $\frac{(n-1)s^2}{\sigma^2}$ assumes that the data in the population from which the sample is drawn, are normally distributed. This χ^2 test statistics is very sensitive to departures from this assumption of normality. If the population is not normally distributed, particularly for small samples, the accuracy of the test can be seriously affected. The procedures for dealing with such a situation are, however, beyond the scope of this text.

2. (Large Samples) If sample size is large, ($n > 30$), we can use Fisher's approximation, namely $\sqrt{2\chi^2} \sim N\left(\sqrt{2n-1}, 1\right)$, and the normal test by considering the statistics

$$Z = \frac{\sqrt{2\chi^2} - \sqrt{2n-1}}{1} \text{ that is, } Z = \sqrt{2\chi^2} - \sqrt{2n-1} \sim N(0, 1)$$

Chapter 7: General Exercises

Testing the Mean μ (σ Known)

In these exercises use the classical and wherever applicable, the *P*-value method.

1. A sample of 100 randomly selected children under the age of 12 showed that an average amount of time a child spends watching television every day is 2.7 hours. Assuming $\sigma = 0.6$ hours,

(a) test the claim that the average time a child spends watching TV is greater than 2.7 hours. Assume a 0.01 significance level.

(b) construct an appropriate confidence interval to validate your conclusion in part(a).

2. The President of a school board claims that public school average attendance rates have dropped below 70% over the past year. Suppose the average rate for 140 randomly selected schools was 73%. Assume $\sigma = 0.09$.

(a) Test the President's claim at a 0.05 significance level. (Note: Even though the data is given in terms of percentages, this test is still about the population average μ.)

(b) Construct an appropriate confidence interval to validate your conclusion in part(a).

3. The following lists the gasoline prices (per gallon) for 8 randomly selected gas stations.

3.55	4.59	3.88	4.08	3.99	4.32	3.69	4.22

(a) Test the claim that average gasoline price per gallon is about $3.99, assuming the population is normally distributed and $\sigma = 0.07$. Use $\alpha = 0.01$.

(b) Construct an appropriate confidence interval to validate your conclusion in part(a).

4. The average length of a movie on a DVD is 1.9 hours. 15 randomly selected DVDs were selected and watched. The average length of a movie was 1.8 hours. Assuming the population is normally distributed and $\sigma = 0.2$ hours,

(a) test the claim that the average length of a movie on a DVD is 1.9 hours.

(b) construct an appropriate confidence interval to validate your conclusion in part(a).

5. The effectiveness of a test preparation course was studied for a simple random sample of 85 subjects who took the test before and after coaching. The differences between the scores resulted in a mean increase of 0.4 with a population standard deviation of 0.6.

(a) Test the claim that the population mean increase is greater than 0, indicating that coaching is effective in raising scores.

(b) Construct an appropriate confidence interval to validate your conclusion in part(a).

6. Forty people are randomly selected and the accuracy of their wrist watches was checked, with positive errors representing watches that are ahead of the correct time and negative errors representing watches that are behind the correct time. The 40 values have a mean error of 12.3 sec. Population standard deviation is known to be 45.0 sec.

(a) Use a 0.1 significance level to test the claim that the population of all watches has a mean error equal to 0 sec.

(b) Construct an appropriate confidence interval to validate your conclusion in part(a).

7. The health of employees is monitored by periodically weighing them in. A sample of 54 employees has a mean weight of 183.9 lb.

(a) Assuming that σ is known to be 31.2 lb, use a 0.10 significance level to test the claim that the mean weight of all such employees is less than 200 lbs.

(b) Construct an appropriate confidence interval to validate your conclusion in part(a).

8. The state Department of Education in U.S. reports that the mean GPA of transfer students in all colleges in Archenland is 2.8. To test this claim, a curious administrator selects a random sample of 900 transfer students and finds that the mean GPA is 2.4. Suppose $\sigma = 0.6$.

(a) At $\alpha = 0.01$, is there enough evidence to conclude that the mean GPA is different from 2.8?

(b) Construct an appropriate confidence interval to validate your conclusion in part(a).

9. A soft drink company claims that the mean caffeine content for a 12-ounce bottle is 40 milligrams. To verify the claim, a quality control officer selects a random sample of twenty-five 12-ounce bottles and calculates a mean caffeine content of 42.2 milligrams. From a previous research study, the population standard deviation is known to be 2.3 milligrams.

(a) At $\alpha = 0.05$, can the quality control officer reject the company's claim? Assume the population is normally distributed.

(b) Construct an appropriate confidence interval to validate your conclusion in part(a).

10. On April 26, 1986, in Chernobyl, Ukraine, a nuclear reactor suffered a catastrophic power increase leading to explosions in the core. This incident resulted in a 2% radiation leak, and approximately 600,000 people were exposed to a high dose of radiation with a standard deviation of 3.97 Gray. 143 randomly selected patients that had been exposed to radiation were tested, and it was found that the mean radiation dose was 12.9 Gray.

(a) Test the claim that the mean radiation exposure was less than 13.4 Gray. Assume a 0.1 significance level.

(b) Construct an appropriate confidence interval to validate your conclusion in part(a).

11. The Tea Drinker's Association estimates that the mean consumption of tea by Americans is more than 7 gallons per year. In a sample of 100 randomly selected Americans, it was found that the mean consumption of tea $\bar{x} = 7.8$ gallons per year with $\sigma = 2.67$ gallons.

(a) At $\alpha = 0.07$, can you support the association's claim?

(b) Construct an appropriate confidence interval to validate your conclusion in part(a).

12. An employment agency reports that the mean annual salary for full-time male workers over the age of 25 with a bachelor's degree is $53,102. Raymond, a small business owner, finds the claim to be suspicious. In order to validate his suspicion, he randomly selects a sample of 12 full time male workers with bachelor's degrees, and finds that the group has a mean annual salary of $52,201 with $\sigma = \$6500$.

(a) Assuming the population is normally distributed, test the agency's claim by using 0.1 significance level.

(b) Construct an appropriate confidence interval to validate your conclusion in part(a).

Testing the Mean μ (σ Unknown)

13. At a university, 346 randomly selected students rated their Physics Professor on a scale of 1-10. The average score was 6.5 with a standard deviation of 1.4.

(a) Using a 0.1 significance level, test the claim that the Professor's mean score is equal to 6.9.

(b) Construct an appropriate confidence interval to validate your conclusion in part(a).

14. The following data gives quarterly rental rates for 2001 and 2002, as given by the U.S. Census Bureau.

8.2	7.8	8.3	7.3	10.2	5.1	8.9	11.3	5.8	8.5	8.2	8.6	7.7	10.2	9.1	9	9.5	10.2
8.3	7.9	8.4	7.4	10.5	5.3	10.3	10.8	5.7	9.1	8.7	9.5	7.8	11	8.4	10	5.8	5.6
8.4	8	8.8	7.1	10.4	5.5	9.3	10.8	6.6	9.4	9.1	9.3	8.9	11	10.3	11.9	7	6.9
8.8	8.4	9	7.8	10.5	5.4	10.3	11.5	6.6	5.6	9.7	10.7	6.7	12.9	7	6	10.3	11.6

(a) Test the claim that $\mu < 9.8$. Use $\alpha = 0.05$.

(b) Construct an appropriate confidence interval to validate your conclusion in part(a).

15. The table below gives the number of floors at 35 randomly selected buildings in the United States.

45	30	24	12	6	37	26
14	4	14	26	18	40	28
36	55	35	17	12	6	8
20	44	23	16	25	27	35
15	10	6	18	4	18	22

(a) Using $\alpha = 0.025$, test the claim that the average number of floors in a building is less than 30.

(b) Construct an appropriate confidence interval to validate your conclusion in part(a).

16. A box of chocolates should weigh 16.4 ounces. A sample of 40 randomly selected boxes had a mean weight of 16.6 ounces, with a standard deviation of 0.4 ounces.

(a) At a 0.05 significance level, test the claim that the mean weight of these boxes is greater than 16.4 ounces.

(b) Construct an appropriate confidence interval to validate your conclusion in part (a).

17. A public official claims that the mean waiting time for bus number 14 during peak hours is less than 10 minutes. Karen randomly took bus number 14 during peak hours on 18 different occasions. Her mean waiting time was 7.8 minutes with a sample standard deviation of 2.5 minutes.

(a) At the 0.01 significance level, test the claim that the mean waiting time for bus number 14 during peak hours is less than 10 minutes, assuming the parent population is normally distributed.

(b) Construct an appropriate confidence interval to validate your conclusion in part (a).

18. Employees in a large accounting firm claim that the company's mean salary is less than $45,000. A random sample of 20 employees revealed a mean salary of $43,500 with a standard deviation of $5200.

(a) Assuming the population is normally distributed, test the employees' claims at a 0.01 significance level.

(b) Construct an appropriate confidence interval to validate your conclusion in part (a).

19. A neurologist measured the brain mass of 22 Baboons and recorded the following results in grams.

131	140	126	133	109	113	132	140	146	123	147
142	142	122	141	111	102	127	139	142	129	132

(a) At a 0.1 significance level, test the claim that the average brain mass of Baboons is equal to 128.1 grams. Assume the population is normally distributed.

(b) Construct an appropriate confidence interval to validate your conclusion in part (a).

20. A social researcher wants to test the claim that convicted burglars spend an average of less than 18.7 months in jail. She takes a random sample of 55 such cases and finds that $\overline{x} = 21.5$ months and $s = 7.4$ months. Assume the population is normally distributed.

(a) Test the claim at the 0.05 significance level.

(b) Construct an appropriate confidence interval to validate your conclusion in part (a).

21. An auto maker claims that the mean gas mileage of its luxury sedans is greater than 25 miles per gallon. A random sample of 8 luxury sedans had a mean of 24 miles per gallon and a standard deviation of 5 miles per gallon.

(a) At $\alpha = 0.05$, can you reject the auto maker's claim that the mean gas mileage of its luxury sedans is greater than 25 miles per gallon, assuming the population is normally distributed?

(b) Construct an appropriate confidence interval to validate your conclusion in part (a).

22. A used car dealer claims that the mean price of a 2007 Ford F-150 Super Cab is more than $16,500. You suspect this claim is incorrect and discover that a random sample of 14 similar vehicles has a mean price of $15,700 and a standard deviation of $1250.

(a) Is there enough evidence to reject the dealer's claim at $\alpha = 0.01$? Assume the population is normally distributed.

(b) Construct an appropriate confidence interval to validate your conclusion in part (a).

23. An industrial company claims that the mean pH level of the water in the nearby river is 6.8 units. You randomly select 26 water samples and measure the pH balance of each one. The sample mean and standard deviation are 6.7 and 0.64, respectively.

(a) Assuming the population is normally distributed, is there enough evidence to reject the company's claim at a 0.05 significance level?

(b) Construct an appropriate confidence interval to validate your conclusion in part (a).

24. A scientist claims that the mean nitrogen dioxide level in India is greater than 43 parts per billion. In order to test this claim, a researcher measures the nitrogen dioxide levels for 46 randomly selected days. Here are the results (in parts per billion).

35	37	61	39	24	55	30	25	21	54	107	23	28	25	36	18	22	17
43	37	40	75	32	39	51	37	20	47	73	102	20	35	21	24	48	70

(a) At $\alpha = 0.1$, can the scientists claim be supported?

(b) Construct an appropriate confidence interval to validate your conclusion in part (a).

25. In Illinois, a random sample of 85 eighth-graders had a mean score of 265 with a standard deviation of 60 on a national mathematics assessment test. This prompted a state school administrator to declare that the mean score for the state's eighth-graders on the examination is greater than 260.

(a) At $\alpha = 0.01$, is there sufficient evidence to support the administrator's claim?

(b) Construct an appropriate confidence interval to validate your conclusion in part (a).

26. A random sample of 6000 people showed that the mean tuna consumption by Americans is 3.2 pounds per year with a standard deviation of 1.13 pounds. After carefully analyzing this information, a nutritionist claims that the true mean tuna consumption by Americans is in fact 3.4 pounds per year.

(a) At $\alpha = 0.1$, can you reject the nutritionist's claim?

(b) Construct an appropriate confidence interval to validate your conclusion in part(a).

27. A university claims that the mean number of class room hours per week for full-time students is more than 14 hours. A random sample of 11 full-time students is selected and the number of classroom hours is listed here.

10.7	9.8	11.6	9.7	7.6	11.3	14.1	8.1	11.5	8.5	6.9

(a) At $\alpha = 0.05$, test the university's claim. Assume that the population is normally distributed.

(b) Construct an appropriate confidence interval to validate your conclusion in part(a).

28. The Citrus Growers Association believes that the mean consumption of fresh citrus fruits by Californians is greater than 94 pounds per year. A random sample of 103 Californians showed a mean consumption of 93.5 pounds per year and a standard deviation of 30 pounds.

(a) At $\alpha = 0.01$, can you reject the association's claim?

(b) Construct an appropriate confidence interval to validate your conclusion in part(a).

29. An automotive battery manufacturer guarantees that the mean reserve capacity of a certain brand of battery is greater than 1.5 hours. To test this claim, you randomly select a sample of 50 batteries and find the mean reserve capacity to be 1.45 hours with a standard deviation of 0.32 hour.

(a) At $\alpha = 0.1$, do you have enough evidence to support the manufacturer's claim?

(b) Construct an appropriate confidence interval to validate your conclusion in part(a).

Test of Hypothesis for Population Proportion p

30. A survey showed that among 785 randomly selected subjects who completed four years of college, 18.3% smoke (based on data from the American Medical Association).

(a) Use a 0.01 significance level to test the claim that the percentage of smokers among those with four year college degrees is less than 20%. (The population has a normal Distribution).

(b) Construct an appropriate confidence interval to validate your conclusion in part(a).

31. A medical researcher claims that less than 20% of American adults are allergic to Penicillin. In a random sample of 100 adults, 18% say they have such an allergy.

(a) Test the researcher's claim at $\alpha = 0.05$. Assume that the original population is normally distributed.

(b) Construct an appropriate confidence interval to validate your conclusion in part(a).

32. The Neuroscience Institute is testing the effects of long-term marijuana use on the brain. From a poll of 2854 individuals examined, 43% were operating on a suboptimal intellectual level.

(a) At a 0.02 significance level, test the claim that among long-term marijuana users, 48% are affected. Assume that the original population is normally distributed.

(b) Construct an appropriate confidence interval to validate your conclusion in part (a).

33. An Organic Chemistry Professor claims that in his class less than 60% of students can accurately demonstrate the classic experiment of an eruption of ammonium dichromate $\left[(NH_4)_2 Cr_2O_7\right]$ volcano. Among 34 randomly selected undergraduate students, 21 accurately develop the reaction.

(a) Using a 0.01 significance level, test the professor's claim. Assume that the original population is normally distributed.

(b) Construct an appropriate confidence interval to validate your conclusion in part (a).

34. Statistics shows that 63% of Americans are in favor of outlawing cigarettes. You decide to test this claim and ask a random sample of 200 Americans whether they are in favor of outlawing cigarettes. Of the 200 Americans selected, 59% are in favor of outlawing cigarettes.

(a) At a 0.01 significance level, is there enough evidence to reject the statistics? Assume that the original population is normally distributed.

(b) Construct an appropriate confidence interval to validate your conclusion in part (a).

35. An independent research center claims that more than 55% of American adults regularly watch CNN. You decide to test this claim and ask a random sample of 530 Americans whether they regularly watch CNN. Of the 530 Americans, 255 respond yes.

(a) At $\alpha = 0.05$, is there enough evidence to support the claim? Assume that the original population is normally distributed.

(b) Construct an appropriate confidence interval to validate your conclusion in part (a).

36. A medical researcher claims that less than 40% of American adults eat breakfast every day. In a random sample of 250 adults, 46% say that they eat breakfast every day.

(a) At $\alpha = 0.03$, is there enough evidence to reject the researcher's claim? Assume that the original population is normally distributed.

(b) Construct an appropriate confidence interval to validate your conclusion in part (a).

37. An environmentalist claims that majority of British consumers want supermarkets to stop selling genetically modified goods. You want to test this claim, and discover that in a random sample of 100 British consumers, 53% say that they want supermarkets to stop selling genetically modified goods.

(a) At a 0.01 significance level, can you support the environmentalist's claim? Assume that the original population is normally distributed.

(b) Construct an appropriate confidence interval to validate your conclusion in part (a).

38. A maker of microwave ovens advertises that less than 10% of its microwaves need repair during the first two years of use. In a random sample of 57 microwaves, 10 needed repairs.

(a) At $\alpha = 0.05$, can you reject the claim that less than 10% of microwaves need repair during the first two years of use? Assume that the original population is normally distributed.

(b) Construct an appropriate confidence interval to validate your conclusion in part (a).

39. A communications industry spokesperson claims that over 80% of Americans either own a cellular phone or have a family member that does. In a random survey of 1036 Americans, 956 said that they or a family member owned a cellular phone.

(a) Test the spokesperson's claim at a 0.1 significance level. Assume that the original population is normally distributed.

(b) Construct an appropriate confidence interval to validate your conclusion in part (a).

Test of Hypothesis for Population Standard Deviation (σ) or Variance (σ^2)

40. Among 91 DUI arrests in a city on a Friday night, the variance in blood alcohol level is 0.021.

(a) At a 0.05 significance level, test the claim that the variance among all DUI arrests is more than 0.019. Assume that the original population is normally distributed.

(b) Construct an appropriate confidence interval to validate your conclusion in part (a).

41. A manufacturer uses a new production method to produce steel rods. A random sample of 17 steel rods resulted in diameters with a standard deviation of 0.2 cm.

(a) Assuming the population is normally distributed with a 0.10 significance level, test the claim that steel rods manufactured using the new production method have diameters with a standard deviation less than 0.15 cm.

(b) Construct an appropriate confidence interval to validate your conclusion in part (a).

42. At a local hospital, 29 randomly selected patients were tested for their glucose level. The variance of the sample was found to be 500 mg/dl.

(a) At a 0.01 significance level, test the claim that the variance of glucose level in the patients' blood is 529 mg/dl. Assume the population is normally distributed.

(b) Construct an appropriate confidence interval to validate your conclusion in part (a).

43. The IQ scores of college students has a normal distribution. The IQ scores of 14 randomly selected college students are listed below.

115	128	107	109	116	124	135
127	115	104	118	126	129	133

(a) Use a 0.05 significance level to test the claim that the standard deviation of IQ scores of college students is greater than 9. Assume the population is normally distributed.

(b) Construct an appropriate confidence interval to validate your conclusion in part (a).

44. The standard deviation of test scores in AP Statistics at a local high school is 16.4. A teacher claims that the standard deviation of test scores is greater than 16.4. A random sample of 22 test scores resulted in scores with a standard deviation of 15.2.

(a) Use a 0.01significance level to test the teacher's claim. Assume the population is normally distributed.

(b) Construct an appropriate confidence interval to validate your conclusion in part (a).

45. A dairy processing company claims that the variance of the amount of fat in the whole milk processed by the company is less than 0.32. You suspect this is wrong and discover that a random sample of 38 milk containers has a variance of 0.36.

(a) At $\alpha = 0.01$, is there enough evidence to reject the company's claim? Assume the population is normally distributed.

(b) Construct an appropriate confidence interval to validate your conclusion in part (a).

46. A restaurant owner claims that the standard deviation of serving times in his restaurant is 2.9 minutes. A random sample of 21 serving times had a standard deviation of 3.1 minutes.

(a) At $\alpha = 0.01$, is there sufficient evidence to support the owner's claim? Assume the population is normally distributed.

(b) Construct an appropriate confidence interval to validate your conclusion in part (a).

47. A sporting goods manufacturer claims that the variance of string tensions for any decent tennis racquet should be about 9 pounds. The string tensions of 18 randomly selected tennis rackets produced a variance of 8.13 pounds.

(a) At $\alpha = 0.05$, is there enough evidence to reject the manufacturer's claim? Assume the population is normally distributed.

(b) Construct an appropriate confidence interval to validate your conclusion in part (a).

48. A school administrator claims that the standard deviation of test scores for 7th -grade students who took a life-science assessment test is less than 28. You have been asked to test this claim. To do so, you randomly select 15 test scores and find that the scores have a standard deviation of 30.

(a) At $\alpha = 0.05$, and assuming that the population is normally distributed, is there enough evidence to support the administrator's claim?

(b) Construct an appropriate confidence interval to validate your conclusion in part (a).

49. The annual salaries in ($) of 16 randomly chosen actuaries are listed below.

45,018	34,952	85,517	45,553	41,900	76,384	48,862	57,615
61,104	96,710	97,875	68,245	39,945	53,582	65,252	72,522

(a) At a 0.01 significance level, can you conclude that the standard deviation of the annual salaries is greater than 20,000? Assume the parent population is normally distributed.

(b) Construct an appropriate confidence interval to validate your conclusion in part(a).

50. The variance in the coolant level in a car radiator is believed to stay at about 2.25 inches. A random sample of the coolant level of 20 radiators showed a variance of 2 inches.

(a) At a 0.05 significant level, is there sufficient evidence to support the belief? Assume the population has a normal distribution.

(b) Construct an appropriate confidence interval to validate your conclusion in part(a).

51. A bottling company claims that the standard deviation of the amount of sports drink in a 12-ounce bottle is less than 0.60. A random sample of 31 bottles has a standard deviation of 0.75.

(a) At 0.01significance level, is there enough evidence to reject the company's claim? Assume the population has a normal distribution.

(b) Construct an appropriate confidence interval to validate your conclusion in part(a).

52. An automotive manufacturer believes that the variation in gas mileage for its luxury sedans is about 25 units. You work for an energy conservation agency and want to test this claim. You find that a random sample of 23 luxury sedans has a standard deviation of 4.5 units.

(a) At $\alpha = 0.05$, do you have enough evidence to reject the manufacturer's claim? Assume the parent population is normally distributed.

(b) Construct an appropriate confidence interval to validate your conclusion in part(a).

53. Shown below are birth weights (in kilograms) of male babies born to mothers taking a special vitamin supplement (based on data from the New York State Department of Health).

3.73	4.37	3.73	4.33	3.39	3.68	4.68	3.52
3.02	4.09	2.47	4.13	4.47	3.22	3.43	2.54

(a) Test the claim that this sample comes from a population with a standard deviation greater than to 0.470 kg, which is the standard deviation for male birth weights in general. The parent population is normally distributed.

(b) Construct an appropriate confidence interval to validate your conclusion in part(a).

54. A state school administrator claims that the standard deviation of SAT verbal test scores is 105. A random sample of 16 SAT verbal test scores has a standard deviation of 113.

(a) At $\alpha = 0.01$, test the administrator's claim. Assume that the population is normally distributed.

(b) Construct an appropriate confidence interval to validate your conclusion in part(a).

8

INFERENCES FROM TWO POPULATIONS

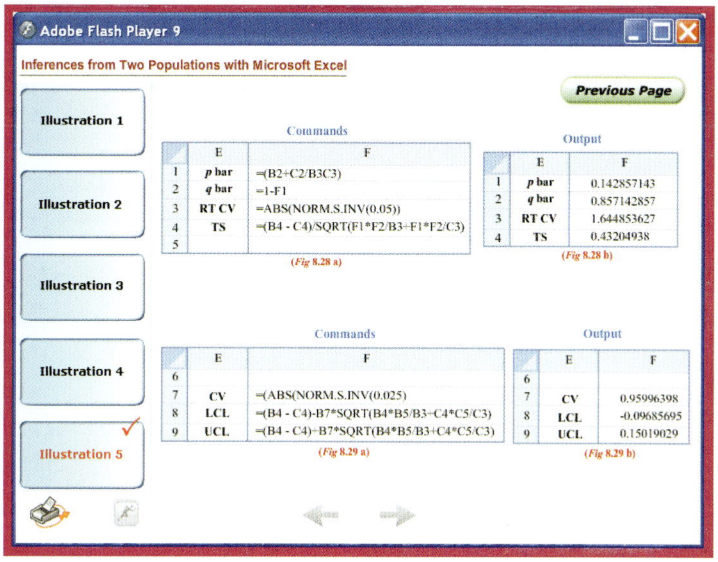

Chapter 8

INFERENCES FROM TWO POPULATIONS

Chapter Overview:

8.1: Introduction

In Chapter 6, Sampling Distributions and Estimation, sample data was used to calculate Point Estimates and to construct Confidence Interval Estimates of the values of the population parameters such as μ, σ, p, and σ^2. In Chapter 7, Hypothesis Testing, we developed methods for testing statistical hypotheses about the population parameters using sample data. In these two chapters all the examples exercises, and illustrations a **single population** and the inferences drawn were based on the analysis of the data obtained from a **single simple random sample**. However, in practice, we come across situations where we want to compare the corresponding parameters of two populations from which the samples are drawn or to compare the data obtained from two simple random samples from the given populations.

The following examples present situations where sample data from two populations is used to draw inferences about those populations.

i) In a survey on smoking, we may want to determine if there is a significant difference between the proportion of smokers with or without college degrees.

ii) When investigating the average salaries of airline pilots, we may want to determine whether there is a significant difference between the average salaries of pilots of two different airlines.

iii) A farmer may be interested in knowing which of the two fertilizers gives a better yield.

iv) We may want to construct a confidence interval for the mean difference between the SAT scores of a group of students before and after taking a prep course.

Earlier in Chapter 7, we compared the population parameters (mean, proportion, standard deviation, variance) with some specified value stated as the original claim in the beginning of the testing process. Furthermore, all inferences in chapter 7 were based on a single samples taken from a single population.

In this chapter we will compare two corresponding parameters of two populations. That is, inferences will be two sample based. In subsequent sections, we will discuss the following in detail.

1. Comparing means of two independent normally distributed populations, when population variances are known (Z-test).
2. Comparing means of two independent normally distributed populations, when population variances are unknown (t-test).
3. Comparing the mean difference for dependent samples (paired t-test).
4. Comparing proportions of a particular attribute in two independent populations (Z-test).
5. Comparing standard deviations or variances of two independent normally distributed populations (F-test).

Note: When deciding to make an inference about population parameters from two samples, it is important to decide whether or not the two samples are **matched (dependent)** or **unmatched (independent)**. In before-after studies, similar to that described in the last example above, there is a natural way to match the data from the two samples. In this case, we would conduct hypothesis testing suitable for matched-pairs. In the other examples above, there is no meaningful way to match the data points from one sample to the other; therefore we would conduct the hypothesis test assuming independent samples (unmatched samples).

8.2: Inferences about Two Means Using the Z-Test (σ_1 and σ_2 Known, Independent Samples)

Objectives

After completing this section you will be able to understand:

A. Sampling Distribution of $\bar{X}_1 - \bar{X}_2$ (σ_1 and σ_2 Known, Independent Samples).
B. Null and Alternative Hypotheses.
C. Test Statistics.
D. Test of Hypothesis.
E. Confidence Interval for the difference between two means.
F. Interpretation of Confidence Interval for $\mu_1 - \mu_2$.

Suppose we want to determine whether there is a significant difference in the average salaries of Airline X pilots and Airline Y pilots. In this study we are not interested in the average salary of all pilots but rather in **comparing** the means of the two groups. Here, the hypothesis can be stated as: "The mean salary of Airline X pilots differs from the mean salary of Airline Y pilots."

Let μ_1 and μ_2 represent the mean salaries for Airline X pilots and Airline Y pilots respectively. Then our hypotheses can be written as follows:

$$H_0 : \mu_1 = \mu_2$$
$$H_1 : \mu_1 \neq \mu_2$$

Here, if we end up rejecting the Null hypothesis, then our conclusion will be interpreted as: "There is a significant difference between the mean salaries of pilots from Airline X and Airline Y." If we fail to reject H_0, then the difference is not significant, which means that the mean salaries of the two groups of pilots are basically the same.

Basic Theory:

Suppose there are two independent populations with means μ_1 and μ_2 and standard deviations σ_1 and σ_2 respectively. A simple random sample of size n_1 from the first population yields an average of $\bar{x}_1$ and a standard deviation of s_1. Another independent simple random sample of size n_2 from the second population yields an average of $\bar{x}_2$ and a standard deviation of s_2. The distribution of the differences between the two sample means in repeated sampling is called the *sampling distribution of* $\bar{X}_1 - \bar{X}_2$.

A. Sampling Distribution of $\bar{X}_1 - \bar{X}_2$ (σ_1 and σ_2 Known, Independent Samples)

If two independent random samples of size n_1 and n_2 are drawn from two populations with means μ_1 and μ_2 and standard deviations σ_1 and σ_2 respectively, then the Central Limit Theorem tells us that the sampling distribution of $\bar{X}_1 - \bar{X}_2$ is approximately normal with

$$\text{Mean} = \mu_{\bar{X}_1-\bar{X}_2} = \mu_1 - \mu_2 \text{ and Standard Deviation } = \sigma_{\bar{X}_1-\bar{X}_2} = \sqrt{\frac{\sigma_1^2}{n_1}+\frac{\sigma_2^2}{n_2}}.$$

Note: For a more detailed discussion of the theory presented here, refer to Section 8.8.

Assumptions for the Z-Test:

1. The samples are **simple random samples**.
2. The samples are **independent** of each other.
3. σ_1^2 and σ_2^2 are known.
4. If the samples are large (n_1 and $n_2 > 30$), there is no normality requirement on the parent populations. If the sample sizes are less than 30, then the populations must be normally distributed.

B. Null and Alternative Hypotheses

Since our claims are always about comparing two population parameters (comparing means in this section), we have three possible alternative hypotheses. Similar to Section 7.2, the null hypothesis is always a statement involving equality. Therefore, when comparing two population means, our null hypothesis always represents the claim that the two population parameters are equal. The alternative hypothesis always contains a strict inequality (<, >, ≠) as shown in Table 8.1.

Left-Tail Test	Right-Tail Test	Two-Tail Test
$H_0 : \mu_1 = \mu_2$ $H_1 : \mu_1 < \mu_2$	$H_0 : \mu_1 = \mu_2$ $H_1 : \mu_1 > \mu_2$	$H_0 : \mu_1 = \mu_2$ $H_1 : \mu_1 \neq \mu_2$

Table 8.1

C. Test Statistics

In order to test a claim about the difference of means ($\mu_1 - \mu_2$), the test statistics used based on the sampling distribution of $\left(\bar{X}_1 - \bar{X}_2\right)$ is

$$Z = \frac{\left(\bar{X}_1-\bar{X}_2\right)-\left(\mu_1-\mu_2\right)}{\sqrt{\frac{\sigma_1^2}{n_1}+\frac{\sigma_2^2}{n_2}}} \qquad (8.1)$$

Note: Given $\left(\bar{X}_1 - \bar{X}_2\right)$ is the observed difference, the expected difference ($\mu_1 - \mu_2$) is zero based on the null hypothesis $H_0 : \mu_1 = \mu_2$ and $\sqrt{\frac{\sigma_1^2}{n_1}+\frac{\sigma_2^2}{n_2}}$ is the standard error of the difference.

Critical Values:

Since the sampling distribution of $\bar{X}_1 - \bar{X}_2$ is approximately normal, the same critical values used in Section 7.2 are used here. They can be found in the z-table.

D. Test of Hypothesis

By following the 4-step hypothesis testing procedure outlined in Chapter 7, we can test the claim about ($\mu_1 - \mu_2$) by any one of the two methods:

- **Classical Method:**

***Step* 1** Set up the **null** and **alternative hypotheses**.
***Step* 2** Calculate the appropriate **Test Statistics**.
***Step* 3** Construct the **Rejection Region** and find the corresponding **Critical Value(s)**.
***Step* 4** State the Decision, and write the conclusion in simple words.

- ***P*-value Method:**

***Step* 1** Set up the **null** and **alternative hypotheses**.
***Step* 2** Calculate the appropriate **Test Statistics**.
***Step* 3** Find the p-value of the computed test statistics. [For details, see Section 7.2H]
***Step* 4** **Decision:** If p-value < α, reject H_0 at α significance level. If p-value > α, fail to reject H_0 at α significance level.

Round off rule: Use one more decimal place than what is present in the sample statistics provided.

E. Confidence Interval for the Difference between Two Means

Another tool that can be used in making inferences about two means is **Interval Estimation** known as **Confidence Intervals**.

Given the null hypothesis $H_0 : \mu_1 = \mu_2$, if the corresponding confidence interval contains zero, the null hypothesis is not rejected, otherwise it is rejected.

Based on the assumption that $\bar{X}_1 - \bar{X}_2 \sim N\left(\mu_1 - \mu_2, \sqrt{\frac{\sigma_1^2}{n_1} + \frac{\sigma_2^2}{n_2}}\right)$ and given that $\bar{x}_1$ and $\bar{x}_2$ are the means of two independent random samples of size n_1 and n_2 from populations with standard deviations σ_1 and σ_2 respectively, then the $100(1 - \alpha)\%$ confidence interval for $\mu_1 - \mu_2$ is given by:

$$\left[\left(\bar{X}_1 - \bar{X}_2\right) - E, \left(\bar{X}_1 - \bar{X}_2\right) + E\right] \quad \text{where } E = z_{\alpha/2}\sigma_{\bar{X}_1 - \bar{X}_2} = z_{\alpha/2}\sqrt{\frac{\sigma_1^2}{n_1} + \frac{\sigma_2^2}{n_2}}$$

$$\left(\bar{x}_1 - \bar{x}_2\right) - z_{\alpha/2}\sqrt{\frac{\sigma_1^2}{n_1} + \frac{\sigma_2^2}{n_2}} < \mu_1 - \mu_2 < \left(\bar{x}_1 - \bar{x}_2\right) + z_{\alpha/2}\sqrt{\frac{\sigma_1^2}{n_1} + \frac{\sigma_2^2}{n_2}} \qquad (8.2)$$

Recall that $z_{\alpha/2}$ is the critical value of the standard normal variable Z for the two-tailed test at "α" significance level.

F. Interpretation of Confidence Interval for $\mu_1 - \mu_2$

When constructing a $100(1 - \alpha)\%$ confidence interval for ($\mu_1 - \mu_2$), we may have any one of the following three situations:

1. The confidence interval contains only positive values (both the lower and upper confidence limits are positive). That is,

$$a < \mu_1 - \mu_2 < b \qquad ; \qquad a > 0,\ b > 0$$

In this case, we conclude with $100(1 - \alpha)\%$ confidence that $\mu_1 - \mu_2 > 0 \Rightarrow \mu_1 > \mu_2$.

2. The confidence interval contains only negative values (both the lower and upper confidence limits are negative). That is,

$$-a < \mu_1 - \mu_2 < -b \quad ; \quad a > 0,\ b > 0$$

In this case, we conclude with $100(1 - \alpha)\%$ confidence that $\mu_1 - \mu_2 < 0 \Rightarrow \mu_1 < \mu_2$.

3. The confidence interval contains both positive and negative values (the lower confidence limit is negative and the upper confidence limit is positive). That is,

$$-a < \mu_1 - \mu_2 < b \quad ; \quad a > 0,\ b > 0$$

In this case we cannot conclude with $100(1 - \alpha)\%$ confidence that $\mu_1 > \mu_2$ or $\mu_1 < \mu_2$. This suggests that it is very likely that $\mu_1 = \mu_2$.

Example 1: The mean height of 50 male students who showed interest in participating in college athletics was 68.2 inches with the population standard deviation of 2.5 inches, while 50 male students who showed no interest in such participation had a mean height of 67.5 inches with a population standard deviation of 2.8 inches.

(a) Test the claim that male students who are interested in participation in college athletics are on the average taller than the other students.

Use **i)** the classical and **ii)** the *p*-value methods.

(b) Construct a 90% confidence interval for $\mu_1 - \mu_2$. Interpret your results.

Solution: **(a) i)** **Classical Method:**

Step 0 The original claim: $\mu_1 > \mu_2 \Rightarrow$ The counter claim: $\mu_1 \le \mu_2$

Interested male students	$\bar{x}_1 = 68.2$	$\sigma_1 = 2.5$	$n_1 = 50$
Not interested male students	$\bar{x}_2 = 67.5$	$\sigma_2 = 2.8$	$n_2 = 50$

Table 8.2

Step 1 Set up the null and alternative hypotheses.

$H_0 : \mu_1 = \mu_2$

$H_1 : \mu_1 > \mu_2$

Step 2 Under $H_0 : \mu_1 = \mu_2 \Rightarrow \mu_1 - \mu_2 = 0$ and sample statistics provided, the test statistics is

$$z = \frac{(\bar{x}_1 - \bar{x}_2) - (\mu_1 - \mu_2)}{\sqrt{\dfrac{\sigma_1^2}{n_1} + \dfrac{\sigma_2^2}{n_2}}} = \frac{(68.2 - 67.5) - 0}{\sqrt{\dfrac{2.5^2}{50} + \dfrac{2.8^2}{50}}} = 1.32$$

Step 3 Since $H_1 : \mu_1 > \mu_2$, the test is right-tailed. Therefore, the rejection region will be in the right-tail of the normal probability curve. The critical value of z at 0.05 significance level, for the right-tail test is 1.645. The rejection region, the critical value, and the calculated value of test statistics are shown in Figure 8.1.

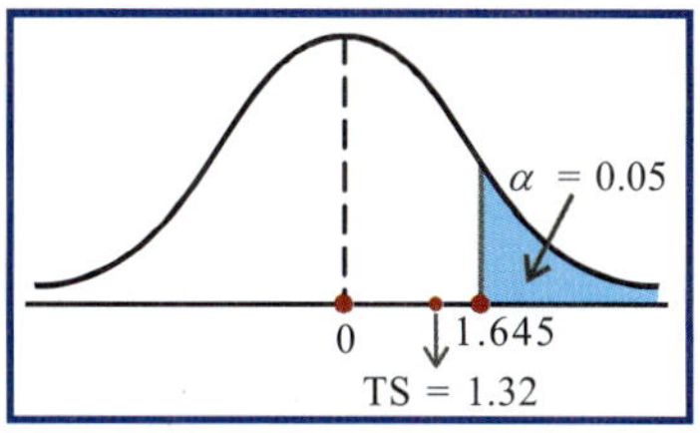

Figure 8.1

Step 4 **Decision:** Since the computed value of the test statistics lies outside the rejection region, we fail to reject H_0 at a 0.05 significance level. *Conclusion:* The data does not provide sufficient evidence against the null hypothesis. Therefore, there does not seem to be a significant difference between the means. The average height of the male students who participate in college athletics does not differ significantly from the average height of the other students.

ii) *P*-value Method:

***Step* 1** Set up the null and alternative hypotheses.

$H_0 : \mu_1 = \mu_2$

$H_1 : \mu_1 > \mu_2$

***Step* 2** Under $H_0 : \mu_1 = \mu_2 \Rightarrow \mu_1 - \mu_2 = 0$ and sample statistics provided, the test statistics is

$$z = \frac{(\bar{x}_1 - \bar{x}_2) - (\mu_1 - \mu_2)}{\sqrt{\frac{\sigma_1^2}{n_1} + \frac{\sigma_2^2}{n_2}}} = \frac{(68.2 - 67.5) - 0}{\sqrt{\frac{2.5^2}{50} + \frac{2.8^2}{50}}} = 1.32$$

***Step* 3** Since $H_1 : \mu_1 > \mu_2$ is right-tailed, the region showing the extreme values is the area to the right of the test statistics. The p-value of the test statistics is given by the shaded area in Figure 8.2.

$$\begin{aligned} p\text{-value} &= P(Z > 1.32) \\ &= 0.5 - P(0 < Z < 1.32) \\ &= 0.5 - 0.4066 \\ &= 0.0934 \end{aligned}$$

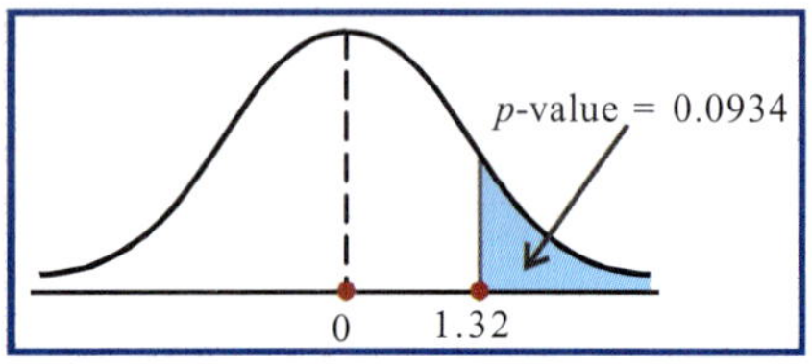

Figure 8.2

***Step* 4** **Decision:** Since the p-value $> \alpha = 0.05$, we fail to reject H_0 at 0.05 significance level.

Hence, there does not seem to be a significant difference between the means.

In short:

Hypotheses	**Test Statistics**
$H_0 : \mu_1 = \mu_2$ $H_1 : \mu_1 > \mu_2$	$z = \frac{(\bar{x}_1 - \bar{x}_2)}{\sqrt{\frac{\sigma_1^2}{n_1} + \frac{\sigma_2^2}{n_2}}} = 1.32$
Rejection Region and Critical Value	**Conclusion**
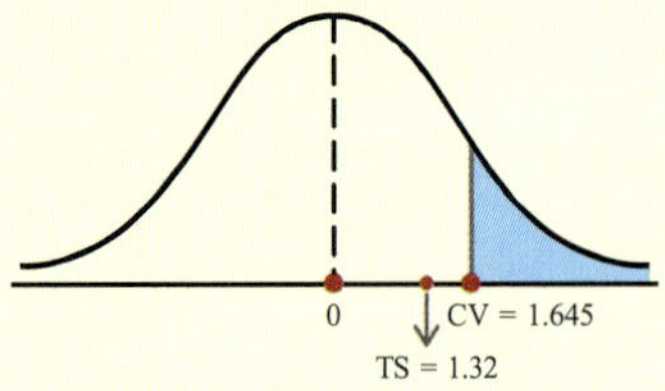	Fail to reject H_0. The average height of the male students who participate in college athletics does not differ significantly from the average height of those who do not.

(b) By using information given in Table 8.2, we have:

$$\sigma_{\bar{X}_1 - \bar{X}_2} = \sqrt{\frac{\sigma_1^2}{n_1} + \frac{\sigma_2^2}{n_2}} = \sqrt{\frac{2.5^2}{50} + \frac{2.8^2}{50}} = 0.53,\ z_{\alpha/2} = \text{Two-tailed CV of } z \text{ at } \alpha = 0.1 = 1.645$$

$$\Rightarrow E = z_{\alpha/2} \cdot \sigma_{\bar{X}_1 - \bar{X}_2} = (1.645)(0.53) = 0.87$$

The 90% confidence interval for the difference of means $(\mu_1 - \mu_2)$ is given by:

$$(\bar{x}_1 - \bar{x}_2) - E < \mu_1 - \mu_2 < (\bar{x}_1 - \bar{x}_2) + E$$

$$(68.2 - 67.5) - 0.87 < \mu_1 - \mu_2 < (68.2 - 67.5) + 0.87$$
$$-0.17 < \mu_1 - \mu_2 < 1.57$$

We can say with 90% confidence that the difference $\mu_1 - \mu_2$ between the population means lies between the limits -0.17 and 1.57. The confidence interval contains both positive and negative values (the lower confidence limit negative and the upper confidence limit positive). This suggests that it is very likely that $\mu_1 = \mu_2$ (the same conclusion obtained in the test of hypothesis).

Example 2: The following table represents 20 randomly selected final scores from two groups of Statistics classes, one using a web-based instructional delivery model and the other using a traditional instructional model.

Web-Based Model				Traditional Model			
83	50	95	76	49	84	55	49
75	84	79	98	77	85	98	27
71	64	89	100	64	63	36	52

From past studies, it is known that $\sigma_1 = 14$ and $\sigma_2 = 21$. Assume the populations are normally distributed.

(a) Is there a significant difference between the means of two populations? Use both the classical and p-value method. Use a 0.01 significance level.

(b) Construct a 99% confidence interval for the difference between the means for the given data.

Solution: **(a)** **Classical Method:**

***Step* 0** The original claim: $\mu_1 \neq \mu_2 \Rightarrow$ The counter claim: $\mu_1 = \mu_2$

Web based model	$\bar{x}_1 = \frac{\sum x}{12} = 80.3$	$\sigma_1 = 14$	$n_1 = 12$
Traditional model	$\bar{x}_2 = \frac{\sum x}{12} = 61.6$	$\sigma_2 = 21$	$n_2 = 12$

Table 8.3

***Step* 1** Set up the null and alternative hypotheses.

$H_0 : \mu_1 = \mu_2$

$H_1 : \mu_1 \neq \mu_2$

***Step* 2** Under $H_0 : \mu_1 = \mu_2 \Rightarrow \mu_1 - \mu_2 = 0$ and sample statistics provided above, the test statistics is

$$z = \frac{(\bar{x}_1 - \bar{x}_2) - (\mu_1 - \mu_2)}{\sqrt{\frac{\sigma_1^2}{n_1} + \frac{\sigma_2^2}{n_2}}} = \frac{(80.3 - 61.6) - 0}{\sqrt{\frac{14^2}{12} + \frac{21^2}{12}}} = 2.567$$

***Step* 3** Since $H_1 : \mu_1 \neq \mu_2$, the test is two-tailed. Therefore, the rejection region will be in the two tails of the normal probability curve. The critical values of z at 0.01 significance level for the two-tailed test are ± 2.575. The rejection region, the critical values, and the calculated value of test statistics are shown in Figure 8.3.

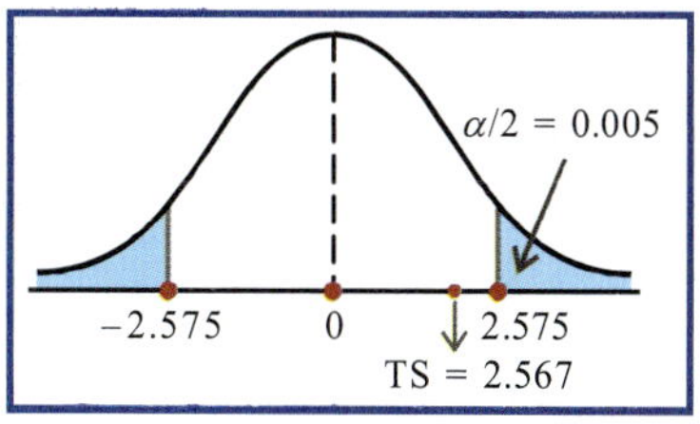

Figure 8.3

***Step* 4** **Decision:** Since the computed value of the test statistics lies outside the rejection region, we fail to reject H_0 at $\alpha = 0.01$. *Conclusion:* The data provides sufficient evidence in favor of the null hypothesis. Therefore, there does not seem to be a significant difference between the means.

P-value Method:

***Step* 1** Set up the null and alternative hypotheses.

$H_0 : \mu_1 = \mu_2$

$H_1 : \mu_1 \neq \mu_2$

***Step* 2** Under $H_0 : \mu_1 = \mu_2 \Rightarrow \mu_1 - \mu_2 = 0$ and sample statistics provided above, the test statistics is

$$z = \frac{(\bar{x}_1 - \bar{x}_2) - (\mu_1 - \mu_2)}{\sqrt{\frac{\sigma_1^2}{n_1} + \frac{\sigma_2^2}{n_2}}} = \frac{(80.3 - 61.6) - 0}{\sqrt{\frac{14^2}{12} + \frac{21^2}{12}}} = 2.567$$

***Step* 3** Since $H_1 : \mu_1 \neq \mu_2$ is two-tailed, the region showing the extreme values is twice the area to the right of the test statistics. The p-value of the test statistics is given by the shaded area in Figure 8.4.

$$\begin{aligned} p\text{-value} &= 2 \times P(Z > 2.567) \\ &= 2 \times [1 - P(Z < 2.567)] \\ &= 2 \times [1 - 0.9949] \\ &= 0.0102 \end{aligned}$$

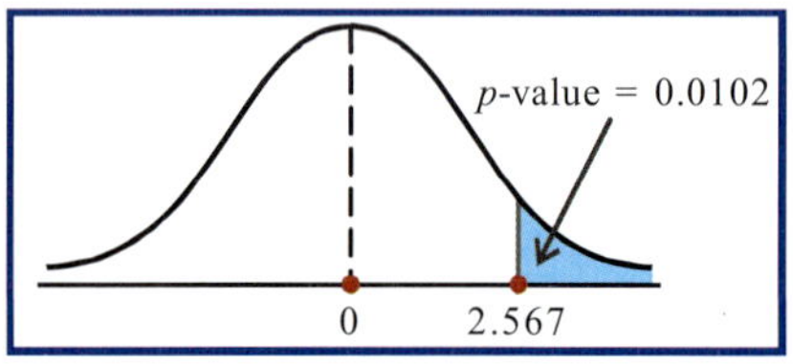

Figure 8.4

***Step* 4** **Decision:** Since the p-value $> \alpha = 0.01$, we fail to reject H_0 at 0.01 significance level.

Hence, there does not seem to be a significant difference between the means.

In short:

Hypotheses	**Test Statistics**
$H_0 : \mu_1 = \mu_2$ $H_1 : \mu_1 \neq \mu_2$	$z = \dfrac{(\bar{x}_1 - \bar{x}_2)}{\sqrt{\frac{\sigma_1^2}{n_1} + \frac{\sigma_2^2}{n_2}}} = 2.567$
Rejection Region and Critical Values	**Conclusion**
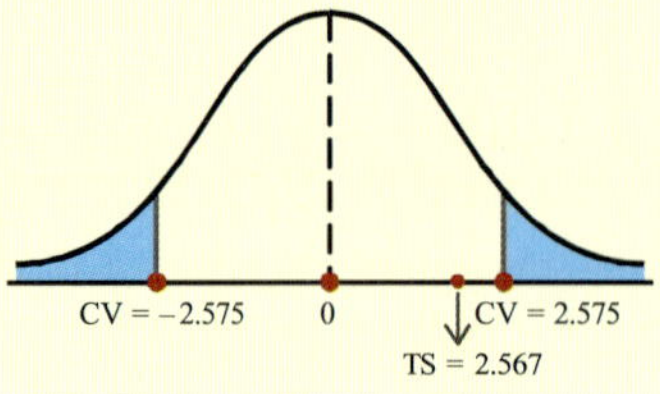	Fail to reject H_0. There does not seem to be a significant difference between the means.

(b) By using the information given in Table 8.3, we have:

$$\sigma_{\bar{X}_1 - \bar{X}_2} = \sqrt{\frac{\sigma_1^2}{n_1} + \frac{\sigma_2^2}{n_2}} = \sqrt{\frac{14^2}{12} + \frac{21^2}{12}} = 7.3,\ z_{\alpha/2} = \text{two-tailed CV of } z \text{ at } \alpha = 0.01 \text{ is } 2.575$$

$$\Rightarrow E = z_{\alpha/2} \cdot \sigma_{\bar{X}_1 - \bar{X}_2} = (2.575)(7.3) = 18.8$$

The 90% confidence interval for the difference of means $(\mu_1 - \mu_2)$ is given by:

$$(\bar{x}_1 - \bar{x}_2) - E < \mu_1 - \mu_2 < (\bar{x}_1 - \bar{x}_2) + E$$

$$(80.3 - 61.6) - 18.8 < \mu_1 - \mu_2 < (80.3 - 61.6) + 18.8$$

$$-0.1 < \mu_1 - \mu_2 < 37.5$$

We can say with 99% confidence that the difference $\mu_1 - \mu_2$ between the population means lies between the limits -0.1 and 37.5. The confidence interval contains both positive and negative values. This suggests that it is very likely that $\mu_1 = \mu_2$, the same conclusion obtained in the test of hypothesis.

Example 3: A study of hospitalization periods of drunk drivers vs. sober drivers as a result of traffic accidents revealed the following. For a group of 150 drunk drivers involved in an accident, the number of days spent in the hospital had a mean of 3.52 days and a population standard deviation of 2.06 days; and for a group of 200 sober drivers who were involved in a car accident, the numbers of days in the hospital had a mean of 1.03 days and a population standard deviation of 0.85 days.

(a) Using $\alpha = 0.01$, is there sufficient evidence to support the claim that the mean length of hospitalization period for drunk drivers is higher than that of the population of sober drivers? Use the classical method.

(b) Construct a 98% confidence interval for $\mu_1 - \mu_2$. Interpret your results.

Solution: **(a)** **Classical Method:**

***Step* 0** The original claim: $\mu_1 > \mu_2 \Rightarrow$ The counter claim: $\mu_1 \leq \mu_2$

Drunk drivers statistics	$\bar{x}_1 = 3.52$	$\sigma_1 = 2.06$	$n_1 = 150$
Sober drivers statistics	$\bar{x}_2 = 1.03$	$\sigma_2 = 0.85$	$n_2 = 200$

Table 8.4

***Step* 1** Set up the null and alternative hypotheses.

$H_0 : \mu_1 = \mu_2$

$H_1 : \mu_1 > \mu_2$

***Step* 2** Under $H_0 : \mu_1 = \mu_2 \Rightarrow \mu_1 - \mu_2 = 0$ and sample statistics provided above, the test statistics is

$$z = \frac{(\bar{x}_1 - \bar{x}_2) - (\mu_1 - \mu_2)}{\sqrt{\dfrac{\sigma_1^2}{n_1} + \dfrac{\sigma_2^2}{n_2}}} = \frac{(3.52 - 1.03) - 0}{\sqrt{\dfrac{2.06^2}{150} + \dfrac{0.85^2}{200}}} = 13.94$$

***Step* 3** Since $H_1 : \mu_1 > \mu_2$, the test is right-tailed. Therefore, the rejection region will be in the right-tail of the normal probability curve. The critical value of z at a 0.01 significance level for the right-tail test is 2.33. The rejection region, the critical value, and the calculated value of test statistics are shown in Figure 8.5.

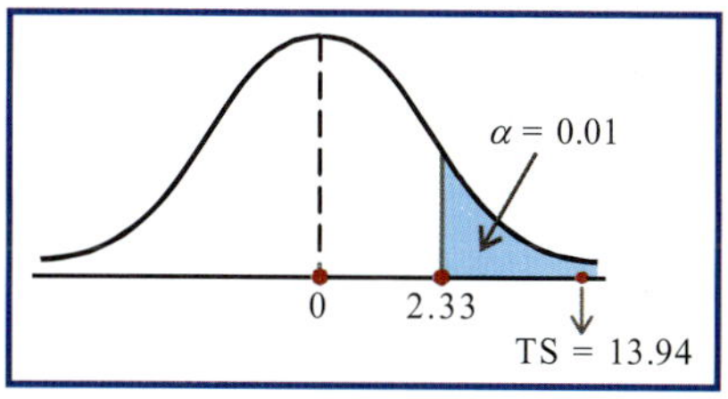

Figure 8.5

Step 4 **Decision:** Since the computed value of the test statistics lies inside the rejection region, we reject H_0 at $\alpha = 0.01$. *Conclusion:* The data provides sufficient evidence against the null hypothesis. Therefore, there seems to be a significant difference between the means.

In short:

Hypotheses	**Test Statistics**
$H_0: \mu_1 = \mu_2$ $H_1: \mu_1 > \mu_2$	$z = \dfrac{(\bar{x}_1 - \bar{x}_2)}{\sqrt{\dfrac{\sigma_1^2}{n_1} + \dfrac{\sigma_2^2}{n_2}}} = 13.94$
Rejection Region and Critical Value	**Conclusion**
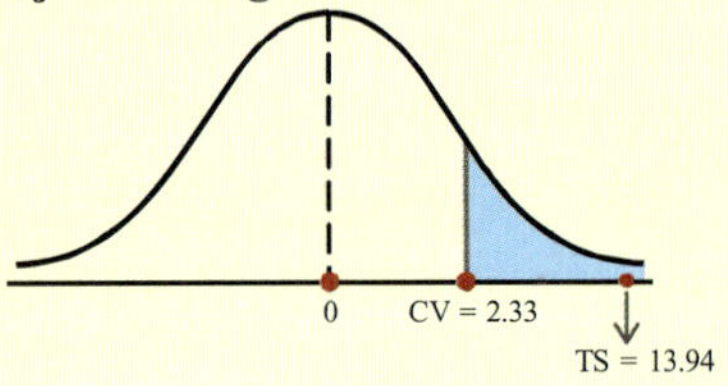	Reject H_0. The mean length of hospitalization period for drunk drivers is higher than that of sober drivers.

(b) By using information given in Table 8.4, we have:

$$\sigma_{\bar{X}_1 - \bar{X}_2} = \sqrt{\frac{\sigma_1^2}{n_1} + \frac{\sigma_2^2}{n_2}} = \sqrt{\frac{2.06^2}{150} + \frac{0.85^2}{200}} = 0.179, \; z_{\alpha/2} = \text{Two-tailed CV of } z \text{ at } \alpha = 0.02 \text{ is } 2.33$$

$$\Rightarrow E = z_{\alpha/2} \cdot \sigma_{\bar{X}_1 - \bar{X}_2} = (2.33)(0.179) = 0.416$$

The 98% confidence interval for the difference of means $(\mu_1 - \mu_2)$ is given by:

$$(\bar{x}_1 - \bar{x}_2) - E < \mu_1 - \mu_2 < (\bar{x}_1 - \bar{x}_2) + E$$

$$(3.52 - 1.03) - 0.416 < \mu_1 - \mu_2 < (3.52 - 1.03) + 0.416$$

$$2.074 < \mu_1 - \mu_2 < 2.906$$

We can say with 98% confidence that the difference $\mu_1 - \mu_2$ between the population means lies between the limits 2.074 and 2.906. Since the confidence interval contains only positive values, we conclude with 98% confidence that $\mu_1 - \mu_2 > 0 \Rightarrow \mu_1 > \mu_2$ which agrees with the result obtained in part (a).

Try it yourself:

1. Find a 95% confidence interval for $\mu_1 - \mu_2$.

Data Set I	**Data Set II**
$\bar{x}_1 = 15.8$	$\bar{x}_2 = 13.5$
$\sigma_1 = 2.5$	$\sigma_2 = 1.9$
$n_1 = 85$	$n_2 = 43$

2. For data given in exercise 1, test the claim that $\mu_1 = \mu_2$. Assume $\alpha = 0.01$. Use both the ***classical*** and the ***p***-value methods.

3. Find a 99% confidence interval for $\mu_1 - \mu_2$.

Data Set I	Data Set II
$\bar{x}_1 = 125.8$	$\bar{x}_2 = 119.3$
$\sigma_1 = 15.8$	$\sigma_2 = 12$
$n_1 = 15$	$n_2 = 10$

4. For data given in exercise 3, test the claim that $\mu_1 = \mu_2$. Assume $\alpha = 0.05$. Use both the ***classical*** and the ***p***-value methods.

5. Given the sample data in the table below:

Group 1	Group 2
$n_1 = 72$	$n_2 = 110$
$\bar{x}_1 = 7.5$	$\bar{x}_2 = 8.7$
$\sigma_1 = 6.2$	$\sigma_2 = 5.4$

Test the claim that $\mu_1 = \mu_2$. Use a 0.05 significance level.

8.3: Inferences about Two Means Using the *t*-Test (σ_1 and σ_2 Unknown, Independent Samples)

Objectives

After completing this section you will be able to understand:

A. *t*-test for difference of Means. (σ_1 and σ_2 Unknown, Independent Samples).

B. Test of Hypothesis for $\mu_1 - \mu_2$ and $100(1-\alpha)\%$ confidence interval for $\mu_1 - \mu_2$. (σ_1 and σ_2 Unknown, $\sigma_1 = \sigma_2$).

C. Test of Hypothesis for $\mu_1 - \mu_2$ and $100(1-\alpha)\%$ confidence interval for $\mu_1 - \mu_2$. (σ_1 and σ_2 Unknown, $\sigma_1 \neq \sigma_2$).

In Section 8.2, we used the *z*-test to test a claim about the difference of two means and to construct a confidence interval for $\mu_1 - \mu_2$. We assumed that the populations standard deviations were known and the parent populations were normally distributed i.e., the samples both had sizes greater than 30. However, in most practical situations, these requirements are not met. Specifically, the values of the populations standard deviations are almost never known in research. In the case where the populations standard deviations are not known, we must use the *t*-test to test the difference between two means and/or to construct a confidence interval for $\mu_1 - \mu_2$. Note that we will still assume that the two samples are **independent** and come from **normally distributed** populations.

A. *t*-test for Difference of Means (σ_1 and σ_2 Unknown, Independent Samples)

- **Assumptions for the *t*-test**

1. The two samples are simple random samples and independent of each other.
2. The two population standard deviations are not known.
3. Both sample sizes are large ($n_1 > 30$, $n_2 > 30$) or both samples come from normally distributed populations.

Remarks:

i) If we are testing inferences about two means using the t-test with independent samples, then the natural assumption about unknown standard deviations is that they are not equal. However, in this textbook we also treat the rare case where we assume that the population standard deviations, although unknown, are equal.

ii) For the validity of the t-test, we should first test the equality of the population variances by applying the so called F-test (Section 8.6.) If the hypothesis $H_0 : \sigma_1 = \sigma_2$ is rejected, then we cannot apply the t-test as such. The discussion of this issue is beyond the scope of this text. However, in applications, the assumptions about the relation between variances or standard deviation are taken to be true.

B. Test of Hypothesis for $\mu_1 - \mu_2$ (σ_1 and σ_2 Unknown, $\sigma_1 = \sigma_2$)

- **Test Statistics**

Under the assumption that $\sigma_1 = \sigma_2 = \sigma$, that is, the population standard deviations are assumed equal but unknown, the test statistics for testing the difference of means is

$$t = \frac{(\bar{X}_1 - \bar{X}_2) - (\mu_1 - \mu_2)}{\sigma_{\bar{X}_1 - \bar{X}_2}} = \frac{(\bar{X}_1 - \bar{X}_2) - (\mu_1 - \mu_2)}{\sqrt{\dfrac{s_p^2}{n_1} + \dfrac{s_p^2}{n_2}}}, \tag{8.3}$$

where $s_p^2 = \dfrac{(n_1 - 1)s_1^2 + (n_2 - 1)s_2^2}{n_1 + n_2 - 2}$ provides an unbiased estimate of the common population variance σ^2 based on the two samples taken together. The suffix p in s_p^2 stands for the **pooled estimate**. The t-statistics defined in (8.3) follows the t-distribution with $(n_1 + n_2 - 2)$ degrees of freedom.

To test the null hypothesis $H_0 : \mu_1 = \mu_2$ or $\mu_1 - \mu_2 = 0$, the test statistics is

$$t = \frac{(\bar{X}_1 - \bar{X}_2)}{\sqrt{\dfrac{s_p^2}{n_1} + \dfrac{s_p^2}{n_2}}} \sim t_{n_1 + n_2 - 2},$$

where $\bar{X}_1 = \dfrac{\sum X_1}{n_1}$, $\bar{X}_2 = \dfrac{\sum X_2}{n_2}$, and $s_p^2 = \dfrac{(n_1 - 1)s_1^2 + (n_2 - 1)s_2^2}{n_1 + n_2 - 2}$.

The rest of the procedure is the same as that of the test of hypothesis for the difference of two means when population standard deviations are known.

- **Confidence Interval for $\mu_1 - \mu_2$**

The $100(1 - \alpha)\%$ confidence interval for $\mu_1 - \mu_2$ in this case is

$$(\bar{x}_1 - \bar{x}_2) - E < \mu_1 - \mu_2 < (\bar{x}_1 - \bar{x}_2) + E, \tag{8.4}$$

where $E = t_v(\alpha/2)\sqrt{\dfrac{s_p^2}{n_1} + \dfrac{s_p^2}{n_2}}$ and $v = n_1 + n_2 - 2$.

Example 4: A researcher gives feed A to a randomly selected sample of 18 pigs and gives feed B to a different randomly selected sample of 15 pigs. The sample of 18 showed an average weight gain of 12 pounds in one week with a standard deviation of 4 pounds. The other sample gave an average weight gain of 10 pounds with a standard deviation of 3 pounds. Assume the population standard deviations are equal $\sigma_1 = \sigma_2 = \sigma$.

(a) Is there sufficient evidence to support the claim that feed A produces a higher average weight gain than feed B? Use the Classical Method and assume the parent populations are normally distributed.

(b) Construct an appropriate confidence interval for the difference of the means. Interpret your results.

Solution: **(a)**

***Step* 0** The original claim: $\mu_A > \mu_B \Rightarrow$ The counter claim: $\mu_A \le \mu_B$

Given sample statistics;

Feed A	$\bar{x}_1 = 12$	$s_1 = 4$	$n_1 = 18$
Feed B	$\bar{x}_2 = 10$	$s_2 = 3$	$n_2 = 15$

Table 8.5

***Step* 1** Set up the null and alternative hypotheses.

$H_0 : \mu_A = \mu_B$

$H_1 : \mu_A > \mu_B$

***Step* 2** Under $H_0 : \mu_A = \mu_B \Rightarrow \mu_A - \mu_B = 0$ and sample statistics provided above, the test statistics is

$$t = \frac{(\bar{x}_1 - \bar{x}_2) - (\mu_A - \mu_B)}{\sqrt{\frac{s_p^2}{n_1} + \frac{s_p^2}{n_2}}} \text{ with } df = n_1 + n_2 - 2 = 31$$

$$\text{where } s_p^2 = \frac{(n_1 - 1)s_1^2 + (n_2 - 1)s_2^2}{n_1 + n_2 - 2} = \frac{(18-1)(4^2) + (15-1)(3^2)}{18+15-2} = 12.8$$

$$\Rightarrow t = \frac{(12-10) - 0}{\sqrt{\frac{12.8}{18} + \frac{12.8}{15}}} = 1.6$$

***Step* 3** Since $H_1 : \mu_A > \mu_B$, the test is right-tailed. Therefore, the rejection region will be in the right tail of the t-curve with $df(v) = 31$. From Table 4, the critical value of t at $\alpha = 0.05$ for 31 df is 1.696. The rejection region, the critical value, and the calculated value of test statistics are shown in Figure 8.6.

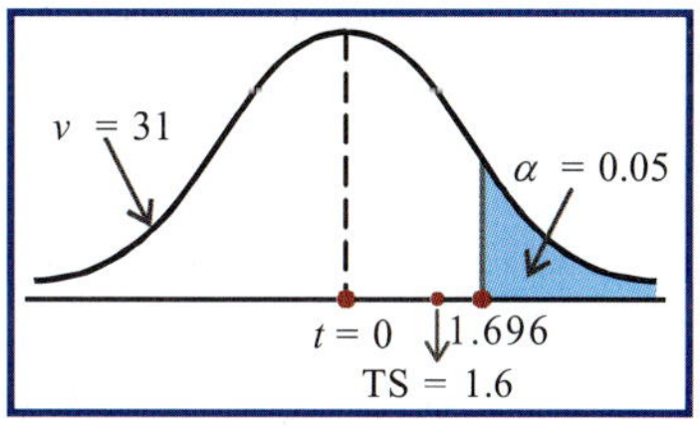

Figure 8.6

***Step* 4** **Decision:** Since the test statistics lies outside the rejection region, we fail to reject H_0 at 5% level of significance. *Conclusion:* The data does not provide sufficient evidence against H_0. There does not seem to be a significant difference between the means. Hence, the two feeds, A and B, seem to be equally effective in producing weight gain.

In short:

Hypotheses	Test Statistics
$H_0: \mu_A = \mu_B$ $H_1: \mu_A > \mu_B$	$t = \dfrac{(\bar{x}_1 - \bar{x}_2)}{\sqrt{\dfrac{s_p^2}{n_1} + \dfrac{s_p^2}{n_2}}} = 1.6$
Rejection Region and Critical Value	**Conclusion**
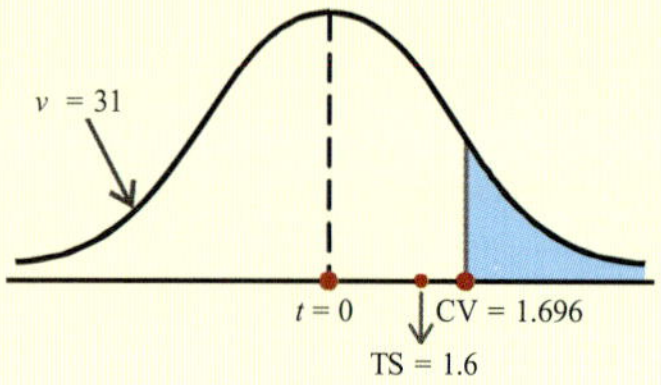	Fail to reject H_0. The two feeds, A and B, seem to be equally effective in producing weight gain.

(b) Since the test of hypothesis in this example is one-tailed at $\alpha = 0.05$, the appropriate confidence level will be $1 - 2\alpha = 1 - 2(0.05) = 0.9 = 90\%$.

By using information given in Table 8.5, we have:

$$s_{\bar{X}_1 - \bar{X}_2} = \sqrt{\frac{s_p^2}{n_1} + \frac{s_p^2}{n_2}} = \sqrt{\frac{12.8}{18} + \frac{12.8}{15}} = 1.3, \quad t_{n_1+n_2-2}(\alpha/2) = t_{31}(0.05) = 1.696$$

$$\Rightarrow E = t_v(\alpha/2) \cdot s_{\bar{X}_1 - \bar{X}_2} = (1.696)(1.3) = 2.2$$

The 90% confidence interval for the difference of means $(\mu_1 - \mu_2)$ is given by:

$$(\bar{x}_1 - \bar{x}_2) - E < \mu_1 - \mu_2 < (\bar{x}_1 - \bar{x}_2) + E$$

$$(12 - 10) - 2.2 < \mu_1 - \mu_2 < (12 - 10) + 2.2$$

$$-0.2 < \mu_1 - \mu_2 < 4.2$$

We can conclude with 90% confidence that the difference $\mu_1 - \mu_2$ between the population means lies between the limits -0.2 and 4.2. The confidence interval contains both positive and negative values. This suggests that it is very likely that $\mu_A - \mu_B = 0 \Rightarrow \mu_A = \mu_B$. The conclusion here agrees with part (a).

Example 5: Given the following sample data from two groups of Statistics classes:

Traditional Based Statistics Group	75	80	45	20	36	78	91	50	64	83	72			
Technology Based Statistics Group	85	77	62	92	74	93	88	65	66	75	85	28	91	73

Assume the populations are normally distributed, and $\sigma_1 = \sigma_2 = \sigma$.

(a) Test the claim that the mean score in a Traditional Based Statistics Group is less than the mean score in the Technology Based Statistics Group. Use the Classical Method.

(b) Construct a 90% confidence interval for $\mu_1 - \mu_2$. Interpret your results.

Solution: **(a)** ***Step 0*** The original claim: $\mu_1 < \mu_2 \Rightarrow$ The counter claim: $\mu_1 \geq \mu_2$

Given the data in this example, we use the descriptive statistics formulas to calculate:

Traditional based Statistics Group	$\bar{x}_1 = 63.1$	$s_1 = 22.3$	$n_1 = 11$
Technology based Statistics Group	$\bar{x}_2 = 75.3$	$s_2 = 17.1$	$n_2 = 14$

Table 8.6

***Step* 1** Set up the null and alternative hypotheses.

$H_0 : \mu_1 = \mu_2$

$H_1 : \mu_1 < \mu_2$

***Step* 2** Under $H_0 : \mu_1 = \mu_2 \Rightarrow \mu_1 - \mu_2 = 0$ and sample statistics calculated above, the test statistics is

$$t = \frac{(\bar{x}_1 - \bar{x}_2) - (\mu_1 - \mu_2)}{\sqrt{\dfrac{s_p^{\ 2}}{n_1} + \dfrac{s_p^{\ 2}}{n_2}}} = \frac{(63.1 - 75.3) - 0}{\sqrt{\dfrac{381.5}{11} + \dfrac{381.5}{14}}} = -1.55$$

$$\text{where } s_p^{\ 2} = \frac{(n_1 - 1)s_1^{\ 2} + (n_2 - 1)s_2^{\ 2}}{n_1 + n_2 - 2} = \frac{(11-1)(22.3^2) + (14-1)(17.1^2)}{11 + 14 - 2} = 381.49$$

***Step* 3** Since $H_1 : \mu_1 < \mu_2$, the test is left-tailed. Therefore, the rejection region will be in the left-tail of the t-curve with $df(v) = 22$. From Table 4, the critical value of t at $\alpha = 0.05$ for 22 df is -1.717. The rejection region, the critical value, and the calculated value of test statistics are shown in Figure 8.7.

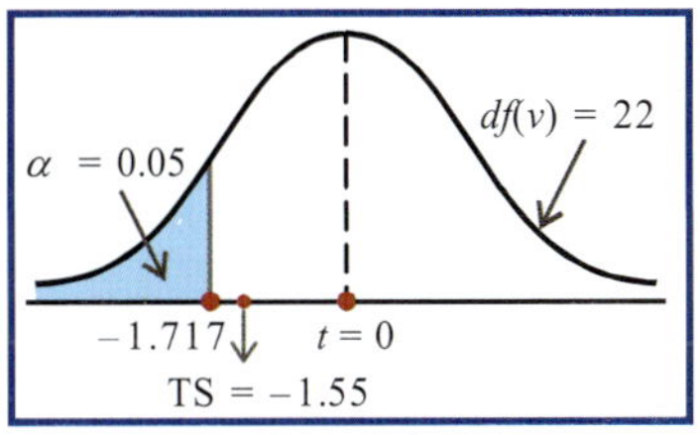

Figure 8.7

***Step* 4** **Decision:** Since the test statistics lies outside the rejection region, we fail to reject H_0 at 5% level of significance. *Conclusion:* The data does not provide sufficient evidence against H_0. There does not seem to be a significant difference between the means. Hence, both instructional delivery models seem to be equally effective.

In short:

Hypotheses	**Test Statistics**
$H_0 : \mu_1 = \mu_2$ $H_1 : \mu_1 < \mu_2$	$t = \dfrac{(\bar{x}_1 - \bar{x}_2)}{\sqrt{\dfrac{s_p^{\ 2}}{n_1} + \dfrac{s_p^{\ 2}}{n_2}}} = -1.55$
Rejection Region and Critical Value 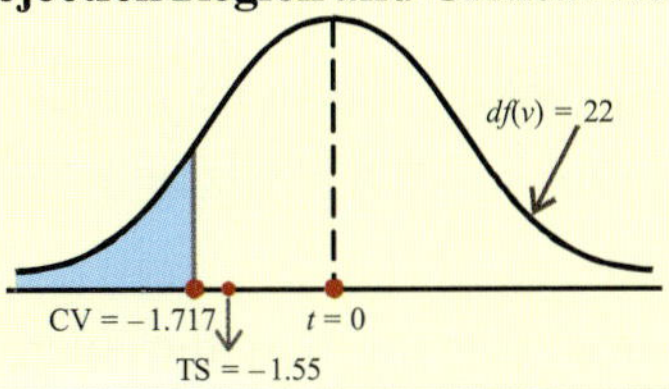 	**Conclusion** Fail to reject H_0. Both instructional delivery models viz. traditional based and technology based seem to be equally effective.

(b) By using information given in Table 8.6, we have:

$$s_{\bar{X}_1 - \bar{X}_2} = \sqrt{\frac{s_p^{\ 2}}{n_1} + \frac{s_p^{\ 2}}{n_2}} = \sqrt{\frac{381.5}{11} + \frac{381.5}{14}} = 7.8,\ t_{n_1 + n_2 - 2}(\alpha/2) = t_{22}(0.05) = 1.717$$

$$\Rightarrow E = t_v(\alpha/2) \cdot s_{\bar{X}_1 - \bar{X}_2} = (7.8)(1.717) = 13.4$$

The 90% confidence interval for the difference of means $(\mu_1 - \mu_2)$ is given by:

$$(\bar{x}_1 - \bar{x}_2) - E < \mu_1 - \mu_2 < (\bar{x}_1 - \bar{x}_2) + E$$

$$(63.1 - 75.3) - 13.4 < \mu_1 - \mu_2 < (63.1 - 75.3) + 13.4$$

$$-25.6 < \mu_1 - \mu_2 < 1.2$$

We can conclude with 90% confidence that the difference between the population means $\mu_1 - \mu_2$ lies between the limits -25.6 and 1.2. The confidence interval contains both positive and negative values. This suggests that it is very likely that $\mu_1 - \mu_2 = 0 \Rightarrow \mu_1 = \mu_2$.

C. Test of Hypothesis for $\mu_1 - \mu_2$ (σ_1 and σ_2 Unknown, $\sigma_1 \neq \sigma_2$)

- **Test of Hypothesis**

To test the equality of means of two normally distributed populations with unknown standard deviations (assumed unequal); that is, to test $H_0 : \mu_1 = \mu_2$, when $\sigma_1 \neq \sigma_2$, various test statistics have been proposed, though none of them completely satisfactory. We give below an **approximate** test for testing the equality of two means when $\sigma_1 \neq \sigma_2$.

- **Test Statistics**

If $\sigma_1 \neq \sigma_2$, then the test statistics for testing the difference of means is

$$t = \frac{(\bar{X}_1 - \bar{X}_2) - (\mu_1 - \mu_2)}{\sqrt{\frac{s_1^2}{n_1} + \frac{s_2^2}{n_2}}} \sim t_v \text{, where } v = \text{smaller of } [(n_1 - 1), (n_2 - 1)] . \qquad (8.5)$$

The rest of the procedure is the same as in the test of hypothesis for the difference of two means when population standard deviations are known.

- **Confidence Interval for $\mu_1 - \mu_2$**

The $100(1 - \alpha)\%$ confidence interval for $\mu_1 - \mu_2$ in this case is:

$$(\bar{x}_1 - \bar{x}_2) - E < \mu_1 - \mu_2 < (\bar{x}_1 - \bar{x}_2) + E, \qquad (8.6)$$

where $E = t_v(\alpha/2)\sqrt{\frac{s_1^2}{n_1} + \frac{s_2^2}{n_2}}$ and v = smaller of $[(n_1 - 1), (n_2 - 1)]$.

Example 6: In the United States, the average class size in public and private institutions in primary education are 23 and 19 respectively. Assume the data were obtained from two samples with standard deviations of 3 and 5, respectively, and sample sizes of 38 and 32, respectively. Assume $\sigma_1 \neq \sigma_2$.

(a) Is there sufficient evidence to show that the average class size in public and private institutions in primary education are different? Use $\alpha = 0.05$.

(b) Construct a 95% confidence interval for $\mu_1 - \mu_2$. Interpret your results.

Solution: **(a)** ***Step 0*** Given sample statistics;

Public	$\bar{x}_1 = 23$	$s_1 = 3$	$n_1 = 38$
Private	$\bar{x}_2 = 19$	$s_2 = 5$	$n_2 = 32$

Table 8.7

The original claim: $\mu_1 \neq \mu_2 \Rightarrow$ The counter claim: $\mu_1 = \mu_2$

Step 1 Set up the null and alternative hypotheses.

$H_0 : \mu_1 = \mu_2$
$H_1 : \mu_1 \neq \mu_2$

Step 2 Under $H_0 : \mu_1 = \mu_2 \Rightarrow \mu_1 - \mu_2 = 0$, sample statistics given above, and the assumption that $\sigma_1 \neq \sigma_2$, the test statistics is

$$t = \frac{(\bar{X}_1 - \bar{X}_2) - (\mu_1 - \mu_2)}{\sqrt{\frac{s_1^2}{n_1} + \frac{s_2^2}{n_2}}} \text{ with } df(v) = \text{smaller of } [(n_1 - 1), (n_2 - 1)] .$$

Therefore, we have:

$$t = \frac{(\bar{x}_1 - \bar{x}_2) - (\mu_1 - \mu_2)}{\sqrt{\frac{s_1^2}{n_1} + \frac{s_2^2}{n_2}}} = \frac{(23 - 19) - 0}{\sqrt{\frac{3^2}{38} + \frac{5^2}{32}}} = \frac{4}{1.009} = 4.0$$

Step 3 Since $H_1 : \mu_1 \neq \mu_2$, the test is two-tailed. Therefore, the rejection region is in both tails of the t-curve with $df(v) = 31$. From Table 4, the critical value of t at $\alpha = 0.05$ for 31 df is 2.0395. The rejection region, the critical values, and the calculated value of test statistics are shown in Figure 8.8.

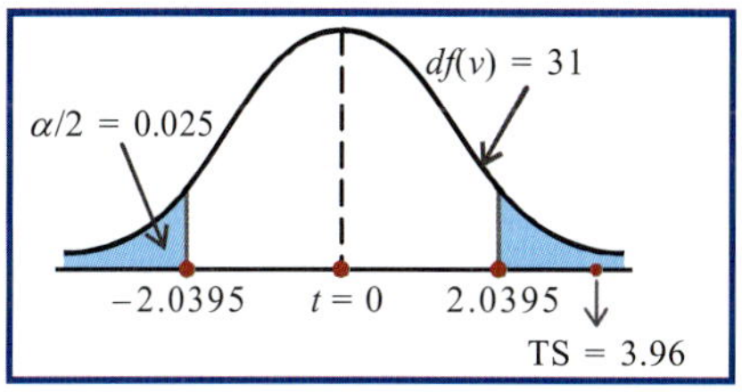

Figure 8.8

Step 4 **Decision:** Since the test statistics lies inside the rejection region, we reject H_0 at 5% level of significance. *Conclusion:* The data provides sufficient evidence against H_0. There seems to be a significant difference between the mean class sizes between public and private schools.

In short:

Hypotheses	**Test Statistics**
$H_0: \mu_1 = \mu_2$ $H_1: \mu_1 \neq \mu_2$	$t = \frac{(\bar{x}_1 - \bar{x}_2)}{\sqrt{\frac{s_1^2}{n_1} + \frac{s_2^2}{n_2}}} = 3.96$
Rejection Region and Critical Values	**Conclusion**
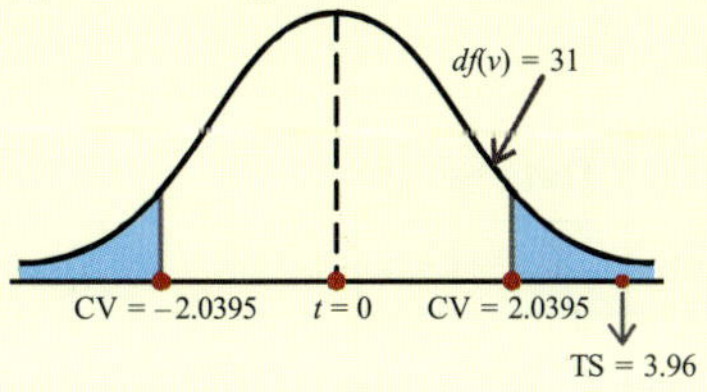	Reject H_0. The mean class sizes in public and private schools are significantly different.

(b) By using information given in Table 8.7, we have:

$$s_{\bar{X}_1 - \bar{X}_2} = \sqrt{\frac{s_1^2}{n_1} + \frac{s_2^2}{n_2}} = \sqrt{\frac{3^2}{38} + \frac{5^2}{32}} = 1.009$$

$$t_v(\alpha/2) = t_{31}(0.025) = 2.0395$$

$$\Rightarrow E = t_{31}(0.025) \cdot s_{\bar{X}_1 - \bar{X}_2} = (2.0395)(1.009) = 2.0579$$

The 95% confidence interval for the difference of means $(\mu_1 - \mu_2)$ is given by:

$$(\bar{x}_1 - \bar{x}_2) - E < \mu_1 - \mu_2 < (\bar{x}_1 - \bar{x}_2) + E$$

$$(23 - 19) - 2.0579 < \mu_1 - \mu_2 < (23 - 19) + 2.0579$$

$$1.94 < \mu_1 - \mu_2 < 6.06$$

We can conclude with 95% confidence that the difference between the population means $\mu_1 - \mu_2$ lies between the limits 1.94 and 6.06. The confidence interval contains only positive values. This suggests that it is very likely that $\mu_1 - \mu_2 > 0 \Rightarrow \mu_1 > \mu_2$.

Try it yourself:

In the following problems, population standard deviations are unknown and samples are independent.

1. Given the following data sets:

Data Set I	Data Set II
$\bar{x}_1 = 125.8$	$\bar{x}_2 = 119.3$
$s_1 = 15.8$	$s_2 = 12$
$n_1 = 15$	$n_2 = 10$

(a) Test the claim that $\mu_1 = \mu_2$ at $\alpha = 0.01$. Assume $\sigma_1^2 = \sigma_2^2$. Parent populations are normally distributed.

(b) Construct a 99% confidence interval for $\mu_1 - \mu_2$.

2. Given the following data sets:

Data Set I	Data Set II
$\bar{x}_1 = 20.4$	$\bar{x}_2 = 25.8$
$s_1 = 7.5$	$s_2 = 4.2$
$n_1 = 18$	$n_2 = 15$

(a) Test the claim that $\mu_1 < \mu_2$ at $\alpha = 0.025$. Assume $\sigma_1^2 \neq \sigma_2^2$. Parent populations are normally distributed.

(b) Construct a 95% confidence interval for $\mu_1 - \mu_2$.

3. Given the following data sets:

Data Set I	Data Set II
$\bar{x}_1 = 15.8$	$\bar{x}_2 = 13.5$
$s_1 = 2.5$	$s_2 = 1.9$
$n_1 = 42$	$n_2 = 38$

(a) Test the claim that $\mu_1 > \mu_2$ at $\alpha = 0.1$. Assume $\sigma_1^2 \neq \sigma_2^2$.

(b) Construct a 98% confidence interval for $\mu_1 - \mu_2$.

8.4: Inferences About Matched Pairs (Dependent Samples)

Objectives

After completing this section you will be able to understand:

A. Matched Pairs t-Test for Difference of Means, Dependent Samples.

B. Confidence Interval for the Mean Difference(μ_d).

In the t-test for difference of means in Section 8.3, the two samples were **independent** of each other. Let us now consider the case where

1. the sample observations $(x_1, x_2, \ldots, x_n)$ and $(y_1, y_2, \ldots, y_n)$ are not independent but they are dependent in pairs (matched samples).
2. the sample sizes are equal; that is, $n_1 = n_2 = n$

In such situations, we use **paired t-test for difference of means**.

A. Matched Pairs t-Test for Difference of Means, Dependent Samples

Suppose a business is interested in finding if advertisement on TV or newspapers is really effective in promoting sales of a product. Let $x_1, x_2, \ldots, x_n$ be the sales (in \$) of a product in n departmental stores for a certain period before advertisement campaign and let $y_1, y_2, \ldots, y_n$ be the corresponding sales for the same product and period in the same departmental stores respectively after the advertisement campaign. (x_i, y_i), $i = 1, 2, \ldots, n$ is the pair of sales in the ith departmental store before and after advertisement. In order to test for a significant difference between the sales x and y, we cannot apply the **difference of the means test** since the samples are not **independent**. In this case we apply the **matched pairs t-test**.

Let $d_i = x_i - y_i$, $(i = 1, 2, \ldots, n)$ denote the differences in the observations for the ith unit.

In this test, the null hypothesis is that the differences are just by chance; that is, the populations **average difference** $\mu_d = 0$. Symbolically, $H_0 : \mu_d = 0$.

The alternative hypothesis H_1 on the other hand, can be written as:

$H_1 : \mu_d < 0$	$H_1 : \mu_d > 0$	$H_1 : \mu_d \neq 0$
Left-Tail Test	**Right-Tail Test**	**Two-Tail Test**

The test statistics is

$$t = \frac{\bar{d} - \mu_d}{\frac{s_d}{\sqrt{n}}} \sim t_{n-1}, \tag{8.7}$$

where $d_i = X_i - Y_i$, $\bar{d} = \frac{\sum d_i}{n}$, and $s_d = \sqrt{\frac{n\sum d_i^2 - \left(\sum d_i\right)^2}{n(n-1)}}$

B. Confidence Interval for the Mean Difference (μ_d)

The $100(1 - \alpha)\%$ confidence interval for the populations mean difference μ_d is

$$\bar{d} - t_{n-1,\alpha/2}\frac{s_d}{\sqrt{n}} < \mu_d < \bar{d} + t_{n-1,\alpha/2}\frac{s_d}{\sqrt{n}} \tag{8.8}$$

Example 7: The table below lists the scores of a group of 13 students who have taken the Graduate Record Exam (GRE) before and after a preparatory course. Assume the population is normally distributed.

(a) Is there sufficient evidence to conclude that the prep course is effective in improving the GRE scores? Use $\alpha = 0.01$.

(b) Construct a 98% confidence interval of the mean difference between the before and after scores (μ_d).

Before	800	950	1240	760	640	990	1000	1080	880	914	730	610	974
After	960	1400	1360	700	990	700	1300	1150	970	1000	700	830	970

Solution: **(a)** Let μ_d be the mean difference (before−after) between the two populations. In this example, effectiveness represents an increase in the GRE score after the prep course has been taken. Therefore, μ_d must be negative in order to verify effectiveness.

Recall: For ungrouped data $\bar{d} = \dfrac{\sum d}{n}$ and $s_d = \sqrt{\dfrac{n\sum d^2 - \left(\sum d\right)^2}{n(n-1)}}$

Before(X)	800	950	1240	760	640	990	1000	1080	880	914	730	610	974	
After(Y)	960	1400	1360	700	990	700	1300	1150	970	1000	700	830	970	
$d = X - Y$	−160	−450	−120	60	−350	290	−300	−70	−90	−86	30	−220	4	$\sum d = -1462$
d^2	25600	202500	14400	3600	122500	84100	90000	4900	8100	7396	900	48400	16	$\sum d^2 = 612412$

Table 8.8

Step 0 The original claim is $\mu_d < 0$ and the counter claim is $\mu_d \geq 0$.
The sample statistics are:

$$\bar{d} = \frac{\sum d}{n} = \frac{-1462}{13} = -112.5$$

$$\text{and } s_d = \sqrt{\frac{n\sum d^2 - \left(\sum d\right)^2}{n(n-1)}} = \sqrt{\frac{13(612412) - (-1462)^2}{(13)(12)}} = 193.2$$

Step 1 Set up the null and alternative hypotheses.

$H_0 : \mu_d = 0$ (cc)

$H_1 : \mu_d < 0$ (oc)

Step 2 Under $H_0 : \mu_d = 0$, the **test statistics** is:

$$t = \frac{\bar{d} - \mu_d}{\frac{s_d}{\sqrt{n}}} = \frac{-112.5 - 0}{\frac{193.2}{\sqrt{13}}} = -2.1$$

Step 3 Since $H_1 : \mu_d < 0$, the test is left-tailed. Therefore, the rejection region will be in the left tail of the t-curve with $df(v) = 12$. From Table 4, the critical value of t at $\alpha = 0.01$ for 12 df is -2.861. The rejection region, the critical value, and the calculated value of test statistics are shown in Figure 8.9.

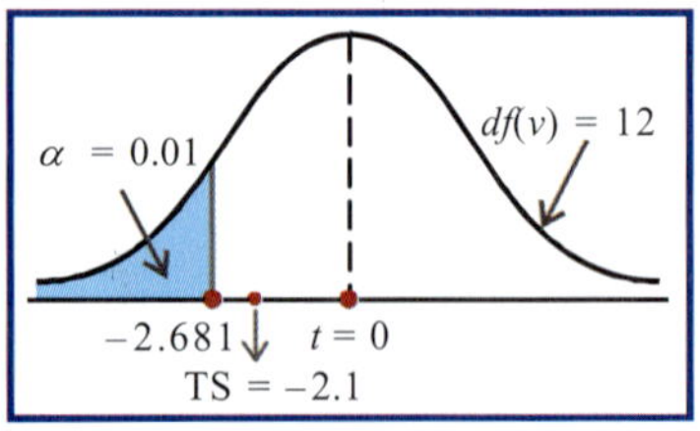

Figure 8.9

Step 4 **Decision:** Since the test statistics lies outside the rejection region, we fail to reject H_0 at 1% level of significance. *Conclusion:* The data does not provide sufficient evidence against H_0. There is not enough evidence to support the claim that the prep course is effective.

In short:

<table>
<tr><td>Hypotheses
$H_0: \mu_d = 0$
$H_1: \mu_d < 0$</td><td>Test Statistics
$t = \dfrac{\bar{d}}{\frac{s_d}{\sqrt{n}}} = -2.1$</td></tr>
<tr><td>Rejection Region and Critical Value
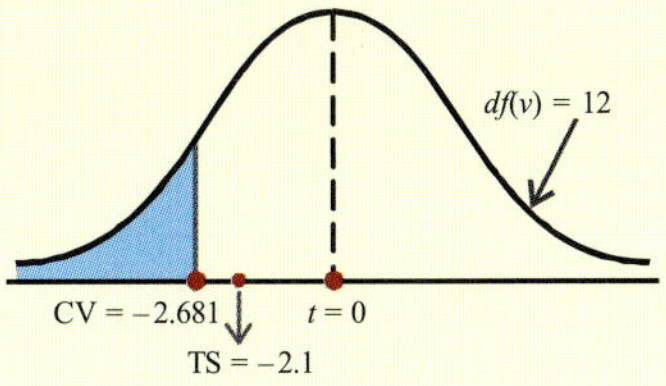
</td><td>Conclusion
Fail to reject H_0. The preparatory course is not effective in improving the GRE scores.</td></tr>
</table>

(b) Calculated sample statistics;

$\bar{d} = -112.5$	$s_d = 193.2$	$n = 13$	$t_{n-1,\,\alpha/2} = t_{12}(0.01) = 2.681$

The 98% confidence interval for the difference of means (μ_d) is given by:

$$\bar{d} - t_{n-1,\,\alpha/2}\frac{s_d}{\sqrt{n}} < \mu_d < \bar{d} + t_{n-1,\,\alpha/2}\frac{s_d}{\sqrt{n}}$$

$$-112.5 - 2.681\frac{193.2}{\sqrt{13}} < \mu_d < -112.5 + 2.681\frac{193.2}{\sqrt{13}}$$

$$-256.2 < \mu_d < 31.2$$

We can conclude with 98% confidence that the mean difference (μ_d) lies between the limits -256.2 and 31.2. The confidence interval contains both positive and negative values. This suggests that it is very likely that $\mu_d = 0$.

Example 8: Memory capacity of students was tested before and after training.

Roll No.	1	2	3	4	5	6	7	8	9	10
Score before training	12	14	11	8	7	10	3	0	5	6
Score after training	15	16	10	7	5	12	10	2	3	8

Assuming the population is normally distributed,

(a) test the belief that training has no effect in increasing the memory capacity of students. Use $\alpha - 0.05$.

(b) construct a 95% confidence interval for the mean difference (μ_d).

Solution: **(a)**

Before(X)	12	14	11	8	7	10	3	0	5	6	
After(Y)	15	16	10	7	5	12	10	2	3	8	
$d = X - Y$	−3	−2	1	1	2	−2	−7	−2	2	−2	$\sum d = -12$
d^2	9	4	1	1	4	4	49	4	4	4	$\sum d^2 = 84$

Table 8.8

***Step* 0** The original claim is $\mu_d = 0$ and the counter claim is $\mu_d \neq 0$.

The sample statistics are

$$\bar{d} = \frac{\sum d}{n} = \frac{-12}{10} = -1.2 \text{ and } s_d = \sqrt{\frac{n\sum d^2 - \left(\sum d\right)^2}{n(n-1)}} = \sqrt{\frac{10(84) - (-12)^2}{(10)(9)}} = 2.8$$

***Step* 1** Set up the null and alternative hypotheses.

$H_0 : \mu_d = 0$ (oc)

$H_1 : \mu_d \neq 0$ (cc)

***Step* 2** Under $H_0 : \mu_d = 0$, the **test statistics** is

$$t = \frac{\bar{d} - \mu_d}{\frac{s_d}{\sqrt{n}}} = \frac{-1.2 - 0}{\frac{2.8}{\sqrt{10}}} = -1.4$$

***Step* 3** Since $H_1 : \mu_d \neq 0$, the test is two-tailed. Therefore, the rejection region will be in both tails of the t-curve with $df(v) = 9$. From Table 4, the critical value of t at $\alpha/2 = 0.025$ for 9 df are ± 2.262. The rejection region, the critical values, and the calculated value of test statistics are shown in Figure 8.10.

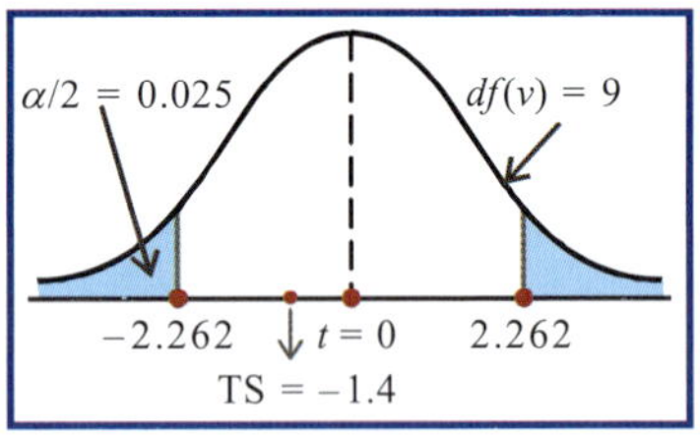

Figure 8.10

***Step* 4** **Decision:** Since the test statistics lies outside the rejection region, we fail to reject H_0 at 5% level of significance. *Conclusion:* The data does not provide sufficient evidence against H_0. There is not enough evidence to support the belief that training course is effective.

In short:

Hypotheses	**Test Statistics**
$H_0 : \mu_d = 0$ $H_1 : \mu_d \neq 0$	$t = \frac{\bar{d}}{\frac{s_d}{\sqrt{n}}} = -1.4$
Rejection Region and Critical Values 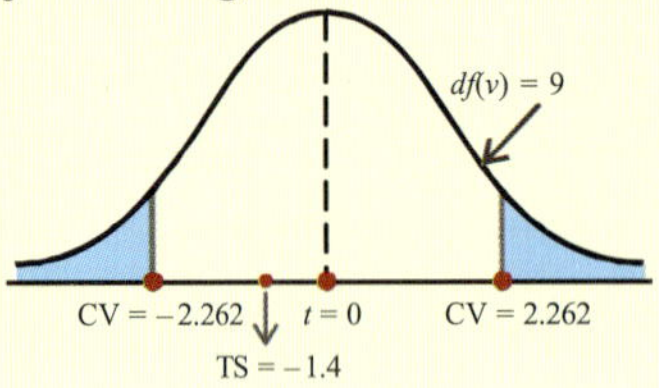	**Conclusion** Fail to reject H_0. The training course has no effect in increasing the memory capacity of students.

(b) Calculated sample statistics;

$\bar{d} = -1.2$	$s_d = 2.8$	$n = 10$	$t_{n-1,\,\alpha/2} = t_9(0.025) = 2.262$

The 95% confidence interval for mean difference (μ_d) is given by:

$$\bar{d} - t_{n-1,\,\alpha/2}\frac{s_d}{\sqrt{n}} < \mu_d < \bar{d} + t_{n-1,\,\alpha/2}\frac{s_d}{\sqrt{n}}$$

$$-1.2 - 2.262\frac{2.8}{\sqrt{10}} < \mu_d < -1.2 + 2.262\frac{2.8}{\sqrt{10}}$$

$$-3.2 < \mu_d < 0.8$$

We can conclude with 95% confidence that the mean difference (μ_d) lies between the limits -3.2 and 0.8. The confidence interval contains both positive and negative values. This suggests that it is very likely that $\mu_d = 0$.

Try it yourself:

In the following problems, assume the population is normally distributed.

1. Construct a 95% confidence interval for μ_d.

$\bar{d} = -2.5$	$s_d = 1.5$	$n = 20$

2. Test the claim that $\mu_d < 0$. Use $\alpha = 0.01$.

$\bar{d} = -2.5$	$s_d = 1.5$	$n = 20$

3. Test the claim that $\mu_d = 0$. Use $\alpha = 0.05$.

$\bar{d} = 8.5$	$s_d = 15.4$	$n = 8$

4. Test the claim that $\mu_d > 0$. Use $\alpha = 0.1$.

$\bar{d} = 5.3$	$s_d = 18.2$	$n = 12$

8.5: Inferences about Two Proportions

A farmer is interested in comparing the proportions of germinating seeds in two brands. A medical researcher is interested in comparing the proportions of smokers with lung cancer and of non-smokers with lung cancer. Such problems deal with the difference of population proportions. For this, we must first describe the sampling distribution of difference of sample proportions $\hat{p}_1 - \hat{p}_2$.

Objectives

After completing this section you will be able to understand:

- **A.** Sampling Distribution of $\hat{p}_1 - \hat{p}_2$.
- **B.** Test of Hypothesis for $\hat{p}_1 - \hat{p}_2$.
- **C.** Confidence Interval for the Difference of Two Proportions.

A. Sampling Distribution of $\hat{p}_1 - \hat{p}_2$

Let us suppose that we want to compare two large populations, population 1 and population 2, with respect to the prevalence of a certain characteristic among their members. We take two independent samples of size n_1 and n_2 from populations 1 and 2 respectively. Let x_1 and x_2 be the observed number of successes in these samples respectively. $\hat{p}_1$ and $\hat{p}_2$ are defined as the following:

$\hat{p}_1 = \frac{x_1}{n_1}$ = Observed proportion (percentage) of successes in the sample from population 1.

$\hat{p}_2 = \frac{x_2}{n_2}$ = Observed proportion (percentage) of successes in the sample from population 2.

Let $\hat{p}_1$ and $\hat{p}_2$ be the corresponding population proportions.

The sampling distribution of the difference of proportions $(\hat{p}_1 - \hat{p}_2)$ is approximately normal with

$$\text{Mean} = \mu_{\hat{p}_1 - \hat{p}_2} = p_1 - p_2 \text{ and Standard Deviation} = \sigma_{\hat{p}_1 - \hat{p}_2} = \sqrt{\frac{p_1 q_1}{n_1} + \frac{p_2 q_2}{n_2}}.$$

Note: $q_1 = 1 - p_1$ and $q_2 = 1 - p_2$.

Therefore, the standard variable for $(\hat{p}_1 - \hat{p}_2)$ is

$$Z = \frac{(\hat{p}_1 - \hat{p}_2) - \mu_{\hat{p}_1 - \hat{p}_2}}{\sigma_{\hat{p}_1 - \hat{p}_2}} = \frac{(\hat{p}_1 - \hat{p}_2) - (p_1 - p_2)}{\sqrt{\frac{p_1 q_1}{n_1} + \frac{p_2 q_2}{n_2}}} \sim N(0,\ 1). \qquad (8.9)$$

To develop the test for two population proportions, we start with setting up the Null Hypothesis and the Test Statistics.

B. Test of Hypothesis for $\hat{p}_1 - \hat{p}_2$

- **Null Hypothesis**

$H_0 : p_1 = p_2 = p$; that is, the population proportions are the same ($p_1 - p_2 = 0$).

- **Test Statistics**

Remark: In general, the common population proportion 'p' is not known and we use its unbiased estimate '$\bar{p}$' taking both samples together.

$$\bar{p} = \frac{x_1 + x_2}{n_1 + n_2} = \frac{n_1 \hat{p}_1 + n_2 \hat{p}_2}{n_1 + n_2}, \text{ and } \bar{q} = 1 - \bar{p} \qquad (8.10)$$

$$Z = \frac{(\hat{p}_1 - \hat{p}_2) - (p_1 - p_2)}{\sqrt{\frac{\bar{p}\bar{q}}{n_1} + \frac{\bar{p}\bar{q}}{n_2}}} \sim N(0,\ 1) \qquad (8.11)$$

From (8.10), under $H_0 : p_1 = p_2$, the **Test Statistics** for the difference of proportions becomes

$$Z = \frac{(\hat{p}_1 - \hat{p}_2)}{\sqrt{\frac{\bar{p}\bar{q}}{n_1} + \frac{\bar{p}\bar{q}}{n_2}}} \sim N(0,\ 1) \qquad (8.12)$$

where $\bar{p}$ and $\bar{q}$ are defined in (8.10).

We can now carry out the test for H_0 (similar to the test for the proportion in Section 7.5) using the 4-step testing procedure by one of the usual (Classical or the p-value) methods.

- **Assumptions for the Normal Test for Two Proportions**

1. The samples are simple random samples.
2. The samples are independent.
3. The conditions of the normal approximation to the binomial distribution are satisfied for each of the samples, $np \geq 5$ and $nq \geq 5$.

C. Confidence Interval for the Difference of Two Proportions

An unbiased estimate of the difference of proportions $p_1 - p_2$ is provided by the difference of sample proportions $\hat{p}_1 - \hat{p}_2$. In Section 8.5A, we saw that the sampling distribution of difference of proportions is given by:

$$\hat{p}_1 - \hat{p}_2 \quad N\left(p_1 - p_2, \sqrt{\frac{p_1 q_1}{n_1} + \frac{p_2 q_2}{n_2}}\right) \qquad (8.13)$$

If p_1 and p_2 are unknown, we use their unbiased estimates provided by the sample proportions, that is

$$p_1 \simeq \hat{p}_1 \,,\quad p_2 \simeq \hat{p}_2 \;\Rightarrow\; q_1 \simeq \hat{q}_1 = 1-\hat{p}_1 \,, \text{ and } \; q_2 \simeq \hat{q}_2 = 1-\hat{p}_2 \,.$$

Therefore, an estimate of the standard error of $(\hat{p}_1 - \hat{p}_2)$ is given by:

$$\hat{\sigma}_{\hat{p}_1 - \hat{p}_2} = \sqrt{\frac{\hat{p}_1\hat{q}_1}{n_1} + \frac{\hat{p}_2\hat{q}_2}{n_2}} = \sqrt{\frac{\hat{p}_1(1-\hat{p}_1)}{n_1} + \frac{\hat{p}_2(1-\hat{p}_2)}{n_2}}$$

The confidence limits for the difference of proportions $p_1 - p_2$, with confidence coefficient $(1 - \alpha)$, are given by:

$$(\hat{p}_1 - \hat{p}_2) \pm E, \text{ where } E = z_{\alpha/2} \cdot \hat{\sigma}_{\hat{p}_1 - \hat{p}_2} = z_{\alpha/2}\sqrt{\frac{\hat{p}_1\hat{q}_1}{n_1} + \frac{\hat{p}_2\hat{q}_2}{n_2}}$$

$z_{\alpha/2}$ is the critical value of the Standard Normal Variable Z for two-tailed test at α significance level. The corresponding confidence interval is given by:

$$\left[(\hat{p}_1 - \hat{p}_2) - E, (\hat{p}_1 - \hat{p}_2) + E\right]$$

$$(\hat{p}_1 - \hat{p}_2) - z_{\alpha/2} \cdot \hat{\sigma}_{\hat{p}_1 - \hat{p}_2} < p_1 - p_2 < (\hat{p}_1 - \hat{p}_2) + z_{\alpha/2} \cdot \hat{\sigma}_{\hat{p}_1 - \hat{p}_2} \qquad (8.14)$$

$$(\hat{p}_1 - \hat{p}_2) - z_{\alpha/2}\sqrt{\frac{\hat{p}_1\hat{q}_1}{n_1} + \frac{\hat{p}_2\hat{q}_2}{n_2}} < p_1 - p_2 < (\hat{p}_1 - \hat{p}_2) + z_{\alpha/2}\sqrt{\frac{\hat{p}_1\hat{q}_1}{n_1} + \frac{\hat{p}_2\hat{q}_2}{n_2}}$$

- **Interpretation of Confidence Interval for $p_1 - p_2$**

100(1 – α)% confidence interval for the difference of proportions can be interpreted exactly as the confidence interval for $\mu_1 - \mu_2$ discussed in Section 8.2.

Round off Rules: Use 3 decimal places for all probabilities (proportions), sample or population. When calculating $x_1 = n_1 \cdot \hat{p}_1$ and $x_2 = n_2 \cdot \hat{p}_2$, round your answers to the nearest whole number. Use 2 decimal places for all values of the Test Statistics; for all other values, use 3 decimal places.

Example 9: To determine if there is a significant difference in the proportion of smokers among the blue collar workers of the United States and Italy, two independent random samples of sizes 1000 and 850 are taken from both populations respectively. If 420 in the U.S. sample are smokers and 350 in the Italy sample are smokers,

(a) test the claim that the proportion of blue collar smokers is higher in the U.S. than Italy. Use $\alpha = 0.05$. Use both the Classical and the p-value methods.

(b) Construct a 90% confidence interval for the difference of proportions $p_1 - p_2$. Interpret your results.

Solution: **(a)** ***Step 0*** In the usual notation we are given:

$$n_1 = 1000, x_1 = 420 \quad ; \quad n_2 = 850, x_2 = 350$$

$\hat{p}_1$ = Sample proportion of smokers among blue collar workers of U.S.

$$= \frac{x_1}{n_1} = \frac{420}{1000} = 0.420$$

$\hat{p}_2$ = Sample proportion of smokers among blue collar workers of Italy

$$= \frac{x_2}{n_2} = \frac{350}{850} = 0.412$$

The stated claim can be represented as $p_1 > p_2$. Hence, the counter claim is $p_1 \leq p_2$.

Classical Method:

***Step* 1** Set up the null and alternative hypotheses.

$H_0 : p_1 = p_2$ (cc)

$H_1 : p_1 > p_2$ (oc)

***Step* 2** Under H_0, the test statistics is

$$Z = \frac{\hat{p}_1 - \hat{p}_2}{\sqrt{\frac{\overline{p}\overline{q}}{n_1} + \frac{\overline{p}\overline{q}}{n_2}}} \sim N(0, 1)$$

where $\overline{p} = \frac{x_1 + x_2}{n_1 + n_2} = \frac{420 + 350}{1000 + 850} = 0.416 \Rightarrow \overline{q} = 1 - \overline{p} = 1 - 0.416 = 0.584$

$$\Rightarrow z = \frac{0.420 - 0.412}{\sqrt{\frac{(0.416)(0.584)}{1000} + \frac{(0.416)(0.584)}{850}}} = \frac{0.008}{0.023} = 0.35$$

***Step* 3** Since $H_1 : p_1 > p_2$, the test is right-tailed. Therefore, the rejection region will be in the right tail of the normal distribution. Hence, the critical value of z at $\alpha = 0.05$ for the right-tailed test is 1.645. The critical value, the rejection region, and the calculated value of test statistics are shown in Figure 8.11.

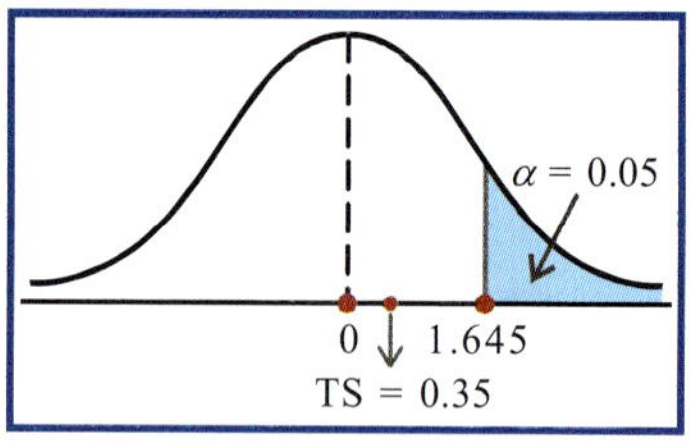

Figure 8.11

***Step* 4** **Decision:** Since the calculated value of test statistics lies outside the rejection region, we fail to reject H_0 at $\alpha = 0.05$. *Conclusion:* The data does not provide sufficient evidence against H_0. There does not seem to be a significant difference between the proportion of smokers among blue collar workers in the U.S. and Italy. Hence, we cannot support the claim that proportion of smokers among blue collar workers is higher in the U.S. than Italy.

***P*-value Method:**

***Step* 1** Set up the null and alternative hypotheses.

$H_0 : p_1 = p_2$ (cc)

$H_1 : p_1 > p_2$ (oc)

***Step* 2** Under H_0, the test statistics is

$$Z = \frac{\hat{p}_1 - \hat{p}_2}{\sqrt{\frac{\overline{p}\overline{q}}{n_1} + \frac{\overline{p}\overline{q}}{n_2}}} \sim N(0, 1)$$

where $\bar{p} = \dfrac{x_1 + x_2}{n_1 + n_2} = \dfrac{420 + 350}{1000 + 850} = 0.416 \quad \Rightarrow \quad \bar{q} = 1 - \bar{p} = 1 - 0.416 = 0.584$

$$\Rightarrow \quad z = \frac{0.420 - 0.412}{\sqrt{\dfrac{(0.416)(0.584)}{1000} + \dfrac{(0.416)(0.584)}{850}}} = \frac{0.008}{0.023} = 0.35$$

Step 3 **P-value:** Since $H_1 : p_1 > p_2$, the test is right-tailed. Therefore, the region showing extreme values is the right tail of the normal probability curve. As shown in Figure 8.12, the p-value of the test statistics is:

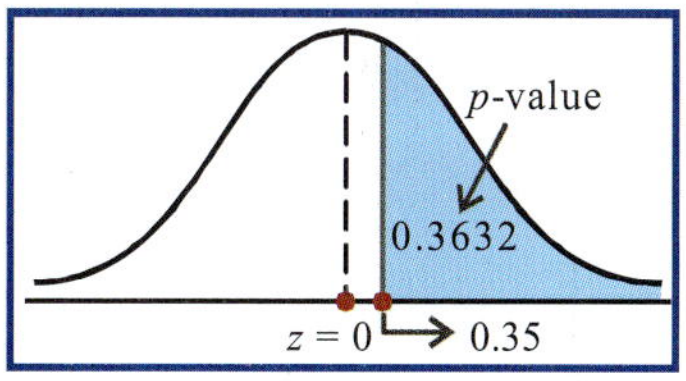

Figure 8.12

$$\begin{aligned} p\text{-value} &= P(Z > 0.35) \\ &= 0.5 - P(0 < Z < 0.35) \\ &= 0.5 - 0.1368 \\ &= 0.3632 \end{aligned}$$

Step 4 **Decision:** Since the p-value $> \alpha = 0.05$, we fail to reject H_0 at 5% level of significance. *Conclusion:* The data does not provide sufficient evidence against H_0.

In short:

Hypotheses	**Test Statistics**
$H_0 : p_1 = p_2$ $H_1 : p_1 > p_2$	$z = \dfrac{\hat{p}_1 - \hat{p}_2}{\sqrt{\dfrac{\bar{p}\bar{q}}{n_1} + \dfrac{\bar{p}\bar{q}}{n_2}}} = 0.35$
Rejection Region and Critical Value	**Conclusion**
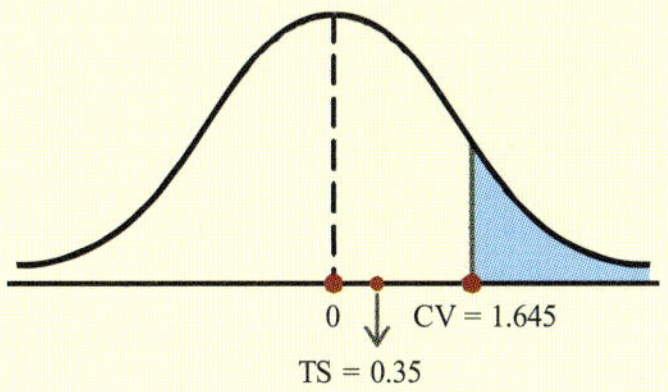	Fail to reject H_0. The proportion of smokers among blue collar workers of the U.S. and Italy are not significantly different.

(b) Let p_1 and p_2 be the proportion of smokers in the populations of blue color workers of the U.S. and Italy, respectively. We have

$$\hat{\sigma}_{\hat{p}_1 - \hat{p}_2} = \sqrt{\frac{\hat{p}_1\hat{q}_1}{n_1} + \frac{\hat{p}_2\hat{q}_2}{n_2}} = \sqrt{\frac{(0.420)(0.580)}{1000} + \frac{(0.412)(0.588)}{850}} = 0.023$$

Computing the z-values for the 90% confidence level gives

$$1 - \alpha = 0.9 \quad \Rightarrow \quad P(0 < Z < z_{\alpha/2}) = 0.45 \quad \Rightarrow \quad z_{\alpha/2} = 1.645$$

Recall:

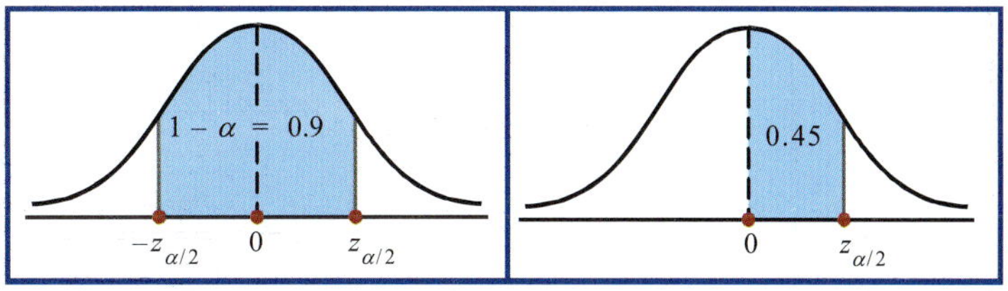

Figure 8.13

For computing E, we have

$$E = z_{\alpha/2} \cdot \hat{\sigma}_{\hat{p}_1 - \hat{p}_2} = (1.645)(0.023) = 0.038.$$

Hence, the 90% confidence interval for the difference of population proportions is:

$$(\hat{p}_1 - \hat{p}_2) - E < p_1 - p_2 < (\hat{p}_1 - \hat{p}_2) + E$$

$$(0.420 - 0.412) - 0.038 < p_1 - p_2 < (0.420 - 0.412) + 0.038$$

$$-0.030 < p_1 - p_2 < 0.046$$

Since, the 90% confidence interval for $p_1 - p_2$ contains both negative and positive limits, it is very likely that $p_1 - p_2 = 0 \Rightarrow p_1 = p_2$.

Example 10: In a random sample of 50 Statistics professors in the Los Angeles Area, 16% indicated that they use technology to teach Statistics. Among a random sample of 90 such professors in the New York area, 13% indicated that they used technology.

(a) Using a 0.01 significance level, test the claim that the proportion of professors using technology in LA is the same as the proportion of Professors using technology in NY. Use the Classical method.

(b) Construct a 99% confidence interval for the difference of proportions $(p_1 - p_2)$.

Solution: **(a)** Let p_1 and p_2 represent the proportion of professors in LA and NY that use technology to teach statistics respectively.

Step 0 We have $n_1 = 50$ and $n_2 = 90$.

$\hat{p}_1$ = Sample proportion of professors in LA who use technology = 0.16, and

$\hat{p}_2$ = Sample proportion of professors in NY who use technology = 0.13.

$$\hat{p}_1 = \frac{x_1}{n_1} \Rightarrow x_1 = \hat{p}_1 \cdot n_1 = (0.16)(50) = 8$$

$$\hat{p}_2 = \frac{x_2}{n_2} \Rightarrow x_2 = \hat{p}_2 \cdot n_2 = (0.13)(90) = 11.7 \approx 12$$

$$\hat{q}_1 = 1 - \hat{p}_1 = 1 - 0.16 = 0.84$$

$$\hat{q}_2 = 1 - \hat{p}_2 = 1 - 0.13 = 0.87$$

The stated claim can be represented as $p_1 = p_2$. Hence the counter claim is $p_1 \neq p_2$.

Classical Method:

Step 1 $H_0 : p_1 = p_2$ (oc)
$H_1 : p_1 \neq p_2$ (cc)

Step 2 Under H_0, the test statistics is

$$Z = \frac{\hat{p}_1 - \hat{p}_2.}{\sqrt{\frac{\bar{p}\bar{q}}{n_1} + \frac{\bar{p}\bar{q}}{n_2}}} \sim N(0, 1)$$

where $\bar{p} = \frac{x_1 + x_2}{n_1 + n_2} = \frac{8+12}{50+90} = 0.143 \Rightarrow \bar{q} = 1 - \bar{p} = 1 - 0.143 = 0.857$

$$\Rightarrow \quad z = \frac{(0.16-0.13)}{\sqrt{\dfrac{(0.143)(0.857)}{50}+\dfrac{(0.143)(0.857)}{90}}} = \frac{0.03}{0.062} = 0.48$$

***Step* 3** Since $H_1: p_1 \neq p_2$, the test is two-tailed. Therefore, the rejection region will be in both tails of the normal distribution. Hence, the critical values of z at $\alpha/2 = 0.005$ for the two-tailed test are $z_{\alpha/2} = 2.575$. The critical values, the rejection regions, and the calculated value of test statistics are shown in Figure 8.14.

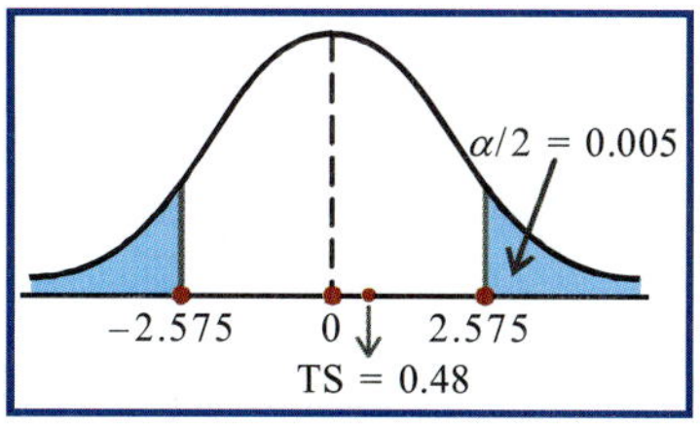

Figure 8.14

***Step* 4** **Decision:** Since the calculated value of test statistics lies outside the rejection region, we fail to reject H_0 at $\alpha = 0.01$. *Conclusion:* The data does not provide sufficient evidence against H_0. There does not seem to be a significance difference between the proportion of professors using technology in LA and NY respectively. Hence, we can support the claim that $p_1 = p_2$.

In short:

Hypotheses $H_0: p_1 = p_2$ $H_1: p_1 \neq p_2$	**Test Statistics** $z = \dfrac{\hat{p}_1 - \hat{p}_2}{\sqrt{\dfrac{\bar{p}\bar{q}}{n_1}+\dfrac{\bar{p}\bar{q}}{n_2}}} = 0.48$
Rejection Region and Critical Values 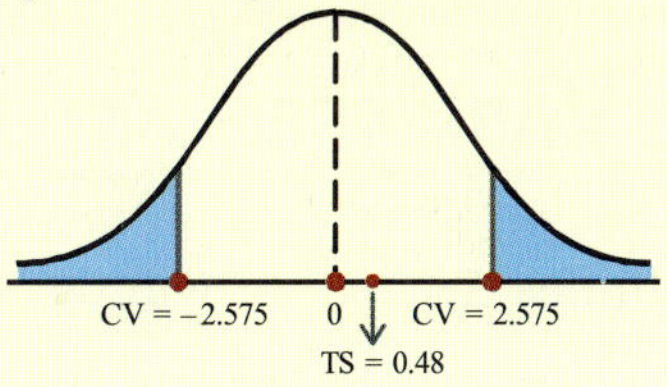	**Conclusion** Fail to reject H_0. The proportion of professors using technology in LA and NY are the same.

(b) By using the information provided in the example, we have

$$\hat{\sigma}_{\hat{p}_1-\hat{p}_2} = \sqrt{\frac{\hat{p}_1\hat{q}_1}{n_1}+\frac{\hat{p}_2\hat{q}_2}{n_2}} = \sqrt{\frac{(0.16)(0.84)}{50}+\frac{(0.13)(0.87)}{90}} = 0.063$$

Computing the z-values for the 99% confidence level gives

$$1-\alpha = 0.99 \Rightarrow P(0 < Z < z_{\alpha/2}) = 0.495 \Rightarrow z_{\alpha/2} = 2.575$$

For computing E, we have

$$E = z_{\alpha/2} \cdot \hat{\sigma}_{\hat{p}_1-\hat{p}_2} = (2.575)(0.063) = 0.162.$$

Hence, the 99% confidence interval for the difference of population proportions is:

$$(\hat{p}_1-\hat{p}_2) - E < p_1 - p_2 < (\hat{p}_1-\hat{p}_2) + E$$

$$(0.16-0.13) - 0.162 < p_1 - p_2 < (0.16-0.13) + 0.162$$

$$-0.132 < p_1 - p_2 < 0.192$$

Since, the 99% confidence interval for $p_1 - p_2$ contains both negative and positive limits, it is very likely that $p_1 - p_2 = 0 \Rightarrow p_1 = p_2$.

Try it yourself:

In the following problems, assume samples are independent and the conditions of the normal approximation to the binomial distribution are satisfied for each of the samples, $np \geq 5$ and $nq \geq 5$.

1. Given the following data sets, construct a 99% confidence interval for $p_1 - p_2$.

Data Set I	Data Set II
$x_1 = 15$	$x_2 = 10$
$n_1 = 45$	$n_2 = 85$

2. For data given in exercise 1, test the claim that $p_1 \neq p_2$. Use $\alpha = 0.04$. Use both the **classical** and the ***p*-value** methods.

3. Given the following data sets, test the claim that $p_1 < p_2$. Use $\alpha = 0.05$. Use both the **classical** and the ***p*-value** methods.

Data Set I	Data Set II
$\hat{p}_1 = 0.34$	$\hat{p}_2 = 0.36$
$n_1 = 100$	$n_2 = 65$

4. For data given in exercise 3, construct a 99% confidence interval for $p_1 - p_2$.

8.6: Inferences About Two Variances and Standard Deviations

Objectives

After completing this section you will be able to understand:

A. *F*-Test or Variance Ratio Test.

B. Test of Hypothesis for $\sigma_1^2 = \sigma_2^2$.

In this Section we will develop the procedure for testing:

i) the equality of two population standard deviation or variances, or

ii) the homogeneity of two independent estimates of the same common population standard deviation or variances.

A. *F*-Test or Variance Ratio Test

Suppose we want to test if two normally distributed populations have the same variance or standard deviation.

Let $X_1, X_2, \ldots, X_{n_1}$ be a simple random sample of size n_1 from the first normally distributed population with variance σ_1^2 and $Y_1, Y_2, \ldots, Y_{n_2}$ be a simple random sample of size n_2 from the second normally distributed population with variance σ_2^2; (assuming that the populations are independent). We set up the null hypothesis $H_0 : \sigma_1^2 = \sigma_2^2$, that is, population variances are the same or that they do not differ significantly.

The test statistics for the *F*-test is

$$F = \frac{s_1^2}{s_2^2} \quad F_{(n_1 - 1,\, n_2 - 1)} \text{ (Snedecor's } F\text{-distribution)} \qquad (8.15)$$

where s_1^2 and s_2^2 are unbiased estimates of the common population variance σ^2 and are given by

$$s_1^2 = \frac{\sum(X_1 - \bar{X}_1)^2}{n_1 - 1}, \text{ where } \bar{X}_1 = \frac{\sum X_1}{n_1}; \text{ and } s_2^2 = \frac{\sum(X_2 - \bar{X}_2)^2}{n_2 - 1}, \text{ where } \bar{X}_2 = \frac{\sum X_2}{n_2}.$$

Remarks: **1.** Since the F-test is based on the ratio of two variances, it is also known as Variance Ratio Test.

2. In $F_{(\nu_1, \nu_2)}$, the first degrees of freedom v_1 corresponds to the numerator and the second degrees of freedom v_2 corresponds to the denominator in

$$F = \frac{s_1^2}{s_2^2} \quad F_{(n_1 - 1,\, n_2 - 1)}.$$

• Assumptions for the *F*-Test

1. The samples are simple random samples.
2. The samples are independent of each other.
3. The parent populations from which the samples are drawn are normally distributed.
4. s_1 is greater than s_2.
5. The F-test will always be a right-tailed test.

• Critical Values of *F*-Distribution

The F-tables (Tables 6A, 6B, and 6C) available in the appendix at the end of the book give the critical values of F for the right-tailed test; that is, the rejection region is determined by the right tail areas. Thus, the critical value $F_{(\nu_1, \nu_2)}(\alpha)$ at level of significance α and (v_1, v_2) degrees of freedom is determined by the equation $P\left(F > F_{(\nu_1, \nu_2)}(\alpha)\right) = \alpha$, as shown in Figure 8.15.

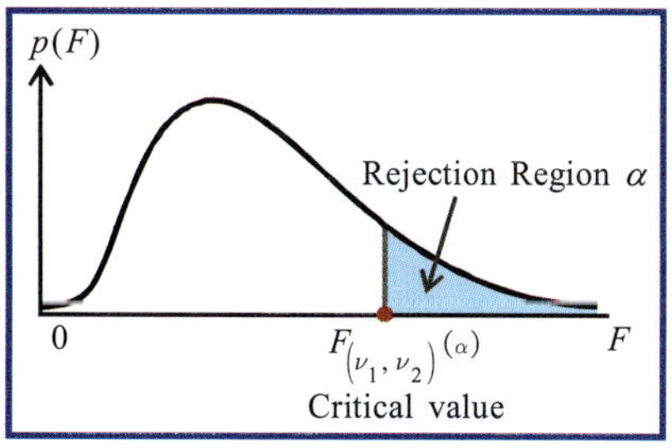

Figure 8.15

Unlike the z, t, and χ^2 distributions where all α values are on the same table, we have separate tables for critical values of $F_{(\nu_1, \nu_2)}(\alpha)$ distribution at different levels of significance. Tables 6A, 6B, and 6C give the critical values of F at $\alpha = 0.01$, 0.025, and 0.05, respectively, for different combinations of (v_1, v_2). Formats of Table 6A, 6B, and 6C are given below. The body of the table consists of the right-tail critical values of F for (v_1, v_2) degrees of freedom and given level of significance α.

Format of Table 6A

(F-Distribution)

Significant Values of the Variance-Ratio $F = \dfrac{s_1^2}{s_2^2}$; $s_1^2 > s_2^2$

Right Tail Areas ($\alpha = 0.01$)

v_2 \ v_1	1	2	3	4	5 ...
1					
2					
14				5.04	

v_1 : Degrees of freedom for the numerator = 4.
v_2 : Degrees of freedom for the denominator = 14.

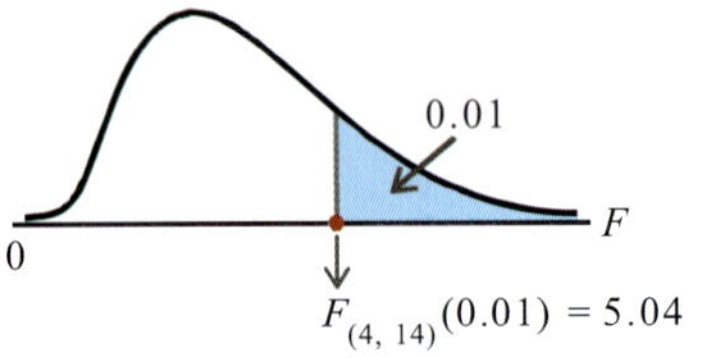

Figure 8.16

B. Test of Hypothesis for $\sigma_1^2 = \sigma_2^2$

- **Null Hypothesis**

$H_0 : \sigma_1^2 = \sigma_2^2$ or $H_0 : \sigma_1 = \sigma_2$; that is, the population variances or standard deviations are the same.

- **Test Statistics**

From (8.16), under $H_0 : \sigma_1^2 = \sigma_2^2$, the test statistics for the equality of two population variances is given by

$$F = \frac{s_1^2}{s_2^2} \quad F_{(n_1 - 1,\, n_2 - 1)} = F_{(\nu_1,\, \nu_2)},\ \text{where } s_1^2 > s_2^2;\ v_1 = n_1 - 1, \text{ and } v_2 = n_2 - 1.$$

By following the usual hypothesis testing procedure, we can test H_0 by any one of the two methods: **(a)** Classical Method and **(b)** P-value Method.

Note: **1.** To obtain exact P-values in the F-Test, we need to utilize technology (Statistical software). Therefore, in this section, we will use the Classical Method to carry out the F-test.

2. A brief discussion on the construction of a confidence interval for the ratio of two variances can be found in Section 8.8.

Example 11: Two random samples of sizes 8 and 11 are drawn from two normally distributed populations. The corresponding variances are 7.14 and 4.85 respectively. Using a 0.05 significance level, test the claim that the two populations have equal variances.

Solution: ***Step* 0** Given statistics: $n_1 = 8$, $n_2 = 11$, $s_1^2 = 7.14$, and $s_2^2 = 4.85$

The stated claim can be represented as $\sigma_1^2 = \sigma_2^2$. Hence the counter claim is $\sigma_1^2 \neq \sigma_2^2$.

***Step* 1** $H_0 : \sigma_1^2 = \sigma_2^2$ (oc)

$H_1 : \sigma_1^2 \neq \sigma_2^2$ (cc)

***Step* 2** Since $s_1^2 > s_2^2$, under H_0, the test statistics is

$$F = \frac{s_1^2}{s_2^2} \quad F_{(n_1-1,\, n_2-1)} = F_{(7,\,10)} = F = \frac{s_1^2}{s_2^2} = \frac{7.14}{4.85} = 1.47$$

***Step* 3** Since $H_1 : \sigma_1^2 \neq \sigma_2^2$, the test is two-tailed. Given that $s_1^2 > s_2^2$, the critical value of F for the two-tailed test at $\alpha = 0.05$ significance level is the same as the right-tailed critical value of F at $\alpha/2 = 0.025$ significance level.

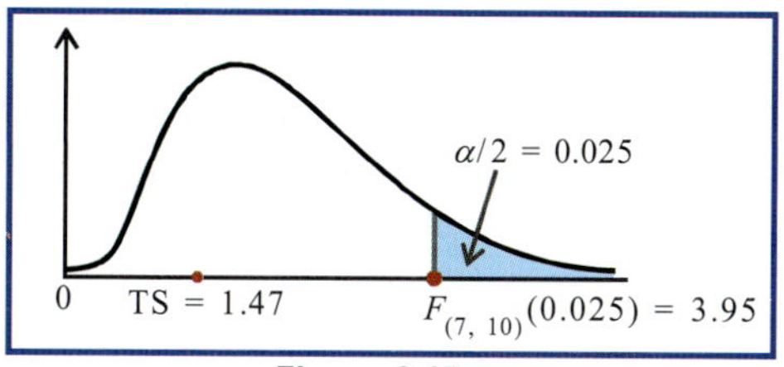

Figure 8.17

$\Rightarrow$ Critical value = $F_{(7,\,10)}(0.025) = 3.95$

The rejection region, the critical values, and the calculated value of test statistics are shown in Figure 8.17.

***Step* 4** **Decision:** Since the calculated value of test statistics lies outside the rejection region, we fail to reject H_0 at $\alpha = 0.05$. *Conclusion:* The data does not provide sufficient evidence against H_0. Hence, we can conclude that there does not seem to be a significant difference between variances– i.e., $\sigma_1^2 = \sigma_2^2$.

In short:

Hypotheses	**Test Statistics**
$H_0: \sigma_1^2 = \sigma_2^2$ $H_1: \sigma_1^2 \neq \sigma_2^2$	$F = \frac{s_1^2}{s_2^2} = 1.47$
Rejection Region and Critical Value	**Conclusion**
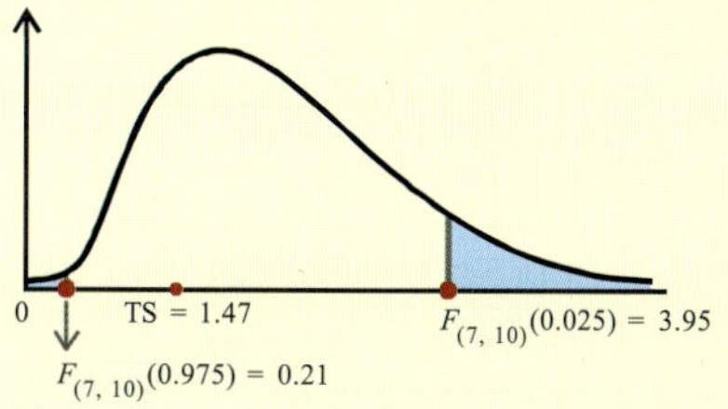	Fail to reject H_0. The two normal populations have equal variances.

Example 12: The weights of 21 green apples were taken and the variance was 9 grams². When 18 red apples were tested, the variance was 7 grams². At the $\alpha = 0.01$ level of significance test to see if the variance of green apples is greater than the variance of red apples.

Solution:

***Step* 0** Given statistics: $n_1 = 21$, $n_2 = 18$, $s_1^2 = 9$, and $s_2^2 = 7$

The stated claim can be represented as $\sigma_1^2 > \sigma_2^2$. Hence the counter claim is $\sigma_1^2 \leq \sigma_2^2$.

***Step* 1** $H_0 : \sigma_1^2 = \sigma_2^2$ (cc)

$H_1 : \sigma_1^2 > \sigma_2^2$ (oc)

***Step* 2** Since $s_1^2 > s_2^2$, under H_0, the test statistics is:

$$F = \frac{s_1^2}{s_2^2} \quad F_{(n_1-1,\, n_2-1)} = F_{(20,\,17)} \Rightarrow F = \frac{s_1^2}{s_2^2} = \frac{9}{7} = 1.29$$

***Step* 3** Since $H_1 : \sigma_1^2 > \sigma_2^2$, the test is one-tailed. Given that $s_1^2 > s_2^2$, the critical value of F for the one-tailed test at $\alpha = 0.01$ significance level is the same as the right-tailed critical value of F at $\alpha = 0.01$ significance level.

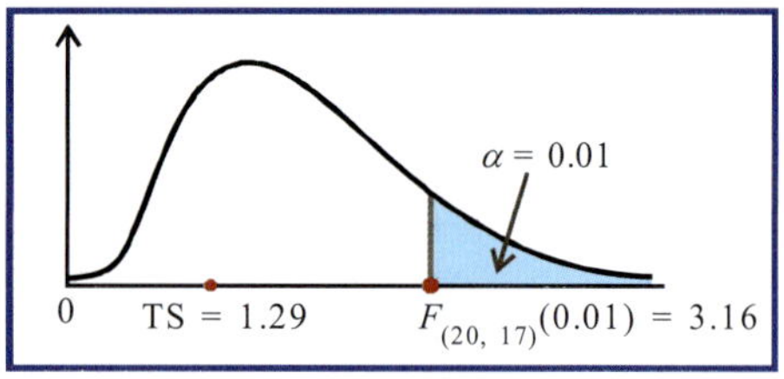

Figure 8.18

$\Rightarrow$ Critical value $= F_{(20,\ 17)}(0.01) = 3.16$

The rejection region, the critical values, and the calculated value of test statistics are shown in Figure 8.18.

***Step* 4** **Decision:** Since the calculated value of test statistics lies outside the rejection region, we fail to reject H_0 at $\alpha = 0.01$. *Conclusion:* The data does not provide sufficient evidence against H_0, hence, we can conclude that there does not seem to be a significant difference between variances, that is $\sigma_1^2 = \sigma_2^2$.

In short:

Hypotheses	**Test Statistics**
$H_0: \sigma_1^2 = \sigma_2^2$ $H_1: \sigma_1^2 > \sigma_2^2$	$F = \dfrac{s_1^2}{s_2^2} = 1.29$
Rejection Region and Critical Value 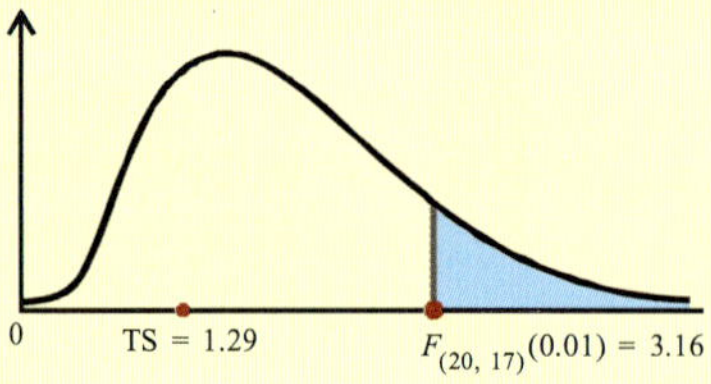	**Conclusion** Fail to reject H_0. The variance of weights of green apples is the same as that of red apples.

Try it yourself:

In the following problems, assume the parent populations are normally distributed.

1. Test the claim that $\sigma_1 > \sigma_2$. Use $\alpha = 0.05$.

Data Set I	Data Set II
$s_1 = 5.8$	$s_2 = 2.8$
$n_1 = 15$	$n_2 = 8$

2. Random samples were drawn from two normally distributed populations and the following results were obtained.

Sample I	16	17	18	19	20	21	22	24	26	27		
Sample II	19	22	23	25	26	28	29	30	31	32	35	35

Test the claim that the two populations have equal standard deviations.

3. Two independent samples are drawn from two normally distributed populations.

Sample I	70	78	73	66	71	64	64	79
Sample II	61	70	66	75	68	79	60	69

Test the claim that $\sigma_1 < \sigma_2$. Use $\alpha = 0.05$.

8.7: Inferences from Two Populations with Microsoft Excel®

Objectives

After completing this section you will be able to carryout the computations in Chapter 8 using Microsoft Excel®.

Illustration 1: Given the following sample statistics:

	A	B	C	D
1	Drunk Drivers		Sober Drivers	
2	n_1	150	n_2	200
3	$\bar{x}_1$	3.52	$\bar{x}_2$	1.03
4	σ_1	2.06	σ_2	0.85
5			α	0.05

(a) Find the Test Statistics for testing $H_0 : \mu_1 = \mu_2$. (Assume Normal Distribution)

(b) Construct a 95% confidence interval for the difference of the means.

Solution: **(a)**

Excel Command	Output
Test Statistics =(B3-D3)/SQRT(B4^2/B2+D4^2/D2)	13.94

Table 8.9

(b)

$$(\bar{x}_1 - \bar{x}_2) - z_{\alpha/2}\sqrt{\frac{\sigma_1^2}{n_1} + \frac{\sigma_2^2}{n_2}} < \mu_1 - \mu_2 < (\bar{x}_1 - \bar{x}_2) + z_{\alpha/2}\sqrt{\frac{\sigma_1^2}{n_1} + \frac{\sigma_2^2}{n_2}}$$

	A	B	C	D
1	Drunk Drivers		Sober Drivers	
2	n_1	150	n_2	200
3	$\bar{x}_1$	3.52	$\bar{x}_2$	1.03
4	σ_1	2.06	σ_2	0.85
5				
6	Critical Value	1.96		

Figure 8.19

Excel Command			Output
	A	B	
7			
8	LCL	=(B3–D3)–B6*SQRT(B4^2/B2+D4^2/D2)	$2.14 < \mu_1 - \mu_2 < 2.84$
9	UCL	=(B3–D3)+B6*SQRT(B4^2/B2+D4^2/D2)	

Table 8.10

Illustration 2: In this demonstration we are going to use the descriptive Statistics Package in Excel® to test the difference between means of two populations with equal variances. The data on the right represents 36 randomly selected final scores from two groups of Algebra classes, one using a technology based instructional delivery system and the other a traditional based one. Is there a significant difference between the means of two populations? Assume $\sigma_1^2 = \sigma_2^2 = 400$. Use $\alpha = 0.05$

Technology Based Class			Traditional Class		
83	84	50	22	84	55
95	74	65	49	34	77
76	97	87	59	48	83
99	100	43	20	76	52
97	63	84	43	100	46
83	89	79	50	77	31
66	70	99	36	27	98
95	85	60	35	27	85
100	79	92	18	63	50
95	71	101	44	18	69
85	46	30	61	82	67
98	56	85	62	68	80

Solution: To find the mean, standard deviation, sample size, and variances of the two populations, select **the Data Tab** and click on the **Data Analysis** icon from the **Analysis** group.

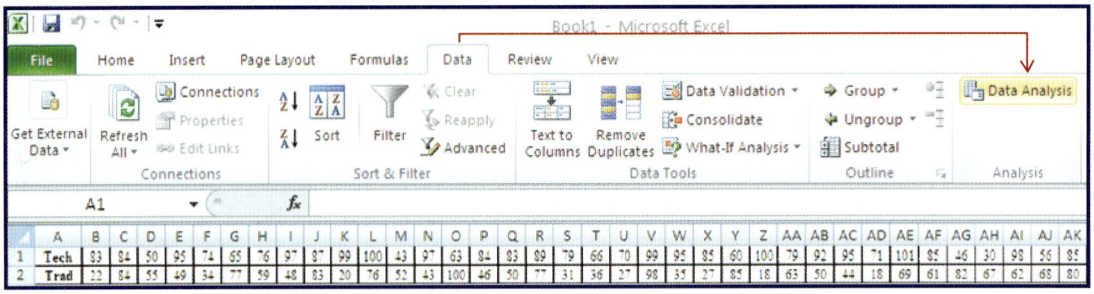

Figure 8.20

Select **z-Test: Two Samples for Means** from the Data Analysis dialog box.

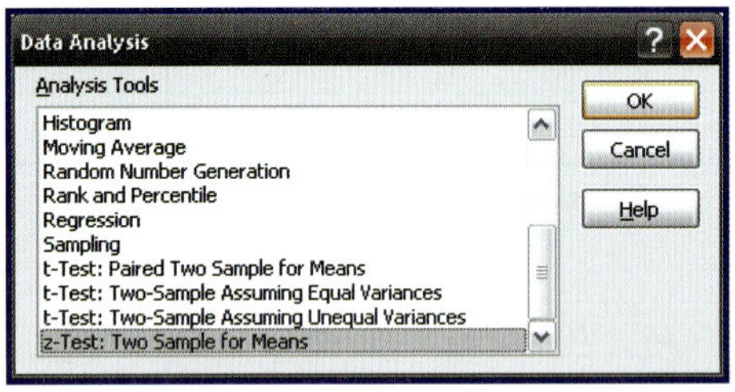

Figure 8.21 a

The following dialog box will appear.

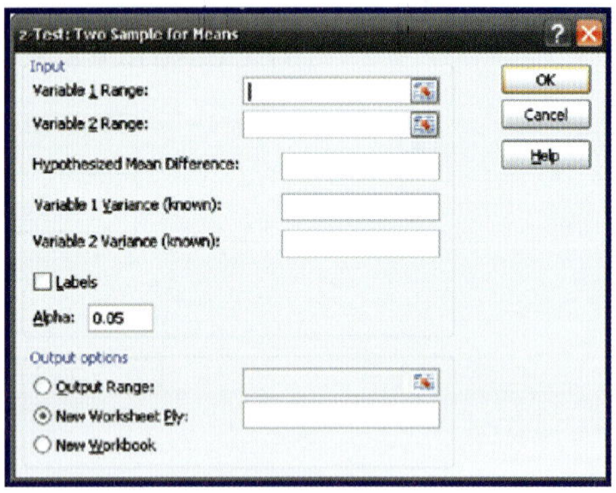

Figure 8.21 b

For the **Variable 1 and 2 Range** select the range of values in Columns A and B and fill out the rest of the boxes appropriately as shown below.

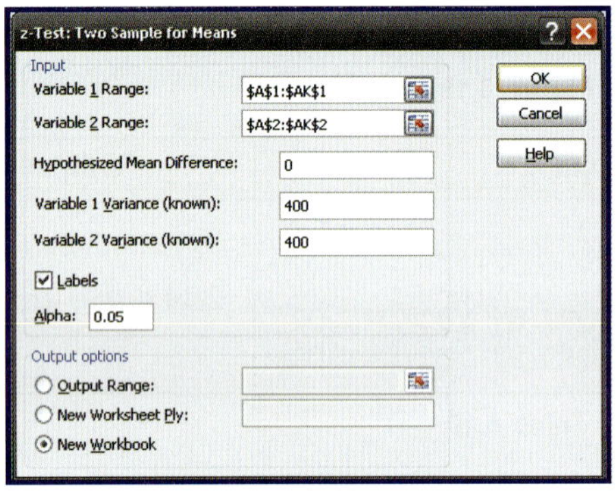

Figure 8.21 c

Press OK.

	A	B	C
1	z-Test: Two Sample for Means		
2			
3		*Tech*	*Trad*
4	Mean	79.47222222	55.44444444
5	Known Variance	400	400
6	Observations	36	36
7	Hypothesized Mean Difference	0	
8	z	5.097061381	
9	P(Z<=z) one-tail	1.72483E-07	
10	z Critical one-tail	1.644853627	
11	P(Z<=z) two-tail	3.44967E-07	
12	z Critical two-tail	1.959963985	

Figure 8.21 d

Now we have all the information we need to carry out the test.

$$H_0 : \mu_1 = \mu_2; \; H_1 : \mu_1 \neq \mu_2$$

From the Figure (8.21 d) we can read the critical value and the test statistics.

CV: 1.96 and TS is 5.097

Since the TS falls inside the Rejection Region, we reject the Null Hypothesis. There seems to be a difference between the means.

Illustration 3: We are given the following sample data from two groups of Statistics classes.

Tech based Statistics class	85	77	62	92	74	45	65	70	93	88
	65	66	25	75	85	91	80	65	69	64
Traditional Statistics class	75	80	45	20	36	78	91	50		

(a) Find the test statistics for the t-value for $\alpha = 0.05$. Assume $\sigma_1^2 = \sigma_2^2$

(b) Construct a 95% confidence interval for $\mu_1 - \mu_2$. Assume $\sigma_1^2 = \sigma_2^2$

Solution: **(a)**

	A	B	C	D	E	F	G	H	I	J	K	L	M	N	O	P	Q	R	S	T	U
1	Tech based Statistics class	85	77	62	92	74	45	65	70	93	88	65	66	25	75	85	91	80	65	69	64
2	Traditional Statistics class	75	80	45	20	36	78	91	50												

Figure 8.22 a

To find the test statistics and the t-value;

1. Click on the Data tab.
2. Click on Data Analysis icon under the Data Analysis group.

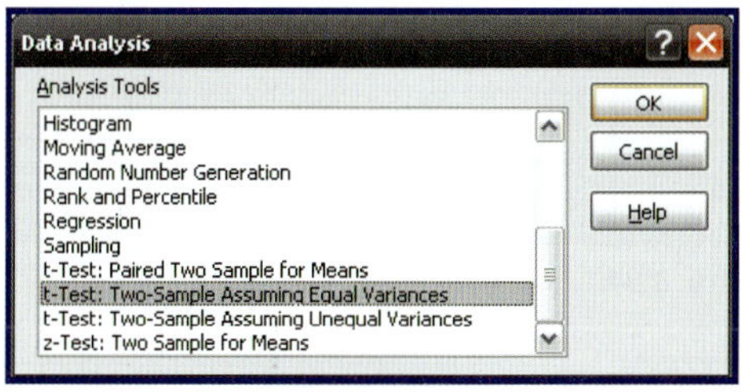

Figure 8.22 a

3. Click on t-test, and press OK.

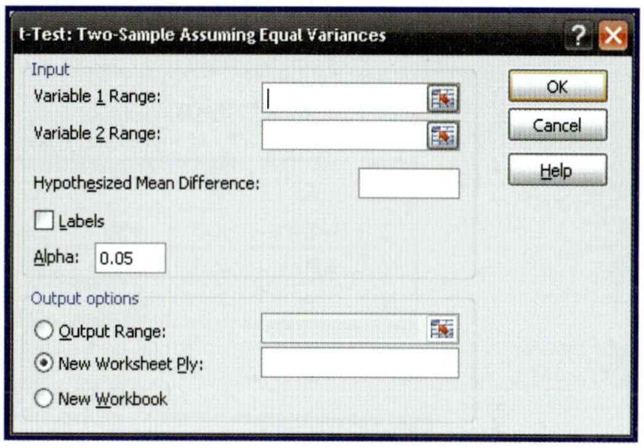

Figure 8.22 b

Fill out the boxes with proper information.

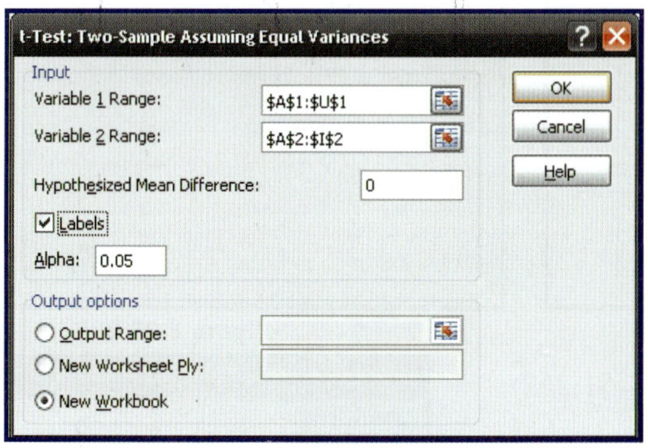

Figure 8.22 c

Press OK.

A	B	C
t-Test: Two-Sample Assuming Equal Variances		
	Tech based Statistics class	*Traditional Statistics class*
Mean	71.8	59.375
Variance	273.4315789	629.6964286
Observations	20	8
Pooled Variance	369.3490385	Pooled Variance
Hypothesized Mean Difference	0	
df	26	
t Stat	1.545463977	Test Statistics
P(T<=t) one-tail	0.06716015	
t Critical one-tail	1.705617901	Critical values for one tail and two tail tests
P(T<=t) two-tail	0.1343203	
t Critical two-tail	2.055529418	

Figure 8.22 d

Since the Test Statistics 1.55 is less than the one-tail Critical Value 1.71 (TS falls outside the Rejection Region), we Fail to Reject the Null Hypothesis. There is insufficient evidence to support the claim that the mean score in a Technology Based Statistics class is more than the mean score in a Traditional Based Statistics class.

(b)

	A	B	C
1		Variable 1	Variable 2
2	Mean	71.8	59.375
3	Variance	273.431579	629.696429
4	Observations	20	8
5	Pooled Variance	369.349038	
6	CV	2.05	

Figure 8.23

Excel Command			Output
	D	E	$-4.056 < \mu_1 - \mu_2 < 28.906$
3	LCL	=(B2-C2)-B6*SQRT(B5/B4+B5/C4)	
4	UCL	=(B2-C2)+B6*SQRT(B5/B4+B5/C4)	

Table 8.11

Illustration 4: The table below lists the before and after GRE scores of a group of 13 students having taken a preparatory course.

Before	800	950	1240	760	640	990	1000	1080	880	914	730	610	974
After	960	1400	1360	700	990	700	1300	1150	970	1000	700	830	970

(a) Construct a 99% confidence interval of the mean difference between the before and after scores (μ_d).

Solution: **(a)**

	A	B	C	D	E	F	G	H	I	J	K	L	M	N
1	**Before**	800	950	1240	760	640	990	1000	1080	880	914	730	610	974
2	**After**	960	1400	1360	700	990	700	1300	1150	970	1000	700	830	970
3	**D**	-160	-450	-120	60	-350	290	-300	-70	-90	-86	30	-220	4

Figure 8.24

Commands

	A	B	C	D	E	F
1	GRE score before the prep course	GRE score after the prep course	d		Parameters	Values
2	800	960	-160		n	=COUNT(C2:C14)
3	950	1400	-450		dbar	=AVERAGE(C2:C14)
4	1240	1360	-120		sd	=STDEV(C2:C14)
5	760	700	60		$t_{n-1}(\alpha/2)$	=T.INV.2T(0.01,12)
6	640	990	-350		E	=F6*F4/SQRT(F2)
7	990	700	290			
8	1000	1300	-300		LCL	=F3–F7
9	1080	1150	-70		UCL	=F3+F7
10	880	970	-90			
11	914	1000	-86			
12	730	700	30			
13	610	830	-220			
14	974	970	4			

Figure 8.25 a

Outputs

	A	B	C	D	E	F
1	GRE score before the prep course	GRE score after the prep course	d		Parameters	Values
2	800	960	-160		n	13
3	950	1400	-450		dbar	-112.4615385
4	1240	1360	-120		sd	193.2168969
5	760	700	60		$t_{n-1}(\alpha/2)$	3.054539586
6	640	990	-350		E	163.6888828
7	990	700	290			
8	1000	1300	-300		LCL	-276.1504213
9	1080	1150	-70		UCL	51.22734433
10	880	970	-90			
11	914	1000	-86			
12	730	700	30			
13	610	830	-220			
14	974	970	4			

Figure 8.25 b

$$-276.175 < \mu_d < 51.252$$

(b) Calculate the test statistics and the critical value using Data Analysis dialog box.

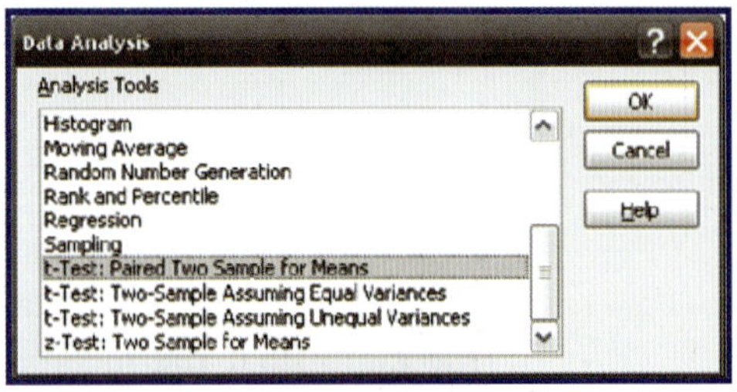

Figure 8.26 a

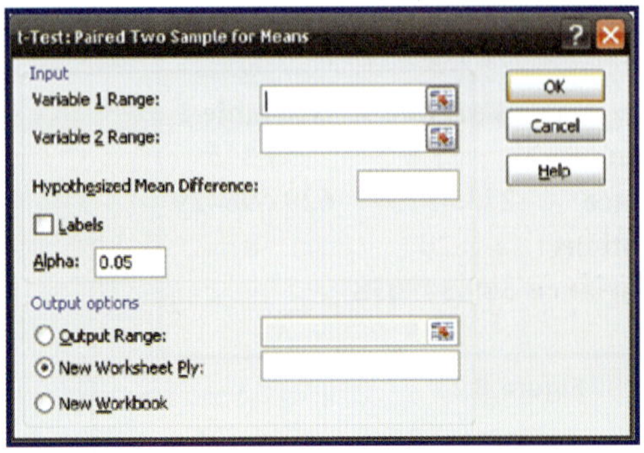

Figure 8.26 b

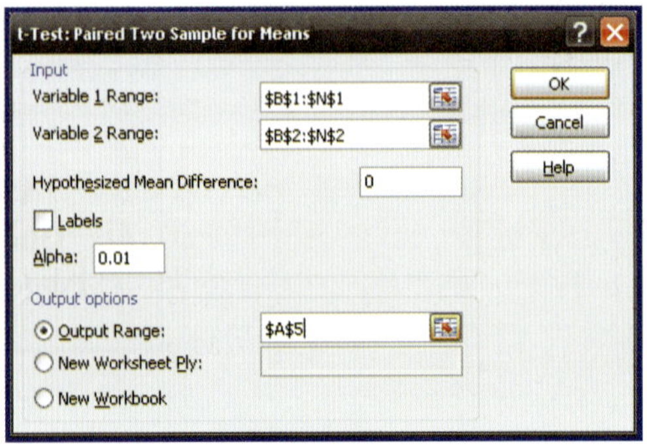

Figure 8.26 c

t-Test: Paired Two Sample for Means		
	Variable 1	*Variable 2*
Mean	889.846154	1002.307692
Variance	31960.9744	58702.5641
Observations	13	13
Pearson Correlation	0.61561517	
Hypothesized Mean Difference	0	
df	12	
t Stat	-2.09860447	← Test Statistics
P(T<=t) one-tail	0.02884385	
t Critical one-tail	2.68099799	← Left tail Critical value
P(T<=t) two-tail	0.0576877	
t Critical two-tail	3.05453959	

Figure 8.26 d

Illustration 5: In a random sample of 50 Statistics professors in the Los Angeles Area, 8 indicated that they use technology to teach Statistics. Among a random sample of 90 such professors in the New York area, 12 indicated that they used technology.

(a) Calculate the Test Statistics for testing $H_0 : p_1 = p_2$ against $H_1 : p_1 > p_2$, by using Microsoft Excel®.

(b) Construct a 95% confidence interval for the difference of proportions $(p_1 - p_2)$ using Microsoft Excel®.

Solution: **(a)** Input the Data:

	A	B	C
1		**LA**	**NY**
2	x	8	12
3	n	50	90
4	***p* hat**	0.16	0.13333333
5	***q* hat**	0.84	0.86666667

Figure 8.27

Commands

	E	F
1	***p* bar**	=(B2+C2)/(B3+C3)
2	***q* bar**	=1–F1
3	**RT CV**	=ABS(NORM.S.INV(0.05))
4	**TS**	=(B4-C4)/SQRT(F1*F2/B3+F1*F2/C3)
5		

Figure 8.28 a

Output

	E	F
1	***p* bar**	0.142857143
2	***q* bar**	0.857142857
3	**RT CV**	1.644853627
4	**TS**	0.43204938

Figure 8.28 b

Commands

	A	B
6		
7	**CV**	=ABS(NORM.S.INV(0.025))
8	**LCL**	=(B4-C4)-B7*SQRT(B4*B5/B3+C4*C5/C3)
9	**UCL**	=(B4-C4)+B7*SQRT(B4*B5/B3+C4*C5/C3)

Figure 8.29 a

Output

	A	B
6		
7	**CV**	1.95996398
8	**LCL**	-0.09685695
9	**UCL**	0.15019029

Figure 8.29 b

8.8: Extended Topics and Proofs for Inferences from Two Populations

Objectives

After completing this section you will be able to understand:

A. Inferences about Two Means.

B. Inferences about Two Proportions.

C. Inferences about Two Standard Deviations or Variances.

A. Inferences About Two Means

- **σ_1 and σ_2 known and samples independent**

If the variances $\sigma_1^{\,2}$ and $\sigma_2^{\,2}$ are known, then the difference of the means $\mu_1 - \mu_2$ follows a normal distribution.

Let $\bar{X}_1$ and $\bar{X}_2$ denote the random variables that represent the sampling distribution of the means from two populations. By Central Limit Theorem, the sampling distributions of $\bar{X}_1$ and $\bar{X}_2$, for large samples with known variances, are given by:

$$\bar{X}_1 \sim N\left(\mu_1, \sigma_1/\sqrt{n_1}\right) \text{ and } \bar{X}_2 \sim N\left(\mu_2, \sigma_2/\sqrt{n_2}\right)$$

$E(\bar{X}_1) = \mu_1,\ E(\bar{X}_2) = \mu_2$	$\Rightarrow E(\bar{X}_1 - \bar{X}_2) = E(\bar{X}_1) - E(\bar{X}_2) = \mu_1 - \mu_2$
$\sigma_{\bar{X}_1} = \dfrac{\sigma_1}{\sqrt{n_1}},\ \sigma_{\bar{X}_2} = \dfrac{\sigma_2}{\sqrt{n_2}}$	$\Rightarrow \sigma_{\bar{X}_1 - \bar{X}_2} = \sqrt{\dfrac{\sigma_1^{\,2}}{n_2} + \dfrac{\sigma_2^{\,2}}{n_2}}$

It follows that

$$Z = \frac{(\bar{X}_1 - \bar{X}_2) - E(\bar{X}_1 - \bar{X}_2)}{\sigma_{\bar{X}_1 - \bar{X}_2}} = \frac{(\bar{X}_1 - \bar{X}_2) - (\mu_1 - \mu_2)}{\sqrt{\dfrac{\sigma_1^{\,2}}{n_1} + \dfrac{\sigma_2^{\,2}}{n_2}}} \sim N(0, 1)$$

We can also obtain a $(1 - \alpha)100\%$ confidence interval from solving $P\left[-z_{\alpha/2} < Z < z_{\alpha/2}\right] = 1 - \alpha$

$$(\bar{x}_1 - \bar{x}_2) - z_{\alpha/2}\sqrt{\frac{\sigma_1^{\,2}}{n_1} + \frac{\sigma_2^{\,2}}{n_2}} < \mu_1 - \mu_2 < (\bar{x}_1 - \bar{x}_2) + z_{\alpha/2}\sqrt{\frac{\sigma_1^{\,2}}{n_1} + \frac{\sigma_2^{\,2}}{n_2}}$$

- **σ_1 and σ_2 unknown but assumed equal and samples independent**

In most applications, the variances are unknown but in some cases it can be assumed that they are equal $\sigma_1^{\,2} = \sigma_2^{\,2} = \sigma^2$. A pooled estimator of the common variance is the weighted average.

$$S_p^{\,2} = \frac{(n_1 - 1)S_1^{\,2} + (n_2 - 1)S_2^{\,2}}{n_1 + n_2 - 2}$$

If we let $W = \dfrac{(n_1 + n_2 - 2)S_p^{\,2}}{\sigma^2}$, then W can be written as the sum of two independent Chi-Square Distributed random variables with $(n_1 - 1)$ and $(n_2 - 1)$ degrees of freedom, respectively.

$$W = \frac{(n_1 - 1)S_1^2}{\sigma^2} + \frac{(n_2 - 1)S_2^2}{\sigma^2} \sim \chi^2_{n_1 + n_2 - 2}$$

We can now write

$$T = \frac{Z}{\sqrt{\dfrac{W}{\nu}}} = \frac{\dfrac{(\bar{X}_1 - \bar{X}_2) - (\mu_1 - \mu_2)}{\sqrt{\dfrac{\sigma_1^2}{n_1} + \dfrac{\sigma_2^2}{n_2}}}}{\sqrt{\dfrac{(n_1 + n_2 - 2)S_p^2}{\sigma^2 (n_1 + n_2 - 2)}}} = \frac{(\bar{X}_1 - \bar{X}_2) - (\mu_1 - \mu_2)}{\sqrt{\dfrac{S_p^2}{n_1} + \dfrac{S_p^2}{n_2}}} \sim t_{n_1 + n_2 - 2}$$

The $(1 - \alpha)100\%$ confidence interval for $(\mu_1 - \mu_2)$ is then given by:

$$(\bar{x}_1 - \bar{x}_2) - t_{n_1 + n_2 - 2, \alpha/2} \sqrt{\frac{s_p^2}{n_1} + \frac{s_p^2}{n_2}} < \mu_1 - \mu_2 < (\bar{x}_1 - \bar{x}_2) + t_{n_1 + n_2 - 2, \alpha/2} \sqrt{\frac{s_p^2}{n_1} + \frac{s_p^2}{n_2}}$$

- **σ_1 and σ_2 unknown but assumed unequal and samples independent**

For the unequal variances, we can use the small sample approximation based on the *t*-distribution.

$$T = \frac{(\bar{X}_1 - \bar{X}_2) - (\mu_1 - \mu_2)}{\sqrt{\dfrac{s_1^2}{n_1} + \dfrac{s_2^2}{n_2}}} = t_v$$

$$\text{where } df\,(v) = \frac{\left(\dfrac{s_1^2}{n_1} + \dfrac{s_2^2}{n_2}\right)^2}{\left[\dfrac{\left(\dfrac{s_1^2}{n_1}\right)^2}{n_1 - 1}\right] + \left[\dfrac{\left(\dfrac{s_2^2}{n_2}\right)^2}{n_2 - 1}\right]}$$

Note: In this text, we have used the following approximation to determine the above degrees of freedom.

v = smaller($n_1 - 1$, $n_2 - 1$)

- **Dependent samples (Matched Pairs)**

Given a random sample of n ordered pairs, (X_i, Y_i), let $D_i = Y_i - X_i$ denote the difference between each ordered pair for $i = 1, 2, \ldots, n$. We assume that D_i are normally distributed with a mean $\mu_D = \mu_2 - \mu_1$ and variance $\sigma_D^2 = \sigma_1^2 + \sigma_2^2 - 2\sigma_{12}$. Hence, $D_i \sim N(\mu_2 - \mu_1, \sigma_D^2)$.

Let $\bar{D} = \dfrac{\sum_{i=1}^{n} D_i}{n} = \bar{Y} - \bar{X}$ and $s_D^2 = \dfrac{n\sum_{i=1}^{n} D_i^2 - \left(\sum_{i=1}^{n} D_i\right)^2}{n(n-1)}$

It follows that $T = \dfrac{\bar{D} - (\mu_2 - \mu_1)}{S_D/\sqrt{n}} \sim t_{n-1}$

The $(1 - \alpha)100\%$ confidence interval for $(\mu_1 - \mu_2)$ is then given by:

$$\bar{d} - t_{n-1,\ \alpha/2}\frac{s_D}{\sqrt{n}} < D_\mu < \bar{d} + t_{n-1,\ \alpha/2}\frac{s_D}{\sqrt{n}}$$

Note: The above theory remains valid even if the samples are independent, except for the degrees of freedom. The degrees of freedom in the matched pair procedure when the samples are independent and variances are assumed equal is $2n - 2$.

B. Inferences about Two Proportions

Assume $X_1 \sim B(n_1, p_1)$ and $X_2 \sim B(n_2, p_2)$. Let $\hat{p}_1 = \frac{x_1}{n_1}$ and $\hat{p}_2 = \frac{x_2}{n_2}$. We know that

$$Z = \frac{(\hat{p}_1 - \hat{p}_2) - (p_1 - p_2)}{\sqrt{\frac{\hat{p}_1\hat{q}_1}{n_1} + \frac{\hat{p}_2\hat{q}_2}{n_2}}} \sim N(0,\ 1).$$

Rationale:

The sampling distributions of $\hat{p}_1$ and $\hat{p}_2$ are given by:

$$\hat{p}_1 \sim N\left(p_1, \sqrt{\frac{p_1q_1}{n_1}}\right) \quad \text{and} \quad \hat{p}_2 \sim N\left(p_2, \sqrt{\frac{p_2q_2}{n_2}}\right)$$

That is,

$$E(\hat{p}_1) = p_1,\ \sigma_{\hat{p}_1} = \sqrt{\frac{p_1 q_1}{n_1}} \quad ; \quad E(\hat{p}_2) = p_2,\ \sigma_{\hat{p}_2} = \sqrt{\frac{p_2 q_2}{n_2}}$$

Therefore,

$$E(\hat{p}_1 - \hat{p}_2) = E(\hat{p}_1) - E(\hat{p}_2) = p_1 - p_2, \text{ and}$$

$$\text{var}(\hat{p}_1 - \hat{p}_2) = \text{var}(\hat{p}_1) + \text{var}(\hat{p}_2) = \frac{p_1 q_1}{n_1} + \frac{p_2 q_2}{n_2} \quad \Rightarrow \quad \sigma_{\hat{p}_1 - \hat{p}_2} = \sqrt{\frac{p_1 q_1}{n_1} + \frac{p_2 q_2}{n_2}}.$$

Hence, the standard variable for $(\hat{p}_1 - \hat{p}_2)$

$$Z = \frac{(\hat{p}_1 - \hat{p}_2) - E(\hat{p}_1 - \hat{p}_2)}{\sigma_{\hat{p}_1 - \hat{p}_2}} = \frac{(\hat{p}_1 - \hat{p}_2) - (p_1 - p_2)}{\sqrt{\frac{p_1q_1}{n_1} + \frac{p_2q_2}{n_2}}} \sim N(0,\ 1)$$

The confidence interval for $p_1 - p_2$ can be derived similar to the one-sample case.

$$(\hat{p}_1 - \hat{p}_2) - z_{\alpha/2}\sqrt{\frac{\hat{p}_1 \hat{q}_1}{n_1} + \frac{\hat{p}_2 \hat{q}_2}{n_2}} < p_1 - p_2 < (\hat{p}_1 - \hat{p}_2) + z_{\alpha/2}\sqrt{\frac{\hat{p}_1 \hat{q}_1}{n_1} + \frac{\hat{p}_2 \hat{q}_2}{n_2}}$$

C. Inferences about Two Standard Deviations or Variances

• Probability Function of *F*-Distribution

The probability density function of *F*-distribution with (v_1, v_2) *df* is

$$p(F) = C\frac{F^{\frac{v_1}{2}-1}}{\left(1+\frac{v_1}{v_2}F\right)^{(v_1+v_2)/2}}; \quad 0 \le F < \infty,$$

where C is a constant depending on *df*'s v_1 and v_2, so that the total area under the probability curve is 1. The graph of the function $p(F)$ varies with the *df* v_1 and v_2.

• Characteristics of $F(v_1, v_2)$ Distribution

1. Mean $= \frac{v_2}{v_2-2}$; $v_2 > 2$; that is, the mean is independent of the first *df* v_1.

2. Mode $= \frac{v_2(v_1-2)}{v_1(v_2+2)} = \left(\frac{v_2}{v_2+2}\right)\left(\frac{v_1-2}{v_1}\right) < 1$ **(Mode is always less than 1)**

3. Skewness $= \frac{\text{Mean} - \text{Mode}}{\text{SD}} > 0$ (*F*-distribution is positively skewed, skewed to the right.)

Assume independent random samples of sizes n_1 and n_2 taken from two normally distributed populations $X \sim N(\mu_1, \sigma_1)$ and $Y \sim N(\mu_2, \sigma_2)$ respectively.

• Confidence Interval for $\frac{\sigma_1^2}{\sigma_2^2}$

We know that $\frac{S_1^2\sigma_2^2}{S_2^2\sigma_1^2} \sim F(n_1 - 1, n_2 - 1)$.

Therefore, $P\left[f_{n_1-1,n_2-1}\left(\frac{\alpha}{2}\right) < \frac{s_1^2\sigma_2^2}{s_2^2\sigma_1^2} < f_{n_1-1,n_2-1}\left(1-\frac{\alpha}{2}\right)\right] = 1 - \alpha.$

Thus, $(1 - \alpha)100\%$ confidence interval for $\frac{\sigma_2^2}{\sigma_1^2}$ will be

$$\frac{s_2^2}{s_1^2}f_{n_1-1,n_2-1}\left(\frac{\alpha}{2}\right) < \frac{\sigma_2^2}{\sigma_1^2} < \frac{s_2^2}{s_1^2}f_{n_1-1,n_2-1}\left(1-\frac{\alpha}{2}\right)$$

Inferences for Two Means-Independent Samples-Known Variances (*Z*-TEST)

In exercises 1-10, use both the *Classical* and the *P-value* methods for testing the hypothesis.

1. Given the sample data in the table below:

Group 1	Group 2
$n_1 = 68$	$n_2 = 104$
$\bar{x}_1 = 8.5$	$\bar{x}_2 = 8.2$
$\sigma_1 = 2.6$	$\sigma_2 = 3.4$

(a) Test the claim that $\mu_1 = \mu_2$. Use a 0.05 significance level.

(b) Construct a 95% confidence interval estimate of the difference between the two Population means.

2. EXPA Media Research reported that adult women watch TV an average of 5.3 hrs. per day, compared to an average of 4.6 hrs. for adult men. Assume that those results are found from a sample of 120 men and 160 women and that the two population standard deviations are minimum 48 and 60 respectively.

(a) Test the belief that women spend more time watching TV than men. Use a 0.05 significance level.

(b) Construct a 90% confidence interval for the difference $\mu_1 - \mu_2$ where μ_1 is the mean time adult women spend watching TV.

3. A study of the socioeconomic effects of education included annual incomes of women with an undergraduate degree and women with a graduate degree.

Undergraduate	$n_1 = 285$	$\bar{x}_1 = \$48,500$	$\sigma_1 = \$8600$
Graduate	$n_2 = 190$	$\bar{x}_2 = \$57,800$	$\sigma_2 = \$12,500$

(a) Using a 0.01 significance level, is there enough evidence to support the claim that women with a graduate degree have incomes with a higher mean?

(b) Construct a 98% confidence interval estimate of the difference between the mean income of women with undergraduate and graduate degrees.

4. Customer waiting times are studied at the Franklin Mutual Bank. When 20 randomly selected customers enter any one of several teller lines for general accounts, their waiting times have a mean of 8.25 min and a population standard deviation of 3.62 min. When 24 randomly selected customers enter the double teller lines that serve the business accounts, their waiting times have a mean of 10.46 min and a population standard deviation of 2.18 min. Assume the parent populations are normally distributed.

(a) Use a 0.05 significance level to test the claim that the mean waiting times for business accounts is longer than the mean waiting times for general accounts.

(b) Construct a 90% confidence interval for the difference of mean waiting times $\mu_1 - \mu_2$.

5. A study of seat belt use involved minors who were hospitalized as a result of traffic accidents. For a group of 45 minors who were not wearing seat belts, the numbers of days spent in the hospital have a mean of 3.42 days and a population standard deviation of 2.16 days. For a group of 52 minors who were wearing seat belts, the numbers of days spent in the hospital have a mean of 1.68 days and a population standard deviation of 0.86.

(a) Using a 0.01 significance level, is there sufficient evidence to support the claim that the population of minors not wearing seat belts has a higher mean number of days spent in the hospital?

(b) Construct a 98% confidence interval for the difference between the means. Does the constructed confidence interval support the conclusion in part (a)?

6. At Pulsar College, 160 randomly selected student cars had ages with a mean of 8.24 years and a population standard deviation of 2.58 years, and 85 randomly selected faculty cars had ages with a mean of 4.68 years and a population standard deviation of 4.32 years.

   (a) Using a 0.005 significance level, is there sufficient evidence to support the claim that student cars are older than faculty cars? Let μ_1 denote the mean age of student cars.

   (b) Construct a 99% confidence interval estimate of the difference between the means $\mu_1 - \mu_2$. Interpret your results and compare it to the conclusion in part (a).

7. As part of a Health Survey, data were collected on the weights of women in two different age groups. For 600 randomly selected women aged 25-35, the mean was 143 pounds and the population standard deviation was 23 pounds. For 860 randomly selected women aged 55-65, the mean and population standard deviation were158 and 28 pounds respectively.

   (a) Test the claim that the mean weights of women in two different age groups are equal. Use a 0.05 significance level.

   (b) Construct a 95% confidence interval for the difference between the mean weights of women in the two age groups.

8. A manufacturer claims that the calling range (in feet) of its 5.8-GHz cordless telephones is greater than that of its leading competitor. You perform a study using 14 randomly selected 900-MHz cordless telephones from the manufacturer and 16 similar phones from its competitor. The results are shown in the accompanying table. Assume the populations are normally distributed.

Manufacturer	Competitor
$n_1 = 14$	$n_2 = 16$
$\bar{x}_1 = 1800$	$\bar{x}_2 = 1500$
$\sigma_1 = 100$	$\sigma_2 = 150$

   (a) At $\alpha = 0.05$, is there enough evidence to support the manufacturer's claim?

   (b) Construct a 90% confidence interval for $\mu_1 - \mu_2$.

9. The mean ACT score for 43 male high school students is 21.2 and the population standard deviation is 4.9. The mean ACT score for 56 female high school students is 20.9 and the population standard deviation is 4.6.

   (a) At $\alpha = 0.01$, do male and female high school students have equal mean ACT scores?

   (b) Construct a 99% confidence interval for $\mu_1 - \mu_2$.

10. Consumers tend to buy more items when there is a sale, but they do not necessarily spend less money on their total purchase because they tend to buy more items. A department store observed 100 randomly selected shoppers during a sale and found that they spent an average of \$70.65, with a population standard deviation of \$32.90. In the same department store, 70 randomly selected shoppers on a day when there was no sale spent an average of \$66.00, with a population standard deviation of \$23.80.

   (a) Using a 0.05 significance level, test the claim that there is no significant difference between the populations means.

   (b) Construct a 95% confidence interval for the difference of the population means.

Inferences for Two Means-Independent Samples-Unknown Variances (*t*-test)

In exercises 11-23, assume the parent populations are normally distributed (if required) and population variances equal.

11. An aptitude test, based on gender was conducted for young adults. The table below lists the results of 15 randomly selected males and females who took the test.

Young Adult	A	B	C	D	E	F	G	H	I	J	K	L	M	N	O
Female	30	19	19	12	29	18	42	20	12	39	14	33	17	31	22
Male	30	6	14	33	14	42	14	22	17	35	11	30	28	17	35

(a) Is there sufficient evidence to support the claim that the mean test scores for females and males are equal? Assume a 0.05 significance level. Use both the *Classical* and the *P-value* methods for testing the hypothesis in the this exercise.

(b) Construct a 95% confidence interval for the difference of the mean test scores.

12. Professor Badalian compared the scores of his morning class with evening class on the same quiz. The 50 morning students' average score was 63.5 with a standard deviation of 4.8, while the 64 evening students' average score was 72.1 with a standard deviation of 5.6.

(a) Test the claim that $\mu_1 = \mu_2$. Use $\alpha = 0.05$.

(b) Construct a 95% confidence interval for the difference between the two population means.

13. Do women understate their weight? The following table gives the reported and the actual weights of 10 randomly selected women.

	1	2	3	4	5	6	7	8	9	10
Reported	130	120	115	124	140	105	112	115	120	125
Actual	132	126	116	127	145	105	120	119	123	127

(a) At the $\alpha = 0.01$, is there sufficient evidence to conclude that the mean of actual weights is higher than the reported?

(b) Construct a 98% confidence interval for $\mu_1 - \mu_2$.

14. One group of Mathematics students took a distance learning class while another group took the same course in a regular classroom. The sample data is available in the table below.

Statistic	Distance Learning	Regular Classroom
Number of students	40	50
Mean scores	30.2	42.5
Sample std. dev.	2.4	2.5

(a) Is there a significance difference in the mean scores? Test the claim using $\alpha = 0.05$ significance level.

(b) Construct a 95% confidence interval for the difference between the two population means.

15. An advertising executive claims that there is a difference in the mean household income for credit card holders of Visa Gold and of American Express Gold. The results of a survey of 38 randomly selected customers from each group are shown in the accompanying table.

Visa Gold	$n_1 = 38$	$\bar{x}_1 = \$64,500$	$s_1 = \$10,600$
American Express Gold	$n_2 = 38$	$\bar{x}_2 = \$68,400$	$s_2 = \$15,200$

(a) Do the results support the executive's claim? Use $\alpha = 0.05$.

(b) Construct a 95% confidence interval for $\mu_1 - \mu_2$.

16. Salaries (in \$) for 10 males and 8 females of low income workers in a certain metropolitan area are given below.

Men ($n = 10$)		**Women ($n = 8$)**	
8,400	8,322	9,200	6,163
9,900	8,041	8,100	8,340
6,352	7,462	8,940	9,600
8,050	9,611	9,930	9,993
8,209	8,814		

(a) Is there a significant difference between the salaries of male and female low income workers? Use $\alpha = 0.01$.

(b) Construct a 95% confidence interval for the difference between the salaries of two different gender workers.

17. Consumer Reports tested several types of snow tires to determine how well each performed under winter conditions. When traveling on ice at 15 mph, 10 *Winter fire* tires had a mean stopping distance of 51 feet with a standard deviation of 8 feet. The mean stopping distance for 12 *XMS Alpine* tires was 55 feet with a standard deviation of 3 feet.

(a) Can you conclude that there is a significant difference in the mean stopping distances of the two types of tires? Use $\alpha = 0.01$.

(b) Construct a 99% confidence interval for $\mu_1 - \mu_2$.

18. In a study of the hypothesis that students who finish an exam later get better scores, Professor Badalian kept track of the order in which papers were handed in. The first 32 papers showed a mean score of 87.5 with a standard deviation of 9.6, while the last 28 papers handed in showed a mean score of 89.5 with a standard deviation of 4.9. Is this a significant difference at $\alpha = 0.05$?

(a) Test the claim that $\mu_1 > \mu_2$.

(b) Construct a 90% confidence interval for the difference between the average scores.

19. At a certain university, the ages of 37 randomly selected male professors have a mean of 43 years with a standard deviation of 8 years, and the 33 randomly selected female professors had a mean age of 39 years with a standard deviation of 10 years.

(a) At a 0.05 significance level, is there enough evidence to conclude that the mean age of male professors at this university is higher than that of female professors?

(b) Construct a 90% confidence interval for the difference of the population means ($\mu_1 - \mu_2$).

20. In a study of the relationship between the salaries of college students before and after they receive their AA degrees, the following data (in dollars) was collected from 10 randomly selected college students.

Before	14254	12657	21646	21315	21321	13065	16156	15164	12131	12154
After	13587	17254	31313	22324	20832	29461	22154	16412	38580	23460

(a) Use 0.01 significance level to test the claim that an AA degree increases the average income of college students. (Hint: $\mu_1 < \mu_2$).

(b) Construct a 99% confidence interval for the difference of the means.

21. Has the cost to outsource a particular job changed from 2010 to 2011? A random sample of 10 companies in spring 2010 showed a sample average cost of \$112 with a sample standard deviation equal to \$28. A random sample of 12 different companies in spring 2011 resulted in a sample average cost of \$125 with a sample standard deviation equal to \$42.

(a) Test the claim that $\mu_1 < \mu_2$. Use $\alpha = 0.01$.

(b) Construct a 98% confidence interval for $\mu_1 - \mu_2$.

22. A safety engineer records the braking distances of two types of tires. 35 tires are randomly selected from each group. The mean braking distance for Tire Type A is 42 feet with a standard deviation of 4.7 feet. The mean braking distance for Tire Type B is 45 feet with a standard deviation of 4.3 feet.

(a) At $\alpha = 0.1$, can the engineer support the claim that the mean braking distances is the same for both types of tires?

(b) Construct a 90% confidence interval for the difference of the population means.

23. In order to test the gasoline consumption (in mpg) of two types of vehicles, 12 cars were randomly selected from each type. The results are listed in the table below.

Low Emission Sedans	33.3	38.7	34.3	36.6	41.2	34.7	33.5	34.2	40.1	39.3	32.4	33.3
Flex Fuel SUV's	26.3	28.9	29.4	27.3	25.4	24.3	25.6	23.3	26.8	25.7	28.9	30.1

(a) Is there sufficient evidence to conclude that there is a significant difference between the average gas consumption of the two types of vehicles?

(b) Construct a 95% confidence interval for the difference of the population means.

Inferences for Two Means-Independent Samples-Unknown Variances (*t*-test)

In exercises 24-38, assume the parent populations are normally distributed (if required) and population variances are unequal. Use both the *Classical* and the *P-value* methods for testing the hypothesis.

24. The following table gives the prices (in $) for several types of services at two independent hair salons.

Services	Salon 1	Salon 2
Child haircut	10	9
Adult haircut	15	20
Shampoo and haircut for a child	20	18
Shampoo and haircut for an adult	25	30
Color	30	40
Styling	25	30

(a) Using $\alpha = 0.01$, test to see whether the mean price for services provided in the second salon is significantly more than the first salon.

(b) Construct a 98% confidence interval for the difference of mean prices between the salons.

25. A BMW car bumper was designed to reduce damage in high-speed collisions. This bumper was installed on an experimental group of cars, but not on a control group. At the end of a trial period, accident data showed 10 repair incidents for the experimental group and 12 repair incidents for the control group. Vehicle downtime (in days per repair incident) is shown below.

Experimental group: 9, 8, 9, 11, 5, 6, 2, 7, 3, 10

Control group: 7, 5, 8, 9, 2, 6, 7, 4, 5, 5, 11, 12

(a) Is the new bumper reducing downtime? Use $\alpha = 0.01$.

(b) Construct a 98% confidence interval for the difference between the downtime of experimental group and control group.

26. A real estate agent claims that there is no significant difference between the mean household incomes of two neighboring cities. The mean income of 12 randomly selected households from the first city was $48,250 with a standard deviation of $8000. In the second city, 10 randomly selected households had a mean income of $53,500 with a standard deviation of $3000.

(a) Test the agent's claim using a 0.01 significance level.

(b) Construct a 99% confidence interval for the difference of the population means.

27. A study of fast food nutrition compared the caloric content of French Fries. Thirty-eight servings of Burger Queen's medium French Fries had a mean of 370 calories and a standard deviation of 50 calories, while 35 servings of Hampton's medium French Fries had a mean of 350 calories and a standard deviation of 45 calories.

(a) At a 0.05 significance level, can you support the claim that the caloric content of the two types of French Fries is different?

(b) Construct a 95% confidence interval for $\mu_1 - \mu_2$.

28. In the study of the hypothesis 'Medicare spending per patient (in \$) in different U.S. metropolitan areas may differ' a researcher collected samples from the Eastern and Western regions. Here is the data summarized in a table.

Statistic	Eastern Region	Western Region
n	14 patients	16 patients
Average	2,120	4,480
SD	1,236	1,123

(a) Is the average spending in the eastern region significantly less than the average spending in the western region? Use $\alpha = 0.01$.

(b) Construct a 98% confidence interval for the difference between the Medicare spending per patient in Eastern and Western regions.

29. In a study of color blindness, 600 men and 500 women are randomly selected and tested. Among the men, 53 have color blindness. Among the women, 15 have color blindness.

(a) Is there sufficient evidence to support the claim that men have a higher rate of color blindness than women? Use a 0.005 significance level.

(b) Construct the 99% confidence interval for the difference between the color blindness rates of men and women.

30. A certain company will purchase the house of any employee who is transferred out of state and will handle all details of reselling the house. The purchase price is based on two assessments, one assessor being chosen by the employee and one by the company. The data given below shows sample of nine assessments.

Assessments of nine Homes (\$ thousands)

Assessed by	H1	H2	H3	H4	H5	H6	H7	H8	H9
Company	328	350	455	278	290	285	535	745	617
Employee	328	335	450	275	350	230	585	795	625

Test at 5% level that two assessors agree. Also construct a 95% confidence interval for the difference between the two assessors chosen one by the employee and the other by the company.

31. In crash tests of five BMW cars, the repair costs (in \$) were computed for a dealership repair center and an independent repair facility. The results are listed in the accompanying table.

Dealership repair center	767	571	904	1,047	418
Independent repair center	513	488	845	971	257

(a) Is there sufficient evidence to support the claim that the independent center has a lower mean repair costs than the dealership? Use a 0.01 significance level.

(b) Construct a 98% confidence interval estimate of the difference of the means between the repair costs of the dealership repair center and the independent repair center. Interpret your answer.

32. Two labs produce LCD display units. 10 displays are chosen at random from each lab. The number of bad pixels are noted for each display units and shown below.

Lab	Number of Bad Pixels									
A	1	0	3	1	1	2	0	2	0	1
B	0	2	2	3	1	0	0	1	2	5

(a) Is there a difference in the pixels defects rate between the two labs? Use $\alpha = 0.05$.

(b) Construct a 95% confidence interval for the difference in the pixels defects rate between the two labs.

33. A researcher wishes to determine whether people with high blood pressure can reduce their blood pressure by following a particular diet. Here is the sample data.

Treatment Group	$n_1 = 48$	$\bar{x}_1 = \$189.1$	$s_1 = \$38.7$
Control Group	$n_2 = 63$	$\bar{x}_2 = \$203.7$	$s_2 = \$42.8$

(a) Test the claim that $\mu_1 < \mu_2$ where μ_1 and μ_2 represent the mean for the treatment group and the control group respectively.

(b) Construct a 90% confidence interval for $\mu_1 - \mu_2$.

34. Rates of return (quarterly) in two investment portfolios A and B are compared over the last 10 quarters. They are weighted similar in safety, but portfolio A is advertised as being "less volatile".

Portfolio A	10.23	8.96	7.54	7.06	9.82	10.93	8.97	12.67	9.81	5.4
Portfolio B	12.49	7.61	8.68	7.68	9.62	11.49	9.91	11.51	7.59	5.1

(a) Is there any significant difference in the average rate of returns in two investment portfolios?

(b) Construct a 90% confidence interval for the difference in the average rate of returns for two investment portfolios.

35. A history teacher claims that the mean score on a history assessment test for nine-year-old boys and girls are equal. The mean score for 13 randomly selected boys is 76.5 with a standard deviation of 11.5, and the mean score for 15 randomly selected girls is 71 with a standard deviation of 13.

(a) Can you reject the teacher's claim? Use $\alpha = 0.02$.

(b) Construct a 98% confidence interval for the difference of the population means.

36. A production manager claims that there is no significant difference in the mean washer diameter manufactured by two different machines. The table below lists 30 randomly selected washer diameters (in inches) produced by the first machine.

0.861	0.864	0.882	0.887	0.858	0.879	0.887	0.876	0.87	0.894	0.884	0.882	0.869	0.859	0.887
0.882	0.862	0.906	0.88	0.887	0.864	0.873	0.86	0.87	0.869	0.877	0.863	0.875	0.883	0.872

The second machine produced the following 30 randomly selected washer diameters.

0.705	0.703	0.715	0.711	0.69	0.72	0.72	0.686	0.7	0.699	0.712	0.718	0.695	0.715	0.691
0.703	0.697	0.694	0.714	0.694	0.672	0.688	0.7	0.71	0.698	0.696	0.7	0.695	0.715	0.696

(a) At 0.01 significance level, can you reject the production manager's claim?

(b) Construct a 99% confidence interval for the difference of the population means.

37. A study finds that the mean annual salary for 33 randomly selected photographers working in North Carolina is \$22,800 and the standard deviation is \$7750. Another study finds that the mean annual salary for 31 randomly selected photographers working in South Carolina is \$43,700 and the standard deviation is \$3695.

(a) At a 0.05 significance level, test the claim that the difference between the mean salaries of photographers in North and South Carolina is significant.

(b) Construct a 95% confidence interval for the difference of the population means. Interpret your results.

38. At two different University laboratories the duration of Exothermic Oxidation Displacement was observed for each student and the following results were recorded.

University A	**University B**
$n_1 = 67$	$n_2 = 95$
$\bar{x}_1 = 36$ min	$\bar{x}_2 = 39$ min
$s_1 = 7$ min	$s_2 = 6$ min

(a) Using a 0.01 significance level, test the claim that the mean displacement times are the same.

(b) Construct a 99% confidence interval for the difference of the mean displacement times.

Inferences for the Mean Difference μ_d, Dependent Samples

In exercises 39-46, let the differences always represent d = Before – After, and assume that normally requirements are satisfied.

39. To evaluate the effectiveness of SAT course, refer to the data in the accompanying table that lists SAT scores before and after the sample of 10 students who took a preparatory course.

Student	A	B	C	D	E	F	G	H	I	J
SAT score before the prep course	700	840	830	860	840	690	830	1180	930	1070
SAT score after the prep course	720	840	820	900	870	700	800	1200	950	1080

(a) Is there sufficient evidence to conclude that the preparatory course is effective in raising scores? Use a 0.05 significance level.

(b) Construct a 90% confidence interval estimate of the mean difference μ_d between the before and after scores.

40. *Tenormin* is a drug designed to regulate the heartbeat and lower the blood pressure. 12 randomly selected subjects were tested with this drug, their blood pressure reading were measured before and after the drug was taken. Here are the results.

Subject	1	2	3	4	5	6	7	8	9	10	11	12
Before	200	174	198	170	179	182	193	209	185	155	169	210
After	191	170	177	167	159	151	176	183	159	145	146	177

(a) Is there sufficient evidence to support the claim that *Tenormin* is effective in lowering blood pressure? Use a 0.01 significance level.

(b) Use the sample data to construct a 98% confidence interval for the mean difference between the before and after readings.

41. A study was conducted to investigate the effectiveness of acupuncture in reducing pain. Results for 6 randomly selected subjects are given in the accompanying table. (The values are before and after acupuncture)

Subject	A	B	C	D	E	F
Before	6.7	6.6	8.9	10.4	8.2	6.5
After	6.9	3.8	7.2	8.3	8.1	6.2

(a) Is there sufficient evidence to conclude that acupuncture has no effect in reducing pain? Use $\alpha = 0.05$.

(b) Construct a 95% confidence interval for the mean of the "before -after" differences. Does the constructed confidence interval support your conclusion in part (a).

42. Ten randomly selected households are surveyed for daily number of gallons of water they use before and after viewing a conservation video.

Before	33	33	38	33	35	35	40	40	42	34
After	34	28	25	28	35	33	31	28	35	36

(a) Use a 0.05 significance level to test the claim that the mean number of gallons consumed is the same before and after the viewing. ***Hint*:** Mean Difference is zero.

(b) Construct a 95% confidence interval for μ_d. Interpret your results.

43. Given $\bar{d} = -4.5$, $s_d = 2.5$ and $n = 25$.

(a) test the claim that $\mu_d \neq 0$. Use a 0.05 significance level.

(b) Construct a 95% confidence interval for μ_d.

44. A Table Tennis paddle manufacturer claims that Table Tennis players can increase their spin during a game using the manufacture's newly designed high friction Yasaka rubbers. Eight players are randomly selected and each is asked to rate their experience using their old rubber on the scale of 0 to 100. After using the new rubber for one month, the same players are again asked to give their rating using the same scale. The before and after scores for each player are given in the table below.

Table Tennis Player	1	2	3	4	5	6	7	8
Rating (old rubber)	89	84	96	82	74	92	85	91
Rating (new rubber)	88	83	98	95	76	91	93	91

(a) Is there enough evidence to support the manufacture's claim at $\alpha = 0.05$?

(b) Construct a 90% confidence interval for the mean difference (μ_d). Interpret your results.

45. A university professor wants to determine whether her student evaluations (on a scale of 0-100) has changed from last year to this year. The following table shows the professor's ratings for the same 16 randomly selected students for last year and this year.

Student	1	2	3	4	5	6	7	8	9	10	11	12	13	14	15	16
Rating (last year)	60	78	89	84	91	60	58	83	75	81	75	70	62	79	88	80
Rating (this year)	78	77	79	75	96	89	88	93	95	84	83	90	77	85	91	74

(a) Is there enough evidence to conclude that the professor's performance rating has not changed? Use a 0.01 significance level.

(b) Construct a 99% confidence interval for the mean difference. Interpret your results.

46. In a study of the relationship between the salaries of college students before and after they receive their AA degrees, the following data (in dollars) was collected from 10 randomly selected college students.

Before	14254	12657	21646	21315	21321	13065	16156	15164	12131	12154
After	13587	17254	31313	22324	20832	29461	22154	16412	38580	23460

(a) Use a 0.01 significance level to show that the mean difference is zero.

(b) Construct a 98% confidence interval for the mean difference.

Inferences for Two Population Proportions

In exercises 47-62, assume the parent populations are normally distributed.

47. In clinical tests of adverse reactions to a certain drug, 7% of the 640 subjects in the treatment group experienced severe headaches, whereas 2% of the 580 subjects in the placebo group experienced severe headaches.

(a) Is there sufficient evidence to support the claim that severe headaches occur at a higher rate among the users of the drug than the non users? Use a 0.01 significance level.

(b) Construct a 98% confidence interval for the difference between the rates ($p_1 - p_2$).

48. In the United States, some fire trucks were painted yellow to heighten their visibility. During a test period, the fleet of red fire trucks made 153,348 runs and had 20 accidents, while the fleet of yellow fire trucks made 135,035 runs and had 4 accidents.

(a) Is there sufficient evidence to support the claim that the yellow fire trucks have a significantly lower accident rate? Use $\alpha = 0.01$ significance level.

(b) Construct a 98% confidence interval for the difference between the two proportions.

49. A marketing survey involves product registration in New York and California. Of 1500 New Yorkers surveyed, 208 registered their product; while 180 out of 700 Californians registered the product.

(a) Test the claim that the two population proportions are the same. Use a 0.05 significance level.

(b) Construct a 90% confidence interval for the difference between the two population proportions.

50. A study of seat belt use involved minors who were hospitalized as a result of traffic accidents. Among 145 minors who were not wearing seat belts, 85 were injured severely. For a group of 152 minors who were wearing seat belts, 56 were injured severely.

(a) Is there sufficient evidence to conclude that the rate (percentage) of severe injuries is lower for minors wearing seat belts? Use a 0.005 significance level.

(b) Construct a 99% confidence interval for the difference of proportions (rates).

51. A study claimed that larger proportion of young children ate cereal at a higher rate than college students. Researchers surveyed both age groups to find the answer. The results are shown in the table below.

Statistic	College Students (Ages 18-28)	Young Children (Ages 6-11)
Number surveyed	1200	1000
Number who eat cereal	840	920

(a) Is there sufficient evidence to conclude that young children eat cereal at a higher rate than college students?

(b) Construct a 99% confidence interval for the difference of proportions.

52. A random sample of 1650 adult males showed that 68% consume 1670 calories per day. A random sample of 2000 adult females showed that 58% consume 1670 calories per day.

(a) Test the claim that there is no significant difference between the percentage of female and male adults consuming 1670 calories per day. Assume a 0.05 significance level.

(b) Construct a 95% confidence interval for the true difference between the consumption rate of male and female adults.

53. 1000 bilingual families that spoke Spanish were surveyed on which language they watch television most of the time. For 600 families that have lived in the United States for less than 5 years, 80% of them watched television in Spanish most of the time; and for the 400 that have lived in the U.S. longer than 5 years, 68% of them had television on in Spanish most of the time.

(a) Using a 0.01 significance level, determine whether there is sufficient evidence to say that the percentage of Spanish-speaking people who have lived in the U.S. for less than 5 years and watch Spanish television is greater than the others.

(b) Construct a 98% confidence interval for the difference of the percentages.

54. Does hormone replacement therapy (HRT) cause breast cancer? Researchers surveyed women in the age group of 52 to 82 who used HRT or a placebo (dummy pill) during the period of last 5-years. Of the 10,000 HRT women, 450 cancers were reported, compared with 320 cancers for the 9,000 women who got the dummy pill. Assume that the participants were randomly assigned to two equal groups.

(a) Is there sufficient evidence to conclude that HRT was associated with increased cancer risk as compared to dummy pill? Use $\alpha = 0.01$ significance level.

(b) Construct a 98% confidence interval for the difference of proportions.

55. 9 of 8,500 people vaccinated against a certain disease later developed the disease. 20 of 10,000 people not vaccinated later developed the disease.

(a) Test the claim that the vaccine is effective in lowering the incidence rate of the disease. Use a significance level of 0.05.

(b) Construct a 90% confidence interval for $p_1 - p_2$.

56. In a study of 1700 teens aged 14-18, half were given written surveys and half were given surveys using an anonymous computer program. Among those given the written surveys, 7.9% indicated that they carried a knife within the last 30 days. Among those given the computer surveys, 10.4% said that they carried a knife within the last 30 days (based on data from the Urban Institute).

(a) The sample percentages of 7.9% and 10.4% are obviously not equal, but is the difference between population percentages significant? Explain. Use a 0.01 significance level.

(b) Construct a 99% confidence interval estimate of the difference between the two population percentages.

57. A sample of college seniors showed that 15 of 50 who owned PlayyStation 3 spent more than an hour a day playing games, compared with 35 of 100 who owned Zoopbox 300.

(a) Test the claim of any significant difference in the population proportions at $\alpha = 0.05$.

(b) Construct a 95% confidence interval for the difference between the two population proportions.

58. A U.S. Department of Justice report included the claim that "in spouse murder cases, wife defendants were less likely to be convicted than husband defendants." Sample data consisted of 184 convictions among 360 husband defendants and 120 convictions among 430 wife defendants.

(a) Test the stated claim using a 0.05 significance level.

(b) Construct a 90% confidence interval for $p_1 - p_2$.

59. A medical research team studied the effects of a new drug on ADD patients. Of 87 patients who used the drug for two years, 20 showed improvements. Of 104 patients who used the drug for six months, 26 showed improvement.

(a) Test the claim that improvement rates are different. Use a 0.01 significance level.

(b) Construct a 98% confidence interval for the difference of the improvement rates.

60. Does Lovastatin reduce the risk of heart attack? A unit of Lovastatin drug bank researchers gave Lovastatin to 1500 people and an inactive substitute to 2000 people of age group 50-60. After 5 years, 52 of the Lovastatin group had suffered a heart attack, as compared with 95 for the inactive pill.

(a) Is there sufficient evidence to support the claim that Lovastatin reduces the risk of heart attack as compared to the inactive pill? Use $\alpha = 0.01$.

(b) Construct a 98% confidence interval for the difference between the proportional.

61. At an international science convention, 20% of the 87 American scientists were women and 18% of the 68 British scientists were women.

(a) Using a 0.01 significance level, test the claim that the population percentages of the female American and British scientists are the same.

(b) Construct a 99% confidence interval for the difference of the population proportions.

62. Among 48 randomly selected students taking an online Basic Math course, 18 passed. Out of 43 randomly selected students taking a traditional Basic Math course, 28 passed.

(a) Using a 0.01 significance level, test the claim that pass rate for a traditional class is higher.

(b) Verify the results in part (a) by constructing a 98% confidence interval for the difference of population pass rates.

Inferences for Two Standard Deviations or Variances

In exercises 63-73, assume the parent populations are normally distributed.

63. Test the claim that populations Sample A and Sample B have different standard deviations. Use a 0.10 significance level.

Sample A	Sample B
$n_1 = 25$	$n_2 = 31$
$\bar{x}_1 = 19.2$	$\bar{x}_2 = 18.6$
$s_1 = 5.6$	$s_2 = 5.2$

64. The weights of 21 randomly selected green apples were taken, and the mean was 40 grams, with a standard deviation of 5 grams. When 18 randomly selected red apples were tested, the mean weight was 38 grams, with a standard deviation of 7 grams. Test the claim that the population standard deviations are different (□$\sigma_1 \neq \sigma_2$).

65. A study is made concerning number of hours spent for reading newspapers. A sample of 13 men showed a standard deviation of 3 hours per week and a sample of 10 women revealed a standard deviation of 6.3 hours per week. Can we conclude that there is more variation in the number of hours women read newspapers than in the number of hours men read newspapers? Use $\alpha = 0.05$.

66. A history teacher claims that the variance of scores on a history assessment test for nine-year-old boys and girls are equal. The mean score for 13 randomly selected boys is 76.5 with a standard deviation of 15.5, and the mean score for 15 randomly selected girls is 71 with a standard deviation of 13. Assuming that the parent populations are normally distributed, can you reject the teacher's claim? Use $\alpha = 0.05$.

67. The data given below shows the weights (in pounds) of randomly selected checked bags for international and domestic flights on the same day.

International (12 bags)	58	29	42	48	32	45	52	21	23	43	36	48
Domestic (9 bags)	39	12	15	17	18	26	28	45	12			

Test the hypothesis that variance of weights of checked bags of international flight passengers is different than domestic flight passengers. Use $\alpha = 0.05$.

68. At a university, 30 students from law school and 36 students from medical school were asked about the number of hours they spend studying. Law students' responses had a standard deviation of 4.6 hours, and medical students' had a standard deviation of 3.8 hours. Use a 0.05 significance level to determine whether the standard deviation for two populations are significantly different.

69. A study involves the number of absences per year among union and nonunion workers. A sample of 24 union workers has a sample standard deviation of 2 days' and a sample of 18 nonunion workers has a standard deviation of 1.5 days. Test at the 0.01 level of significance whether the two populations have the same variations between number of absences among union and nonunion workers.

70. A university professor tries two different approaches in teaching her Calculus classes. In one section she uses technology and in the other section she follows the traditional approach. Both sections were given the same standardized final examination. The group of 36 randomly selected students who were taught with the technology method scored an average of 72.8 with a standard deviation of 22. The traditional lecture based group of 29 randomly selected students scored an average of 68.5 with a standard deviation of 18. Assuming the parent populations are normally distributed, is there sufficient evidence to conclude that the technology based method has a higher variance than the traditional method? Use $\alpha = 0.01$.

71. A manufacturing process drills holes in sheet metal that are supposed to be 0.51 cm in diameter. The hole diameters are carefully measured and recorded (in cm) for 9 randomly chosen sheets before and after the new drill press is installed.

New drill	0.5053	0.4947	0.5195	0.5088	0.4907	0.5031	0.5005	0.5110	0.5107
Old drill	0.5214	0.4998	0.5095	0.4912	0.5040	0.5035	0.4962	0.4956	0.5035

Do these independent random samples prove that the new process has smaller variance? Use $\alpha = 0.05$.

72. Two independent samples are drawn from two normally distributed populations.

Sample I	60	65	71	74	76	82	85	87
Sample II	61	66	67	85	78	63	85	86

Given the sample data, test the claim that the samples come from populations with equal standard deviations. Use a 0.05 significance level.

73. A researcher claims that the variation in the salaries of city college professors is greater than the variation in the salaries of university professors. A sample of 32 city college professors has a standard deviation of \$9228 and a sample of 36 university professors has a standard deviation of \$2103. Test the claim using $\alpha = 0.05$.

9

MULTINOMIAL EXPERIMENTS
CHI-SQUARE TESTS FOR CATEGORICAL DATA

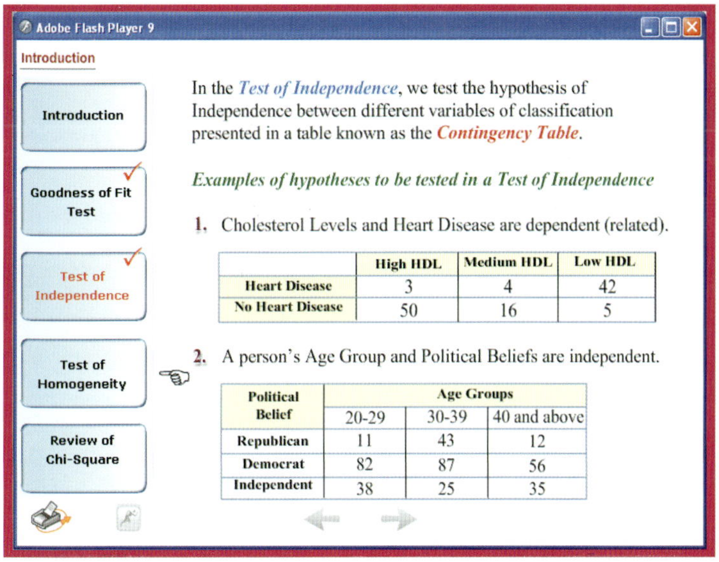

CHAPTER

9

MULTINOMIAL EXPERIMENTS
CHI-SQUARE TESTS FOR CATEGORICAL DATA

Chapter Overview:

9.1: INTRODUCTION

The **Chi-Square Distribution** was earlier used in Chapters 6 and 7 to construct **confidence intervals** as well as to test **hypotheses** concerning the standard deviation or variance of a population. In this chapter, we will use the **Chi-Square Distribution** to test **Categorical Data** that can be organized into tables with distinct cells. In particular, we will discuss Multinomial Experiments such as Goodness of Fit Test, Test of Independence and Homogeneity.

- In the **Goodness of Fit Test**, we determine how well the observed data taken from a single population fits the theoretical distribution or expectation.

Examples of hypotheses to be tested in a Goodness of Fit Test

i) A (six-sided) die is fair. That is, the probability of the top face showing any one number is the same, or uniformly distributed.

$$p_1 = p_2 = p_3 = p_4 = p_5 = p_6 = \frac{1}{6}$$

ii) The probability of a car accident on any given day of the week is uniform.

$$p_M = p_T = p_W = p_{TH} = p_F = p_{SAT} = p_{SUN} = \frac{1}{7}$$

iii) In an auto center, the probability of selling a black car is 40%, the probability of selling a silver car is 30%, the probability of selling a white car is 10%, the probability of selling a yellow car is 10%, and the probability of selling all the other colors is 10%.

$$p_{black} = 0.40,\ p_{silver} = 0.30,\ p_{white} = 0.10,\ p_{yellow} = 0.10,\ p_{other} = 0.10$$

- In the **Test of Independence**, we test the hypothesis of Independence between different variables of classification presented in a table known as the **Contingency Table**.

Examples of hypotheses to be tested in a Test of Independence

i) Cholesterol Levels and Heart Disease are dependent (related).

	High HDL	Medium HDL	Low HDL
Heart Disease	3	4	42
No Heart Disease	50	16	5

Table 9.1

ii) A person's Age Group and Political Beliefs are independent.

Political Belief	**Age Groups**		
	20-29	30-39	40 and above
Republican	11	43	12
Democrat	82	87	56
Independent	38	25	35

Table 9.2

- In the **Test of Homogeneity**, we test the equality of several proportions taken from **different populations**.

Examples of hypotheses to be tested in a Test of Homogeneity

i) The distribution of grades A, B, C, D, and F is the same for high school freshmen, sophomores, juniors, and seniors.

$$P_A = P_B = P_C = P_D = P_F$$

ii) The proportions of DUI arrests are identical in southern, central, and northern California.

$$P_{SC} = P_{CC} = P_{NC}$$

- **Review of Chi-Square Distribution**

All three tests in this chapter (Goodness of Fit Test, Test of Independence, and Test of Homogeneity) use a **right-tail Chi-Square Distribution** with α significance level.

In Figure 9.1, $\chi_v^2(\alpha)$ is known as the right-tailed critical value of χ^2 with v degrees of freedom and α significance level, where $P\left(\chi^2 > \chi_v^2(\alpha)\right) = \alpha$. The critical values of the χ^2-distribution have been tabulated for different values of v (df) and α, and are given in Table 7 at the end of the book.

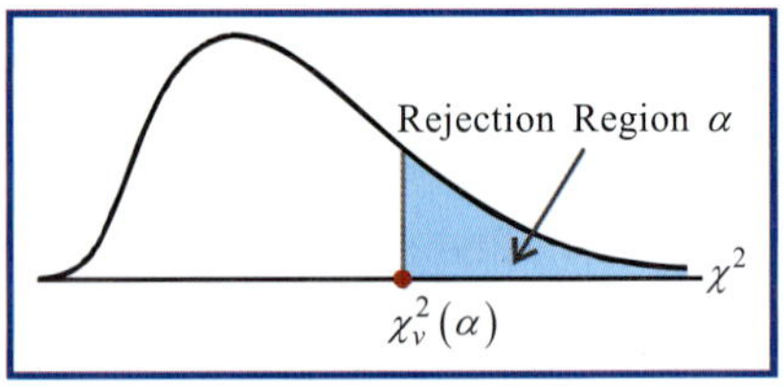

Figure 9.1

9.2: χ^2-Test of Goodness of Fit

A. Multinomial Experiments

Let us consider a random experiment which is repeated a number of times under essentially the same homogeneous conditions. Such experiments are termed as Multinomial Experiments, given the following conditions are satisfied:

i) The number of trials is finite and fixed.

ii) All the trials are independent.

iii) Each trial results in $k > 2$ mutually exclusive outcomes.

iv) The expected outcomes are $E_1, E_2, \ldots, E_k$ with respective probabilities $p_1, p_2, \ldots, p_k$, such that $\sum_{i=1}^{k} p_i = 1$.

Objectives

After completing this section you will be able to understand:

A. Multinomial Experiments.

B. Chi-Square Test of Goodness of Fit.

C. Computing expected Frequencies using Uniform Distribution.

D. Computing expected Frequencies using Non-Uniform Distribution.

The results of any trial can be classified into exactly one of the k-classes, also known as ***categories*** or ***cells***. The respective probabilities of the expected outcomes $E_1, E_2, \ldots, E_k$ remain constant for each trial.

The **Chi-Square Goodness of Fit Test** is used to test the hypothesis in a ***multinomial experiment*** such that an observed frequency distribution fits a claimed theoretical distribution.

Notation:

O: Represents the observed frequency of an outcome (given in the sample data).

E: Represents the expected frequency of an outcome (under the null hypothesis).

k: Represents the number of categories or outcomes.

n: Represents the total number of trials or observed frequencies (also the size of the sample data).

• Conditions for the validity of the Goodness of Fit Test

1. The sample must be random so that the sample observations are independent.
2. For each category, the expected frequency should be at least 5.
3. n, the total frequency, should be reasonably large, say, greater than 50.
4. The data should be given in frequency counts for each of the categories and must not be replaced by relative frequencies or proportions.

B. Chi-Square Test of Goodness of Fit

Suppose we are given a set of observed frequencies (O_i's) under some experiment and we want to test if the experimental results support a particular hypothesis or theory. Karl Pearson, in 1900, developed a procedure for testing the significance of the discrepancy between experimental values and theoretical values obtained under a hypothesis. This test is known as χ^2-Test of Goodness of Fit and is used to test that the deviation between observation (experiment) and theory may be attributed to chance (fluctuations of sampling) or it is really due to the inadequacy of the theory to fit the observed data. That is, to support a particular hypothesis, the observed frequencies must match the expected distribution of probabilities.

• Procedure for χ^2- Test of Goodness of Fit

Classical Method:

***Step* 1 Null and the Alternative Hypotheses:**

In a Goodness of Fit Test, the original claim H_0 always represents the expected probabilities. That is, the observed frequency distribution fits the specified theoretical or hypothesized distribution. The Alternative Hypothesis H_1 always represents the statement that "At least one of the observed probabilities in H_0 is different than expected". That is, the observed frequency distribution does not fit in the specified distribution.

***Step* 2 Test Statistics:**

$$\chi^2 = \sum_{i=1}^{k} \frac{(O_i - E_i)^2}{E_i} = \frac{(O_1 - E_1)^2}{E_1} + \frac{(O_2 - E_2)^2}{E_2} + \ldots + \frac{(O_k - E_k)^2}{E_k}$$

The above test statistics follows a χ^2-distribution with $v = k - 1$ degrees of freedom where $O_1, O_2, \ldots, O_k$ represent the observed frequencies, and $E_1, E_2, \ldots, E_k$ are the corresponding expected or theoretical frequencies obtained from the given hypothesis.

• Computation of Expected Frequencies

Suppose the data from a random sample of size n is divided into k categories labeled 1, 2, … , k; and let n_1, $n_2, \ldots, n_k$ be the respective counts (cell frequencies), with the corresponding cell probabilities $p_1, p_2, \ldots, p_k$ in H_0. These probabilities are constant in each trial. We know from Binomial distribution that if the probability of occurrence of an event is p, then the expected number of events in n trials is np. In a Chi-Square Goodness of Fit Test for a Multinomial Experiment, the expected frequency for any category is the product of the probability assumed in H_0 for that category and the total number of observed frequencies $\sum_{i=1}^{k} O_i = n$.

Therefore, $\quad E_1 = p_1 \times \sum O, \; E_2 = p_2 \times \sum O, \; \ldots, \; E_k = p_k \times \sum O.$

In general, $\quad E_i = p_i \times \sum_{i=1}^{k} O_i \quad$ or $\quad E(n_i) = p_i \times n, \; (i = 1, 2, \ldots, k)$

> ***Remark:*** The sum of observed and expected frequencies must always be the same.
>
> $$\sum O = \sum E$$

***Step* 3 Rejection Region and the Critical Values:**

The rejection region in this chapter will always be in the right-tail of the Chi-Square Distribution, with α significance level and $k - 1$ degrees of freedom. We find the corresponding critical value $\chi^2_{k-1}(\alpha)$ using Table 7 (Chi-Square Table) at the end of the book.

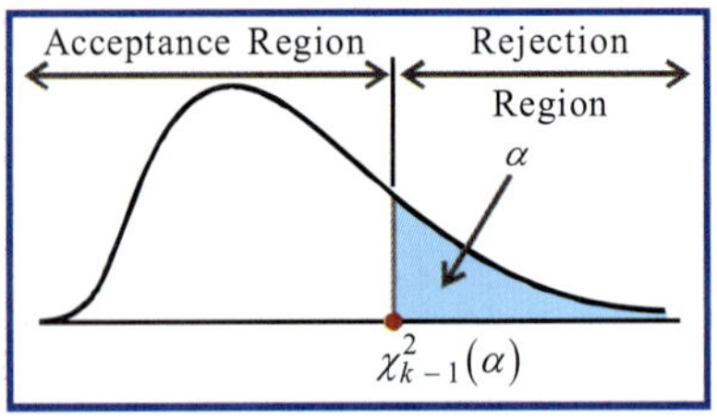

Figure 9.2

***Step* 4 Decision:** If the computed χ^2-value (test statistics) lies in the shaded (rejection) region, then we reject H_0 at α level of significance. If the computed χ^2-value lies in the acceptance region, then we fail to reject H_0 at α level of significance. *Conclusion*: If we reject H_0, then we can conclude that the probabilities are not distributed as expected, and if we fail to reject H_0, we say that the probabilities are distributed as expected.

P-value Method:

***Step* 1 Null and the Alternative Hypotheses:**

In a Goodness of Fit Test, the original claim H_0 always represents the expected probabilities. That is, the observed frequency distribution fits the specified theoretical or hypothesized distribution. The Alternative Hypothesis H_1 always represents the statement that "At least one of the observed probabilities in H_0 is different than expected.". That is, the observed frequency distribution does not fit in the specified distribution.

***Step* 2 Test Statistics:**

$$\chi^2 = \sum_{i=1}^{k} \frac{(O_i - E_i)^2}{E_i} = \frac{(O_1 - E_1)^2}{E_1} + \frac{(O_2 - E_2)^2}{E_2} + \dots + \frac{(O_k - E_k)^2}{E_k}$$

***Step* 3** Compute the *p*-value of the test statistics.

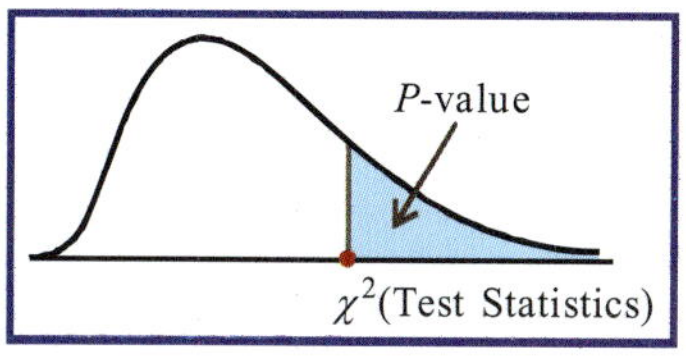

Figure 9.3

$$P\text{-value} = P\left(\chi^2 > \chi^2_{\text{TS}}\right)$$

$$= \left[\begin{array}{l}\text{Area to the right of the computed value of test statistics for } v = k-1 \; df \\ \text{as shown in Figure 9.3.}\end{array}\right]$$

This probability (or area) can be obtained from Table 7 or using a statistical software.

***Step* 4 Decision:** If the *p*-value $\leq \alpha$, then we reject H_0 at α significance level. If the *p*-value $> \alpha$, then we fail to reject H_0 at α significance level. *Conclusion:* Same as the classical method.

C. Computing Expected Frequencies using a Uniform Distribution

Example 1: A die is rolled 120 times and the results are shown in the accompanying table.

Number on the die	1	2	3	4	5	6
Frequency	17	25	17	23	15	23

Use a 0.05 significance level to test the claim that the die is fair (probabilities are uniformly distributed).

Use (a) the Classical method and (b) the *P*-value method.

Solution: **(a)** **Classical Method:**

***Step* 0** Data: In the usual notation, we have $n = 120$, $k = 6$, and $\alpha = 0.05$.

Expected Probabilities:

Let p_1 represent the probability of getting a 1, p_2 getting a 2, and so on.

The original claim is that the die is fair. That is, the probabilities for each of the faces 1, 2, … , 6 are the same so that $p_1 = p_2 = p_3 = p_4 = p_5 = p_6 = \frac{1}{6}$. If the claim is not true, then at least one of the probabilities $p_1, p_2, p_3, p_4, p_5,$ or p_6 must be different from $\frac{1}{6}$.

***Step* 1** Set up the null and the alternative hypotheses.

$$H_0 : p_1 = p_2 = p_3 = p_4 = p_5 = p_6 = \frac{1}{6}$$

H_1 : At least one is different. The die is not fair.

***Step* 2** **Test Statistics:** Under H_0, the expected frequency

$$E_i = p_i \times \sum_{i=1}^{k} O_i$$

for each face $= \frac{1}{6}(120) = 20.$

More precisely,

$E_1 = p_1 \times \sum O = \frac{1}{6} \times 120 = 20,\ E_2 = p_2 \times \sum O = \frac{1}{6} \times 120 = 20$, and so on.

As provided in the example, observed frequencies are the sample values collected in the experiment.

Number on the die	1	2	3	4	5	6
Frequency (*O*'*s*)	17	25	17	23	15	23

Table 9.3

$O_1 = 17,\ O_2 = 25,\ O_3 = 17,\ O_4 = 23,\ O_5 = 15$, and $O_6 = 23$

The calculations for the Chi-square test statistics are given in the following table.

# on the die Categories (k = 6)	**Observed Frequencies (O)**	**Expected Frequencies (E)**	$\frac{(O-E)^2}{E}$
1	17	20	$\frac{(17-20)^2}{20} = 0.45$
2	25	20	$\frac{(25-20)^2}{20} = 1.25$
3	17	20	$\frac{(17-20)^2}{20} = 0.45$
4	23	20	$\frac{(23-20)^2}{20} = 0.45$
5	15	20	$\frac{(15-20)^2}{20} = 1.25$
6	23	20	$\frac{(23-20)^2}{20} = 0.45$
Totals:	$\sum O_i = 120$	$\sum E_i = 120$	$\chi^2 = \sum \frac{(O-E)^2}{E} = 4.3$

Table 9.4

Note: The sum of the observed frequencies must be equal to the sum of the expected frequencies.

$$\sum O_i = \sum E_i = 120$$

Test Statistics is:

$$\chi^2 = \sum \frac{(O-E)^2}{E}$$

$$= \frac{(17-20)^2}{20} + \frac{(25-20)^2}{20} + \frac{(17-20)^2}{20} + \frac{(23-20)^2}{20} + \frac{(15-20)^2}{20} + \frac{(23-20)^2}{20}$$

$$= 4.3$$

***Step* 3** The Rejection Region for a Goodness of Fit Test is a right tail Chi-square distribution with $k - 1 = 5$ degrees of freedom. The critical value for the Chi-square Goodness of Fit Test with 5 *df* at a 0.05 level of significance is $\chi_5^2(0.05) = 11.07$. The calculated value of the test statistics, critical value and the rejection region are shown in Figure 9.4.

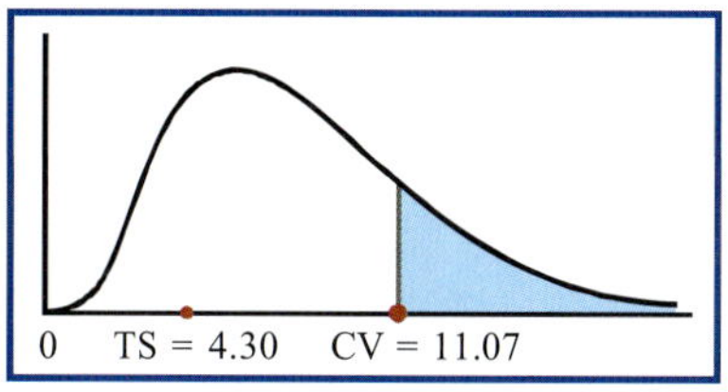

Figure 9.4

***Step* 4** **Decision:** Since the calculated value of χ^2(test statistics) lies outside the rejection region, we fail to reject H_0 at 5% level of significance. *Conclusion:* The data does not provide sufficient evidence against the hypothesis that the die is fair (follows a uniform distribution).

(b) ***P*-value Method:**

***Step* 1** Set up the null and the alternative hypotheses.

$$H_0 : p_1 = p_2 = p_3 = p_4 = p_5 = p_6 = \frac{1}{6}$$

H_1 : At least one is different. The die is not fair.

***Step* 2** **Test Statistics:** Under H_0, the expected frequency

$$E_i = p_i \times \sum_{i=1}^{k} O_i$$

for each face $= \frac{1}{6}(120) = 20.$

More precisely,

$$E_1 = p_1 \times \sum O = \frac{1}{6} \times 120 = 20,\ E_2 = p_2 \times \sum O = \frac{1}{6} \times 120 = 20, \text{ and so on.}$$

As provided in the example, observed frequencies are the sample values collected in the experiment.

Recall:

Number on the die	1	2	3	4	5	6
Frequency (*O*'s)	17	25	17	23	15	23

Table 9.3

$O_1 = 17,\ O_2 = 25,\ O_3 = 17,\ O_4 = 23,\ O_5 = 15$, and $O_6 = 23$

The calculations for the Chi-square test statistics are given in the following table.

Recall:

# on the die Categories ($k = 6$)	Observed Frequencies (O)	Expected Frequencies (E)	$\frac{(O-E)^2}{E}$
1	17	20	$\frac{(17-20)^2}{20} = 0.45$
2	25	20	$\frac{(25-20)^2}{20} = 1.25$
3	17	20	$\frac{(17-20)^2}{20} = 0.45$
4	23	20	$\frac{(23-20)^2}{20} = 0.45$
5	15	20	$\frac{(15-20)^2}{20} = 1.25$
6	23	20	$\frac{(23-20)^2}{20} = 0.45$
Totals:	$\sum O_i = 120$	$\sum E_i = 120$	$\chi^2 = \sum \frac{(O-E)^2}{E} = 4.3$

Table 9.4

Note: The sum of the observed frequencies must be equal to the sum of the expected frequencies.

$$\sum O_i = \sum E_i = 120$$

Test Statistics is:

$$\chi^2 = \sum \frac{(O-E)^2}{E}$$

$$= \frac{(17-20)^2}{20} + \frac{(25-20)^2}{20} + \frac{(17-20)^2}{20} + \frac{(23-20)^2}{20} + \frac{(15-20)^2}{20} + \frac{(23-20)^2}{20}$$

$$= 4.3$$

Step* 3** ***P-value of the test statistics is given by:

P-value = $P(\chi^2 >$ Test Statistics $= 4.3$ with 5 *df*)

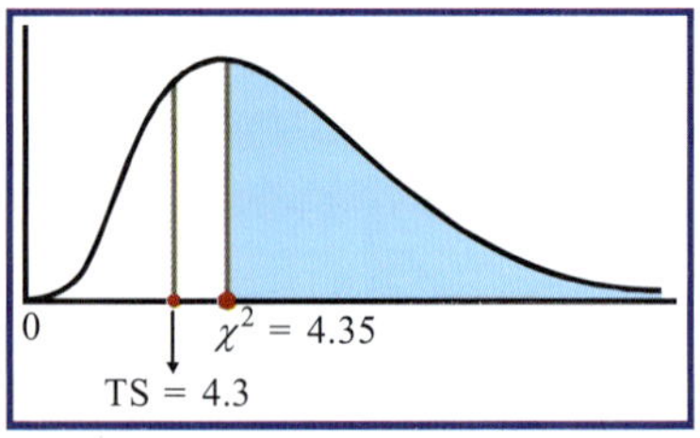

Figure 9.5

From the Chi-square table (Table 7), in the row corresponding to $v = 5$, the value of Chi-square which is closest to 4.3 is 4.35. This corresponds to an α value of 0.5 in the first row of the χ^2-table.

$$\Rightarrow \quad P\left(\chi_5^2 > 4.35\right) = 0.5$$

Since the value of the test statistics $\chi^2 = 4.30$ is to the left of 4.35 (Figure 9.5), $P(\chi^2 > 4.30) > 0.5$.

Hence, the $\boldsymbol{p}$-value of test statistics is greater than 0.5.

***Step* 4** **Decision:** Since the $\boldsymbol{p}$-value > 0.5, we fail to reject H_0 at 0.05 level of significance. *Conclusion*: The data does not provide sufficient evidence against the hypothesis that the die is fair.

Note: We can calculate the exact **p**-value of the test statistics $p(\chi^2 > 4.3)$ using the Microsoft Excel command CHISQ.DIST.RT(x, **deg_freedom**), as shown below.

 =CHISQ.DIST.RT(4.3,5) CHISQ.DIST.RT(x, deg_freedom)	$\Rightarrow P\left(\chi_5^2 > 4.30\right) = 0.5071 \approx 0.51$

In short:

Hypotheses	**Test Statistics**
$H_0: p_1 = p_2 = p_3 = p_4 = p_5 = p_6 = \frac{1}{6}$ H_1: At least one is different. The die is not fair	$\chi^2 = \sum \frac{(O-E)^2}{E} = 4.3$
Rejection Region and Critical Value	**Conclusion**
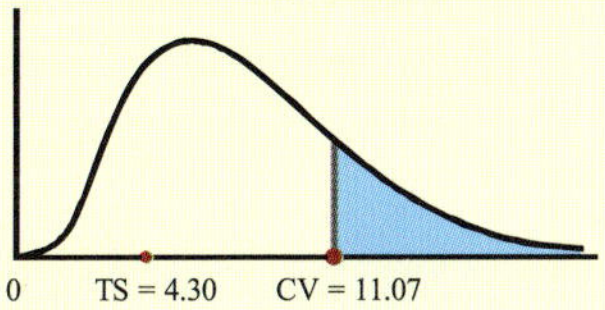	Fail to reject H_0. Data does not provide sufficient evidence against the hypothesis that dice is fair.

D. Computing Expected Frequencies using a Non-Uniform Distribution

Example 2: In a class of size 200, Prof. Mardy gave 22 **A**'s, 33 **B**'s, 95 **C**'s, 35 **D**'s, and 15 **F**'s. Prof. Mardy would like to know whether or not the current semester's grade distribution is significantly different from last semester's distribution for the same course. His records show that last semester 10%, 15%, 60%, 10%, and 5% of the students received the letter grades of A, B, C, D, and F, respectively. Use a 0.05 significance level to check the fit. Use (a) Classical Method and (b) *P*-value Method.

Solution: **(a)** **Classical Method:**

***Step* 0** Data: In the usual notation, we have $n = 200$, $k = 5$, and $\alpha = 0.05$

Expected Probabilities:

$$p_1 = 0.10,\ p_2 = 0.15,\ p_3 = 0.60,\ p_4 = 0.10, \text{ and } p_5 = 0.05$$

Null Hypothesis: There is no significant difference between the distributions of student grades in the two semesters; that is, the grade distribution for this semester **fits** well with last semester's grade distribution.

Alternative Hypothesis: At least one of the probabilities in H_0 is different than expected or the two grade distributions are different.

***Step* 1** Set up the null and alternative hypotheses.

$H_0: p_1 = 0.10,\ p_2 = 0.15,\ p_3 = 0.60,\ p_4 = 0.10, \text{ and } p_5 = 0.05$

H_1 : At least one of the probabilities is different from the stated one. The two grade distributions are different.

***Step* 2** **Test Statistics:** Under H_0, the expected frequency

$$E_i = p_i \times \sum_{i=1}^{k} O_i$$

for each E. More precisely,

$E_1 = p_1 \times \sum O = 0.10 \times 200 = 20$ $\quad$ $E_2 = p_2 \times \sum O = 0.15 \times 200 = 30$

$E_3 = p_3 \times \sum O = 0.60 \times 200 = 120$ $\quad$ $E_4 = p_4 \times \sum O = 0.10 \times 200 = 20$

$E_5 = p_5 \times \sum O = 0.05 \times 200 = 10$

The observed frequencies are the sample values collected in the experiment.

Letter grades	A	B	C	D	F
Frequency (*O*'s)	22	33	95	35	15

Table 9.5

That is: $O_1 = 22$, $O_2 = 33$, $O_3 = 95$, $O_4 = 35$, and $O_5 = 15$.

The calculations for the Chi-square test statistics are given in the following table.

Letter Grade	Observed Frequencies	Expected Frequencies	$\frac{(O-E)^2}{E}$
A	22	20	$\frac{(22-20)^2}{20} = 0.2$
B	33	30	$\frac{(33-30)^2}{30} = 0.3$
C	95	120	$\frac{(95-120)^2}{120} = 5.208$
D	35	20	$\frac{(35-20)^2}{20} = 11.25$
F	15	10	$\frac{(15-10)^2}{10} = 2.5$
Totals:	$\sum O_i = 200$	$\sum E_i = 200$	$\chi^2 = \sum \frac{(O-E)^2}{E} = 19.458$

Table 9.6

Test Statistics is:

$$\chi^2 = \sum \frac{(O-E)^2}{E}$$

$$= \frac{(22-20)^2}{20} + \frac{(33-30)^2}{30} + \frac{(95-120)^2}{120} + \frac{(35-20)^2}{20} + \frac{(15-10)^2}{10}$$

$$= 19.458333 \approx 19.46$$

Step 3 The Rejection Region for a Goodness of Fit Test is a right-tail Chi-square distribution with $k - 1 = 5 - 1 = 4$ degrees of freedom. The critical value for the Chi-square Goodness of Fit Test with 4 *df* at a 0.05 level of significance is $\chi_4^2(0.05) = 9.488$. The calculated value of the test statistics and the rejection region are shown in Figure 9.6.

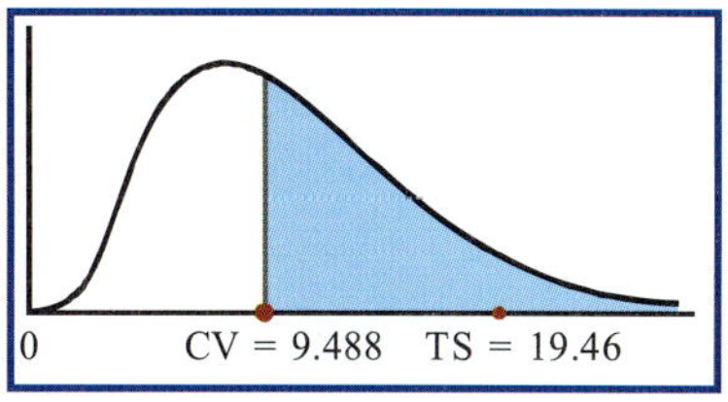

Figure 9.6

Step 4 **Decision:** Since the calculated value of χ^2 (test statistics) lies inside the rejection region, we reject H_0 at 5% level of significance. *Conclusion:* The data does not provide sufficient evidence to conclude that this semester's grade distribution is similar to last semester's grade distribution.

(b) ***P*-value Method:**

Step 1 Set up the null and alternative hypotheses.

$H_0 : p_1 = 0.10, p_2 = 0.15, p_3 = 0.60, p_4 = 0.10$, and $p_5 = 0.05$

H_1 : At least one of the probabilities is different from the stated one. The two grade distributions are different.

Step 2 **Test Statistics:** Under H_0, the expected frequency

$$E_i = p_i \times \sum_{i=1}^{k} O_i$$

for each E. More precisely,

$E_1 = p_1 \times \sum O = 0.10 \times 200 = 20$ $\quad E_2 = p_2 \times \sum O = 0.15 \times 200 = 30$

$E_3 = p_3 \times \sum O = 0.60 \times 200 = 120$ $\quad E_4 = p_4 \times \sum O = 0.10 \times 200 = 20$

$E_5 = p_5 \times \sum O = 0.05 \times 200 = 10$

Recall: The observed frequencies are the sample values collected in the experiment.

Letter grades	A	B	C	D	F
Frequency (*O*'s)	22	33	95	35	15

Table 9.5

That is: $O_1 = 22$, $O_2 = 33$, $O_3 = 95$, $O_4 = 35$, and $O_5 = 15$.

The calculations for the Chi-square test statistics are given in the following table.

Recall:

Letter Grade	**Observed Frequencies**	**Expected Frequencies**	$\frac{(O-E)^2}{E}$
A	22	20	$\frac{(22-20)^2}{20} = 0.2$
B	33	30	$\frac{(33-30)^2}{30} = 0.3$
C	95	120	$\frac{(95-120)^2}{120} = 5.208$

D	35	20	$\frac{(35-20)^2}{20} = 11.25$
F	15	10	$\frac{(15-10)^2}{10} = 2.5$
Totals:	$\sum O_i = 200$	$\sum E_i = 200$	$\chi^2 = \sum \frac{(O-E)^2}{E} = 19.458$

Table 9.6

Test Statistics is:

$$\chi^2 = \sum \frac{(O-E)^2}{E}$$

$$= \frac{(22-20)^2}{20} + \frac{(33-30)^2}{30} + \frac{(95-120)^2}{120} + \frac{(35-20)^2}{20} + \frac{(15-10)^2}{10}$$

$$= 19.458333 \approx 19.46$$

***Step* 3** ***P*-value of the test statistics is given by:**

$$p\text{-value} = P(\chi^2 > \text{test statistics} = 19.46 \text{ with } 4\ df)$$

From table 7, for $v = 4$, value of Chi-square nearest to 19.46 is 14.860. This corresponds to an α value of 0.005 in the first row of the χ^2-table.

$$\Rightarrow P\left(\chi_4^2 > 14.860\right) = 0.005$$

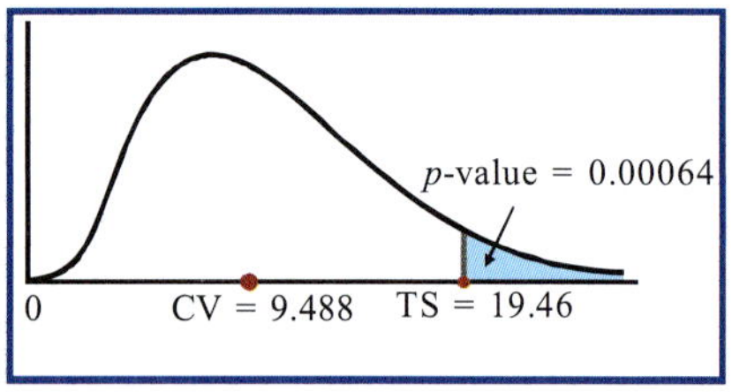

Figure 9.7

Since the value of the test statistics $\chi^2 = 19.46$ is very much to the right of $\chi^2 = 14.860$ (Figure 9.7), the $P(\chi^2 > 19.46) << 0.005$. Hence, ***p*-value** of test statistics is << 0.005.

Note: Double less than sign << means much smaller than.

***Step* 4** **Decision:** Since the ***p*-value** << 0.5, we reject H_0 at 0.05 level of significance. *Conclusion*: The data provides sufficient evidence against the null hypothesis.

Note: We can calculate the exact ***p*-value** of the test statistics $P(\chi^2 > 19.46)$ using the Microsoft Excel command CHISQ.DIST.RT(x, **deg_freedom**) as shown below.

=CHISQ.DIST.RT(19.46,4) CHISQ.DIST.RT(x, **deg_freedom**)	$\Rightarrow P\left(\chi_4^2 > 19.46\right) = 0.000638$

In short:

Hypotheses

$H_0: p_1 = 0.10, p_2 = 0.15, p_3 = 0.60,$
$p_4 = 0.10$ and $p_5 = 0.05$

H_1 : At least one of the probabilities is different than stated.

Test Statistics

$$\chi^2 = \sum \frac{(O-E)^2}{E} = 19.46$$

Rejection Region and Critical Value

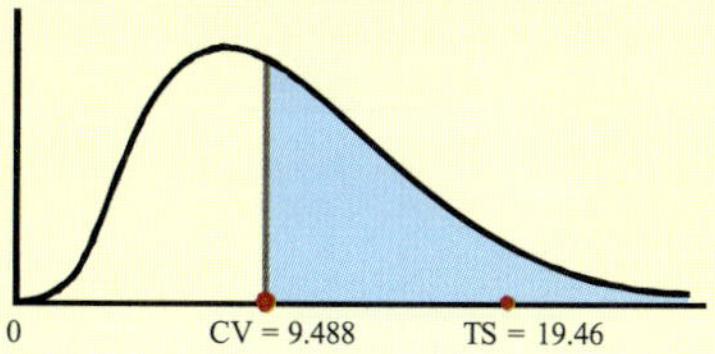

Conclusion

Reject H_0. The current semester's grade distribution is not simiiar to last semester's grade distribution.

Example 3: It is believed that the proportion of people with A, B, O, and AB blood types in the population are respectively 0.3, 0.2, 0.4, and 0.1. When 400 randomly selected people were examined, the observed numbers of each blood type were 118, 86, 136, and 60 respectively. At the 0.01 significance level, test the claim that the probabilities are distributed as believed. Use the Classical method.

Solution: **Classical Method:**

***Step* 0** Data: In the usual notation, we have $n = 400$ and $k = 4$.

Expected Probabilities: $p_A = 0.3$, $p_B = 0.2$, $p_o = 0.4$, $p_{AB} = 0.1$

Null Hypothesis: The probabilities are distributed as believed.

Alternative Hypothesis: At least one of the probabilities in H_0 is different than expected.

***Step* 1** Set up the null and alternative hypotheses.

$H_0 : p_A = 0.3, p_B = 0.2, p_O = 0.4, p_{AB} = 0.1$

H_1 : At least one of the probabilities is different than believed.

***Step* 2** **Test Statistics:** Under H_0, the expected frequency

$$E_i = p_i \times \sum_{i=1}^{k} O_i$$

for each E. More precisely,

$E_1 = p_A \times \sum O = 0.3 \times 400 = 120$ $\quad$ $E_2 = p_B \times \sum O = 0.2 \times 400 = 80$

$E_3 = p_O \times \sum O = 0.4 \times 400 = 160$ $\quad$ $E_4 = p_{AB} \times \sum O = 0.1 \times 400 = 40$

The observed frequencies are the sample values collected in the experiment.

Blood Type	A	B	O	AB
Frequency (*O*'s)	118	86	136	60

Table 9.7

That is: $O_1 = 118$, $O_2 = 86$, $O_3 = 136$, and $O_4 = 60$

The calculations for the Chi-square test statistics are given in the following table.

Blood Type	Observed Frequencies	Expected Frequencies	$\frac{(O-E)^2}{E}$
A	118	120	$\frac{(118-120)^2}{120} = 0.033$
B	86	80	$\frac{(86-80)^2}{80} = 0.45$
O	136	160	$\frac{(136-160)^2}{160} = 3.6$
AB	60	40	$\frac{(60-40)^2}{40} = 10$
Totals:	$\sum O_i = 400$	$\sum E_i = 400$	$\chi^2 = \sum \frac{(O-E)^2}{E} = 14.083$

Table 9.8

Test Statistics is:

$$\chi^2 = \sum \frac{(O-E)^2}{E}$$

$$= \frac{(118-120)^2}{120} + \frac{(86-80)^2}{80} + \frac{(136-160)^2}{160} + \frac{(60-40)^2}{40} = 14.08333 \approx 14.083$$

Step 3 The Rejection Region for a Goodness of Fit Test is a right-tail Chi-square distribution with $k-1 = 4-1 = 3$ degrees of freedom. The critical value for the Chi-square Goodness of Fit Test with 3 *df* at a 0.01 level of significance is $\chi_3^2(0.01) = 11.345$. The calculated value of the test statistics, critical value and the rejection region are shown in Figure 9.8.

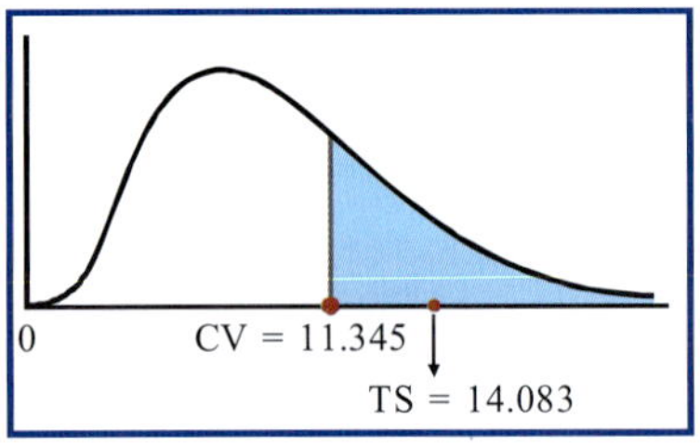

Figure 9.8

Step 4 **Decision:** Since the calculated value of χ^2 (test statistics) lies inside the rejection region, we reject H_0 at 1% level of significance. *Conclusion:* The data does not provide sufficient evidence to support H_0. The distribution of probabilities is not as believed.

In short:

Hypotheses

$H_0: p_A = 0.3, p_B = 0.2, p_O = 0.4$ and $p_{AB} = 0.1$

H_1 : At least one of the probabilities is different than believed.

Test Statistics

$$\chi^2 = \sum \frac{(O-E)^2}{E} = 14.083$$

Rejection Region and Critical Value

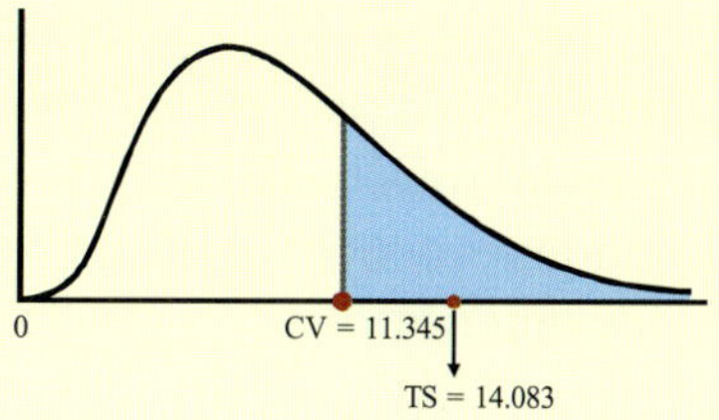

Conclusion

Reject H_0. At least one of the probabilities is different.

Try it yourself:

Use the Goodness of Fit Test to solve the following problems.

1. In a survey in year 2000, married couples were asked the following question: "*Who makes the most money in your household*?" The following table contains the responses in percentages.

Money Maker	Men	Women	Both
Percentages	40	30	30

In a current survey, 500 randomly selected households were asked the same question. The results are given in the accompanying table.

Money Maker	Men	Women	Both
Percentages	200	160	140

Use a 0.05 significance level to test the claim that the current money making pattern is the same as that for 2000.

2. The following table lists the frequency distribution for 100 rolls of a die.

# of dots	1	2	3	4	5	6
Frequency	17	16	18	13	20	16

Use $\alpha = 0.01$, to test the claim that the die is fair.

3. The following table lists the frequency distribution of cars sold at Santiago BMW during the past 12 months.

Month	Jan	Feb	Mar	Apr	May	Jun	Jul	Aug	Sep	Oct	Nov	Dec
# of cars sold	18	43	20	12	24	33	38	27	34	42	21	18

Using a 0.1 significance level, test the claim that the number of cars sold at Santiago each month follows a uniform distribution.

9.3: Test of Independence

Objectives

After completing this section you will be able to understand:

A. Contingency Tables.

B. χ^2-Test of Independence.

A. Contingency Tables

In this section we will study ***contingency tables*** (two-way classification tables). In contingency tables, the frequency count for categorical data is arranged using at least two rows and two columns. The categories of one of the variables are shown along the columns and the categories of the other variable are shown along the rows. Given a contingency table, we will use the χ^2-distribution to develop a test of hypothesis known as **Test of Independence**. This test is used to analyze the association (relationship) between the rows and columns of the given table. In the test of independence, the null hypothesis represents **independence** (no relationship between rows and columns) and the alternative hypothesis represents **dependence** of the classified variables.

To illustrate this, let us consider the data relating LDL's (low-density lipoproteins, also known as bad cholesterol) and heart disease arranged as follows in a contingency table.

	High LDL	Medium LDL	Low LDL
Heart Disease	3	4	42
No Heart Disease	50	16	5

Table 9.9

This is a contingency table of size 2 × 3 (read two by three) because it consists of two rows and three columns. When writing the size of a contingency table, we always list the number of rows first. The two categories are "LDL" and "Heart Disease". The categories of the LDL are listed along the columns of the table and the categories of the heart disease are shown along the rows. The six boxes in the table are called **cells**. Each cell contains the frequency of the contingency.

It is customary to show the row and column totals in a contingency table. Thus, adding a column of row totals and a row of column totals. The above contingency table becomes:

	High LDL	Medium LDL	Low LDL	Row Totals
Heart Disease	3	4	42	**49**
No Heart Disease	50	16	5	**71**
Column Totals	**53**	**20**	**47**	**Grand Total = 120**

Table 9.10

Note: The grand total is total number of observations.

B. χ^2-Test for Independence

- **Assumptions**

1. The sample data are randomly selected so that the sample observations are independent.

2. The total frequency "n" in the contingency table is fixed (predetermined).

3. For every cell in the contingency table, the corresponding expected frequency must be at least 5.

- **Null and Alternative Hypotheses for the Test of Independence**

H_0 : The variables of classification (row and column variables) are ***independent***.

H_1 : The variables of classification are ***dependent***.

- **Test Statistics for the Test of Independence**

Test Statistics = $\chi^2 = \sum \frac{(O-E)^2}{E}$, where the O's are the observed frequencies in each cell and E's are the expected frequencies found using the formula

$$E = \frac{(\text{Row Total})(\text{Column Total})}{(\text{Grand Total})}$$

- **Rejection Region and the Critical Values**

The rejection region in the test of independence is always a right-tailed χ^2- distribution with $(r-1) \times (c-1)$ *df* at an α significance level.

Note: r = the number of rows in the contingency table not including the row and column totals, and

c = the number of columns in the contingency table not including the row and column totals.

Conclusion:

In the test of independence, if H_0 is rejected, we conclude that row and column variables in the contingency table are ***dependent*** (related). If we fail to reject H_0, then row and column variables in the contingency table are ***independent***.

P-Values: In a χ^2-Test, ***p***-values are usually determined using a computer software.

$$p\text{-value} = P(\chi^2 > \text{Test Statistics})$$

If the p-value $< \alpha$, reject the null hypothesis, otherwise fail to reject the null hypothesis.

For a more comprehensive treatment of the theory of test of independence, refer to Section 9.6.

Example 4: The following data presents the cholesterol levels and the corresponding heart disease among a random sample of 120 adults in a city.

	High LDL	Medium LDL	Low LDL	Row Totals
Heart Disease	3	4	42	**49**
No Heart Disease	50	16	5	**71**
Column Totals	**53**	**20**	**47**	**Grand Total = 120**

Test whether or not the cholesterol level and heart disease in the adult population are independent. Use $\alpha = 0.05$. Use the Classical method.

Solution: ***Step 0*** Given: The grand total = n = 120, number of rows = 2, number of columns = 3, and the significance level = 0.05.

Because the data are in the form of a contingency table, we use the **Chi-Square Test for Independence**.

Step 1 Set up the null and the alternative hypotheses.

H_0 : The cholesterol level and heart disease are independent.

H_1 : The cholesterol level and heart disease are dependent (related).

***Step* 2 Test Statistics:** Under the null hypothesis H_0:

The observed frequencies are;

$O_{1,1}$	$O_{1,2}$	$O_{1,3}$	$O_{2,1}$	$O_{2,2}$	$O_{2,3}$
3	4	42	50	16	5

Table 9.11

$$\text{Expected frequency for any cell} = \frac{(\text{row total of that cell})(\text{column total of that cell})}{(\text{grand total})}$$

	Table of Expected Frequencies		
	High LDL	**Medium LDL**	**Low LDL**
Heart Disease	$E_{1,1} = \frac{(49)(53)}{120} = 21.642$	$E_{1,2} = \frac{(49)(20)}{120} = 8.167$	$E_{1,3} = \frac{(49)(47)}{120} = 19.192$
No Heart Disease	$E_{2,1} = \frac{(71)(53)}{120} = 31.358$	$E_{2,2} = \frac{(71)(20)}{120} = 11.833$	$E_{2,3} = \frac{(71)(47)}{120} = 27.808$

Table 9.12

We can now calculate the test statistics.

Remark: Cell (1, 1) refers to the cell in the intersection of the 1st row and 1st column and so on.

Cells	***O***	***E***	$\frac{(O-E)^2}{E}$
(1, 1)	3	21.642	$\frac{(3-21.642)^2}{21.642} = 16.058$
(1, 2)	4	8.167	$\frac{(4-8.167)^2}{8.167} = 2.126$
(1, 3)	42	19.192	$\frac{(42-19.192)^2}{19.192} = 27.107$
(2, 1)	50	31.358	$\frac{(50-31.358)^2}{31.358} = 11.082$
(2, 2)	16	11.833	$\frac{(16-11.833)^2}{11.833} = 1.467$
(2, 3)	5	27.808	$\frac{(5-27.808)^2}{27.808} = 18.707$
Totals:	$\sum O = 120$	$\sum E = 120$	$\chi^2 = \sum\frac{(O-E)^2}{E} = 76.546$

Table 9.13

The **test statistics** is: $\chi^2 = \sum\frac{(O-E)^2}{E} = 76.546$

Step **3** The Rejection Region for a Test of Independence is a right-tail Chi-square distribution with $(c - 1)(r - 1) = (3 - 1)(2 - 1) = 2$ degrees of freedom. The critical value for the Chi-square Test of Independence with 2 *df* at a 0.05 level of significance is $\chi^2_2(0.05) = 5.991$. The calculated value of the test statistics, critical value and the rejection region are shown in Figure 9.9.

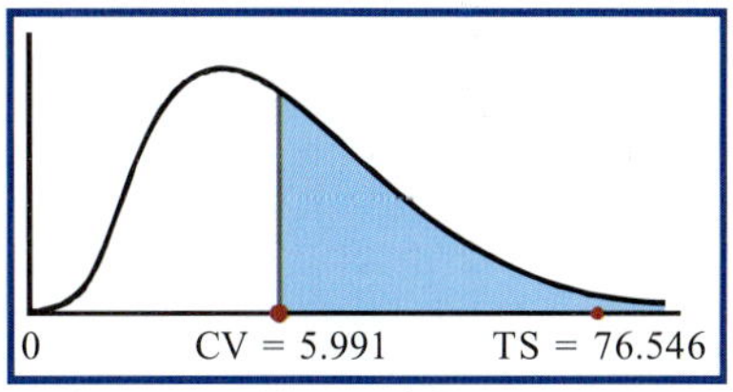

Figure 9.9

Step **4** **Decision:** Since the calculated value of χ^2 (test statistics) lies inside the rejection region, we reject H_0 at 5% level of significance. *Conclusion:* The data does not provide sufficient evidence to support H_0. There is strong evidence to suggest that the cholesterol level and heart disease **are not** independent; that is, they are related (dependent).

In short:

Hypotheses	**Test Statistics**
H_0 : The cholesterol level and heart diseases are independent. H_1 : The cholesterol level and heart diseases are dependent (related).	$\chi^2 = \sum \frac{(O-E)^2}{E} = 76.546$
Rejection Region and Critical Value	**Conclusion**
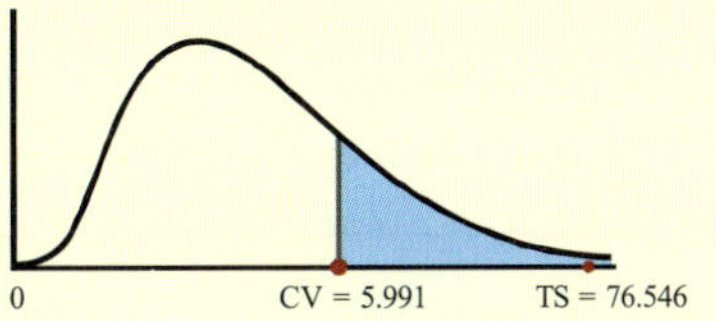	Reject H_0. Cholesterol level and heart diseases are not independent in adults.

Example 5: 344 subjects were classified in the following table according to their blood type and eye color.

Eye color\Blood type	A	B	O	AB
Blue	40	32	38	15
Brown	85	42	70	22

Use a 0.05 significance level to determine if Blood-Type and Eye-Color are independent. Use the Classical method.

Solution: *Step* **0** Given: $n = 344$, number of rows = 2, number of columns = 4, and the significance level = 0.05.

Because the data are in the form of a contingency table, we use the **Chi-Square Test for Independence**.

Step **1** Set up the null and the alternative hypotheses.

H_0 : Blood type and eye color are independent.

H_1 : Blood type and eye color are dependent (related).

Step **2** **Test Statistics: Under** the null hypothesis H_0:

The observed frequencies are:

$O_{1,1}$	$O_{1,2}$	$O_{1,3}$	$O_{1,4}$	$O_{2,1}$	$O_{2,2}$	$O_{2,3}$	$O_{2,4}$
40	32	38	15	85	42	70	22

Table 9.14

The expected frequencies are;

$$\text{Expected frequency for any cell} = \frac{(\text{row total of that cell})(\text{column total of that cell})}{(\text{grand total})}$$

$E_{1,1}$	$E_{1,2}$	$E_{1,3}$	$E_{1,4}$
$\frac{(125)(125)}{344} = 45.422$	$\frac{(125)(74)}{344} = 26.890$	$\frac{(125)(108)}{344} = 39.244$	$\frac{(125)(37)}{344} = 13.445$
$E_{2,1}$	$E_{2,2}$	$E_{2,3}$	$E_{2,4}$
$\frac{(219)(125)}{344} = 79.578$	$\frac{(219)(74)}{344} = 47.110$	$\frac{(219)(108)}{344} = 68.756$	$\frac{(219)(37)}{344} = 23.555$

Table 9.15

We can now calculate the test statistics.

Cells	***O***	***E***	$\frac{(O-E)^2}{E}$
(1, 1)	40	45.422	$\frac{(40-45.422)^2}{45.422} = 0.647$
(1, 2)	32	26.890	$\frac{(32-26.890)^2}{26.890} = 0.971$
(1, 3)	38	39.244	$\frac{(38-39.244)^2}{39.244} = 0.039$
(1, 4)	15	13.455	$\frac{(15-13.455)^2}{13.455} = 0.180$
(2, 1)	85	79.578	$\frac{(85-79.578)^2}{79.578} = 0.369$
(2, 2)	42	47.110	$\frac{(42-47.110)^2}{47.110} = 0.554$
(2, 3)	70	68.756	$\frac{(70-68.756)^2}{68.756} = 0.023$
(2, 4)	22	23.555	$\frac{(22-23.555)^2}{23.555} = 0.103$
Totals:	$\sum O = 344$	$\sum E = 344$	$\chi^2 = \sum\frac{(O-E)^2}{E} = 2.886$

Table 9.16

The **test statistics** is: $\chi^2 = \sum\frac{(O-E)^2}{E} = 2.886$

***Step* 3** The Rejection Region for a Test of Independence is a right-tail Chi-square distribution with $(c - 1)(r - 1) = (4 - 1)(2 - 1) = 3$ degrees of freedom. The critical value for the Chi-square Test of Independence with 3 *df* at a 0.05 level of significance is $\chi_3^2(0.05) = 7.815$. The calculated value of the test statistics and the rejection region are shown in Figure 9.10.

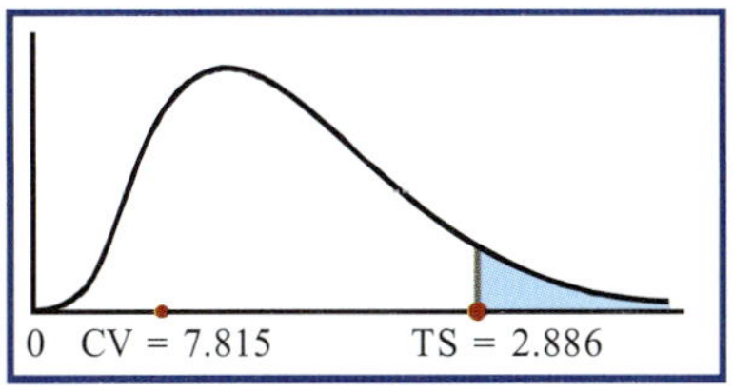

Figure 9.10

***Step* 4** **Decision:** Since the calculated value of χ^2 (test statistics) lies outside the rejection region, we fail to reject H_0 at 5% level of significance. *Conclusion:* The data does not provide sufficient evidence to reject H_0. There is strong evidence to suggest that blood type and eye color are independent.

In short:

Hypotheses	**Test Statistics**
H_0 : Blood type and eye color are independent H_1 : Blood type and eye color are dependent (related)	$\chi^2 = \sum \frac{(O-E)^2}{E} = 2.886$
Rejection Region and Critical Value 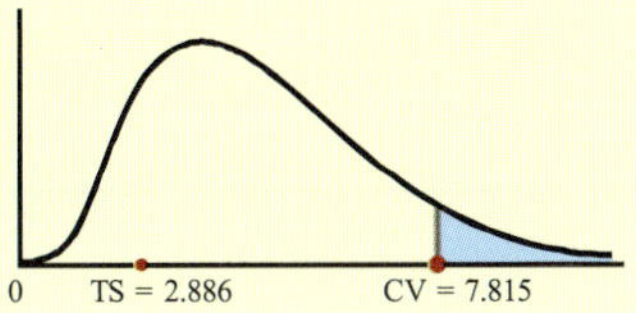	**Conclusion** Fail to reject H_0. Blood type and eye color are independent.

Example 6: Use the ***P***-value method to examine whether a voter's region of residence (city or county) and his or her political party (Republican or Democrat) preference are independent, given the following table. Use $\alpha = 0.01$.

	Republican	**Democrat**
City	470	650
County	550	350

Solution: ***Step* 0** Given: $n = 2020$, number of rows = 2, number of columns = 2, and the significance level = 0.01.

Because the data are in the form of a contingency table, we use the **Chi-Square Test for Independence**.

***Step* 1** Set up the null and the alternative hypotheses.

H_0 : The region of residence and party preference are independent.

H_1 : The region of residence and party preference are dependent.

***Step* 2** **Test Statistics:**Under the null hypothesis H_0:

The observed frequencies are:

$O_{1,1}$	$O_{1,2}$	$O_{2,1}$	$O_{2,2}$
470	650	550	350

Table 9.17

The table provided in this example, with all the row and column totals is shown below.

	Republican	Democrat	Row Total
City	470	650	**1120**
County	550	350	**900**
Column Totals	**1020**	**1000**	**Grand Total = 2020**

Table 9.18

Using Table 9.18, the expected frequencies are;

$E_{1,1}$	$E_{1,2}$	$E_{2,1}$	$E_{2,2}$
565.545	554.455	454.455	445.545

Table 9.19

We can now calculate the test statistics.

Cells	O	E	$\frac{(O-E)^2}{E}$
(1, 1)	470	565.545	16.142
(1, 2)	650	554.455	16.465
(2, 1)	550	454.455	20.087
(2, 2)	350	445.545	20.489
Totals:	$\sum O = 2020$	$\sum E = 2020$	$\chi^2 = \sum \frac{(O-E)^2}{E} = 73.183$

Table 9.20

The **test statistics** is $\chi^2 = \sum \frac{(O-E)^2}{E} = 73.183$

Note: When the degrees of freedom for a contingency table is equal to 1, that is, (The table is a 2 × 2 table.), some researchers suggest using the *Yates correction for continuity* in the test statistics formula.

$$\sum \frac{(|O-E|-0.5)^2}{E}$$

We will not use the Yates Correction in this section because of the already conservative nature of the Chi-square Test.

Step 3 ***P***-value of the test statistics is given by:

p-value = $P(\chi^2 >$ test statistics = 73.183 with 1 df)

From Table 7, the critical value of Chi-square for $v = 1$, and $\alpha = 0.01$ is $\chi_1^2(0.01) = 6.635$. The test statistics and the critical value are shown in Figure 9.11. It is quite clear from this figure that the p-value of the test statistics given by

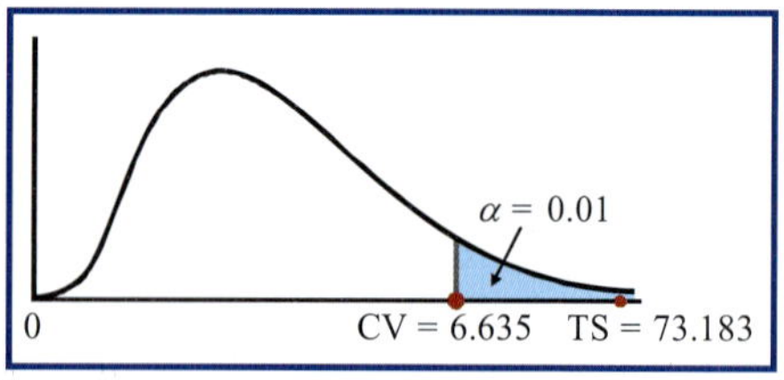

Figure 9.11

$P(\chi^2 > 73.183 \text{ for } df\ v = 1)$ is much smaller than the given significance level $\alpha = 0.01$.

Note: We can find the exact value for $P(\chi^2 > 73.183)$ for dfv = 1 using Microsoft Excel ®.

Command	Output
=CHISQ.DIST.RT(73.183,1)	$1.18169 \times 10^{-17} = 0.000000000000000012 \approx 0$

Step 4 **Decision:** Since the p-value < 0.01, we reject H_0 at 0.01 level of significance. *Conclusion*: The data provides sufficient evidence against the null hypothesis. The region of residence and party preference are certainly related (dependent).

In short:

Hypotheses	**Test Statistics**
H_0: The region of residence and party preference are independent. H_1: The region of residence and party preference are dependent.	$\chi^2 = \sum \frac{(O-E)^2}{E} = 73.183$
Rejection Region and Critical Value	**Conclusion**
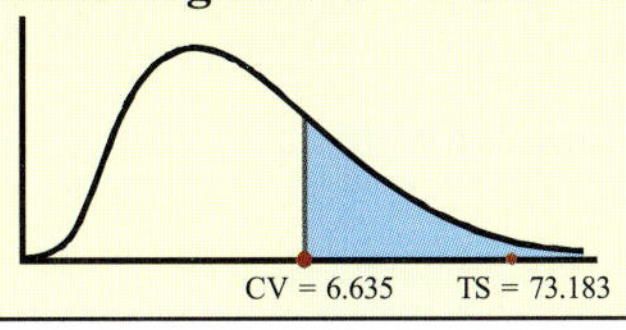	Reject H_0. The region of residence and party preference are not independent.

Try it yourself:

In the following problems, use both the Classical method and the *p*-value method to test the independence between the variables of classification.

1. In a certain community, 389 randomly selected people were classified according to their age group and political belief. The data is given below.

Political Belief	Age Groups		
	20-29	30-39	40 and above
Republican	11	43	12
Democrat	82	87	56
Independent	38	25	35

Using a 0.01 significance level, test the claim that a person's age group and political beliefs are independent.

2. A survey was conducted to investigate the relationship between different delivery methods in teaching statistics and students' pass rates.

	Passing	Failing
Technology based instructions	356	180
Traditional based instructions	240	90

Use a 0.05 significance level, to test the claim that instructional delivery method and pass rate are related.

3. The following data was collected to determine whether the selection of a Premium or a Basic Cable TV programming is independent of the viewer's level of education.

	High School	Bachelor's Degree	Master's Degree
Premium	250	100	120
Basic	40	160	150

Test the above claim using a 0.01 significance level.

9.4: Test of Homogeneity

Objectives

After completing this section you will be able to understand:

A. Test for Equality of Several Proportions (Test of Homogeneity).

In Section 9.2, we used the Chi-square Goodness of Fit Test to determine if an observed frequency distribution fits a claimed distribution (theoretical distribution), assuming the samples are drawn from a **single population**. In Section 9.3, we used the Test of Independence to determine if rows and columns of a contingency table are related or not, once again using the assumption that samples are drawn from a **single population**.

However, we occasionally obtain samples in the form of a contingency table from **different populations** and want to determine if those populations have the same proportions of a particular characteristics.

In this section, we will use the Test of Homogeneity to test claims that different populations have equal proportions of some characteristic.

Note: The fundamental difference between the Goodness of Fit Test and the Test of Independence versus the Test of Homogeneity is the assumption that whether the samples are drawn from a single population or different populations.

A. Test for Equality of Several Proportions (Test of Homogeneity)

Suppose we want to test the equality of several population proportions, that is, $p_1 = p_2 = \ldots = p_c$.

For this test, we draw independent random samples of pre-determined sizes, say, $n_1, n_2, \ldots, n_c$ from c populations respectively and record their responses into one of the r categories for each of these samples. Then we get the data in the form of $r \times c$ contingency table with fixed number of column totals $n_1, n_2, \ldots, n_c$. However, the r row totals will be the result of random sampling.

In the test of homogeneity, we want to test:

H_0: Populations are homogeneous with respect to the cell frequencies (probabilities).

Here, the test of hypothesis for H_0 is the same as that for the ***test of independence*** presented in Section 9.3 with the exception that instead of testing the null hypothesis of independence between rows and columns of a contingency table, we test the null hypothesis that different populations have the same proportions of a certain characteristic.

Example 7: The following data gives the HDL cholesterol levels in random samples of sizes 120, 200, 150, and 130 from the adult population of the cities of Washington D.C., New York, Los Angeles, and Atlanta respectively.

	Washington D.C.	New York	Los Angeles	Atlanta
High HDL	53	80	68	57
Not High HDL	67	120	82	73

Using a 0.025 significance level, test the equality of proportions of adults with high HDL cholesterol levels in these four cities. Use both the Classical and the *P*-value methods.

Solution: **Classical Method:**

***Step* 0** Given: $\alpha = 0.025$ and $n = 600$.

Let p_1, p_2, p_3, and p_4 represent the true proportions of adults with high HDL cholesterol levels in D.C., NY, LA, and ATL respectively.

Since we are testing the equality of proportions of four different populations, we use the Chi-square Test of Homogeneity.

***Step* 1** Set up the null and the alternative hypotheses.

$H_0 : p_1 = p_2 = p_3 = p_4$

H_1 : At least one proportion is different from the others (ALOID)

***Step* 2** **Test Statistics:**

The observed frequencies are;

$O_{1,1}$	$O_{1,2}$	$O_{1,3}$	$O_{1,4}$	$O_{2,1}$	$O_{2,2}$	$O_{2,3}$	$O_{2,4}$
53	80	68	57	67	120	82	73

Table 9.21

Here are all the row and column totals.

	Washington D.C.	New York	Los Angeles	Atlanta	Row Total
High HDL	53	80	68	57	**258**
Not High HDL	67	120	82	73	**342**
Column Totals	**120**	**200**	**150**	**130**	**Grand Total = 600**

Table 9.22

Using Table 9.22, the expected frequencies are:

$E_{1,1}$	$E_{1,2}$	$E_{1,3}$	$E_{1,4}$
$\frac{(258)(120)}{600} = 0.038$	$\frac{(258)(200)}{600} = 0.419$	$\frac{(258)(150)}{600} = 0.190$	$\frac{(258)(130)}{600} = 0.022$
$E_{2,1}$	$E_{2,2}$	$E_{2,3}$	$E_{2,4}$
$\frac{(342)(120)}{600} = 0.029$	$\frac{(342)(200)}{600} = 0.316$	$\frac{(342)(150)}{600} = 0.143$	$\frac{(342)(130)}{600} = 0.016$

Table 9.23

We can now calculate the test statistics.

Cells	O	E	$\frac{(O-E)^2}{E}$
(1, 1)	53	51.6	0.038
(1, 2)	80	86	0.419
(1, 3)	68	64.5	0.19
(1, 4)	57	55.9	0.022
(2, 1)	67	68.4	0.29
(2, 2)	120	114	0.316
(2, 3)	82	85.5	0.143
(2, 4)	73	74.1	0.016
Totals:	$\sum O = 600$	$\sum E = 600$	$\chi^2 = \sum\frac{(O-E)^2}{E} = 1.173$

Table 9.24

The **test statistics** is $\chi^2 = \sum\frac{(O-E)^2}{E} = 1.173$

***Step* 3** The Rejection Region for a Test of Homogeneity is a right-tail Chi-square distribution with $(c - 1)(r - 1) = (4 - 1)(2 - 1) = 3$ degrees of freedom. Using Table 7, the critical value for the Chi-square Test of Homogeneity with 3 *df* at a 0.025 level of significance is $\chi_3^2(0.025) = 9.348$. The calculated value of the test statistics, the rejection region, and the critical value are shown in Figure 9.12.

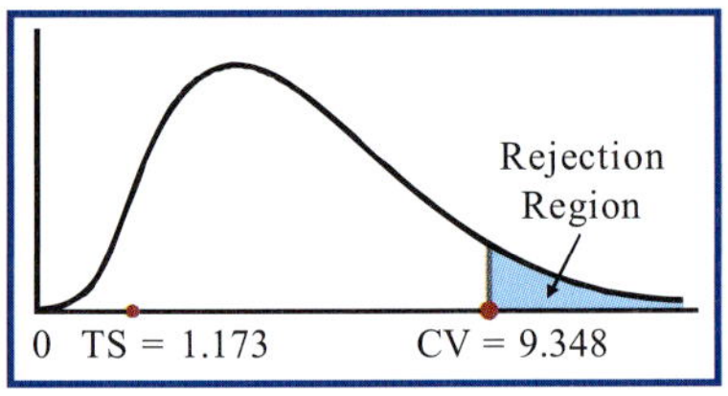

Figure 9.12

***Step* 4** **Decision:** Since the calculated value of χ^2 (test statistics) lies outside the rejection region, we fail to reject H_0 at 0.025 level of significance. *Conclusion:* The data does not provide sufficient evidence to reject H_0. The distribution of probabilities is homogeneous. There is strong evidence to suggest that the proportion of adults with high HDL cholesterol level is the same in all four cities.

P-value Method:

***Step* 1** Set up the null and the alternative hypotheses.

$H_0 : p_1 = p_2 = p_3 = p_4$

H_1 : At least one proportion is different from the others (ALOID)

***Step* 2** **Test Statistics:**

The observed frequencies are;

Recall:

$O_{1,1}$	$O_{1,2}$	$O_{1,3}$	$O_{1,4}$	$O_{2,1}$	$O_{2,2}$	$O_{2,3}$	$O_{2,4}$
53	80	68	57	67	120	82	73

Table 9.21

Here are all the row and column totals.

Recall:

	Washington D.C.	New York	Los Angeles	Atlanta	Row Total
High HDL	53	80	68	57	**258**
Not High HDL	67	120	82	73	**342**
Column Totals	**120**	**200**	**150**	**130**	**Grand Total = 600**

Table 9.22

Using table 9.22, the expected frequencies are:

Recall:

$E_{1,1}$	$E_{1,2}$	$E_{1,3}$	$E_{1,4}$
$\frac{(258)(120)}{600} = 0.038$	$\frac{(258)(200)}{600} = 0.419$	$\frac{(258)(150)}{600} = 0.190$	$\frac{(258)(130)}{600} = 0.022$
$E_{2,1}$	$E_{2,2}$	$E_{2,3}$	$E_{2,4}$
$\frac{(342)(120)}{600} = 0.029$	$\frac{(342)(200)}{600} = 0.316$	$\frac{(342)(150)}{600} = 0.143$	$\frac{(342)(130)}{600} = 0.016$

Table 9.23

We can now calculate the test statistics.

Recall:

Cells	O	E	$\frac{(O-E)^2}{E}$
(1, 1)	53	51.6	0.038
(1, 2)	80	86	0.419
(1, 3)	68	64.5	0.19
(1, 4)	57	55.9	0.022
(2, 1)	67	68.4	0.29
(2, 2)	120	114	0.316
(2, 3)	82	85.5	0.143
(2, 4)	73	74.1	0.016
Totals:	$\sum O = 600$	$\sum E = 600$	$\chi^2 = \sum \frac{(O-E)^2}{E} = 1.173$

Table 9.24

The **test statistics** is $\chi^2 = \sum \frac{(O-E)^2}{E} = 1.173$

Step 3 ***P*-value of the test statistics is given by:**

p-value = $P(\chi^2 >$ test statistics $= 1.173$ with 3 *df*).

From Table 7, the row corresponding to $v = 3$ and the column corresponding to $\alpha = 0.025$ intersect at $\chi^2 = 9.348$. Since the calculated χ^2 value (test statistics) is to the left of 9.348, $P(\chi^2 > 1.173) > 0.025$.

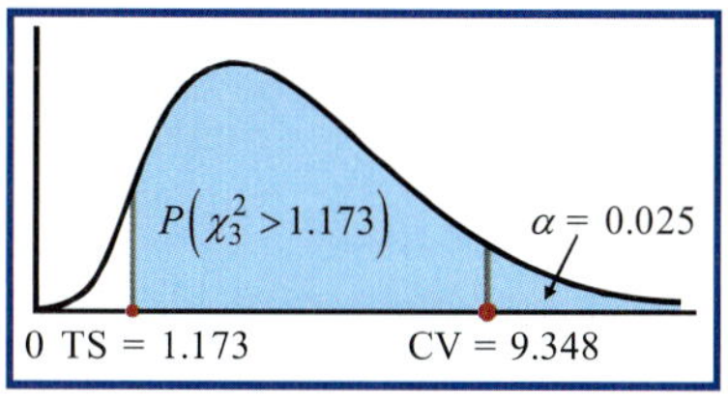

Figure 9.13

__Note:__ We can find the exact value for $P(\chi^2 > 1.173)$ for *df* $v = 3$ using Microsoft Excel®.

Command	Output
=CHISQ.DIST.RT(1.173,3)	0.759

Step 4 **Decision:** Since the *p*-value > $\alpha = 0.025$, we fail to reject H_0 at 0.025 level of significance.

Conclusion: The data does not provide sufficient evidence to reject H_0. The distribution of probabilities is homogeneous. There is strong evidence to suggest that the proportion of adults with high HDL cholesterol level is the same in all four cities.

In short:

Hypotheses	**Test Statistics**
$H_0: p_1 = p_2 = p_3 = p_4$ H_1 : At least one proportion is different from the others (ALOID)	$\chi^2 = \sum \frac{(O-E)^2}{E} = 1.173$
Rejection Region and Critical Value 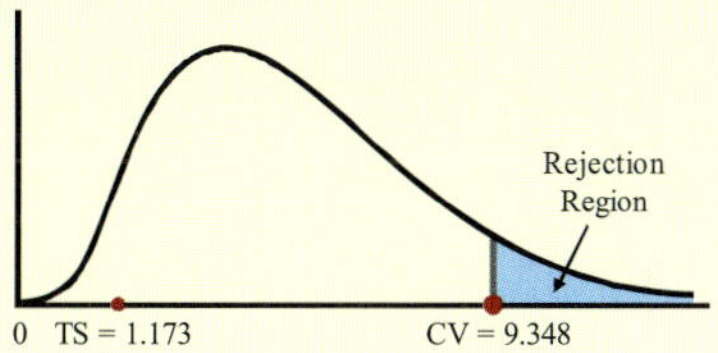 	**Conclusion** Fail to reject H_0. The proportion of adults with high HDL cholesterol level is the same in all four cities.

Example 8: It is found that 35 of 250 housewives in Baltimore, 22 of 220 housewives in Atlanta, and 39 of 300 housewives in New York watch at least one day-time talk show every day. Use a 0.05 significance level to test the claim that there is no difference between the true proportions of housewives who watch day-time talk shows in these cities. Use the Classical method.

Solution: **Classical Method:**

***Step* 0** $\alpha = 0.05$, and the following table summarizes the given data.

	Baltimore	Atlanta	New York	Total
Watch Talk Show	35	22	39	96
Do not watch Talk Show	250 – 35 = 215	220 – 22 = 198	300 – 39 = 261	674
Total	250	220	300	Grand Total = 770

Table 9.25

Let p_1, p_2, and p_3 represent the true proportions of housewives who watch talk shows in the cities of Baltimore, Atlanta, and New York, respectively

Since we are testing the equality of proportions of three different populations, we use the Chi-square Test of Homogeneity.

***Step* 1** Set up the null and the alternative hypotheses.

$H_0 : p_1 = p_2 = p_3$

H_1 : At least one proportion is different from the others

***Step* 2** **Test Statistics:** Under the null hypothesis H_0:

The observed frequencies are:

$O_{1,1}$	$O_{1,2}$	$O_{1,3}$	$O_{2,1}$	$O_{2,2}$	$O_{2,3}$
35	22	39	215	198	261

Table 9.26

The expected frequencies are:

$E_{1,1}$	$E_{1,2}$	· $E_{1,3}$
$\frac{(96)(250)}{770} = 31.169$	$\frac{(96)(220)}{770} = 27.429$	$\frac{(96)(300)}{770} = 37.403$
$E_{2,1}$	$E_{2,2}$	$E_{2,3}$
$\frac{(674)(250)}{770} = 218.831$	$\frac{(674)(220)}{770} = 192.571$	$\frac{(674)(300)}{770} = 262.597$

Table 9.27

We can now calculate the test statistics.

Cells	O	E	$\frac{(O-E)^2}{E}$
(1,1)	35	31.169	0.471
(1,2)	22	27.429	1.075
(1,3)	39	37.403	0.068
(2,1)	215	218.831	0.067
(2,2)	198	192.571	0.153
(2,3)	261	262.597	0.010
Totals:	$\sum O = 770$	$\sum E = 770$	$\chi^2 = \sum \frac{(O-E)^2}{E} = 1.844$

Table 9.28

The **test statistics** is: $\chi^2 = \sum \frac{(O-E)^2}{E} = 1.844$

***Step* 3** The Rejection Region for a Test of Homogeneity is a right-tail Chi-square distribution with $(c-1)(r-1) = (3-1)(2-1) = 2$ degrees of freedom. Using Table 7, the critical value for the Chi-square Test of Homogeneity with 2 df at a 0.05 level of significance is $\chi^2_2(0.05) = 5.991$. The calculated value of the test statistics, the rejection region, and the critical value are shown in Figure 9.14.

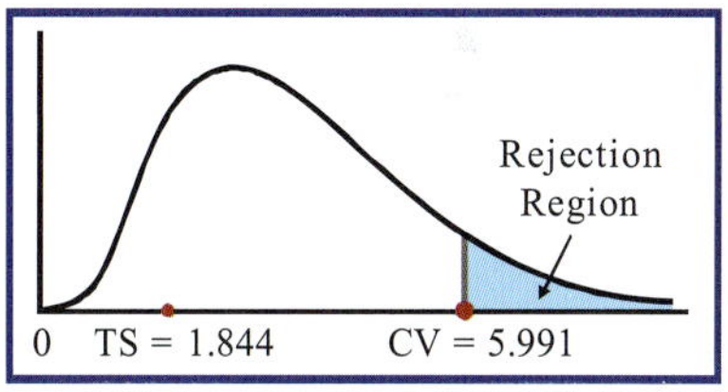

Figure 9.14

***Step* 4** **Decision:** Since the calculated value of χ^2 (test statistics) lies outside the rejection region, we fail to reject H_0 at 5% level of significance. *Conclusion:* The data does not provide sufficient evidence to reject H_0. The distribution of probabilities is homogeneous. There is strong evidence to suggest that the proportion of housewives who watch talk shows is the same in all three cities.

In short:

Hypotheses	**Test Statistics**
$H_0: p_1 = p_2 = p_3$ H_1 : At least one proportion is different from the others	$\chi^2 = \sum \frac{(O-E)^2}{E} = 1.844$
Rejection Region and Critical Value 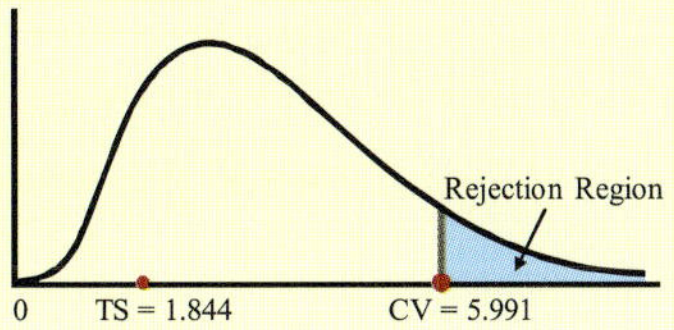	**Conclusion** Fail to reject H_0. The proportion of housewives who watch talk shows is the same in all three cities.

Try it yourself:

In the following problems, use the given significance level to test if the proportions are the same. Use both the Classical and the *P*-value methods.

1. From the adult male population of four large cities, random samples of sizes given below were taken and the number of married and single men were recorded.

	City A	City B	City C	City D	Total
Married	137	164	152	147	**600**
Single	32	57	56	35	**180**
Total	**169**	**221**	**208**	**182**	**Grand Total = 780**

Can it be concluded that, at 5% level of significance, the proportion of married men is the same in all four cities?

2. In a Statistics Common Final Exam, 100 of 125 day-time students on campus passed, 57 of 75 evening students on campus passed, and 36 of 50 off-campus students passed. Use a 0.01 significance level to test the claim that proportions of day-time, evening, and off-campus students that pass the common final are the same.

3. To test whether the proportion of teenage smokers in three school districts are the same, random samples were taken. The following data were obtained.

	District 1	District 2	District 3
Teenage Smokers	17	18	12
Teenage Nonsmokers	68	57	53
Totals	**85**	**75**	**65**

Using a 0.05 significance level, are the proportions of teenage smokers in all three school districts the same?

9.5: Multinomial Experiments with Microsoft Excel®

Objectives

After completing this section you will be able to carryout the computations in Chapter 9 using Microsoft Excel®.

Illustration 1: A die is rolled 120 times and the following data is collected.

Number on the die	1	2	3	4	5	6
Frequency	17	25	17	23	15	23

(a) Calculate the test Statistics for a Goodness of Fit Test.

(b) Find the critical value for this test using a 0.05 significance level.

Solution: **(a)**

Excel Commands

	A	B	C	D
1	# on the die	Observed Frequencies	Expected Frequencies	(O – E)^2/E
2	1	17	=1/6*120	=(B2 – C2)^2/C2
3	2	25	=1/6*120	=(B3 – C3)^2/C3
4	3	17	=1/6*120	=(B4 – C4)^2/C4
5	4	23	=1/6*120	=(B5 – C5)^2/C5
6	5	15	=1/6*120	=(B6 – C6)^2/C6
7	6	23	=1/6*120	=(B7 – C7)^2/C7
8		=SUM(B2:B7)	=SUM(C2:C7)	=SUM(D2:D7)

Figure 9.15 a

Excel Output

	A	B	C	D
1	# on the die	Observed Frequencies	Expected Frequencies	(O – E)^2/E
2	1	17	20	0.45
3	2	25	20	1.25
4	3	17	20	0.45
5	4	23	20	0.45
6	5	15	20	1.25
7	6	23	20	0.45
8		120	120	4.3

Figure 9.15 b

Test Statistics $\chi^2 = \sum \frac{(O-E)^2}{E} = 4.3$

(b)

Original Command	Commands	Output (Critical Value)
=CHISQ.INV.RT(CHISQ.INV.RT(**probability**, deg_freedom)	CHISQ.INV.RT(0.05,5)	11.07049769

Table 9.29

Illustration 2: It is believed that the proportion of people with A, B, O, and AB blood types in the population are respectively 0.3, 0.2, 0.4, and 0.1. When 400 randomly picked people were examined, the observed numbers of each type were 118, 86, 136, and 60.

(a) Calculate the test statistics.

(b) Find the critical value (right-tail Chi-Square value) at a 0.01 significance level.

Solution: **(a)**

Excel Commands

	A	B	C	D
1	Blood Type	O	E	(O – E)^2/E
2	A	118	=0.3*400	=(B2 – C2)^2/C2
3	B	86	=0.2*400	=(B3 – C3)^2/C3
4	O	136	=0.4*400	=(B4 – C4)^2/C4
5	AB	60	=0.1*400	=(B5 – C5)^2/C5
6		=SUM(B2:B5)	=SUM(C2:C5)	=SUM(D2:D5)

Figure 9.16 a

Excel Output

	A	B	C	D
1	Blood Type	O	E	(O – E)^2/E
2	A	118	120	0.033333333
3	B	86	80	0.45
4	O	136	160	3.6
5	AB	60	40	10
6		400	400	14.08333333

Figure 9.16 b

Test Statistics $\chi^2 = \sum \frac{(O-E)^2}{E} = 14.08$

(b)

Original Command	Commands	Output (Critical Value)
=CHISQ.INV.RT(CHISQ.INV.RT(**probability**, deg_freedom)	CHISQ.INV.RT(0.01,3)	11.34486673

Table 9.30

Illustration 3: 344 subjects were classified according to their blood type and eye color.

	A	B	O	AB
Blue	40	32	38	15
Brown	85	42	70	22

(a) Calculate the test statistics for a Test of Independence.

(b) Use a 0.05 significance level, to determine the critical value for a Test of Independence.

Solution: **(a)**

Excel Commands

	A	B	C	D	E	F	G	H	I	J
1	O	E	(O – E)^2/E			A	B	O	AB	Total
2	40	=J2*F4/J4	=(A2 – B2)^2/B2		Blue	40	32	38	15	=SUM(F2:I2)
3	32	=J2*G4/J4	=(A3 – B3)^2/B3		Brown	85	42	70	22	=SUM(F3:I3)
4	38	=J2*H4/J4	=(A4 – B4)^2/B4		Total	=SUM(F2:F3)	=SUM(G2:G3)	=SUM(H2:H3)	=SUM(I2:I3)	=SUM(J2:J3)
5	15	=J2*I4/J4	=(A5 – B5)^2/B5							
6	85	=J3*F4/J4	=(A6 – B6)^2/B6							
7	42	=J3*G4/J4	=(A7 – B7)^2/B7							
8	70	=J3*H4/J4	=(A8 – B8)^2/B8							
9	22	=J3*I4/J4	=(A9 – B9)^2/B9							
10	=SUM(A2:A9)	=SUM(B2:B9)	=SUM(C2:C9)							

Figure 9.17 a

Excel Output

	A	B	C	D	E	F	G	H	I	J
1	O	E	(O – E)^2/E			A	B	O	AB	Total
2	40	45.42151163	0.647111628		Blue	40	32	38	15	125
3	32	26.88953488	0.971264613		Brown	85	42	70	22	219
4	38	39.24418605	0.039445306		Total	125	74	108	37	344
5	15	13.44476744	0.179902577							
6	85	79.57848837	0.369355952							
7	42	47.11046512	0.554374779							
8	70	68.75581395	0.022514444							
9	22	23.55523256	0.102684119							
10	**344**	**344**	**2.886653419**							

Figure 9.17 b

$$\text{Test Statistics } \chi^2 = \sum \frac{(O-E)^2}{E} = 2.89$$

(b)

Original Command	Commands	Output (Critical Value)
=CHISQ.INV.RT(CHISQ.INV.RT(**probability**, deg_freedom)	CHISQ.INV.RT(0.05,3)	7.814727903

Table 9.31

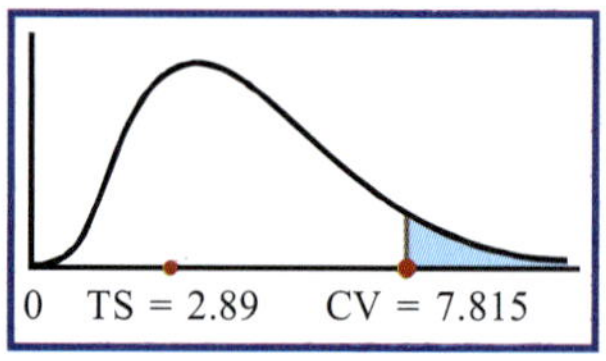

Figure 9.18

We can also use the **CHITEST** built in function in Excel® to obtain the ***p***-value directly from the list of observed and expected frequencies.

Illustration 4: Given:

O	35	78	89	32	20
E	38	62	74	49	12

Use the CHITEST command to calculate the p-value.

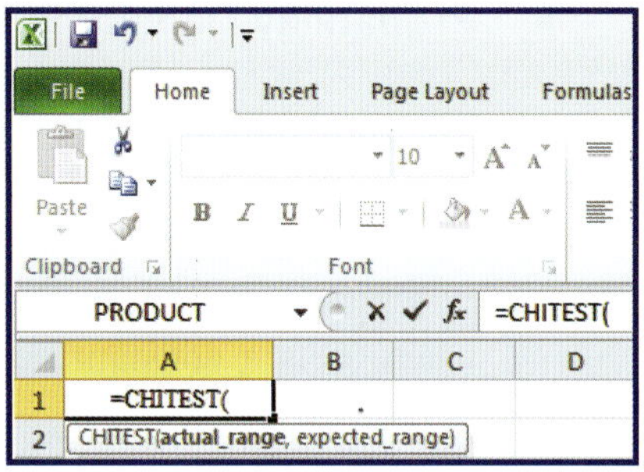

Solution:

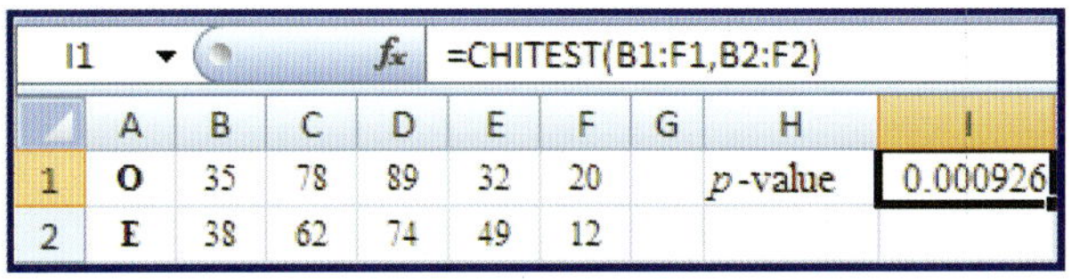

Figure 9.19

Commands	Excel Output (p-value)
=CHITEST(B1:F1,B2:F2)	0.00093

Table 9.32

9.6: Extended Topics and Proofs for Multinomial Experiments: Chi-Square Tests for Categorical Data (Optional)

A. Chi-Square Distribution

Objectives

After completing this section you will be able to understand:

A. Chi-Square Distribution.

B. Chi-Square Test for Independence of Variables of Classification.

C. 2×2 Contingency Table.

D. Yates Correction for Continuity for 2 × 2 Table.

E. Brandt and Snedecor Formula for 2×c Contingency Table.

If $X \sim N(\mu, \sigma)$, then $Z = \dfrac{X-\mu}{\sigma} \sim N(0, 1)$ has the standard normal distribution. If instead we consider the probability distribution of Z^2, the new probability distribution is called the **Chi-square Distribution** with one degree of freedom.

The probability distribution of the sum of squares of n independent standard normal variables is a Chi-square Distribution with n degrees of freedom (df).

If $X_i \sim N(\mu_i, \sigma_i)$ and $i = 1, 2, \ldots, n$ independent normal variables, then $\chi^2 = \sum_{i=1}^{n}\left(\dfrac{X_i-\mu_i}{\sigma_i}\right)^2$ has a Chi-square distribution with n df and we write $\chi^2 \sim \chi^2_n$.

Suppose there are two independent standard normal variables Z_1 and Z_2. The probability distribution of $Z_1^2 + Z_2^2$ follows a Chi-square distribution with two degrees of freedom. Similarly, the probability distribution of the sum of squares of n independent standard normal variables is a Chi-square distribution with n degrees of freedom (df).

• Probability Density Function (pdf) of Chi-Square Distribution

The probability density function (pdf) of χ^2-distribution with n degrees of freedom is given by:

$$p\left(\chi^2\right) = Ce^{\frac{-\chi^2}{2}}\left(\chi^2\right)^{\frac{n}{2}-1} \quad ; \quad 0 \le \chi^2 < \infty \tag{9.1}$$

where C is a constant depending on n.

The Chi-square distribution is a continuous distribution. But unlike the normal and the t-distributions, the Chi-square probability curve is not symmetrical rather skewed to the right. The shape of the Chi-square distribution depends on n, the df.

***Remark* 1:**

i) The observed and expected frequencies have a linear constraint:

$$\sum_{i=1}^{k} O_i = \sum_{i=1}^{k} E_i = n \tag{9.2}$$

where n is the total frequency.

The constraint in (9.2) is termed as linear because it does not involve squares and higher powers of the frequencies. Accordingly, $\sum(O-E) = \sum O - \sum E = n - n = 0$, which provides an important check on our calculations in the computation of χ^2.

ii) Sometimes the simplified form of the formula $\chi^2 = \sum\frac{(O-E)^2}{E}$ is useful for numerical computation of the test statistics χ^2.

$$\chi^2 = \sum_{i=1}^{k}\left(\frac{O_i^2}{E_i}\right) - n \tag{9.3}$$

iii) If the expected frequencies are close to the observed frequencies, we shall get small values of χ^2 and consequently large ***p***-values and hence fail to reject H_0 [Figure 9.20 (a)]. On the other hand if the expected frequencies deviate much from the observed frequencies, then we have large χ^2-values and consequently small ***p***-values leading to rejection of H_0 [Figure 9.20 (b)].

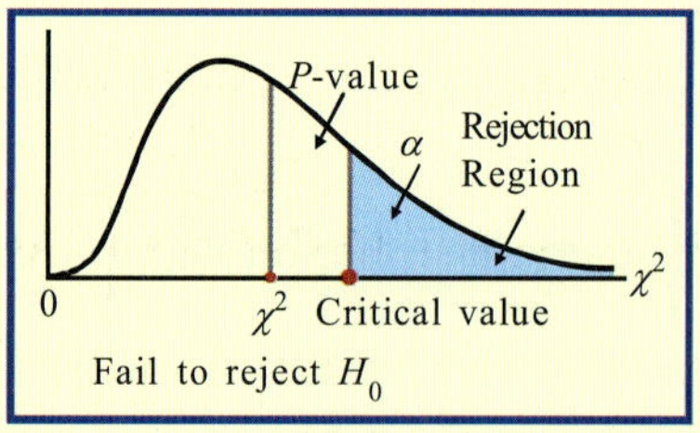

Figure 9.20 a

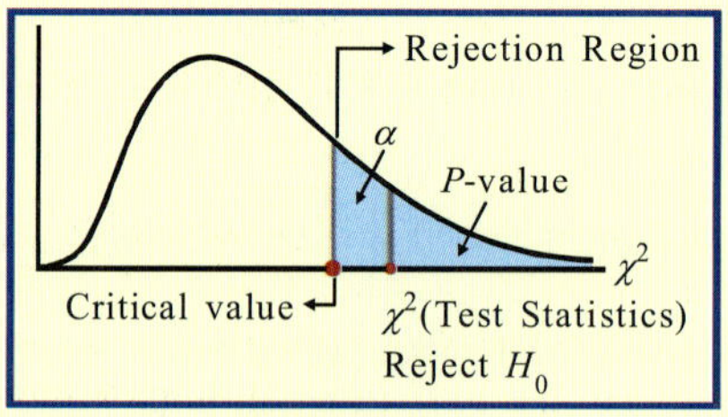

Figure 9.20 b

Thus, in testing goodness of fit, large values of χ^2 indicate the disagreement between observed frequencies and the expected frequencies obtained under the null hypothesis. This means that the critical values and the rejection regions will lie towards the extreme right-tail of the χ^2-distribution. Hence, we shall always be using the right-tailed test for goodness of fit hypothesis.

iv) It should be noted that the χ^2-test depends only on the set of observed and expected frequencies and on degrees of freedom (df) v. It does not make any assumptions regarding the parent population from which the observations are taken. Since χ^2 defined by $\chi^2 = \sum_{i=1}^{k} \frac{(O_i - E_i)^2}{E_i}$ does not involve any population parameters, it is termed statistic and the test is known as ***Non-Parametric test***.

v) Although χ^2-distribution is essentially a continuous distribution, χ^2-test can be applied to discrete random variables whose frequencies can be counted and tabulated with or without grouping.

Remark 2:

In the χ^2-Test for Goodness of Fit, the **Classical method** is much more convenient to use than the ***P*-value method**. Moreover in the Classical method, we can arrive at the conclusions about rejecting (or failing to reject) the null hypothesis without plotting the critical values, rejection region and the test statistic.

- **Goodness of Fit Test, given a Grouped Frequency Table**

Example 9: The data in the given grouped frequency table gives the distribution of the HDL cholesterol level in a sample of 120 adults in a city. Test whether or not the distribution of HDL cholesterol levels in the adult population of the city fits the normal distribution with the mean 39.63 and standard deviation 5.28. Use a 0.05 significance level. Use the Classical method.

HDL Cholesterol Level	Observed Frequency
20 - 25	3
25 - 30	3
30 - 35	15
35 - 40	39
40 - 45	44
45 - 50	13
50 - 55	1
55 - 60	2

Solution: ***Step* 1** H_0 : The distribution of HDL cholesterol levels fits the normal distribution with mean $\mu = 39.63$ and standard deviation $\sigma = 5.28$.

H_1 : The distribution of HDL cholesterol levels does not fit in the normal distribution

***Step* 2** **Test Statistics:**

Computation of Expected Frequencies

For normal distribution with mean $\mu = 39.63$ and standard deviation $\sigma = 5.28$:

$$Z = \frac{X - \mu}{\sigma} = \frac{X - 39.63}{5.28} \sim N(0, 1)$$

The corresponding z-values in various classes and the expected frequencies are obtained in the following table.

Class (X-values)	Class (Z-values)	Class Probability (p)	Expected frequency $np = 120p$
20 - 25	$-3.27 - -2.77$	$0.4999 - 0.4972 = 0.0027$	0.324 0.3
25 - 30	$-2.77 - -1.83$	$0.4972 - 0.4664 = 0.0308$	3.696 3.7
30 - 35	$-1.83 - -0.88$	$0.4664 - 0.3106 = 0.1558$	18.696 18.7
35 - 40	$-0.88 - 0.07$	$0.3106 + 0.0279 = 0.3385$	40.620 40.5
40 - 45	$0.07 - 1.02$	$0.3461 - 0.0279 = 0.3182$	38.184 38.1
45 - 50	$1.02 - 0.01$	$0.4750 - 0.3461 = 0.1289$	15.468 15.6
50 - 55	$1.96 - 2.91$	$0.4982 - 0.4750 = 0.0232$	2.784 2.8
55 - 60	$2.91 - 3.86$	$0.4999 - 0.4982 = 0.0017$	0.204 0.2

Table 9.33

For $z_1 > 0$, $z_2 > 0$, we have:

$$\left.\begin{aligned} P(-z_2 < Z < -z_1) &= P(z_1 < Z < z_2) = P(0 < Z < z_2) - P(0 < Z < z_1) \\ P(-z_2 < Z < z_1) &= P(-z_2 < Z < 0) + P(0 < Z < z_1) \\ &= P(0 < Z < z_2) + P(0 < Z < z_1) \\ P(z_1 < Z < z_2) &= P(0 < Z < z_2) - P(0 < Z < z_1) \end{aligned}\right\} \quad \ldots(4)$$

Note: Using the z table and equation (4), we get the probabilities (areas) as shown in Table 9.33

The sum of the expected frequencies in the first two classes is less than 5; therefore they are to be pooled together with the expected frequency of the third class so that the total is more than 5. Similarly the expected frequencies of the last three classes are to be pooled together so that their sum is more than 5. By pooling the first three and the last three classes, we obtain the table of observed and expected frequencies shown below.

HDL cholesterol Level	Observed Frequency (O)	Expected Frequency (E)
20 - 35	21	22.7
35 - 40	39	40.6
40 - 45	44	38.2
45 - 60	16	18.5

Table 9.34

Therefore, the test statistics is:

$$\chi^2 = \sum\left[\frac{(O-E)^2}{E}\right]$$

$$= \frac{(21-22.7)^2}{22.7} + \frac{(39-40.6)^2}{40.6} + \frac{(44-38.2)^2}{38.2} + \frac{(16-18.5)^2}{18.5}$$

$$= 0.1273 + 0.0631 + 0.8806 + 0.3378 = 1.4088 \quad 1.409$$

Degrees of freedom (v) is given by: $v = k - 1 - 2 - 2 = 8 - 5 = 3$.

1 df is lost because of the linear constraint $\sum O = \sum E$. Two df are lost in pooling the expected frequencies in the first three classes, and Two df are lost in pooling the expected frequencies in the last three classes.

The critical value of Chi-square for 3 *df* and a 0.05 significance level from Table 7 is 7.815.

That is, $\chi^2_v(\alpha) = \chi^2_3(0.05) = 7.815$.

The computed value of test statistics and the critical value of χ^2 along with the rejection region are shown in Figure 9.21.

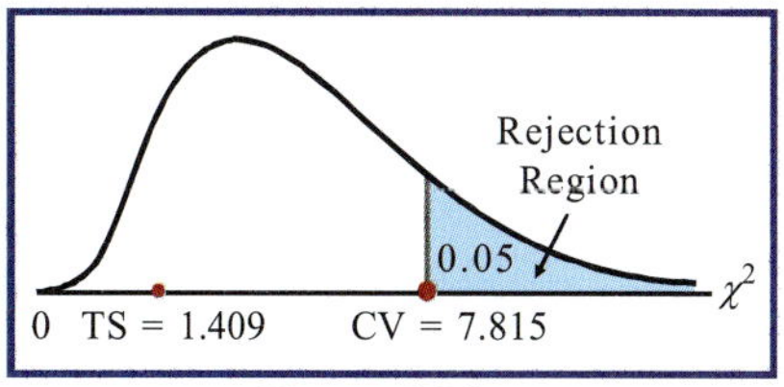

Figure 9.21

Since the computed value of χ^2 falls in the acceptance region, we fail to reject the null hypothesis (H_0) at $\alpha = 0.05$. Hence, the normal distribution provides a good fit to the distribution of HDL cholesterol levels in the adult population of the city.

In short:

Hypotheses	**Test Statistics**
H_0: The distribution of HDL cholesterol levels fits the normal distribution with mean $\mu = 39.63$ and standard deviation $\sigma = 5.28$. H_1: The distribution of HDL cholesterol levels does not fit in the normal distribution.	$\chi^2 = \sum \frac{(O-E)^2}{E} = 1.409$
Rejection Region and Critical Value	**Conclusion**
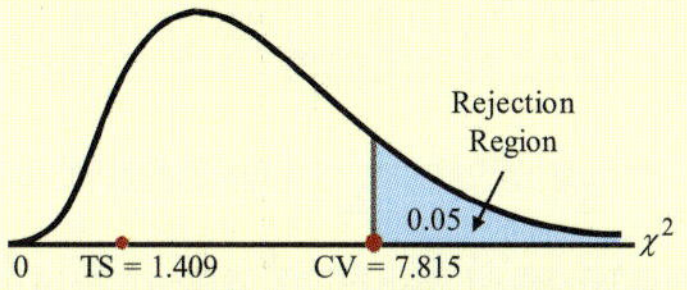	Fail to reject H_0. The normal distribution provides a good fit to the distribution of HDL cholesterol levels in the adult population of the city.

B. Chi-Square Test for Independence of Variables of Classification

Suppose the responses from a random sample of size n are divided into r mutually disjoint (exclusive) and exhaustive classes $A_1, A_2, \ldots, A_r$ with respect to the variable (characteristic) A so that a randomly selected item belongs to one and only one of the categories $A_1, A_2, \ldots, A_r$. Similarly let us suppose that the same sample is divided into "c" mutually disjoint and exhaustive classes or categories, say, $B_1, B_2, \ldots, B_c$ with respect to another variable or characteristic B so that an item selected at random belongs to one and only one of the categories $B_1, B_2, \ldots, B_c$. The frequency distribution of the items belonging to the classes $A_1, A_2, \ldots, A_r$ and $B_1, B_2, \ldots, B_c$ can be represented in the following $r \times c$ **manifold contingency table:**

$r \times c$ manifold contingency table							
$A \backslash B$	B_1	B_2	…	B_j	…	B_c	**Total**
A_1							
A_2							
…							
A_i							
…							
A_r							
Total							

Table 9.35 a

$r \times c$ manifold contingency table							
$A \backslash B$	B_1	B_2	...	B_j	...	B_c	Total
A_1	$(A_1 B_1)$	$(A_1 B_2)$	...	$(A_1 B_j)$	...	$(A_1 B_c)$	(A_1)
A_2	$(A_2 B_1)$	$(A_2 B_2)$	...	$(A_2 B_j)$	...	$(A_2 B_c)$	(A_2)
A_i	$(A_i B_1)$	$(A_i B_2)$	...	$(A_i B_j)$	...	$(A_i B_c)$	(A_i)
A_r	$(A_r B_1)$	$(A_r B_2)$	...	$(A_r B_j)$	...	$(A_r B_c)$	(A_r)
Total	(B_1)	(B_2)	...	(B_j)	...	(B_c)	$\sum_{i=1}^{r}(A_i) = \sum_{j=1}^{c}(B_j) = n$

Table 9.35 b

where: (A_i) is the frequency of the ith category A_i; that is, it is the number of items belonging to the category A_i, $i = 1, 2, \ldots, r$. (B_j) is the number of items belonging to the category B_j, $j = 1, 2, \ldots, c$. And $\left(A_i B_j\right)$ is the number of items belonging to both the categories A_i and B_j.

- **Expected Frequencies**

Under the null hypothesis that the two variables of classification A and B are independent, the expected frequency for the $(i, j)^{\text{th}}$ cell is given by:

$$E\left(A_i B_j\right) = \left(A_i B_j\right)_0 = \frac{(A_i)(B_j)}{n} \quad ; \quad \begin{pmatrix} i = 1, 2, \ldots, r \\ j = 1, 2, \ldots, c \end{pmatrix} \tag{9.5}$$

The rule in (9.5) can be stated in words as follows:

> *Under the hypothesis of independence of variables of classification, the expected frequency for any of the cell frequencies can be obtained by multiplying the row totals and column totals in which the frequency occurs and dividing the product by the total frequency n.*

- **Rationale for Formula (9.5)**

Under the null hypothesis that the two variables (characteristics) A and B are independent, the expected frequency for $\left(A_i B_j\right)$ is given by:

$$\begin{aligned} E\left[\left(A_i B_j\right)\right] &= n \cdot P\left[\left(A_i B_j\right)\right] \\ &= n \cdot P \cdot \left[\left(A_i \text{ and } B_j\right)\right] \\ &= n \cdot P \cdot [A_i] \cdot P\left[B_j\right] \end{aligned}$$

[By multiplication rule of probability, since A and B are independent]

$$= n \times \frac{(A_i)}{n} \times \frac{(B_j)}{n} = \frac{(A_i)(B_j)}{n}$$

If $\left(A_iB_j\right)_0$ denotes the expected frequency of $\left(A_iB_j\right)$ then

$$E\left(A_iB_j\right) = \left(A_iB_j\right)_0 = \frac{(A_i)(B_j)}{n} \;; \qquad \begin{pmatrix} i = 1, 2, \ldots, r \\ j = 1, 2, \ldots, c \end{pmatrix}$$

$$\text{Expected frequency for any cell} = \frac{\text{(Row Total)} \times \text{(Column Total)}}{\text{Grand Total}} \qquad (9.6)$$

- **Application of Chi-Square Test of Goodness of Fit**

$$\text{The statistic is: } \chi^2 = \sum\left[\frac{(O-E)^2}{E}\right] = \sum_i\sum_j\left[\frac{\left\{\left(A_iB_j\right)-\left(A_iB_j\right)_0\right\}^2}{\left(A_iB_j\right)_0}\right] \qquad (9.7)$$

follows χ^2-distribution with $(r-1)\times(c-1)$ degrees of freedom. (See (9.8) below.)

By comparing this calculated value with the tabulated (critical) value of χ^2 for $(r-1)\times(c-1)$ df at a certain level of significance, we reject or fail to reject the null hypothesis of independence of variables at that level of significance.

- **Degrees of Freedom for $r \times c$ Contingency Table**

Since n the total frequency in the contingency table is fixed (known), the sum of r row totals and the sum of c column totals are also fixed, each being equal to n. This means that we can assign the frequencies arbitrarily to any of the $(r-1)$ rows, and the frequency of the remaining r^{th} row can be obtained since the total of all the row frequencies is fixed (n). Similarly, we can assign the frequencies arbitrarily to any of the $(c-1)$ columns.

Hence, by the principle of counting, in $r \times c$ contingency table we can assign $(r-1)(c-1)$ frequencies arbitrarily, which determines the total number of degrees of freedom.

In assigning the frequencies arbitrarily to $(r-1)$ rows and $(c-1)$ columns, we have to keep in mind that

i) No row or column frequency is negative and

ii) No row or column frequency is greater than n, the total frequency.

For r × c contingency table, the degrees of freedom (v) are given by:

$$v = (r-1)(c-1) \qquad (9.8)$$

In particular, for 2×2 contingency table, $df(v) = (2-1)\times(2-1) = 1$

for 3×4 contingency table, $df(v) = (3-1)\times(4-1) = 2\times 3 = 6$

and so on.

C. 2 × 2 Contingency Table

Under the null hypothesis of independence of variables of classification, the value of χ^2 for the 2×2 contingency table

			Total
	a	b	$a+b$
	c	d	$c+d$
Total	$a+c$	$b+d$	$n=a+b+c+d$

Table 9.36

is given by:

$$\chi^2 = \frac{n(ad-bc)^2}{(a+c)(b+d)(a+b)(c+d)} \qquad (9.9)$$

where $n = a + b + c + d$ is the total frequency, and figures in brackets in the denominator are row and column totals.

- **Rationale for (9.9)**

Under the hypothesis of independence of the variables of classification, on using (9.5), we get:

$$E\,(a) = \frac{(a+c)(a+b)}{n};\quad E\,(b) = \frac{(b+d)(a+b)}{n};\quad E\,(c) = \frac{(a+c)(c+d)}{n};\quad E\,(d) = \frac{(b+d)(c+d)}{n}$$

where $n = a + b + c + d$

$$\therefore\ a - E\,(a) = a - \frac{(a+c)(a+b)}{a+b+c+d} = \frac{a(a+b+c+d)-(a+c)(a+b)}{a+b+c+d}$$

$$= \frac{a^2+ab+ac+ad-a^2-ab-ac-bc}{(a+b+c+d)} = \frac{ad-bc}{n}$$

Similarly, we will get:

$$b - E\,(b) = -\frac{(ad-bc)}{n}\quad;\quad c - E\,(c) = -\frac{ad-bc}{n}\quad;\quad d - E\,(d) = \frac{(ad-bc)}{n}$$

$$\therefore \chi^2 = \frac{[a-E(a)]^2}{E(a)} + \frac{[b-E(b)]^2}{E(b)} + \frac{[c-E(c)]^2}{E(c)} + \frac{[d-E(d)]^2}{E(d)}$$

$$= \frac{(ad-bc)^2}{n^2}\left[\frac{1}{E(a)} + \frac{1}{E(b)} + \frac{1}{E(c)} + \frac{1}{E(d)}\right]$$

$$= \frac{(ad-bc)^2}{n^2}\left[\frac{n}{(a+c)(a+b)} + \frac{n}{(b+d)(a+b)} + \frac{n}{(a+c)(c+d)} + \frac{n}{(b+d)(c+d)}\right]$$

$$= \frac{(ad-bc)^2}{n}\left[\left\{\frac{1}{(a+c)(a+b)} + \frac{1}{(b+d)(a+b)}\right\} + \left\{\frac{1}{(a+c)(c+d)} + \frac{1}{(b+d)(c+d)}\right\}\right]$$

$$= \frac{(ad-bc)^2}{n}\left[\frac{b+d+a+c}{(a+c)(a+b)(b+d)} + \frac{b+d+a+c}{(a+c)(c+d)(b+d)}\right]$$

$$= (ad-bc)^2\left[\frac{1}{(a+c)(a+b)(b+d)} + \frac{1}{(a+c)(c+d)(b+d)}\right] \qquad \text{[Because } n = a+b+c+d\text{]}$$

$$= (ad-bc)^2\left[\frac{c+d+a+b}{(a+c)(a+b)(b+d)}\right] = \frac{n(ad-bc)^2}{(a+c)(a+b)(b+d)(c+d)}$$

Formula (9.9) enables us to compute the value of χ^2 directly from the given frequencies without computing the expected frequencies under the hypothesis of independence.

D. Yates Correction for Continuity for 2 × 2 Table

If any expected cell frequency in 2 × 2 table is less than 5, then for the application of χ^2-test it has to be pooled with the preceding or succeeding frequency so that the total is greater than 5. This results in the loss of 1 *df*. Since for 2 × 2 table $df = (2-1) \times (2-1) = 1$, the degrees of freedom left after adjusting for pooling are $v = 1 - 1 = 0$, which is absurd [χ^2 must have at least 1 *df*.]. In such a situation, when any expected cell frequency in a 2 × 2 table is less than 5, we apply the **Yates Correction for 'Continuity'**, creation of Yates (1934). This consists of adding 0.5 to the cell frequency which is less than 5 and adjusting the remaining frequencies accordingly, since row and column totals are fixed and then applying χ^2-test without pooling.

The value of the χ^2 for the 2 × 2 table

a	b
c	d

after applying Yates correction is:

$$\chi^2 = \frac{n\left[\,|ad - bc| - n/2\right]^2}{(a+c)(b+d)(a+b)(c+d)} \tag{9.10}$$

where $|ad - bc|$ is the absolute or modulus value of the deviation $ad - bc$ after ignoring the negative sign, if any, and assuming $n = a + b + c + d$.

> ***Remark:*** In general, Yates correction tends to decrease the value of test statistic. [This is obvious from (9.10)]

E. Brandt and Snedecor Formula for 2 × *c* Contingency Table

Under the hypothesis of independence of variables of classification, the value of χ^2 for $2 \times c$ contingency table

2 × *c* Contingency Table

							Total
	a_1	a_2	a_3	...	...	a_c	$n_1 = \sum a_1$
	b_1	b_2	b_3	...	...	b_c	$n_2 = \sum b_1$
Total	m_1	m_2	m_3	...	...	m_c	$n = n_1 + n_2$

Table 9.37

can be computed from the Brandt and Snedecor formula

$$\chi^2 = \frac{n^2}{n_1 \times n_2}\left[\frac{a_1^2}{m_1} + \frac{a_2^2}{m_2} + \ldots + \frac{a_c^2}{m_c} - \frac{n_1^2}{n}\right] \tag{9.11}$$

This has $v = (2-1) \times (c-1) = (c-1)$ *df*

The proof of (9.11) is beyond the scope of this book.

The formula (9.11) enables us to compute the value of Chi-square directly from the given data (without computing the expected frequencies) under the hypothesis of independence.

Ch 9: General Exercises

Use both the Classical and the *P*-value methods to solve the following exercises.

χ^2-Test of Goodness of Fit

1. In a study of fatal car accidents, 245 cases were randomly selected from the population in which the driver had a blood alcohol level above the legal limit. The results are listed in the accompanying table according to the day of the week.

Day	Mon	Tues	Wed	Thurs	Fri	Sat	Sun
Number	29	30	33	34	38	43	38

At the 0.01 significance level, test the claim that the frequency of such fatal accidents is uniformly distributed throughout the week.

2. The number of overall absences in all the classes Professor Badalian teaches has traditionally had the following distribution.

Days	Mon	Tues	Wed	Thurs	Fri
% of absences	30	10	9	28	23

During the current semester, the number of observed absences showed the following distribution.

Days	Mon	Tues	Wed	Thurs	Fri
% of absences	19	8	6	11	20

Using $\alpha = 0.05$, test the claim that the absences in Professor Badalian's classes are distributed as traditionally expected.

3. The following table lists the age distribution and the number of tickets for speeding for a random sample of 100 drivers.

Age	16-25	26-35	36-45	46-55	56 and older
Arrests	34	22	16	19	9

A previous study discovered that the distribution of drivers ticketed for speeding was as follows:

$P_{16-25} = 0.20$, $P_{26-35} = 0.30$, $P_{36-45} = 0.20$, $P_{46-55} = 0.15$, and $P_{\text{56 and older}} = 0.15$

Using a 5% significance level, test the claim that the distribution of the number of tickets in the given age groups is as expected.

4. The following table gives the percentage of students registered in a specific math class for the Fall semester 2002 in Sun Valley College.

Math Classes	Arithmetic	Algebra	Statistics	Calculus
Percentages	42.3	28.4	21.8	7.5

A recent random sample of 1000 students in Sun Valley College showed that 400 of them were registered in Arithmetic, 290 were registered in Algebra, 220 were registered in Statistics, and 90 were registered in Calculus. Use a 0.05 significance level to test the claim that the current registration percentages are the same as the ones in 2002.

5. Of all students enrolled at W.H. University, 10% are seniors, 25% are juniors, 30% are sophomores, and 35% are freshmen. A random sample of 405 students includes 98 seniors, 87 juniors, 100 sophomores, and 120 freshmen. Use $\alpha = 0.01$ to test the claim that the percent distribution of students is as indicated.

6. A donut shop owner suspects that not all kinds of donuts are equally popular. He has seven kinds of donuts and the number of each kind sold yesterday is given below.

Kind of donut	Plain	Sugar	Chocolate	Cinnamon	Jelly	Chocolate with sprinkles	Custard Filled
Number sold	22	19	30	16	20	12	18

At a 5% significance level, does the owner have enough evidence to conclude that not all kinds of donuts are equally popular?

7. A car dealership wants to determine if the color distribution of cars sold is uniform. During a one week period, these sales were observed.

Color	Silver	Black	White	Red	Blue	Beige
Number sold	15	19	23	15	9	17

At a 0.01 significance level, determine if the color distribution of cars sold is uniform.

8. The U. S. Census Bureau provided data for the number of people in different age groups in Broward County. The following table includes the distribution of the number of people in different age groups for the county.

Age Group	0 - 14	15 - 34	35 - 54	55 - 74	75 and over
Number of people	321,807	404,080	499,559	252,066	136,582

Using a 0.05 significance level, is there enough evidence to suggest a uniform distribution of the number of people in different age groups?

9. An airline suspects that people do not travel uniformly on different days of the week. 3852 randomly selected travelers were surveyed, and the results are summarized in the accompanying table.

Day of the week	Sun	Mon	Tues	Wed	Thurs	Fri	Sat
Number of travelers	490	388	500	473	648	654	699

Using a 0.05 significance level, test the airline's suspicion.

10. According to a scientist's claim, 30% of self-pollinations produce red, 40% pink, and 30% white flowers in a particular plant. A random sample of 230 observations yields 55 red, 110 pink and 65 white flowers. Use a 0.05 significance level to test the scientist's claim.

11. In 2010, the National Center for Education released the following data for the total number, in thousands, of full-time students enrolled in 4-year degree-granting public institutions.

Year	1994	1995	1996	1997	1998	1999	2000	2001	2002	2003	2004	2005	2006	2007
Number of FT students	5432	5546	5694	5848	5905	5900	5852	5825	5815	5806	5835	5893	5970	6055

Using $\alpha = 0.01$, test the claim that distribution of full-time students through the years has remained the same.

12. The U. S. Department of Commerce's Bureau of Economic Analysis provides the current (U. S.) expenditures, in billions of dollars, for 4 quarters in 2001 and three quarters in 2002.

Quarters	Q I in 2001	Q II in 2001	Q III in 2001	Q IV in 2001	Q I in 2002	Q II in 2002	Q III in 2002
Amount	2896.1	2939.3	2976.1	2994.9	3067.3	3117.4	3135.0

At the 0.05 significance level, do you have enough evidence to conclude that the amounts of money spent in the given quarters are uniformly distributed?

13. In an elementary school, it is known that 60% of third-graders live with both parents, 27% live with their mothers, 10% live with their fathers, and the rest live with other guardians. Seventy randomly selected third-graders were surveyed, and the results were: 35 lived with both parents, 22 lived with their mothers, 10 lived with their fathers, and the rest lived with other guardians. Using a 0.01 significance level, test the claim that the observed frequencies fit the expected probabilities.

14. In a class of 200 students, Prof. Mardirosian gave 22 As, 33 Bs, 95 Cs, 35 Ds, and 15 Fs. His records show that last semester 10%, 15%, 60%, 10%, and 5% the students received the letter grades of A, B, C, D, and F, respectively. Use the 0.05 level of significance to check the fit.

15. A sequence of 200 pseudo-random digits is generated on a graphing calculator. The following table gives the frequency of each digit.

Digit	**9**	**8**	**7**	**6**	**5**	**4**	**3**	**2**	**1**	**0**
Frequency	17	18	23	19	21	23	18	22	24	15

Does the data fit a uniform distribution where each digit is equally likely? Use $\alpha = 0.05$.

16. The accompanying table lists the numbers of wining horses starting in different positions in a race.

Starting Position	**1**	**2**	**3**	**4**	**5**	**6**	**7**	**8**
Number of wins	29	19	18	25	17	10	15	11

Test the claim that the probabilities of winning in different starting positions are uniformly distributed.

17. In studying the occurrence of psychological characteristics, the following sample data were obtained.

Characteristics	**A**	**B**	**C**	**D**	**E**	**F**
Frequency	35	30	27	36	42	38

Test the claim that psychological characteristics have a uniform distribution.

18. In a Census, it was found that the total population percentages broke down according to the following geographic regions: Northeast 33%, Midwest 23%, South 22%, and in the West 22%. A statistics student selects a random sample of 2000 subjects and finds that 740 reside in Northeast, 410 in Midwest, 460 in the South, and 390 in the West. At the 0.05 significance level, test the claim that the population distribution is as reported by the Census.

19. A curious Professor asked 70 randomly selected students to identify the day of the week they would prefer to take their exams. The results are listed in the following table. Use a 0.05 significance level to test the claim that the results fit a uniform distribution.

Day	**Mon**	**Tues**	**Wed**	**Thurs**	**Fri**
Number of students	25	18	10	8	9

20. Use a 0.05 significance level and the industrial accident data given here

Days	**Mon**	**Tues**	**Wed**	**Thurs**	**Fri**
% of accidents	24	16	20	14	26

to test the claim that the percent distribution of industrial accidents is as follows.

Days	**Mon**	**Tues**	**Wed**	**Thurs**	**Fri**
% of accidents	18	20	18	20	24

χ^2-Test of Independence

21. The following table lists the number of male and female students registered in 5 randomly selected math classes at a community college.

	Arithmetic	**Algebra**	**Geometry**	**Calculus**	**Statistics**	**Total**
Males	505	690	418	267	355	**2235**
Females	667	654	559	124	465	**2469**
Total	**1172**	**1344**	**977**	**391**	**820**	**4704**

Using a 0.05 significance level, test the claim that the level of mathematics courses and gender are independent from each other.

22. In a community, 389 randomly selected people were classified according to their age group and political beliefs.

Political Belief	20 - 30	30 - 40	40 and above
Republican	11	43	12
Democrat	82	87	56
Independent	38	25	35

Using a 0.01 significance level, test the claim that a person's age group and political belief are dependent.

23. A survey was conducted to investigate the relationship between the use of technology in the classroom and student pass rates.

	Passing	Failing
Using technology	356	180
Not using technology	240	90

Use a 0.05 significance level, to test the claim that the use of technology and pass rates are related.

24. The following chart sorts a sample of 600 randomly selected adults into two categories, eye color and blood type.

Eye Color	Blood Type A	Blood Type B	Blood Type AB	Blood Type O	Total
Blue	99	87	44	15	**245**
Brown	85	102	96	72	**355**
Total	**184**	**189**	**140**	**87**	**600**

Using $\alpha = 0.05$, test the claim that eye color and blood type are independent.

25. Bureau of Labor provides the following information about coal (in QBtu) and total fossil fuel productions (in QBtu) for years 1998-2001.

	1998	1999	2000	2001
Coal	23.935	23.186	22.623	23.574
Fossil Fuel	59.204	57.505	57.054	58.262

Using a 0.1 significance level, determine whether the production of fuels is dependent on the year.

26. In the chart below, you find the amount of carbon dioxide, in millions metric tons of carbon equivalent, emitted by sector in 1999 and 2000.

	Residential	Commercial	Industrial	Transportation	Total
1999	298.8	253.1	465.8	499.4	**1,517.1**
2000	313.4	267.8	465.7	514.8	**1,561.7**
Total	**612.2**	**520.9**	**931.5**	**1014.2**	**3078.8**

Test the claim that the sector that emits carbon dioxide is independent of the year in which data was taken. Use a 0.05 significance level.

27. The following table summarizes the results of a survey released by the Gallup Poll. In May of 2001 and 2002, 1015 randomly selected people were asked what they thought of the frequency of imposing the death penalty. There were four groups: imposed too often, imposed the right amount, imposed not often enough, and had no opinion.

	Too Often	About Right	Not Enough	No Opinion	Total
May 2002	223	244	477	71	**1015**
May 2001	213	345	386	71	**1015**
Total	**436**	**589**	**863**	**142**	**2030**

Test if people's responses are independent of the year in which they were surveyed. Use a 0.1 significance level.

28. The following data was collected to test whether a person's financial status is independent of his or her drinking habits.

	Financial Status		
	Rich	Middle	Poor
Heavy drinker	40	50	35
Moderate drinker	60	55	25
Non drinker	20	45	10

Use $\alpha = 0.05$ to test the claim.

29. The Boeing Commercial Airline Group conducted a survey of 133 planes either being flown or on order from three airline companies based in three different countries. The sample data is summarized in the accompanying table.

	United Air	British Air	Singapore Air
Boeing 767s flying	36	22	14
Boeing 767s ordered	16	23	22

Is there sufficient evidence to support the claim that the distribution of planes flying and planes ordered is independent of the airline company? Use a 0.01 significance level.

30. Use the sample data below to test whether car model is independent of the likelihood of being in an accident. Use a 0.01 significance level.

	Toyota	BMW	Porsche
Has been in an accident	56	38	25
Has not been in an accident	96	57	42

31. The table below shows the age and favorite art museums of 657 randomly selected people.

	Getty	LACMA	Autry
15-24	45	58	92
25-34	67	76	69
35-44	89	94	67

Use a 5 percent level of significance to test the null hypothesis that age-group and preferred museum type are independent.

32. A survey was conducted to determine whether there is a gender gap in the confidence people have in politicians. The sample results are listed in the accompanying table. Use a 0.1 significance level to test the claim that there is such a gender gap.

Confidence in Politicians	Great Deal	Some	Very Little or None
Men	38	115	42
Women	53	175	31

χ^2-Test of Homogeneity

33. In a surgical procedure, 243 out of 270 patients at Hospital A survived, 198 out of 225 patients at Hospital B survived and 161 out of 175 patients at Hospital C survived. Use a 0.01 significance level to test the claim that the proportions of survivals in the three Hospitals are uniform (equal).

34. To test whether the proportion of teenage smokers in three school districts are the same, random samples were taken. The following data were obtained:

17 of 85 teenagers are smokers in District 1,
18 of 75 teenagers are smokers in District 2, and
12 of 65 teenagers are smokers in District 3.

Using a 0.05 significance level, test the claim that the proportions of teenage smokers in all the three school districts are the same.

35. To study whether there is a difference between the proportion of parents who favor a twelve-month school year, random samples from 4 different counties were collected. Here are the results.

	County 1	County 2	County 3	County 4
For	45	85	53	47
Against	35	15	17	13

Using a 0.025 significance level, test the claim that the proportions of parents who favor the 12-month school year are the same in all 4 counties.

36. In a random sample of teachers, lawyers, doctors and businessmen, it is found that 340 of 360 teachers, 120 of 145 lawyers, 80 of 185 doctors and 295 out of 310 businessmen use the internet. Use a 0.1 significance level to test the claim that the proportions of internet users among the selected professions are equal.

37. The same item is manufactured at three locations of a manufacturing company. At location A, 67 out of 1405 items were defective; at location B, 55 out of 1225 items were defective; and 45 out of 1140 items at location C were defective. Use $\alpha = 0.01$ to test the claim that the proportions of defective items in the three locations are equal.

38. In a recent poll, 415 of 540 Republican, 612 of 720 Democrat, and 153 of 170 Independent voters favor the finance reform bill. Use a 0.05 significance level to test the claim that proportions of voters who favor the reform bill are the same.

39. A new branch of a large car rental company is to be opened on an island. The management of this company from previous experiences expects 25% of the car rental contracts to be for small cars, 35% for medium size cars, 10% for large cars, 25% for BMW and the remaining 5% for specialty cars such as vans, pick ups etc. The local manager on this island decides to test whether or not this distribution is actually what they see happening in their office. Out of 200 randomly sampled car rental contracts they note that 37 are for small cars, 81 for medium size cars, 14 for large cars, 61 for BMW's and 7 for specialty cars. Can it, at the 0.01 level of significance, be concluded that the management's expectation is correct?

40. Prof. Badalian's multiple choice exam had 80 questions with the distribution of correct answers as given below: A-16, B-16, C-15, D-22, F-11. At the 0.01 significance level, test the claim that the frequency of answers of multiple choice questions are uniformly distributed.

41. In a four digit lottery, each of four digits is supposed to have the same probability of occurrence. The table given below shows the four digit numbers of 25 randomly selected lottery tickets.

1 4 5 2	4 2 7 8	3 4 7 5	1 8 8 8	6 7 2 2
7 2 3 3	8 8 8 0	3 9 9 8	4 7 0 1	6 9 3 4
6 3 9 4	1 4 0 9	7 7 5 3	3 4 5 4	3 9 8 4
7 7 9 8	9 4 3 2	4 3 0 7	4 6 5 3	1 1 7 2
5 7 5 1	7 5 0 6	6 0 5 1	0 3 8 6	1 2 5 6

Use $\alpha = 0.01$ to test the claim that the digits are from a uniform population.

42. BTW, a chain of restaurants, provide services in food industry, in different parts of the world. In one particular restaurant, the management wants to test the efficiency of public dealing employees in that restaurant, during festival season. A survey was carried out and it was found that the number of customers served at five different counters during Christmas season were 67, 45,35, 30 and 42 the corresponding expected number claimed by the management were 60, 49, 32 ,35 and 43. Test the claim at 5% level of significance.

43. The employees of ABC Corporation were sampled at random from pay records and asked to complete an anonymous job satisfaction survey, yielding the tabulation shown.

Pay Type	Satisfied	Neutral	Dissatisfied	Total
Salaried	60	52	8	**120**
Hourly	203	122	20	**345**
Total	**263**	**174**	**28**	**465**

Using $\alpha = 0.1$, test the claim that job satisfaction is independent of pay categories.

44. A research team in Statistics Department examined parked vehicle in four different suburban shopping malls. Two hundred vehicles were examined in each location.

Vehicle Type	Mall_A	Mall_B	Mall_C	Mall_D	Mall_E	Total
Car	28	52	34	43	47	204
Minivan	47	48	46	49	32	222
Mini truck	33	46	39	25	42	185
Full sized van	37	28	36	46	41	188
Full sized Truck	55	26	45	37	38	201
Total	200	200	200	200	200	1000

At the 0.05 significance level, test the claim that the vehicle type vary by mall location.

45. One hundred twenty five students in an introductory statistics class were asked how many credits they had earned in college and how certain they were about their choice of major.

Credits Earned	Highly Uncertain	Somewhat Certain	Very Certain	Total
0-15	6	6	4	16
15-60	19	13	7	39
60 or more	8	12	5	25
Total	33	31	16	80

At a 0.05 significance level, test the claim that credits earned are independent of choice of major.

46. In an Aircraft, high level of cockpit noise can damage the hearing of pilots. A co-pilot collected 60 noise level measurements using a sound meter under three categories low, medium and high in three flight phases.

Noise Level	Phase-1	Phase-II	Phase-III	Row Total
Low	7	3	5	15
Medium	16	4	6	26
High	2	5	12	19
Column Total	25	12	23	60

Use 0.01 significance level to test whether the noise level and the phases are independent.

47. A chemical extraction plant processes sea water to collect sodium chloride and magnesium. It was observed that sea water contains sodium chloride, magnesium and other compounds in the ratio 54%, 3%, 43%. A sample of 1000 tons of sea water resulted in 620 tons of sodium chloride, 40 tons of magnesium and 340 tons of mixture of other compounds. Are these data consistent with the known composition of sea water at 5% level.

48. A wall paint shop owner is of the opinion that all kinds of branded paints available with him for sale are equally popular. He has ten kinds of branded wall paints and the following table summarizes the number of each kind of wall paint buckets sold by him on a randomly selected day.

W. paint of brand	A	B	C	D	E	F	G	H	I	J
Number of buckets sold	19	18	23	14	17	22	19	21	22	17

At $\alpha = 0.01$ significance level, do we have enough evidence to conclude that all branded wall paints are equally popular.

10

CORRELATION AND REGRESSION ANALYSIS

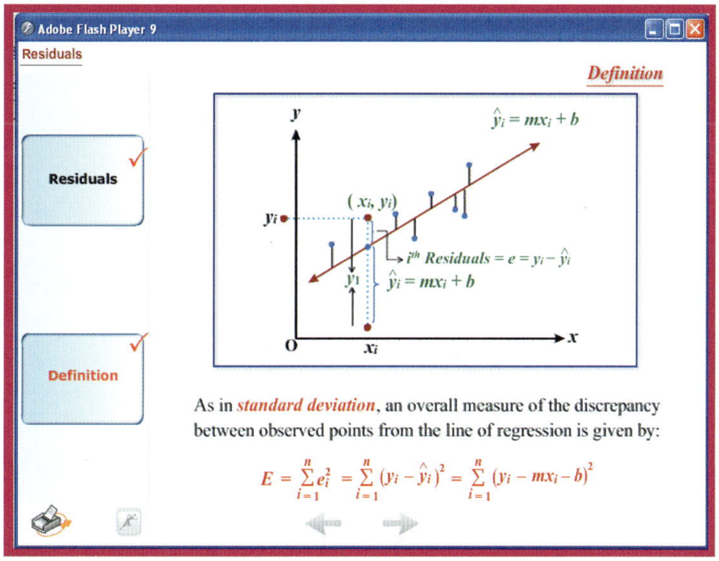

CHAPTER

10

CORRELATION AND REGRESSION ANALYSIS

Chapter Overview:

1. *Introduction*
2. *Linear Correlation*
3. *Regression Analysis and Forecasting*
4. *Variation and Prediction Interval*
5. *Correlation and Regression Analysis at a Glance*
6. *Correlation and Regression Analysis with Microsoft Excel®*
7. *Extended Topics and Proofs for Correlation and Regression Analysis (Optional)*

10.1: INTRODUCTION

Until now, we have used statistical methods for making inferences from one or two samples, using measurements of a single variable (X). For example, for each individual in the random sample, we measured a single characteristic such as the height, weight, or HDL cholesterol level.

Suppose now, for each member in a random sample we measure two variables X and Y, known as Bivariate Data (ordered pairs). In this chapter, we will present methods for determining whether there is a correlation (association/relationship) between the two variables and whether that correlation is linear.

Remark: The study of nonlinear correlation is beyond the scope of this book. Therefore we will only focus on linear correlation in this chapter.

Once a linear correlation is established, we will derive an equation known as the **Regression Equation** or **Line of Best Fit** that best fits the bivariate data and we will use it to predict the value of one variable given the value of the other variable. In addition, we will construct a confidence interval for the predicted value known as the **Prediction Interval**. We conclude with a summary of the Correlation and Regression Analysis given a bivariate data set, and a closer look at the theory behind the concepts presented throughout the chapter.

10.2: Linear Correlation

Objectives

After completing this section you will be able to understand:

A. Correlation and Graphical Representation of Bivariate Data.

B. Karl Pearson's Coefficient of Correlation.

C. Test for Significance of Observed Sample Correlation Coefficient (r).

A. Correlation and Graphical Representation of Bivariate Data

Distributions resulting from experiments where each outcome or unit of the experiment assumes two values are called **bivariate distributions**. Given a bivariate distribution or a bivariate data set (ordered pairs), we are interested to find if there is a relationship between the two variables under study. Correlation is a statistical tool which studies that relationship and Correlation Analysis is the set of methods and techniques for analyzing the relationship between the two variables. The two variables are usually denoted by x and y.

Note: It is customary to represent the independent variable by "x" and the dependent variable by "y".

Definition: Correlation

Two variables are said to be correlated if the change in one variable results in a corresponding change in the other variable.

• Linear and Non-Linear Correlation

The correlation between two variables is said to be *linear* if, corresponding to a unit change in one variable, there is a constant change in the value of the other variable over the entire range of values. That is, a bivariate data set is linearly correlated if the plot of the data set in the xy-plane lies along a linear path. Such phenomena occur frequently in physical sciences.

The relationship between two variables is said to be *non-linear* or *curvilinear* if corresponding to a unit change in one variable, the other variable does not change at a constant rate but it changes at fluctuating rate. In such cases if the data are plotted on the xy-plane we do not get a straight line pattern. Such phenomena are common in the data relating to economics and social sciences.

• Methods of Studying Correlation

We shall confine our discussion to the methods of ascertaining only *linear relationship* between two variables. The commonly used methods for studying the correlation between two variables are:

i) Scatter Diagram method

ii) Karl Pearson's Coefficient of Correlation, also known as the Covariance, method

iii) Two-way frequency table, also known as Bivariate Correlation, method

iv) Rank method

v) Concurrent Deviation method

In this section, we shall discuss the Scatter Diagram method as well as the Pearson's Correlation Coefficient. The Rank method will be discussed in Appendix 3 (Non-Parametric Statistics). The other methods are beyond the scope of this book.

• Scatter Diagram Method

Scatter Diagram (Scatter Plot) is one of the easiest ways of presenting bivariate data and provides one of the simplest tools of visualizing the correlation between two variables. Hence, before developing a formal statistical analysis for correlation, we should use a scatter plot to explore the data visually.

> **Definition:**
>
> *A scatter diagram is a collection of points (x, y) in the xy-plane where the horizontal axis represents the independent variable (X) and the vertical axis represents the dependent variable (Y).*

From scatter plots we can form a fairly good, though rough, idea about the relationship between the two variables. A careful study of the scatter plot provides answers to the following questions.

i) Is there a relationship between X and Y?

ii) Is the relationship positive (increasing from left to right) or negative (decreasing from left to right)?

iii) How strong is the relationship?

i) Is there a relationship between X and Y?

If the points on the scatter plot reveal any trend (either upward or downward), the variables are said to be correlated. On the other hand, if no trend is revealed, the variables are uncorrelated.

If the points cluster around a straight path (Figure 10.1 a), then there is a linear relationship. If the points cluster around a curved path (Figure 10.1 b), then there is a curvilinear (nonlinear) relationship. Otherwise, there is no relationship between the variables (Figure 10.1 c).

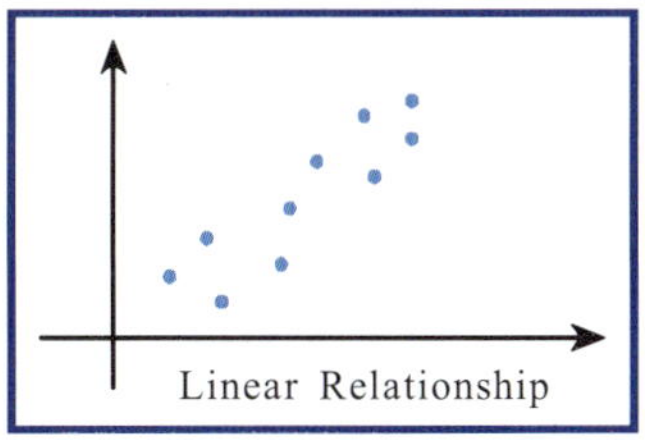

Figure 10.1 a

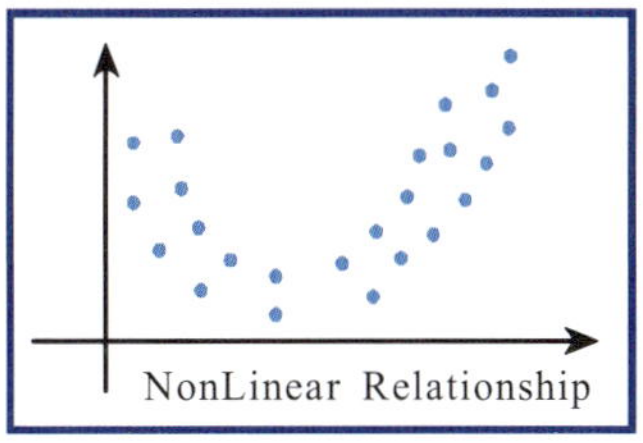

Figure 10.1 b

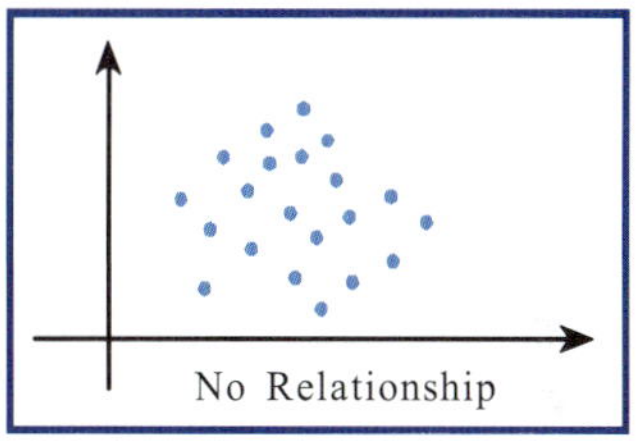

Figure 10.1 c

ii) Is the relationship positive or negative?

The linear relationship between the variables is **positive** if the points in the scatter plot cluster around a line with a positive slope. That is, the line is increasing from left to right (Figure 10.2 a).

On the other hand, if the points depict a downward trend from left to right of the scatter plot, then the correlation is **negative** (Figure 10.2 b).

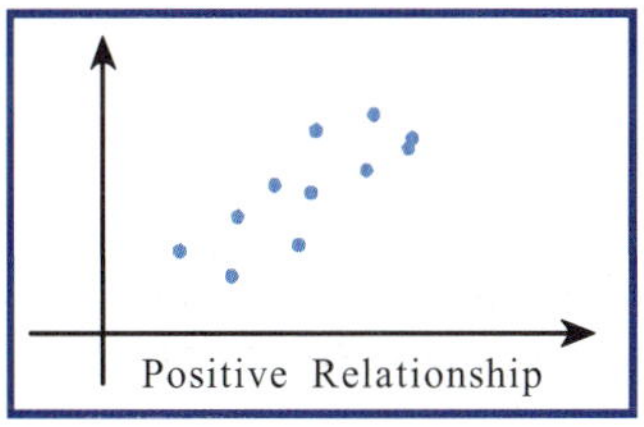

Figure 10.2 a

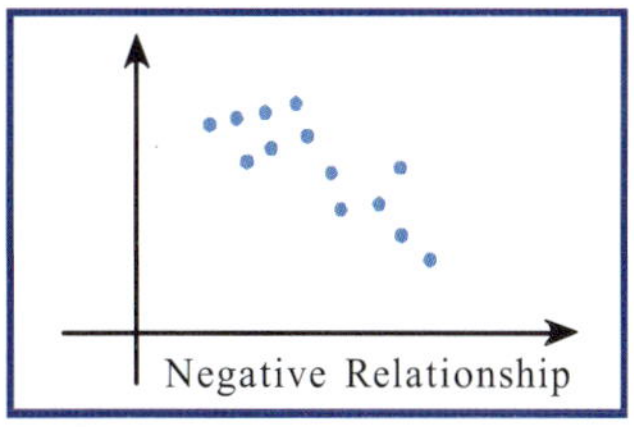

Figure 10.2 b

iii) How strong is the relationship?

A linear relationship between two variables is strong if the points of the scatter plot cluster around a line very closely. On the other hand, if the points are widely scattered around a line, we may suspect a weak or no correlation. The following diagrams (Figure 10.3 a to 10.3 f) of the bivariate data depict the strength of the linear relationship between the variables for different forms of correlation.

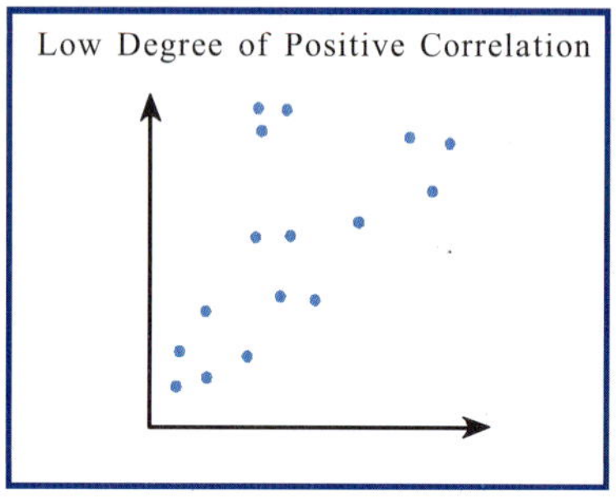

Figure 10.3 a

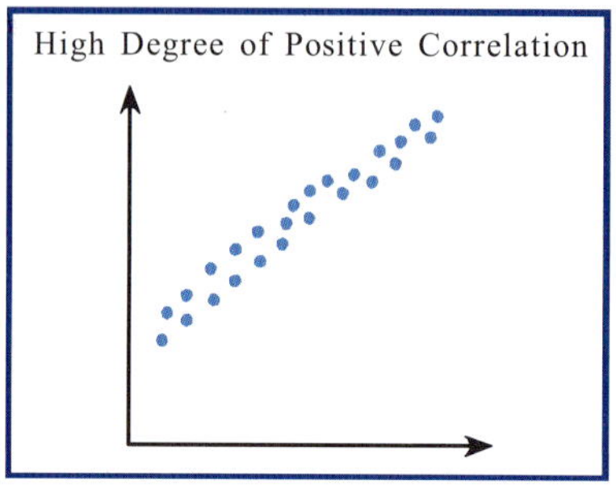

Figure 10.3 b

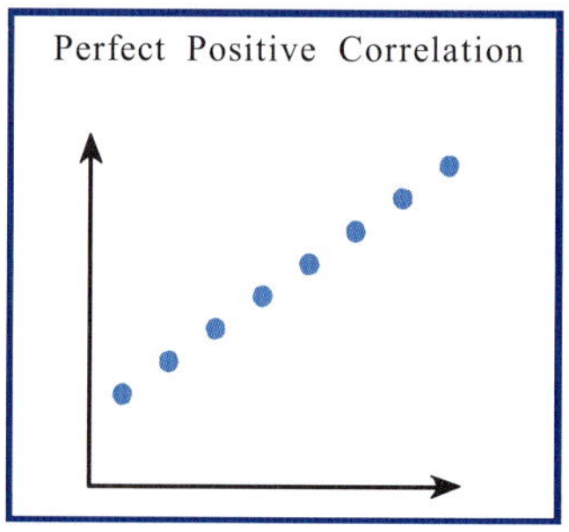

Figure 10.3 c

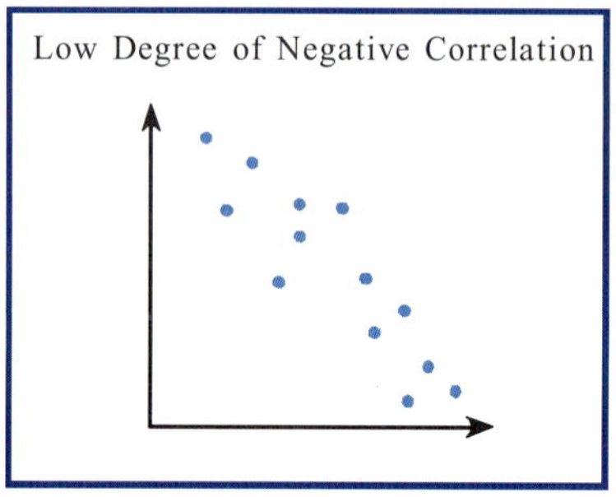

Figure 10.3 d

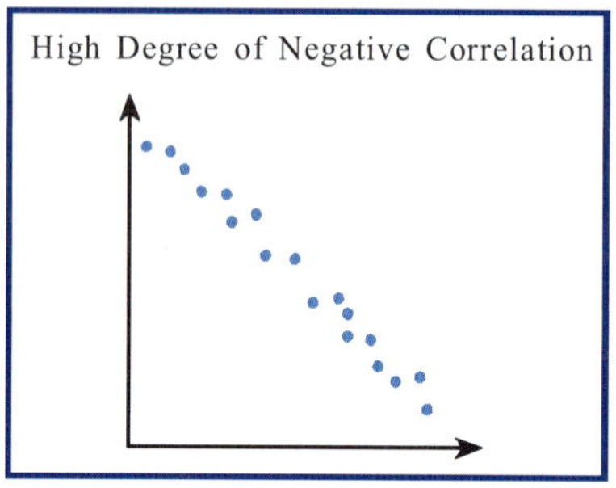

Figure 10.3 e

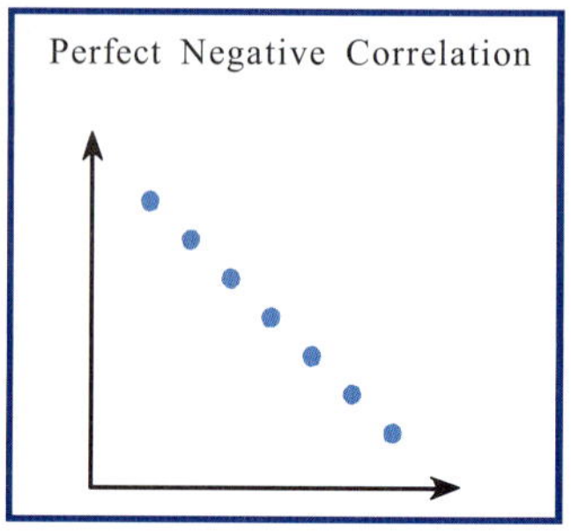

Figure 10.3 f

Example 1: Draw a scatter diagram from the following bivariate data of 10 students in a class.

Heights (inches)	62	72	70	60	67	70	64	65	60	70
Weights (lbs.)	50	65	63	52	56	60	59	58	54	65

Also indicate whether the correlation is positive or negative.

Solution: The scatter plot for the above data is shown below.

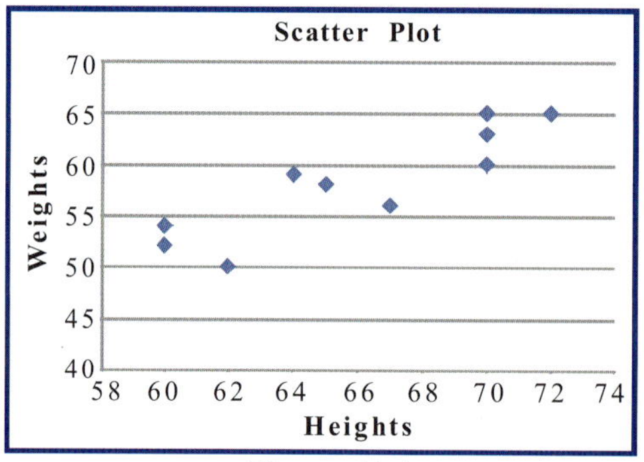

Figure 10.4

Since the points are fairly close to each other, we expect a fairly high degree of correlation between the given bivariate data set. Further, since the points reveal an upward trend from left to right, the correlation is positive. Hence, we may expect a reasonably good degree of positive correlation between the given bivariate data.

Note: Drawing a scatter plot of the bivariate data is usually the first step in investigating the relationship between two variables. However, this method is not suitable if the number of observations is fairly large. Furthermore, the scatter plot only describes the nature of the relationship, not an exact mathematical measure; and at best, it is a subjective visual analysis of the bivariate data.

B. Karl Pearson's Coefficient of Correlation

A mathematical method for measuring the magnitude of linear correlation between a set of bivariate data was suggested by Karl Pearson (1857-1936), a British Biometrician and Statistician, and is by far the most widely used method in practice.

Definition: Karl Pearson's Correlation Coefficient

The correlation coefficient between two variables x and y, usually denoted by $r(x, y)$, r_{xy}, or simply r is a numerical measure of linear relationship between them and is given by:

$$r = \frac{n\sum xy - \sum x \sum y}{\sqrt{n\sum x^2 - \left(\sum x\right)^2}\sqrt{n\sum y^2 - \left(\sum y\right)^2}} \tag{10.1}$$

where

n = number of ordered pairs in sample data

$\sum x$ = sum of all x values

$\sum x^2$ = sum of the square of all x values (We must first square every x value and then add all the squares.)

$\sum xy$ = sum of the products of all x and y values (We must first multiply every x value by its corresponding y value, and then add all such products.)

$\left(\sum x\right)^2$ = square of the sum of all x values

Note: $\sum x^2$ and $\left(\sum x\right)^2$ are quite different and must not be confused.

Alternative formula:

$$r = \frac{\sum(x-\bar{x})(y-\bar{y})}{\sqrt{\sum(x-\bar{x})^2\sum(y-\bar{y})^2}} = \frac{s_{xy}}{\sqrt{s_{xx}\cdot s_{yy}}} \tag{10.2}$$

where

$$s_{xy} = \sum(x-\bar{x})(y-\bar{y}) = \sum xy - \frac{\left(\sum x\right)\left(\sum y\right)}{n}$$

$$s_{xx} = \sum(x-\bar{x})^2 = \sum x^2 - \frac{\left(\sum x\right)^2}{n}$$

$$s_{yy} = \sum(y-\bar{y})^2 = \sum y^2 - \frac{\left(\sum y\right)^2}{n}$$

Dividing the numerator and denominator of (10.2) by $n - 1$, we have:

$$r = \frac{\sum(x-\bar{x})(y-\bar{y})}{(n-1)\cdot s_x \cdot s_y}, \tag{10.3}$$

where s_x and s_y are the sample standard deviations of the variables.

The population correlation coefficient is denoted by ρ (letter Rho of the Greek alphabet), and is given by:

$$\rho = \frac{N\sum xy - \sum x \sum y}{\left[\sqrt{N\sum x^2 - \left(\sum x\right)^2}\right]\left[\sqrt{N\sum y^2 - \left(\sum y\right)^2}\right]}$$

Round off Rule for correlation coefficient: The formula for r is very sensitive to rounding. We should carry as many digits in the decimal places as is reasonable for our problem right from the first step to the last step. Rounding in the middle of calculations will result in substantial errors. The answers given in the book may vary slightly from your answer, depending on the rounding process used. Since the critical values of "r", given in Table 7 at the end of the book, are up to three decimal places, the value of r should be rounded to three decimal places to test its significance.

Note: i) With the advancements in technology, even simple hand held calculators can compute the correlation coefficient with negligible errors. Hence, if you have access to a graphing calculator or a Statistical Software, the round off issues will no longer be significant.

ii) The value of r can significantly be affected by the presence of outliers in the data set.

• Assumptions Underlying Karl Pearson's Correlation Coefficient

1. The bivariate sample data (x_i, y_i), $i = 1, 2, \ldots, n$ is a random sample.
2. The sample observations (x_i, y_i), $i = 1, 2, \ldots, n$ have a joint bivariate normal distribution.
3. All outliers are removed prior to calculating the correlation coefficient.

• Important Properties of Correlation Coefficient

Property 1. Correlation Coefficient cannot exceed 1 numerically.

That is, it lies between -1 and $+1$. Symbolically, $|r| \le 1 \quad \Rightarrow \quad -1 \le r \le 1$

This property provides a check on our calculations. If in any problem, the obtained value of r lies outside the limits ± 1, it would imply that there is some mistake in our calculations. $r = +1$ implies perfect positive correlation between the variables and $r = -1$ implies perfect negative correlation between the variables (variables = bivariate data).

Interpretation of r:

i) A perfect positive correlation $r = +1$ indicates that the scatter diagram will be a straight line increasing from left to right (positive slope).

ii) A perfect negative correlation $r = -1$ indicates that the scatter diagram will be a straight line decreasing from left to right (negative slope).

iii) For other values of r lying between -1 and 1, there are no set guidelines for its interpretation. The best we can conclude is that the closer the value of r is to ± 1, the stronger is the relation between the variables, and the closer the value of r is to 0, the weaker is the relation between the variables. One should be very careful in interpreting the value of sample correlation coefficient r as it is often misinterpreted.

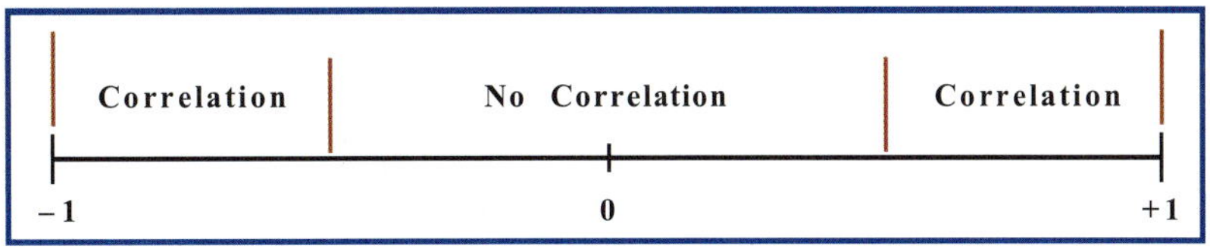

Figure 10.5

iv) Another more useful measure for interpreting the value of r is the coefficient of determination (r^2) discussed in section 10.4.

Property 2. The value of r does not depend on the choice of the variable x or y.

Any one of the two variables can be taken as x and the other as y without affecting the value of r. Thus:

$$r(x, y) = r(y, x)$$

Example 2: The following data give the distance (in miles) between 10 randomly selected towns in Georgia traveling by the roads and taking the shortest distance between them (straight line). Calculate the correlation coefficient for the bivariate data where x represents the road distance and y represents the linear distance.

x	17	29	18	17	11	6	30	20	33	14
y	12	22	11	15	9	5	23	12	28	12

Also indicate whether the correlation is positive or negative.

Solution: We start with calculating the necessary sums for the correlation coefficient formula.

x	17	29	18	17	11	6	30	20	33	14	$\sum x = 195$
y	12	22	11	15	9	5	23	12	28	12	$\sum y = 149$
x^2	289	841	324	289	121	36	900	400	1089	196	$\sum x^2 = 4485$
y^2	144	484	121	225	81	25	529	144	784	144	$\sum y^2 = 2681$
xy	204	638	198	255	99	30	690	240	924	168	$\sum xy = 3446$

Table 10.1

Given $n = 10$, Karl Pearson's correlation coefficient is:

$$r = \frac{n\sum xy - \sum x \sum y}{\sqrt{n\sum x^2 - \left(\sum x\right)^2}\sqrt{n\sum y^2 - \left(\sum y\right)^2}}$$

$$= \frac{10(3446) - (195)(149)}{\sqrt{10(4485) - 195^2}\sqrt{10(2681) - 149^2}}$$

$$= \frac{5405}{\sqrt{(6825)(4609)}} = \frac{5405}{5608.602767} = 0.96369813 \approx 0.964$$

Hence, there seems to be a very strong positive correlation between the given bivariate data.

Example 3: A random sample of size 50 from a bivariate distribution gave the following results:

$$n = 50,\ s_x = 4,\ s_y = 3.5, \text{ and } \sum(x - \bar{x})(y - \bar{y}) = 420$$

Find Karl Pearson's correlation coefficient.

Solution: Using formula (10.3), we have:

$$r = \frac{\sum(x-\bar{x})(y-\bar{y})}{(n-1)\cdot s_x \cdot s_y} = \frac{(420)}{(49)(4)(3.5)} = 0.612$$

Hence, there seems to be a fairly good degree of positive correlation between the bivariate data.

C. Test for Significance of Observed Sample Correlation Coefficient (r)

Sometimes a more precise method for determining linear correlation is necessary. Testing for the significance of observed sample correlation coefficient (r) is one way to determine the existence of linear correlation.

In a Test of Hypotheses for the significance of (r), we will utilize **i)** test statistics t and **ii)** test statistics r.

In the following discussion we will develop a hypotheses testing procedure for determining linear correlation given the following null and alternative hypotheses.

Null Hypotheses: $H_0 : \rho = 0$ (No linear correlation)
Alternative Hypotheses: $H_1 : \rho \neq 0$ (Strong linear correlation)

ρ denotes the population correlation coefficient.

• Method 1: Hypotheses Test for Linear Correlation Using Test Statistics *t* (*t*-test)

Hypotheses: According to R.A. Fisher, the null hypotheses $H_0 : \rho = 0$. That is, the variables are uncorrelated in the population.

$H_0 : \rho = 0$ (No linear correlation)
$H_1 : \rho \neq 0$ (Strong linear correlation)

Test Statistics: The test statistics is $t = r\sqrt{\frac{n-2}{1-r^2}} \sim t_{n-2}$

Rejection Region and Critical Values: The rejection region in this test is a two-tail ***t***-distribution. The critical values are found using the ***t***-table (Table 4) with $n - 2$ degrees of freedom and an $\alpha/2$ significance level.

Criteria	Decision	Conclusion
The test statistics lies inside the rejection region.	Reject H_0	There is sufficient evidence to support the claim of strong linear correlation.
The test statistics lies outside the rejection region.	Fail to Reject H_0	There is no linear correlation.

Table 10.2

• Method 2: Hypotheses Test for Linear Correlation Using Test Statistics *r* (*t*-test)

Hypotheses:

$H_0 : \rho = 0$ (No linear correlation)
$H_1 : \rho \neq 0$ (Strong linear correlation)

Test Statistics: Instead of computing the test statistics from the observed value of sample correlation coefficient r, in this test we use r itself as a test statistics.

Rejection Region and Critical Values: Read the critical value $r_n(\alpha)$ from Table 7, using the sample size n, and an α significance level for a two-tailed test. Thus, the rejection region is:

$$|r| > r_n(\alpha) \quad \Rightarrow \quad r > r_n(\alpha) \text{ or } r < -r_n(\alpha)$$

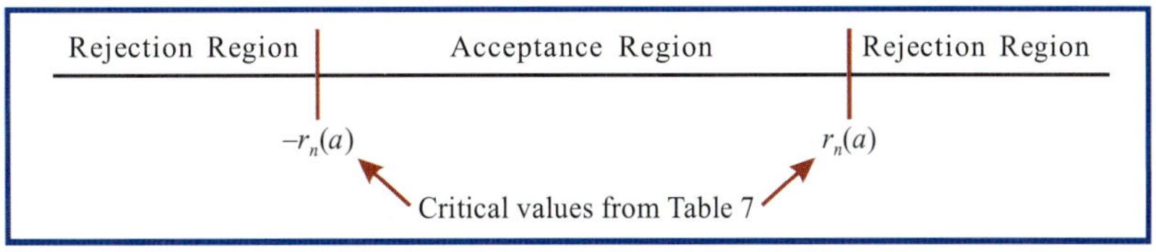

Figure 10.6

Decision/Conclusion: Hence, if the calculated value of r lies in the rejection region ($|r|$ is greater than the tabulated value $r_n(\alpha)$), we reject the null hypotheses, which would imply that there is linear correlation between the variables. If r lies in the acceptance region ($|r|$ is less than $r_n(\alpha)$), we fail to reject the null hypothesis and conclude that there is no linear correlation.

• Confidence Interval for ρ

If the null hypotheses $H_0 : \rho = 0$ is rejected, then we may be interested in obtaining the limits within which the population correlation coefficient (ρ) may be expected to lie. The confidence interval for ρ is:

$$r - t_{n-2}(\alpha/2)\left(\frac{1-r^2}{\sqrt{n}}\right) < \rho < r + t_{n-2}(\alpha/2)\left(\frac{1-r^2}{\sqrt{n}}\right) \qquad (10.4)$$

$t_v(\alpha)$ is the right-tailed critical value of t for v degrees of freedom and an α significance level.

Example 4: A random sample of 27 pairs of observations from a normal population gives a correlation coefficient of 0.42.

(a) Is it likely that the variables in the population are uncorrelated? Use $\alpha = 0.05$.

(b) Construct a 95% confidence interval for ρ.

Solution: **(a)** We are given $n = 27$ and $r = 0.42$.

Null Hypotheses: $H_0 : \rho = 0$ (No linear correlation between the variables)

Alternative Hypotheses: $H_1 : \rho \neq 0$ (Strong linear correlation)

***Method* 1: Test Statistics t**

Under H_0, the test statistics is:

$$t = r\sqrt{\frac{n-2}{1-r^2}} \sim t_{n-2} = t_{25}$$

$$\Rightarrow t = 0.42\sqrt{\frac{27-2}{1-(0.42^2)}} = 2.31398 = 2.314$$

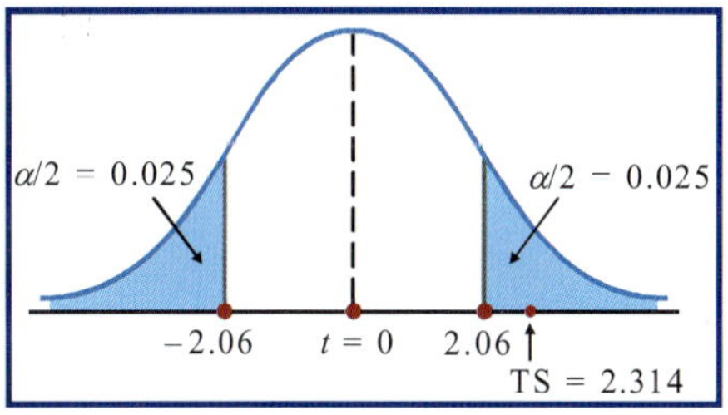

Figure 10.7

From Table 4, the critical value of t for 25 df and a 0.05 significance level, for two-tailed test is 2.060. That is, $t_{25}(0.025) = 2.0595$. The rejection regions, critical values, and the test statistics are shown in Figure 10.7.

Since the test statistics $t = 2.314$ lies inside the rejection region, we reject H_0 at 5% level of significance. There is sufficient evidence to conclude that the given variables are linearly correlated in the population.

Method* 2: Test Statistics *r

From Table 7, the critical value of r at $\alpha = 0.05$ and for $n = 27$ is less than 0.396. Since the value of test statistics $r = 0.42$ is greater than the critical value of r at $\alpha = 0.05$ and for $n = 27$, it lies in the rejection region. Hence $H_0 : \rho = 0$ is rejected at 5% level of significance and we can conclude that the variables are linearly correlated in the population.

(b) 95% confidence interval for ρ

$$\alpha = 0.05 \quad \Rightarrow \quad \frac{\alpha}{2} = 0.025 \quad \Rightarrow \quad t_{25}(0.025) = 2.06$$

$$0.42 - 2.06\left(\frac{1-0.42^2}{\sqrt{27}}\right) < \rho < 0.42 + 2.06\left(\frac{1-0.42^2}{\sqrt{27}}\right) \quad \text{or} \quad 0.093 < \rho < 0.747$$

Example 5: Fuel consumption in *mpg* for 16 randomly selected short and long wheelbase light duty vehicles was recorded by an independent study.

Short wheel fuel consumption in (mpg)	17.5	20.3	21.2	21	20.9	21.1	21.2	21.5
	21.4	22.1	22	22.5	22.1	22.5	23.7	23.8
Long wheel fuel consumption in (mpg)	12.9	16.1	16.8	17.3	16.7	17.3	16.4	17.2
	17	16.2	17.5	17.5	17.7	17.8	17.3	18.3

(a) Draw the scatter plot. **(b)** Calculate the correlation coefficient r.

(c) Using a 0.01 significance level to test the claim that the fuel consumptions in short and long wheelbase light duty vehicles are correlated.

(d) Construct a 99% confidence interval for the population correlation coefficient ρ.

Solution: **(a)**

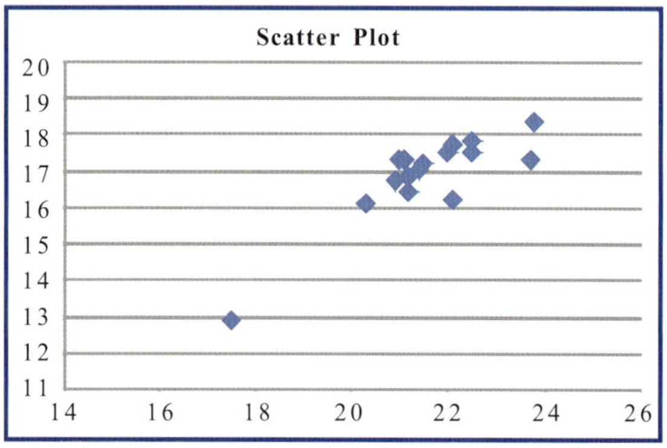

Figure 10.8

(b) In order to calculate the sample correlation coefficient r for the given bivariate data, we must first calculate all the necessary sums shown in Table 10.3.

x	y	x^2	y^2	xy
17.5	12.9	306.25	166.41	225.75
20.3	16.1	412.09	259.21	326.83
21.2	16.8	449.44	282.24	356.16
21	17.3	441	299.29	363.3
20.9	16.7	436.81	278.89	349.03
21.1	17.3	445.21	299.29	365.03
21.2	16.4	449.44	268.96	347.68
21.5	17.2	462.25	295.84	369.8
21.4	17	457.96	289	363.8
22.1	16.2	488.41	262.44	358.02
22	17.5	484	306.25	385
22.5	17.5	506.25	306.25	393.75
22.1	17.7	488.41	313.29	391.17
22.5	17.8	506.25	316.84	400.5
23.7	17.3	561.69	299.29	410.01
23.8	18.3	566.44	334.89	435.54
$\sum x = 343.8$	$\sum y = 270$	$\sum x^2 = 7418.7$	$\sum y^2 = 4578.38$	$\sum xy = 5823.67$

Table 10.3

Hence, the correlation coefficient is:

$$r = \frac{n\sum xy - \sum x \sum y}{\sqrt{n\sum x^2 - \left(\sum x\right)^2}\sqrt{n\sum y^2 - \left(\sum y\right)^2}}$$

$$= \frac{16(5823.67) - (343.8)(270)}{\sqrt{16(7418.7) - 343.8^2}\sqrt{16(4578.38) - 270^2}}$$

$$= \frac{352.72}{(22.377)(18.817)} = 0.838$$

There seems to be a high degree of positive correlation between the variables.

(c) We are given $n = 16$ and $r = 0.838$.

Null Hypotheses: $H_0 : \rho = 0$ (No linear correlation between the variables)
Alternative Hypotheses: $H_1 : \rho \neq 0$ (Strong linear correlation)

***Method* 1: Test Statistics t**

Under H_0, the test statistics is:

$$t = r\sqrt{\frac{n-2}{1-r^2}} \sim t_{n-2} = t_{14}$$

$$\Rightarrow t = 0.838\sqrt{\frac{16-2}{1-(0.838)^2}} = 5.738$$

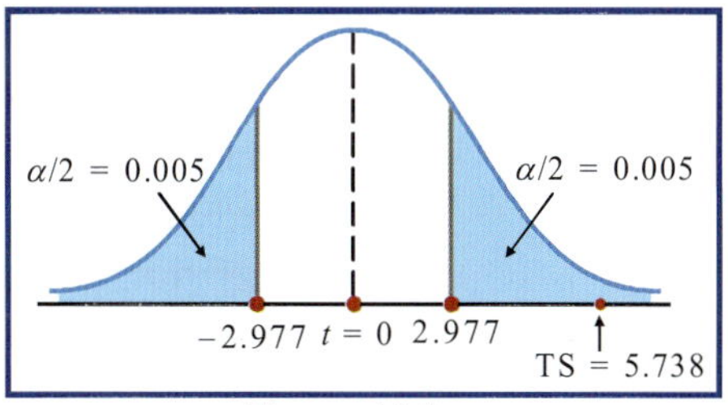

Figure 10.9

From Table 4, the critical value of t for 14 df and a 0.01 significance level for two-tailed test is 2.977. That is, $t_{14}(0.005) = 2.977$. The rejection regions, critical values, and the test statistics are shown in Figure 10.9.

Since the test statistics $t = 5.738$ lies inside the rejection region, we reject H_0 at 1% level of significance. There is sufficient evidence to conclude that the given variables are linearly correlated in the population.

Method* 2: Test Statistics *r

From Table 7, the critical value of r at $\alpha = 0.01$ and for $n = 16$ is 0.623 $\left[r_{16}(0.01) = 0.623\right]$.

Since the value of test statistics $r = 0.838$ is greater than $r_{16}(0.01)$, it lies in the rejection region. Hence $H_0 : \rho = 0$ is rejected at 1% level of significance and we can conclude that the variables are linearly correlated in the population.

(d) 99% confidence interval for ρ

$$\alpha = 0.01 \Rightarrow \frac{\alpha}{2} = 0.005 \Rightarrow t_{14}(0.005) = 2.977$$

$$0.838 - 2.977\left(\frac{1-(0.838)^2}{\sqrt{16}}\right) < \rho < 0.838 + 2.977\left(\frac{1-(0.838)^2}{\sqrt{16}}\right)$$

$$0.616 < \rho < 1.060$$

Try it yourself:

1. Draw the scatter plot and compute the sample correlation coefficient r given the following data sets.

(a)

x	0	2	4	6	8	10
y	12	7	9	4	3	2

(b)

x	3	8	10	2	6	13
y	14	22	24	15	16	20

(c)

x	17.8	18.2	16.7	15.8	17.1	18.6	19.0	20.8	20.9
y	6.8	6.6	6.8	6.5	6.8	6.3	7.0	6.8	6.9

2. Calculate the value of correlation coefficient r, given

$n = 10$, $\sum x = 250$, $\sum y = 300$, $\sum x^2 = 6{,}500$, $\sum y^2 = 10{,}000$, and $\sum xy = 7{,}900$

3. Compute and interpret the correlation coefficient for the grades of 12 students selected at random.

Mathematics Grades:	85	83	87	84	88	82	90	86	88	87	89	90
Physics Grades:	88	86	88	86	90	86	88	85	92	86	88	91

4. The following bivariate data represents the average temperature difference (in °C) between the inside and the outside of a two story house and the average daily gas consumption (in Kwh).

Temp. difference:	10.5	11.5	11.4	12.5	13.5	15.0	15.2	15.6	16.5	17
Daily gas consumption:	70.1	83.0	82.7	80.0	75.5	78.5	81.1	86.5	105.5	86.0

(a) Draw the scatter diagram and interpret the graph.

(b) Compute and interpret the correlation coefficient r.

(c) Test the significance of the correlation coefficient using a 0.05 significance level using both methods presented in this section.

(d) Construct a 95% confidence interval for the population correlation coefficient ρ.

5. Consider the following bivariate data set.

x	23	41	49	50	53	54	56	57	58	60	61
y	27.9	25.9	25.2	31.1	34.7	29.1	32.5	30.3	33.8	41.1	34.5

(a) Use $\alpha = 0.01$ to test the significance of the sample correlation coefficient.

(b) Construct a 99% confidence interval for the population correlation coefficient ρ.

10.3: Regression Analysis and Forecasting

In Section 10.2, we introduced the Scatter Diagram (Plot) and presented methods for finding the value of the linear sample correlation coefficient r and its interpretation. We also introduced a testing procedure for determining whether there is a linear correlation between the variables in the bivariate data. In this section, we will find the equation of a line ($y = mx + b$) that models the linear relationship (describes the relationship between the variables). This equation is known as the **Line of Best Fit** or the **Regression Line**. We will also use the regression equation (regression line) to make **Predictions**.

Objectives

After completing this section you will be able to understand:

- **A.** Regression Analysis.
- **B.** Residuals.
- **C.** Principle of Least Squares.
- **D.** Application of Regression Line for Prediction.

A. Regression Analysis

The literal or dictionary meaning of the word **Regression** is "stepping back or returning to the average value." The term was first used by the British Biometrician Sir Frances Galion (1822-1911) in the later part of the 19th century in connection with some studies he made on estimating the extent to which the stature of the sons of tall parents reverts or regresses back to the mean stature of the population. He studied the relationship between the heights of about one thousand fathers and sons and published the results in 1885 in a paper titled "Regression towards Mediocrity in Hereditary Stature." The interesting features of his study are:

i) The tall fathers have tall sons and short fathers have short sons.

ii) The average height of the sons of group of tall fathers is less than that of the fathers, and the average height of the sons of a group of short fathers is more than that of the fathers.

In other words Galton's studies revealed that the offsprings of abnormally tall or short parents tend to revert or step back to the average height of the population rather than progress to more extremes, a phenomenon which Galton described as **Regression to Mediocrity**.

But today, the word regression, as used in Statistics, has a much wider perspective without any reference to biometry. **Regression Analysis**, in the general sense, means the study of identifying, estimating and validating the relationship between the variables. This relationship is then used for estimation or prediction of the unknown value of one variable from the known value of the other variable. Regression is one of the most important statistical tools which is extensively used in almost all sciences: natural, social and physical. It is specially used in business and economics to study the relationship between two or more variables that are related casually and for estimation of demand and supply curves, cost functions, production and consumption functions, etc.

Prediction and estimation are major problems in almost all spheres of human activity. The estimation or predictions of future production, consumption, prices, investments, sales, profits, income, etc. are of paramount importance to a businessman or an economist. Population estimates and population projections are indispensable for efficient planning of an economy.

Remark:

Regression analysis attempts to establish the *nature of the relationship* between the variables i.e., to study the mathematical relationship between the variables and thereby provide a mechanism for prediction, or forecasting.

Definition: Simple Regression

The study of the relationship between two variables is known as simple regression.

Definition: Multiple Regression

The regression analysis for studying the relationship among three or more variables at a time is known as multiple regression. (Discussed in 10.7)

In simple linear regression analysis there are two types of variables. The variable whose value is influenced or is to be predicted called the **dependent variable** and the variable which influences the values or is used for prediction called the **independent variable**. In regression analysis, the independent variable is also known as **regression**, **predictor,** or **explanatory** variable while the dependent variable is known as **regressed**, **explained**, or **response variable**.

In regression analysis, the **independent variable**, whose values are controlled by the experimenter, is denoted by x and the **dependent variable**, whose values depend on x and a number of unpredictable chance variations or errors, is denoted by y.

Definition: Regression Line

Given a collection of pairs of sample observations on the two variables x and y $\left((x_1, y_1), (x_2, y_2), \ldots, (x_n, y_n)\right)$, the regression equation

$$\hat{y} = mx + b \tag{10.5}$$

*mathematically describes the linear relationship between the two variables x and y. The graph of the linear regression equation is called the **Regression Line** or the **Line of Best Fit**.*

Note: The notation used to describe the regression line in advanced statistics books and journals is $\hat{y} = b_0 + b_1 x$, where b_0 is the y-intercept and b_1 is the slope. However, to simplify the symbolization used for the y-intercept and the slope, in this book, we have used the notation $\hat{y} = mx + b$ which is more familiar to an algebra student.

The following graph is an example of the Regression Line (Line of Best Fit) for 7 pairs of data.

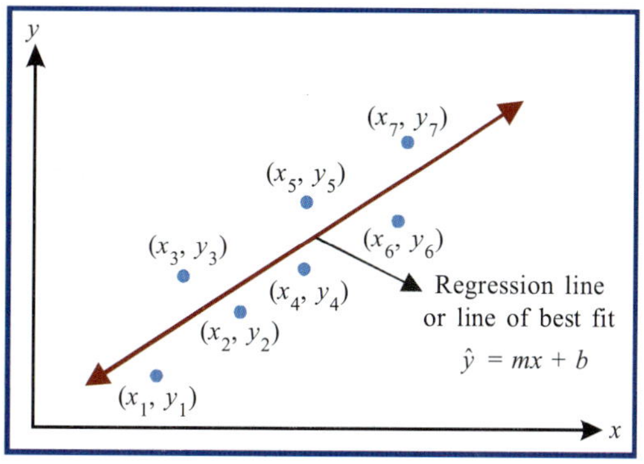

Figure 10.10

B. Residuals

In order to derive the formulas for the slope and the y-intercept of a regression line, we must first introduce the basic concept of **residuals**.

Note: A comprehensive study of the concept of residual requires the knowledge of Calculus and is beyond the scope of this book.

For any given value of $X = x_i$, the estimated value of y given by the regression equation $\hat{y} = mx + b$ is $\hat{y}_i = mx_i + b$. The discrepancy between the observed y_i and the estimated $\hat{y}_i$ is called a **residual** or **error of estimate** for the point (x_i, y_i) and is given by $y_i - \hat{y}_i$.

Definition:

The difference between the observed value of y and its estimated value ŷ given by the regression line is called the residual or error of estimate and is given by:

$$e = y - \hat{y} \qquad (10.6)$$

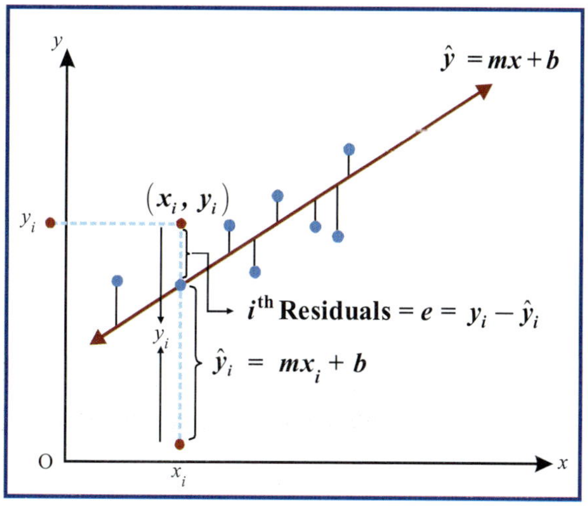

(10.7)

Figure 10.10

As in standard deviation, an overall measure of the discrepancy between observed points from the line of regression is given by:

$$E = \sum_{i=1}^{n} e_i^2 = \sum_{i=1}^{n} (y_i - \hat{y}_i)^2 = \sum_{i=1}^{n} (y_i - mx_i - b)^2 \quad (10.8)$$

C. Principle of Least Squares

The **Principle of Least Squares** consists of minimizing the sum of the squares of the residuals or the errors of estimates (10.8). Therefore, according to the principles of least squares, we shall determine the constants m and b in (10.5) so that the expression E in (10.8) is minimized for variations in the slope (m) and the y-intercept (b).The system of equations which minimizes E in (10.8) for variations in the slope and the y-intercept is given by:

$$\begin{Bmatrix} nb + m\sum x = \sum y \\ b\sum x + m\sum x^2 = \sum xy \end{Bmatrix} \quad (10.9)$$

Note: The derivation of the system in (10.9) requires techniques of Calculus and is beyond the scope of this book.

The solution for the system of equations in (10.9) will give us the y-intercept (b) and the slope (m). Solving the system in (10.9) for m and b gives us:

$$m = \frac{n\sum xy - \sum x \sum y}{n\sum x^2 - \left(\sum x\right)^2} \quad (10.10)$$

$$b = \frac{\sum y \sum x^2 - \sum x \sum xy}{n\sum x^2 - \left(\sum x\right)^2} \quad (10.11)$$

***Proof:* (*Optional*)**
We can use the Gauss-Jordan Method from algebra to solve (10.9).

$$\left[\begin{array}{cc|c} n & \sum x & \sum y \\ \sum x & \sum x^2 & \sum xy \end{array}\right] \Rightarrow \left[\begin{array}{cc|c} 1 & \frac{\sum x}{n} & \frac{\sum y}{n} \\ \sum x & \sum x^2 & \sum xy \end{array}\right] \Rightarrow \left[\begin{array}{cc|c} 1 & \frac{\sum x}{n} & \frac{\sum y}{n} \\ 0 & \frac{-\left(\sum x\right)^2}{n} + \sum x^2 & \frac{-\sum x \sum y}{n} + \sum xy \end{array}\right]$$

$$\left[\begin{array}{cc|c} 1 & \frac{\sum x}{n} & \frac{\sum y}{n} \\ 0 & \frac{-\left(\sum x\right)^2 + n\sum x^2}{n} & \frac{-\sum x \sum y + n\sum xy}{n} \end{array}\right] \Rightarrow \left[\begin{array}{cc|c} 1 & \frac{\sum x}{n} & \frac{\sum y}{n} \\ 0 & 1 & \frac{-\sum x \sum y + n\sum xy}{-\left(\sum x\right)^2 + n\sum x^2} \end{array}\right]$$

$$\left[\begin{array}{cc|c} 1 & 0 & \left(\frac{-\sum x \sum y + n\sum xy}{-\left(\sum x\right)^2 + n\sum x^2}\right)\left(\frac{-\sum x}{n}\right) + \frac{\sum y}{n} \\ 0 & 1 & \frac{-\sum x \sum y + n\sum xy}{-\left(\sum x\right)^2 + n\sum x^2} \end{array}\right] \Rightarrow \left[\begin{array}{cc|c} 1 & 0 & \frac{\sum y \sum x^2 - \sum x \sum xy}{-\left(\sum x\right)^2 + n\sum x^2} \\ 0 & 1 & \frac{-\sum x \sum y + n\sum xy}{-\left(\sum x\right)^2 + n\sum x^2} \end{array}\right]$$

Therefore,

y-intercept	Slope
$\dfrac{\sum y \sum x^2 - \sum x \sum xy}{n\sum x^2 - \left(\sum x\right)^2}$	$\dfrac{n\sum xy - \sum x \sum y}{n\sum x^2 - \left(\sum x\right)^2}$

Table 10.4

Remarks:

i) Recall $r(x, y) = \dfrac{n\sum xy - \sum x \sum y}{\sqrt{n\sum x^2 - \left(\sum x\right)^2}\sqrt{n\sum y^2 - \left(\sum y\right)^2}} = \dfrac{s_{xy}}{\sqrt{s_{xx}}\sqrt{s_{yy}}}$

Hence, we can write the formula for the slope and y-intercept of the regression line as:

$$m = \frac{n\sum xy - \sum x \sum y}{n\sum x^2 - \left(\sum x\right)^2} = \frac{s_{xy}}{s_{xx}} = r\frac{s_y}{s_x}, \text{ and } b = \frac{\sum y}{n} - m \cdot \frac{\sum x}{n} = \overline{y} - m \cdot \overline{x}$$

Notice that the numerical calculations required for computing m have already been done in computing r; and to compute b, we can simply use the slope and the average x and y values.

ii) $\sum_{i=1}^{n} e_i = \sum_{i=1}^{n}\left(y_i - \hat{y}_i\right) = 0$

Round off Rule for the slope and the y-intercept: Round the slope and the y-intercept to three decimal places.

Example 6: Consider the following bivariate data set.

x	27	22	45	41	39	50	53	57	58	60
y	8.1	9.4	25.5	23.2	18.8	20.5	22	25	31.5	30

(a) Find the slope and y-intercept of the Regression Line.

(b) Construct the Scatter Plot and draw the Regression Line.

Solution: **(a)** To find the slope and the y-intercept, we must first calculate all the sums.

x	27	22	45	41	39	50	53	57	58	60	$\sum x = 452$
y	8.1	9.4	25.5	23.2	18.8	20.5	22	25	31.5	30	$\sum y = 214$
x^2	729	484	2025	1681	1521	2500	2809	3249	3364	3600	$\sum x^2 = 21962$
y^2	65.61	88.36	650.25	538.24	353.44	420.25	484	625	992.25	900	$\sum y^2 = 5117.4$
xy	218.7	206.8	1147.5	951.2	733.2	1025	1166	1425	1827	1800	$\sum xy = 10500.4$

Table 10.5

We have $n = 10$, $\sum x = 452$, $\sum y = 214$, $\sum x^2 = 21962$, $\sum y^2 = 5117.4$ and $\sum xy = 10500.4$

Therefore, the slope of the regression line is

$$m = \frac{n\sum xy - \sum x \sum y}{n\sum x^2 - \left(\sum x\right)^2} = \frac{10(10500.4) - (452)(214)}{10(21962) - (452)^2} = \frac{8276}{15316} = 0.540$$

and the y-intercept is

$$b = \frac{\sum y \sum x^2 - \sum x \sum xy}{n\sum x^2 - \left(\sum x\right)^2} = \frac{(214)(21962)-(452)(10500.4)}{10(21962)-(452)^2} = \frac{-46312.8}{15316} = -3.024$$

Hence, the regression equation for the given sample bivariate data is $\hat{y} = 0.540x - 3.024$

(b)

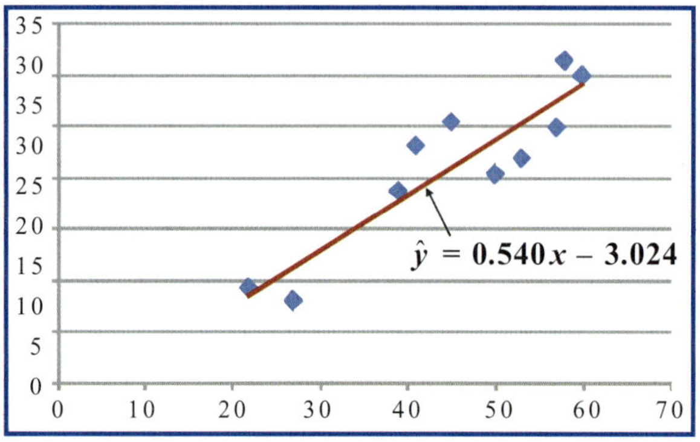

Figure 10.10

D. Application of Regression Line for Prediction

The equation of the regression line is commonly used to predict the value of the dependent variable y for a given value of the independent variable x known as x_0.

Notation:

In this book, we will use $\hat{y}$ to represent the predicted value of the dependent variable y, and x_0 to represent the value of the independent variable x used for prediction.

For example, the predicted value of y, written as $\hat{y}$ when $x = x_0$, is given by:

$$\hat{y}_{x=x_0} = m \cdot x_0 + b \qquad (10.12)$$

However, the regression equation should be used for prediction and estimation with utmost care and caution. It only makes sense to use the regression equation for estimation if it fits the data well. Hence, technically, before using the line of regression, we should first test for **Goodness of Fit**. The following points should be kept in mind while using the regression line for prediction.

1. Test the significance of the observed sample correlation coefficient $r = r(x, y)$ as discussed in Section 10.2. If the value of r is significant, we can use the regression line for estimation and prediction.
2. If r is not significant, then the linear model is not a good fit and hence the regression line should not be used for prediction. In this case (r not significant), the best predicted value of y for any given x is the mean value of y denoted as $\bar{y}$.

The above results can be summarized as:

i) If r is significant, then the best predicted value of y, denoted by $\hat{y}$ for a given x denoted by x_0, is obtained by substituting x_0 into the regression equation. Thus:

$$\hat{y}_{x=x_0} = m \cdot x_0 + b \qquad (10.13)$$

ii) If r is not significant, then the best predicted value for any given x is $\hat{y} = \bar{y}$.

3. If the linear regression is a good fit to the given data (r is significant), it is worthwhile to use the regression line to estimate y for any given x. However, in doing so, we should not go beyond the available sample data on the independent variable x.

   For example, suppose the regression line of y (blood pressure) on x (age of women), say, is a good fit for the given data. It would be futile to estimate the blood pressure of a woman whose age is 270 years ($x_0 = 270$). As another example, suppose the correlation coefficient between x (the dose of a drug in milligrams, ranging from 3 mg to 10 mg) and y (the number of days of relief from allergy) on a group of n patients is significant. It would be nonsensical to use the regression line to estimate the value of y for a given dose of, say, 50 mg. In fact such a high dose may result in the death of the patient.

4. We should not use the linear regression model to make predictions for y corresponding to far distant values of x. At a distant value of x, there may be drastic change in the pattern of the relationship between the variables, not exhibited in the current available data. Hence, the predicted values of y for far distant values of x may be extremely unreliable.

Example 7: The following table gives the age and blood pressure of 10 randomly selected women

Age (x)	56	42	36	47	49	42	60	72	63	55
Blood Pressure (y)	147	125	118	128	145	140	155	160	149	150

(a) Find the sample correlation coefficient r.

(b) Test the significance of r at $\alpha = 0.01$.

(c) Determine the Regression Line for the given bivariate data.

(d) Estimate the blood pressure of a woman whose age is 45 years. That is, $x_0 = 45$.

Solution:

Calculations for the Regression Line and the Correlation Coefficient				
x	y	x^2	y^2	xy
56	147	3136	21609	8232
42	125	1764	15625	5250
36	118	1296	13924	4248
47	128	2209	16384	6016
49	145	2401	21025	7105
42	140	1764	19600	5880
60	155	3600	24025	9300
72	160	5184	25600	11520
63	149	3969	22201	9387
55	150	3025	22500	8250
$\sum x = 522$	$\sum y = 1417$	$\sum x^2 = 28348$	$\sum y^2 = 202493$	$\sum xy = 75188$

Table 10.6

(a) The correlation coefficient $r = r(x, y)$ is:

$$r = \frac{n\sum xy - \sum x \sum y}{\sqrt{n\sum x^2 - \left(\sum x\right)^2}\sqrt{n\sum y^2 - \left(\sum y\right)^2}}$$

$$= \frac{10(75188) - (522)(1417)}{\sqrt{10(28348) - (522)^2}\sqrt{10(202493) - (1417)^2}} = \frac{12206}{(104.862)(130.841)} = 0.892$$

(b) We have $n = 10$ and Test Statistics $r = 0.892$.

$H_0 : \rho = 0$
$H_1 : \rho \neq 0$

From Table 7, the critical value of r for $n = 10$ and $\alpha = 0.01$ for two-tailed test is [0.765]. Since the computed value of test statistics $r = 0.892$ is greater than critical value $\left[r_{10}(0.01) = 0.765\right]$, it lies in the rejection region. Hence, we reject H_0 at 1% level of significance and conclude that the variables are correlated in the population.

(c) Regression Equation: $\hat{y} = mx + b$

Using formulas (10.10) and (10.11) respectively, we get

$$m = \frac{n\sum xy - \sum x \sum y}{n\sum x^2 - \left(\sum x\right)^2} = \frac{12206}{10996} = 1.110;\ b = \bar{y} - m \cdot \bar{x} = 141.7 - (1.110)(52.2) = 83.756$$

Hence, the equation of the regression line (line of best fit) is:

$$\hat{y} = 1.110x + 83.756$$

(d) Since the value of r is significant, we can use the regression equation obtained in part **(c)** to estimate y for any given x. Therefore, the most likely estimate of the blood pressure (y) of a woman whose age(x) is 45 years is obtained from substituting $x_0 = 45$ into $y = 1.110x + 83.756$, and is given by:

$$\hat{y}_{x=45} = 1.110(45) + 83.756 = 133.708 \approx 134$$

Important Remark:

It should be clearly understood that the regression equation obtained from the given sample data is an estimate of the true regression equation $y = Mx + B$ for the entire population from which the sample is drawn. A different independent sample from the given population will result in another estimate of $y = Mx + B$, which will possibly be different from $\hat{y} = 1.110x + 83.758$.

Note:

	Population	**Sample**
Equation of the regression line	$y = Mx + B$	$\hat{y} = mx + b$
Slope of the regression line	M	m
***y*-intercept of the regression line**	B	b

Table 10.7

Example 8: The weights of a calf taken at weekly intervals are:

Age (weeks)	1	2	3	4	5	6	7	8	9	10
Weight (lbs.)	52.5	58.7	65.0	70.2	75.4	81.1	87.2	95.5	102.2	108.4

(a) Fit a linear regression equation to the given bivariate data set by the principle of least squares.

(b) Obtain an estimate of the weight of the calf at the age of 13 weeks.

(c) Estimate the weights of the calf at ages 1, 2, 3, ..., 10 weeks respectively using the regression equation obtained in (a).

(d) Find the error $\left(e = y - \hat{y}\right)$ in each case, and verify that $\sum e = 0$.

Solution:

Calculations for the Regression Line			
x	y	x^2	xy
1	52.5	1	52.5
2	58.7	4	117.4
3	65	9	195
4	70.2	16	280.8
5	75.4	25	377
6	81.1	36	486.6
7	87.2	49	610.4
8	95.5	64	764
9	102.2	81	919.8
10	108.4	100	1084
$\sum x = 55$	$\sum y = 796.2$	$\sum x^2 = 385$	$\sum xy = 4887.5$

Table 10.8

(a) Regression Equation: $y = mx + b$

Using formulas (10.10) and (10.11) respectively, we get

$$m = \frac{n\sum xy - \sum x \sum y}{n\sum x^2 - \left(\sum x\right)^2} = \frac{5084}{825} = 6.16242 \approx 6.162;$$

$$b = \overline{y} - m \cdot \overline{x} = 79.62 - (6.162)(5.5) = 45.729.$$

Hence, the equation of the regression line (line of best fit) is: $\hat{y} = 6.162x + 45.729$

(b) An estimate of calf's weight at the age of 13 weeks is obtained on taking $x_0 = 13$ and is given by:

$$\hat{y}_{x=13} = (6.162)(13) + 45.729 = 125.835$$

(c) $\hat{y}_{x=1} = 6.162(1) + 45.729 = 51.891$, $\hat{y}_{x=2} = 6.162(2) + 45.729 = 58.053$, etc. The results for all given x values are summarized in the following table.

x	1	2	3	4	5	6	7	8	9	10
$\hat{y}$	51.891	58.053	64.215	70.377	76.539	82.701	88.863	95.025	101.187	107.349

Table 10.9

(d)

x	y	$\hat{y}$	$e = y - \hat{y}$
1	52.5	51.891	0.609
2	58.7	58.053	0.647
3	65	64.215	0.785
4	70.2	70.377	−0.177
5	75.4	76.539	−1.139
6	81.1	82.701	−1.601
7	87.2	88.863	−1.663
8	95.5	95.025	0.475
9	102.2	101.187	1.013
10	108.4	107.349	1.051
Sum			3.55271E-14

Table 10.10

$$\sum e = 3.5527 \times 10^{-14} \approx 0.$$

Note: We are using more decimal places for computing errors to show that the sum of the errors is not always exactly zero. The reason for this is due to the rounding of the value of slope and the *y*-intercept to three decimal places.

Try it yourself:

1. Find the equation of the regression line when $\bar{x} = 100$, $\bar{y} = 35$, $s_x = 5.7$, $s_y = 7.3$, and $r = -0.85$.

2. Predict the *y*-value when the equation of the regression line is $\hat{y} = -0.453x + 24.814$ given $x_0 = 8$.

3. Predict the *y*-value when the equation of the regression line is $\hat{y} = 1.364x - 54.028$ given $x_0 = 25.6$.

4. Given the following bivariate data, find the equation of the regression line, predict the *y*-value for each *x*, compute the residuals $(y_i - \hat{y}_i)$, and show that $\sum e = 0$.

x	55	65	75	95	105	115
y	60	47	35	20	18	19

5. Given the following bivariate data set, (a) find the equation of the regression line and (b) predict the value of *y* given $x_0 = 100$.

Age	23	41	49	50	53	54	56	57	58	60	61
% Body Fat	20.2	25.9	25.2	31.1	34.7	29.1	32.5	30.3	30.4	41.1	34.5

6. Given the following bivariate data set, (a) find the equation of the regression line and (b) predict the value of *y* given $x_0 = 14.2$.

Temp. difference	10.5	11.5	11.4	12.5	13.5	15	15.2	15.6	16.5	17
Daily gas consumption	70.1	83	82.7	80	75.5	83.2	81.1	86.5	99.4	86

10.4: Variation and Prediction Interval

A. Explained and Unexplained Variation

In the previous section we used the regression line to predict an individual value (point estimate), of y for a given x. However, the predicted value $\hat{y}$ may not be the true value of y that corresponds to the given value of $x = x_0$. (Note that $\hat{y}$ is only an estimate of the true "y" value. Its accuracy at best, is as good as our derived model from the collected data.) In this section we will construct the prediction interval, which is a confidence interval estimate for the true value of y using the point estimate $\hat{y}$. We will also briefly describe **Explained** and **Unexplained Variations**, and the **Coefficient of Determination**. For a more detailed analysis of these topics, refer to Section 10.7.

Objectives

After completing this section you will be able to understand:

- **A.** Explained and Unexplained Variation.
- **B.** Standard Error of Estimate.
- **C.** Prediction Interval.
- **D.** Coefficient of Determination $\left(r^2\right)$.

(Optional Content)

Definition: *The* ***residual sum of squares*** *or the* ***sum of squares due to error (SSE)*** *is given by*:

$$SSE = \sum_{i=1}^{n}\left(y_i - \hat{y}_i\right)^2 = \sum_{i=1}^{n} e_i^2 = s_{yy} - \frac{s_{xy}^2}{s_{xx}} \qquad (10.14)$$

Remark:

$s_{yy} = \sum_{i=1}^{n}\left(y_i - \bar{y}\right)^2$ is called the total variation in the y-values.

The total variation in the y-values can be split into two parts.

i) The **explained variation**, the variation that is explained by the regression line is the variability which is explained by the linear regression of y. It is also known as the **sum of squares** due to linear regression.

$$\text{Explained Variation} = \sum_{i=1}^{n}\left(\hat{y}_i - \bar{y}\right)^2$$

$$= \sum_{i=1}^{n}\left(y_i - \bar{y}\right)^2 - \sum_{i=1}^{n}\left(y_i - \hat{y}\right)^2 = s_{yy} - SSE = \frac{s_{xy}^2}{s_{xx}}$$

ii) The **un-explained variation**, the variation which is not explained by the linear regression model, due to other factors affecting y such as random factors beyond the human control and other characteristics not included in the study. It is the residual variation (also known as the **error sum of squares**).

$$\text{Un-explained Variation} = SSE = \sum_{i=1}^{n}\left(y_i - \hat{y}_i\right)^2 = \sum_{i=1}^{n} e_i^2$$

B. Standard Error of Estimate

The regression equations enable us to estimate (predict) the value of the dependent variable y for any given value of the independent variable x. The estimates so obtained are, however, not perfect. A measure of the precision or reliability of the estimates obtained from the fitted regression equation is provided by the **standard error of the estimate**. **Standard Error** is an expression analogous to standard deviation and gives us a measure of the scatter of the observations about the line of regression.

The standard error of estimate is a measure of the variability or scatter of the observed values around the regression line. Thus, for the regression line, the standard error of estimate, denoted by s_e, is given as:

$$s_e = \sqrt{\frac{1}{n-2}\sum_{i=1}^{n}\left(y_i - \hat{y}_i\right)^2} = \sqrt{\frac{1}{n-2}\sum_{i=1}^{n} e_i^2} \qquad (10.15)$$

We can also write the standard error of estimate in terms of the y-intercept and the slope of the estimated regression line.

$$s_e = \sqrt{\frac{\sum y^2 - b\sum y - m\sum xy}{n-2}} \qquad (10.16)$$

Example 9: Consider the following bivariate data set.

x	1	2	3	4	5	6	7	8	9	10
y	52.5	58.7	65.0	70.2	75.4	81.1	87.2	95.5	102.2	108.4

Obtain the standard error of estimate of y by using both the formulas (10.15) and (10.16), and compare the results.

Solution: **Using formula (10.15):**

From the given data, we get:

x	y	$\hat{y}$	e	e^2
1	52.5	51.889	0.611	0.373
2	58.7	58.051	0.649	0.421
3	65	64.213	0.787	0.619
4	70.2	70.375	−0.175	0.031
5	75.4	76.537	−1.137	1.293
6	81.1	82.699	−1.599	2.557
7	87.2	88.861	−1.661	2.759
8	95.5	95.023	0.477	0.228
9	102.2	101.185	1.015	1.03
10	108.4	107.347	1.053	1.109

Table 10.11

Hence, $\sum e^2 = 10.42$.

$$s_e = \sqrt{\frac{1}{n-2}\sum_{i=1}^{n} e_i^2} = \sqrt{\frac{10.42}{10-2}} = 1.141$$

Using formula (10.16):

From the given data, we get:

x	y	x^2	y^2	xy
1	52.5	1	2756.25	52.5
2	58.7	4	3445.69	117.4
3	65	9	4225	195
4	70.2	16	4928.04	280.8
5	75.4	25	5685.16	377
6	81.1	36	6577.21	486.6
7	87.2	49	7603.84	610.4
8	95.5	64	9120.25	764
9	102.2	81	10444.84	919.8
10	108.4	100	11750.56	1084

Table 10.12

$\sum x = 55$, $\sum y = 796.2$, $\sum y^2 = 66536.84$, $\sum xy = 4887.5$, $m = 6.1624$, and $b = 45.7267$

Note: We are using more decimal places for better precision.

$$s_e = \sqrt{\frac{\sum y^2 - b\sum y - m\sum xy}{n-2}} = \sqrt{\frac{66536.84 - 45.7267(796.2) - 6.1624(4887.5)}{10-2}} = \sqrt{\frac{10.511}{8}} = 1.146$$

This is approximately the same as the result from using formula (10.15).

C. Prediction Interval

We can use the regression equation $y = mx + b$ to estimate the value of $y = \hat{y}$ for any given value of $x = x_0 \cdot \hat{y}$ is the best **point estimate** of y based on the principle of least squares. However, we cannot say anything about the accuracy of this estimate. In this section we will obtain the confidence interval estimate for the true value of y using its point estimate $\hat{y}$ with desired confidence level $100(1 - \alpha)\%$. We call such confidence interval the **prediction interval** since it refers to a future or unknown observation.

Assuming that the y-values are normally distributed as $N(\hat{y}, s_e)$ about the regression line, the prediction interval for y, when $x = x_0$, with a confidence level $100(1 - \alpha)\%$ is:

$$\hat{y} - E < y < \hat{y} + E \qquad (10.17)$$

The margin of error E is given by:

$$E = t_{n-2,\,\alpha/2} \cdot s_e \cdot \sqrt{1 + \frac{1}{n} + \frac{n(x_0 - \bar{x})^2}{n\sum x^2 - \left(\sum x\right)^2}}$$

$$= t_{n-2,\,\alpha/2} \sqrt{\frac{\sum y^2 - b\sum y - m\sum xy}{n-2}} \sqrt{1 + \frac{1}{n} + \frac{n(x_0 - \bar{x})^2}{n\sum x^2 - \left(\sum x\right)^2}} \qquad (10.18)$$

s_e is the standard error of estimate of y, and $t_v(\alpha)$ is the right-tailed critical value of t-distribution for v degrees of freedom at an α significance level.

Example 10: Consider the bivariate data relating to the age and blood pressure of a woman.

Age (x)	56	42	36	47	49	42	60	72	63	55
Blood Pressure (y)	147	125	118	128	145	140	155	160	149	150

Predict the blood pressure of a woman whose age (x) is 52 years and construct the 95% prediction interval for the blood pressure.

Solution: From the given bivariate data we obtain the following results:

$n = 10,\ \sum x = 522,\ \sum y = 1417,\ \sum x^2 = 28348,\ \sum y^2 = 202493,\ \sum xy = 75188,$

$\bar{x} = 52.5,\ \bar{y} = 141.7,\ m = 1.110,$ and $b = 83.758$

Hence, the **regression equation** is:

$$\hat{y} = 1.110x + 83.758$$

The **best estimate** of the blood pressure (y) of a woman whose age (x) is 52 years is obtained by substituting $x = x_0 = 52$ in $\hat{y} = 1.110x + 83.758$.

$$\hat{y} = 1.110(52) + 83.758$$

Prediction Interval for y:

The critical value of t for 8 degrees of freedom and 95% confidence level is $t_8(0.025) = 2.306$.

The standard error of estimate is:

$$s_e = \sqrt{\frac{\sum y^2 - b\sum y - m\sum xy}{n-2}} = \sqrt{\frac{202493-(83.758)(1417)-(1.110)(75188)}{10-2}}$$

$$= \sqrt{\frac{349.234}{8}} = 6.607$$

The margin of error is:

$$E = t_8(0.025)\cdot s_e \cdot \sqrt{1+\frac{1}{n}+\frac{n(x_0-\bar{x})^2}{n\sum x^2-\left(\sum x\right)^2}}$$

$$= (2.306)\cdot(6.607)\cdot\sqrt{1+\frac{1}{10}+\frac{10(52-52.2)^2}{10(28348)-(522)^2}}$$

$$= 15.236\cdot\sqrt{1+0.1+0.0000364} = (15.236)(1.049) = 15.983$$

Therefore, the 95% prediction interval for the blood pressure of a 52-year-old woman is:

$$\hat{y} - E < y < \hat{y} + E$$

$$141.478 - 15.983 < y < 141.478 + 15.983$$

$$125.495 < y\ 157.461$$

Thus, we can conclude with a 95% confidence that the blood pressure of a 52-year-old woman will lie between 125.495 and 157.461.

Optional:

D. Coefficient of Determination $\left(r^2\right)$

The coefficient of correlation between two variables is the measure of the linear relationship between them and indicates the amount of variation of one variable which is accounted for by another variable. A more useful and readily comprehensible measure for this purpose is the **coefficient of determination** which gives the percentage variation in the dependent variable (y) that is accounted for by the independent variable (x). That is, the coefficient of determination gives the ratio of the variability which is explained by the linear regression of y to the total variability in y. The coefficient of determination is given by the square of the correlation coefficient r^2.

$$\textbf{Coefficient of Determination} = r^2 = \frac{\text{Explained Variation}}{\text{Total Variation}} = \frac{\sum(\hat{y}-\bar{y})^2}{\sum(y-\bar{y})^2} \qquad (10.19)$$

The coefficient of determination is a much more useful and better measure for interpreting the value of r. According to **Tuttle**:

> "*The coefficient of correlation has been grossly overrated and is used entirely too much. Its square, the coefficient of determination, is a much more useful measure of the linear covariation of two variables. The reader should develop the habit of squaring every correlation coefficient he finds cited or stated before coming to any final conclusion about the extent of the linear relationship between the two correlated variables.*"

For example, if the value of $r = 0.8$, we cannot conclude that 80% of the variation in y is due to variation in x. However, the coefficient of determination in this case is $r^2 = 0.64$, which implies that only 64% of the variation in y has been explained by x and the remaining 36% of the variation is due to other factors affecting y, but not included in the study.

By the same argument, while comparing two correlation coefficients, one of which is 0.4 and the other is 0.8, it is misleading to conclude that the correlation in the second case is twice as high as correlation in the first case. The coefficient of determination clearly explains this viewpoint, since in the case r is 0.4, the coefficient of determination is 0.16 and in the case r is 0.8, the coefficient of determination is 0.64. From this we conclude that the correlation in the second case is four times as high as the correlation in the first case.

The above discussion implies that:

> "The closeness of the relationship between two variables as determined by correlation coefficient r is not proportional."

Example 11: The following table gives the age and blood pressure of 10 randomly selected women.

Age(x)	56	42	36	47	49	42	60	72	63	55
Blood Pressure(y)	147	125	118	128	145	140	155	160	149	150

What percentage of the variation in blood pressure (y) is explained by the linear relationship between age and blood pressure?

Solution: The correlation coefficient r for the above bivariate data set is 0.892. Hence, the coefficient of determination $r^2 = (0.892)^2 = 0.796$. This means that 79.6% of the total variation in the blood pressure of women is explained by the variation in their age. The remaining 20.4% of the variation in blood pressure can be explained by factors other than age. Some of these factors may be weight, diet, working conditions, physical exercise, heredity or chance factors.

Try it yourself:

Use the Goodness of Fit Test to solve the following problems.

1. Given the following bivariate data set

x	90	50	85	70	60	80	75	110
y	15	23	15	13	19	18	19	10

compute the standard error of estimate using the formulas

(a) $s_e = \sqrt{\frac{1}{n-2}\sum_{i=1}^{n} e_i^2}$, and **(b)** $s_e = \sqrt{\frac{\sum y^2 - b\sum y - m\sum xy}{n-2}}$

2. Find the 95% prediction interval for the above data set when $x_0 = 100$.

3. Construct the 99% prediction interval for the following bivariate data set when $x_0 = 35$.

x	20	45	40	43	50	60	29
y	18	40	34	36	35	42	21

10.5: Correlation and Regression Analysis at a Glance

In this section, we present an example which includes all the topics discussed in this chapter except the optional ones.

- **Summary of the Procedure for Regression Analysis**

1. Determine if a Linear Correlation (linear relationship) exists between the variables by
   i) a Scatter Plot,
   ii) finding "r"(The sample Correlation Coefficient), and
   iii) a Hypothesis Testing Procedure.
2. Once a Linear Relationship is established, we then find the equation of the **Regression Line** ($y = mx + b$), also known as the **Line of Best Fit**.
3. We use the equation of the regression line (our linear model) and an x_0 value to compute (predict) $\hat{y}$.
4. Once a $\hat{y}$ value is found, we can proceed to calculate a **Prediction Interval** for the true value of y.

Example 12: A correlation analysis was performed by using the weekly earnings (in \$) of 15 college students in Atlanta University and the number of hours they worked per week.

Hours Worked (X)	35	37	37	34	31	27	28	28	33	26	28	33	32	39	36
Weekly Earning (Y)	159	162	168	160	159	159	162	162	166	152	161	167	166	165	169

(a) Find Karl Pearson's correlation coefficient.

(b) Draw the scatter plot.

(c) Using $\alpha = 0.05$ significance level to test the claim that the weekly earnings and hours worked by college students are correlated.

(d) Construct a 95% confidence interval for the population correlation coefficient ρ.

(e) Find the slope and y-intercept of the regression line.

(f) Fit a linear regression equation for the given bivariate data.

(g) Estimate the weekly earning of a student, who works 42 hours in a week.

(h) Estimate the weekly earning for all values of X.

(i) Find the error $(e = y - \hat{y})$ in each case, and verify $\sum e = 0$.

(j) Obtain the standard error of estimate of y, by using both the formulas (10.15) and (10.16).

(k) Find the 95% prediction interval from the above data set when $x_0 = 45$ hours.

(l) What percentage of the variation in weekly earning (y) is explained by the linear relationship between weekly earning and hours worked per week?

Solution: **(a)** We start with calculating the necessary sums for the correlation coefficient formula.

x	y	x^2	y^2	xy
35	159	1225	25281	5565
37	161	1369	25921	5957
37	167	1369	27889	6179
34	160	1156	25600	5440
31	158	961	24964	4898
27	158	729	24964	4266
28	161	784	25921	4508
28	161	784	25921	4508
33	166	1089	27556	5478
26	151	676	22801	3926
28	160	784	25600	4480
33	166	1089	27556	5478
32	165	1024	27225	5280
39	165	1521	27225	6435
36	168	1296	28224	6048
$\sum x = 484$	$\sum y = 2426$	$\sum x^2 = 15856$	$\sum y^2 = 392648$	$\sum xy = 78446$

Table 10.13

Given $n = 15$, Karl Pearson's correlation coefficient is

$$r = \frac{n\sum xy - \sum x \sum y}{\sqrt{n\sum x^2 - \left(\sum x\right)^2}\sqrt{n\sum y^2 - \left(\sum y\right)^2}}$$

$$= \frac{15(78446) - (484)(2426)}{\sqrt{15(15856) - (484)^2}\sqrt{15(392648) - (2426)^2}}$$

$$= \frac{2506}{\sqrt{(3584)(4244)}} = \frac{2506}{3900.064} = 0.642554 \approx 0.643$$

Hence, there seems to be a very strong positive correlation between the given bivariate data.

(b)

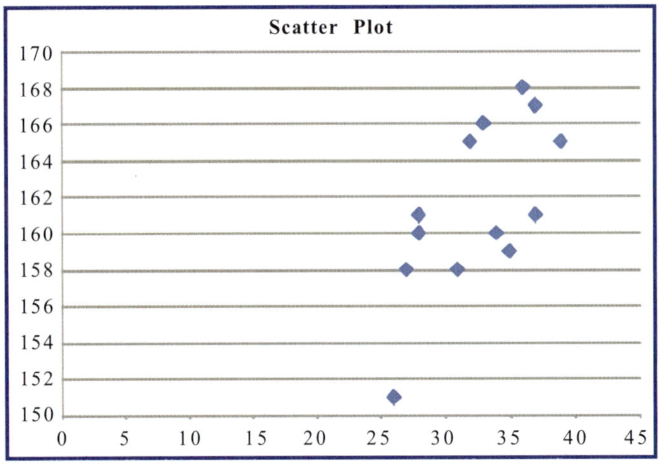

Figure 10.11

(c) We are given $n = 15$ and $r = 0.643$.

Null Hypothesis: $H_0 : \rho = 0$ (No linear correlation between the variables)

Alternative Hypothesis: $H_1 : \rho \neq 0$ (Strong linear correlation)

Method* 1: Test Statistics *t

Under H_0, the test statistics is

$$t = r\sqrt{\frac{n-2}{1-r^2}} \sim t_{n-2} = t_{13}$$

$$\Rightarrow t = 0.643\sqrt{\frac{15-2}{1-(0.643^2)}} = 3.0271$$

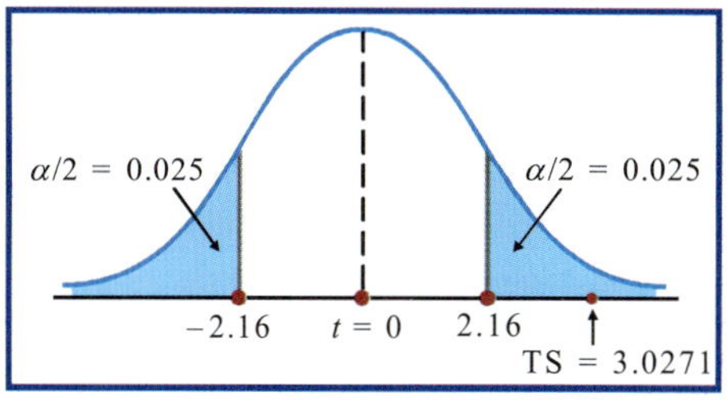

Figure 10.12

For a two-tailed test, using Table 4, the t-value with 13 df and a 0.05 significance level is 2.16. That is, $t_{13}(0.025) = 2.160368652$. The rejection regions, critical values, and the test statistics are shown in Figure 10.12.

Since the test statistics $t = 3.0271$ lies inside the rejection region, we reject H_0 at 5% level of significance. There is sufficient evidence to conclude that the given variables are linearly correlated in the population.

Method* 2: Test Statistics *r

From Table 8, the critical value of r at $\alpha = 0.05$ and for $n = 15$ is equal to 0.514. Since the value of test statistics $r = 0.643$ is greater than the critical value of r at $\alpha = 0.05$ and for $n = 15$, it lies in the rejection region. Hence $H_0 : \rho = 0$, is rejected at 5% level of significance and we can conclude that the variables are linearly correlated in the population.

(d) 95% confidence interval for ρ

$$\alpha = 0.05 \Rightarrow \frac{\alpha}{2} = 0.025 \Rightarrow t_{13}(0.025) = 2.16$$

$$r - t_{n-2}(\alpha/2)\left(\frac{1-r^2}{\sqrt{n}}\right) < \rho < r + t_{n-2}(\alpha/2)\left(\frac{1-r^2}{\sqrt{n}}\right)$$

$$0.643 - 2.16\left(\frac{1-0.643^2}{\sqrt{15}}\right) < \rho < 0.643 + 2.16\left(\frac{1-0.643^2}{\sqrt{15}}\right)$$

$$0.3158 < \rho < 0.9702$$

(e) We have, $n = 15$, $\sum x = 484$, $\sum y = 2426$, $\sum x^2 = 15856$, $\sum y^2 = 392648$ and $\sum xy = 78446$

Therefore, the slope of the regression line is

$$m = \frac{n\sum xy - \sum x \sum y}{n\sum x^2 - \left(\sum x\right)^2} = \frac{15(78446)-(484)(2426)}{15(15856)-(484)^2} = \frac{2506}{3584} = 0.699$$

and the y-intercept is

$$b = \frac{\sum y \sum x^2 - \sum x \sum xy}{n\sum x^2 - \left(\sum x\right)^2} = \frac{(2426)(15856)-(484)(78446)}{15(15856)-(484)^2} = \frac{498792}{3584} = 139.172$$

(f) The regression equation for the given sample bivariate data is $\hat{y} = 0.699x + 139.172$.

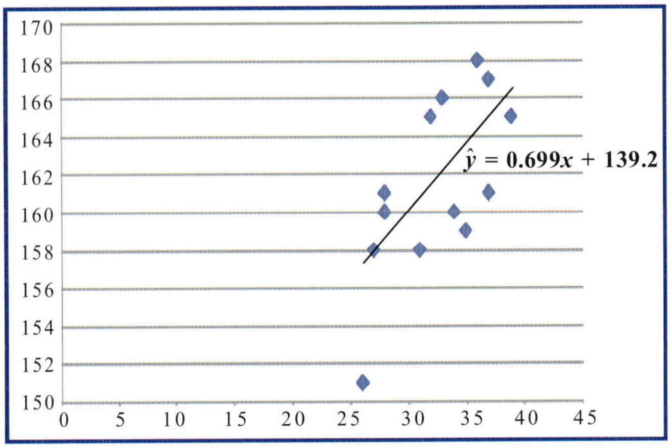

Figure 10.13

(g)

$$\hat{y} = 0.699x + 139.172$$

$$\hat{y}_{x=42} = (0.699)\cdot 42 + 139.172 = 168.53$$

(h)

x	y	$\hat{y}$
35	159	163.6445
37	161	165.043
37	167	165.043
34	160	162.9453
31	158	160.8477
27	158	158.0508
28	161	158.75
28	161	158.75
33	166	162.2461
26	151	157.3516
28	160	158.75
33	166	162.2461
32	165	161.5469
39	165	166.4414
36	168	164.3438

Table 10.14

(i)

x	y	$\hat{y}$	$e = y - \hat{y}$
35	159	163.6445	−4.64454
37	161	165.043	−4.04298
37	167	165.043	1.95702
34	160	162.9453	−.94532
31	158	160.8477	−2.84766
27	158	158.0508	−0.05079
28	161	158.75	2.24999

Table cont. next page

28	161	158.75	2.24999
33	166	162.2461	3.75390
26	151	157.3516	−6.35157
28	160	158.75	1.24999
33	166	162.2461	3.75390
32	165	161.5469	3.45312
39	165	166.4414	−1.44142
36	168	164.3438	3.65624
Total			−0.00013

Table 10.15

$$\sum e = -0.00013 \approx 0$$

(j) **Using formula (10.15):**

From the given data, we get:

x	y	$\hat{y}$	e	e^2
35	159	163.6445	−4.64454	21.57175
37	161	165.043	−4.04298	16.34567
37	167	165.043	1.95702	3.829935
34	160	162.9453	−2.94532	8.674916
31	158	160.8477	−2.84766	8.10919
27	158	158.0508	−0.05079	0.002579
28	161	158.75	2.24999	5.062469
28	161	158.75	2.24999	5.062469
33	166	162.2461	3.75390	14.09175
26	151	157.3516	−6.35157	40.34243
28	160	158.75	1.24999	1.562483
33	166	162.2461	3.75390	14.09175
32	165	161.5469	3.45312	11.92402
39	165	166.4414	−1.44142	2.07768
36	168	164.3438	3.65624	13.3681

Table 10.16

Hence, $\sum e^2 = 166.1171875$.

$$s_e = \sqrt{\frac{1}{n-2}\sum_{i=1}^{n} e_i^2} = \sqrt{\frac{166.1171875}{15-2}} = 3.575$$

Using formula (10.16):

From the given data, we get:

$n = 15$, $\sum y = 2426$, $\sum y^2 = 392648$, $\sum xy = 78446$, $m = 0.699219$ and $b = 139.1719$.

$$s_e = \sqrt{\frac{\sum y^2 - b\sum y - m\sum xy}{n-2}}$$

$$= \sqrt{\frac{392648 - 139.1719(2426) - 0.699219(78446)}{15-2}} = \sqrt{\frac{166.0369}{13}} = 3.575$$

This is approximately the same as the result from using formula (10.15).

(k) From the given bivariate data we obtain the following results:

$n = 15$, $\sum x = 484$, $\sum y = 2426$, $\sum x^2 = 15856$, $\sum y^2 = 392648$, $\sum xy = 78446$,

$x_0 = 45$, $\bar{x} = 32.267$, $\bar{y} = 161.733$, $m = 0.699219$ and $b = 139.1719$

Hence, the **regression equation** is

$$\hat{y} = 0.699219x + 139.1719$$

The **best estimate** of the weekly earning (y) of a student whose working hours/week is 45 hours is obtained by substituting $x = x_0 = 45$ into $y = 0.699219x + 139.1719$ and is:

$$\hat{y} = 0.699219(45) + 139.1719 = 170.637$$

Prediction Interval for *y*:

The critical value of t for 13 degrees of freedom and 95% confidence level is $t_{13}(0.025) = 2.16$.

The standard error of estimate is $s_e = 3.575$.

The margin of error is

$$E = t_{13}(0.025) \cdot s_e \cdot \sqrt{1 + \frac{1}{n} + \frac{n(x_0 - \bar{x})^2}{n\sum x^2 - \left(\sum x\right)^2}}$$

$$= (2.16) \cdot (3.575) \cdot \sqrt{1 + \frac{1}{15} + \frac{15(45 - 32.267)^2}{15(15856) - (484)^2}}$$

$$= 7.772 \cdot \sqrt{1 + 0.066667 + 0.678555} = (7.722)(1.321) = 10.20076 \approx 10.2.$$

Therefore, the 95% prediction interval for weekly earning of a student working 45 hours per week is:

$$\hat{y} - E < y < \hat{y} + E$$

$$170.638 - 10.202 < y < 170.638 + 10.202$$

$$160.435 < y\ 180.839$$

Thus, we can conclude with a 95% confidence that weekly earning of a student who works 45 hours in a week will lie between \$160.435 and \$180.839.

(l) The correlation coefficient r for the above bivariate data set is 0.643. Hence, the coefficient of determination $r^2 = (0.643)^2 = 0.413$. This means that 41.3% of the total variation in the weekly earning of students was explained by the variation in their working hours per week. The remaining 58.7% of the variation in the weekly earning can be explained by factors other than working hours per week. Some of these factors may be qualification, working experience, work quality, or chance factors.

10.6: Correlation and Regression Analysis with Microsoft Excel®

Objectives

After completing this section you will be able to carryout the computations in this chapter using Microsoft Excel®.

Illustration 1: Construct the scatter plot. Insert the regression line and its equation on the scatter plot.

x	5	20	18	26	14	30	28	19	23
y	10	25	30	30	20	38	31	23	26

Solution: **Step 1** Type the data into a worksheet. Then, select the data including the labels. (Figure 10.14 a)

Step 2 Click on the Scatter with Markers option in the Charts Group under the Insert Tab. (Figure 10.14 b)

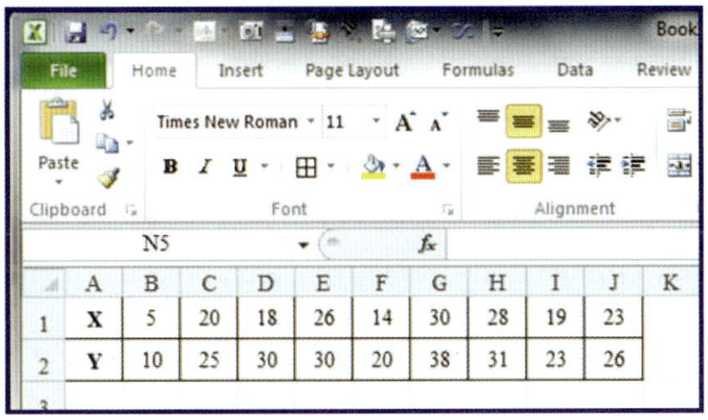

Figure 10.14 a

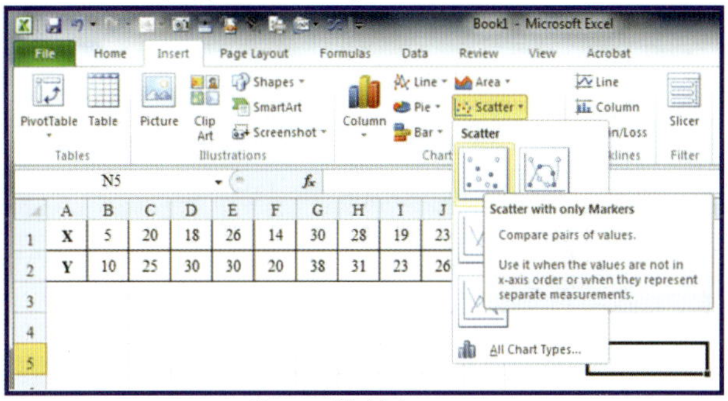

Figure 10.14 b

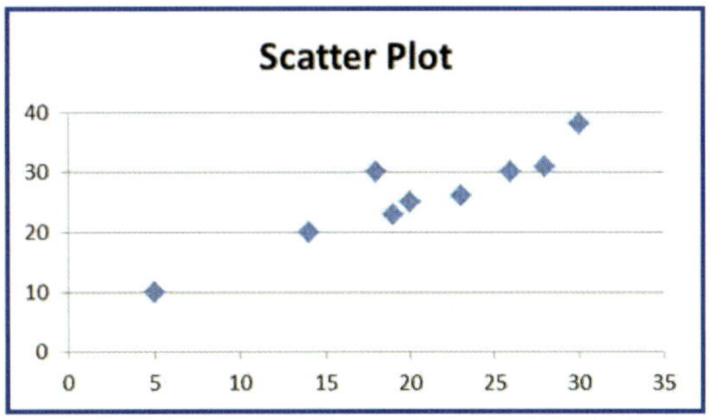

Figure 10.14 c

Right click on the any dot of the scatter plot, choose option **Add Trendline** and select options **Linear** and **Display Equation on chart**. Close the window. The following figure will Displayed.

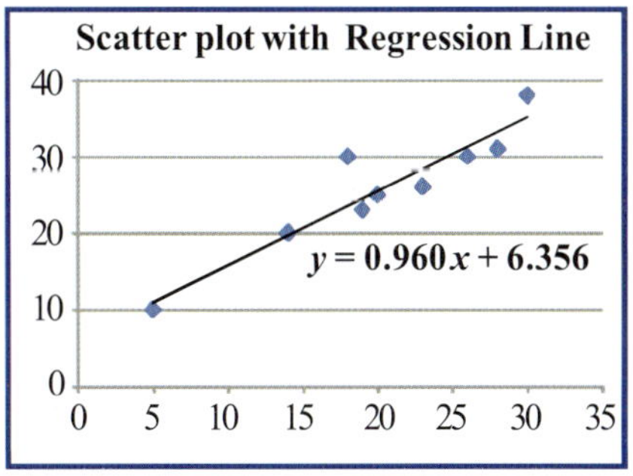

Figure 10.14 d

Illustration 2: Calculate the correlation coefficient for the following bivariate data.

x	5	20	18	26	14	30	28	19	23
y	10	25	30	30	20	38	31	23	26

Solution:

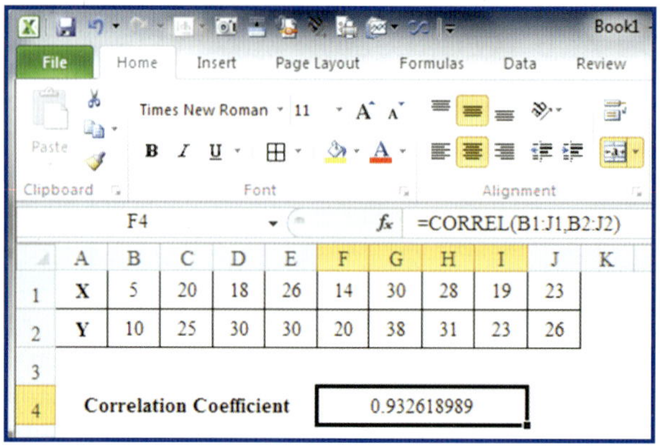

Figure 10.15 a

We can also calculate the correlation coefficient using the correlation command under the Data Analysis group.

***Step* 1** Click on the Data tab.

***Step* 2** Click on Data Analysis in the Analysis Group. If Data Analysis icon does not appear in your ribben, follow the steps to create it in Chapter "0".

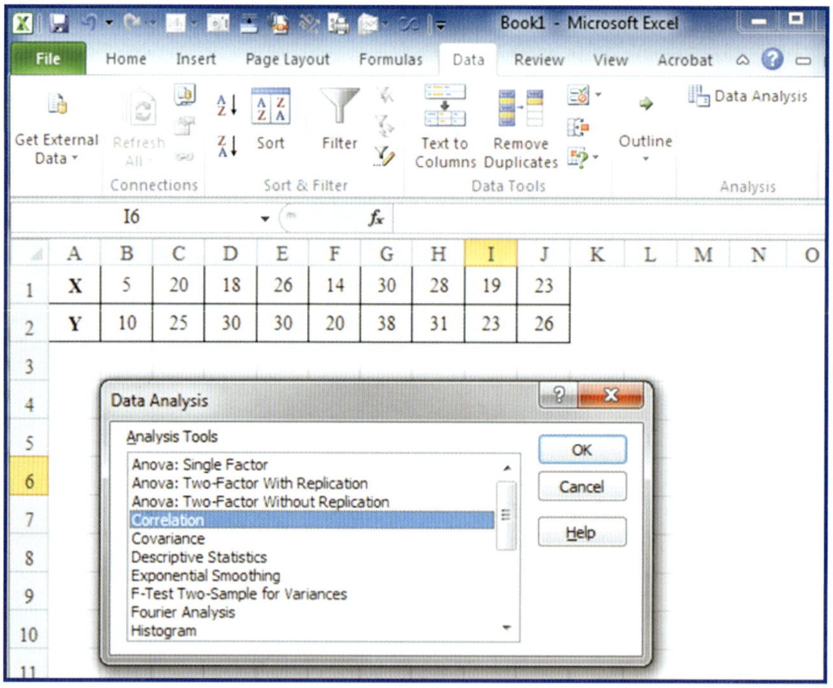

Figure 10.15 b

***Step* 3** Click on correlation in the Data Analysis dialog box and press OK. The following dialog box will appear.

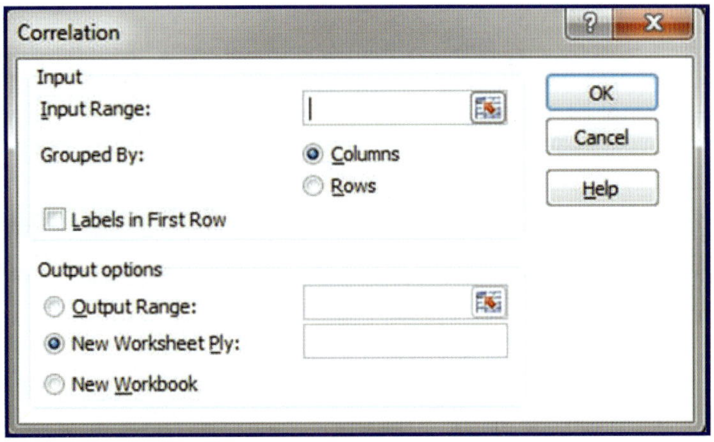

Figure 10.15 c

***Step* 4** In the Input Range box, select the given bivariate data with or without labels. Select Columns or Rows option depending if the bivariate data is inputted in rows or columns. Check the Labels in First Row, if you have selected the labels in the Input Range. Select New Workbook (Figure 10.16 d).

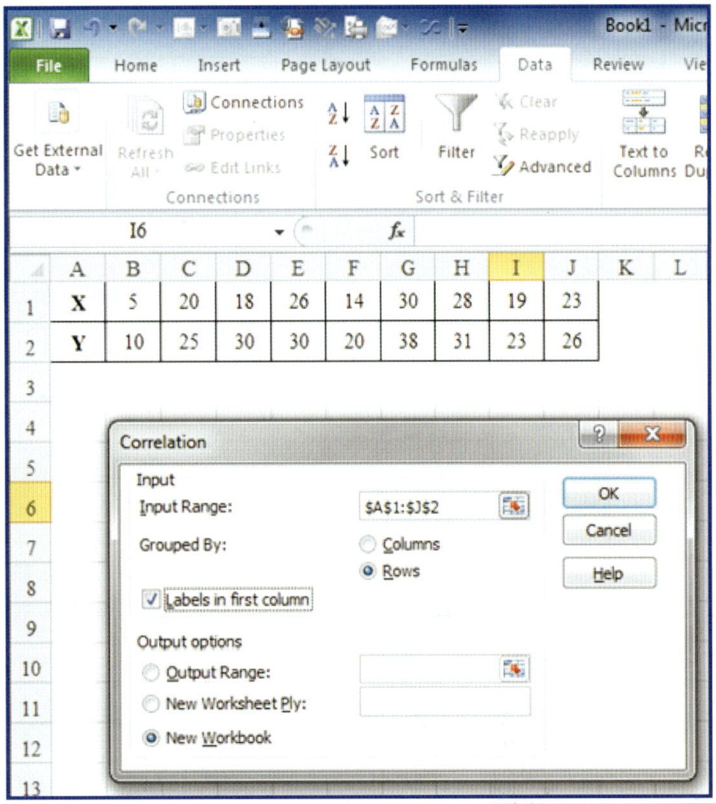

Figure 10.15 d

***Step* 5** Click on OK.

$r = 0.932619$

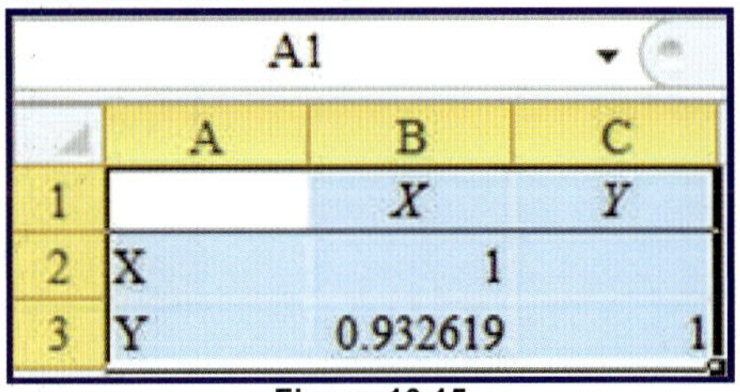

Figure 10.15 e

Illustration 3: Calculate the slope and the y-intercept of the regression line for the following bivariate data.

x	5	20	18	26	14	30	28	19	23
y	10	25	30	30	20	38	31	23	26

Solution:

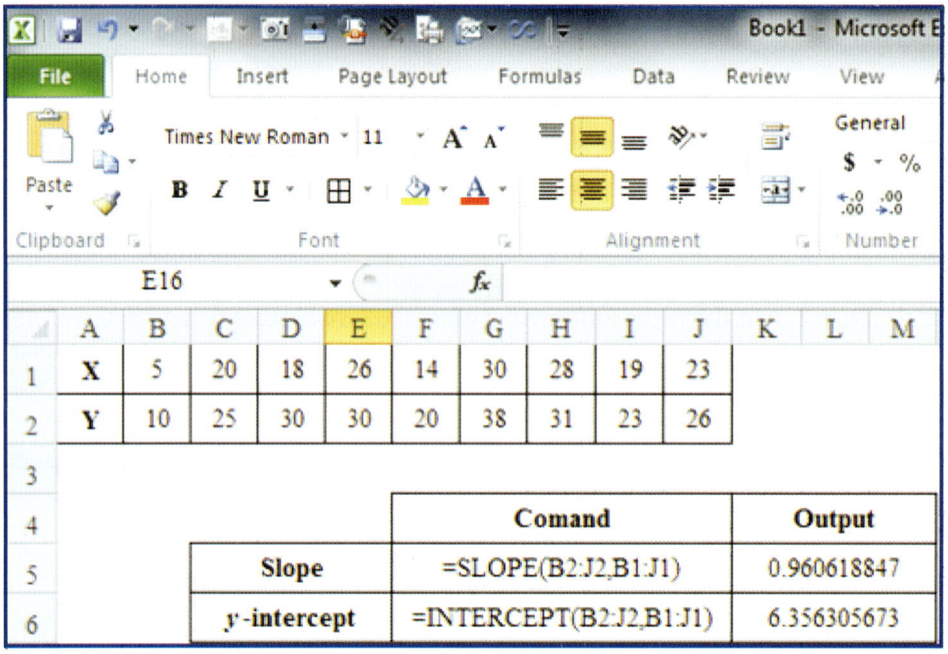

	A	B	C	D	E	F	G	H	I	J
1	X	5	20	18	26	14	30	28	19	23
2	Y	10	25	30	30	20	38	31	23	26

	Comand	Output
Slope	=SLOPE(B2:J2,B1:J1)	0.960618847
y-intercept	=INTERCEPT(B2:J2,B1:J1)	6.356305673

Figure 10.16

Illustration 4: Calculate $\hat{y}$ when $x_0 = 24$ given the following bivariate data.

x	5	20	18	26	14	30	28	19	23
y	10	25	30	30	20	38	31	23	26

Solution:

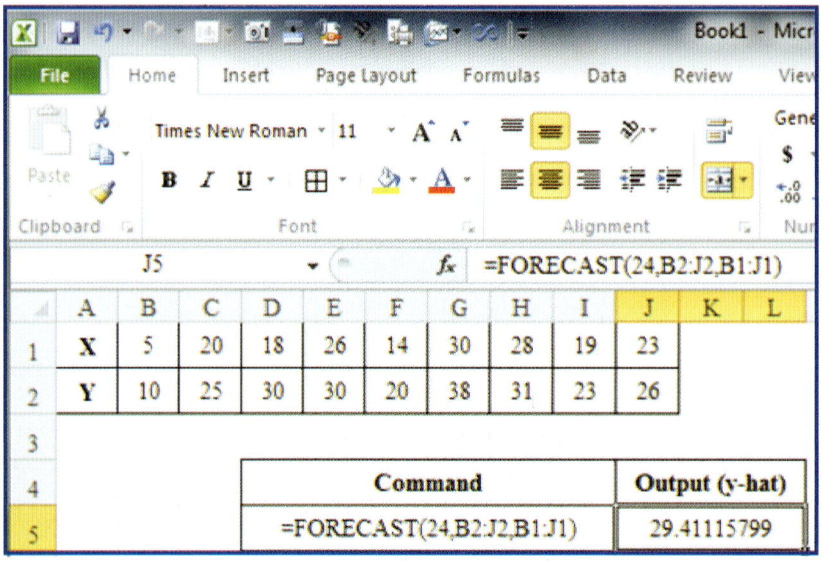

	A	B	C	D	E	F	G	H	I	J
1	X	5	20	18	26	14	30	28	19	23
2	Y	10	25	30	30	20	38	31	23	26

Command	Output (y-hat)
=FORECAST(24,B2:J2,B1:J1)	29.41115799

Figure 10.17

10.7: Extended Topics and Proofs for Correlation and Regression Analysis (Optional)

The study of correlation is easy and quite useful in physical sciences since, on the basis of experimental results, it is easy to establish mathematical relationship between two or more variables under study. However, in social and economic sciences, it becomes very difficult to establish mathematical relationship between the variables under study since in such phenomena the values of the variables under study are affected simultaneously by multiplicity of factors separately. Hence, in the data relating to social and economic phenomena, the study of correlation cannot be as accurate and precise.

Objectives

After completing this section you will be able to understand:

A. Correlation and Causation.

B. Properties of Sample Correlation Coefficient.

C. One-tailed Test for the Significance of Linear Correlation.

D. Regression Line.

E. Analysis of the Regression Equation.

A. Correlation and Causation

Correlation analysis enables us to have an idea about the degree and direction of the relationship between the two variables under study. However, it fails to reflect upon the cause and effect relationship between the variables. In a bivariate distribution, if the variables have the cause and effect relationship, they are bound to vary in sympathy with each other; therefore, there is bound to be a high degree of correlation between them. In other words, **causation always implies correlation**. However, the converse is not true; that is, even a fairly high degree of correlation between the two variables need not imply a cause and effect relationship between them. The high degree of correlation between the variables may be due to the following reasons.

1. Mutual dependence. The phenomena under study may inter-influence each other. Such situations are usually observed in data relating to economic and business situations. For instance, it is well known principle in economics that prices of a commodity are influenced by the forces of supply and demand. For instance, if the price of a commodity increases, its demand generally decreases, other factors remaining constant. Here increased price is the cause, and reduction in demand is the effect. However, a decrease in the demand of a commodity due to emigration of the people or due to fashion or some other factors like changes in the tastes and habits of people may result in decrease in its price. Here, the cause is the reduced demand, and the effect is the reduced price. Accordingly the two variables may show a good degree of correlation due to interaction of one on the other; yet it becomes very difficult to isolate the exact cause from the effect.

2. Both the variables being influenced by the same external factors. A high degree of correlation between the two variables may be due to the effect or interaction of a third variable or a number of variables on each of these two variables. For example a fairly high degree of correlation may be observed between the yield per hectare of two crops, say, rice and potato, due to the effect of a number of factors like favorable weather conditions, fertilizers used, irrigation facilities, etc. But none of these two variables is the cause of the other.

3. Pure chance. A small randomly selected sample from a bivariate distribution may show a fairly high degree of correlation although, in actuality, the variables may not be correlated in the population. Such correlation may be attributed to chance fluctuations. Moreover, the conscious or unconscious bias on the part of the investigator, in the selection of the sample, may also result in high degree of correlation in the sample. In this connection, it may be worthwhile to mention the two phenomena where a fairly high degree of correlation may be observed, even if it is not possible to conceive them as being causally related. For example we may observe a high degree of correlation between the size of shoes and the intelligence of a group of people. Such correlation is called **spurious or nonsense correlation**.

• Remarks about the Correlation Coefficient

1. Karl Pearson's coefficient is also known as **Product Moment Correlation Coefficient**. [The justification for this name is beyond the scope of the book.]

2. **Rounding (r):** The formulas for r are sensitive to rounding. As in other calculations, we should carry as many digits in the decimal places as is reasonable for our problem right from the first step to the last step. Rounding in the middle of the calculations will result in substantial errors. The answers given in the book may vary slightly, depending on the rounding process used. Since the critical values of r, given in Table 7 in the Appendix, are up to three decimal places, the value of r should be rounded to three decimal places to test its significance.

3. **Assumptions Underlying Karl Pearson's Correlation Coefficient:** Pearson's correlation coefficient r is based on the following assumptions.

   i) The variables x and y under study are linearly related. In other words, the scatter diagram of the bivariate data will give approximately a straight line.

   ii) The bivariate sample data (x_i, y_i), $i = 1, 2, \ldots, n$ is a random sample.

   iii) Each of the variables (series) is being affected by a large number of independent contributory causes of such a nature as to produce **normal distribution**. More rigorously, this assumption can be restated as follows: **The sample observations (x, y) have a joint bivariate normal distribution.**

   This assumption is not easy to verify. However, from practical point of view this implies that

   (a) for any given value of x, the conditional distribution of y is normal; and

   (b) for any given value of y, the conditional distribution of x is normal.

   For example, the variables (series) relating to ages, heights, weights supply, price etc. conform to this assumption.

   iv) The forces so operating on each side of the variable series are not independent of each other but are related in a causal fashion. In other words, cause and effect relationship exists between different forces operating on the items of the two variable series. These forces must be common to both series. If the operating forces are entirely independent of each other and not related in any fashion, then there cannot be any correlation between the variables under study. For example, the correlation coefficient between

   (a) the series of heights and income of individuals over a period of time,

   (b) the series of marriage rate and the rate of agricultural production in a country over a period of time and

   (c) the series relating to the size of the shoes and intelligence of a group of individuals, should each be zero since the forces affecting the two variable series in each of the above cases are entirely independent of each other. However, if in any of the above cases the value of r for a given set of data is not zero, then such correlation is termed as **chance correlation** or **spurious/nonsense correlation**.

B. Properties of Sample Correlation Coefficient

1. Limits for Correlation Coefficient

Pearson's correlation coefficient cannot exceed 1 numerically. In other words, it lies between –1 and +1. Symbolically, $|r| \leq 1 \quad \Rightarrow \quad -1 \leq r \leq +1$.

2. The value of r does not depend on the choice of the variable x and y.

Any one of the two variables can be taken as x and the other as y without affecting the value of r. Thus:

$$r(x, y) = r(y, x)$$

3. Correlation Coefficient is independent of the change of origin and scale.

Mathematically, if x and y are the given variables and they are transformed to the new variables u and v by the change of origin and scale, namely, $u = \dfrac{x-a}{h}$ and $v = \dfrac{y-b}{k}$ where a, b, h, and k are constants, $h > 0$, $k > 0$; then the correlation coefficient between x and y is the same as the correlation coefficient between u and v.

$$r(x, y) = r(u, v) \Rightarrow r_{xy} = r_{uv} = \frac{n\sum uv - \sum u \sum v}{\sqrt{n\sum u^2 - \left(\sum u\right)^2}\sqrt{n\sum v^2 - \left(\sum v\right)^2}}$$

4. Two independent variables are uncorrelated but the converse is not true.

If x and y are independent variables, then $r(x, y) = r_{xy} = 0$.

> ***Remark:***
> One should not be confused with the words **uncorrelation** and **independence**. When $r(x, y) = 0$, the uncorrelation between the variables simply implies the absence of any linear relationship between them. They may, however, be related in some other form (other than a straight line), such as, quadratic, logarithmic or trigonometric form.

5. If the variables x and y are connected by the linear equation $ax + by + c = 0$, then

$r(x, y) = +1$ if a and b are opposite signs.
$r(x, y) = -1$ if a and b are the same signs.

6. If x and y are two variables and a, b, c and d are any numbers so that $a \neq 0$, $c \neq 0$, then

$$r(ax + b, cy + d) = r(ax, cy) = \frac{ac}{|ac|} r(x, y)$$

$$= r(x, y) \text{ if } a \text{ and } c \text{ are the same signs.}$$

$$= -r(x, y) \text{ if } a \text{ and } c \text{ are opposite signs.}$$

C. One-tailed Test for the Significance of Linear Correlation

The probability density function of the sample correlation coefficient r, based on a sample size n, under the null hypothesis $H_0 : \rho = 0$ is given by:

$$p(r) = C\left(1 - r^2\right)^{\frac{n-4}{2}} \quad ; -1 \leq r \leq 1$$

Since the pdf of "r"is symmetric about $r = 0$, the single tailed tests for

$H_1 : \rho > 0$ (Right-tailed) or

$H_1 : \rho < 0$ (Left-tailed)

can be similarly developed on using the following result:

The significant (critical) values of r or (t) at level of significance α for a single-tailed test are the same as the critical values of r or (t) for two-tailed test at level of significance (2α).

D. Regression Line

Notation:

In this chapter we have used the letter b to represent the ***y*-intercept** of the regression equation and the letter m to represent the **slope**. In more advanced statistics books, the ***y*-intercept** is denoted by b_0 and the slope by b_1. Therefore, the regression equation is written as $y = b_0 + b_1 x$.

Notation for the Regression Equation		
	Population Parameter	**Sample Statistics**
Equation of the regression line	$y = \beta_0 + \beta_1 x$	$y = b_0 + b_1 x$
y-intercept of the regression line	β_0	b_0
Slope of the regression line	β_1	b_1

E. Analysis of the Regression Equation

Expected value of the slope for sample data is $E(\hat{\beta}_1) = E(\hat{m}) = m.$

Expected value of the y-intercept for sample data is $E(\hat{\beta}_0) = E(\hat{b}) = b.$

Variance of the slope for sample data is $V(\hat{m}) = V(\hat{\beta}_1) = \dfrac{\sigma^2}{s_{xx}}.$

Variance of the y-intercept for sample data is $V(\hat{b}) = V(\hat{\beta}_0) = \sigma^2\left[\dfrac{1}{n} + \dfrac{\bar{x}^2}{s_{xx}}\right].$

Confidence interval for the slope of the regression line (β_1) using the *t*-distribution is

$$P\left(-t_{n-2,\,\alpha/2} < \frac{\hat{\beta}_1 - \beta_1}{s_e / \sqrt{s_{xx}}} < t_{n-2,\,1-\alpha/2}\right) = 1 - \alpha.$$

Hence, a $100(1 - \alpha)\%$ confidence interval for β_1 is

$$\hat{\beta}_1 - t_{n-2,\,\alpha/2}\frac{s_e}{\sqrt{s_{xx}}} < \beta_1 < \hat{\beta}_1 + t_{n-2,\,\alpha/2}\frac{s_e}{\sqrt{s_{xx}}}.$$

Confidence interval for the *y*-intercept of the regression line (β_0) using the *t*-distribution

A $100(1 - \alpha)\%$ confidence interval for β_0 is

$$\hat{\beta}_0 - t_{n-2,\,\alpha/2}\sqrt{s_e^2\left(\frac{1}{n} + \frac{\bar{x}^2}{s_{xx}}\right)} < \beta_0 < \hat{\beta}_0 + t_{n-2,\,\alpha/2}\sqrt{s_e^2\left(\frac{1}{n} + \frac{\bar{x}^2}{s_{xx}}\right)}.$$

Ch 10: General Exercises

In exercises 1-5, draw the scatter diagram, and compute $r = r(x, y)$ for the given bivariate data sets. Interpret the results.

1.

x	15	23	31	14	20	12	20	18
y	39	49	67	35	44	25	41	42

2.

x	25	30	28	32	35	38
y	18	22	25	24	29	30

3.

x	35	37	32	34	30	38	40
y	4	5	6	5	10	3	2

4.

x	1	2	5	3	7	9	8	10	11
y	12	14	7	9	6	5	17	20	25

5.

x	8	1	3	6	12	7	10
y	13	30	23	35	24	18	15

In exercises 6-9, obtain the value of the correlation coefficient $r = r(x, y)$ for the information given.

6. $n = 7$, $\sum x = 41$, $\sum y = 14.4$, $\sum x^2 = 357$, $\sum y^2 = 32.7$, and $\sum xy = 74$.

7. $n = 8$, $\sum x = 609$, $\sum y = 504$, $\sum x^2 = 47995$, $\sum y^2 = 32954$, and $\sum xy = 39565$

8. $n = 9$, $\sum x = 183$, $\sum y = 257$, $\sum x^2 = 3855$, $\sum y^2 = 7437$, and $\sum xy = 5267$

9. $n = 15$, $\sum x = 2688$, $\sum y = 2426$, $\sum x^2 = 502672$, $\sum y^2 = 392648$, and $\sum xy = 434324$

Given the bivariate data sets in exercises 10-15, (a) draw the scatter diagram, (b) compute and interpret the sample correlation coefficient, (c) test the significance of the correlation coefficient at a 0.05 significance level using both methods outlined in the chapter, and (d) construct a 95% prediction interval for the population correlation coefficient ρ.

10.

x	0	2	4	6	8	10
y	12	7	9	4	3	2

11.

x	3	8	10	2	6	13
y	14	22	24	15	16	20

12.

x	8	1	3	6	12	7	10
y	13	30	23	35	24	18	15

13.

x	5	15	10	20	41	26
y	20	13	29	8	43	36

14.

x	20	16	15	18	12	10	8	12	5	14
y	12	10	14	16	9	10	7	10	3	13

15.

x	70	67	63	64	65	68	64	70	72	68	71	69	73
y	68	68	66	65	66	70	65	67	73	72	69	68	70

In exercises 16-20, find the equation of the regression line using the given bivariate data sets.

16.

x	50	20	30	60	15	40
y	75	30	45	80	35	60

17.

x	8	7	9.8	9	7.1	5	6.5	8.5
y	7	4.8	8.5	6.8	6	5.2	2.1	7.2

18.

x	15	10	12	9	5	14	7
y	8	23	18	27	39	11	33

19.

x	46	42	44	40	43	41	45
y	40	38	36	35	39	37	41

20.

x	70	67	63	64	65	68	64	70	72	68	71	69	73
y	68	68	66	65	66	70	65	67	73	72	69	68	70

In exercises 21-24, predict the y-value $(\hat{y})$ for the given ~~the~~ regression equation and the value of x_0.

21. $\hat{y} = 2.35x - 125.4$, $x_0 = 5.6$.

22. $\hat{y} = -12.5x - 18.63$, $x_0 = 4$.

23. $\hat{y} = -1.52x - 30.07$, $x_0 = 3.1$.

24. $\hat{y} = 6.34x - 9.55$, $x_0 = 2.3$.

In exercises 25-27, given the bivariate data sets, (a) find the equation of the regression line, (b) predict the y-value for each, (c) compute the residuals $(y_i - \hat{y}_i)$, and (d) show that $\sum e = 0$.

25.

x	14	19	24	21	26	22	15	20	19
y	31	36	48	37	50	45	33	41	39

26.

x	72	69	67	68	71	72	71	74	73	66	67
y	69	65	66	64	66	63	66	65	67	60	63

27.

x	10	8	13	9	11	14	6	4	12	7	5
y	9.1	8.1	8.7	8.8	9.3	8.1	6.1	3.1	9.1	7.3	4.7

In exercises 28-31, construct a prediction interval for the given bivariate data susing the indicated confidence level and x_0 value.

28. Confidence level: 95%, $x_0 = 18$

x	14	19	24	21	26	22	15	20	19
y	31	36	48	37	50	45	33	41	39

29. Confidence level: 99%, $x_0 = 48$

x	46	42	44	40	43	41	45
y	40	38	36	35	39	37	41

30. Confidence level: 98%, $x_0 = 150$

x	237	162	121	88	28	10	4
y	20	25	30	35	40	45	50

31. Confidence level: 95%, $x_0 = 100$

x	71	79	105	115	76	83	114	57	114	94	75	76
y	71	82	99	114	70	82	113	44	113	91	83	72

Comprehensive Exercises

32. A study was conducted to determine the relationship between a teacher's monthly income and the amount of money he or she spends on entertainment per month. The data from a random sample of 10 teachers is shown below.

x (\$)	3654	4860	3025	10400	8970	7800	9200	5300	4900	7800
y (\$)	310	400	200	780	600	568	648	600	392	600

(a) Draw a scatter diagram. Label the chart and the axis.

(b) Find the correlation coefficient.

(c) Test for linear correlation using $\alpha = 0.01$. (Use both methods outlined in this chapter.)

(d) Find the equation of the regression line.

(e) Predict the value of y if $x_0 = \$6500$.

(f) Construct a 99% prediction interval for y.

33. The following table represents the heights (in inches) of 11 randomly selected siblings.

Height of Brothers	72	69	67	68	71	72	71	74	73	66	67
Heights of Sisters	69	65	66	64	66	63	66	65	67	60	63

(a) Draw a scatter diagram. Label the chart and the axis.

(b) Find the correlation coefficient.

(c) Test for linear correlation using $\alpha = 0.05$. (Use both methods outlined in this chapter.)

(d) Find the equation of the regression line.

(e) Predict the value of y if $x_0 = 70$.

(f) Construct a 95% prediction interval for y.

34. The following table shows the amounts of advertising budgets and annual sales of 9 randomly selected companies.

Advertising (thousands \$)	45	52	46	48	50	47	53	49	51
Sales (millions \$)	9.5	11	9.7	9.8	10	9.7	11	9.9	10

(a) Draw a scatter diagram. Label the chart and the axis.

(b) Find the correlation coefficient.

(c) Test for linear correlation using $\alpha = 0.05$. (Use both methods outlined in this chapter.)

(d) Find the equation of the regression line.

(e) Predict the value of y if $x_0 = 66$.

(f) Construct a 95% prediction interval for y.

35. Consider the following bivariate data.

x	3	4	5	6	7	8	9	10
y	16	19	26	29	30	38	40	51

(a) Draw a scatter diagram. Label the chart and the axis.

(b) Find the correlation coefficient.

(c) Test for linear correlation using $\alpha = 0.05$. (Use both methods outlined in this chapter.)

(d) Find the equation of the regression line.

(e) Predict the value of y if $x_0 = 15$.

(f) Construct a 95% prediction interval for y.

36. Consider the following bivariate data set.

x	1	2	3	4	5	6	7	8	9	10	11
y	45487	45430	45741	46451	47322	48145	48813	49609	50502	51375	51968

(a) Draw a scatter diagram. Label the chart and the axis.

(b) Find the correlation coefficient.

(c) Test for linear correlation using $\alpha = 0.05$. (Use both methods outlined in this chapter.)

(d) Find the equation of the regression line.

(e) Predict the value of y if $x_0 = 15$.

(f) Construct a 95% prediction interval for y.

37. The table below gives the percent change from April 1, 1990, to April 1, 2000 of the residential population of the top 12 most populated states as provided by the U.S. Department of Commerce.

Rank	1	2	3	4	5	6	7	8	9	10	11	12
%	3.4	4.7	5.5	8.6	8.6	9.8	10.8	11.4	13.2	15	21	26

(a) Draw a scatter diagram. Label the chart and the axis.

(b) Find the correlation coefficient.

(c) Test for linear correlation using $\alpha = 0.01$. (Use both methods outlined in this chapter.)

(d) Find the equation of the regression line.

(e) Predict the value of y if $x_0 = 15$.

(f) Construct a 95% prediction interval for y.

38. The chart below summarizes the data provided by the U.S. Department of Labor on the unemployment rate versus the change in payroll employment for May-October 2002.

Unemployment Rate	5.8	5.9	5.9	5.7	5.6	5.7
Change in Payroll Employment	31	38	41	27	22	28

(a) Draw a scatter diagram. Label the chart and the axis.

(b) Find the correlation coefficient.

(c) Test for linear correlation using $\alpha = 0.05$. (Use both methods outlined in this chapter.)

(d) Find the equation of the regression line.

(e) Predict the value of y if $x_0 = 4.6$.

(f) Construct a 95% prediction interval for y.

39. The Federal Aviation Administration provided the data of the percentage of public and private airports with lit runways for the years1990 to 2001.

% Private	7	6.8	6.6	6.3	6.2	6.4	6.4	6.4	6.3	6.7	7.2	7.3
% Public	73	74	74	69	68	69	69	70	69	76	77	84

(a) Draw a scatter diagram. Label the chart and the axis.

(b) Find the correlation coefficient.

(c) Test for linear correlation using $\alpha = 0.01$. (Use both methods outlined in this chapter.)

(d) Find the equation of the regression line.

(e) Predict the value of y if $x_0 = 3.6$.

(f) Construct a 95% prediction interval for y.

40. A department store gives in-service training to its salesmen followed by a test. The manager is considering whether she should terminate the services of any salesman who does not do well on the test. The following data give the test scores and sales made by nine randomly selected salesmen during a certain period.

Test Scores	14	19	24	21	26	22	15	20	19
Sales in (thousands)	31	36	48	37	50	45	33	41	39

(a) Draw a scatter diagram. Label the chart and the axis.

(b) Find the correlation coefficient.

(c) Test for linear correlation using $\alpha = 0.01$. (Use both methods outlined in this chapter.)

(d) Find the equation of the regression line.

(e) Predict the value of y if $x_0 = 3.6$.

(f) Construct a 95% prediction interval for y.

APPENDICES

Appendix 1: A Brief Introduction to Microsoft Excel®
Appendix 2: Analysis of Variance
Appendix 3: Non-Parametric Statistics
Appendix 5: Formulas
Appendix 4: Statistical Tables

APPENDIX 1

A BRIEF INTRODUCTION TO MICROSOFT EXCEL®

Appendix Overview:

1. *Introduction*
2. *The Layout of Excel®*
3. *Introduction to Formulas in Excel®*
4. *Creating Graphs Using Excel®*
5. *The Analysis ToolPak*

A1.1: INTRODUCTION

Microsoft Excel is a spreadsheet application that features calculation and graphing tools. It has gone from being a very difficult tool to a very user-friendly application. This appendix will take you through Excel's general layout, formulas, graphing tools and Analysis ToolPak. Unless otherwise noted, the contents of this appendix have been limited to what is used directly with the textbook.

This tutorial has made the assumption that the reader has had very little experience with Excel, be it an old or new version; however, it does assume that the reader has had some prior experience with computers. For example, the reader should know how to open/save/print/close documents. The instructions in this appendix are tailored to Microsoft Excel 2013 operating on a Microsoft Windows operating system. If you are using a Macintosh operating system, the commands and locations of menus may be different than what is discussed in this appendix. In this event, consult the help menu in Excel on your Macintosh computer.

The use of Excel for the coursework in this text has been left to the discretion of the instructor. If your instructor chose to not utilize this appendix, it is still strongly encouraged that you learn this material. Although the concepts have been tailored to the statistics discipline, they are applicable across all fields.

A1.2: THE LAYOUT OF EXCEL®

Objectives

After completing this section, you will understand:

- **A.** the Quick Access Toolbar
- **B.** the workbook
- **C.** the Ribbon
- **D.** drag-and-drop editing

The layout of Excel has been through many changes since its inception in 1985. The most notable change occurred in Office 2007 with the creation of the Ribbon. This is part of the Microsoft Office Fluent user interface, and it replaced toolbars and menus with icons and pop-up description windows.

This section of the appendix will cover the most significant components of the layout of Excel (with respect to this course). We will begin by discussing the Quick Access Toolbar, workbook and Ribbon. We will end this section with a brief tutorial on how to use drag-and-drop editing via a detailed example.

A. The Quick Access Toolbar

The Quick Access Toolbar is a customizable toolbar which consists of commands that are independent of any of the tabs you are currently working in. It is located at the very top of Excel and is pictured below. The commands in the toolbar by default are save, undo, and redo. Directly to the right of the redo command is an icon that leads to a drop-down menu of other commands that you can add. A few examples of commands you can add to this toolbar are create new workbook, open workbook, sort ascending and sort descending.

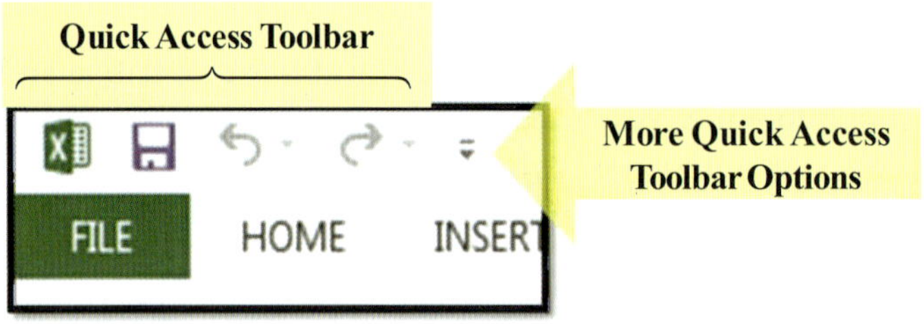

B. The Workbook

The workbook consists of all of the worksheets saved within the file. There is at least one worksheet in every workbook. A worksheet (or spreadsheet) consists of rows and columns ready for data entry. All of the data in an Excel workbook is saved in the worksheets. A screenshot of a worksheet is below.

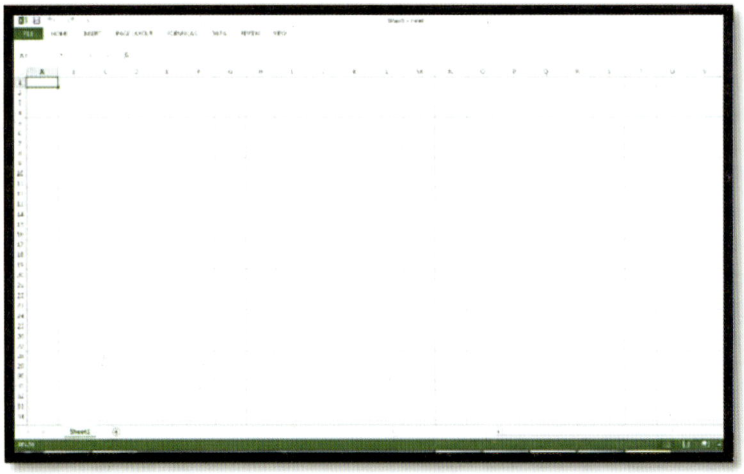

A workbook must have at least one worksheet. By default, a worksheet is automatically loaded with each new workbook. Some versions have more than one worksheet launched when Excel is opened. If yours only has one and you would to like to add more, simply click on the plus sign next to the worksheet name which is located at the bottom of the worksheet (pictured below).

The name of a worksheet is customizable. To change the name of a worksheet, simply double-click on the worksheet name. For example, if you would like to save a new worksheet as "STATS", double-click on the worksheet name "Sheet1" and start typing STATS. When finished typing STATS, press enter. To save the workbook, click on the floppy disk icon in the Quick Access Toolbar. Save the workbook as you would any other document. The screenshot below shows a workbook name of "Spring Semester" with worksheets "STATS", "ENG", and "CHEM."

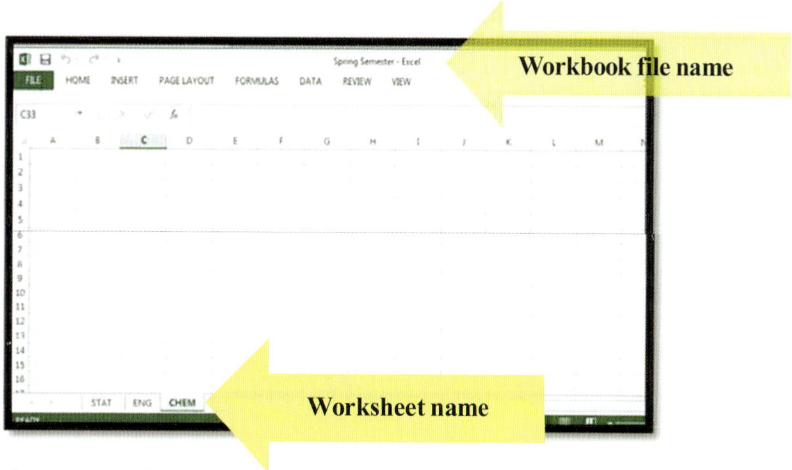

The File tab houses the new workbook, open, save, save as, print, add-ins and options commands. Notice that the Quick Access Toolbar already gave you accessibility to three of these commands.

C. The Ribbon

The next major section of the layout is the Ribbon. This is the most obvious difference between older and newer versions of Excel. The Ribbon replaced toolbars and menus with icons and pop-up menus, and cannot be deleted or replaced with toolbars or menus from previous versions. The Ribbon consists of tabs and groups that have been named according to their function, and are shown in the picture below.

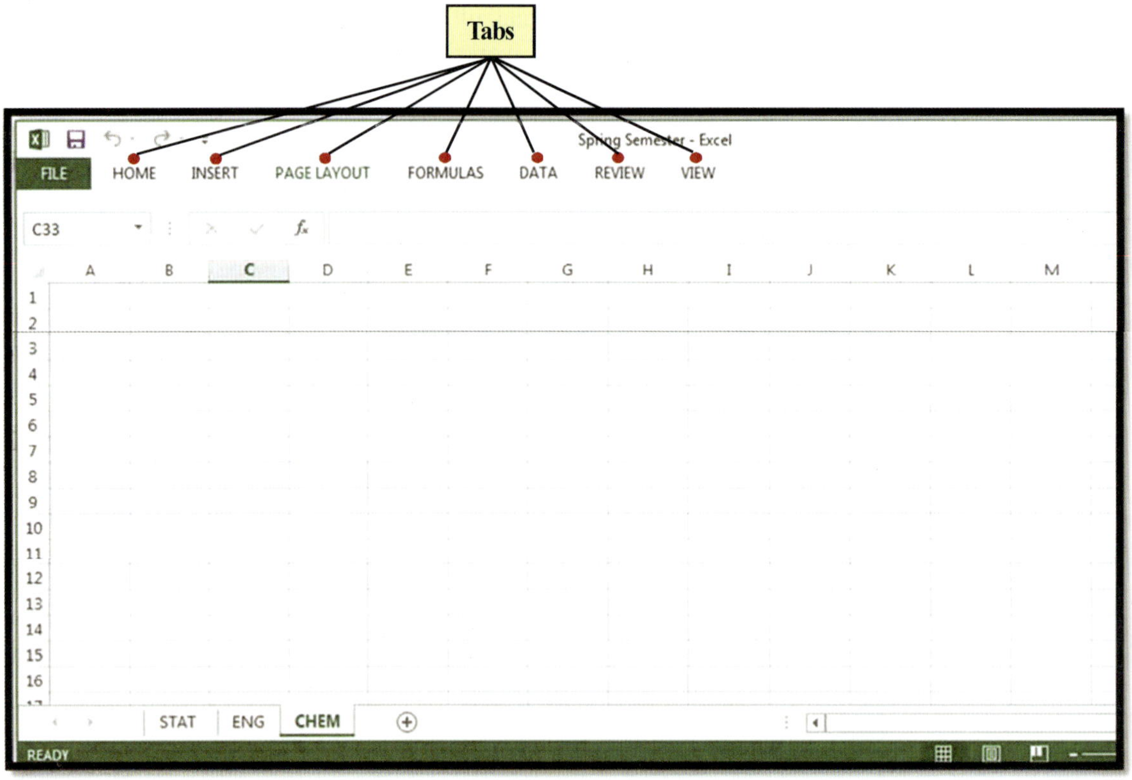

The contents of each tab are hidden by default. To view the contents of the tab, simply click on the tab whose contents you want to see. When you click on the tab, the content will appear and the formula bar/cell reference area will disappear. If you click the Home tab, your screen will look similar to the screenshot below. The Groups within the Home tab have also been mapped in this picture.

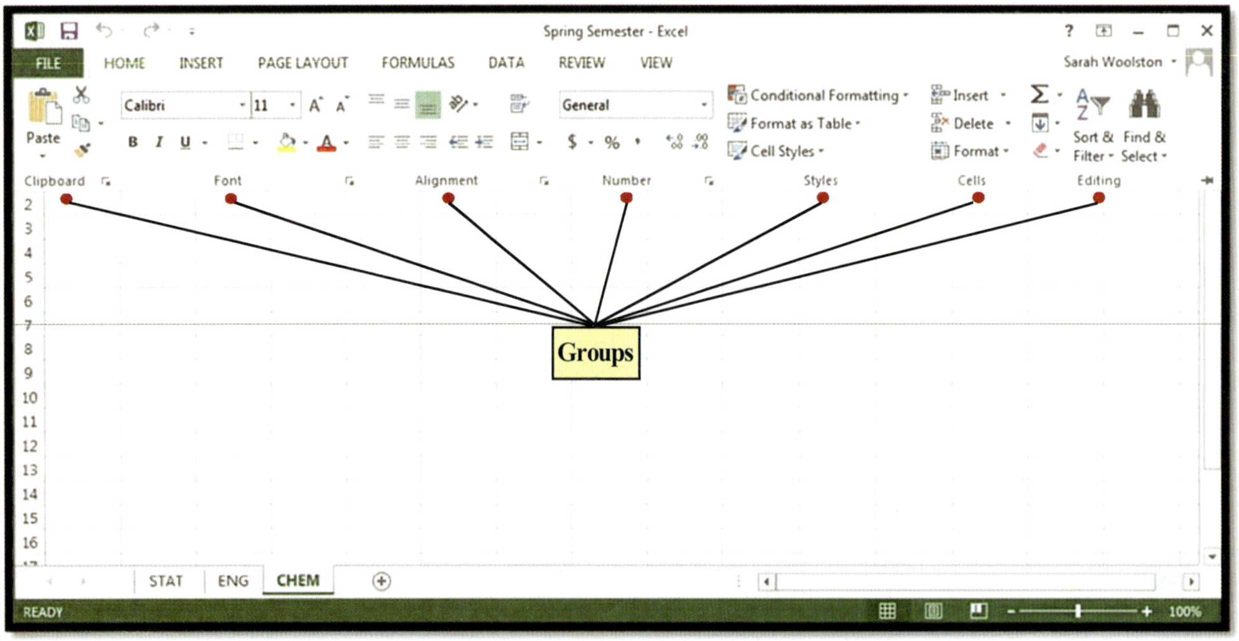

The contents within each tab are organized by groups. The groups are named based on the operations they perform. For example, the Font group contains commands that pertain specifically to the font in the cells of your worksheet. Likewise, the Alignment tab contains commands that pertain specifically to the alignment of the cells in your worksheet. If you would like to view other available commands within the group that do not have icons, click on the dialog launcher icon (pictured), and a dialog box containing all of the respective commands will appear. This is demonstrated below.

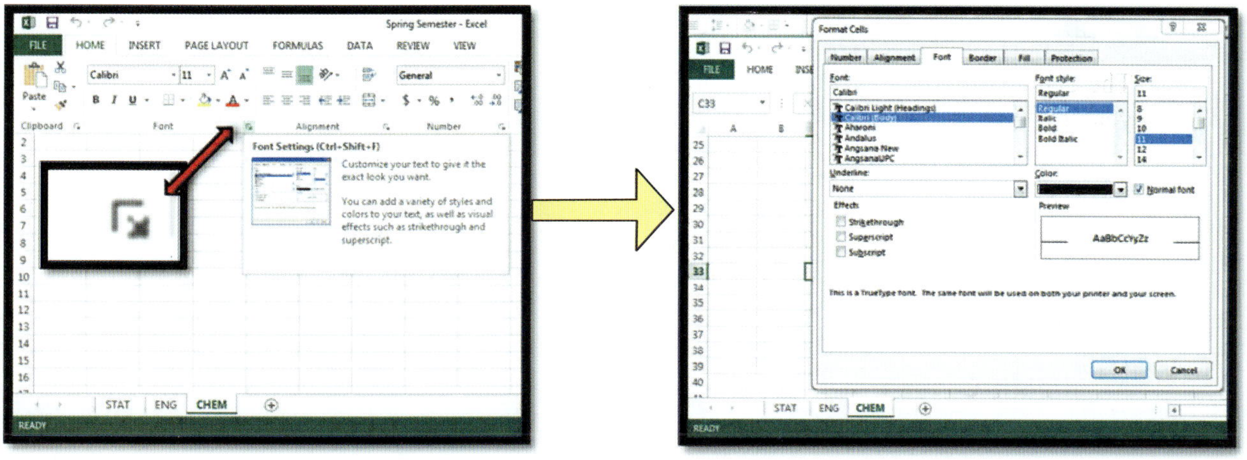

The dialog box layout is probably the most familiar amongst experienced Excel users. The look and design of this dialog box resembles the versions of Excel prior to 2007. If you have experience with Excel, but it was in versions prior to 2007, then the dialog launchers may be easier for you to work with.

In the picture above on the left, there is a pop-up window displayed that says "Font Settings." This pop-up menu is another benefit of the Ribbon. Each icon has a corresponding pop-up window that explains exactly what that feature does without actually having to click on the icon (which would consequently perform the operation or open the dialog box). To access this window, scroll your mouse over the icon, and the pop-up window that appears will explain what that feature will do to your cell/ worksheet.

The following table summarizes each tab and the groups contained within them. The tabs and groups listed are only those which are used directly with this course. A brief discussion of the tab groups follows after the table.

Tab Name	Groups
HOME	Clipboard, Font, Alignment, Numbers, Cells and Editing
INSERT	Charts and Illustrations.
PAGE LAYOUT	Page Setup and Scale to Fit.
FORMULAS	Function Library
DATA	Analysis, Sort & Filter
REVIEW	Proofing

In order to successfully use Excel in this course, you will need to familiarize yourself with each of these tabs and their groups. Each of the tab's groups are explained in detail below. To optimize your understanding of this section, have Excel open and follow along with text by looking for each tab and group as they are discussed.

Home Tab Groups

The Home tab houses the most commonly used commands across all Microsoft Office® applications. Their significance in this course is quite grand since it controls the overall presentation of your spreadsheet. If you have to turn-in assignments, you will want your fonts and alignments consistent. When working with numbers in cells, you will want the correct formatting of the numbers in the cell (decimal places, percent signs, etc.). More details for each of the groups are stated below.

Clipboard Group

The Clipboard group contains the commands for cut, copy and paste. You can also copy elements to your clipboard for later use. All of these features are also available via right-clicking on the cell.

Font Group

The Font group houses all of the commands for formatting the font of your document. In this group, you can change the color and size of your text. You can also choose borders and colors for the cells in your spreadsheet.

Alignment Group

The Alignment group commands will change the alignment of your cell(s). If you want the text in the center of the cell, click the Center icon. If you want text in one cell to be centered evenly over a table, select a row of cells above the table that encompasses the width of the table, and click the Merge & Center icon. If you want to type on multiple lines in the same cell, then click the Wrap Text icon.

Number Group

The Number group commands consist of various ways to format numbers in your cell(s). If you would like the numbers to have a percent sign (%), then click the % icon. If you would like the numbers to have a certain number of decimal places, then use the increase/decrease decimal icon to achieve the desired decimal places. If there is a specific format you want but do not see an icon for it, then click on the drop-down menu (which has General selected by default) or the dialog launcher in the Number group.

Cell Group

The Cell group contains commands for inserting cells, rows, columns and additional sheets. Previously, you learned that the shortcut for adding additional sheets to your workbook is found at the bottom of the worksheet (the plus sign). The same command can be found in this group. You can also add cells, rows and columns by right-clicking on the row/column of which you would like to insert the new row/column next to, and then select insert new row/column from the right-click drop-down menu.

Editing Group

The Editing group contains commands for clearing the contents of a cell, sorting/filtering and finding/selecting contents in formulas. You can clear the formatting or contents of a cell by selecting the Clear icon. To sort or filter data, select the Sort & Filter icon. If you select the icon, a drop-down menu will appear giving you options on how to sort your data. The Find & Select icon allows you to find or replace content through the entire worksheet. The AutoSum icon is also located here, and an example utilizing this AutoSum icon follows at the end of this section.

Insert Tab Groups

The Insert tab houses the commands for inserting charts and illustrations. These features will be used quite often in this course, especially inserting charts. The Charts group will probably be visited as often as the Home and Formulas tabs.

Charts Group

The Charts group houses the available charts in Excel. You can turn your data set into many different graphical representations. The types of graphical representations that are used most often in this course are column (bar) charts, histograms, pie charts and line graphs. Detailed instructions on how to insert these for a data set are in Section 1.4 of this appendix.

Illustrations Group

The Illustrations group contains the commands for inserting pictures and graphics into your spreadsheet. Illustrations will take your spreadsheet to a whole new level. If there is a particular picture stored on your computer that you want to add, click on the Pictures icon. The Browse window will appear. Find and select the picture you want to add, then click the Insert button. If you would like to add shapes, such as arrows or callout graphics (e.g. the colored arrows found earlier in this section), click on the Shapes icon.

Page Layout Tab Groups

The Page Layout groups that are most important for our purposes are the Page Setup and Scale to Fit groups. Although this tab will not be used as often as the Home, Insert and Formulas tabs, it is useful to know. You will probably only access this tab when you need to print your spreadsheet for an assignment.

Page Setup Group

The Page Setup group allows you to set margins and print areas. A margin is the border area of your document. You can control this by clicking on the Margins icon. After clicking the icon, you can choose a pre-programmed margin setting, or you can customize your margins by selecting the "Custom Margins..." option at the bottom of the drop-down menu." The Print Area icon allows you to choose which part of the document you would like to print. Sometimes printing the entire spreadsheet is unnecessary, and this option allows you to print only the essential parts of your spreadsheet. To do this, select the area you wish to print with your mouse, then click the Print Area icon and select "Set Print Area."

Scale to Fit Group

The Scale to Fit group allows you to fit your spreadsheet on a specified number of pages. Quite often in Excel, your spreadsheet will not fit on a single page, and your printed spreadsheet will span several pages. To avoid this, visit the Scale to Fit group. There are two drop-down menu choices here, namely "Height" and "Width." Click on each of the drop-down menus and select "1 page" to make your entire spreadsheet print on one page. By doing this, Excel re-sizes all of your rows and columns symmetrically to fit on one page.

Formulas Tab Groups

In this course, the Formulas tab groups will probably be the most frequently visited tab besides the Home tab. It has the Function Library group which contains all of the predefined formulas in Excel, along with the Insert Function tool. A more detailed discussion of how to use this tab and its formulas is in Section 1.3.

Function Library Group

The Function Library group has a variety of icons. To the very left is the Insert Function tool icon, and when this is clicked a dialog box appears. This dialog box allows you to search Excel's database of predefined functions. Simply type in a keyword or description of the formula you want, and Excel will yield results pertaining to your search. You may also find formulas without using the Insert Function tool by checking the categories of formulas labeled with graphical icons. For example, if you are looking for the formula of a math function, click the Math & Trig icon. A drop-down menu will appear, and it will list several math functions. If you still cannot find it, search for it using the Insert Function tool.

Data Tab Groups

The Data tab will only be used in this course when you need to sort data or use the Analysis ToolPak. The Analysis ToolPak is an add-in for Excel which provides additional formulas and charts that are specific to statistics. A more detailed discussion of the Analysis ToolPak follows in Appendix 1.5.

Analysis Group

The Analysis group will contain the icon for the Analysis ToolPak. This add-in does not come pre-loaded in Excel, so you will have to load it. Detailed instructions on how to load this add-in is in Appendix 1.5 Once the Analysis ToolPak has been loaded, the icon for it will be housed in the Analysis group.

Sort & Filter Group

You may notice that the command for Sort & Filter lies in both the Home and the Data tabs. The only difference is the Sort from Smallest to Largest/Sort from Largest to Smallest icons in the Data tab are listed in the drop-down menu of the Sort & Filter icon in the Home tab. Regardless of which tab this command is in, it operates exactly the same.

Review Tab Groups

The only reason you will probably use the Review tab in this course is to check for spelling and grammar mistakes. There are many other commands in this tab, such as password protecting worksheets, translating words, or adding comments to spreadsheets, but they will not be discussed in this appendix.

Proofing Group

The Proofing group contains the commands for checking spelling and grammar mistakes in your spreadsheet. To use it, click on the Spelling icon. A dialog box will appear with the mistakes Excel has found. If the word is truly misspelled, click the correct spelling of the word from the selection labeled "Suggestions." If Excel does not recognize the word but it is spelled correctly, then click "Ignore Once". If you are using that same word throughout your entire spreadsheet, then click "Ignore All." If your workbook contains multiple sheets, then you must check for spelling errors in each worksheet individually. Just repeat the spell-check process for each worksheet.

D. Drag-and-Drop Editing

The last topic to be covered in this section is drag-and-drop editing. This editing technique makes copying formulas to additional cells easy and fast. To use drag-and-drop editing, select the cell you wish to copy and place your cursor in the lower right corner of the selected cell. A plus sign will appear. Once the plus sign appears, left-click the mouse and drag the cursor over the cells you want the formula copied to. An example of this technique is below.

Example 1: Type the following data set into two rows (as shown) of an Excel worksheet and sum the rows individually.

	A	B	C	D	E	F	G	H
1	1	2	3	4	5	6	7	
2	8	9	10	11	12	13	14	
3	15	16	17	18	19	20	21	
4								

Solution :

Steps 1. After copying the distribution into Excel, select cell H1 (see image 1).

2. In the Home tab, find the Editing group and click the AutoSum icon (see image 2).

   If a drop-down menu appears, click the Σ Sum option. You will notice that when you use the AutoSum shortcut, Excel will predict which cells you want to sum by highlighting a range of cells (see image 3). If the range Excel selected was from A1:G1, press enter. If not, adjust your cell range.

3. Select cell H1 by left-clicking the cell. Move your cursor to the bottom right corner of the cell. Your cursor should look like a plus sign. Once it looks like a plus sign, left-click and drag over cells H2 and H3 (see image 4). Once you have dragged your cursor cells H2 and H3, let go out of mouse.

4. The formula from cell H1 should have been copied to cells H2 and H3. The numerical values will be displayed in the cell, but the formula will be displayed in the formula bar (see image 5).

1.

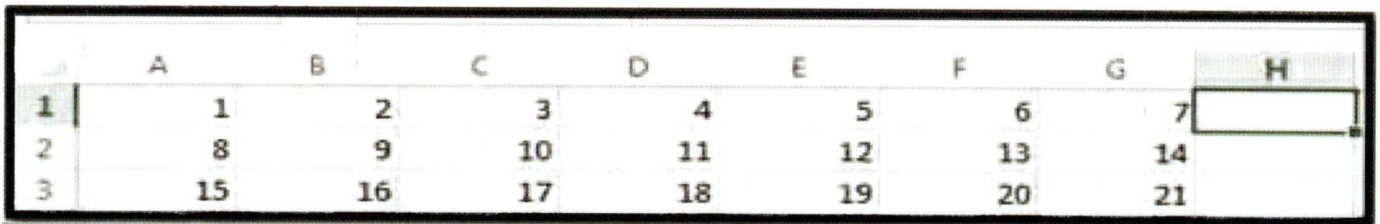

	A	B	C	D	E	F	G	H
1	1	2	3	4	5	6	7	
2	8	9	10	11	12	13	14	
3	15	16	17	18	19	20	21	

2.

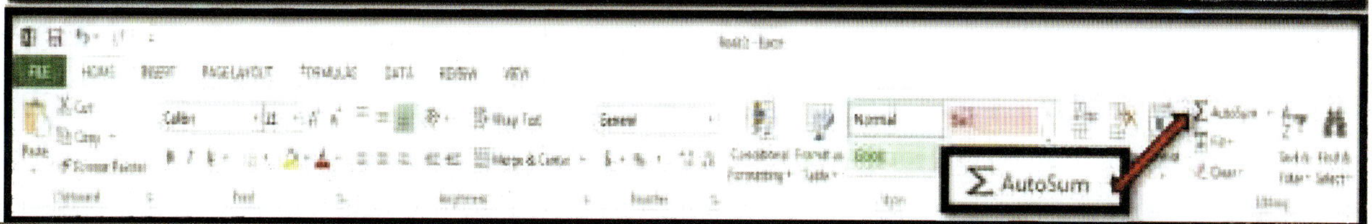

3.

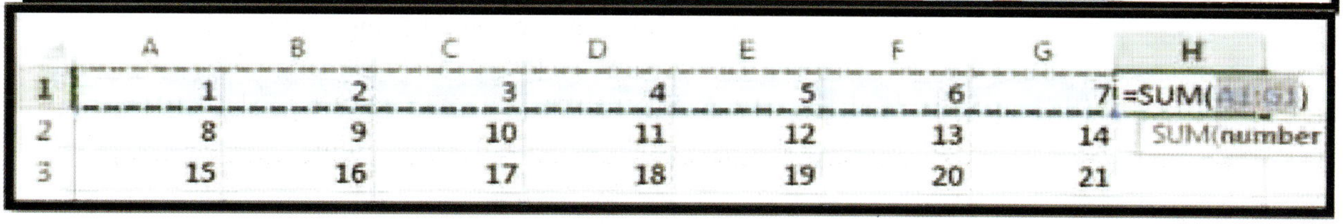

	A	B	C	D	E	F	G	H
1	1	2	3	4	5	6	7	=SUM(A1:G1)
2	8	9	10	11	12	13	14	SUM(number
3	15	16	17	18	19	20	21	

4.

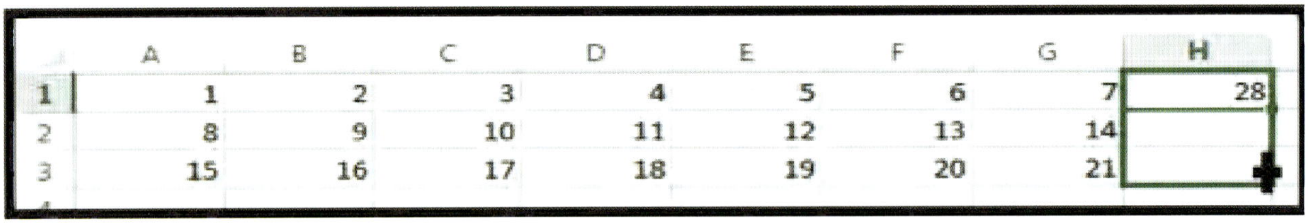

5.

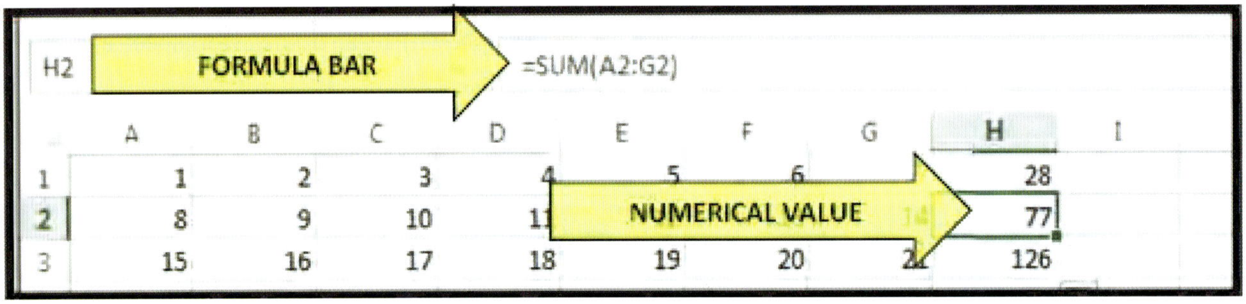

Note: The AutoSum icon is a shortcut for the SUM formula in Excel.

A1.3: Introduction to Formulas in Excel®

Objectives

After completing this section, you will understand:

- **A.** the components of a formula
- **B.** formulas of common functions
- **C.** the Insert Function tool
- **D.** basic formula creation

Formulas are equations that perform calculations or operations on values in a worksheet. A formula can consist of any or all of the following components: functions, references, operators and constants. Each of these components is essential to understanding and utilizing the predefined functions in Excel. The first part of this section is dedicated to these concepts. The rest of this section will discuss the order of operations that Excel follows when performing calculations. We will use this along with our knowledge of formulas to create customized formulas in Excel.

A. The Components of a Formula

Functions

A function is a predefined equation that performs calculations on values in worksheets, known as **arguments**. These arguments may be numbers, cell locations or an array of numbers or cell locations. In Excel, an array is defined as a collection of items. The general structure of a function is demonstrated in the figure below :

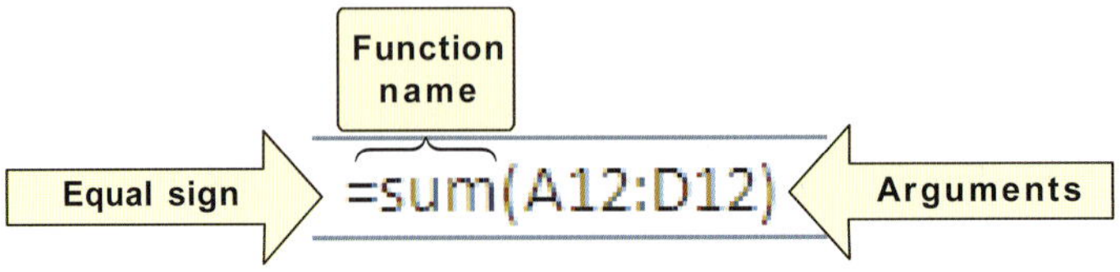

All functions in Excel follow this format. There is an equal sign and a function name with arguments in parentheses. If an equal sign is not typed at the beginning of the formula, then no calculation will be performed. The *range* operator (colon) will be explained in detail later in this section. In Excel, we often use cell locations as arguments in place of numerical values to help relieve the stress of iterative calculations. When we use a cell location in a formula, it allows us to change the value in the referenced cell without having to perform a new calculation.

References

A reference identifies a cell or a selection of cells in a worksheet. Furthermore, it instructs Excel which values to use in a formula. When using references, you can use multiple cells in a single formula, or use one cell in multiple formulas. You can also use one cell from a worksheet in other cells or formulas throughout the entire workbook. In the diagram above, our arguments were the cell references A12 through D12. This formula told Excel to calculate the sum of the 12th row from column A to column D.

Operators

Operators are used to specify the type of calculation you want to perform on the cells in a formula. The four types of operators in Excel are arithmetic, comparison, text and reference. We will not be discussing the text operator (&), since it is not used in this course.

Arithmetic

Arithmetic operators are used to perform basic mathematical operations. In Excel, this terminology is used to denote the following operations:

+	Addition
-	Subtraction
*	Multiplication
/	Division
%	Percent
^	Exponentiation

These symbols can be found either on the numeric keypad, the number row, or both places. Some characters share the same key. If the operator you need shares the same location as another character, and it is above the other character, then press and hold the shift key first before pressing that key to produce the symbol. If it does not lie above the other character, then just press the key. The following diagram below maps where the number row, numeric keypad and shift keys lie on the keyboard.

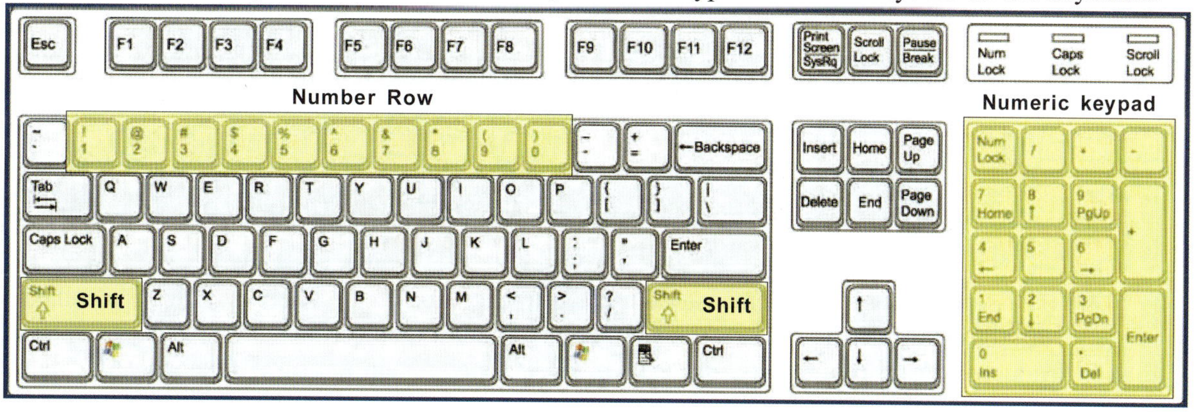

Comparison

In mathematics, operators are used to define a relationship between two entities. In Excel, these same operators are called comparison operators. They are as follows:

=	Equal to (equal sign)
>	Greater than
<	Less than
>=	Greater than or equal to
<=	Less than or equal to
< >	Not equal to

The *equal to* operator is located on the same key as the *addition* operator in the number row. The *greater than* and *lesser than* operators are located towards the bottom of the keyboard, next to the question mark key. These operators do shares the same keys as other characters, so be sure to press the shift key before pressing those keys. The following is another diagram showing where these can be found.

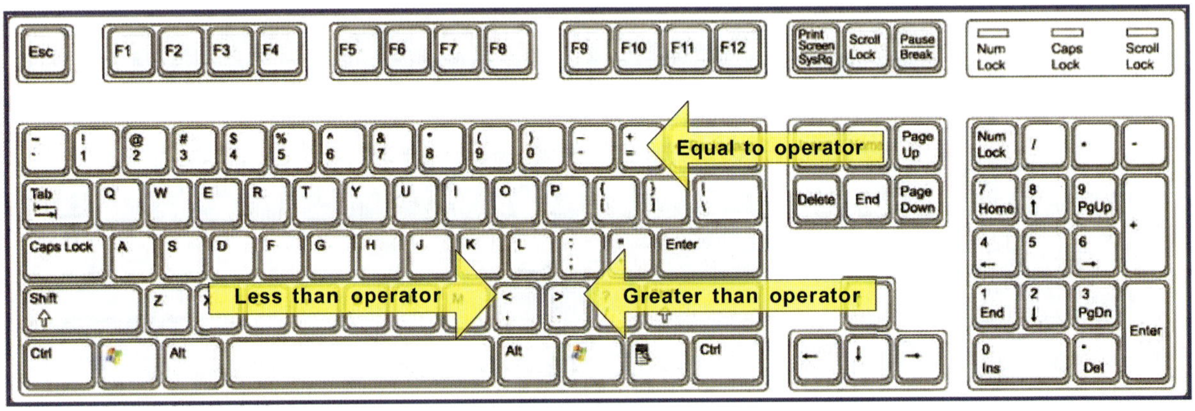

Reference

Reference operators combine ranges of cells for formulas. There are only three, and they are defined below:

:	*The range operator* requires two reference cells. The chosen formula will be performed on all the cells that lie in between the range, including the two reference cells. Ex: [STDEV(A1:F1)].
,	*The union operator* combines multiple reference cells into one formula. Ex: [AVERAGE(C6:D6,H7:K7)]. This will perform the calculation (C6+D6+H7+I7+J7+K7)/6.
(SPACE)	*The intersection operator* performs the calculation only on the cells that each data set shares. Ex: [SUM(A2:D4 B1:B10) – only values from B2 and B3 will be summed.

The screenshots below demonstrate these reference operators. The first row of pictures demonstrates the formulas and the cells referenced in the formulas. The second row of pictures shows the numeric value obtained from using a different reference operator on the same function and data set.

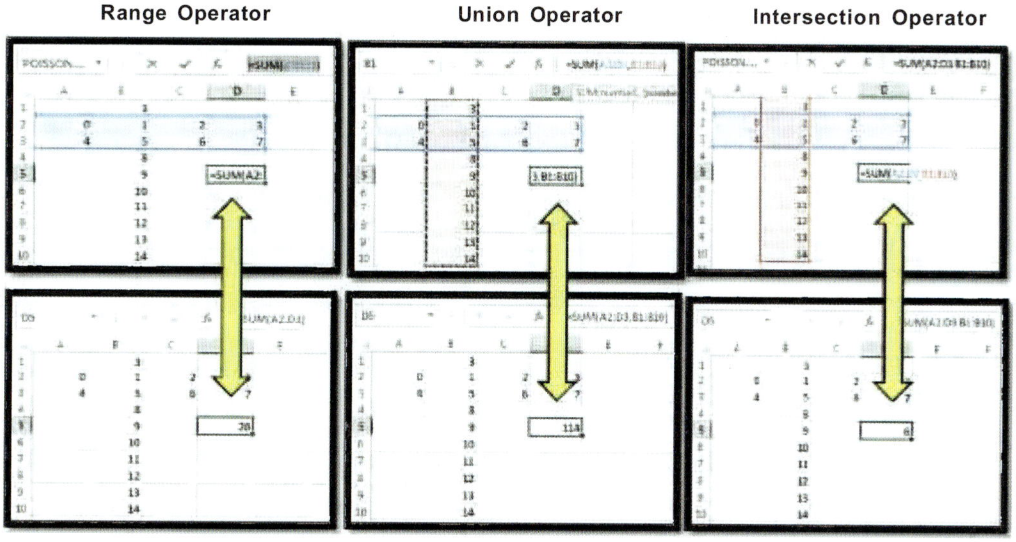

Each of these operators have only one location on the keyboard. The *range* operator is located on the same key as the *less than* operator; however, you do not need to press the shift button to produce it. Notice the symbol is located beneath the *less than* symbol. The *union* operator is located on the same key as the semicolon, but the shift key must be pressed prior to pressing the semicolon key to produce the *range* operator. The diagram below maps their locations.

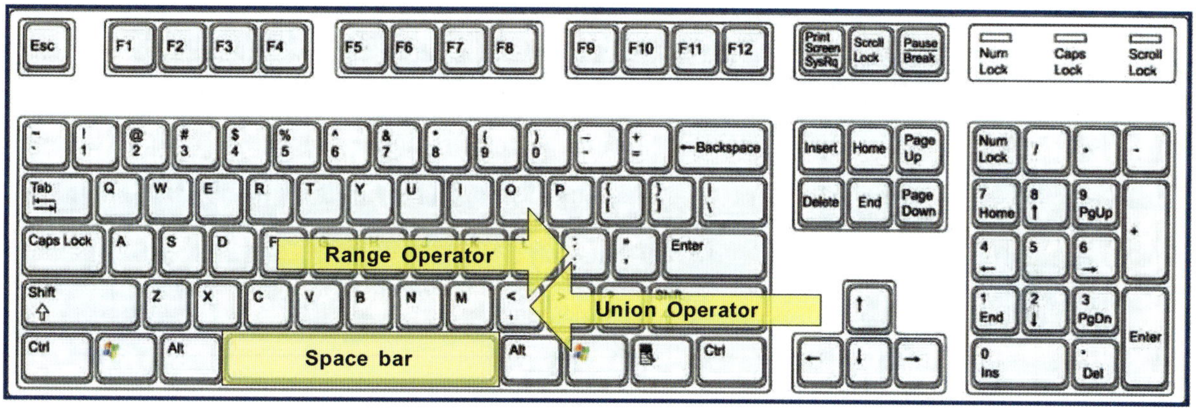

Constants

In mathematics, a constant is a fixed number that does not vary. They are used quite frequently in Excel formulas. Furthermore, it is sometimes easier to use actual numbers as opposed to cell locations in certain situations. For example, if the square root of 16 only needs to be calculated once, the formula =SQRT(16) could be written in any cell. Conversely, if you typed 16 in cell A3 you would have to select another cell and write =SQRT(A3). In other words, an extra step was added by using a cell location. The use of cell locations is most beneficial when tedious iterative calculations need to be performed. If these kind of calculations do not need to me made, then using a constant in the formula is oftentimes faster and easier.

B. Formulas of Common Functions in Excel

The following is a compilation of common functions and operations (with their respective formulas) in Excel. These common formulas are categorized into two tables – mathematical and statistical. The function names are listed in alphabetical order, and a brief description has been provided for each of them.

As stated above, you must type an *equal to* operator first (=), then type the name of the function, select the function name from the pop-up drop-down menu, and finally insert data into the formula by using either a number, cell location, or an array.

MATHEMATICAL

Function Name	Description
ABS	Returns the absolute value of a number
DEGREES	Returns degrees when the original argument was given in radians
EXP	Returns the number e raised to any number (e^x)
FACT	Returns the product of positive integers $\leq n$ ($n!$)
FACTDOUBLE	Returns the double factorial of a number
GCD	Returns the greatest common divisor
INT	Returns the nearest integer by rounding down
LCM	Returns the least common multiple
LN	Returns the natural logarithm of a number
LOG10	Returns the logarithm to base 10 of a number
PI	Returns the value of π
PRODUCT	Returns the product of all numbers/cells given as arguments
RAND	Returns a random number between 0 and 1
RANDBETWEEN	Returns a random number between a range you choose
SLOPE	Returns the slope of a linear regression line through known *x's* and *y's*
SQRT	Returns the square root of a number
SUM	Returns the sum of all the numbers in a selection of cells

STATISTICAL

Function Name	Description
AVERAGE	Returns the expected value (arithmetic mean) of a data set
COMBIN	Number of combinations taken n at a time $C_k^n = \dfrac{n!}{k!(n-k)!}$
COMBINA	Returns the number of combinations taken n at a time with replacement
CORREL	Returns the correlation coefficient between two data sets
COUNT	Counts the number of data points in a data set
MAX	Returns the maximum value in a data set
MEDIAN	Returns the value that lies in the middle of an ordered data set

MIN	Returns the minimum value in a data set
MODE	Returns the most occurring value in a data set
NORMAL.DIST	Returns the normal distribution for a specified mean and standard deviation
PERMUT	Returns the number of arrangements taken n at a time from m objects $A_k^n = P_k^n = \dfrac{n!}{(n-k)!}$
PERMUTATIONA	Returns the number of arrangements taken n at a time with replacement [P_k^n]
POISSON.DIST	Returns the Poisson distribution
STDEV.P	Returns the standard deviation based on a population given as arguments
STDEV.S	Returns the standard deviation based on a sample of the population used as arguments
T.DIST	Returns the left-tailed Student's t-distribution
VAR.P	Returns the variance based on a population given as arguments
VAR.S	Returns the variance based on a sample of the population given as arguments

C. The Insert Function Tool

The Insert Function tool allows you to search Excel's function library. It can also directly insert the formula of a function from the library to the cell. The functions cover a wide array of subjects and disciplines including finance, statistics, mathematics, trigonometry and engineering. There are also logical, date and time, and information functions. The Insert Function tool is helpful when you do not know the function name in Excel. To access the Insert Function tool, you can click on the Insert Function icon in the Function Library group of the Formulas tab. There are other ways to access this feature, but only this method will be discussed.

Note: The shortcut command for accessing the Insert Function window is holding down shift and pressing F3.

Example 2: The following is the theoretical probability distribution of a fair die. Find the formula for the expected value ($\Sigma\ x \cdot p$) of the variable by using the Insert Function tool, and use it to calculate E(x).

x	1	2	3	4	5	6
$p(x)$	1/6	1/6	1/6	1/6	1/6	1/6

Solution :

Steps

1. After copying the distribution into Excel, select cell A9 (see image 1).
2. Click on the Formulas tab and select the Insert Function icon (see image 2).

3. When this dialog box appears, type "mean" in the "Search for a function" text box, and click GO. The search in Excel will yield several results for the mean. Since we are interested in the expected value of our distribution, we will click AVERAGE. Once it is selected, click OK (see image 3).

4. When the Function Arguments dialog box appears, you will type in the rows or columns in which your data is stored. In this case, the expected value of the distribution is the arithmetic mean. Consequently, we will only need the values from column A to compute the expected value. These values are located in cells A2:A7, and we will type these cell locations into the Number1 box. Notice that to the right of your Number1 input, you will see the numerical values you selected in brackets. Also notice that below the Number1 and Number2 inputs, there is the statement "= 3.5". This is the result Excel will give you in the cell you originally selected for the formula (A9). Now click OK (see image 4).

5. Your screen should look similar to the referenced below. Notice that in cell A9, the numerical value is displayed, while in the formula bar, the formula is displayed. Once you know the formula, you will not need to access the Lookup & Reference tool to use the formula again. To use the average function in the future, simply type =AVERAGE, then press tab (or double click AVERAGE from the pop-up drop-down menu) and select the data set to be averaged (see image 5).

1.

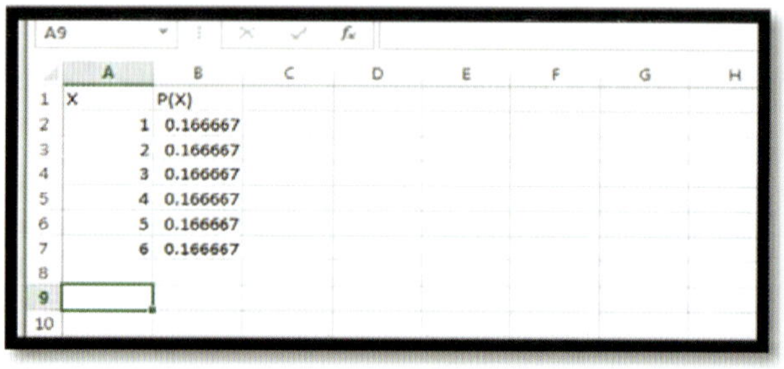

2.

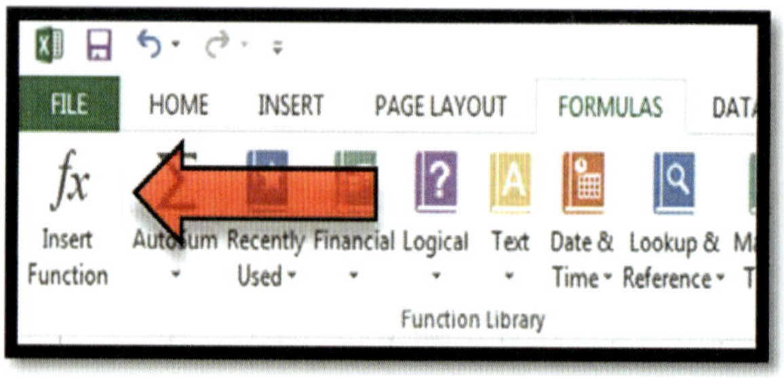

3.

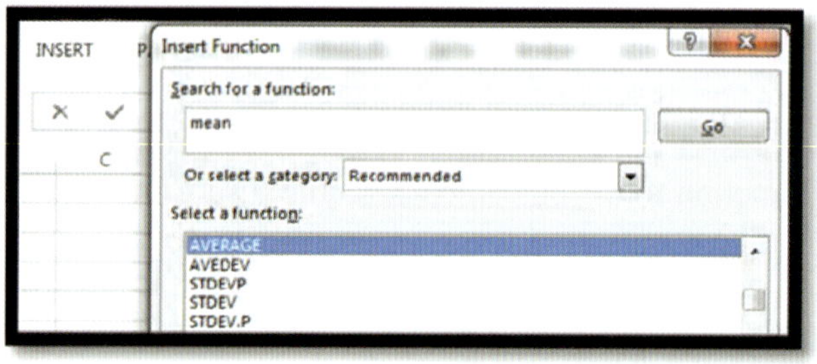

4.

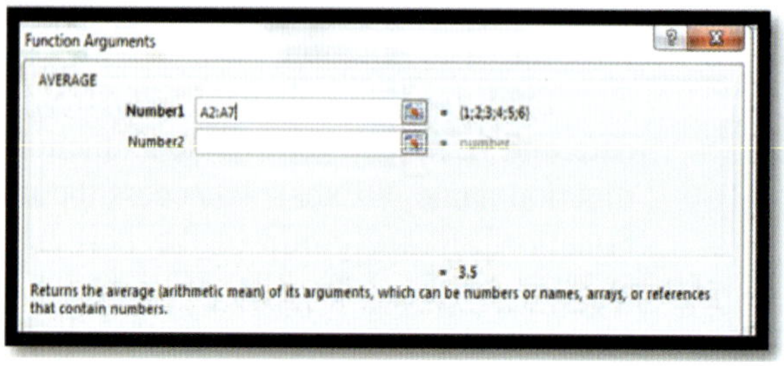

5.

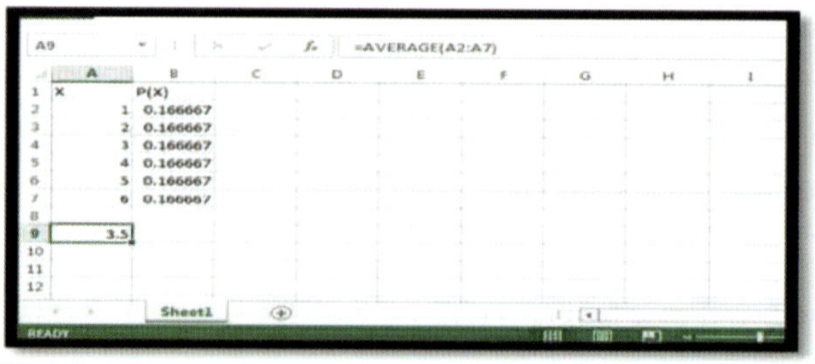

Note: For more information on the expected value of a random variable with a discrete probability distribution, please see Section 4.4.

Note: If you are preparing a spreadsheet that is to be opened by another party, be sure that you are using formulas that are compatible with their version of Excel. When looking up a formula via the "Insert Function tool", be sure to read the formula description carefully. Excel will let you know which formula is compatible with previous versions.

The predefined functions in Excel are very useful and are frequently used in business, academia, research and many other fields. However, there are many formulas that do not have built-in functions in Excel. That being said, this software will let you create formulas that suit your needs. The next subsection explains how to create basic formulas.

D. Basic Formula Creation

Before you start creating your own formulas in Excel, it is imperative that you understand the order of operations Excel follows when evaluating equations. After you understand this, you will be ready to start creating your own formulas.

Order of Operations in Excel

When it comes to interpreting and evaluating equations, Excel follows the same protocol as mathematics. In mathematics and Excel, we call this protocol the **order of operations**. The acronym *Please Excuse My Dear Aunt Sally* may help you remember this. The order is:

1. Please = Parentheses
2. Excuse = Exponents
3. My = Multiplication
4. Dear = Division
5. Aunt = Addition
6. Sally = Subtraction

This list begins with the operation that is performed first, and ends with the operation that is performed last.

Note: Using parentheses are essential to creating accurate formulas. By using parentheses, you are telling Excel which part of the formula to perform first. For example, A1-A2/B2, (A1-A2)/B2. The left-side formula will tell Excel to subtract the quotient of A2 divided by B2 from A1. Whereas the right-side formula will tell Excel to divide the difference of A1 and B2 by B2.

Example 3: Enter the following numbers into a column, and use another column to square each number. Finally, sum the two columns individually.

32	54	111	12	514	67	95	88	40

Solution

Steps

1. Enter the given data into a column, and label it "x" (see image 1).
2. In cell C2, type: =B2^2 and in cell A2, type x^2 (see image 2).
3. Using the drag-and-drop technique, copy the formula from cell C2 to C10 (see image 3).
4. In cell B11, sum the B column by using the SUM formula. After inserting the formula, press enter (see image 4).

5. Drag-and-drop the formula from cell B11 to cell C11 (see image 5).

1.

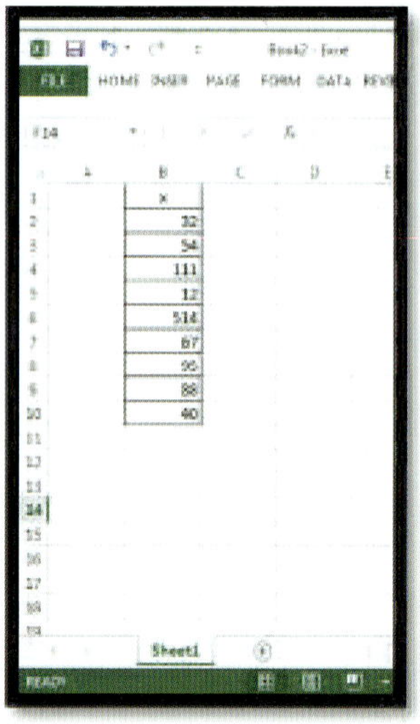

2.

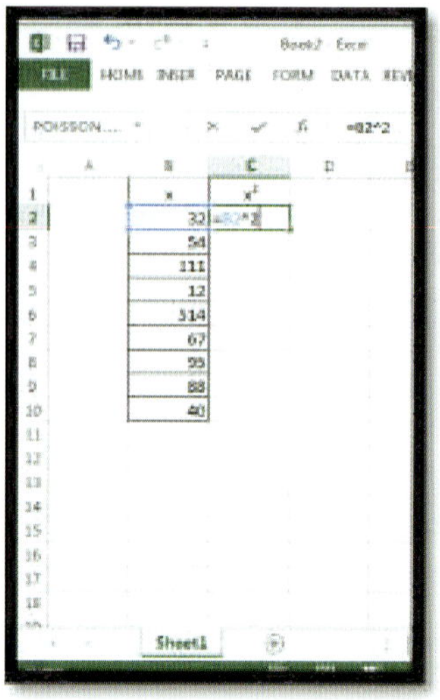

3.

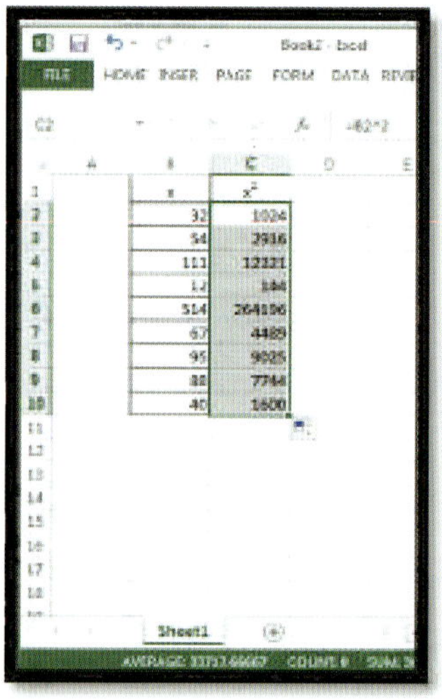

4.

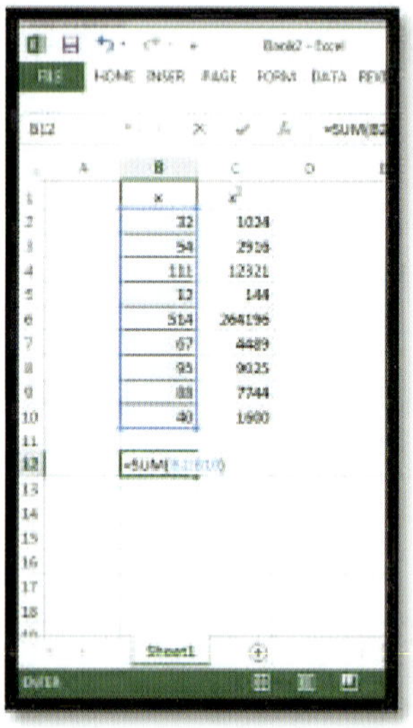

5.

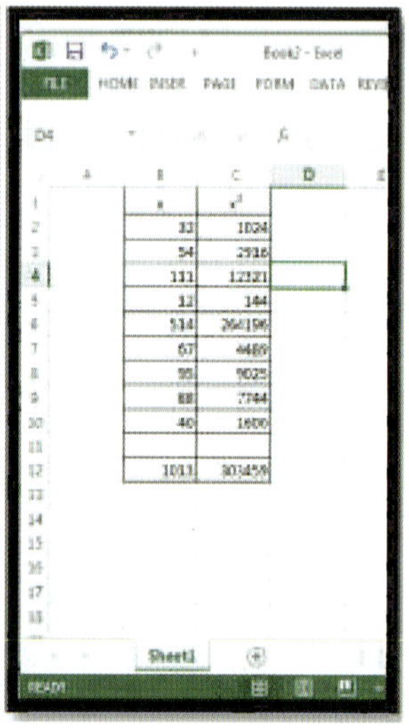

Note: If you are not familiar with the drag-and-drop technique, please see Appendix 1.2.

Example 4: Using the values you obtained from the previous example, evaluate the standard deviation

$$\left(\sqrt{\frac{23\left(\sum x^2\right)-\left(\sum x\right)^2}{(12)(14)}}\right) \text{ of } x.$$

Solution :

Steps 1. Let $y = \sqrt{\dfrac{23\left(\sum x^2\right)-\left(\sum x\right)^2}{(12)(14)}}$.

2. Using the values obtained from example 2, create a formula in Excel that will perform the calculation for y.

3. The numerical value that is returned after the calculation is performed is the value of y.

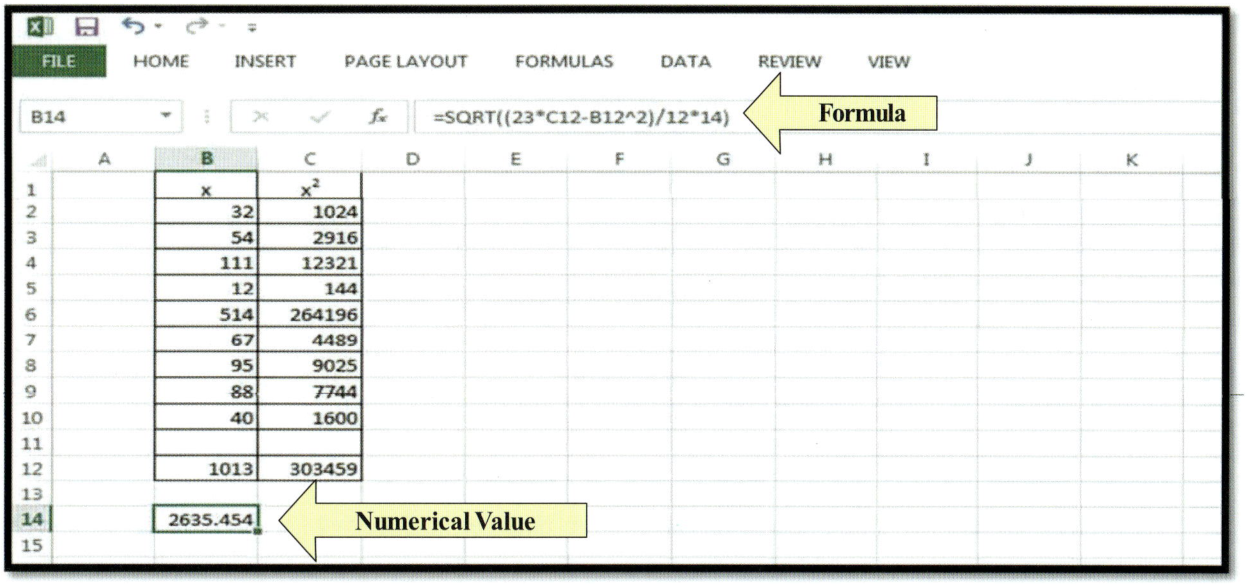

	A	B	C
1		x	x^2
2		32	1024
3		54	2916
4		111	12321
5		12	144
6		514	264196
7		67	4489
8		95	9025
9		88	7744
10		40	1600
11			
12		1013	303459
13			
14		2635.454	
15			

Note: For more information on the standard deviation of a probability distribution, please see Section 4.4.

Example 5: Given the following table:

x	0	1	2	3	4	5	6
p	0.122	0.087	0.017	0.069	0.117	0.114	0.475

(a) Evaluate: $\sum(x \cdot p)$ **(b)** Evaluate: $\sqrt{\sum\left(x^2 \cdot p\right) - \left(\sum x \cdot p\right)^2}$

Solution The following tables summarize both of these calculations. The first table has the formulas displayed. The second table shows the numerical values produced by the formulas.

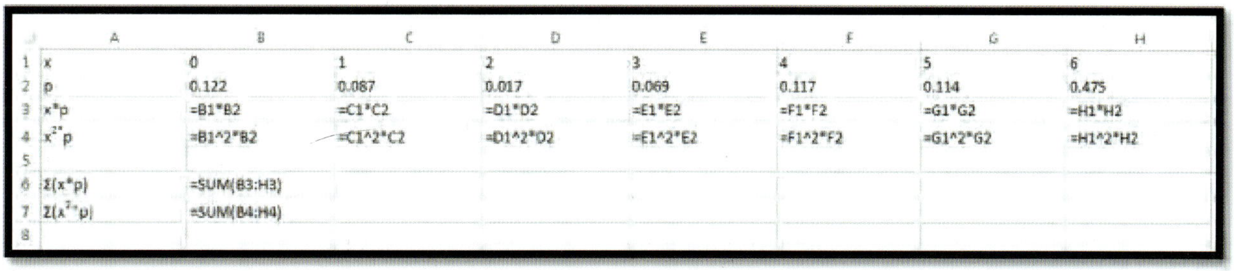

	A	B	C	D	E	F	G	H
1	x	0	1	2	3	4	5	6
2	p	0.122	0.087	0.017	0.069	0.117	0.114	0.475
3	x*p	=B1*B2	=C1*C2	=D1*D2	=E1*E2	=F1*F2	=G1*G2	=H1*H2
4	x^{2}*p	=B1^2*B2	=C1^2*C2	=D1^2*D2	=E1^2*E2	=F1^2*F2	=G1^2*G2	=H1^2*H2
5								
6	Σ(x*p)	=SUM(B3:H3)						
7	Σ(x^2*p)	=SUM(B4:H4)						
8								

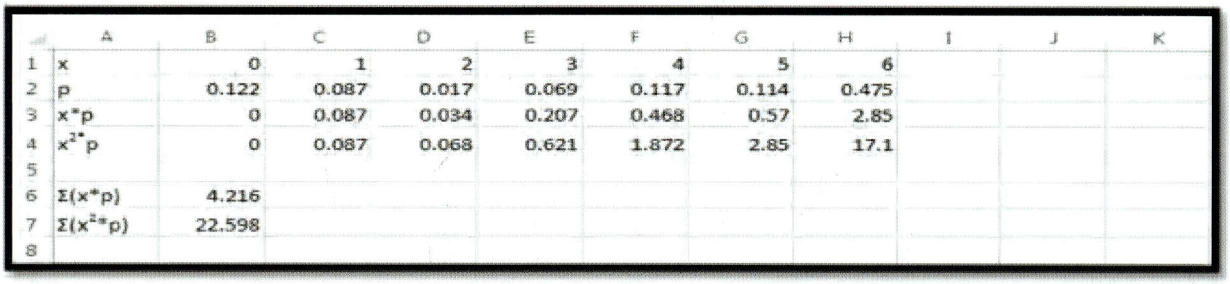

	A	B	C	D	E	F	G	H	I	J	K
1	x	0	1	2	3	4	5	6			
2	p	0.122	0.087	0.017	0.069	0.117	0.114	0.475			
3	x*p	0	0.087	0.034	0.207	0.468	0.57	2.85			
4	x^2*p	0	0.087	0.068	0.621	1.872	2.85	17.1			
5											
6	Σ(x*p)	4.216									
7	Σ(x^2*p)	22.598									
8											

A1.4: Creating Graphs Using Excel®

Objectives

After completing this section, you will know how to create:

A. bar graphs
B. pie charts
C. line graphs

Although organized data charts can be very useful in performing calculations, they will never be as powerful as graphical representations. Primarily, visuals are always easier to digest and understand than text alone. To that end, it is just as essential for you to learn how to create various forms of charts as it is to collect and organize data. In this section, we will discuss bar charts, pie charts and line graphs.

A. Bar Graphs

Bar graphs, also known as bar charts or column graphs, are a popular form of visual representation for a data set. They primarily consist of rectangles whose heights represent the frequency of the variable being studied. A close relative of the bar graph is the histogram. The only difference between them is the histogram has no gaps in between each class interval (bar). The following two examples will demonstrate how to generate a bar graph and histogram from a data set in Excel.

Example 6: Given the following frequency table, draw a 3-D column graph representing this data in Excel.

x	1	2	3	4	5	6
Frequency	15	10	5	18	26	9

Solution :

Steps

1. Select both rows of data and include an extra blank column to the left of the data set (see image 1).
2. With the data selected, click on the Insert tab and go to the Charts group. Look for the "Insert Column Chart" icon (see image 2). Once you have located it, click on the icon.
3. When you click on the icon, a drop-down menu will appear. Select the first option under the 3-D column in the Charts group (see image 3).
4. Your chart should look like the accompanying picture (see image 4).

1.

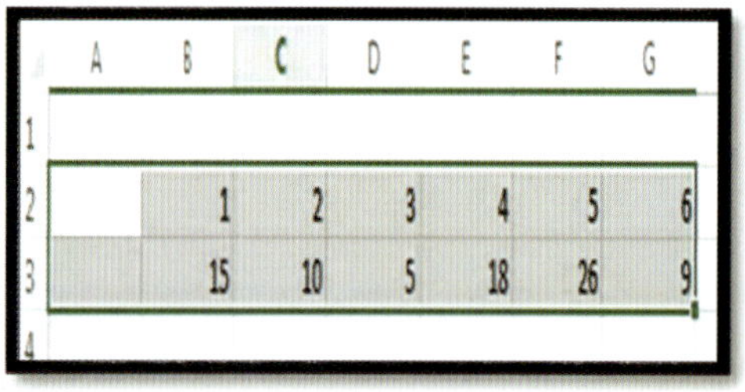

2.

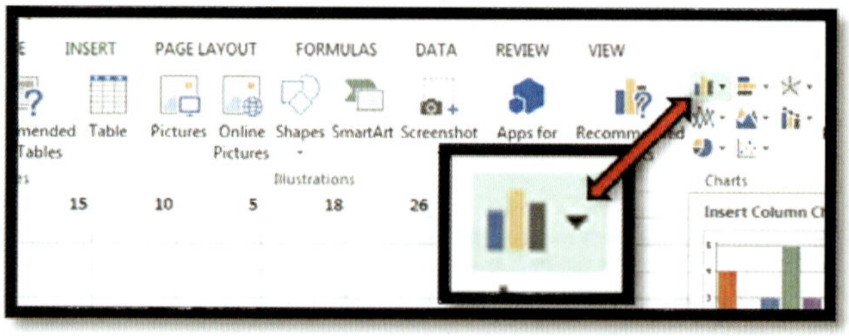

3.

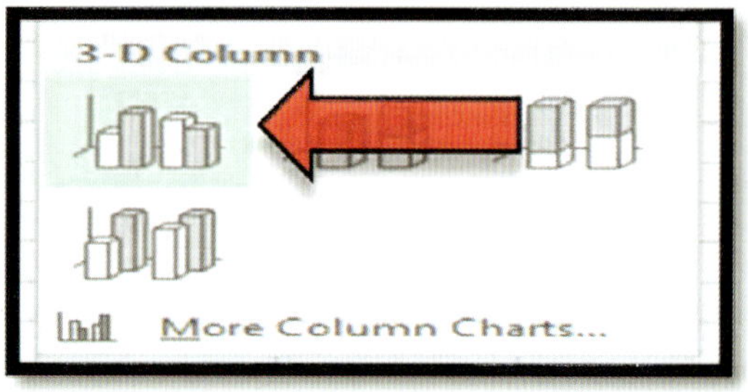

4.

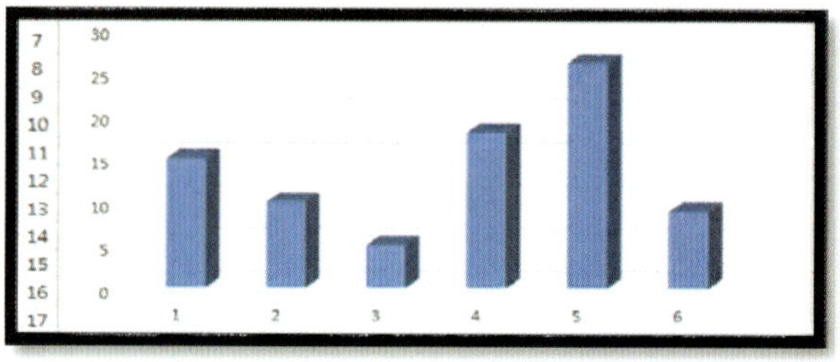

Example 7: Create a histogram from the bar graph in example 5.

Solution :

Steps

1. Click on any of the bars in the graph. Once they are selected, right-click the mouse and click on "Format Data Series" from the drop-down menu (see image 1).
2. In the SERIES OPTIONS group, select the column icon (see image 2).
3. In the next SERIES OPTIONS subgroup is an option called Gap Width. Change it to 0% (see image 3).
4. This graph is now a histogram (see image 4).

1.

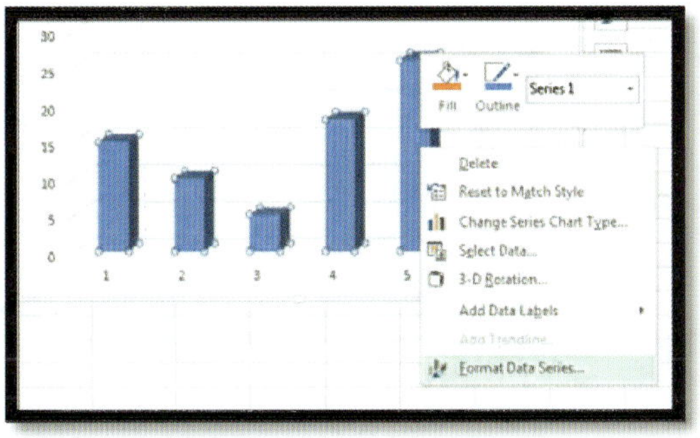

2.

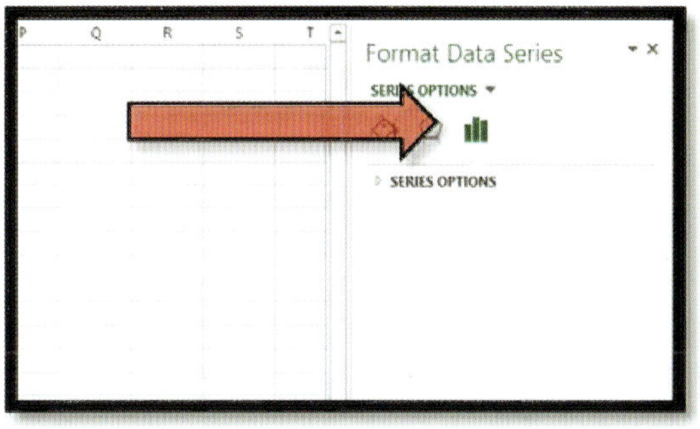

3.

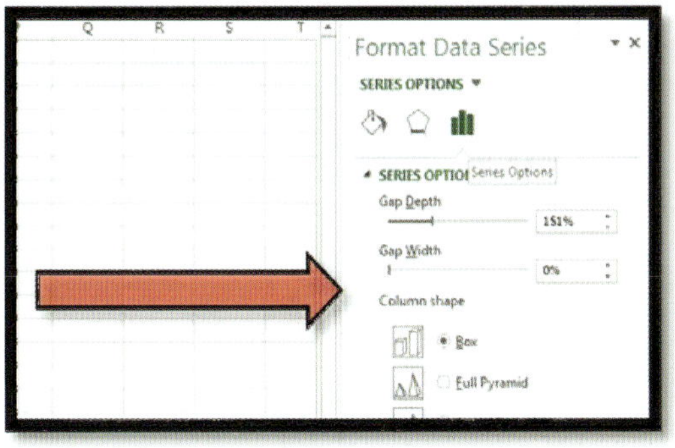

4.

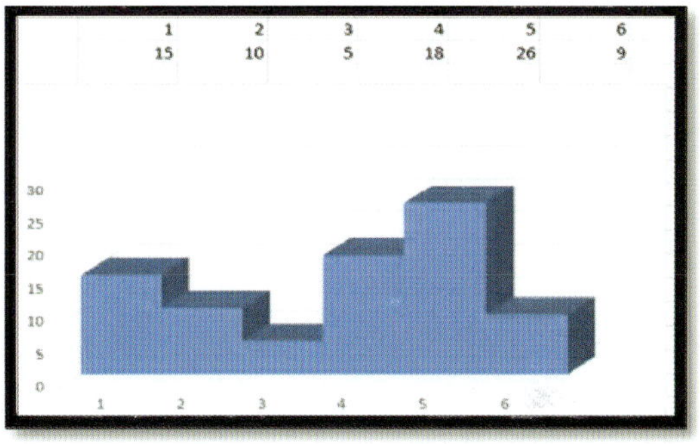

B. Pie Charts

A pie chart is a circular chart that has been divided into sectors representing numerical proportion. Even more precise, the arc length of each sector is proportional to the quantity it represents. These are perhaps the most widely used and recognized charts in business and the media. To do these by hand can be quite tedious, but Excel significantly simplifies the process. The following example illustrates how to produce a pie chart in Excel.

Example 8: A student conducted a simple study at a children's day care center. He was interested in which vegetables were favored the most amongst 8-year-old children. He obtained the following data.

Broccoli	Corn	Carrots	Green Beans	Potatoes	Mushrooms
3	6	4	5	12	2

Generate a pie chart in Excel that represents this data.

Solution :

Steps 1. Select both rows of data (see image 1).

2. Click on the Insert tab and select the "Insert Pie or Doughnut Chart" icon in the Charts group (see image 2).

3. Under the 3-D Pie section, there is only one option to select. Select it (see image 3).

4. You have created a pie chart (see image 4). You may see a "Chart Title" text box at the head of your chart. You can delete it by clicking on it and pressing delete, or you can change the name by double-clicking on it.

1.

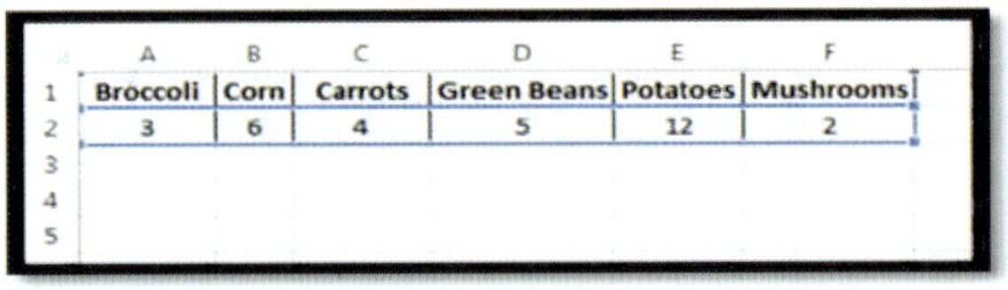

2.

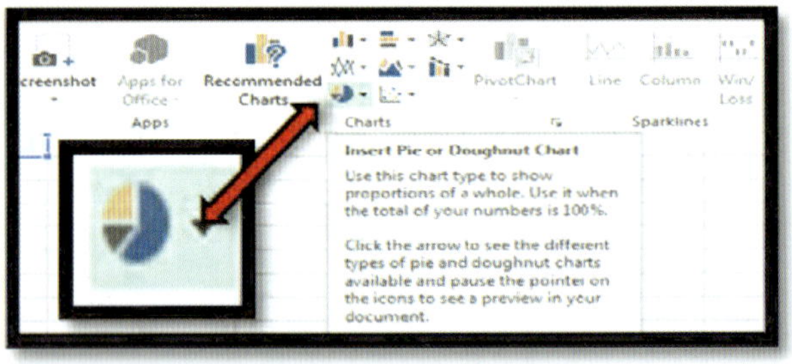

3.

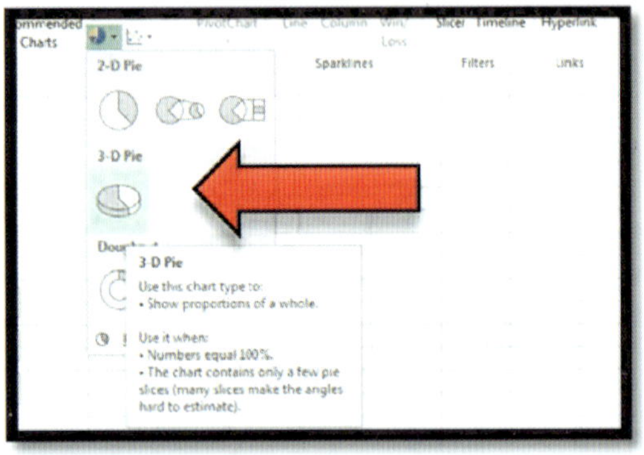

4.

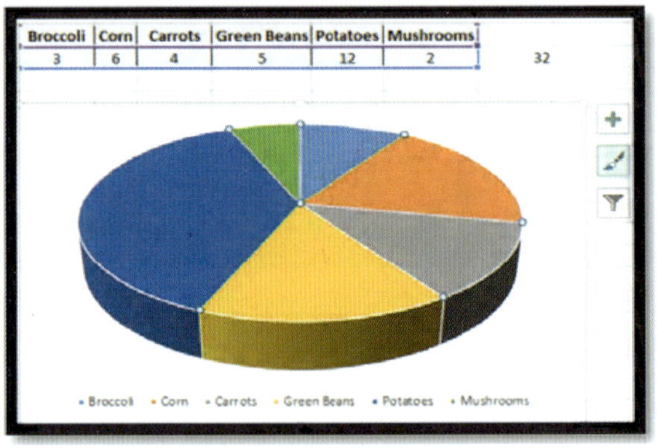

C. Line Graphs

A line graph is a type of graph that displays data points which are connected by line segments. Most typically, a line graph is used to show trends of data over a specified interval. The most commonly used interval for these graphs is time. In the following example, we will obtain the line graph of a data set and then we will "smooth out"the line segments of the graph which will turn it into a curve.

Example 9: Draw the graph of the function $f(x) = x^3$ using the following values for x:

$$x = \{-3, -2, -1, 0, 1, 2, 3\}$$

Solution :

Steps 1. Input the data into Excel and exponentiate each x value as indicated (see image 1).

2. As in example 6, select both rows of data with an additional blank column (see image 2).

3. Go to the Charts group under the Insert tab as before, and click on the Insert Line Chart icon (see image 3).

4. Select the first option in the 2-D Line group (see image 4). The graph produced has points which connect through line segments.

5. Make this into a smooth curve by right-clicking on the line of the graph and selecting “Format Data Series” from the drop-down menu (see image 5).

6. When the Format Data Series window appears, select the fill icon. Under the fill icon, there will be other options for the line. Scroll down all the way to bottom and check the box “Smoothed line” (see image 6).

7. Your graph should look like the one pictured (see image 7).

1.

-3	-2	-1	0	1	2	3
-27	-8	-1	0	1	8	27

2.

	-3	-2	-1	0	1	2	3
	-27	-8	-1	0	1	8	27

3.

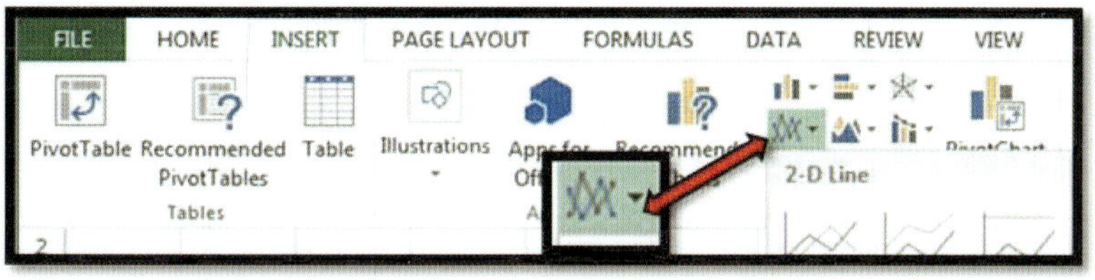

4.

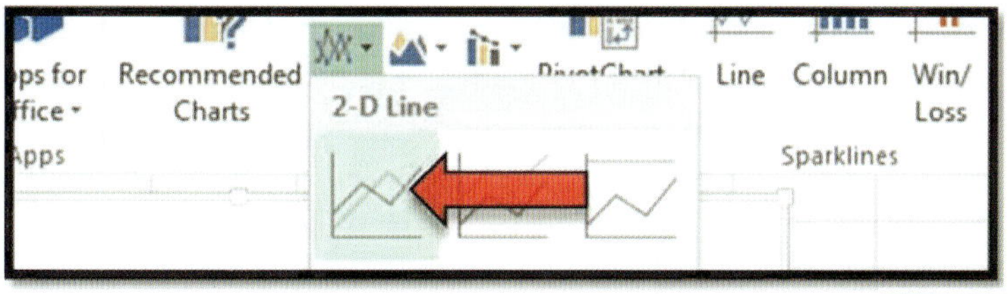

5.

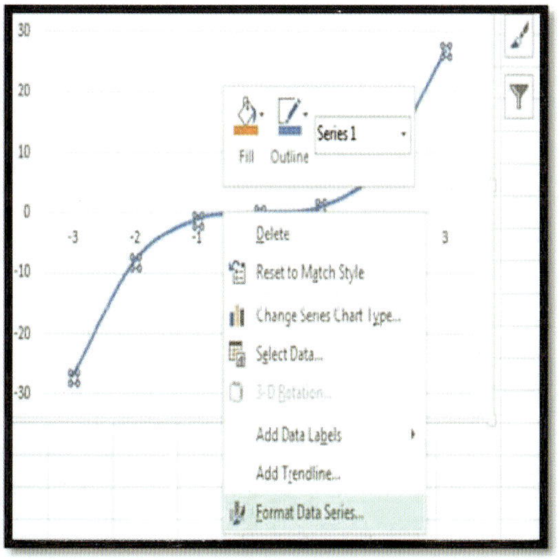

6.

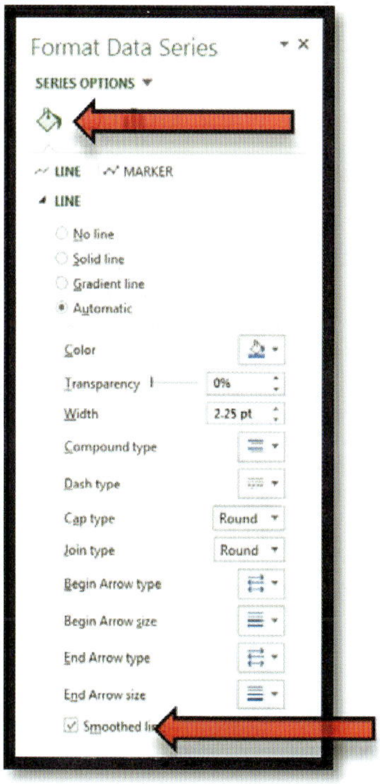

7.

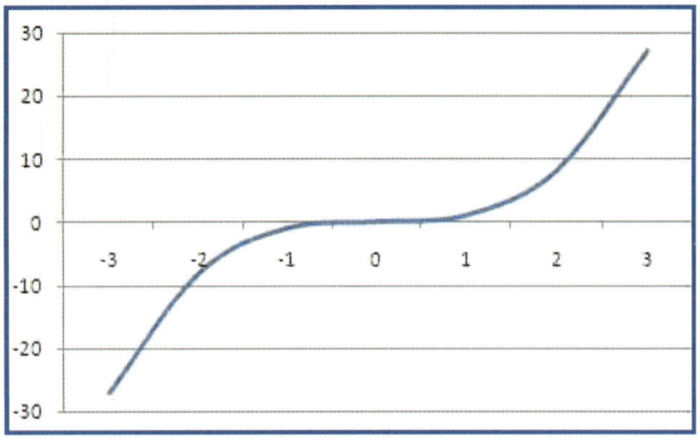

A1.5: The Analysis ToolPak

Objectives

After completing this section, you will be comfortable with:

A. loading the Analysis ToolPak

B. accessing the Analysis ToolPak

An additional component that can be installed on a computer or other device in order to increase its capabilities or functionalities is called an **add-in**. Excel has several add-ins that are available for download. The only one that we are primarily concerned with in this text is the Analysis ToolPak.

With this add-in, there is nothing to download; however, we will need to activate it. This section is dedicated to showing you this process.

A. Loading the Analysis ToolPak

As stated above, you will not need to download anything additional from Microsoft Office® in order to utilize the Analysis ToolPak. All we need to do is load it. The following steps, also made available at www.office.microsoft.com, outline this process. Step-by-step screenshots for steps 2-6 follow immediately after.

***Step* 1** Open Microsoft Excel.

***Step* 2** Create a new blank workbook. (see image 2)

***Step* 3** Click on the **File** tab and select **Options**. (see image 3)

***Step* 4** Click **Add-Ins**, and then in the Manage box, select Excel Add-ins and click Go.(see image 4 & 5)

***Step* 5** In the **Add-Ins** available box, select the **Analysis ToolPak** check box, and then click **OK**. If the Analysis ToolPak is not listed in the Add-Ins available box, click Browse to locate it. If you get prompted that the Analysis ToolPak is not currently installed on your computer, click Yes to install it. (see image 4 & 5).

***Step* 6** The Data Analysis icon should appear in the Analysis group in Excel. If it does not appear, then close and re-open Excel.

2.

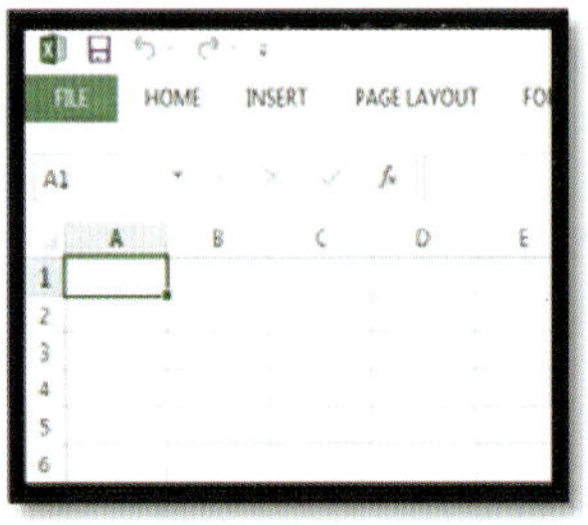

3.

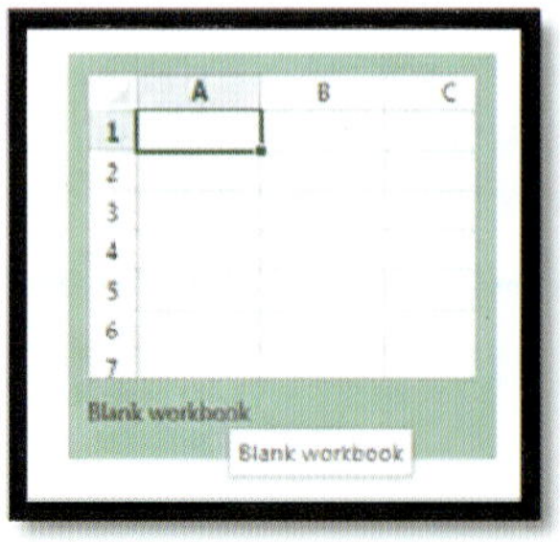

3. *cont'd*

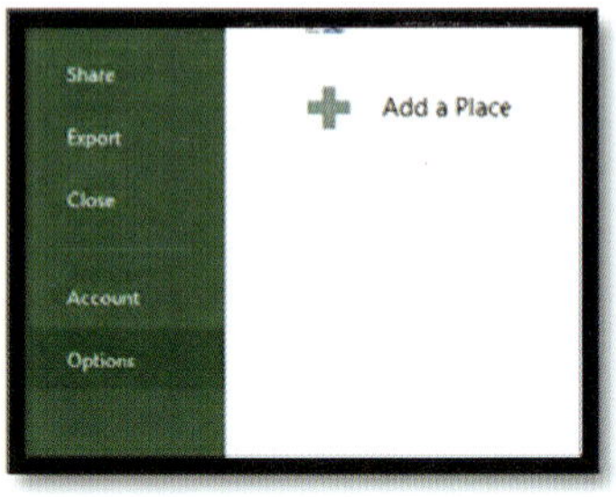

4. & 5.

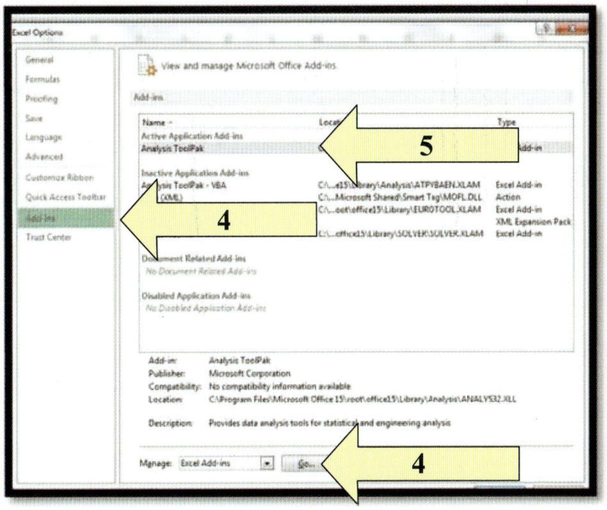

6.

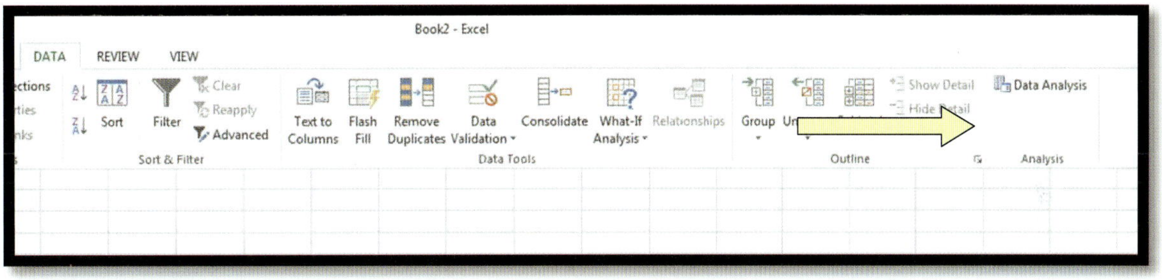

The Analysis ToolPak is one of the most important packages in Excel. It contains several built-in statistical procedures that can facilitate calculations for you in future chapters. With this ToolPak, you can calculate analysis of variance, correlations, descriptive statistics, sampling and regression. To start doing these calculations, click on the Data Analysis icon. In case you cannot find it, refer to Steps 5 and 6 above. The following screenshot displays the Data Analysis dialog box, and it is launched when you click on the icon.

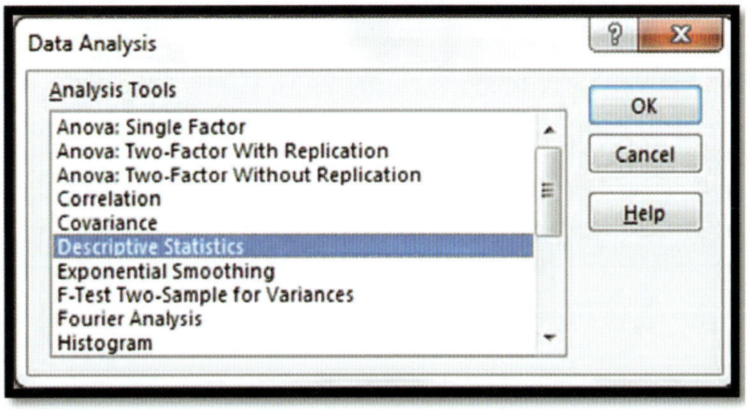

Microsoft Excel is a very powerful and comprehensive software. With this brief introduction, you have only been exposed to some of its basic features. A good amount of practice is required to increase your comfort level with the software.

Appendix 1: General Exercises

1. Given the data sets in columns A and B, calculate the following expressions in the indicated cells or columns.

A	12	40	12	5	6	7	86	2	34	4	5	6	7	665
B	18	28	70	50	12	98	74	18	45	36	65	82	100	17

(a) $\frac{(A-B)^2}{B}$ in column C. **(b)** $\sum \frac{(A-B)^2}{B}$ in cell C15. **(c)** $\sum A's$ in cell A15.

(d) $\sum B's$ in cell B15. **(e)** $A^2's$ in column D. **(f)** $\sum A^2's$ in cell D15.

(g) $\sqrt{\frac{14\left(\sum A^2\right)-\left(\sum A\right)^2}{14(13)}}$ in cell H8.

2. Evaluate $\frac{x-y}{w}$, given $x = 130$, $y = 105$, and $w = 12$.

3. Given $a = 23.5$, $b = 20.3$, $c = 1.96$, $d = 2.4$, $e = 45$, $f = 1.8$, and $g = 34$.

Evaluate $(a-b) - c\sqrt{\frac{d^2}{e}+\frac{f^2}{g}}$

4. Write a command in Excel ® for the formula $(a-b) - c\sqrt{\frac{d^2}{e}+\frac{f^2}{g}}$ in a general format, so that when the values of the given variables change, so will the result obtained from $(a-b) - c\sqrt{\frac{d^2}{e}+\frac{f^2}{g}}$. Try using the values given in problem #3, and verify that your general command produces the same answer as obtained in exercise 3. Now, use the following new set of parameters to evaluate $(a-b) - c\sqrt{\frac{d^2}{e}+\frac{f^2}{g}}$.

$a = 140.6$, $b = 117.5$, $c = 2.58$, $d = 22.7$, $e = 105$, $f = 18.3$ and $g = 84$.

5. Given the following table;

X	0	1	2	3	4	5	6	7	8	9	10
P	0.0282	0.1211	0.2335	0.2668	0.2001	0.1029	0.0368	0.0090	0.0014	0.0001	0.0000

(a) Evaluate $K = \sum(X \cdot P) = x_1p_1 + x_2p_2 + \ldots + x_np_n$ **(b)** Evaluate $\sqrt{\sum(X^2 \cdot P) - K^2}$

6. Write a general command in Excel® that will solve a Quadratic Equation $ax^2 + bx + c = 0$ for any given a, b and c. **Recall:** The Quadratic formula is $\frac{-b \pm \sqrt{b^2-4ac}}{2a}$.

7. Draw the graph for the function $f(x) = x^2$ using the following values for x. $x = \{-3, -2, -1, 0, 1, 2, 3\}$.

APPENDIX 2

ANALYSIS OF VARIANCE

Appendix Overview:

1. *Introduction*
2. *Meaning and Importance of ANOVA*
3. *One-Way Analysis of Variance*
4. *Two-Way Analysis of Variance*

A2.1: INTRODUCTION

In Sections 8.2 and 8.3, we discussed the procedure for testing two means. In practice, however, there are ample cases where we must deal with and test three or mores means. For instance, a farmer may be interested in comparing the average yield of more than two fertilizers; a hospital might want to compare three brands of pain killers before adopting one; a consumer interest group might be interested in testing the hypothesis that the average fuel consumption per mile for four popular midsize automobiles would be the same. Can statistics be of any help? If yes, how do we go about comparing three or more means.

One obvious method is to compare two means at a time by using the ***Z*** or ***t***-Tests illustrated in Sections 8.2 and 8.3. However, these procedures are very time consuming, since as few as 4 sample means would require $C_2^4 = 6$, different tests to compare 6 possible pairs of sample means. Therefore, there must be a procedure that can compare all the means simultaneously. One such procedure is known as the **analysis of variance** (ANOVA), a term introduced by R. A. Fisher in 1920's to deal with problems in the analysis of **agronomic data**.

ANOVA is the most commonly used statistical procedure in Natural and Social Science Research today, and can be a very effective tool in testing the effects of Treatments and/or Products.

Note: A detailed discussion of ANOVA is beyond the scope of this book, and it is usually caried out using statistical software.

In our basic discussion of ANOVA we will focus on investigating questions such as:

1. Are all the means from more than two populations equal?
2. Are all the means from more than two treatments on one population equal? (Do the treatments have a common overall effect?)

A2.2: Meaning and Importance of ANOVA

Objectives

After completing this section you will be able to understand:

A. Assumptions for the ANOVA Test.

B. *F*-Distribution.

Variation is inherent in nature. The basic purpose of analysis of variance is to test the homogeneity of several means based on variation in the data. This variation in any set of numerical data is due to a number of causes which may be classified as "assignable" and "chance."

The variation due to assignable causes can be detected and measured, whereas the variation due to chance causes is beyond the human control.

Definition

According to Fisher, Analysis of Variance (ANOVA) is the "separation of variance ascribable to one group of causes from the variance ascribable to the other group."

In analysis of variance, the total variation in the sample data is expressed as the sum of its non-negative components where each of these components is a measure of the variation due to some specific independent source or factor or cause. The ANOVA consists of estimation of the amount of variation due to each of the independent factors (causes) separately and their estimation due to assignable factors (causes) with the estimate due to chance factor (causes), the latter being known as experimental error or simply error.

A. Assumptions for the ANOVA Test

ANOVA test is based on the test statistic $\boldsymbol{F}$ (or Variance Ratio). For the validity of the $\boldsymbol{F}$-test in ANOVA, the following assumptions are made.

1. The observations are independent.
2. Parent populations from which observations are taken are normal.
3. Various treatment and environmental effects are additive in nature.

In the following sections we will discuss the analysis of variance for **(a)** One-Way classification known as One-Way ANOVA and **(b)** Two-Way classification known as Two-Way ANOVA.

Remarks:

1. ANOVA technique enables us to compare several population means simultaneously which results in a great deal of savings in terms of time and resources as compared to several experiments required for comparing two populations means at a time.
2. As pointed out earlier, the origin of the ANOVA technique lies in agricultural experiments and as such its language is loaded with such terms as treatments, blocks, plots etc. However, ANOVA technique is so versatile that it finds applications in the design of all types of experiments in diverse fields such as industry, education, psychology, business etc.
3. It should be clearly understood that ANOVA technique is not designed to test the equality of several population variances. Rather, its objective is to test the equality of several population means or the homogeneity of several independent sample means.
4. In addition to testing the homogeneity of several sample means, the ANOVA technique is now frequently applied in testing the linearity of the fitted regression line or the significance of the correlation ratio. The discussion of these topics is, however, beyond the scope of this book.

5. Because the calculations of the test statistics and p-value in ANOVA are quite complicated and cumbersome, we will emphasize the interpretation of the results obtained using the Data Analysis package in Microsoft Excel®.
6. Manual calculations of the test statistics and the p-value will be presented as optional content.

B. *F*-Distribution

ANOVA technique is based on the ***F***-distribution, discussed in Section 8.6. There, we explained how to read the critical values of ***F*** for the **numerator** and **denominator** degrees of freedom at levels of significance $\alpha = 0.01$, 0.025, 0.05, and 0.1 given in the **Tables 6(A)**, **(B)**, **(C)**, and **(D)** at the end of the book.

Review of the *F*-Distribution:

1. The shape of the F-distribution depends on the number of degrees of freedom. The F-distribution has two degrees of freedom $F(v_1, v_2)$, where v_1 is known as the numerator degrees of freedom and v_2 the denominator degrees of freedom.
2. The critical values of the ***F***-distribution are non-negative.

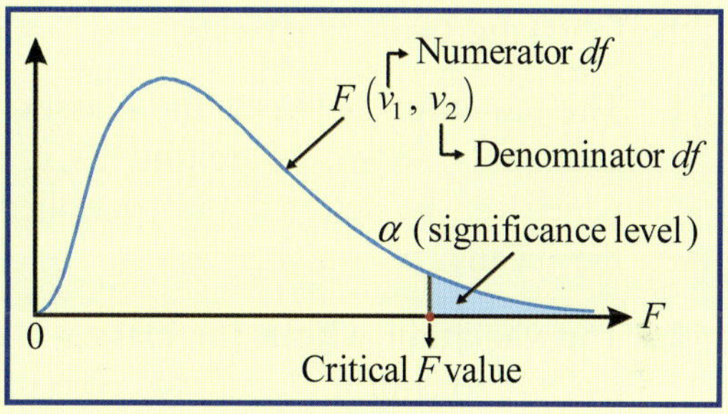

Figure A2.1

3. The ***F***-distribution is a continuous distribution skewed to the right.
4. In this chapter, the rejection region is always in the right-tail of the ***F***-distribution curve.

Example 1: Find the ***F*** value for 5 degrees of freedom for the numerator, 14 degrees of freedom for the denominator, and a 0.025 area (significance level) in the right tail of the ***F***-distribution.

v_2 \ v_1	1	2	3	4	5	6	. . .	30	40
1	647.789	799.500	864.163	899.583	921.848	937.111	. . .	1001.414	1005.598
2	38.506	39.000	39.165	39.248	39.298	39.331		39.465	39.473
3	17.443	16.044	15.439	15.101	14.885	14.735		14.081	14.037
4	12.218	10.649	9.979	9.605	9.364	9.197	. . .	8.461	8.411
5	10.007	8.434	7.764	7.388	7.146	6.978		6.227	6.175
6	8.813	7.260	6.599	6.227	5.988	5.820		5.065	5.012
7	8.073	6.542	5.890	5.523	5.285	5.119	. . .	4.362	4.309
8	7.571	6.059	5.416	5.053	4.817	4.652		3.894	3.840
9	7.209	5.715	5.078	4.718	4.484	4.320		3.560	3.505
10	6.937	5.456	4.826	4.468	4.236	4.072	. . .	3.311	3.255
11	6.724	5.256	4.630	4.275	4.044	3.881		3.118	3.061
12	6.554	5.096	4.474	4.121	3.891	3.728	. . .	2.963	2.906
13	6.414	4.965	4.347	3.996	3.767	3.604		2.837	2.780
14	6.298	4.857	4.242	3.892	3.663	3.501	. . .	2.732	2.674
15	6.200	4.765	4.153	3.804	3.576	3.415		[illegible]	[illegible]
⋮	⋮	⋮	⋮	⋮	⋮	⋮	. . .	⋮	⋮
24	5.717	4.319	3.721	3.379	3.155	2.995		2.209	2.146
25	5.686	4.291	3.694	3.353	3.129	2.969	. . .	2.182	2.118
30	5.568	4.182	3.589	3.250	3.026	2.867		2.074	2.009
40	5.424	4.051	3.463	3.126	2.904	2.744		1.943	1.875
60	5.286	3.925	3.343	3.008	2.786	2.627	. . .	1.815	1.744
120	5.152	3.805	3.227	2.894	2.674	2.515		1.690	1.614
∞	5.024	3.689	3.116	2.786	2.567	2.408	. . .	1.566	1.484

Figure A2.2 a

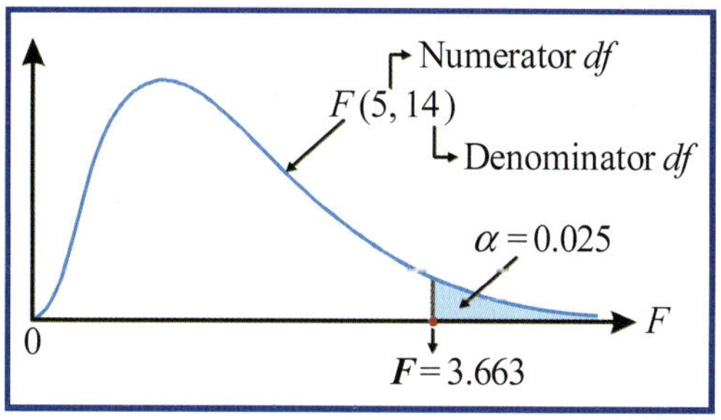

Figure A2.2 b

A2.3: One-Way Analysis of Variance

Objectives

After completing this section you will be able to understand:

A. Partition of Total Variation.

B. Hypothesis Testing for three or more means (One-Way ANOVA).

One-Way analysis of variance technique is used to test the equality of three or more population means based on classification of "n" sample observations divided into "k" classes on the basis of a single characteristic or criterion.

Note: Even though the ANOVA test is used to compare three or more means, the corresponding F-Test uses variances to calculate the test statistics.

A. Partition of Total Variation

In the One-Way analysis of variance procedure, the total variation in the observations can be split into the following two components.

1. The **variation between the classes-** *i.e.*, the variation due to different bases of classification commonly known as **treatments or factors**. This type of variation is due to **assignable causes** which can be detected and controlled by human endeavor and it involves finding the variance of the means.

2. The **variation within the classes-** *i.e.*, the inherent variation of the random variable within the observations of a class. This type of variation is due to chance causes beyond human control. Variation within the classes is calculated using all the sample data and is not affected by the differences in the means.

The main objective of analysis of variance technique is to examine if there is significant difference between the class means in view of the inherent variability within the separate classes. If there is no difference, then the variation between the classes and the variation within the classes will be approximately equal to 1 and the null hypothesis will not be rejected. On the other hand, if the difference between the means is significant, then the variation between the classes will be greater than the variation within classes. Hence the F value will be greater than 1 and H_0 will be rejected.

B. Hypothesis Testing for three or more means (One-Way ANOVA)

We can carry out the test for the equality of several (k) population means by using the Classical Method. P-value Method will be discussed in A2.5.

Method 1: Classical Method

***Step* 1** Set up the hypotheses

Null Hypothesis: $H_0 : \mu_1 = \mu_2 = \ldots = \mu_k$, that is, all the means are equal.

Alternative Hypothesis: H_1 : At least one mean is different from the others.

***Step* 2** Test Statistics for One-Way ANOVA

Test Statistics formula for unequal sample sizes

$$F = \frac{\text{variance between the groups}}{\text{variance within the groups}} = \frac{s_B^2}{s_W^2} = \frac{\dfrac{\sum n_i\left(\overline{x}_i - \overline{\overline{x}}\right)^2}{k-1}}{\dfrac{\sum s_i^2\left(n_i - 1\right)}{\sum\left(n_i - 1\right)}}$$

$$= \frac{\dfrac{n_1(\bar{x}_1 - \bar{\bar{x}})^2 + n_2(\bar{x}_2 - \bar{\bar{x}})^2 + \ldots + n_k(\bar{x}_k - \bar{\bar{x}})^2}{k-1}}{\dfrac{s_1^2(n_1 - 1) + s_2^2(n_2 - 1) + \ldots + s_k^2(n_k - 1)}{N-k}} = \frac{\dfrac{SS_B}{k-1}}{\dfrac{SS_W}{N-k}} = \frac{MS_B}{MS_W}$$

$\bar{x}_i$	$\dfrac{\sum x}{n}$ for each category.
s_i^2	variance for each category.
$\bar{\bar{x}}$	$\dfrac{\sum \bar{x}_i \cdot n_i}{\sum n_i}$ average of averages called x double bar.
k	number of categories.
$k - 1$	numerator degrees of freedom.
$N - k$	denominator degrees of freedom.
SS_B	$\sum n_i(\bar{x}_i - \bar{\bar{x}})^2$: Sum of squares between the groups.
SS_W	$\sum s_i^2(n_i - 1)$: Sum of squares within the groups.
MS_B	$\dfrac{SS_B}{k-1}$ mean squares between groups.
MS_W	$\dfrac{SS_W}{N-k}$ mean squares within groups.

Table A2.1 summarizes the steps in calculating the above test statistics.

Source	***SS* (Sum of squares)**	***df***	***MS* (Mean square)**	***F***
Between Groups	$SS_B = \sum_{i=1}^{k} n_i(\bar{x}_i - \bar{\bar{x}})^2$	$k - 1$	$MS_B = \dfrac{SS_B}{k-1}$	$F = \dfrac{MS_B}{MS_W}$ with $df = (k - 1, n - k)$
Within Groups	$SS_W = \sum_{i=1}^{k} s_i^2(n_i - 1)$	$n - k$	$MS_W = \dfrac{SS_W}{N-k}$	

Table A2.1

Special Case for the Test Statistics (Equal Sample Sizes)

Test Statistics: $F = \dfrac{ns_{\bar{x}}^2}{s_p^2}$ where

$s_{\bar{x}}^2$ = variance between sample means (the variance for $\bar{x}_1, \bar{x}_2, \bar{x}_3, \ldots, \bar{x}_k$),

$\overline{s^2}$ = the average of variances for each category,

$\overline{s^2} = \dfrac{\sum s_i^2}{k}$, and

n is the number of observations in each group.

Step 3 Rejection Region and Critical Values:

The rejection region for One-Way ANOVA test is a right-tailed F distribution, with α significance level, $k-1$ df known as **numerator degrees of freedom**, and $N-k$ df known as **denominator degrees of freedom**.

Note: The numerator degrees of freedom is found in the top row of the F table, and the denominator degrees of freedom is found in the first column of the F table.

Step 4 Decision and Conclusion:

If the computed value of the test statistic lies in the rejection region, we **Reject** H_0 at level of significance α. If the computed value of the test statistic lies outside the rejection region, we **Fail to Reject** H_0. Write an appropriate conclusion in simple words.

Example 2: A school administrator wants to test the consistence of scores in several Statistics classes. To study the equality of average scores, data from three randomly selected Statistics classes was collected and is displayed in the following table.

Sample grades from each class							
Statistics Class 1	87	65	80	90	34		
Statistics Class 2	65	78	98	54	34	34	66
Statistics Class 3	90	100	34	65	66	78	

Using a 0.05 significance level, test the claim that average scores are equal throughout different Statistics classes.

Solution : **Step 1** Set up the hypotheses

Null Hypothesis: $H_0 : \mu_1 = \mu_2 = \mu_3$, that is, all the means are equal.

Alternative Hypothesis: H_1 : At least one mean is different from the others.

Step 2 Test Statistics for One-Way ANOVA

$$k-1 = 3-1 = 2$$

$$N-k = 18-3 = 15$$

$$\bar{\bar{x}} = 68.217$$

	$\bar{x}_i$	s_i^2	n_i
Statistics Class 1	71.2	525.7	5
Statistics Class 2	61.286	534.238	7
Statistics Class 3	72.167	534.567	6

Table A2.2

$$SS_B = \sum_{i=1}^{3} n_i\left(\bar{x}_i - \bar{\bar{x}}\right)^2$$

$$= 5(71.2-68.217)^2 + 7(61.268-68.217)^2 + 6(72.167-68.217)^2 = 474.378$$

$$MS_B = \frac{SS_B}{k-1} = \frac{474.378}{2} = 237.189$$

$$SS_W = \sum_{i=1}^{3} s_i^2\left(n_i - 1\right) = 525.7(4) + 534.238(6) + 534.567(5) = 7981.063$$

$$MS_W = \frac{SS_W}{N-k} = \frac{7981.063}{15} = 532.071$$

$$\text{Test Statistics} = F = \frac{MS_B}{MS_W} = \frac{237.189}{532.071} = 0.446$$

Summary of the results from above

Source	*SS* (Sum of squares)	*df*	*MS* (Mean square)	Variance Ratio
Between Groups	474.378	2	237.189	$F_{(2,15)} = 0.446$
Within Groups	7981.063	15	532.071	

Table A2.3

***Step* 3** Rejection Region and Critical Values:

The critical value for the right-tailed ***F***-test at α = 0.05 significance level with 2 numerator *df* and 15 denominator *df* is 3.682. That is, $F_{(2,15)}(0.05) = 3.682$.

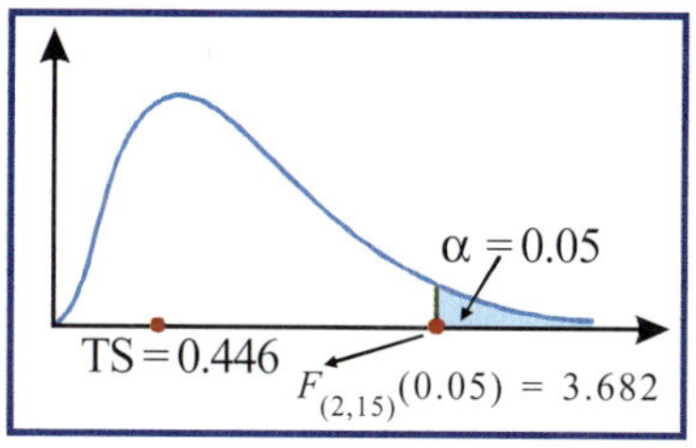

Figure A2.3

***Step* 4** **Decision:** Since the calculated value of test statistics lies outside the rejection region, we fail to reject H_0 at α = 0.05. *Conclusion:* The data does not provide sufficient evidence against H_0. There does not seem to be significant difference between the means.

In short:

Hypotheses	**Test Statistics**
$H_0: \mu_1 = \mu_2 = \mu_3$, that is, all the means are equal. H_1: At least one mean is different from the others.	$F = \frac{MS_B}{MS_W} = \frac{237.189}{532.071} = 0.446$ with $df = (2, 15)$
Rejection Region and Critical Value	**Conclusion**
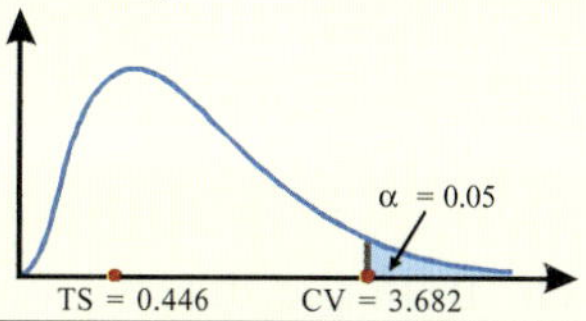	Fail to reject H_0. There does not seem to be significant difference between the means.

Example 3: A researcher plans to test the hypothesis that the average number of days a given patient is kept in the three local hospitals- St. Vincent Hospital, St. Joseph Hospital, and St. Stephens Hospital- is the same. An examination of the hospital records reveals the number of days 21 randomly selected patients, in three groups of seven, stayed in each hospital. Test the hypothesis at α = 0.025.

Hospital	**Number of days spent by seven patients at the hospital**						
St. Vincent	8	5	9	2	7	8	2
St. Joseph	4	3	8	7	7	1	5
St. Stephens	1	4	9	8	7	2	3

Using a 0.05 significance level, test the claim that the average number of days a patient is kept in the hospital is the same.

Solution : ***Step* 1** Set up the hypotheses

Null Hypothesis: $H_0 : \mu_1 = \mu_2 = \mu_3$, that is, all the means are equal.

Alternative Hypothesis: H_1 : At least one mean is different from the others.

***Step* 2** Test Statistics for One-Way ANOVA:

	$\bar{x}_i$	s_i^2	n_i
St. Vincent	5.857	8.476	7
St. Joseph	5	6.333	7
St. Stephens	4.857	9.810	7

Table A2.4

$n_1 = n_2 = n_3 = n = 7$

$s_{\bar{x}}^2 = 0.292$

$\overline{s^2} = 8.206$

$$\text{Test Statistics} = F = \frac{ns_{\bar{x}}^2}{\overline{s^2}} = \frac{(7)(0.292)}{8.206} = 0.249$$

***Step* 3** Rejection Region and Critical Values:

The critical value for the right-tailed ***F***-test at $\alpha = 0.025$ significance level with 2 numerator *df* and 18 denominator *df* is 4.560. That is, $F_{(2,18)}(0.025) = 4.560$.

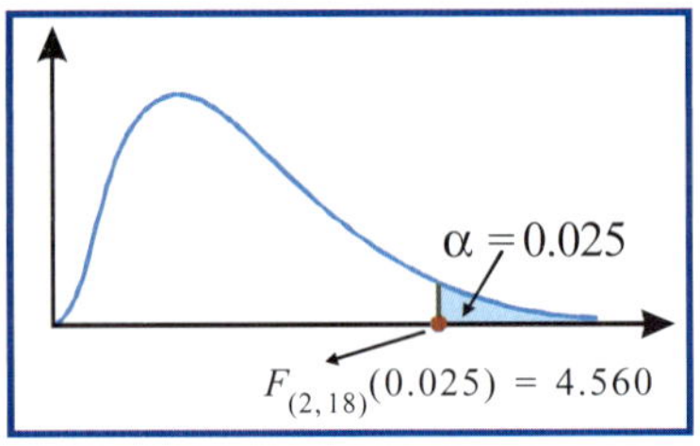

Figure A2.4

***Step* 4** **Decision:** Since the calculated value of test statistics lies outside the rejection region, we fail to reject H_0 at $\alpha = 0.025$. *Conclusion:* The data does not provide sufficient evidence against H_0. There does not seem to be significant difference between the means.

Example 4: The following data is given.

$\bar{x}_1 = 27, s_1 = 8, n_1 = 4; \bar{x}_2 = 25, s_2 = 9, n_2 = 7; \bar{x}_3 = 28, s_3 = 5, n_3 = 5$; and $\alpha = 0.05$.

Test the hypotheses:

H_0 : All the means are equal.

H_1 : At least one mean is different from the others.

Solution : Since we are given the values of sample statistics, we can directly compute SS_B, SS_W, and degrees of freedom.

***Step* 1** The overall mean (mean of the means) $\bar{\bar{x}}$ is given by

$$\bar{\bar{x}} = \frac{n_1\bar{x}_1 + n_2\bar{x}_2 + n_3\bar{x}_3}{n_1 + n_2 + n_3} = \frac{(4)(27)+(7)(25)+(5)(28)}{4+7+5} = 26.438$$

Calculation of SS_B and SS_W

Between Groups Sum of Squares	Within Groups Sum of Squares
$SS_B = \sum_{i=1}^{3} n_i(\bar{x}_i - \bar{\bar{x}})^2$ $= 4(27-26.438)^2 + 7(25-26.438)^2 + 5(28-26.438)^2$ $\approx 4(0.3158) + 7(2.068) + 5(2.440) = 27.939$	$SS_W = \sum_{i=1}^{3} s_i^2(n_i - 1)$ $= 8^2\cdot(3) + 9^2\cdot(6) + 5^2\cdot(4)$ $= 778$

Table A2.5

Degrees of Freedom:

For $SS_B = k - 1 = 3 - 1 = 2$
For $SS_W = N - k = 16 - 3 = 13$

Source	*SS* (Sum of squares)	*df*	*MS* (Mean square)	*F*
Between Groups	27.939	2	$\frac{28.939}{2} = 13.970$	$F_{(2,13)}\ \frac{13.970}{59.846} = 0.233$
Within Groups	778	13	$\frac{778}{13} = 59.846$	

Table A2.6

Step 2 Critical Value and the Rejection Region

In a One-Way ANOVA, the rejection region is a right-tailed ***F***-distribution. The critical value of ***F*** for (2, 13) degrees of freedom at 5% level of significance is $F_{(2,\,13)}(0.05) = 3.806$.

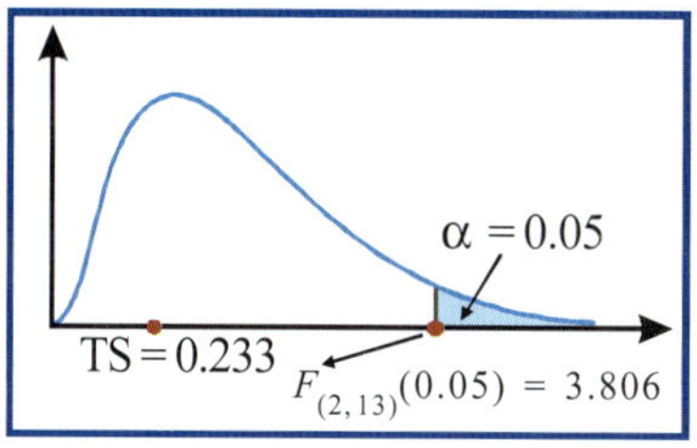

Figure A2.5

Step 3 Decision and conclusion

Since the calculated value of $F = 0.233$ is less than the critical value $F_{(2,\,13)}(0.05) = 3.806$, it lies outside the rejection region. Hence, we fail to reject the null hypothesis. There does not seem to be a significant difference between the means.

In short:

Hypotheses	**Test Statistics**
H_0 : All the means are equal. H_1 : At least one mean is different from the others.	$F = \frac{MS_B}{MS_W} = \frac{13.970}{59.846} = 0.233$ with $df = (2, 13)$
Rejection Region and Critical Value 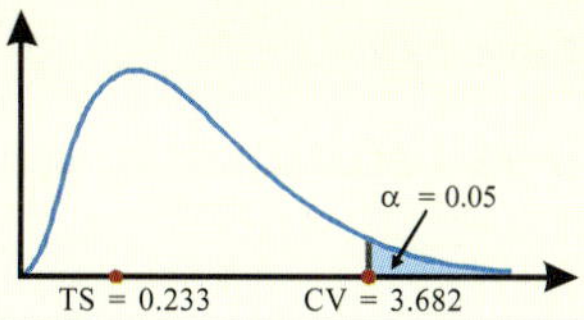	**Conclusion** Fail to reject H_0 at a = 0.05. There does not seem to be significant difference between the means.

A2.4: Two-Way Analysis of Variance

Objectives

After completing this section you will be able to understand:

A. Assumptions for the Two-Way ANOVA.

B. Hypothesis Testing for Two-Factor ANOVA.

In a One-Way ANOVA, the data is categorized into groups according to a single factor (or treatment). Now, suppose the n observations are classified into k categories (or classes) $A_1, A_2, \ldots, A_k$ according to some criterion A and into h categories $B_1, B_2, \ldots, B_k$ according to some criterion B, having kh combinations (A_i, B_j) $i = 1, 2, \ldots, k$; $j = 1, 2, \ldots, h$, called cells. This scheme of classification according to two factors or criteria is called ***Two-Way classification*** and its analysis is called ***Two-Way analysis of variance***. The number of observations in each cell may be equal or different, but we will consider one observation per cell so that $n = hk$, that is, the total number of cells is $n = hk$.

In the Two-Way classification, the values of the response variable are affected by two factors. For example, the yield of milk may be affected by differences in **treatments**, such as **rations** as well as the differences in **variety**, such as **breed and stock** of the cows. Let us now suppose that the cows are divided into h different groups or classes according to their breed and stock, each group containing k cows and then let us consider the effect of k treatments (such as rations given at random to cows in each group) on the yield of milk.

Let $x_{ij} = \left[\begin{array}{c}\text{Yield of milk from the cow of} \\ j^{\text{th}} \text{ breed or stock, fed on the ration } i\end{array}\right]$; $i = 1, 2, \ldots, k; j = 1, 2, \ldots, h.$

The suffix i refers to the treatments (rations) and the suffix j refers to the breed and stock of the cow. The yields can be expressed as variable values in the following $k \times h$ Two-Way table. One factor of variation (breed and stock of cows) is represented along the columns and the other factor of variation (treatments) is represented along the rows of table.

		Varieties of Cows						
		1	2	…	j	…	h	Row Totals
Treatments (Rations)	1	x_{11}	x_{12}	…	x_{1j}	…	x_{1h}	$x_{1\cdot}$
	2	x_{21}	x_{22}	…	x_{2j}	…	x_{2h}	$x_{2\cdot}$
	⋮	⋮	⋮		⋮		⋮	⋮
	i	x_{i1}	x_{i2}	…	x_{ij}	…	x_{ih}	$x_{i\cdot}$
	⋮	⋮	⋮		⋮		⋮	⋮
	k	x_{k1}	x_{k2}	…	x_{kj}	…	x_{kh}	$x_{k\cdot}$
	Column Totals	$x_{\cdot 1}$	$x_{\cdot 2}$	…	$x_{\cdot j}$	…	$x_{\cdot h}$	$x_{\cdot\cdot}$

Table A2.7

In Two-Way classification, the total variation in the observation x_{ij} can be split into the following three components.

1. The variation between the classes due to factor one represented along the k rows of the table, that is the **variation between the treatments (rations)**.

2. The variation between the classes due to the other factor represented along the h columns of the table, that is, the **variation between varieties** (breed or stock) of the cows.

3. The inherent variation within the observations of each class due to a combination of a large number of uncontrolled or extraneous factors of random nature, **chance variation**.

Our objective in Two-Way ANOVA is to examine if the row means or column means differ significantly in view of inherent variability within the different classes.

A. Assumptions for the Two-Way ANOVA

1. The parent population from which the samples are drawn must be normally distributed.
2. The observations (samples) must be independent.
3. The population variances from which the samples were drawn must be equal.
4. Sample sizes must be equal for each group.

B. Hypothesis Testing for Two-Factor ANOVA

Null and Alternative Hypothesis

We set up the **null hypothesis** that the treatments as well as varieties are homogenous (equal). In other words, the null hypothesis for treatments and varieties are respectively:

$$\begin{cases} H_{0T} : \mu_{1T} = \mu_{2T} = \ldots = \mu_{kT} \\ H_{0V} : \mu_{1V} = \mu_{2V} = \ldots = \mu_{hV} \end{cases}$$

Alternative Hypotheses:

$$\begin{cases} H_{1T} : \text{At least one of the } \mu_{iT} \text{'s is different} \\ H_{1V} : \text{At least one of the } \mu_{jV} \text{'s is different} \end{cases}$$

- **Test Statistics**

Under the null hypothesis $H_{0T} : \mu_{1T} = \mu_{2T} = \ldots = \mu_{kT}$, the test statistics is:

$$F_T = \frac{MS_T}{MS_e} \sim F_{k-1,\,(k-1)(h-1)}(\alpha)$$

Under the null hypothesis $H_{0V} : \mu_{1V} = \mu_{2V} = \ldots = \mu_{hV}$, the test statistics is:

$$F_V = \frac{MS_V}{MS_e} \sim F_{h-1,\,(k-1)(h-1)}(\alpha)$$

Summary of Two-Way ANOVA Calculations for test statistics

Source of Variation	Sum of squares (SS)	df	Mean Sum of squares (MS)	Variation Ratio(F)
Treatments Factor A (Rows)	$SS_T = h\sum_i \left(\bar{x}_{i.} - \bar{\bar{x}}\right)^2$	$k-1$	$MS_T = \dfrac{SS_T}{k-1}$	$F_T = \dfrac{MS_T}{MS_e}$, $df = (k-1, (h-1)(k-1))$
Varieties Factor B (Columns)	$SS_W = k\sum_j \left(\bar{x}_{.j} - \bar{\bar{x}}\right)^2$	$h-1$	$MS_V = \dfrac{SS_V}{h-1}$	$F_V = \dfrac{MS_V}{MS_e}$, $df = (h-1, (h-1)(k-1))$
Residuals (Error)	SS_e	$(k-1) \times (h-1)$	$MS_e = \dfrac{SS_e}{(k-1)(h-1)}$	

Table A2.8

Note: Since the computational procedure for the test statistics' in Two-Way ANOVA is quite complicated and lengthy, we will use the **Anova Two-Factor without Replication** package from the Data Analysis Group in Excel® to carry out the computations for **Sum of Squares, Mean Sum of Squares,** and the **F values**.

- ### Critical Values and Conclusion

By comparing the test statistics values with the **tabulated (critical) values** of F for respective df and at the given level of significance, the null hypotheses of the homogeneity of various treatments and various varieties may be rejected or not rejected.

Example 5: Five doctors each test five treatments for a certain disease and observe the number of days each patient takes to recover. The results (recovery time in days) are given bellow.

		Treatments				
Doctors		1	2	3	4	5
	1	10	14	23	18	20
	2	11	15	24	17	21
	3	9	12	20	16	19
	4	8	13	17	17	20
	5	12	15	19	15	22

Test the significance of variation between the doctors and treatments at a 5 % level of significance.

Solution : Here the factors of variation are

A: Doctors, represented along the rows of the given table; and

B: Treatments, represented along the columns of the given table.

Null Hypotheses: The doctors as well as the treatments are homogeneous.

$H_{0A}: \mu_{1A} = \mu_{2A} = \mu_{3A} = \mu_{4A} = \mu_{5A}$ that is,

there is no significant difference between the doctors.

$H_{0B}: \mu_{1B} = \mu_{2B} = \mu_{3B} = \mu_{4B} = \mu_{5B}$ that is,

there is no significant difference between the treatments.

Alternative Hypotheses:

H_{1A} : At least one of the doctors differs significantly.

H_{1B} : At least one of the treatments differs significantly.

Test Statistics:

From the Data Analysis Dialog Box select ANOVA: Two-Factor Without Replication and fill in the required cells in the dialog box and press OK.

Anova: Two-factor without Replication using Data Analysis Dialog box

SUMMARY	Count	Sum	Average	Variance		
1	5	85	17	26		
2	5	88	17.6	25.8		
3	5	76	15.2	21.7		
4	5	75	15	21.5		
5	5	83	16.6	15.3		
1	5	50	10	2.5		
2	5	69	13.8	1.7		
3	5	103	20.6	8.3		
4	5	83	16.6	1.3		
5	5	102	20.4	1.3		
ANOVA						
Source of Variation	SS	df	MS	F	P-value	F crit
Rows	25.84	4	6.46	2.9907407	0.0508117	3.0069173
Columns	406.64	4	101.66	47.064815	1.187E-08	3.0069173
Error	34.56	16	2.16			
Total	467.04	24				

Figure A2.6

Using the information from the above table, under the null hypotheses H_{0A} and H_{0B} we get

$$F_A = \frac{MS_T}{MS_e} = 2.991 \sim F_{(4,\ 16)} \quad \text{and} \quad F_B = \frac{MS_V}{MS_e} = 47.065 \sim F_{(4,\ 16)}.$$

Critical Values: From table 6(c), the critical (tabulated) value of F for (4, 16) df at 5% level of significance is 3.007.

Decision and Conclusion:

Since the calculated value of test statistics $F_A = 2.991$ is less than the critical value (3.007), we fail to reject the null hypothesis H_{0A} at 5% level of significance. This means that the data does not provide sufficient evidence against H_{0A}; that is, the difference between the means is not significant. Hence, we may conclude that there is no significant difference between the doctors; that is, all five doctors may be regarded as equally good.

Since the calculated value of test statistics $F_B = 47.065$ is very much greater than the critical value of 3.007, we reject the null hypothesis H_{0B} at a 5% level of significance and conclude that difference between the treatment is highly significant.

A2.5: Analysis of Variance with Microsoft Excel®

> **Objectives**
>
> After completing this section you will be able to carry out the computations in this chapter using Microsoft Excel®.

One-Way ANOVA using Data Analysis package in Excel®:

Illustration 1: A school administrator wants to test the consistency of scores in several Statistics classes. To study the equality of average scores, data from 3 randomly selected Statistics classes was collected.

	Sample Test Scores In Each Class						
Statistics Class A	87	65	80	90	34	65	88
Statistics Class B	65	78	98	54	34	34	66
Statistics Class C	90	100	34	65	66	78	86

Test the claim that the averages are the same. Use a 0.05 significance level.

Solution : $H_0 : \mu_A = \mu_B = \mu_C$ (there is no significant difference between the means)

H_1 : At least one mean significantly differs from the others

Test Statistics and the Critical Value:

Input the data in the given table into a worksheet, and from the Data Tab, click on the Data Analysis icon in the Analysis Group as shown below.

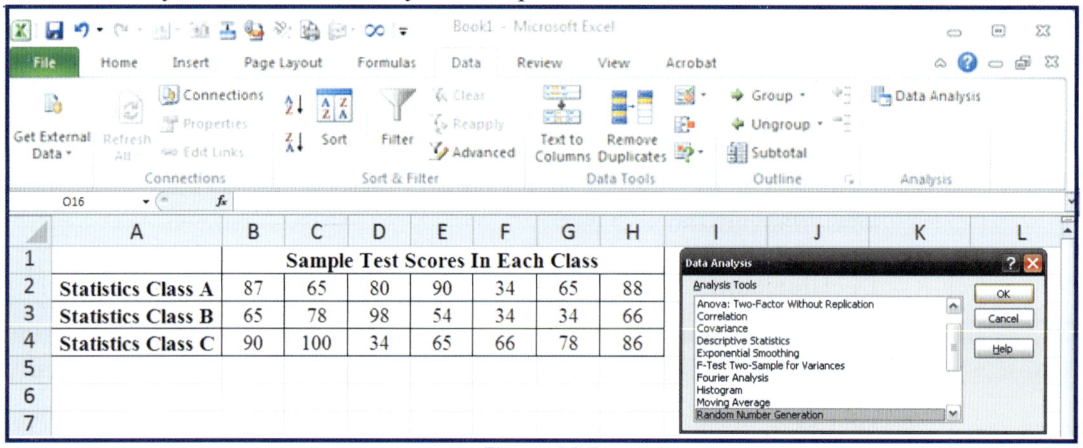

Figure A2.7 a

Select **Anova: Single Factor** and press OK. The following dialog box will appear.

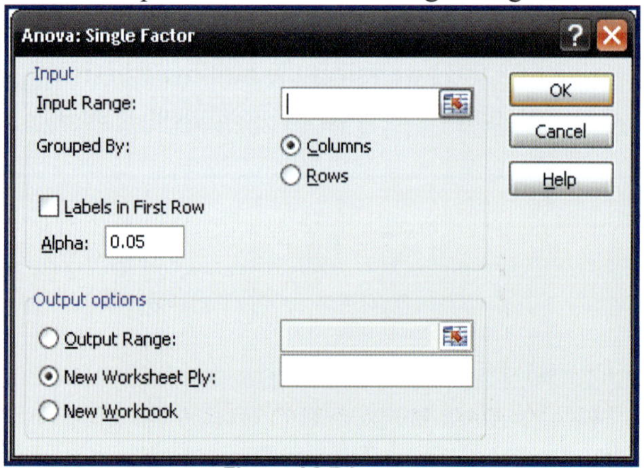

Figure A2.7 b

Fill out the input boxes as shown below:

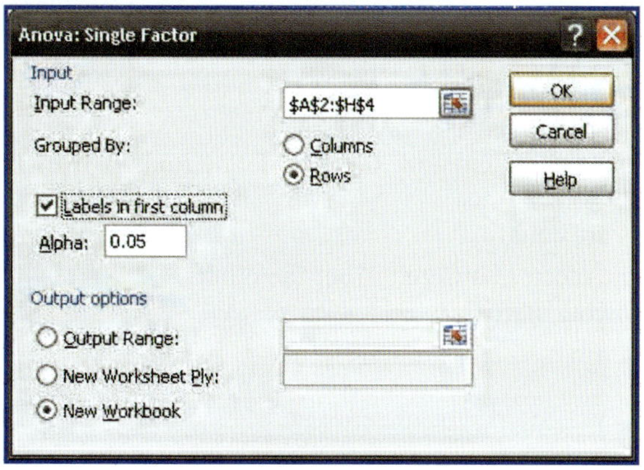

Figure A2.7 c

Press OK. The following results will display and show all the necessary information.

	A	B	C	D	E	F	G
1	**Anova: Single Factor**						
2							
3	**SUMMARY**						
4	***Groups***	***Count***	***Sum***	***Average***	***Variance***		
5	**Statistics Class A**	7	509	72.71429	401.2381		
6	**Statistics Class B**	7	429	61.28571	534.2381		
7	**Statistics Class C**	7	519	74.14286	472.8095		
8							
9	**ANOVA**						
10	***Source of Variation***	***SS***	***df***	***MS***	***F***	***P-value***	***F crit***
11	**Between Groups**	695.2381	2	347.619	0.740515	0.490846	3.554557
12	**Within Groups**	8449.714	18	469.4286			
13							
14	**Total**	9144.952	20				

Figure A2.7 d

In the display above, ***F** crit* represents the critical value (***F*** value = 3.555), ***F*** represents the test statistics = 0.741, and the ***p***-value = 0.491.

Decision and conclusion:

Since the test statistics lies outside the rejection region, we fail to reject the null hypothesis. There does not seem to be a significant difference between the means.

P-value Method

$H_0 : \mu_A = \mu_B = \mu_C$ (there is no significant difference between the means)

H_1 : At least one mean significantly differs from the others

_P_-value

The ***p***-value of the computed test statistics $F = 0.741$ is given by

$$P\text{-value} = P\left(F > F_{(2,18)}(0.05)\right) = 0.491$$

Decision and conclusion:

Since the ***p***-value is greater than $\alpha = 0.05$, we fail to reject H_0. There does not seem to be a significant difference between the means.

One-Way Analysis of Variance

In exercises 1-6, use the Classical and the *P*-value methods to test the following hypotheses.

1. H_0 : All means are equal
H_1 : At least one mean is different from the others.
Given the following statistics:
$\bar{x}_1 = 5.75, s_1 = 1.3, n_1 = 4;\ \bar{x}_2 = 3.5, s_2 = 1.5, n_2 = 6;\ \bar{x}_3 = 7.2, s_3 = 1.6, n_3 = 8;$ and $\alpha = 0.05$.

2. H_0 : All means are equal
H_1 : At least one mean is different from the others.
Given the following statistics:
$\bar{x}_1 = 20.4, s_1 = 18.2, n_1 = 11;\ \bar{x}_2 = 38.3, s_2 = 15.1, n_2 = 10;\ \bar{x}_3 = 44.1, s_3 = 15.5, n_3 = 9;$ and $\alpha = 0.01$.

3. H_0 : All means are equal
H_1 : At least one mean is different from the others.
Given the following statistics:
$\bar{x}_1 = 165.25, s_1 = 18.98, n_1 = 8;\ \bar{x}_2 = 157.57, s_2 = 3.82, n_2 = 7;\ \bar{x}_3 = 180.11, s_3 = 14.32, n_3 = 9;$ and $\alpha = 0.025$.

4. H_0 : All means are equal
H_1 : At least one mean is different from the others.
Given the following statistics:
$\bar{x}_1 = 58, s_1 = 11, n_1 = 15;\ \bar{x}_2 = 62, s_2 = 17, n_2 = 15;\ \bar{x}_3 = 58, s_3 = 14, n_3 = 15;$ and $\alpha = 0.05$.

5. H_0 : All means are equal
H_1 : At least one mean is different from the others.

Sample A	Sample B	Sample C
30	32	30
28	31	26
24	28	24
26	36	32

6. H_0 : All means are equal
H_1 : At least one mean is different from the others.

Sample A	Sample B	Sample C
52	78	74
81	93	68
61	58	73
89	87	50
82		89
		90

Use both the Classical and *P*-value methods to test the hypotheses in the following exercises.

7. The following table contains the prices of 6 different model cars in 4 different dealerships. Using a 0.05 significance level, test the claim that the mean price is the same for 4 different dealerships.

	Dealership I	Dealership II	Dealership III	Dealership IV
Prices in $	54000	50000	44000	50000
	28000	33000	40000	31000
	43000	50000	55000	40000
	78000	82000	85000	80000
	93000	88000	80000	90000
	38000	40000	33000	42000

8. A hospital administrator plans to test the hypothesis that the average number of days a patient stays in 3 neighboring hospitals is the same. The hospitals tested are St. Joseph, Memorial, and Cedars- Sinai. An examination of the hospital records reveals that 8 patients who stayed in each hospital spent the number of days outlined in the following table. Test the hypothesis using a 0.05 significance level.

	Day Spent in the Hospital							
St. Joseph Hospital	9	5	7	9	7	8	2	3
Memorial Hospital	4	7	6	8	9	4	5	2
Cedars- Sinai Hospital	2	1	6	5	4	9	7	8

9. 16 college students were randomly assigned to three groups in order to experiment with three different methods of teaching Algebra. At the end of the semester, the same test was given to all students. The table below gives the scores of students in the three groups.

Method I	Method II	Method III
48	55	84
73	85	68
51	70	95
65	69	74
87	67	
95		
69		

At a 1% significance level, can we reject the null hypothesis that the mean algebra score of all students taught by each of these three methods is the same?

10. A researcher wishes to see whether there is any difference in the weight gains of athletes following one of three special diets. Athletes are randomly assigned to three groups and placed on the diet for six week. The weight gains (in pounds) are shown here. At $\alpha = 0.05$, can the researcher conclude that there is a difference in the mean diets.

Diet A	Diet B	Diet C
3	10	8
6	12	3
7	11	2
4	14	5
	8	
	6	

11. The ages of top executives in three major corporations are listed below.

Corporation 1	49	58	55	60	62
Corporation 2	61	59	58	54	
Corporation 3	56	63	60	48	

Use the analysis of variance at the 0.01 level of significance to determine whether or not the average ages of top executives at these corporations differ significantly.

12. The same size houses in the same area are heated with gas, electricity, and heat pump. The following table lists random samples taken from each group.

Heat Pump	220	170	190	180	
Electricity	200	185	210	195	
Gas	191	165	170	150	160

Use the analysis of variance at the 0.01 level of significance to indicate whether or not the average monthly heating bills using various forms of heating differ significantly.

13. Random samples of same size deluxe room air-conditioners from three manufacturers showed the following life expectancies (in months of operation).

Manufacturer 1	34	28	32	29	31	
Manufacturer 2	23	34	37	32		
Manufacturer 3	30	39	35	37	32	31

Use the analysis of variance at the 0.05 level of significance to indicate whether or not the average life expectancies of three different brands of room air-conditioners differ significantly.

14. Use the following data to compare four treatments.

Treatment 1	3	5	1	6	
Treatment 2	2	4	6	5	3
Treatment 3	1	7	5	4	6
Treatment 4	3	2	4		

Use the analysis of variance at the 0.05 level of significance to determine whether or not the treatment means differ significantly.

15. A manufacturer of machine parts is considering one of the 4 machines currently in the market. The following is the daily output on 5 randomly selected days for each machine.

Machine I	72	56	68	65	60
Machine II	62	70	66	64	78
Machine III	68	72	74	70	66
Machine IV	64	72	68	68	58

Do the machines have an equal average output rate? Use a 0.01 significance level.

16. A manufacturing company has purchased three new machines of different makes and wishes to determine whether one of them is faster than the others in producing a certain product. Five hourly production figures are observed at random from each machine and the results are given in the following table.

	Machine A_1	**Machine A_2**	**Machine A_3**
Observations	25	31	24
	30	39	30
	36	38	28
	38	42	25
	31	35	28

Using a 0.05 significance level, test the claim that the means are the same.

17. Weight gain (in pounds) in pigs subjected to four different treatments are given in the following table.

TREATMENTS			
1	2	3	4
16.5	16.8	16.4	18.5
15.9	18.0	15.6	19.5
15.9	18.0	16.9	18.6
16.7	16.6	13.8	20.1
17.0	17.0	15.3	16.5
14.6	16.1	19.0	17.5
13.0	17.0	16.0	18.7
16.4	17.9	14.2	16.6
15.8	19.1	15.5	16.5

Use a One-Way ANOVA to test the equality of the means.

Two-Way ANOVA Exercises

18. To study the performance of three detergents and three different water temperatures, the following "Whiteness" readings were obtained with specially designed equipment.

Water Temperature	Detergent A	Detergent B	Detergent C
Cold water	55	51	68
Warm water	52	49	67
Hot water	51	49	56

Perform a two-way analysis of variance, using 5% level of significance.

19. Three different Algebra classes were taught using three different methods of teaching. The following table lists the average scores obtained at the end of 3 semesters.

Classes	Detergent A	Detergent B	Detergent C
Class A	65	82	73
Class B	79	60	68
Class C	70	71	70

Perform a two-way analysis of variance, using 1% level of significance.

20. The following are the defective pieces produced by four operators working in turn on four different machines.

	Operator			
Machine	B_1	B_2	B_3	B_4
A_1	34	28	33	29
A_2	31	24	35	22
A_3	27	20	43	72
A_4	28	28	29	26

Perform analysis of variance at 0.05 level of significance to as certain whether variability in production is due to variability in operator's performance or variability in machine's performance.

21. In an experiment on the effects of temperature conditions in human performance, 8 practiced subjects were given a sensorimotor test in each of 4 temperature conditions. Since the subjects were all practiced, the order in which the tests were done was unimportant. The tests were randomized amongst the subjects, so that for each condition there were equal numbers of first testing, second testing, third testing and fourth testing. The scores in the tests are shown in the following table.

	Subjects							
Temperature	**1**	**2**	**3**	**4**	**5**	**6**	**7**	**8**
1	76	80	79	90	85	101	94	83
2	75	81	77	90	86	98	93	85
3	76	78	66	91	82	98	92	83
4	68	75	72	85	82	90	82	77

Perform a Two-Way ANOVA to show whether there is any significant difference between the scores of subjects due to temperature conditions.

22. The following table gives quality rating of ten service stations by five professional raters.

	Service Station									
Rater	**1**	**2**	**3**	**4**	**5**	**6**	**7**	**8**	**9**	**10**
A	99	70	90	99	65	85	75	70	85	92
B	96	65	80	95	70	88	70	51	84	91
C	95	60	48	87	48	75	71	93	80	93
D	98	65	70	95	67	82	73	94	86	80
E	97	65	62	99	60	80	76	92	90	89

Use a Two-Way ANOVA to analyze the data and discuss whether there is a significant difference between raters or between service stations.

23. Four experimenters determine the moisture content of samples of a powder, each by taking a random sample from each of the six consignments. Their assessment are given in the table below.

	Consignments					
Observers	**1**	**2**	**3**	**4**	**5**	**6**
1	9	10	9	10	11	11
2	12	11	9	11	10	10
3	11	10	10	12	11	10
4	12	13	11	14	12	10

(a) Perform an analysis of variance on the given data and discuss whether there is a significant difference between consignments or between observers.

(b) If there is a significant difference between the levels of any of the factor(s) of variation, then determine the corresponding pairs of sample means which differ significantly. Use a 0.01 significance level.

24. Perform a Two-Way ANOVA on the data given below. (X_{ij}'s are the given observations)

	Treatment I				
Treatment II	**i**	**ii**	**iii**	**iv**	**v**
i	30	24	33	36	27
ii	26	29	24	31	35
iii	38	28	35	30	35

25. Four varieties of potato are planted, each on five plots of the same size and shape. Each variety is treated with five different fertilizers. The yields in tons are as follows:

Variety \ Plots	1	2	3	4	5
1	19	22	26	18	21
2	25	19	23	26	22
3	17	19	22	20	21
4	21	18	25	23	24

Perform an Analysis of Variance and show whether there is any significant difference between the yields of different varieties due to different fertilizers.

26. The following data represent the number of units of production per day turned out by 5 different workmen using different types of machines.

Workmen	Machine Types			
	A	B	C	D
1	46	40	49	38
2	48	42	54	45
3	36	38	46	34
4	35	40	48	35
5	40	44	51	41

(a) Test whether the mean productivity is the same for the four brands of machine types.

(b) Test whether 5 different workmen differ with respect to productivity.

APPENDIX 3

NON-PARAMETRIC STATISTICS

Appendix Overview:

A3.1: INTRODUCTION

Statistical tests that make inferences about population parameters are called Parametric Statistical Tests. Most parametric tests rely on the fact that the distribution of the random variable being sampled is known and the population from which the samples are drawn is normally distributed. This has been mostly, if not completely, the method of approach in the materials presented so far. In practice, however, the distribution of the samples drawn from a population is not known. In such situation, we need statistical methods which do not depend on the distribution of the sampled population. That is, we need statistical tests which do not make any assumption regarding the distribution of the parent population. Tests that satisfy this criterion are called distribution free or Non-Parametric tests. In this appendix, we will discuss several such methods.

A. Parametric and Non-Parametric Tests

In previous chapters, almost all the **tests of significance** were based on the assumption that the parent populations were **normally distributed**. Such tests, which deal with the parameters of the population, are known as Parametric Tests. The Z, t, χ^2, and F-tests are examples of such tests.

- **Parametric Test**

  A Parametric statistical test is a test whose model specifies certain conditions about the parameters of the population from which the samples are drawn.

- **Non-Parametric Test**

  A Non-Parametric statistical test is a test that does not depend on the particular distribution of the target population from which the samples are drawn. In other words, Non-Parametric tests do not make any assumption regarding the distribution of the population.

B. Advantages and Limitations of Non-Parametric Methods Over Parametric Methods

Below we will give a brief and comparative study of Parametric and Non-Parametric methods and their relative merits and demerits.

Advantages of Non-Parametric Methods

1. Non-Parametric methods are readily comprehensible, very simple and easy to apply and do not require complicated sampling theory. Moreover, most Non-Parametric tests do not require lengthy and tedious calculations, consequently less time consuming.

2. The Non-Parametric methods require fewer (assumptions) and less stringent restrictions about the parent population than their parametric counterparts.

3. Non-Parametric techniques will apply to and account for data which are mere classifications (measured in nominal scale), while Non-Parametric methods exists to deal with such data. Non-Parametric methods are specially suitable for studies in behavioral sciences like psychology, education, social sciences etc. where the data can generally be classified or ranked but not exactly measured. For instance, no parametric test can be applied if the scores of a college student are given in grades such as A^+, A, A^-, B^+, B, B^- and so on. In fact Non-Parametric tests can be used for all types of data:

   i) Qualitative data (Nominal scaling like "increase, decrease or no change"; "less than, greater than, equal to" etc.)

   ii) Ranked data (ordinal scaling)

   iii) Quantitative data, the data which has been measured more precisely (Interval or Ratio scaling.)

4. Non-Parametric methods can be used in situations where the normality assumption is not true. Since, in general the socio-economic data are not normally distributed, Non-Parametric tests have found applications in psychometry, sociology and educational statistics.

Limitations of Non-Parametric Tests

1. Non-Parametric tests can be used if the measurements are nominal or ordinal. Even in that case, if a parametric test exists, it is more powerful than the Non-Parametric test. In other words, if all the assumptions of a statistical model are satisfied by the data and if the measurements are of required strength, then the Non-Parametric tests are unnecessary.

2. In Parametric methods we compute the sample statistics and use them to obtain the point estimates or interval estimates of the corresponding population parameters. On the other hand, Non-Parametric methods are based on the data which are nominal and ordinal in nature and hence are designed to test statistical hypothesis only and not estimating the parameters.

3. The Non-Parametric methods result in loss of information as they are based on the signs (plus or minus) of the deviations or the ranks of the variable values and not on the exact magnitude of the variable values. However, Non-Parametric statistics are, generally, only slightly less efficient than their parametric counterparts.

C. Four Main Non-Parametric Tests

Here is the list of Non-Parametric methods discussed in this appendix.

1. **Sign Test**

   i) The Single Sample Sign Test (Large Sample)

   ii) The Single Sample Sign Test (Small Sample)

   iii) The Single Sample Sign Test (Paired Sample)

2. **Wilcoxon Signed-Rank Test**

   i) Wilcoxon Rank Sum Test (Small Sample)

   ii) Wilcoxon Rank Sum Test (Large Sample)

   iii) Wilcoxon Rank Sum Test (Matched Pairs)

3. **Mann-Whitney Test**

   i) Mann-Whitney Test For Independent Samples (Small Sample)

   ii) Mann-Whitney Test For Independent Samples (Large Sample)

4. **Spearman's Test for Rank Correlation Coefficient**

Note: The hypothesis testing procedure in this chapter remains the same as previous chapters.

i) State the null and alternative hypotheses.

ii) Find the test statistics.

iii) Find the rejection region, and critical value(s).

iv) Write the decision rule, and state the conclusion in simple words.

A3.2: Sign Test

Sign test is the simplest of all the Non-Parametric tests. As the name suggests, it is based on the signs (plus or minus) of the deviations rather than the exact magnitudes of the variable values. In this section, we will discuss:

- Single sample sign test.
- Paired samples sign test.

Objectives

After completing this section you will be able to understand:

A. Single Sample Sign Test.

B. Paired Samples Sign Test.

A. Single Sample Sign Test

Single sample sign test is used to test the hypothesis concerning the **median** for one population. Suppose we want to test the hypothesis that the median (η) of a population has a specified value, η_0; that is, $H_0 : \eta = \eta_0$. As usual, the alternative hypothesis H_1 may be one-tailed or two-tailed.

Left-tailed	Right-tailed	Two-tailed
$H_1 : \eta < \eta_0$	$H_1 : \eta > \eta_0$	$H_1 : \eta \neq \eta_0$

Table A3.1

- **Testing Procedure**

Let $x_1, x_2, \ldots, x_n$ be a random sample of size n from the given population with median $\eta = \eta_0$ under H_0. Subtract the value η_0 from each of the sample observations $(x_i - \eta_0)$ and write:

i) a plus sign (+), if the deviation is positive. That is, $x_i - \eta_0 > 0$.

ii) a minus sign (–), if the deviation is negative. That is, $x_i - \eta_0 < 0$.

iii) zero (0), if the deviation is zero. That is, $x_i - \eta_0 = 0$.

Alternatively, replace each sample value exceeding η_0 with a plus (+) sign, each sample value less than η_0 with a minus (–) sign, and each value equal to η_0 with zero (0).

By the definition of median, and based on the continuity assumption,

$$P(X > \text{Median}) = P(X < \text{Median}) = 0.5,$$

that is, under $H_0 : P(X > \eta_0) = P(X < \eta_0) = 0.5$.

Hence, if the null hypothesis is true, then the number of plus signs should be approximately equal to the number of minus signs. If one sign is more frequent than the other sign, then we reject the null hypothesis. If the difference in the number of plus and minus signs is due to chance variations (or fluctuations of sampling), then we fail to reject the null hypothesis. To determine whether or not the difference between the number of plus and minus signs is due to chance variations, we use the usual hypothesis testing procedure discussed in chapter 8. We describe the sign test in two situations:

i) When the sample size is large.

ii) When the sample size is small.

> ***Remark:***
>
> A zero value is neither positive nor negative. Hence, in case of ties, that is, if any deviation is zero, we discard these tied observations (or zeros) and reduce the sample size by the number of zeros. In other words, the effective sample size becomes the total number of plus and minus signs. Since this results in loss of the information contained in the sample, this method is not recommended if the sample size is small.

- ## Single Sample Sign Test (Large Sample)

Let us define X = number of plus signs and n = Total number of non zero deviations.

By the above discussion, the given sample data is dichotomized into two classes of plus and minus signs and each sign is the result of an independent trial.

Under $H_0 : P(X > \eta_0) = P(X < \eta_0) = 0.5$ implies that p = probability of positive sign = 0.5, which is constant for each trial. Hence, under H_0, the variable X has a binomial distribution with parameters n and $p = 0.5 \Rightarrow q = 1 - p = 0.5$. Therefore, the mean and standard deviation are as follows:

$$\text{mean } (\mu) = np = \frac{n}{2} \;;\; \text{Standard deviation } (\sigma) = \sqrt{npq} = \sqrt{n\left(\frac{1}{2}\right)\left(\frac{1}{2}\right)} = \frac{1}{2}\sqrt{n}\,.$$

If n is large, we use normal approximation to the binomial distribution after applying the continuity correction and then apply the normal test.

Hence, for large samples ($n > 25$), the test statistic is

$$Z = \frac{(X + 0.5) - \frac{n}{2}}{\frac{1}{2}\sqrt{n}} \quad N(0,1)\text{ , if } X < \frac{n}{2} \qquad (1)$$

or

$$Z = \frac{(X - 0.5) - \frac{n}{2}}{\frac{1}{2}\sqrt{n}} \quad N(0,1)\text{ , if } X > \frac{n}{2} \qquad (2)$$

For testing the significance of Z, we can use either the Classical Method, P-value Method or, Method 3 as an alternative.

Method 1: Classical Method

***Step* 1** Set up the null and alternative hypotheses as follows:

$H_0: \eta = \eta_0$

$$H_1: \begin{cases} \eta < \eta_0 & \text{or} \\ \eta > \eta_0 & \text{or} \\ \eta \neq \eta_0 \end{cases}$$

***Step* 2 Test Statistic**

Subtract η_0 from each sample observation and replace each deviation by its sign (+ or –) or zero (if the deviation is zero.) Find X = number of positive deviations and n = Total number of plus and minus signs after dropping out zeros, if any.

Compute the **test statistic** Z by using equation (1) or (2).

***Step* 3 Critical Values**

Obtain the critical value(s) of Z for left-tailed, right-tailed, or two-tailed tests at an α significance level, from Table 3.

***Step* 4 Decision and Conclusion**

If the calculated value of the test statistics falls in the critical region, then we reject H_0, otherwise fail to reject H_0. Write the conclusion of the test in simple language.

***Method* 2: *P*-Value Method**

***Step* 1** Set up the null and alternative hypotheses.

$H_0: \eta = \eta_0$

$$H_1: \begin{cases} \eta < \eta_0 & \text{or} \\ \eta > \eta_0 & \text{or} \\ \eta \neq \eta_0 \end{cases}$$

***Step* 2 Test Statistic**

Subtract η_0 from each sample observation and replace each deviation by its sign (+ or –) or zero (if the deviation is zero.) Find X = number of positive deviations and n = Total number of plus and minus signs after dropping out zeros, if any.

Compute the **test statistic** Z by using equation (1) or (2).

***Step* 3 *P*-value**

Calculate the ***p***-value of the test statistics obtained in Step 2 as discussed in Chapter 7.

***Step* 4 Decision and Conclusion**

If ***p***-value < α, we reject H_0 at α significance level, otherwise we fail to reject H_0.

***Method* 3:**

The significance of the difference between positive and negative signs can also be tested by using the following test statistic.

Test Statistic

X = Smaller of [Number of plus signs, Number of minus signs]

Critical Value

The critical value of X for one-tailed and two-tailed tests are given in Table 8 at the end of the book. Since the test statistic is the number of less frequent signs, the critical region is in the left hand tail.

Decision

Reject H_0 if X is less than or equal to the critical value given in Table 8, otherwise we fail to reject H_0.

Example 1: To test the claim that the median age of mathematics faculty in California Community Colleges is less than 42 years, the results from a random sample of 33 mathematics faculty gave the following ages (in years).

56	62	61	54	52	32	24	35	50	42	52
49	26	31	31	54	38	36	45	53	37	40
38	31	29	25	45	52	48	39	30	38	64

Use the sign test at the 0.05 level of significance to test the claim by Classical and ***P***-value Method.

Solution : ***Method* 1: Classical Method**

***Step* 1 Null and Alternative Hypotheses**

$H_0 : \eta = 42$

$H_1 : \eta < 42$

***Step* 2 Test Statistic**

Subtract 42 from each sample value and record the differences in a table.

14	20	19	12	10	−10	−18	−7	8	0	10
7	−16	−11	−11	12	−4	−6	3	11	−5	−2
−4	−11	−13	−17	3	10	6	−3	−12	−4	22

Table A3.2

Use the Excel command "=COUNTIF(range, criteria)" to count the number of plus signs in the table.

Note: The range in the countif command represents the range of the original data in Excel®, and criteria in the sign test means ">0".

	A	B	C	D	E	F	G	H	I	J	K
1	56	62	61	54	52	32	24	35	50	42	52
2	49	26	31	31	54	38	36	45	53	37	40
3	38	31	29	25	45	52	48	39	30	38	64
4											
5	14	20	19	12	10	-10	-18	-7	8	0	10
6	7	-16	-11	-11	12	-4	-6	3	11	-5	-2
7	-4	-11	-13	-17	3	10	6	-3	-12	-4	22
8											
9	*X*	=COUNTIF(A5:K7,">0")									
10		COUNTIF(range, criteria)									

Figure A3.1

X = number of plus signs = 15, and n = 33 − 1 = 32.

Since $n > 25$, we use the normal test. After applying the continuity correction for normal approximation to the binomial distribution, given that $X < \frac{n}{2}$, the test statistic is

$$Z = \frac{(X+0.5)-\frac{n}{2}}{\frac{1}{2}\sqrt{n}} = \frac{(15+0.5)-\frac{32}{2}}{\frac{1}{2}\sqrt{32}} = -0.18.$$

Step 3 **Critical Values**

The critical value of Z for a left-tailed test at 5% level of significance is -1.645. The critical value, the rejection region and the test statistic are shown in Figure A3.2.

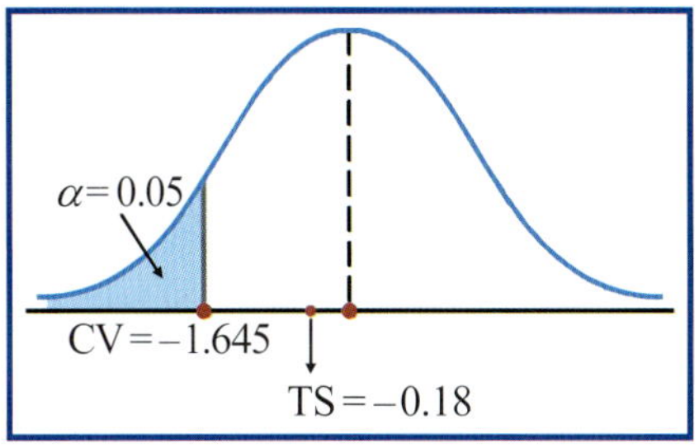

Figure A3.2

Step 4 **Decision and Conclusion**

Since $Z = -0.18$ falls outside of the rejection region, we fail to reject H_0. There is sufficient evidence against the claim that the median age of the mathematics faculty is less than 42.

Method 2: ***P*-Value Method**

Step 1 **Null and Alternative Hypotheses**

$H_0 : \eta = 42$

$H_1 : \eta < 42$

Step 2 **Test Statistic** $Z = -0.18$

Step 3 ***P*-value**

The ***p***-value of the test statistics for the left-tailed test is given by

$$p\text{-value} = P(Z < -0.18)$$
$$= 0.5 - 0.0714 = 0.4286.$$

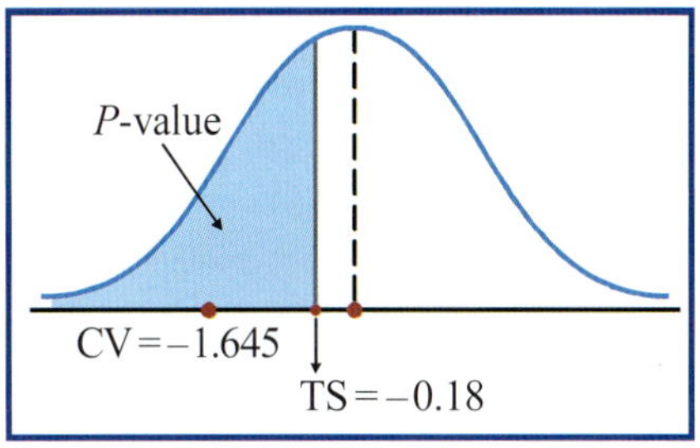

Figure A3.3

Step 4 **Decision and Conclusion**

Since the ***p***-value > α, we fail to reject H_0 at $\alpha = 0.05$. There is sufficient evidence against the claim that the median age of the mathematics faculty is less than 42.

In short:

Hypotheses	**Test Statistics**
$H_0: \eta = 42$ $H_1: \eta < 42$	$Z = \frac{(X+0.5)-\frac{n}{2}}{\frac{1}{2}\sqrt{n}} = -0.18$
Rejection Region and Critical Value 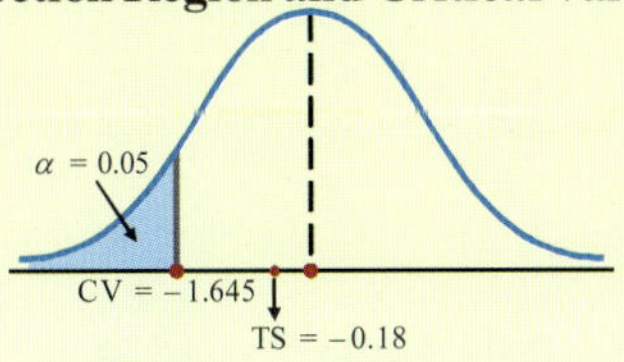	**Conclusion** Fail to reject H_0. There is sufficient evidence against the claim that the median age of the mathematics faculty is less than 42.

- ## Single Sample Sign Test (Small Sample)

This test is used to test the hypothesis concerning the median of a single population when the sample size is small. If the sample size is less than or equal to 25, then we cannot use the normal test as discussed earlier in this section. Here we proceed as follows.

Method 1: Classical Method

Step 1 Null and Alternative Hypotheses

$H_0: \eta = \eta_0$

$$H_1: \begin{cases} \eta < \eta_0 & \text{or} \\ \eta > \eta_0 & \text{or} \\ \eta \neq \eta_0 \end{cases}$$

Step 2 Test Statistic

Subtract median (η_0) from each data value and write '+'for positive difference and '–' for negative difference. The test statistics X = number of plus signs.

Step 3 Critical Value(s) and the Rejection Region

(a) The test criterion for the right-tailed test at α significance level is:

Reject H_0 if $X \geq x_\alpha$, where x_α is the smallest integer so that $P(X \geq x_\alpha)$ is as close to α as possible.

(b) The test criterion for the left-tailed test at α significance level is:

Reject H_0 if $X \leq x_\alpha$, where x_α is the largest integer so that $P(X \leq x_\alpha)$ is as close to α as possible.

(c) The test criterion for the two-tailed test at α significance level is:

Reject H_0 if $X > x_{\alpha/2}$ or $X < x'_{\alpha/2}$, where $x_{\alpha/2}$ is the smallest integer and $x'_{\alpha/2}$ is the largest integer so that $P(X \geq x_{\alpha/2})$ and $P(X \leq x'_{\alpha/2})$ are each as close to $(\alpha/2)$, as possible.

For a given value of n as obtained earlier in this section and $p = 0.5$, compute the critical values as explained above on using Table 2(a) of the binomial probabilities.

Step 4 Decision and Conclusion

If the value of test statistics X falls in the critical region then we reject H_0, otherwise fail to reject H_0. Write the conclusion in simple language.

Method 2: *P*-Value Method

Step 1 Null and Alternative Hypotheses

$H_0: \eta = \eta_0$

$$H_1: \begin{cases} \eta < \eta_0 & \text{or} \\ \eta > \eta_0 & \text{or} \\ \eta \neq \eta_0 \end{cases}$$

Step 2 Test Statistic

The test statistics X = number of plus signs = k (k is an arbitrary constant)

Step 3 *P*-value

P-value of the test statistic $X = k$ obtained in Step 2 is given by:

For a right-tailed test	For a left-tailed test	For a two-tailed test
P-value = $P(X \geq k)$	*P*-value = $P(X \leq n - k)$	*P*-value = $2P(X \geq k)$

Table A3.3

Step 4 **Decision and Conclusion**

If ***p***-value < α, we reject H_0 at α significance level, otherwise we fail to reject H_0. Write the conslusion in simple language.

Example 2: Following are responses to the question: "How many hours do you study before a major statistics exam?"

6	5	1	2	2	5	7	5	3	7	4	7

Use the sign test at a 0.05 significance level to test the hypothesis that the median number of hours a student studies before a statistics test is 3. Use the Classical and ***P***-value Methods.

Solution : ***Method* 1: Classical Method**

Step 1 **Null and Alternative Hypotheses**

$H_0 : \eta = 3$

$H_1 : \eta \neq 3$

Step 2 **Test Statistics**

Subtracting 3 from each observation and writing the sign of the difference.

x	6	5	1	2	2	5	7	5	3	7	4	7
$x - \eta$	3	2	−2	−1	−1	2	4	2	0	4	1	4
Sign of the difference	+	+	−	−	−	+	+	+	0	+	+	+

Table A3.4

By discarding 0, we have X = number of (+) signs = 8, and n = 12 – 1 = 11.

Step 3 **Critical Value(s) and the Rejection Region**

From Table 2(a) of the binomial probabilities, for n = 11 and p = 0.5, we have

$$\left.\begin{array}{l} P(X=0)=P(X=11)=0.000 \\ P(X=1)=P(X=10)=0.005 \\ P(X=2)=P(X=9)=0.027 \end{array}\right\} \Rightarrow \left\{\begin{array}{l} P(X\leq 1) = P(X\geq 10)=0.000+0.005 = 0.005 \\ P(X\leq 2)=P(X\geq 9) = 0.005+0.027 = 0.032 \end{array}\right. \quad (3)$$

For a two-tailed test, the rejection region has a probability 0.025 in each tail (because of symmetry of the binomial distribution under $H_0 : p = 0.5$). Of the two cumulative probabilities in (3), 0.032 is much closer to 0.025. Hence, the two-tailed rejection region that has a probability as close to α = 0.05 is given by

$$(X \geq 9 \text{ or } X \leq 2), \text{ that is; } (X = 9, 10, 11 \text{ or } X = 0, 1, 2).$$

The rejection region and the test statistics are shown in the following diagram.

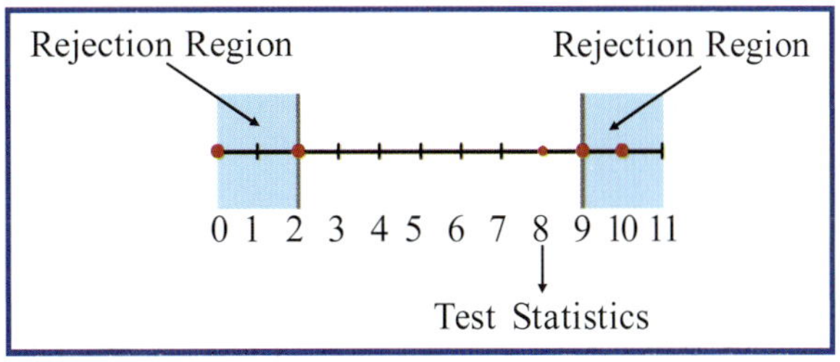

Figure A3.4

Step 4 Decision and Conclusion

The calculated value of test statistic $X = 8$, lies outside the rejection region. Therefore, we fail to reject H_0 at a 0.05 significance level. Hence, there is not sufficient evidence to reject the hypothesis that the median number of hours of study before a statistics test is 3 hours.

Method 2: P-Value Method

Step 1 Null and Alternative Hypotheses

$H_0: \eta = 3$

$H_1: \eta \neq 3$

Step 2 Test Statistic

The test statistics $X = 8$

Step 3 P-value

P-value of the test statistic $X = 8$ obtained in Step 2 is given by:

$$
\begin{aligned}
p\text{-value} &= 2P(X \geq 8) \\
&= 2[P(X = 8) + P(X = 9) + P(X = 10) + P(X = 11)] \\
&= 2(0.081 + 0.027 + 0.005 + 0.000) = 0.226
\end{aligned}
$$

Step 4 Decision and Conclusion

Since ***p***-value $> \alpha$, we fail to reject H_0 at a 0.05 significance level. Hence, there is not sufficient evidence to reject the hypothesis that the median number of hours of study before a statistics test is 3 hours.

In short:

Hypotheses	**Test Statistics**
$H_0: \eta = 3$ $H_1: \eta \neq 3$	X = number of (+) signs = 8, and $n = 11$.
Rejection Region and Critical Value 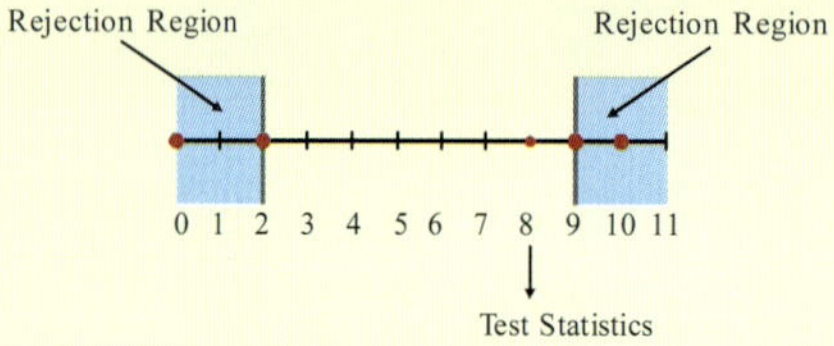	**Conclusion** Fail to reject H_0. There is not sufficient evidence to reject the hypothesis that the median number of hours of study before a statistics test is 3 hours.

Method 3:

$n = 11$, Number of plus signs = 8, Number of minus signs = 3.

Test Statistic:

X = Smaller of [Number of plus signs, Number of minus signs]

= Min[8, 3] = 3

Critical Value: From Table 8, the critical value of X for $n = 11$ at a 0.05 significance level, for two-tailed test is 1.

Decision: Since the computed value of test statistic $X = 3$ is greater than the critical value 1, we fail to reject H_0 at a 0.05 significance level. Hence, there is not sufficient evidence to reject the hypothesis that the median number of hours of study before a statistics test is 3 hours.

> ***Remark:***
>
> We may conclude that out of the three methods Classical Method, P-value Method and Method 3, the method based on test statistic X = Smaller of [Number of plus signs, Number of minus signs] is the most convenient to use, followed by the P-value Method.

B. Paired Samples Sign Test

The procedure of the single sample sign test can be extended to the paired sample problems. Let $(x_i, y_i) : i = 1, 2, \ldots, n$ be a random sample of n-paired observations. This test is similar to paired t-test in Chapter 8, where we assume that the samples were drawn from a normally distributed population. However, in the paired samples sign test, we will not make any assumptions regarding the distribution of the parent population. The only assumptions are:

1. Measurements are such that the deviations $d_i = x_i - y_i;\ i = 1, 2, \ldots, n$ can be expressed in terms of positive or negative signs.
2. Variables have continuous distributions.
3. d_i's ; $i = 1, 2, \ldots, n$ are independent.

***Step* 1 Null and Alternative Hypotheses**

$$H_0 : \eta_1 = \eta_2$$

$$H_1 : \begin{cases} \eta_1 < \eta_2 & \text{or} \\ \eta_1 > \eta_2 & \text{or} \\ \eta_1 \neq \eta_2 \end{cases}$$

***Step* 2 Test Statistic**

Assuming x-values are the first sample and y-values the second sample, find the deviations $d_i = x_i - y_i;\ i = 1, 2, \ldots, n$. Replace each of these deviations with their signs (+, –, 0). In case $x_i = y_i \Rightarrow d_i = 0$, discard these deviations and reduce the sample size by the number of such zeros. Under $H_0 : \eta_1 = \eta_2$, we would expect the number of plus signs to be approximately equal to the number of minus signs. Hence $d \sim B(n, p = 0.5)$.

Now, we can use any one of the three methods discussed in single sample sign test (small samples) to test H_0.

Example 3: Two women are randomly selected and their weights (in lbs.), before and after they are put on a new diet, are recorded. The data are:

Weight before diet	180	178	165	200	160	145	170	210	185	155
Weight after diet	174	181	157	198	152	150	160	205	178	160

Use the paired samples sign test at a 0.05 level of significance to test the claim that the new weight loss diet is effective.

Solution: Since the sample size is small, we will use the binomial distribution.

Method 1: Classical Method

***Step* 1** **Null and Alternative Hypotheses**

$H_0 : \eta_1 = \eta_2$

$H_1 : \eta_1 > \eta_2$

***Step* 2** **Test Statistic**

Subtracting the "weight after the diet" from "weight before the diet" and writing the sign of the differences, we get

Weight before diet (x)	180	178	165	200	160	145	170	210	185	155
Weight after diet (y)	174	181	157	198	152	150	160	205	178	160
Sign of the difference (d_i's)	+	–	+	+	+	–	+	+	+	–

Table A3.5

X = number of (+) signs = 7.

***Step* 3** **Critical Value(s) and the Rejection Region**

From Table 2(a) of the binomial probabilities, for $n = 10$ and $p = 0.5$, we obtain the one-tailed rejection region that has a binomial probability as close to 0.05 as possible. The region contains $X = 8$, 9, and 10, because

$$P(X \geq 8) = P(X = 8) + P(X = 9) + P(X = 10)$$
$$= 0.044 + 0.010 + 0.001 = 0.055$$

The rejection region and the test statistics are shown in the following diagram.

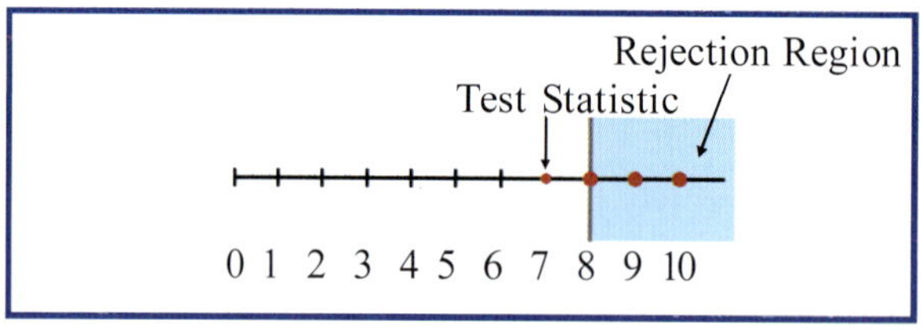

Figure A3.5

***Step* 4** **Decision and Conclusion**

The calculated value of test statistic $X = 7$, lies outside the rejection region. Therefore, we fail to reject H_0 at a 0.05 significance level. The data does not provide sufficient evidence against the null hypothesis which may be regarded as true and we may conclude that both diets are equally good.

Method 2: *P*-Value Method

***Step* 1** **Null and Alternative Hypotheses**

$H_0 : \eta_1 = \eta_2$

$H_1 : \eta_1 > \eta_2$

***Step* 2** **Test Statistic**

X = number of (+) signs = 7.

***Step* 3** ***P*-value**

***P*-value of the test statistic $X = 8$ obtained in Step 2 is given by:

$$\begin{aligned} p\text{-value} &= P(X \geq 7) \\ &= P(X = 7) + P(X = 8) + P(X = 9) + P(X = 10) \\ &= 0.117 + 0.044 + 0.010 + 0.001 \;=\; 0.0172 \end{aligned}$$

***Step* 4** **Decision and Conclusion**

Since ***p***-value > α, we fail to reject H_0 at a 0.05 significance level. The data does not provide sufficient evidence against the null hypothesis which may be regarded as true and we may conclude that both diets are equally good.

***Method* 3:**

***Step* 1** **Null and Alternative Hypotheses**

$H_0 : \eta_1 = \eta_2$

$H_1 : \eta_1 > \eta_2$

***Step* 2** **Test Statistic** X = Smaller of [Number of plus signs, Number of minus signs]

$= \text{Min}[7, 3] = 3$

***Step* 3** **Critical Value:** From Table 8, the critical value of X for $n = 10$ at a 0.05 significance level for one-tailed test is 1, therefore, the rejection region is $X \leq 1$.

***Step* 4** **Decision and Conclusion**

Since the computed value of test statistic $X = 3$ lies outside the rejection region, we fail to reject H_0 at a 0.05 significance level. The data does not provide sufficient evidence against the null hypothesis which may be regarded as true and we may conclude that both diets are equally good.

In short:

Hypotheses	**Test Statistics**
$H_0 : \eta_1 = \eta_2$ $H_1 : \eta_1 > \eta_2$	X = number of (+) signs = 7 and $n = 10$.
Rejection Region and Critical Value	**Conclusion**
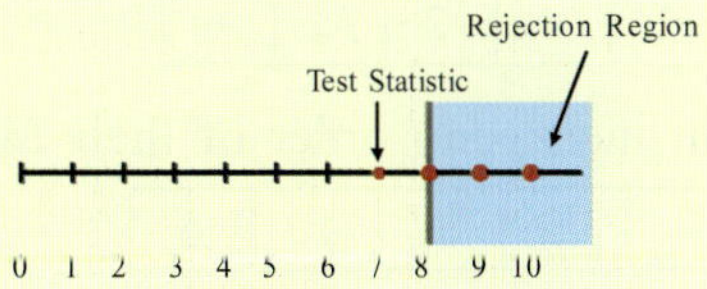	Fail to reject H_0. The data does not provide sufficient evidence against the null hypothesis and we conclude that both diets are equally good.

A3.3: Wilcoxon Signed-Rank Test

Objectives

After completing this section you will be able to understand:

- **A.** Signed Ranks
- **B.** Single Sample Wilcoxon Signed-Rank Test (Small Samples: $n \leq 30$)
- **C.** Single Sample Wilcoxon Signed-Rank Test For Matched Pairs
- **D.** Wilcoxon Signed-Rank Test (Large Sample)

The sign test is not very sensitive to the amount of variation of a data value from the claimed value of the median. It uses only the information whether the value is above or below the claimed value. Consequently, it may fail in rejecting a false null hypothesis. In this section we introduce another test called the **Wilcoxon Signed-Rank Test** that, in addition to the signs, also uses the amount of variation of data values from the claimed value. Therefore it is considered better than the sign test.

A. Signed Ranks

In testing the null hypothesis that $\eta = \eta_0$, we subtract η_0 from each sample observation. If the difference is zero for any observation, then we omit that observation and reduce the sample size (n) accordingly. The remaining differences are arranged in increasing order of their absolute values. The ordering determines the rank (position) of each data value. To each rank we attach the sign (plus or minus) of the corresponding difference. These ranks are called the **signed-ranks**.

These concepts are explained in Examples 4 and 5.

Notation:

T^+: Sum of the positive ranks

T^-: The absolute value of the sum of negative ranks

T_0: Smaller of (T^+, T^-)

Example 4: For the null hypothesis $\eta = 5$, compute the values of T^+, T^- and T for the following observations:

8	9	3	5	4	11

Solution : ***Step 1*** Subtracting 5 from each observation, we get

3	4	−2	0	−1	6

By discarding zero, we are left with 5 data values

3	4	−2	−1	6

Step 2 Arranging the above numbers in increasing order of their **absolute value**, we get

$\lvert -1 \rvert$	$\lvert -2 \rvert$	3	4	6

Step 3 Assigning rank numbers to these absolute values, we get

Values →	$\lvert -1 \rvert$	$\lvert -2 \rvert$	3	4	6
Ranks →	1	2	3	4	5

Table A3.6

Step 4 Assigning rank numbers to these absolute values, we get

Arranged Data	→	−1	−2	3	4	6
Rank	→	1	2	3	4	5
Signed-Rank	→	−1	−2	3	4	5

Table A3.7

Step 5 Computing T^+, T^- and T, we get

T^+ : The sum of the positive ranks = 3 + 4 + 5 = 12

T^- : The absolute value of the sum of negative ranks = |–1 – 2| = |–3| = 3

T_0 : Smaller of (T^+, T^-) = smaller of (12, 3) = 3.

Remark:

Repeated Ranks or Ties: In the above example, all the absolute values obtained in **Step 2** were distinct, that is, different. If the absolute values of two or more differences are the same, we call it a tie. In case of ties, common ranks are assigned to the repeated items. These common ranks are the arithmetic mean of the ranks which these items would have got if they were different from each other and the next item will get the rank next to the rank used in computing the common rank. For example, suppose an item is repeated at rank 4, that is, the items at places 4 and 5 are the same. Then the common rank to be assigned to each item is (4 + 5)/2, that is 4.5 which is the average of 4 and 5, the ranks which these observations would have assumed if they were different. The next item will be assigned the rank 6. If an item is repeated 3 times at rank 7, then the common rank to be assigned to each value will be (7 + 8 + 9)/3, that is, 8 which is the arithmetic mean of 7, 8 and 9, namely, the ranks these observations would have got if they were different from each other. The next rank to be assigned will be 10.

Example 5: For the null hypothesis $\eta = 8$, compute T^+, T^- and T for the following observations:

11	3	8	10	2	5	9	15	16	13

Solution : **Step 1** Subtracting 8 from each observation, we get

3	–5	0	2	–6	–3	1	7	8	5

By discarding zero, we are left with 9 data values.

3	–5	2	–6	–3	1	7	8	5

Step 2 Arranging the above numbers in increasing order of their **absolute value**, we get

1	2	|–3|	3	|–5|	5	|–6|	7	8

Step 3 Assigning rank numbers to these absolute values, we get

Values →	1	2	|–3|	3	|–5|	5	|–6|	7	8
Ranks →	1	2	3.5	3.5	5.5	5.5	7	8	9

Table A3.8

Note: Since observation 3 is repeated at the 3rd and 4th places, the common rank given to each is $\frac{1}{2}(3+4)$ = 3.5. Similarly, the observation 5 is repeated at the 5th and 6th places, hence the common rank given to each is $\frac{1}{2}(5+6)$ = 5.5.

Step 4 Assigning the signed-ranks, we get

Arranged Data	1	2	|–3|	3	|–5|	5	|–6|	7	8
Rank	1	2	3.5	3.5	5.5	5.5	7	8	9
Signed-Rank	1	2	–3.5	3.5	–5.5	5.5	–7	8	9

Table A3.9

Step 5 Computing T^+, T^- and T, we get

$T^+ = 1 + 2 + 3.5 + 5.5 + 8 + 9 = 29$

$T^- = |-3.5 - 5.5 - 7| = |-16| = 16$

T_0 = Smaller of (T^+, T^-) = smaller of (29, 16) = 16.

B. Single Sample Wilcoxon Signed-Rank Test (Small Samples: $n \leq 30$)

In testing the null hypothesis that the population has a specified median value, $\eta = \eta_0$, we expect the two values T^+ and T^- to be of comparable size. If $T^+ < T^-$, then the observed values are farther below η_0 than above η_0. Consequently, $\eta < \eta_0$. If $T^+ > T^-$, then this would imply that the observed values are farther above η_0 than below η_0. Thus, $\eta > \eta_0$. Keeping this in mind, we present the following procedure for the **Wilcoxon Signed-Rank Test**.

Method 1: Classical Method

Step 1 **Null and Alternative Hypotheses**

$H_0 : \eta = \eta_0$

$$H_1 : \begin{cases} \eta < \eta_0 & \text{or} \\ \eta > \eta_0 & \text{or} \\ \eta \neq \eta_0 \end{cases}$$

Step 2 **Test Statistic**

Subtract η_0 from each observed value, discard the zeros, and assign the signed-ranks to each of the remaining differences, as explained in Part A above.

Compute

T^+ = The sum of the positive ranks.

T^- = The absolute value of the sum of negative ranks.

T_0 = Smaller of (T^+, T^-)

Step 3 **Critical Value(s) and the Rejection Region**

Let n be the number of observations left after discarding the zeros in Step 2. For **small samples** ($n \leq 30$), the critical value of T at α significance level for one-tailed or two-tailed test can be obtained from Table 9 at the end of the book. (We denote the critical value of T as T_c.)

Since the test statistic T_0 is the smaller of T^+ and T^-, the rejection region is in the left tail. The rejection region is given by $\{T \leq T_c\}$, as shown in Figure A3.6.

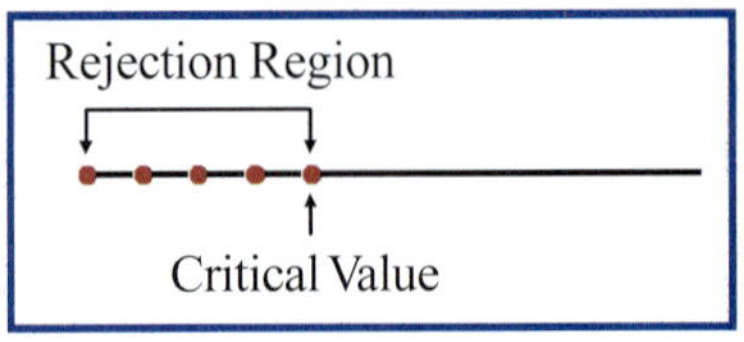

Figure A3.6

Step 4 **Decision and Conclusion**

If the computed value of the test statistic T_0 is less than or equal to the critical value obtained from Table 9 (that is, if $T_0 \leq T_c$), reject H_0; otherwise H_0 may be regarded as true. Write the conclusion in simple language.

Method 2: *P*-Value Method

***Step* 1** **Null and Alternative Hypotheses**

$H_0 : \eta = \eta_0$

$$H_1 : \begin{cases} \eta < \eta_0 & \text{or} \\ \eta > \eta_0 & \text{or} \\ \eta \neq \eta_0 \end{cases}$$

***Step* 2** **Test Statistic**

T_0 = smaller of (T^+, T^-).

***Step* 3** ***P*-value**

***P*-value of the test statistic T_0 is given by p-value = $P(T \le T_0)$ and can be obtained from Table 9.

***Step* 4** **Decision and Conclusion**

If the ***p***-value $< \alpha$, we reject H_0 at α significance level, otherwise we fail to reject H_0. Write the conslusion in simple language.

Example 6: A teacher claims that the median time to do a particular type of mathematics problem should be at most 3 minutes, but her students believe that the median time is more than 3 minutes. A random sample of 10 students completed the problem in the following times in minutes:

2.5	2	4	4.5	4	2.5	4.5	3	3.5	5

Use the Wilcoxon Signed-Rank Test with 0.05 level of significance to test the teacher's claim by the Classical and ***P***-value Methods.

Solution : ***Method* 1: Classical Method**

***Step* 1** **Null and Alternative Hypotheses**

$H_0 : \eta = 3$

$H_1 : \eta > 3$

***Step* 2** **Test Statistics**

$x - \eta$	–0.5	–1	1	1.5	1	–0.5	1.5	0	0.5	2
Arranged Data	|–0.5|	|–0.5|	0.5	|–1|	1	1	1.5	1.5	2	
Rank Calculations	$\frac{1+2+3}{3} = 2$			$\frac{4+5+6}{3} = 5$			$\frac{7+8}{2} = 7.5$			
Rank	2	2	2	5	5	5	7.5	7.5	9	
Signed-Rank	–2	–2	2	–5	5	5	7.5	7.5	9	

Table A3.10

Computing T^+ and T^-, we get

$$T^+ = 2 + 5 + 5 + 7.5 + 7.5 + 9 = 36$$

$$T^- = |-2 - 2 - 5| = 9$$

$$\Rightarrow T_0 = \text{Smaller of } (36, 9) = 9.$$

Step 3 Critical Value (s) and the Rejection Region

Using Table 9, the critical value of T (T_c) for $n = 9$ and $\alpha = 0.05$ for a single-tailed test is $T_c = 8$.

The critical value, the rejection region and the test statistic are shown in Figure A3.7.

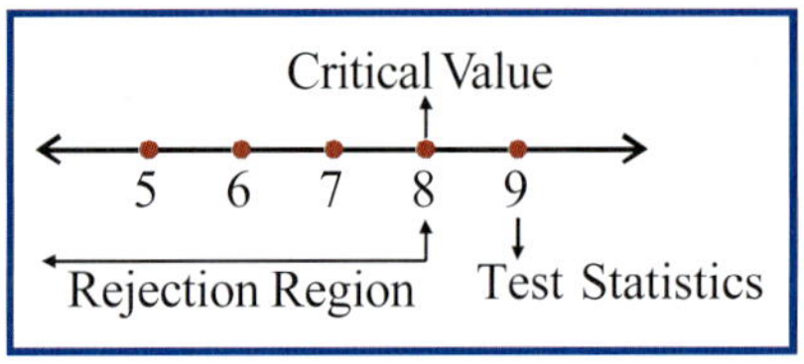

Figure A3.7

Step 4 Decision and Conclusion

Since the computed value of test statistic $T_0 = 9$, lies outside the rejection region, we fail to reject H_0 at 5% level of significance. There is not sufficient evidence to reject the teacher's claim.

Method 2: *P*-Value Method

Step 1 Null and Alternative Hypotheses

$H_0 : \eta = 3$

$H_1 : \eta > 3$

Step 2 Test Statistic

Test statistics $= T_0 = 9$

Step 3 *P*-value

P-value of the test statistic T_0 is given by p-value $= P(T \leq 9)$

From Table 9, the critical value of T for $n = 9$ and $\alpha = 0.05$, for a single-tailed test is $T_c = 8$, so that $P(T \leq 8) = 0.05$.

Therefore,

$$p\text{-value} = P(T \leq 9) > P(T \leq 8) = 0.05$$
$$> 0.05 = \alpha$$

Step 4 Decision and Conclusion

Since, the ***P***-value $> \alpha$, we fail to reject H_0 at a 0.05 significance level. There is not sufficient evidence to reject the teacher's claim.

In short:

Hypotheses $H_0 : \eta = 3$ $H_1 : \eta > 3$	**Test Statistics** $T_0 = \text{Smaller of } (T^+, T^-) = 9$
Rejection Region and Critical Value 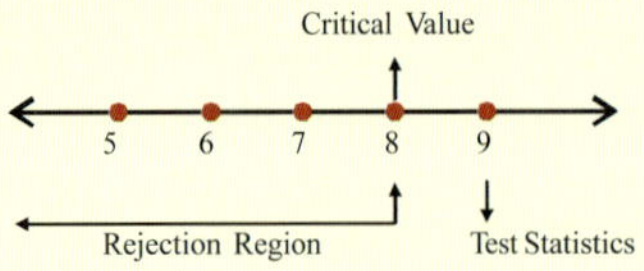 	**Conclusion** Fail to reject H_0. There is not sufficient evidence to reject the teacher's claim.

Remark:

If we apply both the **Sign Test** and **Wilcoxon Signed-Rank Test** to the same problem, both tests will not necessarily lead to the same conclusion. Since the **Wilcoxon Test** uses more information contained in the sample as compared with the **Sign Test**, it may be regarded as the better choice.

C. Single Sample Wilcoxon Signed-Rank Test for Matched Pairs

The Wilcoxon Signed-Rank Test can also be used to compare two non-normal populations when n pairs (x_i, y_i) ; $i = 1, 2, \ldots, n$ of sample observations are given. In testing the null hypothesis that $\eta_1 = \eta_2$, we examine the differences $d_i = x_i - y_i$; $i = 1, 2, \ldots, n$ for each pair and proceed exactly similarly as in Wilcoxon Signed-Rank Test discussed in Part B. The following example illustrates the procedure.

Example 7: A random sample of 10 women is selected and their weights (in lbs.), before and after they are put on a new diet, are recorded.

Weight before diet	180	178	165	200	160	145	170	210	185	155
Weight after diet	174	181	157	198	152	150	160	205	178	160

Use the Wilcoxon Singed-Rank Test at a 0.05 level of significance to test the claim that the new diet is effective in reducing the weight. Use the Classical Method.

Solution : ***Method* 1: Classical Method**

***Step* 1 Null and Alternative Hypotheses**

$H_0 : \eta_1 = \eta_2$

$H_1 : \eta_1 > \eta_2$ (new diet is effective)

***Step* 2 Test Statistics**

$d = x - y$	6	−3	8	2	8	−5	10	5	7	−5
Arranged Data	2	|−3|	|−5|	|−5|	5	6	7	8	8	10
Ranks	1	2	4	4	4	6	7	8.5	8.5	10
Signed-Rank	1	−2	−4	−4	4	6	7	8.5	8.5	10

Table A3.11

Computing T^+ and T^-, we get

$$T^+ = 1 + 4 + 6 + 7 + 8.5 + 8.5 + 10 = 45$$

$$T^- = |-2 - 4 - 4| = 10$$

$$\Rightarrow \; T_0 = 10.$$

***Step* 3 Critical Value (s) and the Rejection Region**

Using Table 9, the critical value of T (T_c) for $n = 10$ and $\alpha = 0.05$ for a single-tailed test is $T_c = 11$.

The critical value, the rejection region and the test statistic are shown in Figure A3.8.

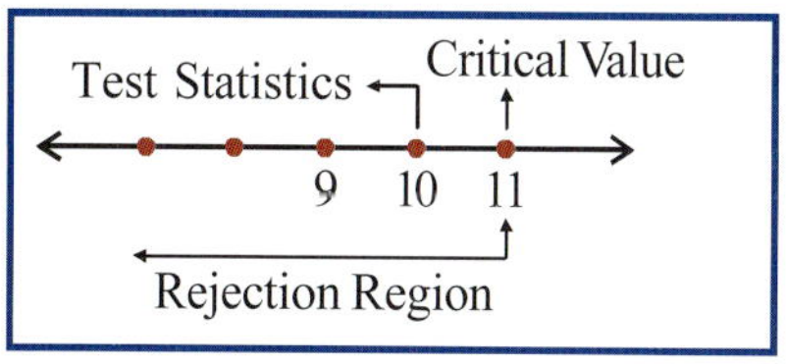

Figure A3.8

***Step* 4 Decision and Conclusion**

Since the computed value of test statistic $T_0 = 10$ lies inside the rejection region, we reject H_0 at 5% level of significance. We conclude that the new diet is definitely effective in reducing weight.

In short:

Hypotheses	**Test Statistics**
$H_0 : \eta_1 = \eta_2$ $H_1 : \eta_1 > \eta_2$	T_0 = Smaller of $(T^+, T^-) = 10$
Rejection Region and Critical Value 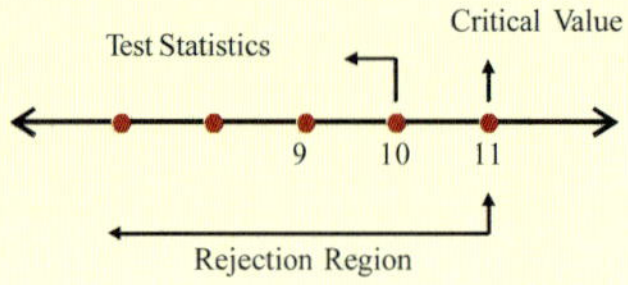	**Conclusion** Reject H_0. The new diet is effective in reducing the weight.

D. Wilcoxon Signed-Rank Test (Large Sample)

In the Wilcoxon Signed-Rank Test, if the sample size is large ($n > 30$), then we can use the **normal test**. For large values of n, we proceed exactly as in the case of Wilcoxon Signed-Rank Test [discussed earlier] up to the computation of the statistics T^+, T^-, and T_0.

Let n be the number of observations left after discarding the zero(0) differences. For large values of n ($n > 30$), the statistic T_0 is approximately normal with

Mean $\left(\mu_{T_0}\right)$	**Standard Deviation** $\left(\sigma_{T_0}\right)$
$\dfrac{n(n+1)}{4}$	$\sqrt{\dfrac{n(n+1)(2n+1)}{24}}$

Table A3.12

Hence, for large n, the **test statistic** is

$$Z = \frac{T_0 - \mu_{T_0}}{\sigma_{T_0}} \sim N(0, 1)$$

The critical values of Z at given level of significance (α) and for the one-tailed or two-tailed test can be obtained from Table 3 of standard normal probabilities.

Example 8: To test the claim that the median age of mathematics faculty in California community colleges is less than 42 years, a survey from a random sample of 32 mathematics faculty yields the following ages (in years):

56	62	61	54	52	32	24	35	50	42	52
49	26	31	31	54	38	36	45	53	37	40
38	31	29	25	45	52	48	39	30	38	

Use the Wilcoxon Signed-Rank Test at the 0.05 level of significance using the Classical Method.

Solution : ***Method* 1: Classical Method**

***Step* 1** **Null and Alternative Hypotheses**

$H_0 : \eta = 42$

$H_1 : \eta < 42$

***Step* 2** **Test Statistics**

Subtracting $\eta = 42$ from each observed value, we get

14	20	19	12	10	−10	−18	−7	8	0	10
7	−16	−11	−11	12	−4	−6	3	11	−5	−2
−4	−11	−13	−17	3	10	6	−3	−12	4	

Table A3.13

Discarding zero and arranging the remaining 31 numbers in increasing order of their absolute values, we get

|−2|	|−3|	3	3	|−4|	|−4|	|−4|	|−5|	|−6|	6	|−7|
7	8	|−10|	10	10	10	|−11|	|−11|	|−11|	11	|−12|
12	12	|−13|	14	|−16|	|−17|	|−18|	19	20		

Table A3.14

Assigning the signed-rank to each of these absolute numbers respectively, we get

−1	−3	3	3	−6	−6	−6	−8	−9.5	9.5	−11.5
11.5	13	−15.5	15.5	15.5	15.5	−19.5	−19.5	−19.5	19.5	−23
23	23	−25	26	−27	−28	−29	30	31		

Table A3.15

Computing T^+ and T^-, we get

$T^+ = 3 + 3 + 9.5 + 11.5 + 13 + 15.5 + 15.5 + 15.5 + 19.5 + 23 + 23 + 26 + 30 + 31 = 239$

$T^- = |-1 -3 -6 -6 -6 -8 -9.5 -9.5 -11.5 -15.5 -19.5 -19.5 -19.5 -23 -25 - 27 -28 -29| = 257$

$\Rightarrow T_0 = 239.$

n = Number of deviations after discarding (0) = 31. Since $n > 30$, we use the normal test.

The test statistic for large sample is

$$Z = \frac{T_0 - \mu_{T_0}}{\sigma_{T_0}} = \frac{239 - \frac{(31)(32)}{4}}{\sqrt{\frac{(31)(32)(63)}{24}}} = -0.18$$

***Step* 3 Critical Value(s) and the Rejection Region**

The critical value of Z for the left-tailed test at 5% level of significance is −1.645. Therefore, the rejection region is $Z < -1.645$. The critical value, the rejection region and the test statistic are shown in Figure A3.9.

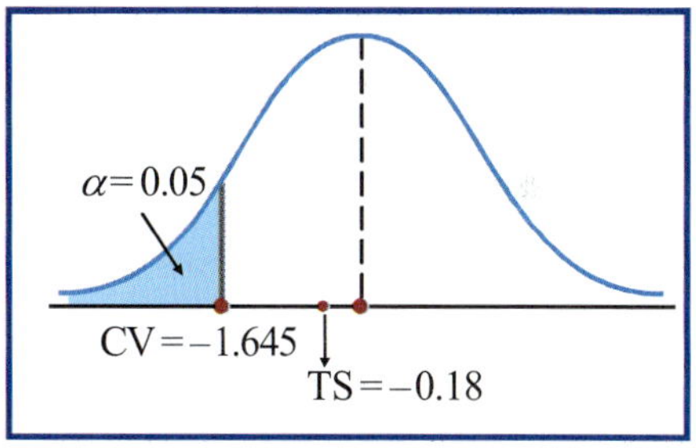

Figure A3.9

***Step* 4 Decision and Conclusion**

Since the test statistic lies outside the rejection region, we fail to reject H_0 at a 0.05 significance level. The data does not provide sufficient evidence against the null hypothesis. The median age of mathematics faculty in California community colleges does not seem to be less than 42 years.

In short:

Hypotheses	**Test Statistics**
$H_0: \eta = 42$ $H_1: \eta < 42$	T_0 = Smaller of $(T^+, T^-) = 10$
Rejection Region and Critical Value	**Conclusion**
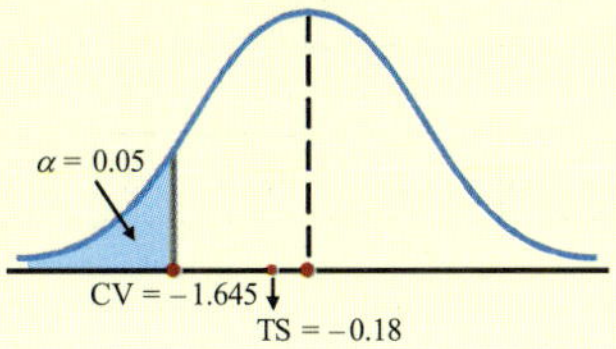	Fail to reject H_0 at $\alpha = 0.05$. There does not seem to be significant difference between the means.

A3.4: Mann-Whitney Test

Objectives

After completing this section you will be able to understand:

A. *U*-Statistics.
B. Mann-Whitney Test for Independent Samples (Small Samples).
C. Mann-Whitney Test for Independent Samples (Large Sample).

The Wilcoxon signed rank test can be used to compare two populations when the two sets of sample data are paired. Suppose there is no natural way to pair up the data values; that is, the samples are independent. In this section we will introduce a test, called the **Mann-Whitney Test**, that enables us to compare two independent samples by testing a hypothesis on the two population medians. This Non-Parametric test for two independent samples was described by Wilcoxon and studied by Mann and Whitney. The Mann-Whitney Test is the most widely used test as an alternative to the *t*-test for two independent samples when we do not make the *t*-test assumption about the population. The only assumption is that the random variable(s) is (are) continuous.

A. *U*-Statistics

The Mann-Whitney Test uses the sum of the ranks of each sample. To assign ranks to each observation of the two samples, we combine the two samples into a single sample and arrange the observations of the combined sample in ascending order of magnitude. Then ranks are assigned to the combined sample data.

T_x: Sum of the ranks of the observations of the first sample (X values) in the combined sample.

T_y: Sum of the ranks of the observations of the second sample (Y values) in the combined sample.

Mann-Whitney U-Statistics are defined as follows.

$$U_x = T_x - \frac{n_1(n_1+1)}{2}$$

$$U_y = T_y - \frac{n_2(n_2+1)}{2}$$

where n_1 and n_2 are the number of observations in the first and second samples respectively.

$$U = \text{Smaller of } (U_x, U_y)$$

B. Mann-Whitney Test for Independent Samples (Small Samples)

The Mann-Whitney Test uses the test statistic U to compare the population medians η_x and η_y. The various steps for Mann-Whitney Test are outlined below. The variable X refers to the first sample observations and the variable Y refers to the second sample observations.

Method 1: Classical Method

Step 1 Null and Alternative Hypotheses

$$H_0 : \eta_x = \eta_y$$

$$H_1 : \begin{cases} \eta_x < \eta_y & \text{or} \\ \eta_x > \eta_y & \text{or} \\ \eta_x \neq \eta_y \end{cases}$$

Step 2 **Test Statistic**

Combine the observations of both samples to obtain a single sample and arrange the observations of the combined samples in ascending order of their magnitude. Assign ranks to the combined sample data and find the values of the statistics T_x, T_y, U_x and U_y.

The test statistic U is defined by: U = Smaller of (U_x and U_y) = U_0

Step 3 **Critical Value(s) and the Rejection Region**

Since the test statistic U is the smaller of U_x and U_y, the rejection region is in the left hand tail. The critical values of U for different values of (n_1 and n_2) and desired level of significance $\alpha = 0.05$ and 0.01 for single-tailed and two-tailed tests are given in Table 10(A) and 10(B) in the Appendix 3.

The rejection region is given by: $U \leq$ Critical value of U obtained from Tables 10(B) and 10(B)

Step 4 **Decision and Conclusion**

Reject H_0 if U lies in the rejection region, otherwise we fail to reject H_0. Write the conclusion of the test in simple language.

Method **2:** ***P*-Value Method**

Step 1 **Null and Alternative Hypotheses**

$$H_0 : \eta_x = \eta_y$$

$$H_1 : \begin{cases} \eta_x < \eta_y & \text{or} \\ \eta_x > \eta_y & \text{or} \\ \eta_x \neq \eta_y & \end{cases}$$

Step 2 **Test Statistic**

U = smaller of (U_x and U_y) = U_0.

Step 3 ***P*-value**

Since the test criterion is left-tailed, the ***p***-value of the test statistic obtained in Step 2 is given by:

$$p\text{-value} = P(U \leq U_0)$$

Step 4 **Decision and Conclusion**

If ***p***-value < α, reject H_0 at α significance level, otherwise we fail to reject H_0. Write the conclusion of the test in simple language.

Example 9: For the following data, test the hypothesis that the median measure in the population X is less than the median measure in the population Y using Mann-Whitney U-test. Use $\alpha = 0.05$.

X	60	45	23	32			
Y	10	25	20	54	32	65	8

Solution :Step 1 **Null and Alternative Hypotheses**

$H_0 : \eta_x = \eta_y$

$H_1 : \eta_x < \eta_y$

***Step* 2 Test Statistic**

To find the rank number for each observation, we combine the samples into a single sample and arrange the numbers in the combined sample in increasing order of their magnitudes. Here is the combined arranged data.

Sample	Y	Y	Y	X	Y	X	Y	X	Y	X	Y
Combined data	8	10	20	23	25	32	32	45	54	60	65
Ranks	1	2	3	4	5	6.5	6.5	8	9	10	11

Table A3.16

$T_x = 4 + 6.5 + 8 + 10 = 28.5$, $T_y = 1 + 2 + 3 + 5 + 6.5 + 9 + 11 = 37.5$, $n_1 = 4$, and $n_2 = 7$.

$$U_x = T_x - \frac{n_1(n_1+1)}{2} = 28.5 - \frac{(4)(5)}{2} = 18.5$$

$$U_y = T_y - \frac{n_2(n_2+1)}{2} = 37.5 - \frac{(7)(8)}{2} = 9.5$$

$\Rightarrow U_x = 18.5$, and $U_y = 9.5$.

Test Statistic $= U =$ Smaller of $(U_x, U_y) =$ Smaller of $(18.5, 9.5) = 9.5$

***Step* 3 Critical Value (s) and the Rejection Region**

The critical value of U for $n_1 = 4$, $n_2 = 7$ and $\alpha = 0.05$ for single-tailed test is 4 (Table 10(A)). The rejection region is $U \le 4$.

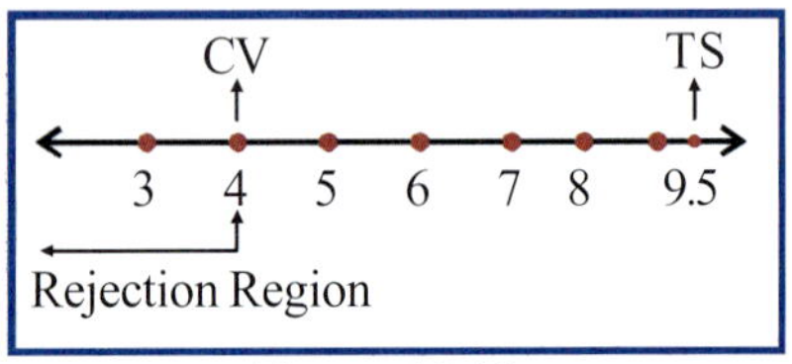

Figure A3.10

***Step* 4 Decision and Conclusion**

Since the test statistic $U = 9.5$ lies outside the rejection region, we fail to reject H_0 at a 0.05 significance level. The data does not provide sufficient evidence against the null hypothesis which may be regarded as true. Hence, we may conclude that the median measures of the populations X and Y are the same.

In short:

<table>
<tr><td>Hypotheses
$H_0: \eta_x = \eta_y$
$H_1: \eta_x < \eta_y$</td><td>Test Statistics
$U =$ Smaller of $(U_x, U_y) = 9.5$</td></tr>
<tr><td>Rejection Region and Critical Value
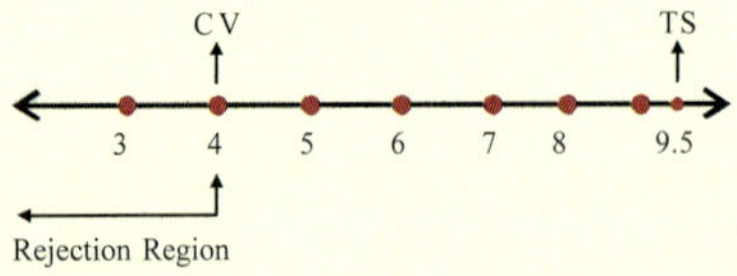
</td><td>Conclusion
Fail to reject H_0. The data does not provide sufficient evidence against the null hypothesis and we may conclude that the median measures of the populations X and Y are the same.</td></tr>
</table>

C. Mann-Whitney Test for Independent Samples (Large Sample)

If the samples are larger than size 20, then we may use the normal test for testing $H_0 : \eta_x = \eta_y$. Mann-Whitney proved that under H_0 the sampling distribution of U is asymptotically normal with mean and standard deviation.

Mean (μ_u)	Standard Deviation (σ_u)
$\frac{n_1 n_2}{2}$	$\sqrt{\frac{n_1 n_2 (n_1 + n_2 + 1)}{12}}$

Table A3.17

Hence, for large samples, $U \sim N(\mu_u, \sigma_u)$. This implies that $Z = \frac{U - \mu_u}{\sigma_u} \sim N(0, 1)$ for large samples, and we can use the z-test in the usual manner. The normal approximation of U is fairly accurate if both n_1 and n_2 are greater than 10.

Assumptions for the Test

- **i)** The two samples are random.
- **ii)** The two samples are independent.
- **iii)** The sizes of both samples are more than 10.
- **iv)** No assumption is made about the nature of the distributions from which the samples are drawn.

Example 10: The data below shows the salaries (in \$1000) in randomly selected advertisements in two different occupations (Education & Marketing).

Education (X)	22	40	18	25	15	23	16	19	21	30	
Marketing (Y)	28	24	20	45	50	39	26	55	48	41	42

Use the Mann-Whitney test at 0.01 level of significance to test whether the median salary in the field of education is lower than the median salary in the field of marketing.

Solution : ***Step* 1** **Null and Alternative Hypotheses**

$H_0 : \eta_x = \eta_y$

$H_1 : \eta_x < \eta_y$

***Step* 2** **Test Statistic**

Combine the two samples into a single sample, arrange the combined observations in ascending order, and assign the rank number to each observation in the combined sample.

Sample	X	X	X	X	Y	X	X	X	Y	X	Y	Y	X	Y	X	Y	Y	Y	Y	Y	Y
Observations	15	16	18	19	20	21	22	23	24	25	26	28	30	39	40	41	42	45	48	50	55
Rank	1	2	3	4	5	6	7	8	9	10	11	12	13	14	15	16	17	18	19	20	21

Table A3.18

Computation of statistic U_x, U_y, and U

Here $n_1 = 10$, $n_2 = 11$

$T_x = 1 + 2 + 3 + 4 + 6 + 7 + 8 + 10 + 13 + 15 = 69$

$T_y = 5 + 9 + 11 + 12 + 14 + 16 + 17 + 18 + 19 + 20 + 21 = 162$

$$U_x = T_x - \frac{n_1(n_1+1)}{2} = 69 - \frac{(10)(11)}{2} = 14;$$

$$U_y = T_y - \frac{n_2(n_2+1)}{2} = 162 - \frac{(11)(12)}{2} = 96.$$

U = Smaller of (14, 96) = 14.

Since $n_1 = 10$ and $n_2 = 11$, we use the normal approximation to U and apply the Z-test. Under H_0, the test statistic is

$$Z = \frac{U - \mu_u}{\sigma_u} \sim N(0, 1)$$

where $\mu_u = \frac{n_1 n_2}{2} = \frac{(10)(11)}{2} = 55$ and $\sigma_u = \sqrt{\frac{n_1 n_2 (n_1 + n_2 + 1)}{12}} \sqrt{\frac{(10)(11)(22)}{12}} = 14.2.$

Substituting the values of μ_u and σ_u into $Z = \frac{U - \mu_u}{\sigma_u}$, we get

$$Z = \frac{14 - 55}{14.2} = -2.89$$

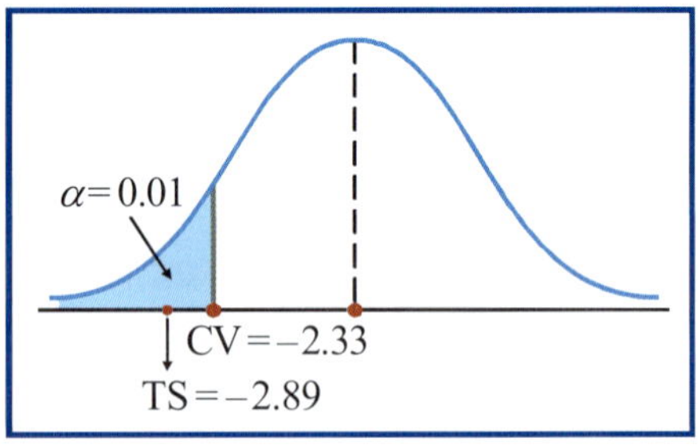

Figure A3.11

***Step* 3 Critical Value(s) and the Rejection Region**

The critical value of Z for the left-tailed test at 1% level of significance is −2.33, so the rejection region is $Z \leq -2.33$.

***Step* 4 Decision and Conclusion**

Since the test statistic $Z = -2.89$ lies in the rejection region, we reject H_0 at 1% level of significance. Hence, we can conclude that $\eta_x < \eta_y$; that is, the median salary in the field of education is less than the median salary in the field of marketing.

In short:

Hypotheses	**Test Statistics**
$H_0: \eta_x = \eta_y$ $H_1: \eta_x < \eta_y$	$Z = \frac{U - \mu_u}{\sigma_u} = -2.89$
Rejection Region and Critical Value 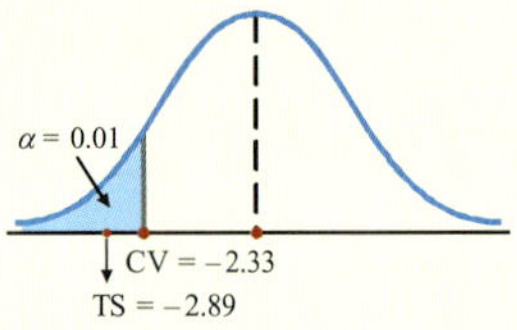	**Conclusion** Reject H_0. The median salary in the field of education is less than the median salary in the field of marketing.

A3.5: Rank Correlation Coefficient

In Chapter 10, we used a parametric method (Karl Pearson's Correlation Coefficient 'r') for studying the linear relationship between the two variables in the sample. In this section we will study a Non-Parametric alternative to r, which is based on the ranks of the observations of the independent variable X and the dependent variable Y.

Objectives

After completing this section you will be able to understand:

A. Spearman's Rank Correlation Coefficient.

B. Properties of Spearman's Rank Correlation Coefficient (r_{sp})

C. Test for Significance of Sample Rank Correlation Coefficient (r_{sp})

A. Spearman's Rank Correlation Coefficient

Sometimes we come across statistical series in which the variables under consideration cannot under go of quantitative measurement but can be arranged in serial order. This happens when we are dealing with qualitative characteristics (attributes) such as honesty, beauty, character, morality, etc. These cannot be measured quantitatively but can be arranged serially. In such situations Karl Pearson's coefficient of correlation cannot be used as such. **Charles Edward Spearman**, a British psychologist, developed a formula in 1904 which consists of obtaining the correlation coefficient using the rankings only.

Note: For the quantitative data, each of the variables X and Y must be ranked separately.

Since we are only using ranks, which are not normally distributed, Spearman's formula (will be shown later) cannot be used to study the existence of a linear relationship between the variables X and Y, as was done by Karl Pearson's correlation coefficient. Spearman's rank correlation coefficient gives a measure of the existence of a monotone relationship between the variables. By a monotone relationship we mean a relationship in which:

i) As X increases, Y also increases for all values of (X, Y). This is known as **monotone increasing relationship**. [See Figures A3.12]

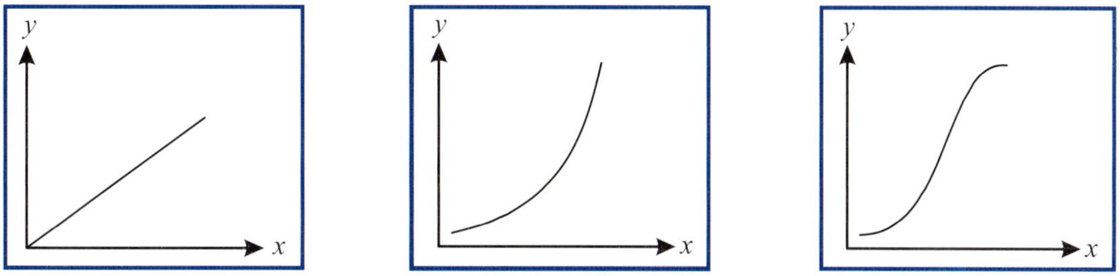

(a) monotone increasing **(b) monotone increasing** **(c) monotone increasing**

Figure A3.12

ii) As X increases, Y deceases for all values of (X, Y). This is known as **monotone decreasing relationship**. [See Figures A3.13]

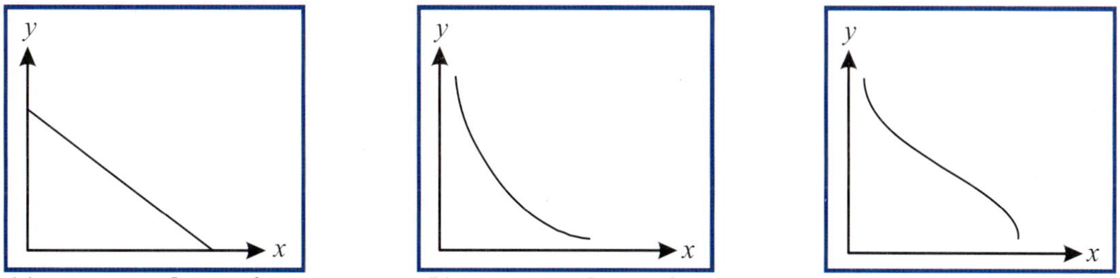

(a) monotone decreasing **(b) monotone decreasing** **(c) monotone decreasing**

Figure A3.13

However if, as X increases, Y increases (decreases) at first but later starts to decrease (increase) and so on, then the relationship is known as **non-monotone relationship**. [See Figures A3.14].

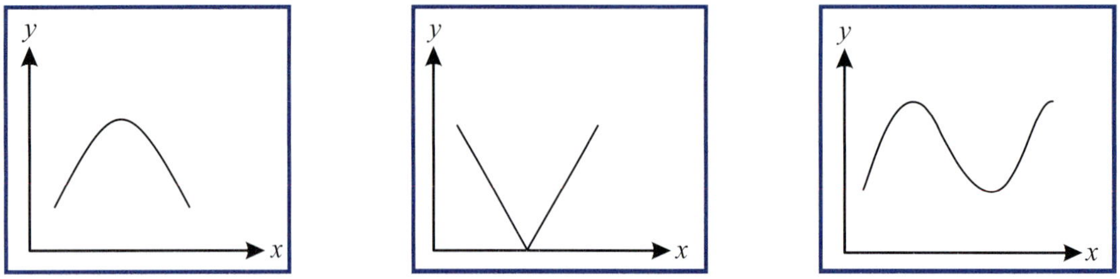

(a) Non-monotone decreasing **(b) Non-monotone decreasing** **(c) Non-monotone decreasing**

Figure A3.14

Spearman's rank correlation coefficient, usually denoted by r_{sp}, is given by the formula

$$r_{sp} = 1 - \frac{6\sum d^2}{n\left(n^2 - 1\right)}, \tag{4}$$

where

d = Difference between the ranks for the two (X, Y) values within a pair

n = Number of pairs in the sampled data

- ## Assumptions

Karl Pearson's correlation coefficient assumes that the parent population from which sample observations are drawn is a bivariate normal distribution. However, the rank correlation coefficient r_{sp} does not make any strict assumptions about the form of the parent distribution from which the sample observations are drawn. It simply assumes that

1. **The sample data have been randomly selected.**
2. **The *n* pairs of sample observations are independent.**
3. **Each pair has the same continuous bivariate distribution.**

- ## Computation of Rank Correlation Coefficient

We discuss below the method of computing the Spearman's rank correlation coefficient r_{sp} under the following situations:

i) When actual ranks are given
ii) When ranks are not given

Case i When actual ranks are given

The following steps are involved

***Step* 1** Compute d, the difference of ranks of the two values within a pair.

***Step* 2** Compute d^2.

***Step* 3** Obtain the sum $\sum d^2$.

***Step* 4** Compute the rank correlation coefficient by the Spearman's formula (4).

Case ii When actual ranks are not given

Spearman's rank correlation formula (4) can also be used even if we are dealing with variables which are measured quantitatively, that is, when the actual data but not the ranks relating to two variables are given. In such cases, we will have to convert the data into ranks. Ranks are given to the X-values and the Y-values separately in the usual manner and then we proceed as in case (i).

Remarks:

1. The subscript sp in r_{sp} is used in honor of Spearman, who derived this formula.
2. The rank correlation coefficient for the entire population data is denoted by ρ_{sp}.
3. In formula (4), we always have $\sum d = 0$, which provides a check for numerical calculations.
4. Interpretation of r_{sp}

   Rank correlation coefficient r_{sp} is a measure of any monotone (increasing or decreasing) relationship and not merely a linear relationship between the variables X and Y. Hence r_{sp} is more meaningful than Karl Pearson's correlation coefficient r.

   If $r_{sp} = 1$, then the relationship between X and Y is perfectly monotone increasing.

   If $r_{sp} = -1$, then the relationship between X and Y is perfectly monotone decreasing.

   If $r_{sp} = 0$, then there is no monotone relationship between X and Y.

   The values of r_{sp} close to ± 1 indicate a strong tendency between X and Y to have a monotone increasing (decreasing) relationship and the values of r_{sp} close to 0 indicate a very weak or practically no monotone relationship between the variables.

   $r_{sp} > 0$ implies that larger (smaller) values of Y. Similarly, $r_{sp} < 0$ implies that larger (smaller) values of X are associated with smaller (larger) values of Y.
5. Spearman's formula is easy to understand and apply, compared with Karl Person's formula. The values obtained by the two formulae, Pearsonian r and Spearman's r_{sp}, are generally different. The difference arises due to the fact that when ranking is used instead of full set of observations, there is always some loss of information. Unless many ties exist in the data, the coefficient of rank correlation should be only slightly different from the Pearsonian coefficient.
6. Spearman's formula is the only formula to be used for finding correlation coefficient if we are dealing with qualitative characteristics which cannot be measured quantitatively but can be arranged serially. However can also be used even if actual data are given.
7. **Effect of Outliers:** In computing the rank correlation coefficient r_{sp}, instead of the exact magnitudes of the paired observations, we use their corresponding ranks which are not affected very much by outliers (that is, very large or very small observations). Therefore, r_{sp} is less sensitive to outliers or extreme observations. Hence, in case of extreme observations, Spearman's formula is preferred to Pearson's formula.
8. The Spearman's rank correlation coefficient r_{sp} provides a sample estimate for the Spearman's rank correlation coefficient ρ_{sp} in the population from which the sample observations are taken.

B. Properties of Spearman's Rank Correlation Coefficient (r_{sp})

1. Limits for r_{sp}

Spearman's rank correlation coefficient lies between –1 and 1.

$$|r_{sp}| \leq 1 \quad \text{or} \quad -1 \leq r_{sp} \leq 1$$

Attainment of Limits

i) $r_{sp} = 1$, if the ranks of the values (X, Y) within a pair are the same, as shown in Table A3.19.

Rank (X)	1	2	3	...	...	n
Rank (Y)	1	2	3	...	...	n

Table A3.19

ii) $r_{sp} = -1$, if the ranks of X and Y values are in the opposite (reverse) directions, as shown in Table A3.20.

Rank (X)	1	2	3	...	...	$n-2$	$n-1$	n
Rank (Y)	n	$n-1$	$n-2$	...	...	3	2	1

Table A3.20

2. The Spearman's rank correlation coefficient is independent of change of origin and scale. If we transform the variables (X, Y) to the variables (U, V) by the transformation

$$U = \frac{X-A}{h}\ ;\ V = \frac{Y-B}{k}\ ;\ h > 0,\ k > 0$$

then $$r_{sp}(X, Y) = r_{sp}(U, V) \qquad (5)$$

The relation in (5) is obvious because, as a result of change of origin and scale in (X, Y) to (U, V), the ranks of the pairs (U_i, V_i) corresponding to the pairs (X_i, Y_i) are not affected.

3. The rank correlation coefficient is not affected if we interchange the variables X and Y.

$$r_{sp}(X, Y) = r_{sp}(Y, X)$$

Interchanging the variables X and Y will result in change of signs of the deviations d, while d^2 and consequently r_{sp} will not be affected.

Remarks:

In computing the ranks for X and Y, we should use the same method of ranking for each of the variables X and Y

i) either from high to low, or

ii) from low to high.

In each case, we get the same value of r_{sp}.

However, if one variable is ranked from low to high and the other variable is ranked from high to low, then the value of r_{sp} will remain the same in magnitude but change in sign.

Example 11: Find Spearman's rank correlation coefficient for the following data

(a)

Rank (X)	1	2	3	4	5	6	7	8
Rank (Y)	1	2	3	4	5	6	7	8

(b)

Rank (X)	1	2	3	4	5	6	7	8
Rank (Y)	8	7	6	5	4	3	2	1

Solution: Let Rank $(X) = R_x$ and Rank $(Y) = R_y$; $n = 8$ (in each case).

(a)

R_x	R_y	$d = R_x - R_y$	d^2
1	1	0	0
2	2	0	0
3	3	0	0
4	4	0	0
5	5	0	0
6	6	0	0
7	7	0	0
8	8	0	0
Total		$\sum d = 0$	$\sum d^2 = 0$

Table A3.21

$$r_{sp} = 1 - \frac{6\sum d^2}{n(n^2-1)} = 1 - \frac{6(0)}{8(63)} = 1 - 0 = 1$$

(b)

R_x	R_y	$d = R_x - R_y$	d^2
1	8	−7	49
2	7	−5	25
3	6	−3	9
4	5	−1	1
5	4	1	1
6	3	3	9
7	2	5	25
8	1	7	49
Total		$\sum d = 0$	$\sum d^2 = 168$

Table A3.22

$$r_{sp} = 1 - \frac{6\sum d^2}{n(n^2-1)} = 1 - \frac{6(168)}{8(63)} = 1 - 2 = -1$$

Example 12: Ten competitors in a beauty contest are ranked by three judges in the following order:

1st Judge	1	6	5	10	3	2	4	9	7	8
2nd Judge	3	5	8	4	7	10	2	1	6	9
3rd Judge	6	4	9	8	1	2	3	10	5	7

Use the rank correlation coefficient to determine which pair of judges has the nearest approach to common tastes in beauty.

Solution: Let R_1, R_2 and R_3 denote the ranks given by the first, second and third judges respectively and let $(r_{ij})_{sp}$ be the rank correlation coefficient between the ranks given by i^{th} and j^{th} judges. $i \neq j = 1, 2, 3$. Let $d_{ij} = R_i - R_j$ be the difference of ranks of an individual given by the ith and j^{th} judges.

Calculation of Rank Correlation Coefficient

R_1	R_2	R_3	$d_{12} = R_1 - R_2$	$d_{13} = R_1 - R_3$	$d_{23} = R_2 - R_3$	d_{12}^2	d_{13}^2	d_{23}^2
1	3	6	−2	−5	−3	4	25	9
6	5	4	1	2	1	1	4	1
5	8	9	−3	−4	−1	9	16	1
10	4	8	6	2	−4	36	4	16
3	7	1	−4	2	6	16	4	36
2	10	2	−8	0	8	64	0	64
4	2	3	2	1	−1	4	1	1
9	1	10	8	−1	−9	64	1	81
7	6	5	1	2	1	1	4	1
8	9	7	−1	1	2	1	1	4
			$\sum d_{12} = 0$	$\sum d_{13} = 0$	$\sum d_{23} = 0$	$\sum d_{12}^2 = 200$	$\sum d_{13}^2 = 60$	$\sum d_{23}^2 = 214$

Table A3.23

We have $n = 10$.

Spearman's rank correlation coefficients are given by:

$$(r_{12})_{sp} = 1 - \frac{6\sum d_{12}^2}{n(n^2-1)} = 1 - \frac{6(200)}{10(99)} = -0.2121$$

$$(r_{13})_{sp} = 1 - \frac{6\sum d_{13}^2}{n(n^2-1)} = 1 - \frac{6(60)}{10(99)} = \frac{7}{11} = 0.6363$$

$$(r_{23})_{sp} = 1 - \frac{6\sum d_{23}^2}{n(n^2-1)} = 1 - \frac{6(214)}{10(99)} = -\frac{49}{165} = -0.2970$$

Since $(r_{13})_{sp}$ is maximum, the pair of 1st and 3rd judges has the nearest approach to common tastes in beauty.

Remark:

Since $(r_{12})_{sp}$ and $(r_{23})_{sp}$ are negative, the pair of judges (1, 2) and (2, 3) have opposite tastes for beauty.

C. Test for Significance of Sample Rank Correlation Coefficient (r_{sp})

Suppose a random sample (X_i, Y_i) ; ($i = 1, 2, \ldots, n$), of n pairs of observations is drawn from a bivariate distribution and let r_{sp} be the observed value of the rank correlation coefficient in the sample. The problem is to find if the sample correlation coefficient (r_{sp}) is significant of any correlation between the variables in the population or is it just due to fluctuations of sampling?

Hypothesis Testing Procedure for ρ_{sp}

***Step* 1** **Null and Alternative Hypotheses**

$H_0 : \rho_{sp} = 0$ (There is no monotone relation)

$H_1 : \rho_{sp} \neq 0$ (There is monotone relation)

ρ_{sp} is the population rank correlation coefficient.

***Step* 2** **Test Statistic = r_{sp}**

***Step* 3** **Critical Value(s) and the Rejection Region**

The critical values $r_{sp}(n, \alpha)$ of r_{sp} for a two-tailed test, α significance level and sample size n are given by:

$$P\left[\left|r_{sp}\right| > r_{sp}(n, \alpha)\right] = \alpha$$

These values are tabulated for different combinations of n and α in Table 11, at the end of the book. Thus the rejection region is

$$\left|r_{sp}\right| > r_{sp}(n, \alpha) \quad \Rightarrow \quad r_{sp} > r_{sp}(n, \alpha) \quad \text{or} \quad r_{sp} < -r_{sp}(n, \alpha)$$

Hence, if the calculated value of $\left|r_{sp}\right|$ is greater than the tabulated value, we reject the null hypothesis $H_0 : \rho_{sp} = 0$, of no monotone relation; otherwise we fail to reject H_0.

Remarks:

1. One-Tailed Tests.

Under H_0, the distribution of the test statistic r_{sp} is symmetric about $r_{sp} = 0$ (like the standard normal distribution or t-distribution). Hence, the single tailed tests for

$$\begin{cases} H_1 : \rho_{sp} > 0 & \text{(Right-tailed)} \\ H_1 : \rho_{sp} < 0 & \text{(Left-tailed)} \end{cases}$$

can be similarly developed with the following result.

The critical values of r_{sp} at an α significance level for a single-tailed test are the same as the critical values of r_{sp} for a two-tailed test at a 2α significance level.

2. Large Samples (n > 30)

If n is large ($n > 30$), then under the null hypothesis $H_0 : \rho_{sp} = 0$ (no correlation) the statistic $Z = r_{sp}\sqrt{n-1}$ has a standard normal distribution.

That is,

$$Z = r_{sp}\sqrt{n-1} \sim N(0, 1) \text{ for large } n$$

Example 13: The following data gives the distance (in miles) by road and straight line (shortest distance), between two towns in New York. Obtain Spearman's rank correlation coefficient r_{sp} and test its significance at $\alpha = 0.05$.

Road Distance (X)	17	29	18	17	11	6	30	20	33	14
Road Distance (Y)	12	22	11	15	9	5	23	12	28	12

Solution: ***Step* 1** Arranging the X-values in ascending order, we get

(X)	6	11	14	17	17	18	20	29	30	33
Rank (X)	1	2	3	4.5	4.5	6	7	8	9	10

Table A3.24

***Step* 2** Arranging the Y-values in ascending order, we get

(Y)	5	9	11	12	12	12	15	22	23	28
Rank (Y)	1	2	3	5	5	5	7	8	9	10

Table A3.25

The data along with their ranks is given in the following table.

R_x = Rank of X-values ; R_y = Rank of Y-values ; $n = 10$.

(a)

X	R_x	Y	R_y	$d = R_x - R_y$	d^2
17	4.5	12	5	−0.5	0.25
29	8	22	8	0	0
18	6	11	3	3	9
17	4.5	15	7	−2.5	6.25
11	2	9	2	0	0
6	1	5	1	0	0
30	9	23	9	0	0
20	7	12	5	2	4
33	10	28	10	0	0
14	3	12	5	−2	4
				$\sum d = 0$	$\sum d^2 = 23.5$

Table A3.26

The Spearman's rank correlation coefficient is given by

$$r_{sp} = 1 - \frac{6\sum d^2}{n(n^2-1)} = 1 - \frac{6(23.5)}{10(99)} = 0.86$$

Test of Significance for r_{sp}

***Step* 1** **Null and Alternative Hypotheses**

$H_0 : \rho_{sp} = 0$ (There is no monotone relation)

$H_1 : \rho_{sp} \neq 0$ (There is monotone relation)

***Step* 2** **Test Statistic = r_{sp} = 0.86**

***Step* 3** **Critical Value(s) and the Rejection Region**

The critical values r_{sp} for $n = 10$ and at $\alpha = 0.05$, for a two-tailed test is 0.648 (From Table 11) so that the rejection region is $r_{sp} > 0.648$ and $r_{sp} < -0.648$.

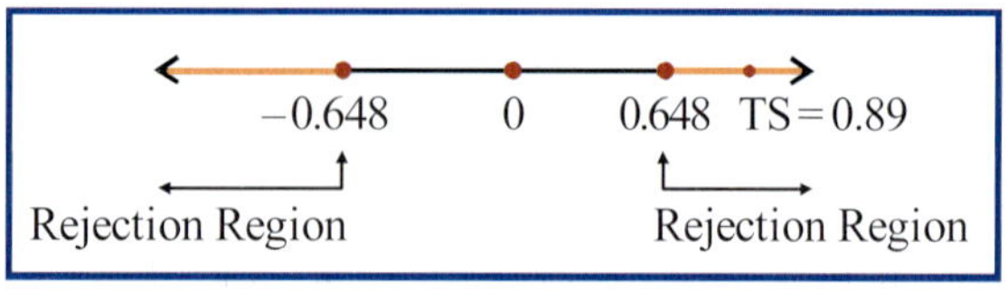

Figure A3.15

***Step* 4** **Decision and Conclusion**

Since the computed value of the test statistic r_{sp} = 0.86 lies in the rejection region, we reject $H_0 : \rho_{sp} = 0$ at 5% level of significance. Hence, we conclude that there is monotone increasing (because $r_{sp} > 0$) relation between the variables X and Y in the population.

In short:

Hypotheses	Test Statistics
$H_0: \rho_{sp} = 0$ $H_1: \rho_{sp} \neq 0$	$r_{sp} = 1 - \dfrac{6\sum d^2}{n(n^2-1)} = 0.86$
Rejection Region and Critical Value	**Conclusion**
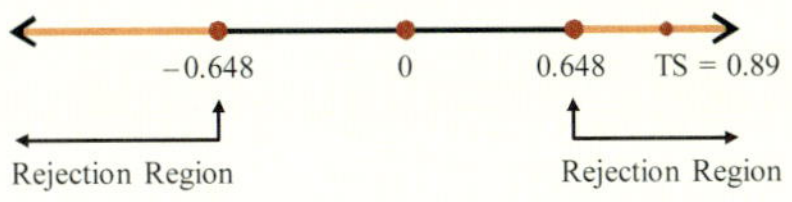	Reject H_0. We conclude that there is monotone increasing relation between the variables X and Y in the population.

Example 14: The readings of HDL and LDL cholesterol levels of 35 randomly selected adults in a city produced the rank correlation coefficient of 0.64. Test the significance of the correlation between the HDL and LDL cholesterol levels in the population of the city? Use $\alpha = 0.01$.

Solution: We are given $n = 35$, $r_{sp} = 0.64$, and $\alpha = 0.01$. Since $n > 30$, we will use the normal test.

***Step* 1 Null and Alternative Hypotheses**

$H_0: \rho_{sp} = 0$

$H_1: \rho_{sp} \neq 0$

***Step* 2 Test Statistic**

Under H_0, the test statistic is

$$Z = r_{sp}\sqrt{n-1} = (0.64)\sqrt{35-1} = 3.73$$

***Step* 3 Critical Value(s) and the Rejection Region**

The critical value of Z for two-tailed test at a 0.01 significance level is 2.575. Hence, the rejection region is

$$Z > 2.575 \quad \text{or} \quad Z < -2.575$$

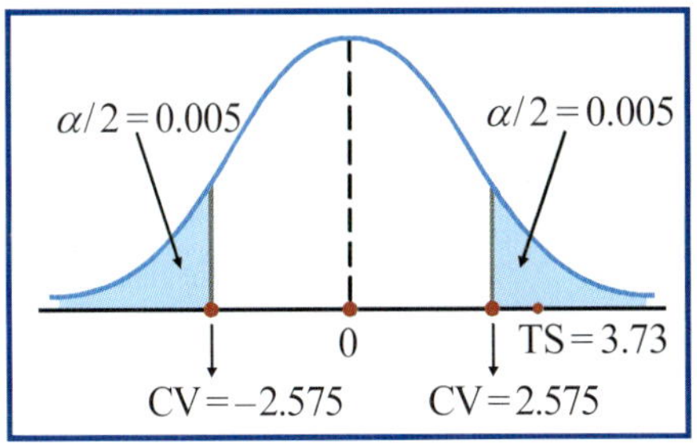

Figure A3.16

***Step* 4 Decision and Conclusion**

Since the computed value of the test statistic $Z = 3.73$ lies in the rejection region, we reject $H_0: \rho_{sp} = 0$ at 1% level of significance. Hence, we conclude that there is monotone relationship between the HDL and LDL cholesterol levels in the population.

In short:

Hypotheses	Test Statistics
$H_0: \rho_{sp} = 0$ $H_1: \rho_{sp} \neq 0$	$Z = r_{sp}\sqrt{n-1} = 3.73$
Rejection Region and Critical Value	**Conclusion**
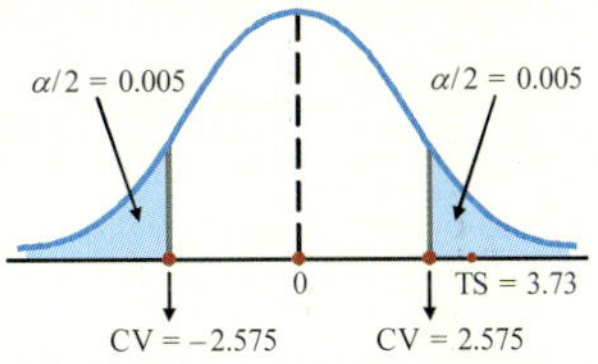	Reject H_0. We conclude that there is monotone relationship between the HDL and LDL cholesterol levels in the population.

Non-Parametric Tests

In exercises 1-7, use the Classical and the *P*-value methods for testing the hypothesis.

Sign Test (Large Samples)

1. The following data represents the final scores of 30 randomly selected students from Prof. Badalian's Statistics classes. Test the claim that the median score is less than 60. Use the sign test at a 0.05 significance level.

13	70	50	66	82	98
44	20	36	8	77	65
90	60	44	86	15	75
37	48	66	72	68	66
61	75	70	40	94	59

2. Test the hypothesis that the median life (in thousand miles) of a tire is more than 48 months. A random sample of 33 tires revealed the following tread lives in months.

39	42	37	53	55	49	48	47	56	60	63
65	46	42	59	58	56	40	46	45	42	59
57	52	49	53	62	63	45	57	58	52	48

Use $\alpha = 0.1$.

3. A random sample of 32 checking accounts at First State Bank gives the following monthly balances. Use the sign test at 0.05 level of significance and test the hypothesis that the median monthly balance is less than \$200.

185	210	324	150	165	134	165	195	245	164	155
320	175	146	148	152	126	159	164	158	211	215
249	168	146	164	157	231	194	182	168	154	

4. The following table lists a random set of SAT scores.

553	535	556	547	565	555	565	558
521	504	510	502	521	513	521	513
539	526	519	519	524	520	525	521
556	540	552	537	556	542	566	550
500	507	489	506	492	509	495	511
542	535	532	534	538	538	536	538
515	499	502	497	507	502	507	504
517	496	505	491	505	494	508	495
482	462	479	468	485	471	489	473
501	497	490	492	497	496	498	496
478	470	474	474	483	477	484	477

Using the sign test with a 0.05 significance level, test the claim that the SAT scores have a median greater than 500.

5. The median height of 15-year old male population of a state is claimed to be 67 inches. Thirty Five 15-year old males are chosen at random. Their heights are recorded (in inches). Use the sign test at 0.01 level of significance to test the hypothesis.

69	72	68	69	71	70	65	64	66	61	66	65
70	69	68	69	72	68	66	65	64	66	67	65
66	72	69	65	67	68	69	68	66	65	64	

6. Use the sign test at 0.05 level of significance to test the hypothesis that the median breaking strength of the springs is 575 lb. A random sample of 36 springs yielded the following breaking strengths:

578	610	601	598	578	546	568	591	579
549	568	572	570	572	571	574	613	615
617	604	608	569	572	575	576	578	574
582	586	583	580	574	578	574	572	573

7. Use the sign test at 0.02 level of significance to test the hypothesis that the median weight of 8-to-10-year-old boys is less than 65 lb. A random sample of 36, 8-to-10-year-old boys yielded the following weights (in lbs.):

64	63	68	67	66	67	70	66	64	63	67	70
66	67	69	68	63	62	64	59	62	64	65	63
64	70	67	63	65	66	67	66	64	63	62	60

In exercises 8-13, use the Classical, *P*-value and the methods 3 for testing the required claim.

Sign Test (Small Samples)

8. A travel agency that promotes a particular holiday package advertises that the median cost (in dollars) per day of the hotels in this travel package is $230. A curious traveler selects a random sample of 10 hotels in the list of this travel package and records the cost (in dollars) per day,

85	250	340	100	92	70	190	210	280	100

Use the sign test for a small sample to test the agency's claim. ($\alpha = 0.01$)

9. Use the sign test, at 0.05 level of significance, to test the hypothesis that the median waiting time for a student who wishes to use a computer at the Academic Computing Center of the university is more than 20 minutes. A random survey yielded the following waiting time (in minutes):

24	26	15	14	10	10	13	22	27	28	20	27	24	27

10. Use the sign test, at 0.05 level of significance, to test Big Burger's claim that the median number of hours (per week) put in by their senior citizen employees is less than 16 hours. A random sample of ten senior citizen employees of their Big Burger was chosen and the following number of hours per week that each worked were recorded.

15	18	16	14	20	16	17	17	12	15

11. Use the sign test, at 0.05 level of significance, to test manufacturer's hypothesis that the median mpg on highway driving of their small cars is more than 40. A random sample of 9 small cars of the manufacturer showed the following miles per gallon on highway driving:

38	42	45	40	37	36	41	40	39

12. A leading tire manufacturer introduces a new tire with a median life expectancy of more than 60,000 miles. An independent agency tested a random sample of 8 such tires and recorded the number of miles (in thousands) the tires last.

62	63	59	56	61	60	64	57

Use the sign test, at 5% level of significance to test the manufacturer's claim.

13. A travel agency, which promotes a particular holiday resort, advertises that the median cost per day of a motel in the city is $50. A cautious traveler selects a random sample of 8 motels in that city and records the cost (in dollar) per day.

52	51	49	50	53	52	48	47

Use the sign test, at 1% level of significance to test the travel agency's claim.

In exercises 14-16, use the Classical, *P*-value and the methods 3 for testing the stated claim.

Sign Test (Paired Samples)

14. Parents of freshmen have often claimed that their sons and daughters lose weight while staying and eating in the dormitories. A random sample of freshmen is selected and their weights are recorded at the beginning and at the end of the first semester.

Before	150	143	180	110	160	175	140
After	145	155	170	120	170	163	125

Use the paired sign test to test the claim at a 0.01 significance level.

15. To test whether students perform equally well on the verbal and non-verbal parts of the SAT test, the scores from a random sample of 7 students are recorded (in pairs).

SAT (verbal)	560	510	620	480	590	610	420
SAT (non-verbal)	620	650	700	620	450	640	400

Use the sign test at 5% level of significance.

16. ABC Coaching institute claims that their coaching helps raising the SAT scores. The SAT scores of students who took the SAT test before and after coaching are as follows:

Before coaching	950	1100	875	1200	1280	1300	1410
After coaching	1050	1120	900	1190	1300	1310	1390

Use the paired samples sign test at a 0.01 significance level to test the coaching institute's claim.

In exercises 17-21, use (a) the Classical Method and (b) the *P*-value Method for testing the hypothesis.

Wilcoxon Signed-Rank Test

17. Use the sign test at 0.05 level of significance to test the hypothesis that the median waiting time for a student who wishes to use a computer at Math Academia is more than 20 minutes. A random survey yielded the following waiting times (in minutes).

24	26	15	14	10	10	13	22	27	28	20	27	24	27

18. Use the Wilcoxon signed-rank test at 0.05 level of significance to test Big Burger's claim that the median number of hours (per week) put in by their senior citizen employees is less than 16 hours. A random sample of ten senior citizen employees of Big Burger was chosen and the following number of hours per week that each worked were recorded.

15	18	16	14	20	16	17	17	12	15

19. Test the hypothesis of Exercise no. 11 using the Wilcoxon signed-rank test, at 5 % level of significance.

20. A leading tire manufacturer introduces a new tire with a median life expectancy of more than 60,000 miles. An independent agency tested a random sample of 8 such tires and recorded the number of miles (in thousands) the tires last.

62	63	59	56	61	60	64	57

Use a 5% level of significance to test the manufacturer's claim.

21. Test the hypothesis of Exercise no. 13 using the Wilcoxon signed-rank test, at 10 % level of significance.

In exercises 22-24, use (a) the Classical Method and (b) the *P*-value Method for testing the stated claim.

Wilcoxon Signed-Rank Test (Matched Pairs)

22. To test whether students perform equally well on the verbal and non-verbal parts of the SAT test, the scores from a random sample of 7 pairs of students are recorded and summarized in the following table.

SAT(verbal)	560	510	620	480	590	610	420
SAT (non-verbal)	620	650	700	620	450	640	400

Use the Wilcoxon Signed-Rank test using a 0.05 significance level.

23. A researcher gives a new feed to a randomly selected sample of 6 cows and records their weights before and after the experimental period.

Weight before	1250	1175	1400	1380	1280	1510
Weight after	1400	1350	1390	1250	1310	1450

Use the Wilcoxon Signed-Rank Test at a 0.5 level of significance to test the researchers claim that the new feed is effective in increasing the weight.

24. Test the hypothesis of Exercise no. 16 using the Wilcoxon signed-rank test, at 0.01 significance level.

In exercises 25-29, use (a) the Classical Method and (b) the *P*-value Method for testing the desired claim.

Wilcoxon Signed-Rank Test (Large Samples)

25. To test the hypothesis that the median number of days automobiles remain in the long-term parking lot at BWI is 3 days. The results of a random sample of 32 automobiles indicated the following days:

4	3	5	3	4	7	3	4	6	4	2
3	4	2	3	4	5	7	8	4	4	2
5	3	4	5	7	9	5	2	3	5	

Use $\alpha = 0.02$ for this test.

26. Test the hypothesis that the median life (in thousand miles) of a tire is more than 48 months. A random sample of 33 tires revealed the following tread lives in months:

39	42	37	53	55	49	48	47	56	60	63
65	46	42	59	58	56	40	46	45	42	59
57	52	49	53	62	63	45	57	58	52	48

Use $\alpha = 0.01$ for this test.

27. A random sample of 32 checking accounts at First State Bank gives the following monthly balances (in $). Use the Wilcoxon Signed-Rank Test at 0.05 level of significance to test the hypothesis that the median monthly balance is less than $200.

185	210	324	150	165	134	165	195	245	164	155
320	175	146	148	152	126	159	164	158	211	215
249	168	146	164	157	231	194	182	168	154	

28. The results in minutes of a random sample of 35 long distance calls is given in the following table. Use the Wilcoxon Signed-Rank Test at 0.01 level of significance to test the claim that the median length of a long distance telephone call is less than 15 minutes.

12	18	23	28	15	10	13	3	6	16	25	35
23	25	28	23	27	28	5	9	37	27	32	19
13	19	18	23	35	4	18	25	31	24	21	

29. Use a 0.05 level of significance to test the hypothesis that the median breaking strength of the springs of a certain brand is 575 lb. A random sample of 36 springs yielded the following breaking strengths:

578	610	601	598	578	546	568	591	579	549	568	572
570	572	571	574	613	615	617	604	608	569	572	575
576	578	574	582	586	583	580	574	578	574	572	573

In exercises 30-35, use the Classical and the *P*-value Methods for testing the hypothesis.

Mann Whitney Test (Small Samples)

30. Given the following table, test the hypothesis that the median income per day in population **A** is less than the median income per day in population **B**. Use $\alpha = 0.05$.

A	50	75	40	85	60	90	50	45	80
B	100	85	70	65	45	70	80	95	

31. Given the following table, test the hypothesis that the median measure in population **A** is less than the median measure in population **B**. Use $\alpha = 0.05$.

A	15	7	9	18			
B	10	8	21	22	5	17	18

32. Given the following table, test the hypothesis that the median measure in population **A** is equal to the median measure in population **B**. Use $\alpha = 0.1$.

A	50	52	41	56	38	25	48
B	28	38	56	47	40		

33. Given the following table, test the hypothesis that the median measure in population **A** is greater than the median measure in population **B**. Use $\alpha = 0.05$.

A	70	78	65	51	41	30
B	49	57	30	78	72	75

34. Two different foods are tested for the difference in weight gains. A random sample of 6 mice was given food A and the other sample of 8 mice was given food B. The weight gains were recorded.

Food A	17	14	15	18	12	13		
Food B	15	16	13	16	18	19	17	19

Test the claim that the median of the two food groups are equal. Use a 0.1 significance level.

35. A researcher wants to determine whether or not there is a difference in the median weight loss under two different weight reducing diets. The weight losses (in pounds) in 5 randomly selected individuals under each diet for a month are recorded.

Diet A	5	0	1	3	4
Diet B	8	2	6	6	1

Test the claim using a 0.1 significance level.

In exercises 36-39, use the Classical and the *P*-value Methods for testing the stated claim.

Mann-Whitney Test (Large Samples).

36. Test the claim that $\eta_A > \eta_B$. Use $\alpha = 0.05$.

A	32	50	28	25	26	29	31	40	50	35
B	34	38	30	55	60	49	36	65	58	50

37. Test the claim that $\eta_A < \eta_B$. Use $\alpha = 0.05$.

A	50	53	42	56	38	25	48	35	26	40
B	28	38	56	47	40	51	20	22	32	52

38. Test the claim that $\eta_A = \eta_B$. Use $\alpha = 0.1$.

A	54	14	32	18	14	44	63	42	54	47
B	41	28	43	37	52	52	23	20	52	58

39. To test whether or not there is a difference in the median yields of two varieties of corn, 10 acres of each variety are planted and grown under similar conditions. Yields (in bushels) of variety A and variety B are recorded.

A	110	71	64	102	104	95	90	85	100	75
B	88	76	79	60	105	115	98	85	90	78

Use a 0.05 significance level.

In exercises 40-44, compute Spearman's rank correlation coefficient r_{sp}.

40.

X	1	2	3	4	5	6	7	8	9	10
Y	3	9	8	14	17	21	23	25	27	25

41.

X	0	2	4	6	8	10
Y	12	7	9	4	3	2

42.

X	3	8	10	2	6	13
Y	14	22	24	15	16	20

43.

X	8	1	3	6	12	7	10
Y	13	30	23	35	24	18	15

44.

X	25	30	28	32	35	38
Y	18	22	25	24	29	30

45. Compute and interpret the rank correlation coefficient for the following grades of 12 students selected at random: Use $\alpha = 0.05$.

Mathematics Grade	85	83	87	84	88	82	90	86	88	87	89	90
Physics Grade	88	86	88	86	90	86	88	85	92	86	88	91

46. Compute and interpret the rank correlation coefficient for the following ages of 8 couples selected at random: Use $\alpha = 0.05$.

Husband's Age	20	45	40	43	50	60	29	70
Wife's Age	18	40	34	36	35	42	21	48

47. Compute and interpret the rank correlation coefficient for the following amounts of advertising budgets and annual sales. Use $\alpha = 0.01$.

Advertising (thousands \$)	45	52	46	48	50	47	53	49	51
Sales (millions \$)	9.5	10.5	9.7	9.8	10.0	9.7	10.5	9.9	10.3

48. Compute and interpret the rank correlation coefficient for the following SAT-verbal scores and the first year college performance (GPA) of 8 randomly selected students. Use $\alpha = 0.05$.

Mathematics Grade	560	620	495	610	650	710	640	680
Physics Grade	2.6	2.8	3.1	2.5	3.2	3.9	3.0	3.8

49. Test the significance of the correlation between two randomly selected samples where $n = 45$ and $r_{sp} = 0.74$. Use $\alpha = 0.01$.

50. Test the significance of the correlation between two randomly selected samples where $n = 38$ and $r_{sp} = 0.58$. Use $\alpha = 0.05$.

51. Test the significance of the correlation between two randomly selected samples where $n = 86$ and $r_{sp} = 0.64$. Use $\alpha = 0.01$.

FORMULA SHEET

FORMULA SHEET

Chapter 2: Data Analysis

$$\textbf{Class Width} = \frac{\text{largest value} - \text{smallest value}}{\text{number of classes}}$$

$$\textbf{Mid-Point} = \text{class mark} = \frac{\text{upper limit} + \text{lower limit}}{2}$$

$$\textbf{Relative Frequency} = \frac{\text{class frequency}}{\text{sample size}} = \frac{f}{n}$$

$$\textbf{Central Angle for Pie-Charts} = \text{relative frequency} \times 360$$

$$\textbf{Sample Mean } (\bar{x}) = \frac{\sum x_i}{n} = \frac{x_1 + x_2 + \ldots + x_n}{n}$$

$$\textbf{Population Mean } (\mu) = \frac{\sum x_i}{N} = \frac{x_1 + x_2 + \ldots + x_N}{N}$$

$$\textbf{Weighted Mean } (\bar{x}) = \frac{\sum x_i \cdot w_i}{\sum w_i} = \frac{x_1 w_1 + x_2 w_2 + \ldots + x_n w_n}{w_1 + w_2 + \ldots + w_n}$$

$$\textbf{Range} = \text{maximum data value} - \text{minimum data value}$$

$$\textbf{Population Variance } \left(\sigma^2\right) = \frac{\sum (x_i - \mu)^2}{N}$$

$$\textbf{Sample Variance } \left(s^2\right) = \frac{\sum (x_i - \bar{x})^2}{n - 1} = \frac{n \sum x_i^2 - \left(\sum x_i\right)^2}{n(n-1)}$$

$$\textbf{Population Standard Deviation } (\sigma) = \sqrt{\sigma^2} = \sqrt{\frac{\sum (x_i - \mu)^2}{N}}$$

$$\textbf{Sample Standard Deviation } (s) = \sqrt{s^2} = \sqrt{\frac{n \sum x_i^2 - \left(\sum x_i\right)^2}{n(n-1)}}$$

$$\textbf{Range Rule of Thumb } (\sigma) \approx \frac{\text{range}}{4}$$

$$\textbf{Mean of Grouped Data} = \frac{\sum x_i \cdot f_i}{\sum f_i}$$

$$\textbf{Standard Deviation of Grouped Data } (s) = \sqrt{\frac{n \sum x_i^2 \cdot f_i - \left(\sum x_i \cdot f_i\right)^2}{n(n-1)}}$$

$$\textbf{Median of Grouped Data } (s) = L + \frac{h}{f_m}\left(\frac{n+1}{2} - c\right)$$

Empirical Rule: (Assume the distribution of data is bell-shaped)

i) Approximately 68.3% of data falls within one standard deviation of the mean.

ii) Approximately 95.4% of data falls within two standard deviation of the mean.

iii) Approximately 99.7% of data falls within three standard deviation of the mean.

Chebyshev's Theorem: The proportion of any data set lying within k standard deviation of the mean is at least $1 - \frac{1}{k^2}$, for $k > 1$.

***Z*-Score :** $Z\text{-score} = \dfrac{X - \mu}{\sigma}$ = Standard scores

Rank of Quartiles : The rank of $Q_1 = 0.25(n + 1)$

The rank of $Q_2 = 0.50(n + 1)$

The rank of $Q_3 = 0.75(n + 1)$

Interquartile Range : $IQR = Q_3 - Q_1$

Percentile : $$\text{Percentile of } X = \frac{\text{number of values below } X + 0.5}{\text{total number of given values in the data set}} \times 100\%$$

Position of Percentile : $$\text{Position of the } p^{\text{th}} \text{ percentile} = \frac{\text{Percentile}}{100} \times n = \frac{P}{100} \cdot n$$

Chapter 3: Classical Probability

Factorial : $n! = n \times (n-1) \times (n-2) \times \ldots \times 3 \times 2 \times 1$

Combination : $C_k^n = \dfrac{n!}{k!(n-k)!}$

Arrangement(Permutation) : $A_k^n = P_k^n = \dfrac{n!}{(n-k)!}$

Empirical Probability : $P(E) = \dfrac{f}{n}$

Classical Probability : $P(E) = \dfrac{n(E)}{n(S)}$

Complement Rule : $P(\bar{E}) = 1 - P(E)$

Addition Rule : $P(E \text{ or } F) = P(E) + P(F) - P(E \cap F)$

Multiplication Rule for Dependent events : $P(E \text{ and } F) = P(E) \cdot P(F|E) = P(F) \cdot P(E|F)$

Multiplication Rule for Independent events : $P(E \text{ and } F) = P(E) \cdot P(F)$

Chapter 4: Discrete Probability Distribution

Expected Value : $E(X)=\mu=\sum x_i \cdot P(X=x_i)=\sum(Xp)$

Variance : $\sigma^2=E(X-\mu)^2=\sum\left[x_i^2\cdot P(X=x_i)\right]-\mu^2=\sum\left(X^2p\right)-\mu^2$

Standard Deviation : $\sigma=\sqrt{\sigma^2}=\sqrt{\sum\left[x_i^2\cdot P(X=x_i)\right]-\mu^2}=\sqrt{\sum\left(X^2p\right)-\mu^2}$

Binomial Probability Distribution (BPD) : $P(X=x)=C_x^n p^x q^{n-x}=\binom{n}{x}p^x q^{n-x},\quad q=1-p\ ;$

$x=0,\ 1,\ 2,\ldots,\ n.$

Mean of a (BPD) : $\mu=n\cdot p$

Variance of a (BPD) : $\sigma^2=npq=np(1-p)$

Standard Deviation of a (BPD) : $\sigma=\sqrt{npq}$

Poisson Distribution : $P(X=x)=\dfrac{e^{-\mu}\mu^x}{x!}\ ;\ x=0,\ 1,\ 2,\ \ldots$

Mean of a Poisson Distribution : $\text{Mean}=\mu$

Variance of a Poisson Distribution : $\sigma^2=\mu$

Multinomial Probability Distribution : $p(x_1,x_2,\ldots,x_k)=\dfrac{n!}{x_1!x_2!\ldots x_k!}\cdot p_1^{x_1}\cdot p_2^{x_2}\ldots\cdot p_k^{x_k}$,

$x_1+x_2+\ldots+x_k=n\ ,\ 0<p_i<1\ ,\ x_i\in\{0,\ 1,\ 2,\ \ldots,\ n\}\ ;\ i=1,\ 2,\ \ldots,\ k$

Hypergeometric Probability Distribution : $p(x)=\dfrac{\binom{k}{x}\binom{N-k}{n-x}}{\binom{N}{n}}$, $\max(0,\ n+k-N)\le x\le\min(n,\ k)$

Chapter 5: Normal Probability Distribution

Standard Score: $Z=\dfrac{X-\mu}{\sigma}$

Continuity Corrections: $P(X=a)\approx P(a-0.5<X<a+0.5)$

$P(X\le a)\approx P(X<a+0.5)$

$P(X\ge a)\approx P(X>a-0.5)$

$P(X<a)\approx P(X<a-0.5)$

$P(X>a)\approx P(X>a+0.5)$

$P(a\le X\le b)\approx P(a-0.5<X<b+0.5)$

Chapter 6: Point and Interval Estimation

Central Limit Theorem (Mean of the Sampling Distribution) : $\mu_{\bar{X}} = \mu$

Central Limit Theorem (Standard Deviation of the Sampling Distribution) : $\sigma_{\bar{X}} = \dfrac{\sigma}{\sqrt{n}}$

Standardization of $\bar{X}$: $Z_{\bar{X}} = \dfrac{\bar{X} - \mu_{\bar{X}}}{\sigma_{\bar{X}}} = \dfrac{\bar{X} - \mu}{\sigma / \sqrt{n}} \sim N(0,1)$

Expected Value of the Sampling Distribution of $\hat{p}$: $E(\hat{p}) = p$

Variance of the Sampling Distribution of $\hat{p}$: $\text{var}(\hat{p}) = \sigma_{\hat{p}}^2 = \dfrac{pq}{n}$, $q = 1 - p$

Standardization of $\hat{p}$: $Z = \dfrac{\hat{p} - E(\hat{p})}{\sigma_{\hat{p}}} = \dfrac{\hat{p} - p}{\sqrt{\dfrac{pq}{n}}} \sim N(0,1)$ for large values of n.

Flow Charts to Estimate σ :

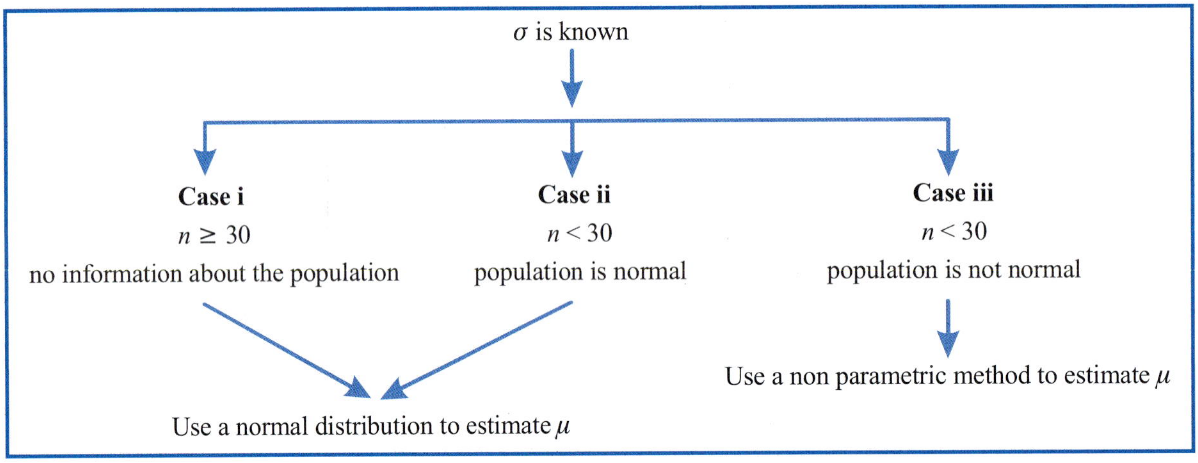

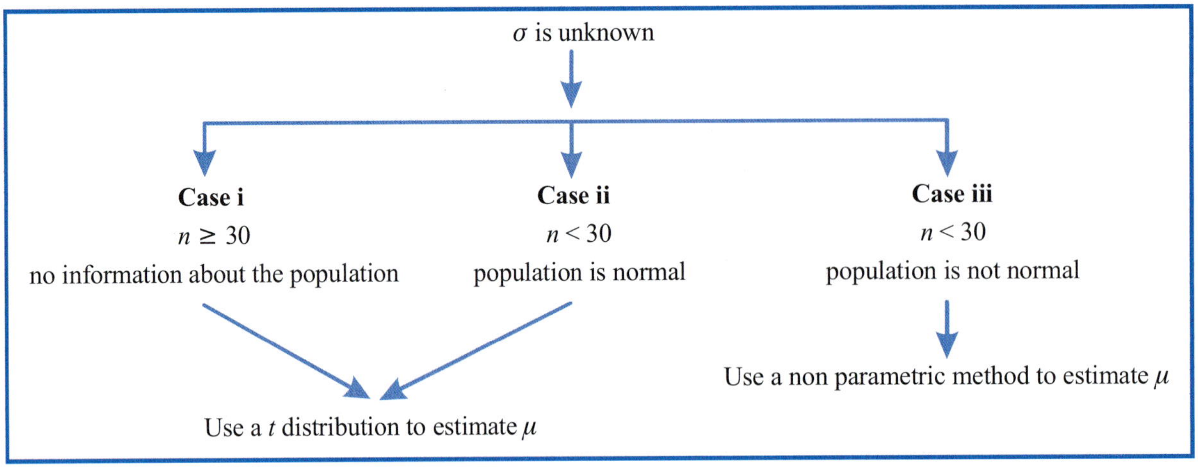

100(1 − α)% Confidence Interval for Population Mean (σ known) :

$$\bar{x} - z_{\alpha/2}\frac{\sigma}{\sqrt{n}} < \mu < \bar{x} + z_{\alpha/2}\frac{\sigma}{\sqrt{n}}$$

Minimum Sample Size for Means : $n = \left(\frac{z_{\alpha/2}\sigma}{E}\right)^2$

100(1 − α)% Confidence Interval for Population Mean (σ unknown):

$$\bar{x} - t_{\alpha/2,n-1}\frac{s}{\sqrt{n}} < \mu < \bar{x} + t_{\alpha/2,n-1}\frac{s}{\sqrt{n}}$$

100(1 − α)% Confidence Interval for Population Proportion :

$$\hat{p} - z_{\alpha/2}\sqrt{\frac{\hat{p}\hat{q}}{n}} < p < \hat{p} + z_{\alpha/2}\sqrt{\frac{\hat{p}\hat{q}}{n}}$$

Minimum Sample Size for Proportion ($\hat{p}$ given) : $n = \frac{z_{\alpha/2}^2\hat{p}\hat{q}}{E^2}$

Minimum Sample Size for Proportion ($\hat{p}$ not given) : $n = \frac{z_{\alpha/2}^2(0.5)(0.5)}{E^2} = \frac{z_{\alpha/2}^2(0.25)}{E^2}$

100(1 − α)% Confidence Interval for Population Variance (Large Sample):

$$s^2\left(1 - z_{\alpha/2}\sqrt{\frac{2}{n}}\right) < \sigma^2 < s^2\left(1 + z_{\alpha/2}\sqrt{\frac{2}{n}}\right)$$

100(1 − α)% Confidence Interval for Population Standard Deviation (Large Sample) :

$$s\sqrt{1 - z_{\alpha/2}\sqrt{\frac{2}{n}}} < \sigma < s\sqrt{1 + z_{\alpha/2}\sqrt{\frac{2}{n}}}$$

100(1 − α)% Confidence Interval for Population Variance (Small Sample):

$$\frac{(n-1)s^2}{\chi_R^2} < \sigma^2 < \frac{(n-1)s^2}{\chi_L^2}$$

100(1 − α)% Confidence Interval for Population Standard Deviation (Small Sample):

$$\sqrt{\frac{(n-1)s^2}{\chi_R^2}} < \sigma < \sqrt{\frac{(n-1)s^2}{\chi_L^2}}$$

Chapter 7: Hypothesis Testing

Flow Charts for all Possible Conclusions Based on Our Decision in Testing of Hypothesis:

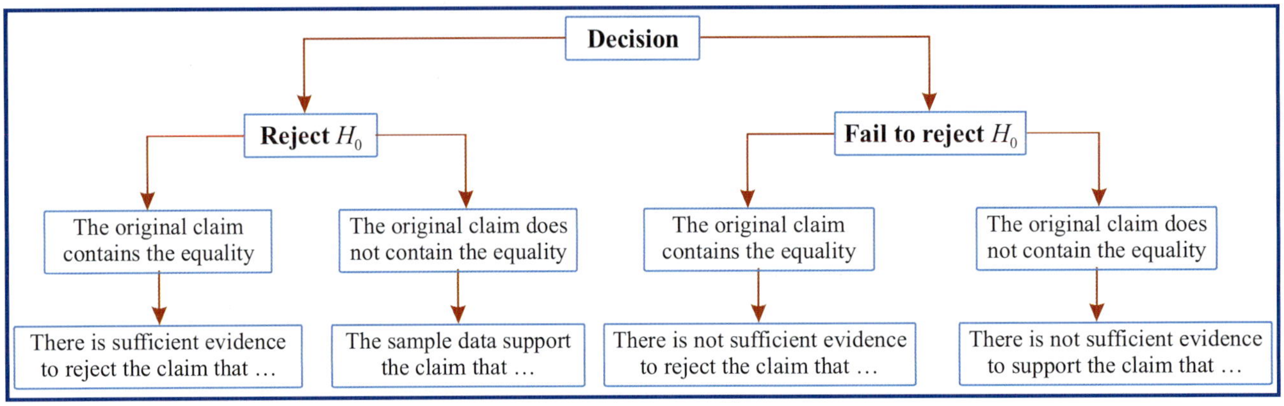

Test Statistics for Population Mean (σ known): $Z = \dfrac{\bar{X} - \mu}{\sigma / \sqrt{n}}$

Test Statistics for Population Mean (σ unknown): $t = \dfrac{\bar{X} - \mu}{s / \sqrt{n}}$, degrees of freedom for $t = n - 1$

Summary for the t-test: σ unknown

Sample Size (n)	Population Distribution	Test Statistics and its sampling distribution
$n \le 30$	Normal	$t = \dfrac{\bar{X} - \mu}{s / \sqrt{n}} \sim t_{n-1}$; σ unknown
$n > 30$	Any	$t = \dfrac{\bar{X} - \mu}{s / \sqrt{n}} \sim t_{n-1}$; σ unknown
$n \le 30$	Unknown	Use non-parametric tests (Chapter 12)

Test Statistics for Population Proportion: $Z = \dfrac{\hat{p} - p}{\sqrt{\dfrac{p(1-p)}{n}}}$

Test Statistics for Population Variance or Standard Deviation: $\chi^2 = \dfrac{(n-1)s^2}{\sigma^2}$

Chapter 8: Inferences from Two Populations

100(1 – α)% Confidence Interval Estimation for $\mu_1 - \mu_2$ (Independent Samples, σ_1 and σ_2 are known):

$$(\bar{x}_1 - \bar{x}_2) - z_{\alpha/2}\sqrt{\frac{\sigma_1^2}{n_1} + \frac{\sigma_2^2}{n_2}} < \mu_1 - \mu_2 < (\bar{x}_1 - \bar{x}_2) + z_{\alpha/2}\sqrt{\frac{\sigma_1^2}{n_1} + \frac{\sigma_2^2}{n_2}}$$

100(1 – α)% Confidence Interval Estimation for $\mu_1 - \mu_2$ (Independent Samples, σ_1 and σ_2 are unknown and assumed unequal):

$$(\bar{x}_1 - \bar{x}_2) - t_{\alpha/2}\sqrt{\frac{s_1^2}{n_1} + \frac{s_2^2}{n_2}} < \mu_1 - \mu_2 < (\bar{x}_1 - \bar{x}_2) + t_{\alpha/2}\sqrt{\frac{s_1^2}{n_1} + \frac{s_2^2}{n_2}}$$

degrees of freedom for $t_{\alpha/2}$ = smaller of $n_1 - 1$ and $n_2 - 1$.

100(1 – α)% Confidence Interval Estimation for $\mu_1 - \mu_2$ (Independent Samples, σ_1 and σ_2 are unknown and assumed equal):

$$(\bar{x}_1 - \bar{x}_2) - t_{\alpha/2}\sqrt{\frac{s_p^2}{n_1} + \frac{s_p^2}{n_2}} < \mu_1 - \mu_2 < (\bar{x}_1 - \bar{x}_2) + t_{\alpha/2}\sqrt{\frac{s_p^2}{n_1} + \frac{s_p^2}{n_2}}$$

$s_p^2 = \dfrac{(n_1 - 1)s_1^2 + (n_2 - 1)s_2^2}{n_1 + n_2 - 2}$,and degrees of freedom for $t_{\alpha/2} = n_1 + n_2 - 2$

100(1 – α)% Confidence Interval for Matched Pairs (Dependent Samples):

$$\bar{d} - t_{\alpha/2}\frac{s_d}{\sqrt{n}} < \mu_d < \bar{d} + t_{\alpha/2}\frac{s_d}{\sqrt{n}}$$, degrees of freedom for $t_{\alpha/2} = n - 1$

$$\bar{d} = \frac{\sum d}{n}, \text{ and } s_d = \sqrt{\frac{n\sum d^2 - \left(\sum d\right)^2}{n(n-1)}}$$

100(1 – α)% Confidence Interval Estimation for two population Proportions:

$$(\hat{p}_1 - \hat{p}_2) - z_{\alpha/2}\sqrt{\frac{\hat{p}_1\hat{q}_1}{n_1} + \frac{\hat{p}_2\hat{q}_2}{n_2}} < p_1 - p_2 < (\hat{p}_1 - \hat{p}_2) + z_{\alpha/2}\sqrt{\frac{\hat{p}_1\hat{q}_1}{n_1} + \frac{\hat{p}_2\hat{q}_2}{n_2}}$$

100(1 – α)% Confidence Interval Estimation for Ratio of Variances:

$\dfrac{s_1^2}{s_2^2}\cdot\dfrac{1}{F_{\alpha/2}} < \dfrac{\sigma_1^2}{\sigma_2^2} < \dfrac{s_1^2}{s_2^2}\cdot\dfrac{1}{F_{1-\alpha/2}}$, degrees of freedom of $F_{a/2}$ and $F_{1-a/2}$ are $(n_1 - 1, n_2 - 1)$

100(1 – α)% Confidence Interval Estimation for Ratio of SD's of two independent Normal populations:

$\dfrac{s_1}{s_2}\cdot\sqrt{\dfrac{1}{F_{\alpha/2}}} < \dfrac{\sigma_1}{\sigma_2} < \dfrac{s_1}{s_2}\cdot\sqrt{\dfrac{1}{F_{1-\alpha/2}}}$, degrees of freedom of $F_{a/2}$ and $F_{1-a/2}$ are $(n_1 - 1, n_2 - 1)$

Test Statistics for Testing differences of two Population Means (Independent Samples, σ_1 and σ_2 are known):

$$Z = \frac{(\bar{x}_1 - \bar{x}_2) - (\mu_1 - \mu_2)}{\sqrt{\dfrac{\sigma_1^2}{n_1} + \dfrac{\sigma_2^2}{n_2}}}$$

Test Statistics for Testing differences of two Population Means (Independent Samples, σ_1 and σ_2 are unknown and assumed unequal):

$$t = \frac{(\bar{x}_1 - \bar{x}_2) - (\mu_1 - \mu_2)}{\sqrt{\frac{s_1^2}{n_1} + \frac{s_2^2}{n_2}}}$$,degrees of freedom for $t_{\alpha/2} =$ smaller of $n_1 - 1$ and $n_2 - 1$

Test Statistics for Testing differences of two Population Means (Independent Samples, σ_1 and σ_2 are unknown and assumed equal):

$$t = \frac{(\bar{x}_1 - \bar{x}_2) - (\mu_1 - \mu_2)}{\sqrt{\frac{s_p^2}{n_1} + \frac{s_p^2}{n_2}}}$$, degrees of freedom for $t_{\alpha/2} = n_1 + n_2 - 2$

Test Statistics for Testing Matched Pairs (Dependent Samples):

$$t = \frac{\bar{d} - \mu_d}{\frac{s_d}{\sqrt{n}}}$$

Test Statistics for Testing differences of two Population Proportions:

$$Z = \frac{(\hat{p}_1 - \hat{p}_2) - (p_1 - p_2)}{\sqrt{\frac{\bar{p}\bar{q}}{n_1} + \frac{\bar{p}\bar{q}}{n_2}}}$$, where $\bar{p} = \frac{x_1 + x_2}{n_1 + n_2}$, and $\bar{q} = 1 - \bar{p}$

Test Statistics for Testing equality of two Population Variances or Standard Deviations:

$$F = \frac{s_1^2}{s_2^2}$$, where $s_1 > s_2$, df of F = $(n_1 - 1, n_2 - 1)$

Chapter 9: Multinomial Experiments

Test Statistics for Goodness of Fit Tests in Multinomial Experiments:

$$\chi^2 = \sum_i \frac{(O_i - E_i)^2}{E_i}$$, where $\sum O_i = \sum E_i$, df of χ^2: $k - 1$

Test Statistics for testing independence of attributes.

$$\chi^2 = \sum_i \frac{(O_i - E_i)^2}{E_i}$$, where $E = \frac{(\text{row total}) \times (\text{column total})}{\text{grand total}}$, and df of χ^2: $(r - 1)(c - 1)$

Chapter 10: Correlation and Regression

Sample Correlation Coefficient : $r = \dfrac{n\sum xy - \sum x \sum y}{\sqrt{n\sum x^2 - \left(\sum x\right)^2}\sqrt{n\sum y^2 - \left(\sum y\right)^2}}$

Test Statistics for Testing Linear Correlation between two variables in population :

$$t = r\sqrt{\frac{n-2}{1-r^2}}\ ,\ df = n - 2$$

Slope of the Regression Line : $m = \dfrac{n\sum xy - \sum x \sum y}{n\sum x^2 - \left(\sum x\right)^2}$

Y-intercept of the Regression Line : $b = \dfrac{\sum y \sum x^2 - \sum x \sum xy}{n\sum x^2 - \left(\sum x\right)^2}$

Prediction Interval : $\hat{y} - E < y < \hat{y} + E$

$$E = t_{\alpha/2, n-2}\sqrt{\frac{\sum y^2 - b\sum y - m\sum xy}{n-2}}\sqrt{1 + \frac{1}{n} + \frac{n\left(x - \bar{x}\right)^2}{n\sum x^2 - \left(\sum x\right)^2}}$$

Multiple Regression Model : $\hat{y} = b + b_1x_1 + b_2x_2 + \cdots + b_kx_k$

Appendix 2: Analysis of Variance

Test Statistics for One Way ANOVA : $F = \dfrac{\text{variance between samples}}{\text{variance within samples}}$

Test Statistics for Equal Sample Sizes:

Variance between Samples : $ns_{\bar{x}}^2$ where $s_{\bar{x}}^2$ is the variance of sample means.

Variance within Samples : s_p^2 (pooled variance)

$$F = \frac{ns_{\bar{x}}^2}{s_p^2}$$

Degrees of Freedom:

Numerator $df = k - 1$, where k is the number of samples, and n is the sample size

Denominator $df = k\ (n - 1)$.

Test Statistics for Unequal Sample Sizes:

$$F = \frac{\text{variance between samples}}{\text{variance within samples}} = \frac{\left[\dfrac{\sum n_i\left(\overline{x}_i - \overline{\overline{x}}\right)}{k-1}\right]}{\left[\dfrac{\sum(n_i-1)s_i^2}{\sum(n_i-1)}\right]}, \text{ where } \quad \overline{x} = \frac{\sum n_i\,\overline{x}_i}{\sum n_i}$$

Degrees of Freedom:

Numerator $df = k - 1$, where k is the number of samples, and n is the sample size

Denominator $df = N - k$

Chapter 12: Non-Parametric Statistics

Single Sample Sign Test (Large Sample):

Test Statistics: $$Z = \frac{(X+0.5)-\frac{n}{2}}{\frac{1}{2}\sqrt{n}}, \quad \text{if } X < \frac{n}{2}$$

or $$Z = \frac{(X-0.5)-\frac{n}{2}}{\frac{1}{2}\sqrt{n}}, \quad \text{if } X > \frac{n}{2}$$

Wilcoxon Signed-Rank Test:

Test Statistics: T = smaller of (T^+, T^-)

Wilcoxon Signed-Rank Test (Large Sample *n* > 30):

Mean: $$\mu_T = \frac{n(n+1)}{4}$$

Standard Deviation: $$\sigma_T = \sqrt{\frac{n(n+1)(2n+1)}{24}}$$

The Test Statistics is: $$Z = \frac{T-\mu_T}{\sigma_T} = \frac{T - \frac{n(n+1)}{4}}{\sqrt{\frac{n(n+1)(2n+1)}{24}}}$$

Mann-Whitney Test: $$U_x = T_x - \frac{n_1(n_1+1)}{2}$$

$$U_y = T_y - \frac{n_2(n_2+1)}{2}$$

U = smaller of (U_x, U_y)

TABLES

TABLE 1

RANDOM NUMBERS

Row \ Column	5	10	15	20	25	30	35	40	45	50
1	02711	08182	75997	79866	58095	83319	80295	79741	74599	84379
	94873	90935	31684	63952	09865	14491	99518	93394	34691	14985
	54921	78680	06635	98689	17306	25170	65928	87709	30533	89736
	77640	97636	37397	93379	56454	59818	45827	74164	71666	46977
5	61545	00835	93251	87203	36759	49197	85967	01704	19634	21898
	17147	19519	22497	16857	42426	84822	92598	49186	88247	39967
	13748	04742	92460	85801	53444	65626	58710	55406	17173	69776
	87455	14813	50373	28037	91182	32786	65261	11173	34376	36408
	08999	57409	91185	10200	61411	23392	47797	56377	71635	08601
10	78804	81333	53809	32471	46034	36306	22498	19239	85428	55721
	82173	26921	28472	98958	07960	66124	89731	95069	18625	92405
	97594	25168	89178	68109	05043	17407	48201	83917	11413	72920
	73881	67176	93504	42636	38233	16154	96451	5792 5	29667	30859
	46071	22912	90326	42453	88108	72064	58601	32357	90610	32921
15	44492	19686	12495	93135	95185	77799	52441	88272	22024	80631
	31864	72170	37722	55794	14636	05148	54505	50113	21119	25228
	51574	90692	43339	65689	76539	27909	05467	21727	51141	72949
	35350	76132	92925	92124	92634	35681	43690	89136	35599	84138
	46943	36502	01172	46045	46991	33804	80006	35542	61056	75666
20	22665	87226	33304	57975	03985	21566	65796	72915	81466	89205
	39437	97957	11838	10433	21564	51570	73558	27495	34533	57808
	77082	47784	40098	97962	89845	28392	78187	06112	08169	11261
	24544	25649	43370	28007	06779	72402	62632	53956	24709	06978
	27503	15558	37738	24849	70722	71859	83736	06016	94397	12529
25	24590	24545	06435	52758	45685	90151	46516	49644	92686	84870
	48155	86226	40359	28723	15364	69125	12609	57171	86857	31702
	20226	53752	90648	24362	83314	00014	19207	69413	97016	86290
	70178	73444	38790	53626	93780	18629	68766	24371	74639	30782
	10169	41465	51935	05711	09799	79077	88159	33437	68519	03040
30	81084	03701	28598	70013	63794	53169	97054	60303	23259	96196
	69202	20777	21727	81511	51887	16175	53746	46516	70339	62727
	80561	95787	89426	93325	86412	57479	54194	52153	19197	81877
	08199	26703	95128	48599	09333	12584	24374	31232	61782	44032
	98883	28220	39358	53720	80161	83371	15181	11131	12219	55920
35	84568	69286	76054	21615	80883	36797	82845	39139	90900	18172
	04269	35173	95745	53893	86022	77722	52498	84193	22448	22571
	10538	13124	36099	13140	37706	44562	57179	11693	67877	01549
	77843	24955	25900	63843	95029	93859	93634	20205	66294	41218
	12034	94636	49455	76362	83532	31062	69903	91186	65768	55949
40	10524	72829	47641	93315	80875	28090	97728	52560	34937	79548
	68935	76632	46984	61772	92786	22651	07086	89754	44143	97687
	89450	65665	29190	43709	11172	34481	95977	47535	25658	73898
	90696	20451	24211	97310	60446	73530	62865	96574	13829	72226
	49006	32047	93086	00112	20470	17136	28255	86328	07293	38809
45	74591	87025	52368	59416	34417	70557	86746	55809	53628	12000
	06315	17012	77103	00968	07235	10728	42189	33292	51487	64443
	62386	09184	62092	46617	99419	64230	95034	85481	07857	42510
	86848	82122	04028	36959	87827	12813	08627	80699	13345	51695
	65643	69480	46598	04501	40403	91408	32343	48130	49303	90689
50	11084	46534	78957	77353	39578	77868	22970	84349	09184	70603

TABLE 2(A)
BINOMIAL PROBABILITIES

$n = 1$

Probability

x \ p	0.05	0.10	0.20	0.30	0.40	0.50	0.60	0.70	0.80	0.90	0.95
0	0.950	0.900	0.800	0.700	0.600	0.500	0.400	0.300	0.200	0.100	0.050
1	0.050	0.100	0.200	0.300	0.400	0.500	0.600	0.700	0.800	0.900	0.950

$n = 2$

Probability

x \ p	0.05	0.10	0.20	0.30	0.40	0.50	0.60	0.70	0.80	0.90	0.95
0	0.903	0.810	0.640	0.490	0.360	0.250	0.160	0.090	0.040	0.010	0.003
1	0.095	0.180	0.320	0.420	0.480	0.500	0.480	0.420	0.320	0.180	0.095
2	0.003	0.010	0.040	0.090	0.160	0.250	0.360	0.490	0.640	0.810	0.903

$n = 3$

Probability

x \ p	0.05	0.10	0.20	0.30	0.40	0.50	0.60	0.70	0.80	0.90	0.95
0	0.857	0.729	0.512	0.343	0.216	0.125	0.064	0.027	0.008	0.001	0.000
1	0.135	0.243	0.384	0.441	0.432	0.375	0.288	0.189	0.096	0.027	0.007
2	0.007	0.027	0.096	0.189	0.288	0.375	0.432	0.441	0.384	0.243	0.135
3	0.000	0.001	0.008	0.027	0.064	0.125	0.216	0.343	0.512	0.729	0.857

$n = 4$

Probability

x \ p	0.05	0.10	0.20	0.30	0.40	0.50	0.60	0.70	0.80	0.90	0.95
0	0.815	0.656	0.410	0.240	0.130	0.063	0.026	0.008	0.002	0.000	0.000
1	0.171	0.292	0.410	0.412	0.346	0.250	0.154	0.076	0.026	0.004	0.000
2	0.014	0.049	0.154	0.265	0.346	0.375	0.346	0.265	0.154	0.049	0.014
3	0.000	0.004	0.026	0.076	0.154	0.250	0.346	0.412	0.410	0.292	0.171
4	0.000	0.000	0.002	0.008	0.026	0.063	0.130	0.240	0.410	0.656	0.815

$n = 5$

Probability

x \ p	0.05	0.10	0.20	0.30	0.40	0.50	0.60	0.70	0.80	0.90	0.95
0	0.774	0.590	0.328	0.168	0.078	0.031	0.010	0.002	0.000	0.000	0.000
1	0.204	0.328	0.410	0.360	0.259	0.156	0.077	0.028	0.006	0.000	0.000
2	0.021	0.073	0.205	0.309	0.346	0.313	0.230	0.132	0.051	0.008	0.001
3	0.001	0.008	0.051	0.132	0.230	0.313	0.346	0.309	0.205	0.073	0.021
4	0.000	0.000	0.006	0.028	0.077	0.156	0.259	0.360	0.410	0.328	0.204
5	0.000	0.000	0.000	0.002	0.010	0.031	0.078	0.168	0.328	0.590	0.774

***n* = 6**

Probability

x \ p	0.05	0.10	0.20	0.30	0.40	0.50	0.60	0.70	0.80	0.90	0.95
0	0.735	0.531	0.262	0.118	0.047	0.016	0.004	0.001	0.000	0.000	0.000
1	0.232	0.354	0.393	0.303	0.187	0.094	0.037	0.010	0.002	0.000	0.000
2	0.031	0.098	0.246	0.324	0.311	0.234	0.138	0.060	0.015	0.001	0.000
3	0.002	0.015	0.082	0.185	0.276	0.313	0.276	0.185	0.082	0.015	0.002
4	0.000	0.001	0.015	0.060	0.138	0.234	0.311	0.324	0.246	0.098	0.031
5	0.000	0.000	0.002	0.010	0.037	0.094	0.187	0.303	0.393	0.354	0.232
6	0.000	0.000	0.000	0.001	0.004	0.016	0.047	0.118	0.262	0.531	0.735

***n* = 7**

Probability

x \ p	0.05	0.10	0.20	0.30	0.40	0.50	0.60	0.70	0.80	0.90	0.95
0	0.698	0.478	0.210	0.082	0.028	0.008	0.002	0.000	0.000	0.000	0.000
1	0.257	0.372	0.367	0.247	0.131	0.055	0.017	0.004	0.000	0.000	0.000
2	0.041	0.124	0.275	0.318	0.261	0.164	0.077	0.025	0.004	0.000	0.000
3	0.004	0.023	0.115	0.227	0.290	0.273	0.194	0.097	0.029	0.003	0.000
4	0.000	0.003	0.029	0.097	0.194	0.273	0.290	0.227	0.115	0.023	0.004
5	0.000	0.000	0.004	0.025	0.077	0.164	0.261	0.318	0.275	0.124	0.041
6	0.000	0.000	0.000	0.004	0.017	0.055	0.131	0.247	0.367	0.372	0.257
7	0.000	0.000	0.000	0.000	0.002	0.008	0.028	0.082	0.210	0.478	0.698

***n* = 8**

Probability

x \ p	0.05	0.10	0.20	0.30	0.40	0.50	0.60	0.70	0.80	0.90	0.95
0	0.663	0.430	0.168	0.058	0.017	0.004	0.001	0.000	0.000	0.000	0.000
1	0.279	0.383	0.336	0.198	0.090	0.031	0.008	0.001	0.000	0.000	0.000
2	0.051	0.149	0.294	0.296	0.209	0.109	0.041	0.010	0.001	0.000	0.000
3	0.005	0.033	0.147	0.254	0.279	0.219	0.124	0.047	0.009	0.000	0.000
4	0.000	0.005	0.046	0.136	0.232	0.273	0.232	0.136	0.046	0.005	0.000
5	0.000	0.000	0.009	0.047	0.124	0.219	0.279	0.254	0.147	0.033	0.005
6	0.000	0.000	0.001	0.010	0.041	0.109	0.209	0.296	0.294	0.149	0.051
7	0.000	0.000	0.000	0.001	0.008	0.031	0.090	0.198	0.336	0.383	0.279
8	0.000	0.000	0.000	0.000	0.001	0.004	0.017	0.058	0.168	0.430	0.663

***n* = 9**

Probability

x \ p	0.05	0.10	0.20	0.30	0.40	0.50	0.60	0.70	0.80	0.90	0.95
0	0.630	0.387	0.134	0.040	0.010	0.002	0.000	0.000	0.000	0.000	0.000
1	0.299	0.387	0.302	0.156	0.060	0.018	0.004	0.000	0.000	0.000	0.000
2	0.063	0.172	0.302	0.267	0.161	0.070	0.021	0.004	0.000	0.000	0.000
3	0.008	0.045	0.176	0.267	0.251	0.164	0.074	0.021	0.003	0.000	0.000
4	0.001	0.007	0.066	0.172	0.251	0.246	0.167	0.074	0.017	0.001	0.000
5	0.000	0.001	0.017	0.074	0.167	0.246	0.251	0.172	0.066	0.007	0.001
6	0.000	0.000	0.003	0.021	0.074	0.164	0.251	0.267	0.176	0.045	0.008
7	0.000	0.000	0.000	0.004	0.021	0.070	0.161	0.267	0.302	0.172	0.063
8	0.000	0.000	0.000	0.000	0.004	0.018	0.060	0.156	0.302	0.387	0.299
9	0.000	0.000	0.000	0.000	0.000	0.002	0.010	0.040	0.134	0.387	0.630

***n* = 10**
Probability

x p	0.05	0.10	0.20	0.30	0.40	0.50	0.60	0.70	0.80	0.90	0.95
0	0.599	0.349	0.107	0.028	0.006	0.001	0.000	0.000	0.000	0.000	0.000
1	0.315	0.387	0.268	0.121	0.040	0.010	0.002	0.000	0.000	0.000	0.000
2	0.075	0.194	0.302	0.233	0.121	0.044	0.011	0.001	0.000	0.000	0.000
3	0.010	0.057	0.201	0.267	0.215	0.117	0.042	0.009	0.001	0.000	0.000
4	0.001	0.011	0.088	0.200	0.251	0.205	0.111	0.037	0.006	0.000	0.000
5	0.000	0.001	0.026	0.103	0.201	0.246	0.201	0.103	0.026	0.001	0.000
6	0.000	0.000	0.006	0.037	0.111	0.205	0.251	0.200	0.088	0.011	0.001
7	0.000	0.000	0.001	0.009	0.042	0.117	0.215	0.267	0.201	0.057	0.010
8	0.000	0.000	0.000	0.001	0.011	0.044	0.121	0.233	0.302	0.194	0.075
9	0.000	0.000	0.000	0.000	0.002	0.010	0.040	0.121	0.268	0.387	0.315
10	0.000	0.000	0.000	0.000	0.000	0.001	0.006	0.028	0.107	0.349	0.599

***n* = 11**
Probability

x p	0.05	0.10	0.20	0.30	0.40	0.50	0.60	0.70	0.80	0.90	0.95
0	0.569	0.314	0.086	0.020	0.004	0.000	0.000	0.000	0.000	0.000	0.000
1	0.329	0.384	0.236	0.093	0.027	0.005	0.001	0.000	0.000	0.000	
2	0.087	0.213	0.295	0.200	0.089	0.027	0.005	0.001	0.000	0.000	0.000
3	0.014	0.071	0.221	0.257	0.177	0.081	0.023	0.004	0.000	0.000	0.000
4	0.001	0.016	0.111	0.220	0.236	0.161	0.070	0.017	0.002	0.000	0.000
5	0.000	0.002	0.039	0.132	0.221	0.226	0.147	0.057	0.010	0.000	0.000
6	0.000	0.000	0.010	0.057	0.147	0.226	0.221	0.132	0.039	0.002	0.000
7	0.000	0.000	0.002	0.017	0.070	0.161	0.236	0.220	0.111	0.016	0.001
8	0.000	0.000	0.000	0.004	0.023	0.081	0.177	0.257	0.221	0.071	0.014
9	0.000	0.000	0.000	0.001	0.005	0.027	0.089	0.200	0.295	0.213	0.087
10	0.000	0.000	0.000	0.000	0.001	0.005	0.027	0.093	0.236	0.384	0.329
11	0.000	0.000	0.000	0.000	0.000	0.000	0.004	0.020	0.086	0.314	0.569

***n* = 12**
Probability

x p	0.05	0.10	0.20	0.30	0.40	0.50	0.60	0.70	0.80	0.90	0.95
0	0.540	0.282	0.069	0.014	0.002	0.000	0.000	0.000	0.000	0.000	0.000
1	0.341	0.377	0.206	0.071	0.017	0.003	0.000	0.000	0.000	0.000	0.000
2	0.099	0.230	0.283	0.168	0.064	0.016	0.002	0.000	0.000	0.000	0.000
3	0.017	0.085	0.236	0.240	0.142	0.054	0.012	0.001	0.000	0.000	0.000
4	0.002	0.021	0.133	0.231	0.213	0.121	0.042	0.008	0.001	0.000	0.000
5	0.000	0.004	0.053	0.158	0.227	0.193	0.101	0.029	0.003	0.000	0.000
6	0.000	0.000	0.016	0.079	0.177	0.226	0.177	0.079	0.016	0.000	0.000
7	0.000	0.000	0.003	0.029	0.101	0.193	0.227	0.158	0.053	0.004	0.000
8	0.000	0.000	0.001	0.008	0.042	0.121	0.213	0.231	0.133	0.021	0.002
9	0.000	0.000	0.000	0.001	0.012	0.054	0.142	0.240	0.236	0.085	0.017
10	0.000	0.000	0.000	0.000	0.002	0.016	0.064	0.168	0.283	0.230	0.099
11	0.000	0.000	0.000	0.000	0.000	0.003	0.017	0.071	0.206	0.377	0.341
12	0.000	0.000	0.000	0.000	0.000	0.000	0.002	0.014	0.069	0.282	0.540

***n* = 13**

Probability

x \ p	0.05	0.10	0.20	0.30	0.40	0.50	0.60	0.70	0.80	0.90	0.95
0	0.513	0.254	0.055	0.010	0.001	0.000	0.000	0.000	0.000	0.000	0.000
1	0.351	0.367	0.179	0.054	0.011	0.002	0.000	0.000	0.000	0.000	0.000
2	0.111	0.245	0.268	0.139	0.045	0.010	0.001	0.000	0.000	0.000	0.000
3	0.021	0.100	0.246	0.218	0.111	0.035	0.006	0.001	0.000	0.000	0.000
4	0.003	0.028	0.154	0.234	0.184	0.087	0.024	0.003	0.000	0.000	0.000
5	0.000	0.006	0.069	0.180	0.221	0.157	0.066	0.014	0.001	0.000	0.000
6	0.000	0.001	0.023	0.103	0.197	0.209	0.131	0.044	0.006	0.000	0.000
7	0.000	0.000	0.006	0.044	0.131	0.209	0.197	0.103	0.023	0.001	0.000
8	0.000	0.000	0.001	0.014	0.066	0.157	0.221	0.180	0.069	0.006	0.000
9	0.000	0.000	0.000	0.003	0.024	0.087	0.184	0.234	0.154	0.028	0.003
10	0.000	0.000	0.000	0.001	0.006	0.035	0.111	0.218	0.246	0.100	0.021
11	0.000	0.000	0.000	0.000	0.001	0.010	0.045	0.139	0.268	0.245	0.111
12	0.000	0.000	0.000	0.000	0.000	0.002	0.011	0.054	0.179	0.367	0.351
13	0.000	0.000	0.000	0.000	0.000	0.000	0.001	0.010	0.055	0.254	0.513

***n* = 14**

Probability

x \ p	0.05	0.10	0.20	0.30	0.40	0.50	0.60	0.70	0.80	0.90	0.95
0	0.488	0.229	0.044	0.007	0.001	0.000	0.000	0.000	0.000	0.000	0.000
1	0.359	0.356	0.154	0.041	0.007	0.001	0.000	0.000	0.000	0.000	0.000
2	0.123	0.257	0.250	0.113	0.032	0.006	0.001	0.000	0.000	0.000	0.000
3	0.026	0.114	0.250	0.194	0.085	0.022	0.003	0.000	0.000	0.000	0.000
4	0.004	0.035	0.172	0.229	0.155	0.061	0.014	0.001	0.000	0.000	0.000
5	0.000	0.008	0.086	0.196	0.207	0.122	0.041	0.007	0.000	0.000	0.000
6	0.000	0.001	0.032	0.126	0.207	0.183	0.092	0.023	0.002	0.000	0.000
7	0.000	0.000	0.009	0.062	0.157	0.209	0.157	0.062	0.009	0.000	0.000
8	0.000	0.000	0.002	0.023	0.092	0.183	0.207	0.126	0.032	0.001	0.000
9	0.000	0.000	0.000	0.007	0.041	0.122	0.207	0.196	0.086	0.008	0.000
10	0.000	0.000	0.000	0.001	0.014	0.061	0.155	0.229	0.172	0.035	0.004
11	0.000	0.000	0.000	0.000	0.003	0.022	0.085	0.194	0.250	0.114	0.026
12	0.000	0.000	0.000	0.000	0.001	0.006	0.032	0.113	0.250	0.257	0.123
13	0.000	0.000	0.000	0.000	0.000	0.001	0.007	0.041	0.154	0.356	0.359
14	0.000	0.000	0.000	0.000	0.000	0.000	0.001	0.007	0.044	0.229	0.488

$n = 15$

Probability

x \ p	0.05	0.10	0.20	0.30	0.40	0.50	0.60	0.70	0.80	0.90	0.95
0	0.463	0.206	0.035	0.005	0.000	0.000	0.000	0.000	0.000	0.000	0.000
1	0.366	0.343	0.132	0.031	0.005	0.000	0.000	0.000	0.000	0.000	0.000
2	0.135	0.267	0.231	0.092	0.022	0.003	0.000	0.000	0.000	0.000	0.000
3	0.031	0.129	0.250	0.170	0.063	0.014	0.002	0.000	0.000	0.000	0.000
4	0.005	0.043	0.188	0.219	0.127	0.042	0.007	0.001	0.000	0.000	0.000
5	0.001	0.010	0.103	0.206	0.186	0.092	0.024	0.003	0.000	0.000	0.000
6	0.000	0.002	0.043	0.147	0.207	0.153	0.061	0.012	0.001	0.000	0.000
7	0.000	0.000	0.014	0.081	0.177	0.196	0.118	0.035	0.003	0.000	0.000
8	0.000	0.000	0.003	0.035	0.188	0.196	0.177	0.081	0.014	0.000	0.000
9	0.000	0.000	0.001	0.012	0.061	0.153	0.207	0.147	0.043	0.002	0.000
10	0.000	0.000	0.000	0.003	0.024	0.092	0.186	0.206	0.103	0.010	0.001
11	0.000	0.000	0.000	0.001	0.007	0.042	0.127	0.219	0.188	0.043	0.005
12	0.000	0.000	0.000	0.000	0.002	0.014	0.063	0.170	0.250	0.129	0.031
13	0.000	0.000	0.000	0.000	0.000	0.003	0.022	0.092	0.231	0.267	0.135
14	0.000	0.000	0.000	0.000	0.000	0.000	0.005	0.031	0.132	0.343	0.366
15	0.000	0.000	0.000	0.000	0.000	0.000	0.000	0.005	0.035	0.206	0.463

$n = 16$

Probability

x \ p	0.05	0.10	0.20	0.30	0.40	0.50	0.60	0.70	0.80	0.90	0.95
0	0.440	0.185	0.028	0.003	0.000	0.000	0.000	0.000	0.000	0.000	0.000
1	0.371	0.329	0.113	0.023	0.003	0.000	0.000	0.000	0.000	0.000	0.000
2	0.146	0.275	0.211	0.073	0.015	0.002	0.000	0.000	0.000	0.000	0.000
3	0.036	0.142	0.246	0.146	0.047	0.009	0.001	0.000	0.000	0.000	0.000
4	0.006	0.051	0.200	0.204	0.101	0.028	0.004	0.000	0.000	0.000	0.000
5	0.001	0.014	0.120	0.210	0.162	0.067	0.014	0.001	0.000	0.000	0.000
6	0.000	0.003	0.055	0.165	0.198	0.122	0.039	0.006	0.000	0.000	0.000
7	0.000	0.000	0.020	0.101	0.189	0.175	0.084	0.019	0.001	0.000	0.000
8	0.000	0.000	0.006	0.049	0.142	0.196	0.142	0.049	0.006	0.000	0.000
9	0.000	0.000	0.001	0.019	0.084	0.175	0.189	0.101	0.020	0.000	0.000
10	0.000	0.000	0.000	0.006	0.039	0.122	0.198	0.165	0.055	0.003	0.000
11	0.000	0.000	0.000	0.001	0.014	0.067	0.162	0.210	0.120	0.014	0.001
12	0.000	0.000	0.000	0.000	0.004	0.028	0.101	0.204	0.200	0.051	0.006
13	0.000	0.000	0.000	0.000	0.001	0.009	0.047	0.146	0.246	0.142	0.036
14	0.000	0.000	0.000	0.000	0.000	0.002	0.015	0.073	0.211	0.275	0.146
15	0.000	0.000	0.000	0.000	0.000	0.000	0.003	0.023	0.113	0.329	0.371
16	0.000	0.000	0.000	0.000	0.000	0.000	0.000	0.003	0.028	0.185	0.440

n = 17

Probability

x \ p	0.05	0.10	0.20	0.30	0.40	0.50	0.60	0.70	0.80	0.90	0.95
0	0.418	0.167	0.023	0.002	0.000	0.000	0.000	0.000	0.000	0.000	0.000
1	0.374	0.315	0.096	0.017	0.002	0.000	0.000	0.000	0.000	0.000	0.000
2	0.158	0.280	0.191	0.058	0.010	0.001	0.000	0.000	0.000	0.000	0.000
3	0.041	0.156	0.239	0.125	0.034	0.005	0.000	0.000	0.000	0.000	0.000
4	0.008	0.060	0.209	0.187	0.080	0.018	0.002	0.000	0.000	0.000	0.000
5	0.001	0.017	0.136	0.208	0.138	0.047	0.008	0.001	0.000	0.000	0.000
6	0.000	0.004	0.068	0.178	0.184	0.094	0.024	0.003	0.000	0.000	0.000
7	0.000	0.001	0.027	0.120	0.193	0.148	0.057	0.009	0.000	0.000	0.000
8	0.000	0.000	0.008	0.064	0.161	0.185	0.107	0.028	0.002	0.000	0.000
9	0.000	0.000	0.002	0.028	0.107	0.185	0.161	0.064	0.008	0.000	0.000
10	0.000	0.000	0.000	0.009	0.057	0.148	0.193	0.120	0.027	0.001	0.000
11	0.000	0.000	0.000	0.003	0.024	0.094	0.184	0.178	0.068	0.004	0.000
12	0.000	0.000	0.000	0.001	0.008	0.047	0.138	0.208	0.136	0.017	0.001
13	0.000	0.000	0.000	0.000	0.002	0.018	0.080	0.187	0.209	0.060	0.008
14	0.000	0.000	0.000	0.000	0.000	0.005	0.034	0.125	0.239	0.156	0.041
15	0.000	0.000	0.000	0.000	0.000	0.001	0.010	0.058	0.191	0.280	0.158
16	0.000	0.000	0.000	0.000	0.000	0.000	0.002	0.017	0.096	0.315	0.374
17	0.000	0.000	0.000	0.000	0.000	0.000	0.000	0.002	0.023	0.167	0.418

n = 18

Probability

x \ p	0.05	0.10	0.20	0.30	0.40	0.50	0.60	0.70	0.80	0.90	0.95
0	0.397	0.150	0.018	0.002	0.000	0.000	0.000	0.000	0.000	0.000	0.000
1	0.376	0.300	0.081	0.013	0.001	0.000	0.000	0.000	0.000	0.000	0.000
2	0.168	0.284	0.172	0.046	0.007	0.001	0.000	0.000	0.000	0.000	0.000
3	0.047	0.168	0.230	0.105	0.025	0.003	0.000	0.000	0.000	0.000	0.000
4	0.009	0.070	0.215	0.168	0.061	0.012	0.001	0.000	0.000	0.000	0.000
5	0.001	0.022	0.151	0.202	0.115	0.033	0.004	0.000	0.000	0.000	0.000
6	0.000	0.005	0.082	0.187	0.166	0.071	0.015	0.001	0.000	0.000	0.000
7	0.000	0.001	0.035	0.138	0.189	0.121	0.037	0.005	0.000	0.000	0.000
8	0.000	0.000	0.012	0.081	0.173	0.167	0.077	0.015	0.001	0.000	0.000
9	0.000	0.000	0.003	0.039	0.128	0.185	0.128	0.039	0.003	0.000	0.000
10	0.000	0.000	0.001	0.015	0.077	0.167	0.173	0.081	0.012	0.000	0.000
11	0.000	0.000	0.000	0.005	0.037	0.121	0.189	0.138	0.035	0.001	0.000
12	0.000	0.000	0.000	0.001	0.015	0.071	0.166	0.187	0.082	0.005	0.000
13	0.000	0.000	0.000	0.000	0.004	0.033	0.115	0.202	0.151	0.022	0.001
14	0.000	0.000	0.000	0.000	0.001	0.012	0.061	0.168	0.215	0.070	0.009
15	0.000	0.000	0.000	0.000	0.000	0.003	0.025	0.105	0.230	0.168	0.047
16	0.000	0.000	0.000	0.000	0.000	0.001	0.007	0.046	0.172	0.284	0.168
17	0.000	0.000	0.000	0.000	0.000	0.000	0.001	0.013	0.081	0.300	0.376
18	0.000	0.000	0.000	0.000	0.000	0.000	0.000	0.002	0.018	0.150	0.397

$n = 19$

Probability

x \ p	0.05	0.10	0.20	0.30	0.40	0.50	0.60	0.70	0.80	0.90	0.95
0	0.377	0.135	0.014	0.001	0.000	0.000	0.000	0.000	0.000	0.000	0.000
1	0.377	0.285	0.068	0.009	0.001	0.000	0.000	0.000	0.000	0.000	0.000
2	0.179	0.285	0.154	0.036	0.005	0.000	0.000	0.000	0.000	0.000	0.000
3	0.053	0.180	0.218	0.087	0.017	0.002	0.000	0.000	0.000	0.000	0.000
4	0.011	0.080	0.218	0.149	0.047	0.007	0.001	0.000	0.000	0.000	0.000
5	0.002	0.027	0.164	0.192	0.093	0.022	0.002	0.000	0.000	0.000	0.000
6	0.000	0.007	0.095	0.192	0.145	0.052	0.008	0.001	0.000	0.000	0.000
7	0.000	0.001	0.044	0.153	0.180	0.096	0.024	0.002	0.000	0.000	0.000
8	0.000	0.000	0.017	0.098	0.180	0.144	0.053	0.008	0.000	0.000	0.000
9	0.000	0.000	0.005	0.051	0.146	0.176	0.098	0.022	0.001	0.000	0.000
10	0.000	0.000	0.001	0.022	0.098	0.176	0.146	0.051	0.005	0.000	0.000
11	0.000	0.000	0.000	0.008	0.053	0.144	0.180	0.098	0.017	0.000	0.000
12	0.000	0.000	0.000	0.002	0.024	0.096	0.180	0.153	0.044	0.001	0.000
13	0.000	0.000	0.000	0.001	0.008	0.052	0.145	0.192	0.095	0.007	0.000
14	0.000	0.000	0.000	0.000	0.002	0.022	0.093	0.192	0.164	0.027	0.002
15	0.000	0.000	0.000	0.000	0.001	0.007	0.047	0.149	0.218	0.080	0.011
16	0.000	0.000	0.000	0.000	0.000	0.002	0.017	0.087	0.218	0.180	0.053
17	0.000	0.000	0.000	0.000	0.000	0.000	0.005	0.036	0.154	0.285	0.179
18	0.000	0.000	0.000	0.000	0.000	0.000	0.001	0.009	0.068	0.285	0.377
19	0.000	0.000	0.000	0.000	0.000	0.000	0.000	0.001	0.014	0.135	0.377

$n = 20$

Probability

x \ p	0.05	0.10	0.20	0.30	0.40	0.50	0.60	0.70	0.80	0.90	0.95
0	0.358	0.122	0.012	0.001	0.000	0.000	0.000	0.000	0.000	0.000	0.000
1	0.377	0.270	0.058	0.007	0.000	0.000	0.000	0.000	0.000	0.000	0.000
2	0.189	0.285	0.137	0.028	0.003	0.000	0.000	0.000	0.000	0.000	0.000
3	0.060	0.190	0.205	0.072	0.012	0.001	0.000	0.000	0.000	0.000	0.000
4	0.013	0.090	0.218	0.130	0.035	0.005	0.000	0.000	0.000	0.000	0.000
5	0.002	0.032	0.175	0.179	0.075	0.015	0.001	0.000	0.000	0.000	0.000
6	0.000	0.009	0.109	0.192	0.124	0.037	0.005	0.000	0.000	0.000	0.000
7	0.000	0.002	0.055	0.164	0.166	0.074	0.015	0.001	0.000	0.000	0.000
8	0.000	0.000	0.022	0.114	0.180	0.120	0.035	0.004	0.000	0.000	0.000
9	0.000	0.000	0.007	0.065	0.160	0.160	0.071	0.012	0.000	0.000	0.000
10	0.000	0.000	0.002	0.031	0.117	0.176	0.117	0.031	0.002	0.000	0.000
11	0.000	0.000	0.000	0.012	0.071	0.160	0.160	0.065	0.007	0.000	0.000
12	0.000	0.000	0.000	0.004	0.035	0.120	0.180	0.114	0.022	0.000	0.000
13	0.000	0.000	0.000	0.001	0.015	0.074	0.166	0.164	0.055	0.002	0.000
14	0.000	0.000	0.000	0.000	0.005	0.037	0.124	0.192	0.109	0.009	0.000
15	0.000	0.000	0.000	0.000	0.001	0.015	0.075	0.179	0.175	0.032	0.002
16	0.000	0.000	0.000	0.000	0.000	0.005	0.035	0.130	0.218	0.090	0.013
17	0.000	0.000	0.000	0.000	0.000	0.001	0.012	0.072	0.205	0.190	0.060
18	0.000	0.000	0.000	0.000	0.000	0.000	0.003	0.028	0.137	0.285	0.189
19	0.000	0.000	0.000	0.000	0.000	0.000	0.000	0.007	0.058	0.270	0.377
20	0.000	0.000	0.000	0.000	0.000	0.000	0.000	0.001	0.012	0.122	0.358

n = 25

Probability

x \ p	0.05	0.10	0.20	0.30	0.40	0.50	0.60	0.70	0.80	0.90	0.95
0	0.277	0.072	0.004	0.000	0.000	0.000	0.000	0.000	0.000	0.000	0.000
1	0.365	0.199	0.024	0.001	0.000	0.000	0.000	0.000	0.000	0.000	0.000
2	0.231	0.266	0.071	0.007	0.000	0.000	0.000	0.000	0.000	0.000	0.000
3	0.093	0.226	0.136	0.024	0.002	0.000	0.000	0.000	0.000	0.000	0.000
4	0.027	0.138	0.187	0.057	0.007	0.000	0.000	0.000	0.000	0.000	0.000
5	0.006	0.065	0.196	0.103	0.020	0.002	0.000	0.000	0.000	0.000	0.000
6	0.001	0.024	0.163	0.147	0.044	0.005	0.000	0.000	0.000	0.000	0.000
7	0.000	0.007	0.111	0.171	0.080	0.014	0.001	0.000	0.000	0.000	0.000
8	0.000	0.002	0.062	0.165	0.120	0.032	0.003	0.000	0.000	0.000	0.000
9	0.000	0.000	0.029	0.134	0.151	0.061	0.009	0.000	0.000	0.000	0.000
10	0.000	0.000	0.012	0.092	0.161	0.097	0.021	0.001	0.000	0.000	0.000
11	0.000	0.000	0.004	0.054	0.147	0.133	0.043	0.004	0.000	0.000	0.000
12	0.000	0.000	0.001	0.027	0.114	0.155	0.076	0.011	0.000	0.000	0.000
13	0.000	0.000	0.000	0.011	0.076	0.155	0.114	0.027	0.001	0.000	0.000
14	0.000	0.000	0.000	0.004	0.043	0.133	0.147	0.054	0.004	0.000	0.000
15	0.000	0.000	0.000	0.001	0.021	0.097	0.161	0.092	0.012	0.000	0.000
16	0.000	0.000	0.000	0.000	0.009	0.061	0.151	0.134	0.029	0.000	0.000
17	0.000	0.000	0.000	0.000	0.003	0.032	0.120	0.165	0.062	0.002	0.000
18	0.000	0.000	0.000	0.000	0.001	0.014	0.080	0.171	0.111	0.007	0.000
19	0.000	0.000	0.000	0.000	0.000	0.005	0.044	0.147	0.163	0.024	0.001
20	0.000	0.000	0.000	0.000	0.000	0.002	0.020	0.103	0.196	0.065	0.006
21	0.000	0.000	0.000	0.000	0.000	0.000	0.007	0.057	0.187	0.138	0.027
22	0.000	0.000	0.000	0.000	0.000	0.000	0.002	0.024	0.136	0.226	0.093
23	0.000	0.000	0.000	0.000	0.000	0.000	0.000	0.007	0.071	0.266	0.231
24	0.000	0.000	0.000	0.000	0.000	0.000	0.000	0.001	0.024	0.199	0.365
25	0.000	0.000	0.000	0.000	0.000	0.000	0.000	0.000	0.004	0.072	0.277

TABLE 2(B)

POISSON PROBABILITIES

For a given value of μ, entry indicates the probability of obtaining a specific value of X

x \ μ	0.1	0.2	0.3	0.4	0.5	0.6	0.7	0.8	0.9	1.0
0	0.9048	0.8187	0.7408	0.6703	0.6065	0.5488	0.4966	0.4493	0.4066	0.3679
1	0.0905	0.1637	0.2222	0.2681	0.3033	0.3293	0.3476	0.3595	0.3659	0.3679
2	0.0045	0.0164	0.0333	0.0536	0.0758	0.0988	0.1217	0.1438	0.1647	0.1839
3	0.0002	0.0011	0.0033	0.0072	0.0126	0.0198	0.0284	0.0383	0.0494	0.0613
4	0.0000	0.0001	0.0003	0.0007	0.0016	0.0030	0.0050	0.0077	0.0111	0.0153
5	0.0000	0.0000	0.0000	0.0001	0.0002	0.0004	0.0007	0.0012	0.0020	0.0031
6	0.0000	0.0000	0.0000	0.0000	0.0000	0.0000	0.0001	0.0002	0.0003	0.0005
7	0.0000	0.0000	0.0000	0.0000	0.0000	0.0000	0.0000	0.0000	0.0000	0.0001

x \ μ	1.1	1.2	1.3	1.4	1.5	1.6	1.7	1.8	1.9	2.0
0	0.3329	0.3012	0.2725	0.2466	0.2231	0.2019	0.1827	0.1653	0.1496	0.1353
1	0.3662	0.3614	0.3543	0.3452	0.3347	0.3230	0.3106	0.2975	0.2842	0.2707
2	0.2014	0.2169	0.2303	0.2417	0.2510	0.2584	0.2640	0.2678	0.2700	0.2707
3	0.0738	0.0867	0.0998	0.1128	0.1255	0.1378	0.1496	0.1607	0.1710	0.1804
4	0.0203	0.0260	0.0324	0.0395	0.0471	0.0551	0.0636	0.0723	0.0812	0.0902
5	0.0045	0.0062	0.0084	0.0111	0.0141	0.0176	0.0216	0.0260	0.0309	0.0361
6	0.0008	0.0012	0.0018	0.0026	0.0035	0.0047	0.0061	0.0078	0.0098	0.0120
7	0.0001	0.0002	0.0003	0.0005	0.0008	0.0011	0.0015	0.0020	0.0027	0.0034
8	0.0000	0.0000	0.0001	0.0001	0.0001	0.0002	0.0003	0.0005	0.0006	0.0009
9	0.0000	0.0000	0.0000	0.0000	0.0000	0.0000	0.0001	0.0001	0.0001	0.0002

x \ μ	2.1	2.2	2.3	2.4	2.5	2.6	2.7	2.8	2.9	3.0
0	0.1225	0.1108	0.1003	0.0907	0.0821	0.0743	0.0672	0.0608	0.0550	0.0498
1	0.2572	0.2438	0.2306	0.2177	0.2052	0.1931	0.1815	0.1703	0.1596	0.1494
2	0.2700	0.2681	0.2652	0.2613	0.2565	0.2510	0.2450	0.2384	0.2314	0.2240
3	0.1890	0.1966	0.2033	0.2090	0.2138	0.2176	0.2205	0.2225	0.2237	0.2240
4	0.0992	0.1082	0.1169	0.1254	0.1336	0.1414	0.1488	0.1557	0.1622	0.1680
5	0.0417	0.0476	0.0538	0.0602	0.0668	0.0735	0.0804	0.0872	0.0940	0.1008
6	0.0146	0.0174	0.0206	0.0241	0.0278	0.0319	0.0362	0.0407	0.0455	0.0504
7	0.0044	0.0055	0.0068	0.0083	0.0099	0.0118	0.0139	0.0163	0.0188	0.0216
8	0.0011	0.0015	0.0019	0.0025	0.0031	0.0038	0.0047	0.0057	0.0068	0.0081
9	0.0003	0.0004	0.0005	0.0007	0.0009	0.0011	0.0014	0.0018	0.0022	0.0027
10	0.0001	0.0001	0.0001	0.0002	0.0002	0.0003	0.0004	0.0005	0.0006	0.0008
11	0.0000	0.0000	0.0000	0.0000	0.0000	0.0001	0.0001	0.0001	0.0002	0.0002
12	0.0000	0.0000	0.0000	0.0000	0.0000	0.0000	0.0000	0.0000	0.0000	0.0001

x \ μ	3.1	3.2	3.3	3.4	3.5	3.6	3.7	3.8	3.9	4.0
0	0.0450	0.0408	0.0369	0.0334	0.0302	0.0273	0.0247	0.0224	0.0202	0.0183
1	0.1397	0.1304	0.1217	0.1135	0.1057	0.0984	0.0915	0.0850	0.0789	0.0733
2	0.2165	0.2087	0.2008	0.1929	0.1850	0.1771	0.1692	0.1615	0.1539	0.1465
3	0.2237	0.2226	0.2209	0.2186	0.2158	0.2125	0.2087	0.2046	0.2001	0.1954
4	0.1733	0.1781	0.1823	0.1858	0.1888	0.1912	0.1931	0.1944	0.1951	0.1954
5	0.1075	0.1140	0.1203	0.1264	0.1322	0.1377	0.1429	0.1477	0.1522	0.1563
6	0.0555	0.0608	0.0662	0.0716	0.0771	0.0826	0.0881	0.0936	0.0989	0.1042
7	0.0246	0.0278	0.0312	0.0348	0.0385	0.0425	0.0466	0.0508	0.0551	0.0595
8	0.0095	0.0111	0.0129	0.0148	0.0169	0.0191	0.0215	0.0241	0.0269	0.0298
9	0.0033	0.0040	0.0047	0.0056	0.0066	0.0076	0.0089	0.0102	0.0116	0.0132
10	0.0010	0.0013	0.0016	0.0019	0.0023	0.0028	0.0033	0.0039	0.0045	0.0053
11	0.0003	0.0004	0.0005	0.0006	0.0007	0.0009	0.0011	0.0013	0.0016	0.0019
12	0.0001	0.0001	0.0001	0.0002	0.0002	0.0003	0.0003	0.0004	0.0005	0.0006
13	0.0000	0.0000	0.0000	0.0000	0.0001	0.0001	0.0001	0.0001	0.0002	0.0002
14	0.0000	0.0000	0.0000	0.0000	0.0000	0.0000	0.0000	0.0000	0.0000	0.0001

x \ μ	4.1	4.2	4.3	4.4	4.5	4.6	4.7	4.8	4.9	5.0
0	0.0166	0.0150	0.0136	0.0123	0.0111	0.0101	0.0091	0.0082	0.0074	0.0067
1	0.0679	0.0630	0.0583	0.0540	0.0500	0.0462	0.0427	0.0395	0.0365	0.0337
2	0.1393	0.1323	0.1254	0.1188	0.1125	0.1063	0.1005	0.0948	0.0894	0.0842
3	0.1904	0.1852	0.1798	0.1743	0.1687	0.1631	0.1574	0.1517	0.1460	0.1404
4	0.1951	0.1944	0.1933	0.1917	0.1898	0.1875	0.1849	0.1820	0.1789	0.1755
5	0.1600	0.1633	0.1662	0.1687	0.1708	0.1725	0.1738	0.1747	0.1753	0.1755
6	0.1093	0.1143	0.1191	0.1237	0.1281	0.1323	0.1362	0.1398	0.1432	0.1462
7	0.0640	0.0686	0.0732	0.0778	0.0824	0.0869	0.0914	0.0959	0.1002	0.1044
8	0.0328	0.0360	0.0393	0.0428	0.0463	0.0500	0.0537	0.0575	0.0614	0.0653
9	0.0150	0.0168	0.0188	0.0209	0.0232	0.0255	0.0281	0.0307	0.0334	0.0363
10	0.0061	0.0071	0.0081	0.0092	0.0104	0.0118	0.0132	0.0147	0.0164	0.0181
11	0.0023	0.0027	0.0032	0.0037	0.0043	0.0049	0.0056	0.0064	0.0073	0.0082
12	0.0008	0.0009	0.0011	0.0013	0.0016	0.0019	0.0022	0.0026	0.0030	0.0034
13	0.0002	0.0003	0.0004	0.0005	0.0006	0.0007	0.0008	0.0009	0.0011	0.0013
14	0.0001	0.0001	0.0001	0.0001	0.0002	0.0002	0.0003	0.0003	0.0004	0.0005
15	0.0000	0.0000	0.0000	0.0000	0.0001	0.0001	0.0001	0.0001	0.0001	0.0002

x \ μ	5.1	5.2	5.3	5.4	5.5	5.6	5.7	5.8	5.9	6.0
0	0.0061	0.0055	0.0050	0.0045	0.0041	0.0037	0.0033	0.0030	0.0027	0.0025
1	0.0311	0.0287	0.0265	0.0244	0.0225	0.0207	0.0191	0.0176	0.0162	0.0149
2	0.0793	0.0746	0.0701	0.0659	0.0618	0.0580	0.0544	0.0509	0.0477	0.0446
3	0.1348	0.1293	0.1239	0.1185	0.1133	0.1082	0.1033	0.0985	0.0938	0.0892
4	0.1719	0.1681	0.1641	0.1600	0.1558	0.1515	0.1472	0.1428	0.1383	0.1339
5	0.1753	0.1748	0.1740	0.1728	0.1714	0.1697	0.1678	0.1656	0.1632	0.1606
6	0.1490	0.1515	0.1537	0.1555	0.1571	0.1584	0.1594	0.1601	0.1605	0.1606
7	0.1086	0.1125	0.1163	0.1200	0.1234	0.1267	0.1298	0.1326	0.1353	0.1377
8	0.0692	0.0731	0.0771	0.0810	0.0849	0.0887	0.0925	0.0962	0.0998	0.1033
9	0.0392	0.0423	0.0454	0.0486	0.0519	0.0552	0.0586	0.0620	0.0654	0.0688
10	0.0200	0.0220	0.0241	0.0262	0.0285	0.0309	0.0334	0.0359	0.0386	0.0413
11	0.0093	0.0104	0.0116	0.0129	0.0143	0.0157	0.0173	0.0190	0.0207	0.0225
12	0.0039	0.0045	0.0051	0.0058	0.0065	0.0073	0.0082	0.0092	0.0102	0.0113
13	0.0015	0.0018	0.0021	0.0024	0.0028	0.0032	0.0036	0.0041	0.0046	0.0052
14	0.0006	0.0007	0.0008	0.0009	0.0011	0.0013	0.0015	0.0017	0.0019	0.0022
15	0.0002	0.0002	0.0003	0.0003	0.0004	0.0005	0.0006	0.0007	0.0008	0.0009
16	0.0001	0.0001	0.0001	0.0001	0.0001	0.0002	0.0002	0.0002	0.0003	0.0003
17	0.0000	0.0000	0.0000	0.0000	0.0000	0.0001	0.0001	0.0001	0.0001	0.0001

x \ μ	6.1	6.2	6.3	6.4	6.5	6.6	6.7	6.8	6.9	7.0
0	0.0022	0.0020	0.0018	0.0017	0.0015	0.0014	0.0012	0.0011	0.0010	0.0009
1	0.0137	0.0126	0.0116	0.0106	0.0098	0.0090	0.0082	0.0076	0.0070	0.0064
2	0.0417	0.0390	0.0364	0.0340	0.0318	0.0296	0.0276	0.0258	0.0240	0.0223
3	0.0848	0.0806	0.0765	0.0726	0.0688	0.0652	0.0617	0.0584	0.0552	0.0521
4	0.1294	0.1249	0.1205	0.1162	0.1118	0.1076	0.1034	0.0992	0.0952	0.0912
5	0.1579	0.1549	0.1519	0.1487	0.1454	0.1420	0.1385	0.1349	0.1314	0.1277
6	0.1605	0.1601	0.1595	0.1586	0.1575	0.1562	0.1546	0.1529	0.1511	0.1490
7	0.1399	0.1418	0.1435	0.1450	0.1462	0.1472	0.1480	0.1486	0.1489	0.1490
8	0.1066	0.1099	0.1130	0.1160	0.1188	0.1215	0.1240	0.1263	0.1284	0.1304
9	0.0723	0.0757	0.0791	0.0825	0.0858	0.0891	0.0923	0.0954	0.0985	0.1014
10	0.0441	0.0469	0.0498	0.0528	0.0558	0.0588	0.0618	0.0649	0.0679	0.0710
11	0.0244	0.0265	0.0285	0.0307	0.0330	0.0353	0.0377	0.0401	0.0426	0.0452
12	0.0124	0.0137	0.0150	0.0164	0.0179	0.0194	0.0210	0.0227	0.0245	0.0263
13	0.0058	0.0065	0.0073	0.0081	0.0089	0.0099	0.0108	0.0119	0.0130	0.0142
14	0.0025	0.0029	0.0033	0.0037	0.0041	0.0046	0.0052	0.0058	0.0064	0.0071
15	0.0010	0.0012	0.0014	0.0016	0.0018	0.0020	0.0023	0.0026	0.0029	0.0033
16	0.0004	0.0005	0.0005	0.0006	0.0007	0.0008	0.0010	0.0011	0.0013	0.0014
17	0.0001	0.0002	0.0002	0.0002	0.0003	0.0003	0.0004	0.0004	0.0005	0.0006
18	0.0000	0.0001	0.0001	0.0001	0.0001	0.0001	0.0001	0.0002	0.0002	0.0002
19	0.0000	0.0000	0.0000	0.0000	0.0000	0.0000	0.0001	0.0001	0.0001	0.0001

x \ μ	7.1	7.2	7.3	7.4	7.5	7.6	7.7	7.8	7.9	8.0
0	0.0008	0.0007	0.0007	0.0006	0.0006	0.0005	0.0005	0.0004	0.0004	0.0003
1	0.0059	0.0054	0.0049	0.0045	0.0041	0.0038	0.0035	0.0032	0.0029	0.0027
2	0.0208	0.0194	0.0180	0.0167	0.0156	0.0145	0.0134	0.0125	0.0116	0.0107
3	0.0492	0.0464	0.0438	0.0413	0.0389	0.0366	0.0345	0.0324	0.0305	0.0286
4	0.0874	0.0836	0.0799	0.0764	0.0729	0.0696	0.0663	0.0632	0.0602	0.0573
5	0.1241	0.1204	0.1167	0.1130	0.1094	0.1057	0.1021	0.0986	0.0951	0.0916
6	0.1468	0.1445	0.1420	0.1394	0.1367	0.1339	0.1311	0.1282	0.1252	0.1221
7	0.1489	0.1486	0.1481	0.1474	0.1465	0.1454	0.1442	0.1428	0.1413	0.1396
8	0.1321	0.1337	0.1351	0.1363	0.1373	0.1381	0.1388	0.1392	0.1395	0.1396
9	0.1042	0.1070	0.1096	0.1121	0.1144	0.1167	0.1187	0.1207	0.1224	0.1241
10	0.0740	0.0770	0.0800	0.0829	0.0858	0.0887	0.0914	0.0941	0.0967	0.0993
11	0.0478	0.0504	0.0531	0.0558	0.0585	0.0613	0.0640	0.0667	0.0695	0.0722
12	0.0283	0.0303	0.0323	0.0344	0.0366	0.0388	0.0411	0.0434	0.0457	0.0481
13	0.0154	0.0168	0.0181	0.0196	0.0211	0.0227	0.0243	0.0260	0.0278	0.0296
14	0.0078	0.0086	0.0095	0.0104	0.0113	0.0123	0.0134	0.0145	0.0157	0.0169
15	0.0037	0.0041	0.0046	0.0051	0.0057	0.0062	0.0069	0.0075	0.0083	0.0090
16	0.0016	0.0019	0.0021	0.0024	0.0026	0.0030	0.0033	0.0037	0.0041	0.0045
17	0.0007	0.0008	0.0009	0.0010	0.0012	0.0013	0.0015	0.0017	0.0019	0.0021
18	0.0003	0.0003	0.0004	0.0004	0.0005	0.0006	0.0006	0.0007	0.0008	0.0009
19	0.0001	0.0001	0.0001	0.0002	0.0002	0.0002	0.0003	0.0003	0.0003	0.0004
20	0.0000	0.0000	0.0001	0.0001	0.0001	0.0001	0.0001	0.0001	0.0001	0.0002
21	0.0000	0.0000	0.0000	0.0000	0.0000	0.0000	0.0000	0.0000	0.0001	0.0001

x \ μ	8.1	8.2	8.3	8.4	8.5	8.6	8.7	8.8	8.9	9.0
0	0.0003	0.0003	0.0002	0.0002	0.0002	0.0002	0.0002	0.0002	0.0001	0.0001
1	0.0025	0.0023	0.0021	0.0019	0.0017	0.0016	0.0014	0.0013	0.0012	0.0011
2	0.0100	0.0092	0.0086	0.0079	0.0074	0.0068	0.0063	0.0058	0.0054	0.0050
3	0.0269	0.0252	0.0237	0.0222	0.0208	0.0195	0.0183	0.0171	0.0160	0.0150
4	0.0544	0.0517	0.0491	0.0466	0.0443	0.0420	0.0398	0.0377	0.0357	0.0337
5	0.0882	0.0849	0.0816	0.0784	0.0752	0.0722	0.0692	0.0663	0.0635	0.0607
6	0.1191	0.1160	0.1128	0.1097	0.1066	0.1034	0.1003	0.0972	0.0941	0.0911
7	0.1378	0.1358	0.1338	0.1317	0.1294	0.1271	0.1247	0.1222	0.1197	0.1171
8	0.1395	0.1392	0.1388	0.1382	0.1375	0.1366	0.1356	0.1344	0.1332	0.1318
9	0.1256	0.1269	0.1280	0.1290	0.1299	0.1306	0.1311	0.1315	0.1317	0.1318
10	0.1017	0.1040	0.1063	0.1084	0.1104	0.1123	0.1140	0.1157	0.1172	0.1186
11	0.0749	0.0776	0.0802	0.0828	0.0853	0.0878	0.0902	0.0925	0.0948	0.0970
12	0.0505	0.0530	0.0555	0.0579	0.0604	0.0629	0.0654	0.0679	0.0703	0.0728
13	0.0315	0.0334	0.0354	0.0374	0.0395	0.0416	0.0438	0.0459	0.0481	0.0504
14	0.0182	0.0196	0.0210	0.0225	0.0240	0.0256	0.0272	0.0289	0.0306	0.0324
15	0.0098	0.0107	0.0116	0.0126	0.0136	0.0147	0.0158	0.0169	0.0182	0.0194
16	0.0050	0.0055	0.0060	0.0066	0.0072	0.0079	0.0086	0.0093	0.0101	0.0109
17	0.0024	0.0026	0.0029	0.0033	0.0036	0.0040	0.0044	0.0048	0.0053	0.0058
18	0.0011	0.0012	0.0014	0.0015	0.0017	0.0019	0.0021	0.0024	0.0026	0.0029
19	0.0005	0.0005	0.0006	0.0007	0.0008	0.0009	0.0010	0.0011	0.0012	0.0014
20	0.0002	0.0002	0.0002	0.0003	0.0003	0.0004	0.0004	0.0005	0.0005	0.0006
21	0.0001	0.0001	0.0001	0.0001	0.0001	0.0002	0.0002	0.0002	0.0002	0.0003
22	0.0000	0.0000	0.0000	0.0000	0.0001	0.0001	0.0001	0.0001	0.0001	0.0001

x \ μ	9.1	9.2	9.3	9.4	9.5	9.6	9.7	9.8	9.9	10.0
0	0.0001	0.0001	0.0001	0.0001	0.0001	0.0001	0.0001	0.0001	0.0001	0.0000
1	0.0010	0.0009	0.0009	0.0008	0.0007	0.0007	0.0006	0.0005	0.0005	0.0005
2	0.0046	0.0043	0.0040	0.0037	0.0034	0.0031	0.0029	0.0027	0.0025	0.0023
3	0.0140	0.0131	0.0123	0.0115	0.0107	0.0100	0.0093	0.0087	0.0081	0.0076
4	0.0319	0.0302	0.0285	0.0269	0.0254	0.0240	0.0226	0.0213	0.0201	0.0189
5	0.0581	0.0555	0.0530	0.0506	0.0483	0.0460	0.0439	0.0418	0.0398	0.0378
6	0.0881	0.0851	0.0822	0.0793	0.0764	0.0736	0.0709	0.0682	0.0656	0.0631
7	0.1145	0.1118	0.1091	0.1064	0.1037	0.1010	0.0982	0.0955	0.0928	0.0901
8	0.1302	0.1286	0.1269	0.1251	0.1232	0.1212	0.1191	0.1170	0.1148	0.1126
9	0.1317	0.1315	0.1311	0.1306	0.1300	0.1293	0.1284	0.1274	0.1263	0.1251
10	0.1198	0.1210	0.1219	0.1228	0.1235	0.1241	0.1245	0.1249	0.1250	0.1251
11	0.0991	0.1012	0.1031	0.1049	0.1067	0.1083	0.1098	0.1112	0.1125	0.1137
12	0.0752	0.0776	0.0799	0.0822	0.0844	0.0866	0.0888	0.0908	0.0928	0.0948
13	0.0526	0.0549	0.0572	0.0594	0.0617	0.0640	0.0662	0.0685	0.0707	0.0729
14	0.0342	0.0361	0.0380	0.0399	0.0419	0.0439	0.0459	0.0479	0.0500	0.0521
15	0.0208	0.0221	0.0235	0.0250	0.0265	0.0281	0.0297	0.0313	0.0330	0.0347
16	0.0118	0.0127	0.0137	0.0147	0.0157	0.0168	0.0180	0.0192	0.0204	0.0217
17	0.0063	0.0069	0.0075	0.0081	0.0088	0.0095	0.0103	0.0111	0.0119	0.0128
18	0.0032	0.0035	0.0039	0.0042	0.0046	0.0051	0.0055	0.0060	0.0065	0.0071
19	0.0015	0.0017	0.0019	0.0021	0.0023	0.0026	0.0028	0.0031	0.0034	0.0037
20	0.0007	0.0008	0.0009	0.0010	0.0011	0.0012	0.0014	0.0015	0.0017	0.0019
21	0.0003	0.0003	0.0004	0.0004	0.0005	0.0006	0.0006	0.0007	0.0008	0.0009
22	0.0001	0.0001	0.0002	0.0002	0.0002	0.0002	0.0003	0.0003	0.0004	0.0004
23	0.0000	0.0001	0.0001	0.0001	0.0001	0.0001	0.0001	0.0001	0.0002	0.0002
24	0.0000	0.0000	0.0000	0.0000	0.0000	0.0000	0.0000	0.0001	0.0001	0.0001

x	μ = 20	x	μ = 20	x	μ = 20
0	0.0000	14	0.0387	28	0.0181
1	0.0000	15	0.0516	29	0.0125
2	0.0000	16	0.0646	30	0.0083
3	0.0000	17	0.0760	31	0.0054
4	0.0000	18	0.0844	32	0.0034
5	0.0001	19	0.0888	33	0.0020
6	0.0002	20	0.0888	34	0.0012
7	0.0005	21	0.0846	35	0.0007
8	0.0013	22	0.0769	36	0.0004
9	0.0029	23	0.0669	37	0.0002
10	0.0058	24	0.0557	38	0.0001
11	0.0106	25	0.0446	39	0.0001
12	0.0176	26	0.0343		
13	0.0271	27	0.0254		

TABLE 3

STANDARD NORMAL DISTRIBUTION TABLE (*Z*-TABLE)

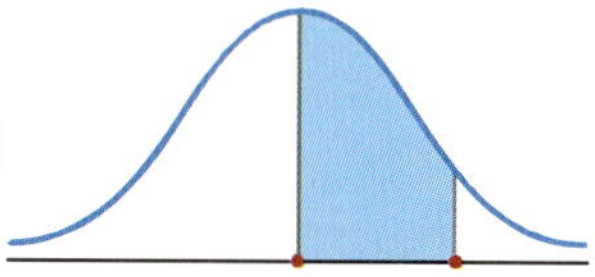

$$Z = \frac{X - \mu}{\sigma} \sim N(0, 1)$$

The following table gives the shaded area in the diagram, namely, $P(0 < Z < z)$ for different values of z.

z	0	0.01	0.02	0.03	0.04	0.05	0.06	0.07	0.08	0.09
0	0.0000	0.0040	0.0080	0.0120	0.0160	0.0199	0.0239	0.0279	0.0319	0.0359
0.1	0.0398	0.0438	0.0478	0.0517	0.0557	0.0596	0.0636	0.0675	0.0714	0.0753
0.2	0.0793	0.0832	0.0871	0.0910	0.0948	0.0987	0.1026	0.1064	0.1103	0.1141
0.3	0.1179	0.1217	0.1255	0.1293	0.1331	0.1368	0.1406	0.1443	0.1480	0.1517
0.4	0.1554	0.1591	0.1628	0.1664	0.1700	0.1736	0.1772	0.1808	0.1844	0.1879
0.5	0.1915	0.1950	0.1985	0.2019	0.2054	0.2088	0.2123	0.2157	0.2190	0.2224
0.6	0.2257	0.2291	0.2324	0.2357	0.2389	0.2422	0.2454	0.2486	0.2517	0.2549
0.7	0.2580	0.2611	0.2642	0.2673	0.2704	0.2734	0.2764	0.2794	0.2823	0.2852
0.8	0.2881	0.2910	0.2939	0.2967	0.2995	0.3023	0.3051	0.3078	0.3106	0.3133
0.9	0.3159	0.3186	0.3212	0.3238	0.3264	0.3289	0.3315	0.3340	0.3365	0.3389
1	0.3413	0.3438	0.3461	0.3485	0.3508	0.3531	0.3554	0.3577	0.3599	0.3621
1.1	0.3643	0.3665	0.3686	0.3708	0.3729	0.3749	0.3770	0.3790	0.3810	0.3830
1.2	0.3849	0.3869	0.3888	0.3907	0.3925	0.3944	0.3962	0.3980	0.3997	0.4015
1.3	0.4032	0.4049	0.4066	0.4082	0.4099	0.4115	0.4131	0.4147	0.4162	0.4177
1.4	0.4192	0.4207	0.4222	0.4236	0.4251	0.4265	0.4279	0.4292	0.4306	0.4319
1.5	0.4332	0.4345	0.4357	0.4370	0.4382	0.4394	0.4406	0.4418	0.4429	0.4441
1.6	0.4452	0.4463	0.4474	0.4484	0.4495	0.4505	0.4515	0.4525	0.4535	0.4545
1.7	0.4554	0.4564	0.4573	0.4582	0.4591	0.4599	0.4608	0.4616	0.4625	0.4633
1.8	0.4641	0.4649	0.4656	0.4664	0.4671	0.4678	0.4686	0.4693	0.4699	0.4706
1.9	0.4713	0.4719	0.4726	0.4732	0.4738	0.4744	0.4750	0.4756	0.4761	0.4767
2	0.4772	0.4778	0.4783	0.4788	0.4793	0.4798	0.4803	0.4808	0.4812	0.4817
2.1	0.4821	0.4826	0.4830	0.4834	0.4838	0.4842	0.4846	0.4850	0.4854	0.4857
2.2	0.4861	0.4864	0.4868	0.4871	0.4875	0.4878	0.4881	0.4884	0.4887	0.4890
2.3	0.4893	0.4896	0.4898	0.4901	0.4904	0.4906	0.4909	0.4911	0.4913	0.4916
2.4	0.4918	0.4920	0.4922	0.4925	0.4927	0.4929	0.4931	0.4932	0.4934	0.4936
2.5	0.4938	0.4940	0.4941	0.4943	0.4945	0.4946	0.4948	0.4949	0.4951	0.4952
2.6	0.4953	0.4955	0.4956	0.4957	0.4959	0.4960	0.4961	0.4962	0.4963	0.4964
2.7	0.4965	0.4966	0.4967	0.4968	0.4969	0.4970	0.4971	0.4972	0.4973	0.4974
2.8	0.4974	0.4975	0.4976	0.4977	0.4977	0.4978	0.4979	0.4979	0.4980	0.4981
2.9	0.4981	0.4982	0.4982	0.4983	0.4984	0.4984	0.4985	0.4985	0.4986	0.4986
3	0.4987	0.4987	0.4987	0.4988	0.4988	0.4989	0.4989	0.4989	0.4990	0.4990
3.1	0.4990	0.4991	0.4991	0.4991	0.4992	0.4992	0.4992	0.4992	0.4993	0.4993
3.2	0.4993	0.4993	0.4994	0.4994	0.4994	0.4994	0.4994	0.4995	0.4995	0.4995
3.3	0.4995	0.4995	0.4995	0.4996	0.4996	0.4996	0.4996	0.4996	0.4996	0.4997
3.4	0.4997	0.4997	0.4997	0.4997	0.4997	0.4997	0.4997	0.4997	0.4997	0.4998
3.5	0.4998	0.4998	0.4998	0.4998	0.4998	0.4998	0.4998	0.4998	0.4998	0.4998
3.6	0.4998	0.4998	0.4999	0.4999	0.4999	0.4999	0.4999	0.4999	0.4999	0.4999
3.7	0.4999	0.4999	0.4999	0.4999	0.4999	0.4999	0.4999	0.4999	0.4999	0.4999

TABLE 4

AREA IN THE RIGHT TAIL UNDER THE *t*-DISTRIBUTION CURVE

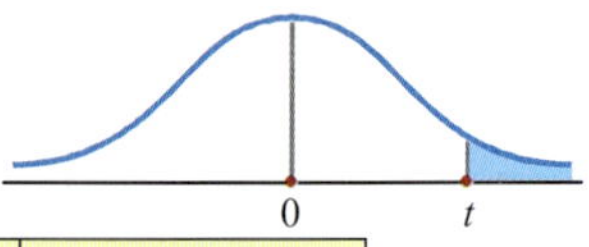

df \ α	0.10	0.05	0.025	0.01	0.005	0.001
1	3.0777	6.3138	12.7062	31.8205	63.6567	318.3088
2	1.8856	2.9200	4.3027	6.9646	9.9248	22.3271
3	1.6377	2.3534	3.1824	4.5407	5.8409	10.2145
4	1.5332	2.1318	2.7764	3.7469	4.6041	7.1732
5	1.4759	2.0150	2.5706	3.3649	4.0321	5.8934
6	1.4398	1.9432	2.4469	3.1427	3.7074	5.2076
7	1.4149	1.8946	2.3646	2.9980	3.4995	4.7853
8	1.3968	1.8595	2.3060	2.8965	3.3554	4.5008
9	1.3830	1.8331	2.2622	2.8214	3.2498	4.2968
10	1.3722	1.8125	2.2281	2.7638	3.1693	4.1437
11	1.3634	1.7959	2.2010	2.7181	3.1058	4.0247
12	1.3562	1.7823	2.1788	2.6810	3.0545	3.9296
13	1.3502	1.7709	2.1604	2.6503	3.0123	3.8520
14	1.3450	1.7613	2.1448	2.6245	2.9768	3.7874
15	1.3406	1.7531	2.1314	2.6025	2.9467	3.7328
16	1.3368	1.7459	2.1199	2.5835	2.9208	3.6862
17	1.3334	1.7396	2.1098	2.5669	2.8982	3.6458
18	1.3304	1.7341	2.1009	2.5524	2.8784	3.6105
19	1.3277	1.7291	2.0930	2.5395	2.8609	3.5794
20	1.3253	1.7247	2.0860	2.5280	2.8453	3.5518
21	1.3232	1.7207	2.0796	2.5176	2.8314	3.5272
22	1.3212	1.7171	2.0739	2.5083	2.8188	3.5050
23	1.3195	1.7139	2.0687	2.4999	2.8073	3.4850
24	1.3178	1.7109	2.0639	2.4922	2.7969	3.4668
25	1.3163	1.7081	2.0595	2.4851	2.7874	3.4502
26	1.3150	1.7056	2.0555	2.4786	2.7787	3.4350
27	1.3137	1.7033	2.0518	2.4727	2.7707	3.4210
28	1.3125	1.7011	2.0484	2.4671	2.7633	3.4082
29	1.3114	1.6991	2.0452	2.4620	2.7564	3.3962
30	1.3104	1.6973	2.0423	2.4573	2.7500	3.3852
31	1.3095	1.6955	2.0395	2.4528	2.7440	3.3749
32	1.3086	1.6939	2.0369	2.4487	2.7385	3.3653
33	1.3077	1.6924	2.0345	2.4448	2.7333	3.3563
34	1.3070	1.6909	2.0322	2.4411	2.7284	3.3479
35	1.3062	1.6896	2.0301	2.4377	2.7238	3.3400
36	1.3055	1.6883	2.0281	2.4345	2.7195	3.3326
37	1.3049	1.6871	2.0262	2.4314	2.7154	3.3256
38	1.3042	1.6860	2.0244	2.4286	2.7116	3.3190
39	1.3036	1.6849	2.0227	2.4258	2.7079	3.3128

df \ α	0.10	0.05	0.025	0.01	0.005	0.001
40	1.3031	1.6839	2.0211	2.4233	2.7045	3.3069
41	1.3025	1.6829	2.0195	2.4208	2.7012	3.3013
42	1.3020	1.6820	2.0181	2.4185	2.6981	3.2960
43	1.3016	1.6811	2.0167	2.4163	2.6951	3.2909
44	1.3011	1.6802	2.0154	2.4141	2.6923	3.2861
45	1.3006	1.6794	2.0141	2.4121	2.6896	3.2815
46	1.3002	1.6787	2.0129	2.4102	2.6870	3.2771
47	1.2998	1.6779	2.0117	2.4083	2.6846	3.2729
48	1.2994	1.6772	2.0106	2.4066	2.6822	3.2689
49	1.2991	1.6766	2.0096	2.4049	2.6800	3.2651
50	1.2987	1.6759	2.0086	2.4033	2.6778	3.2614
51	1.2984	1.6753	2.0076	2.4017	2.6757	3.2579
52	1.2980	1.6747	2.0066	2.4002	2.6737	3.2545
53	1.2977	1.6741	2.0057	2.3988	2.6718	3.2513
54	1.2974	1.6736	2.0049	2.3974	2.6700	3.2481
55	1.2971	1.6730	2.0040	2.3961	2.6682	3.2451
56	1.2969	1.6725	2.0032	2.3948	2.6665	3.2423
57	1.2966	1.6720	2.0025	2.3936	2.6649	3.2395
58	1.2963	1.6716	2.0017	2.3924	2.6633	3.2368
59	1.2961	1.6711	2.0010	2.3912	2.6618	3.2342
60	1.2958	1.6706	2.0003	2.3901	2.6603	3.2317
61	1.2956	1.6702	1.9996	2.3890	2.6589	3.2293
62	1.2954	1.6698	1.9990	2.3880	2.6575	3.2270
63	1.2951	1.6694	1.9983	2.3870	2.6561	3.2247
64	1.2949	1.6690	1.9977	2.3860	2.6549	3.2225
65	1.2947	1.6686	1.9971	2.3851	2.6536	3.2204
66	1.2945	1.6683	1.9966	2.3842	2.6524	3.2184
67	1.2943	1.6679	1.9960	2.3833	2.6512	3.2164
68	1.2941	1.6676	1.9955	2.3824	2.6501	3.2145
69	1.2939	1.6672	1.9949	2.3816	2.6490	3.2126
70	1.2938	1.6669	1.9944	2.3808	2.6479	3.2108
71	1.2936	1.6666	1.9939	2.3800	2.6469	3.2090
72	1.2934	1.6663	1.9935	2.3793	2.6459	3.2073
73	1.2933	1.6660	1.9930	2.3785	2.6449	3.2057
74	1.2931	1.6657	1.9925	2.3778	2.6439	3.2041
75	1.2929	1.6654	1.9921	2.3771	2.6430	3.2025
76	1.2928	1.6652	1.9917	2.3764	2.6421	3.2010
77	1.2926	1.6649	1.9913	2.3758	2.6412	3.1995
78	1.2925	1.6646	1.9908	2.3751	2.6403	3.1980
79	1.2924	1.6644	1.9905	2.3745	2.6395	3.1966
80	1.2922	1.6641	1.9901	2.3739	2.6387	3.1953
∞	1.2816	1.6449	1.9600	2.3263	2.5758	3.0902

TABLE 5

CRITICAL VALUES $\chi_\upsilon^2(\alpha)$ OF CHI-SQUARE DISTRIBUTION

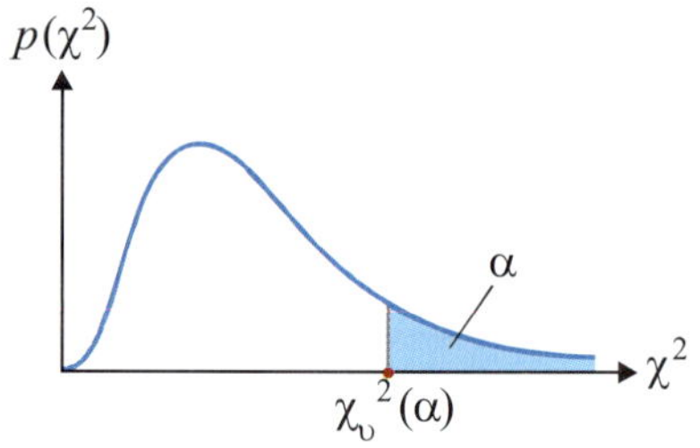

df \ α	0.995	0.99	0.975	0.95	0.90	0.10	0.05	0.025	0.01	0.005
1	0.000	0.000	0.001	0.004	0.016	2.706	3.841	5.024	6.635	7.879
2	0.010	0.020	0.051	0. 103	0.211	4.605	5.991	7.378	9.210	10.597
3	0.072	0.115	0.216	0.352	0.584	6.251	7.815	9.348	11.345	12.838
4	0.207	0.297	0.484	0.711	1.064	7.779	9.488	11.143	13.277	14.860
5	0.412	0.554	0.831	1.145	1.610	9.236	11.070	12.833	15.086	16.750
6	0.676	0.872	1.237	1.634	2.204	10.645	12.592	14.449	16.812	18.548
7	0.989	1.239	1.690	2.167	2.833	12.017	14.067	16.013	18.475	20.278
8	1.344	1.646	2.180	2.733	3.490	13.362	15.507	17.535	20.090	21.955
9	1.735	2.088	2.700	3.325	4.168	14.684	16.919	19.023	21.666	23.589
10	2.156	2.558	3.247	3.940	4.865	15.987	18.307	20.483	23.209	25.188
11	2.603	3.053	3.816	4.575	5.578	17.275	19.675	21.920	24.725	26.757
12	3.074	3.571	4.404	5.226	6.304	18.549	21.026	23.337	26.217	28.999
13	3.565	4.107	5.009	5.892	7.042	19.812	22.362	24.736	27.688	29.819
14	4.075	4.660	5.629	6.571	7.790	21.064	23.685	26.119	29.141	31.319
15	4.601	5.229	6.262	7.261	8.547	22.307	24.996	27.488	30.578	32.801
16	5.142	5.812	6.908	7.962	9.312	23.542	26.296	28.845	32.000	34.267
17	5.697	6.408	7.564	8.672	10.085	24.769	27.587	30.191	33.409	35.715
18	6.265	7.015	8.231	8.231	10.865	25.989	28.869	31.526	34.805	37.156
19	6.844	7.633	8.907	10.117	11.651	27.204	30.144	32.852	36.191	38.582
20	7.434	8.260	9.591	10.851	12.443	28.412	31.410	34.170	37.566	39.997
21	8.034	8.897	10.283	11.591	13.240	29.615	32.671	35.479	38.932	41.401
22	8.643	9.542	10.982	12.338	14.042	30.813	33.924	36.781	40.289	42.796
23	9.260	10.196	11.689	13.091	14.848	32.007	35.172	38.076	41.638	44.181
24	9.886	10.856	12.401	13.848	15.659	33.196	36.415	39.364	42.980	45.559
25	10.520	11.524	13.120	14.611	16.473	34.382	37.652	40.646	44.314	46.928
26	11.160	12.198	13.844	15.379	17.292	35.563	38.885	41.923	45.642	48.290
27	11.808	12.879	14.573	16.151	18.114	36.741	40.113	43.194	46.963	49.645
28	12.461	13.565	15.308	16.928	18.939	37.916	41.337	44.461	48.278	50.993
29	13.121	14.256	16.047	17.708	19.768	39.087	42.557	45.722	49.588	52.336
30	13.787	14.953	16.791	18.493	20.599	40.256	43.773	46.979	50.892	53.672
40	20.707	22.164	24.433	26.509	29.051	51.805	55.758	59.342	63.691	66.766
50	27.991	29.707	32.357	34.764	37.689	63.167	67.505	71.420	76.154	79.490
60	35.534	37.485	40.482	43.188	46.459	74.397	79.082	83.298	88.379	91.952
70	43.275	45.442	48.758	51.739	55.329	85.527	90.531	95.023	100.425	104.215
80	51.172	53.540	57.153	60.391	64.278	96.578	101.879	106.629	112.329	116.321
90	59.196	61.754	65.647	69.126	73.291	107.565	113.145	118.136	124.116	128.299
100	67.328	70.065	74.222	77.929	82.358	118.498	124.342	129.561	135.807	140.169

For larger values of υ, the quantity $\sqrt{2\chi^2} - \sqrt{2\nu - 1}$ may be used as a standard normal variable.

TABLE 6(A)

CRITICAL VALUES OF THE VARIANCE RATIO
F-DISTRIBUTION (RIGHT-TAIL AREAS) – $\alpha = 0.01$

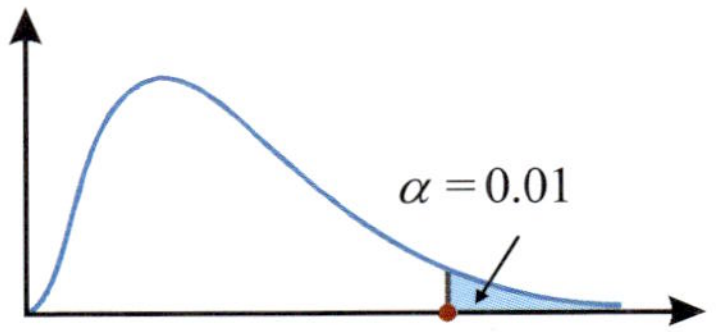

υ_2 \ υ_1	1	2	3	4	5	6	7	8	9	10	12	15	20	24	30	40
1	4052.181	4999.500	5403.352	5624.583	5763.650	5858.986	5928.356	5981.070	6022.473	6055.847	6106.321	6157.285	6208.730	6234.631	6260.649	6286.782
2	98.503	99.000	99.166	99.249	99.299	99.333	99.356	99.374	99.388	99.399	99.416	99.433	99.449	99.458	99.466	99.474
3	34.116	30.817	29.457	28.710	28.237	27.911	27.672	27.489	27.345	27.229	27.052	26.872	26.690	26.598	26.505	26.411
4	21.198	18.000	16.694	15.977	15.522	15.207	14.976	14.799	14.659	14.546	14.374	14.198	14.020	13.929	13.838	13.745
5	16.258	13.274	12.060	11.392	10.967	10.672	10.456	10.289	10.158	10.051	9.888	9.722	9.553	9.466	9.379	9.291
6	13.745	10.925	9.780	9.148	8.746	8.466	8.260	8.102	7.976	7.874	7.718	7.559	7.396	7.313	7.229	7.143
7	12.246	9.547	8.451	7.847	7.460	7.191	6.993	6.840	6.719	6.620	6.469	6.314	6.155	6.074	5.992	5.908
8	11.259	8.649	7.591	7.006	6.632	6.371	6.178	6.029	5.911	5.814	5.667	5.515	5.359	5.279	5.198	5.116
9	10.561	8.022	6.992	6.422	6.057	5.802	5.613	5.467	5.351	5.257	5.111	4.962	4.808	4.729	4.649	4.567
10	10.044	7.559	6.552	5.994	5.636	5.386	5.200	5.057	4.942	4.849	4.706	4.558	4.405	4.327	4.247	4.165
11	9.646	7.206	6.217	5.668	5.316	5.069	4.886	4.744	4.632	4.539	4.397	4.251	4.099	4.021	3.941	3.860
12	9.330	6.927	5.953	5.412	5.064	4.821	4.640	4.499	4.388	4.296	4.155	4.010	3.858	3.780	3.701	3.619
13	9.074	6.701	5.739	5.205	4.862	4.620	4.441	4.302	4.191	4.100	3.960	3.815	3.665	3.587	3.507	3.425
14	8.862	6.515	5.564	5.035	4.695	4.456	4.278	4.140	4.030	3.939	3.800	3.656	3.505	3.427	3.348	3.266
15	8.683	6.359	5.417	4.893	4.556	4.318	4.142	4.004	3.895	3.805	3.666	3.522	3.372	3.294	3.214	3.132
16	8.531	6.226	5.292	4.773	4.437	4.202	4.026	3.890	3.780	3.691	3.553	3.409	3.259	3.181	3.101	3.018
17	8.400	6.112	5.185	4.669	4.336	4.102	3.927	3.791	3.682	3.593	3.455	3.312	3.162	3.084	3.003	2.920
18	8.285	6.013	5.092	4.579	4.248	4.015	3.841	3.705	3.597	3.508	3.371	3.227	3.077	2.999	2.919	2.835
19	8.185	5.926	5.010	4.500	4.171	3.939	3.765	3.631	3.523	3.434	3.297	3.153	3.003	2.925	2.844	2.761
20	8.096	5.849	4.938	4.431	4.103	3.871	3.699	3.564	3.457	3.368	3.231	3.088	2.938	2.859	2.778	2.695
21	8.017	5.780	4.874	4.369	4.042	3.812	3.640	3.506	3.398	3.310	3.173	3.030	2.880	2.801	2.720	2.636
22	7.945	5.719	4.817	4.313	3.988	3.758	3.587	3.453	3.346	3.258	3.121	2.978	2.827	2.749	2.667	2.583
23	7.881	5.664	4.765	4.264	3.939	3.710	3.539	3.406	3.299	3.211	3.074	2.931	2.781	2.702	2.620	2.535
24	7.823	5.614	4.718	4.218	3.895	3.667	3.496	3.363	3.256	3.168	3.032	2.889	2.738	2.659	2.577	2.492
25	7.770	5.568	4.675	4.177	3.855	3.627	3.457	3.324	3.217	3.129	2.993	2.850	2.699	2.620	2.538	2.453
30	7.562	5.390	4.510	4.018	3.699	3.473	3.304	3.173	3.067	2.979	2.843	2.700	2.549	2.469	2.386	2.299
40	7.314	5.179	4.313	3.828	3.514	3.291	3.124	2.993	2.888	2.801	2.665	2.522	2.369	2.288	2.203	2.114
60	7.077	4.977	4.126	3.649	3.339	3.119	2.953	2.823	2.718	2.632	2.496	2.352	2.198	2.115	2.028	1.936
120	6.851	4.787	3.949	3.480	3.174	2.956	2.792	2.663	2.559	2.472	2.336	2.192	2.035	1.950	1.860	1.763
∞	6.635	4.605	3.782	3.319	3.017	2.802	2.639	2.511	2.407	2.321	2.185	2.039	1.878	1.791	1.696	1.592

TABLE 6(B)

CRITICAL VALUES OF THE VARIANCE RATIO
F-DISTRIBUTION (RIGHT-TAIL AREAS) $\alpha = 0.025$

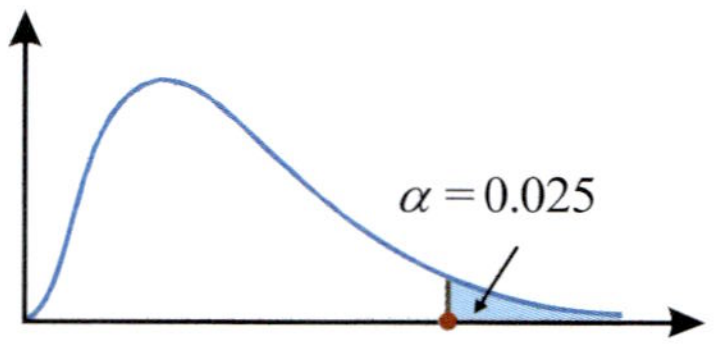

υ_2 \ υ_1	1	2	3	4	5	6	7	8	9	10	12	15	20	24	30	40
1	647.789	799.500	864.163	899.583	921.848	937.111	948.217	956.656	963.285	968.627	976.708	984.867	993.103	997.249	1001.414	1005.598
2	38.506	39.000	39.165	39.248	39.298	39.331	39.355	39.373	39.387	39.398	39.415	39.431	39.448	39.456	39.465	39.473
3	17.443	16.044	15.439	15.101	14.885	14.735	14.624	14.540	14.473	14.419	14.337	14.253	14.167	14.124	14.081	14.037
4	12.218	10.649	9.979	9.605	9.364	9.197	9.074	8.980	8.905	8.844	8.751	8.657	8.560	8.511	8.461	8.411
5	10.007	8.434	7.764	7.388	7.146	6.978	6.853	6.757	6.681	6.619	6.525	6.428	6.329	6.278	6.227	6.175
6	8.813	7.260	6.599	6.227	5.988	5.820	5.695	5.600	5.523	5.461	5.366	5.269	5.168	5.117	5.065	5.012
7	8.073	6.542	5.890	5.523	5.285	5.119	4.995	4.899	4.823	4.761	4.666	4.568	4.467	4.415	4.362	4.309
8	7.571	6.059	5.416	5.053	4.817	4.652	4.529	4.433	4.357	4.295	4.200	4.101	3.999	3.947	3.894	3.840
9	7.209	5.715	5.078	4.718	4.484	4.320	4.197	4.102	4.026	3.964	3.868	3.769	3.667	3.614	3.560	3.505
10	6.937	5.456	4.826	4.468	4.236	4.072	3.950	3.855	3.779	3.717	3.621	3.522	3.419	3.365	3.311	3.255
11	6.724	5.256	4.630	4.275	4.044	3.881	3.759	3.664	3.588	3.526	3.430	3.330	3.226	3.173	3.118	3.061
12	6.554	5.096	4.474	4.121	3.891	3.728	3.607	3.512	3.436	3.374	3.277	3.177	3.073	3.019	2.963	2.906
13	6.414	4.965	4.347	3.996	3.767	3.604	3.483	3.388	3.312	3.250	3.153	3.053	2.948	2.893	2.837	2.780
14	6.298	4.857	4.242	3.892	3.663	3.501	3.380	3.285	3.209	3.147	3.050	2.949	2.844	2.789	2.732	2.674
15	6.200	4.765	4.153	3.804	3.576	3.415	3.293	3.199	3.123	3.060	2.963	2.862	2.756	2.701	2.644	2.585
16	6.115	4.687	4.077	3.729	3.502	3.341	3.219	3.125	3.049	2.986	2.889	2.788	2.681	2.625	2.568	2.509
17	6.042	4.619	4.011	3.665	3.438	3.277	3.156	3.061	2.985	2.922	2.825	2.723	2.616	2.560	2.502	2.442
18	5.978	4.560	3.954	3.608	3.382	3.221	3.100	3.005	2.929	2.866	2.769	2.667	2.559	2.503	2.445	2.384
19	5.922	4.508	3.903	3.559	3.333	3.172	3.051	2.956	2.880	2.817	2.720	2.617	2.509	2.452	2.394	2.333
20	5.871	4.461	3.859	3.515	3.289	3.128	3.007	2.913	2.837	2.774	2.676	2.573	2.464	2.408	2.349	2.287
21	5.827	4.420	3.819	3.475	3.250	3.090	2.969	2.874	2.798	2.735	2.637	2.534	2.425	2.368	2.308	2.246
22	5.786	4.383	3.783	3.440	3.215	3.055	2.934	2.839	2.763	2.700	2.602	2.498	2.389	2.331	2.272	2.210
23	5.750	4.349	3.750	3.408	3.183	3.023	2.902	2.808	2.731	2.668	2.570	2.466	2.357	2.299	2.239	2.176
24	5.717	4.319	3.721	3.379	3.155	2.995	2.874	2.779	2.703	2.640	2.541	2.437	2.327	2.269	2.209	2.146
25	5.686	4.291	3.694	3.353	3.129	2.969	2.848	2.753	2.677	2.613	2.515	2.411	2.300	2.242	2.182	2.118
30	5.568	4.182	3.589	3.250	3.026	2.867	2.746	2.651	2.575	2.511	2.412	2.307	2.195	2.136	2.074	2.009
40	5.424	4.051	3.463	3.126	2.904	2.744	2.624	2.529	2.452	2.388	2.288	2.182	2.068	2.007	1.943	1.875
60	5.286	3.925	3.343	3.008	2.786	2.627	2.507	2.412	2.334	2.270	2.169	2.061	1.944	1.882	1.815	1.744
120	5.152	3.805	3.227	2.894	2.674	2.515	2.395	2.299	2.222	2.157	2.055	1.945	1.825	1.760	1.690	1.614
∞	5.024	3.689	3.116	2.786	2.567	2.408	2.288	2.192	2.114	2.048	1.945	1.833	1.708	1.640	1.566	1.484

TABLE 6(C)

CRITICAL VALUES OF THE VARIANCE RATIO
F-DISTRIBUTION (RIGHT-TAIL AREAS) $\alpha = 0.05$

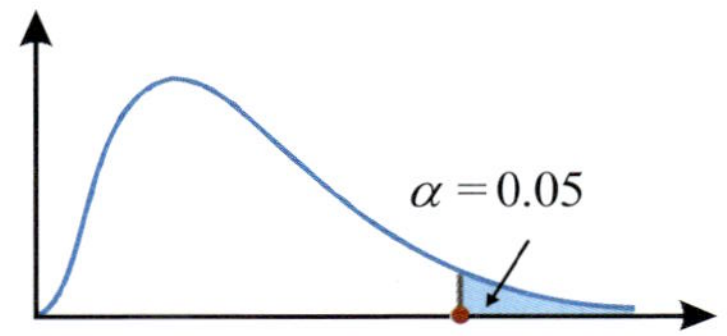

υ_2 \ υ_1	1	2	3	4	5	6	7	8	9	10	12	15	20	24	30	40
1	161.448	199.500	215.707	224.583	230.162	233.986	236.768	238.883	240.543	241.882	243.906	245 950	248.013	249.052	250.095	251.143
2	18.513	19.000	19.164	19.247	19.296	19.330	19.353	19.371	19.385	19.396	19.413	19 429	19.446	19.454	19.462	19.471
3	10.128	9.552	9.277	9.117	9.013	8.941	8.887	8.845	8.812	8.786	8.745	8 703	8.660	8.639	8.617	8.594
4	7.709	6.944	6.591	6.388	6.256	6.163	6.094	6.041	5.999	5.964	5.912	5 858	5.803	5.774	5.746	5.717
5	6.608	5.786	5.409	5.192	5.050	4.950	4.876	4.818	4.772	4.735	4.678	4 619	4.558	4.527	4.496	4.464
6	5.987	5.143	4.757	4.534	4.387	4.284	4.207	4.147	4.099	4.060	4.000	3.938	3.874	3.841	3.808	3.774
7	5.591	4.737	4.347	4.120	3.972	3.866	3.787	3.726	3.677	3.637	3.575	3.511	3.445	3.410	3.376	3.340
8	5.318	4.459	4.066	3.838	3.687	3.581	3.500	3.438	3.388	3.347	3.284	3.218	3.150	3.115	3.079	3.043
9	5.117	4.256	3.863	3.633	3.482	3.374	3.293	3.230	3.179	3.137	3.073	3.006	2.936	2.900	2.864	2.826
10	4.965	4.103	3.708	3.478	3.326	3.217	3.135	3.072	3.020	2.978	2.913	2.845	2.774	2.737	2.700	2.661
11	4.844	3.982	3.587	3.357	3.204	3.095	3.012	2.948	2.896	2.854	2.788	2.719	2.646	2.609	2.570	2.531
12	4.747	3.885	3.490	3.259	3.106	2.996	2.913	2.849	2.796	2.753	2.687	2.617	2.544	2.505	2.466	2.426
13	4.667	3.806	3.411	3.179	3.025	2.915	2.832	2.767	2.714	2.671	2.604	2.533	2.459	2.420	2.380	2.339
14	4.600	3.739	3.344	3.112	2.958	2.848	2.764	2.699	2.646	2.602	2.534	2.463	2.388	2.349	2.308	2.266
15	4.543	3.682	3.287	3.056	2.901	2.790	2.707	2.641	2.588	2.544	2.475	2.403	2.328	2.288	2.247	2.204
16	4.494	3.634	3.239	3.007	2.852	2.741	2.657	2.591	2.538	2.494	2.425	2.352	2.276	2.235	2.194	2.151
17	4.451	3.592	3.197	2.965	2.810	2.699	2.614	2.548	2.494	2.450	2.381	2.308	2.230	2.190	2.148	2.104
18	4.414	3.555	3.160	2.928	2.773	2.661	2.577	2.510	2.456	2.412	2.342	2.269	2.191	2.150	2.107	2.063
19	4.381	3.522	3.127	2.895	2.740	2.628	2.544	2.477	2.423	2.378	2.308	2.234	2.155	2.114	2.071	2.026
20	4.351	3.493	3.098	2.866	2.711	2.599	2.514	2.447	2.393	2.348	2.278	2.203	2.124	2.082	2.039	1.994
21	4.325	3.467	3.072	2.840	2.685	2.573	2.488	2.420	2.366	2.321	2.250	2.176	2.096	2.054	2.010	1.965
22	4.301	3.443	3.049	2.817	2.661	2.549	2.464	2.397	2.342	2.297	2.226	2.151	2.071	2.028	1.984	1.938
23	4.279	3.422	3.028	2.796	2.640	2.528	2.442	2.375	2.320	2.275	2.204	2.128	2.048	2.005	1.961	1.914
24	4.260	3.403	3.009	2.776	2.621	2.508	2.423	2.355	2.300	2.255	2.183	2.108	2.027	1.984	1.939	1.892
25	4.242	3.385	2.991	2.759	2.603	2.490	2.405	2.337	2.282	2.236	2.165	2.089	2.007	1.964	1.919	1.872
30	4.171	3.316	2.922	2.690	2.534	2.421	2.334	2.266	2.211	2.165	2.092	2.015	1.932	1.887	1.841	1.792
40	4.085	3.232	2.839	2.606	2.449	2.336	2.249	2.180	2.124	2.077	2.003	1.924	1.839	1.793	1.744	1.693
60	4.001	3.150	2.758	2.525	2.368	2.254	2.167	2.097	2.040	1.993	1.917	1.836	1.748	1.700	1.649	1.594
120	3.920	3.072	2.680	2.447	2.290	2.175	2.087	2.016	1.959	1.910	1.834	1.750	1.659	1.608	1.554	1.495
∞	3.841	2.996	2.605	2.372	2.214	2.099	2.010	1.938	1.880	1.831	1.752	1.666	1.571	1.517	1.459	1.394

TABLE 6(D)

CRITICAL VALUES OF THE VARIANCE RATIO *F*-DISTRIBUTION (RIGHT-TAIL AREAS) $\alpha = 0.10$

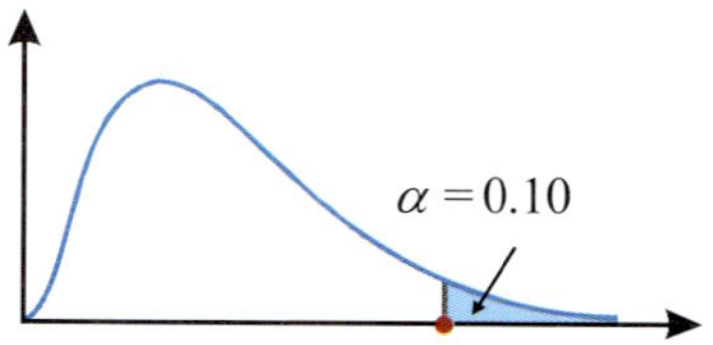

υ_2 \ υ_1	1	2	3	4	5	6	7	8	9	10	12	15	20	24	30	40
1	39.863	49.500	53.593	55.833	57.240	58.204	58.906	59.439	59.858	60.195	60.705	61.220	61.740	62.002	62.265	62.529
2	8.526	9.000	9.162	9.243	9.293	9.326	9.349	9.367	9.381	9.392	9.408	9.425	9.441	9.450	9.458	9.466
3	5.538	5.462	5.391	5.343	5.309	5.285	5.266	5.252	5.240	5.230	5.216	5.200	5.184	5.176	5.168	5.160
4	4.545	4.325	4.191	4.107	4.051	4.010	3.979	3.955	3.936	3.920	3.896	3.870	3.844	3.831	3.817	3.804
5	4.060	3.780	3.619	3.520	3.453	3.405	3.368	3.339	3.316	3.297	3.268	3.238	3.207	3.191	3.174	3.157
6	3.776	3.463	3.289	3.181	3.108	3.055	3.014	2.983	2.958	2.937	2.905	2.871	2.836	2.818	2.800	2.781
7	3.589	3.257	3.074	2.961	2.883	2.827	2.785	2.752	2.725	2.703	2.668	2.632	2.595	2.575	2.555	2.535
8	3.458	3.113	2.924	2.806	2.726	2.668	2.624	2.589	2.561	2.538	2.502	2.464	2.425	2.404	2.383	2.361
9	3.360	3.006	2.813	2.693	2.611	2.551	2.505	2.469	2.440	2.416	2.379	2.340	2.298	2.277	2.255	2.232
10	3.285	2.924	2.728	2.605	2.522	2.461	2.414	2.377	2.347	2.323	2.284	2.244	2.201	2.178	2.155	2.132
11	3.225	2.860	2.660	2.536	2.451	2.389	2.342	2.304	2.274	2.248	2.209	2.167	2.123	2.100	2.076	2.052
12	3.177	2.807	2.606	2.480	2.394	2.331	2.283	2.245	2.214	2.188	2.147	2.105	2.060	2.036	2.011	1.986
13	3.136	2.763	2.560	2.434	2.347	2.283	2.234	2.195	2.164	2.138	2.097	2.053	2.007	1.983	1.958	1.931
14	3.102	2.726	2.522	2.395	2.307	2.243	2.193	2.154	2.122	2.095	2.054	2.010	1.962	1.938	1.912	1.885
15	3.073	2.695	2.490	2.361	2.273	2.208	2.158	2.119	2.086	2.059	2.017	1.972	1.924	1.899	1.873	1.845
16	3.048	2.668	2.462	2.333	2.244	2.178	2.128	2.088	2.055	2.028	1.985	1.940	1.891	1.866	1.839	1.811
17	3.026	2.645	2.437	2.308	2.218	2.152	2.102	2.061	2.028	2.001	1.958	1.912	1.862	1.836	1.809	1.781
18	3.007	2.624	2.416	2.286	2.196	2.130	2.079	2.038	2.005	1.977	1.933	1.887	1.837	1.810	1.783	1.754
19	2.990	2.606	2.397	2.266	2.176	2.109	2.058	2.017	1.984	1.956	1.912	1.865	1.814	1.787	1.759	1.730
20	2.975	2.589	2.380	2.249	2.158	2.091	2.040	1.999	1.965	1.937	1.892	1.845	1.794	1.767	1.738	1.708
21	2.961	2.575	2.365	2.233	2.142	2.075	2.023	1.982	1.948	1.920	1.875	1.827	1.776	1.748	1.719	1.689
22	2.949	2.561	2.351	2.219	2.128	2.060	2.008	1.967	1.933	1.904	1.859	1.811	1.759	1.731	1.702	1.671
23	2.937	2.549	2.339	2.207	2.115	2.047	1.995	1.953	1.919	1.890	1.845	1.796	1.744	1.716	1.686	1.655
24	2.927	2.538	2.327	2.195	2.103	2.035	1.983	1.941	1.906	1.877	1.832	1.783	1.730	1.702	1.672	1.641
25	2.918	2.528	2.317	2.184	2.092	2.024	1.971	1.929	1.895	1.866	1.820	1.771	1.718	1.689	1.659	1.627
30	2.881	2.489	2.276	2.142	2.049	1.980	1.927	1.884	1.849	1.819	1.773	1.722	1.667	1.638	1.606	1.573
40	2.835	2.440	2.226	2.091	1.997	1.927	1.873	1.829	1.793	1.763	1.715	1.662	1.605	1.574	1.541	1.506
60	2.791	2.393	2.177	2.041	1.946	1.875	1.819	1.775	1.738	1.707	1.657	1.603	1.543	1.511	1.476	1.437
120	2.748	2.347	2.130	1.992	1.896	1.824	1.767	1.722	1.684	1.652	1.601	1.545	1.482	1.447	1.409	1.368
∞	2.706	2.303	2.084	1.945	1.847	1.774	1.717	1.670	1.632	1.599	1.546	1.487	1.421	1.383	1.342	1.295

TABLE 7

CRITICAL VALUES OF KARL PEARSON'S CORRELATION COEFFICIENT *r* FOR TWO-TAILED TEST

n \ α	**0.05**	**0.01**
4	0.950	0.999
5	0.878	0.959
6	0.811	0.917
7	0.754	0.875
8	0.707	0.834
9	0.666	0.798
10	0.632	0.765
11	0.602	0.735
12	0.576	0.708
13	0.553	0.684
14	0.532	0.661
15	0.514	0.641
16	0.497	0.623
17	0.482	0.606
18	0.468	0.590
19	0.456	0.575
20	0.444	0.561
25	0.396	0.505
30	0.361	0.463
35	0.335	0.430
40	0.312	0.402
45	0.294	0.378
50	0.279	0.361
60	0.254	0.330
70	0.236	0.305
80	0.220	0.286
90	0.207	0.269
100	0.196	0.256

TABLE 8

CRITICAL VALUES FOR THE SIGN TEST

α / (n)	0.005 (one tail)	0.01 (one tail)	0.025 (one tail)	0.05 (one tail)
	0.01 (two tails)	0.02 (two tails)	0.05 (two tails)	0.10 (two tails)
1	*	*	*	*
2	*	*	*	*
3	*	*	*	*
4	*	*	*	*
5	*	*	*	0
6	*	*	0	0
7	*	0	0	0
8	0	0	0	1
9	0	0	1	1
10	0	0	1	1
11	0	1	1	2
12	1	1	2	2
13	1	1	2	3
14	1	2	2	3
15	2	2	3	3
16	2	2	3	4
17	2	3	4	4
18	3	3	4	5
19	3	4	4	5
20	3	4	5	5
21	4	4	5	6
22	4	5	5	6
23	4	5	6	7
24	5	5	6	7
25	5	6	7	7

Test statistic X = Number of the less frequent sign.

Reject H_0 if X is less than or equal to the critical value in the above table.

TABLE 9

CRITICAL VALUES OF *T* FOR THE WILCOXON SIGNED-RANK TEST

α / (n)	0.005 (one tail)	0.01 (one tail)	0.025 (one tail)	0.05 (one tail)
	0.01 (two tails)	0.02 (two tails)	0.05 (two tails)	0.10 (two tails)
5	*	*	*	1
6	*	*	1	2
7	*	0	2	4
8	0	2	4	6
9	2	3	6	8
10	3	5	8	11
11	5	7	11	14
12	7	10	14	17
13	10	13	17	21
14	13	16	21	26
15	16	20	25	30
16	19	24	30	36
17	23	28	35	41
18	28	33	40	47
19	32	38	46	54
20	37	43	52	60
21	43	49	59	68
22	49	56	66	75
23	55	62	73	83
24	61	69	81	92
25	68	77	90	101
26	76	85	98	110
27	84	93	107	120
28	92	102	117	130
29	100	111	127	141
30	109	120	137	152

Notes:

1. * Indicates that it is not possible to get a value in the critical region.

2. Reject H_0 if the test statistic T is less than or equal to the critical value given in the above table.

TABLE 10 (A)

CRITICAL VALUES OF *U* IN THE MANN-WHITNEY TEST

A. The entries are the critical values of U for a one-tailed test at $\alpha = 0.025$ or for a two-tailed test at $\alpha = 0.05$.

n_2 \ n_1	1	2	3	4	5	6	7	8	9	10	11	12	13	14	15	16	17	18	19	20
1																				
2								0	0	0	0	1	1	1	1	1	2	2	2	2
3					0	1	1	2	2	3	3	4	4	5	5	6	6	7	7	8
4				0	1	2	3	4	4	5	6	7	8	9	10	11	11	12	13	13
5			0	1	2	3	5	6	7	8	9	11	12	13	14	15	17	18	19	20
6			1	2	3	5	6	8	10	11	13	14	16	17	19	21	22	24	25	27
7			1	3	5	6	8	10	12	14	16	18	20	22	24	26	28	30	32	34
8		0	2	4	6	8	10	13	15	17	19	22	24	26	29	31	34	36	38	41
9		0	2	4	7	10	12	15	17	20	23	26	28	31	34	37	39	42	45	48
10		0	3	5	8	11	14	17	20	23	26	29	33	36	39	42	45	48	52	55
11		0	3	6	9	13	16	19	23	26	30	33	37	40	44	47	51	55	58	62
12		1	4	7	11	14	18	22	26	29	33	37	41	45	49	53	57	61	65	69
13		1	4	8	12	16	20	24	28	33	37	41	45	50	54	59	63	67	72	76
14		1	5	9	13	17	22	26	31	36	40	45	50	55	59	64	67	74	78	83
15		1	5	10	14	19	24	29	34	39	44	49	54	59	64	70	75	80	85	90
16		1	6	11	15	21	26	31	37	42	47	53	59	64	70	75	81	86	92	98
17		2	6	11	17	22	28	34	39	45	51	57	63	67	75	81	87	93	99	105
18		2	7	12	18	24	30	36	42	48	55	61	67	74	80	86	93	99	106	112
19		2	7	13	19	25	32	38	45	52	58	65	72	78	85	92	99	106	113	119
20		2	8	13	20	27	34	41	48	55	62	69	76	83	90	98	105	112	119	127

TABLE 10(B)

CRITICAL VALUES OF *U* IN THE MANN-WHITNEY TEST

B. The entries are the critical values of U for a one-tailed test at α = 0.05 or for a two-tailed test at α = 0.10.

n_2 \ n_1	1	2	3	4	5	6	7	8	9	10	11	12	13	14	15	16	17	18	19	20
1																			0	0
2					0	0	0	1	1	1	1	2	2	2	3	3	3	4	4	4
3			0	0	1	2	2	3	3	4	5	5	6	7	7	8	9	9	10	11
4			0	1	2	3	4	5	6	7	8	9	10	11	12	14	15	16	17	18
5		0	1	2	4	5	6	8	9	11	12	13	15	16	18	19	20	22	23	25
6		0	2	3	5	7	8	10	12	14	16	17	19	21	23	25	26	28	30	32
7		0	2	4	6	8	11	13	15	17	19	21	24	26	28	30	33	35	37	39
8		1	3	5	8	10	13	15	18	20	23	26	28	31	33	36	39	41	44	47
9		1	3	6	9	12	15	18	21	24	27	30	33	36	39	42	45	48	51	54
10		1	4	7	11	14	17	20	24	27	31	34	37	41	44	48	51	55	58	62
11		1	5	8	12	16	19	23	27	31	34	38	42	46	50	54	57	61	65	69
12		2	5	9	13	17	21	26	30	34	38	42	47	51	55	60	64	68	72	77
13		2	6	10	15	19	24	28	33	37	42	47	51	56	61	65	70	75	80	84
14		2	7	11	16	21	26	31	36	41	46	51	56	61	66	71	77	82	87	92
15		3	7	12	18	23	28	33	39	44	50	55	61	66	72	77	83	88	94	100
16		3	8	14	19	25	30	36	42	48	54	60	65	71	77	83	89	95	101	107
17		3	9	15	20	26	33	39	45	51	57	64	70	77	83	89	96	102	109	115
18		4	9	16	22	28	35	41	48	55	61	68	75	82	88	95	102	109	116	123
19	0	4	10	17	23	30	37	44	51	58	65	72	80	87	94	101	109	116	123	130
20	0	4	11	18	25	32	39	47	54	62	69	77	84	92	100	107	115	123	130	138

TABLE 11

CRITICAL VALUES OF SPEARMAN'S RANK CORRELATION COEFFICIENT r_{sp}

α / (n)	0.05 (one tail)	0.025 (one tail)	0.01 (one tail)	0.005 (one tail)
	0.10 (two tails)	0.05 (two tails)	0.02 (two tails)	0.01 (two tails)
5	0.900	*	*	*
6	0.829	0.886	0.943	*
7	0.714	0.786	0.893	*
8	0.643	0.738	0.833	0.881
9	0.600	0.683	0.783	0.833
10	0.564	0.648	0.745	0.794
11	0.523	0.623	0.736	0.818
12	0.497	0.591	0.703	0.780
13	0.475	0.566	0.673	0.745
14	0.457	0.545	0.646	0.716
15	0.441	0.525	0.623	0.689
16	0.425	0.507	0.601	0.666
17	0.412	0.492	0.582	0.645
18	0.399	0.476	0.564	0.625
19	0.388	0.462	0.549	0.608
20	0.377	0.450	0.534	0.591
21	0.368	0.438	0.521	0.576
22	0.359	0.428	0.508	0.562
23	0.351	0.418	0.496	0.549
24	0.343	0.409	0.485	0.537
25	0.336	0.400	0.475	0.526
26	0.329	0.392	0.465	0.515
27	0.323	0.385	0.456	0.505
28	0.317	0.377	0.448	0.496
29	0.311	0.370	0.440	0.487
30	0.305	0.364	0.432	0.478

ANSWERS

ANSWERS

Chapter 1: General Exercises

5. **(a)** N = 281, n = 23 **(b)** N = 214, n = 15 **(c)** N = 32, n = 4 **(d)** N = 512, n = 57 **(e)** N = 341, n = 19

7. **(a)** Quantitative **(b)** Qualitative **(c)** Quantitative **(d)** Qualitative **(e)** Quantitative

Chapter 2: Try it yourself

Section 2.2 A

1.

X	Tally bars	f
0	|||	3
1	𝍸 𝍸 𝍸 |	16
2	𝍸 𝍸 ||||	14
3	𝍸 |||	8
4	𝍸 ||||	9

3.

Classes	Frequency
0	8
1	14
2	8
3	8
4	7
5	3
6	2

Section 2.2 B

1.

Classes	Frequency
0-20	13
20-40	11
40-60	44
60-80	64
80-101	78

3.

Classes	Frequency
0-15	18
15-30	14
30-45	12
45-60	6

5.

Classes	Frequency
0-6	3
6-12	13
12-18	8
18-24	9
24-30	3
30-36	9
36-42	10
42-48	5

Section 2.2 C

1.(a)

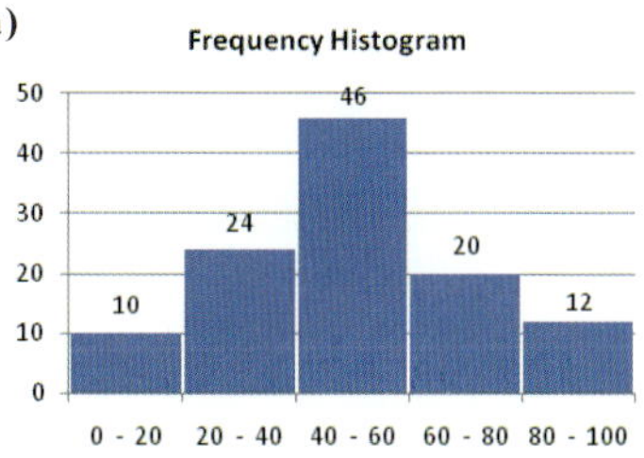

(b)

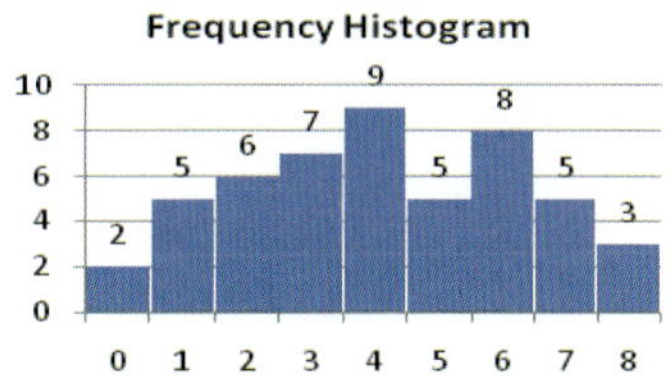

(c)

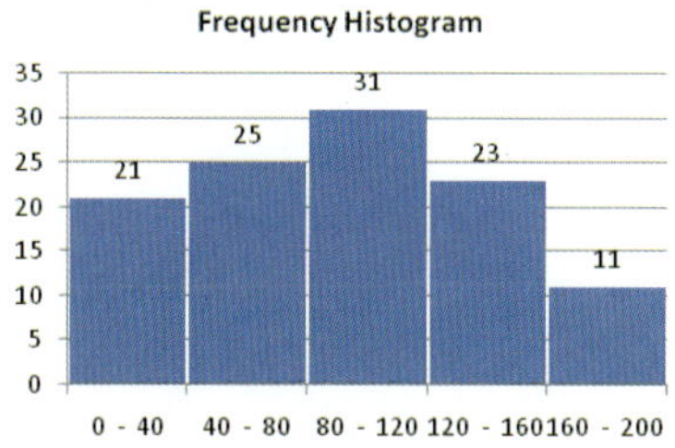

Section 2.2 D

1.(a)

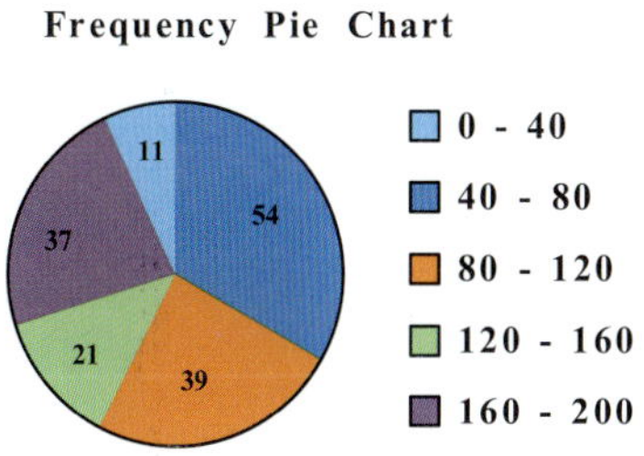

(b)

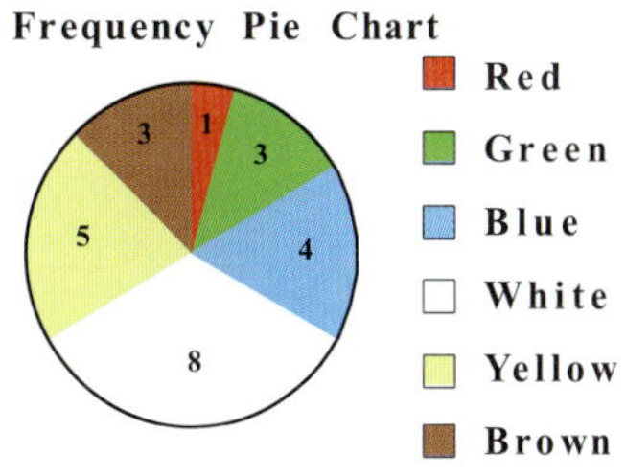

(c)

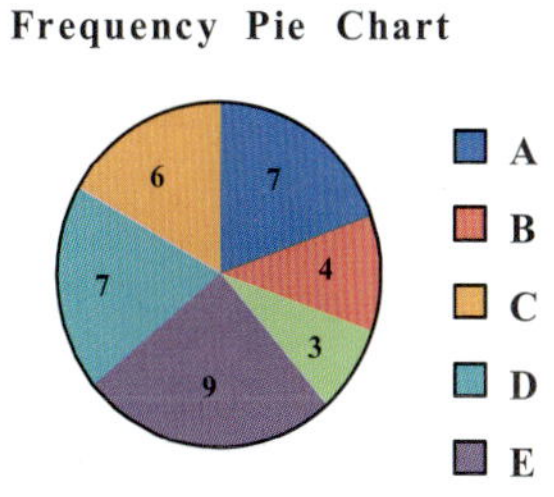

Section 2.3 A

1.

Class	f	Relative Frequency	Cumulative Frequency
0 - 40	27	0.25	27
40 - 80	10	0.9	37
80 - 120	21	0.20	58
120 - 160	37	0.35	95
160 - 200	11	0.10	106

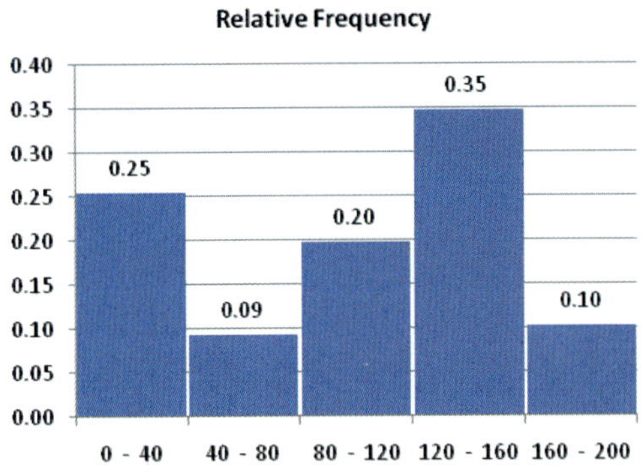

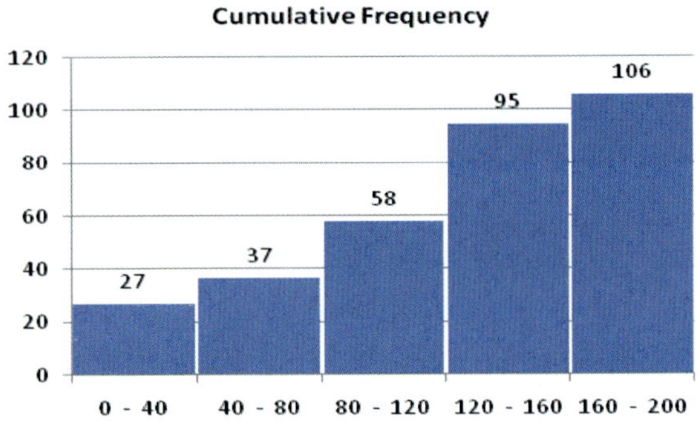

3. (a)

Class	f	Relative Frequency	Cumulative Frequency
0 - 10	12	0.20	12
10 - 20	17	0.28	29
20 - 30	20	0.33	49
30 - 40	8	0.13	57
40 - 50	3	0.05	60

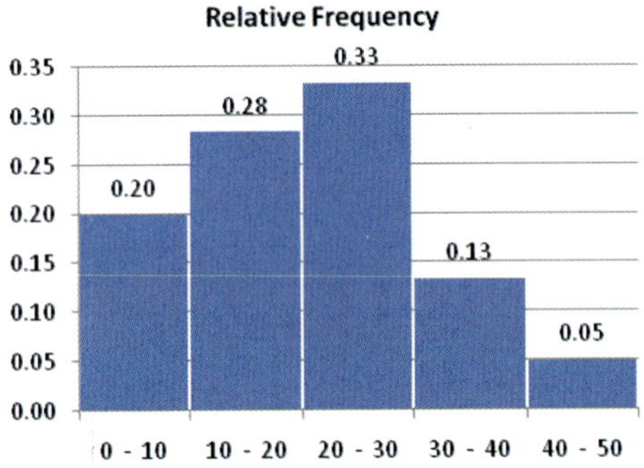

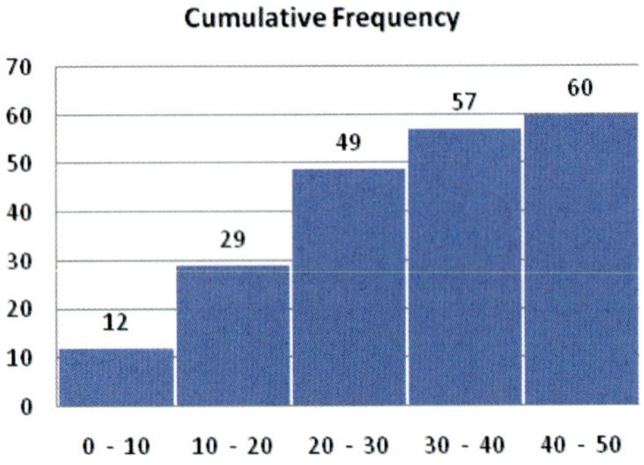

3. (b)

Class	f	Relative Frequency	Cumulative Frequency
30 - 40	10	0.10	10
40 - 50	17	0.17	27
50 - 60	47	0.46	74
60 - 70	20	0.19	94
70 - 80	9	0.09	103

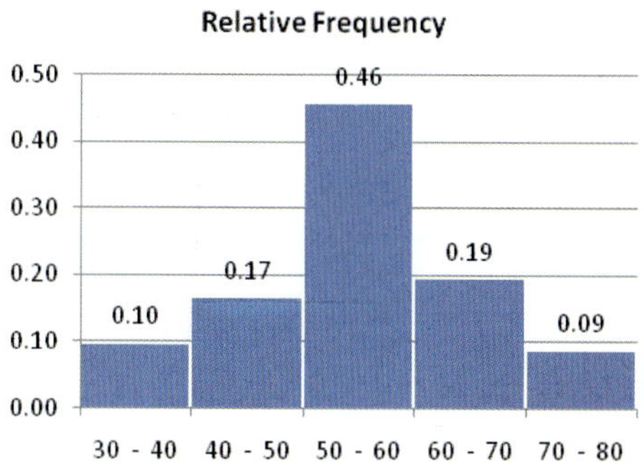

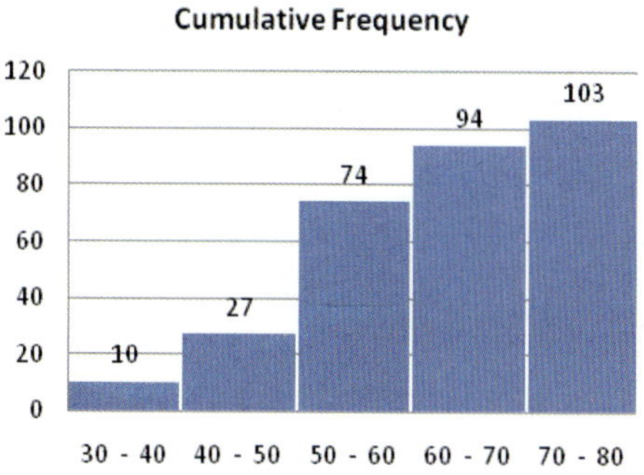

Section 2.4 A

1. (a)

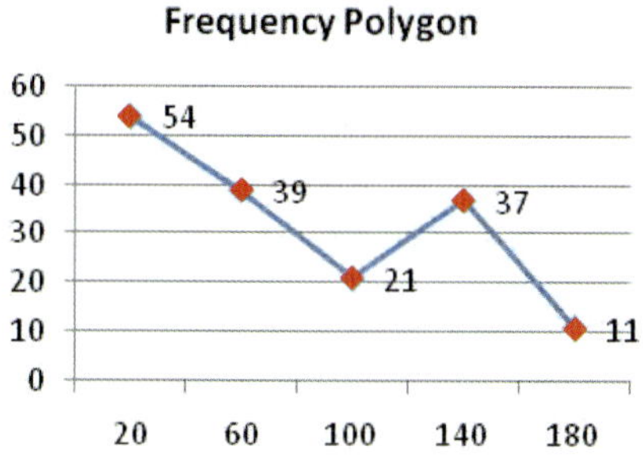

(b)

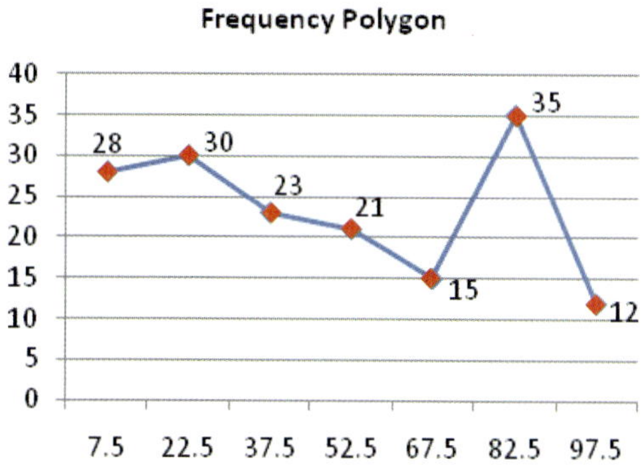

(c)

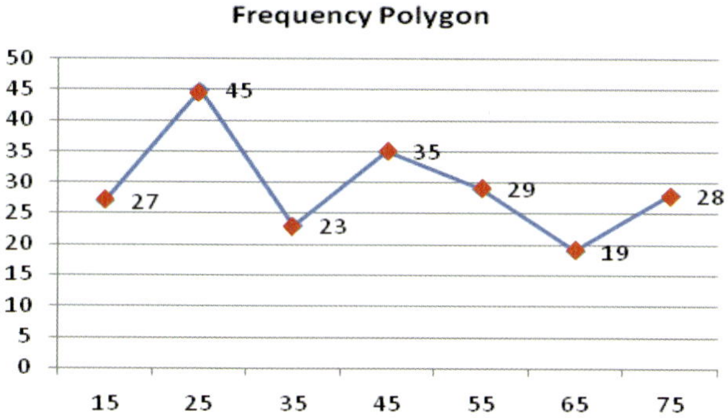

Section 2.4 B

1. (a)

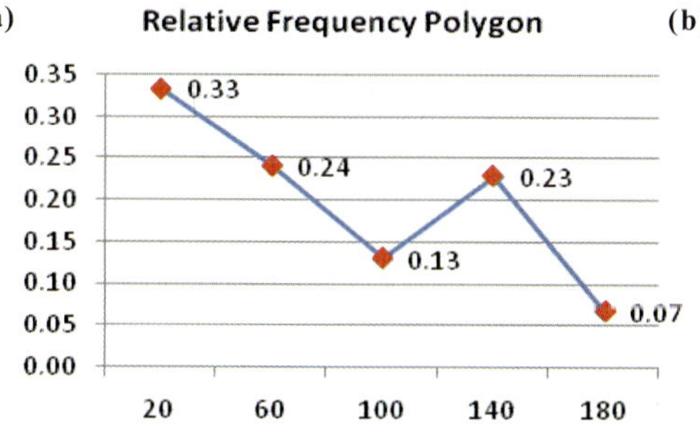

(b)

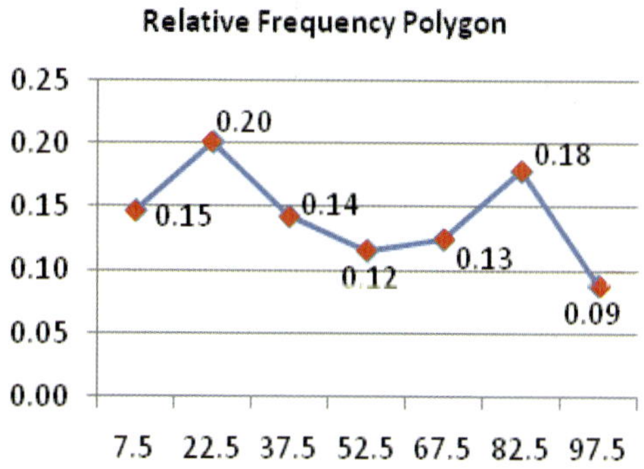

(c)

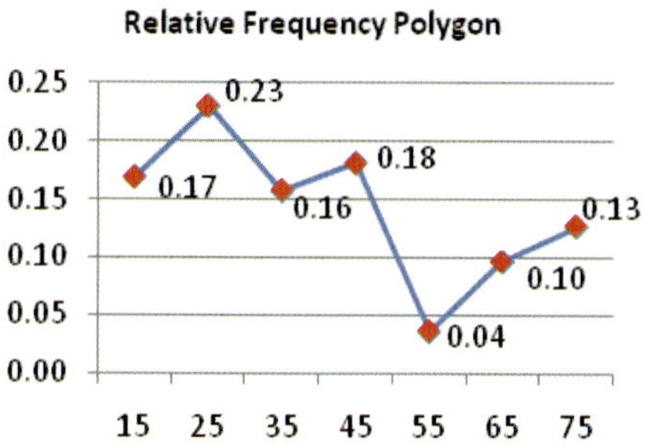

Section 2.4 C

1. (a)

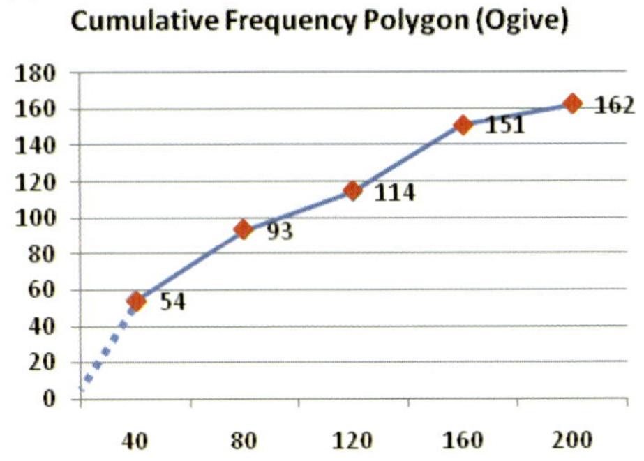

(b)

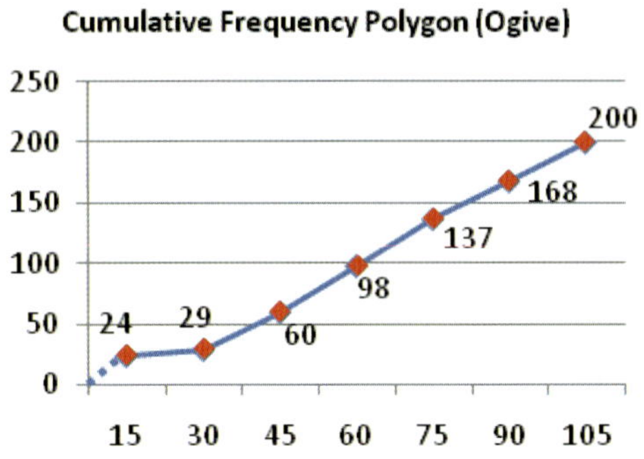

(c)

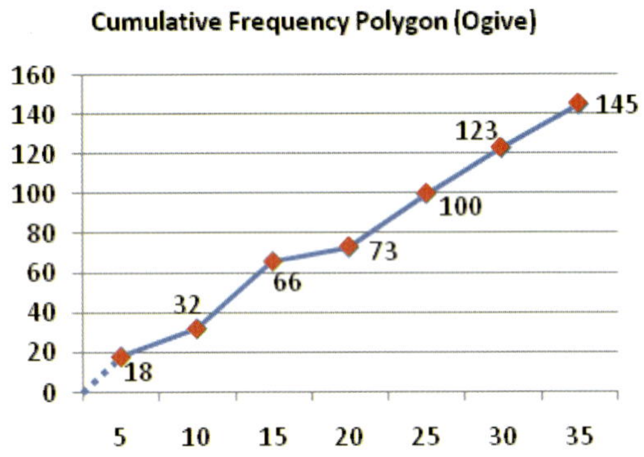

Section 2.5 A

1. (a)

Stem	Leaf
0	7
1	0
3	5
4	5 6
5	0 0 2 4
6	3 5 5 8 8
7	0 0 5 5 8
8	0 0 1 3 5 7 8
9	0 0 3 5 5 8
10	0 0
11	5 7

(b)

Stem	Leaf
0	0 2 3 6 8 8 8
1	0 3 3 5 7 7
2	0 3 6 7 8 9
3	2 3 4 6 8 9 9 9 9
4	1 5 9
5	1 1 2 5 6
6	0 2 4 6
7	0 6 6 7 9 6 7
8	0 0 1 3 8

(c)

Stem	Leaf
0	4 5 6 7
1	3 9
2	0 0 1 9
3	2 5 5 7 7
4	0 1 3 3 4 4 5
5	0 2 2 4 4 5
6	5 7
7	0 4 4 5 6 6 7
8	2 3 5

Section 2.6 B

1. 23.5 **3.** 5

Section 2.6 C

1. (a) Variance = 125.05 ; SD = 11.18 **(b)** Variance = 138.94 ; SD = 11.79

Section 2.6 D

1. (a) Range = (12.02, 33.18) **(b)** Range = (−0.76, 37.48) **(c)** Range = (−3.97, 108.23)

Section 2.7 A

1. **(a)** Mean = 19.38 ; Median = 21.15 ; Mode = 7.5 **(b)** Mean = 36.30 ; Median = 35.9 ; Mode = 15

3. **(a)** Mean = 54.94 ; Median = 55.48 ; Mode = 55 ; Variance = 672.99 ; SD = 25.94

(b) Mean = 41.55 ; Median = 41.42 ; Mode = 42.5 ; Variance = 55.47 ; SD = 7.45

Section 2.8 B

1. **(a)** 75% **(b)** 95.4% **3.** **(a)** 84% **(b)** 75% **(c)** 88.89% **(d)** 30.56%

Section 2.8 C

1. 1.5 ; usual

Section 2.8 E

1. **(a)** Q_1 = 17.5 ; Q_2 = 27.5 ; Q_3 = 38.5 **(b)** Q_1 = 17 ; Q_2 = 25 ; Q_3 = 34

Section 2.8 F

1. **(a)** 4 **(b)** 46 **(c)** 91th percentile

Section 2.9 A

1. **(a)** Min = 0 ; Q_1 = 6.5 ; Median = 19 ; Q_3 = 36 ; Max = 93

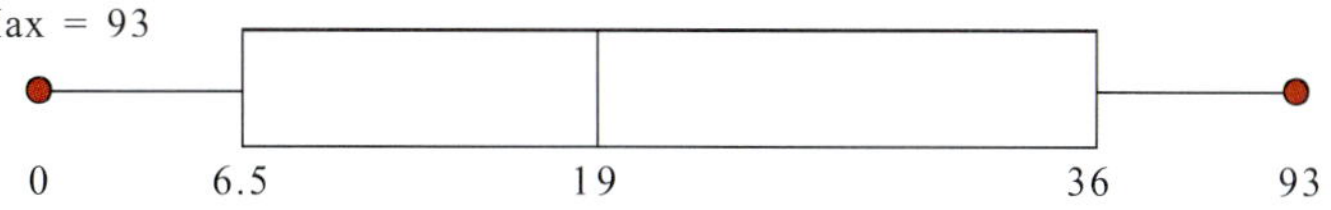

(b) Min = 10 ; Q_1 = 18 ; Median = 29 ; Q_3 = 35 ; Max = 42

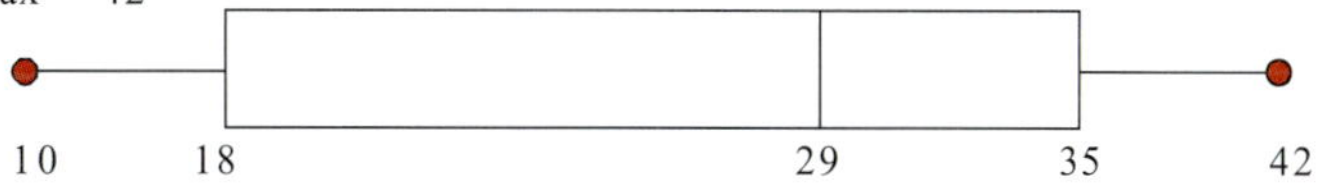

(c) Min = 10 ; Q_1 = 18 ; Median = 29 ; Q_3 = 37 ; Max = 44

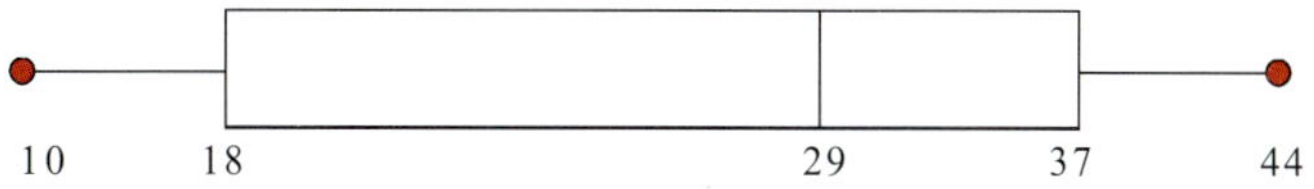

1.

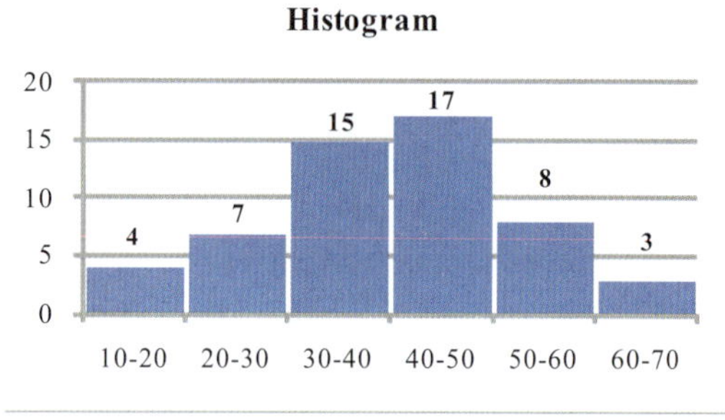

Classes	Frequency	RF	CF
10-20	4	0.07	4
20-30	7	0.13	11
30-40	15	0.28	26
40-50	17	0.31	43
50-60	8	0.15	51
60-70	3	0.06	54
	54		

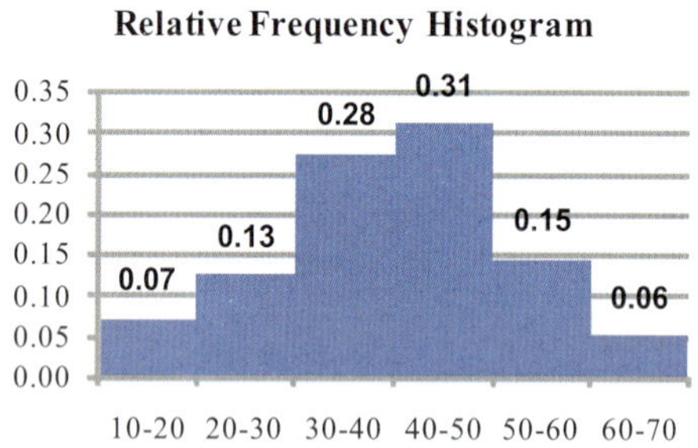

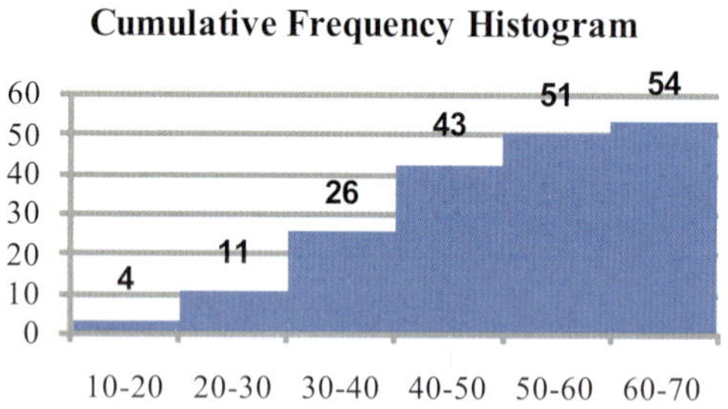

3.

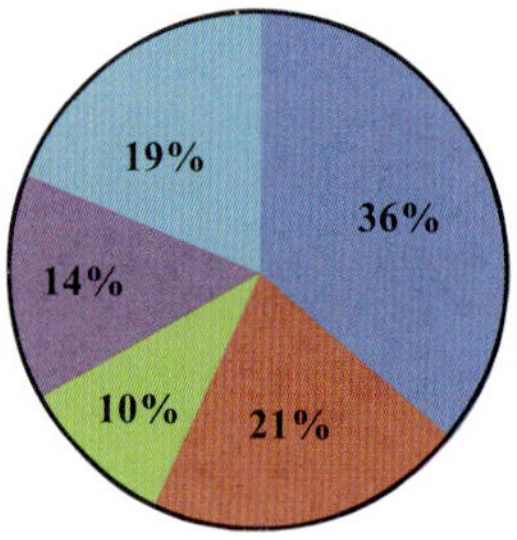

5.

Classes	Frequency
20-50	2
50-80	4
80-110	22
110-140	28
140-170	4

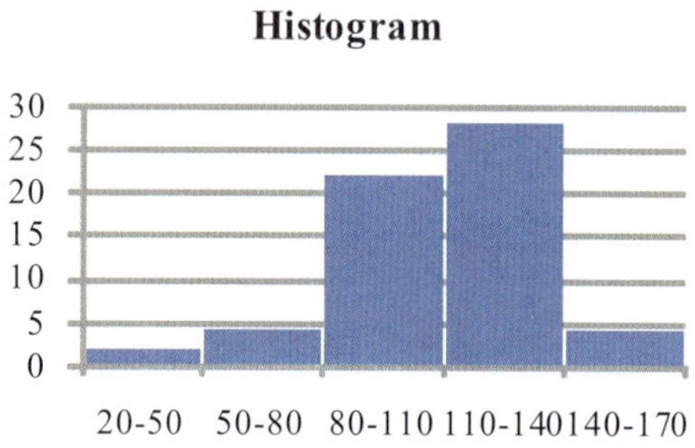

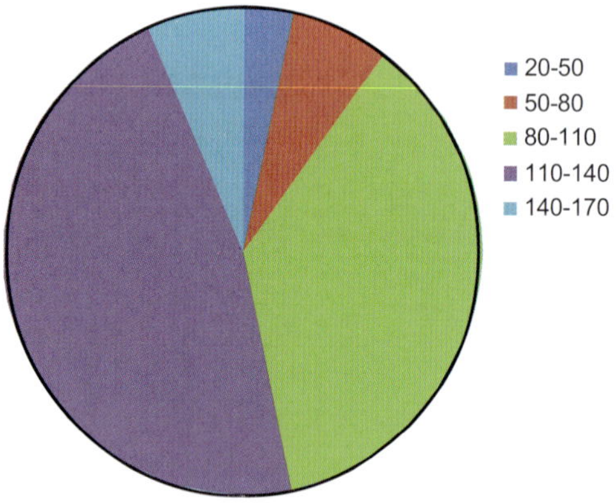

7.

Classes	Frequency
20-50	2
50-80	6
80-110	21
110-140	27
140-170	30

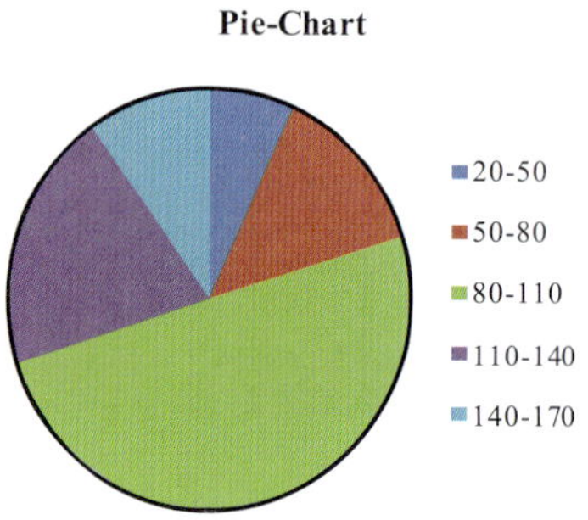

9. (a)

Classes	Frequency
0-20	20
20-40	12
40-60	5
60-80	1
80-100	2
	40

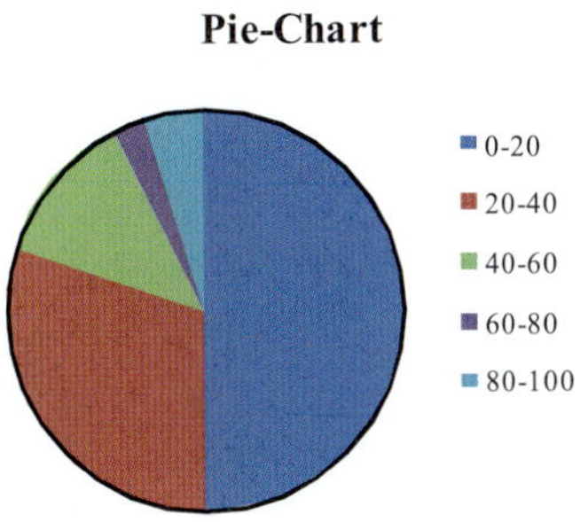

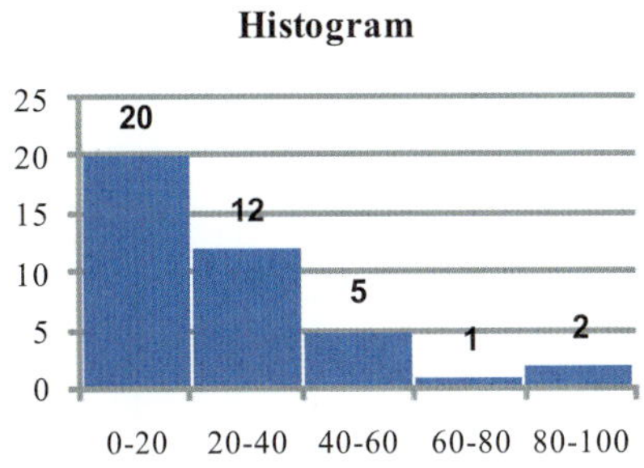

(b)

Classes	RF
0-20	0.5
20-40	0.3
40-60	0.125
60-80	0.025
80-100	0.05
	1

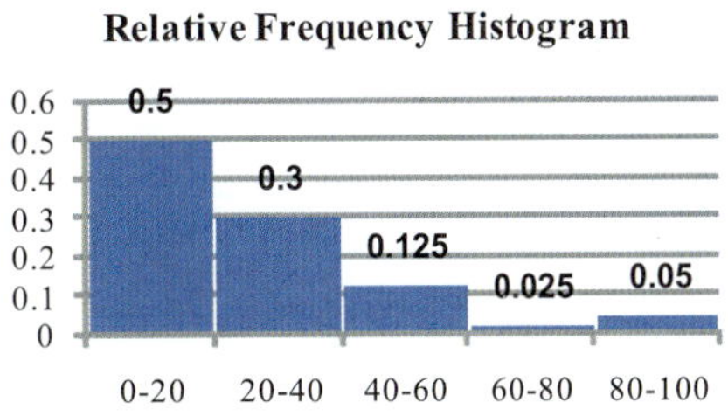

(c)

Classes	CF
0-20	20
20-40	32
40-60	37
60-80	38
80-100	40

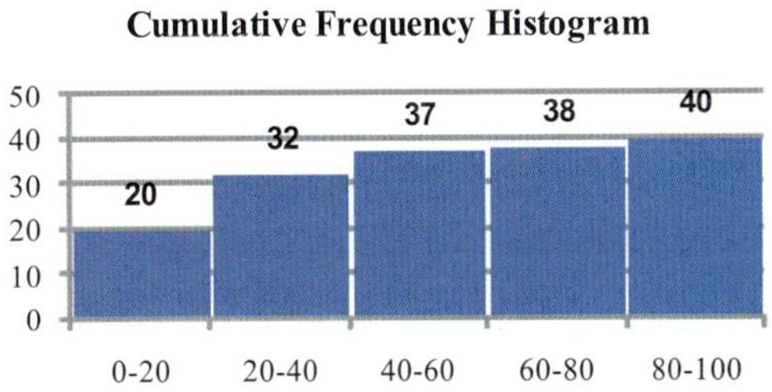

(d)

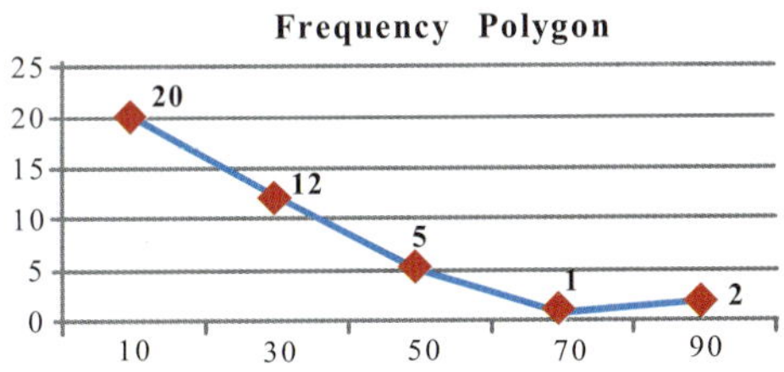

(e)

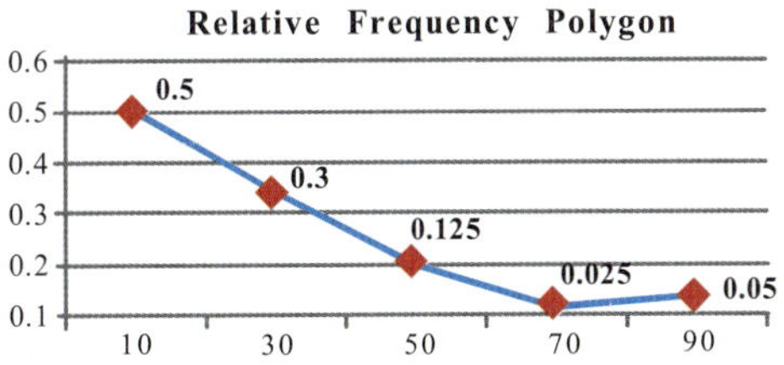

(f)

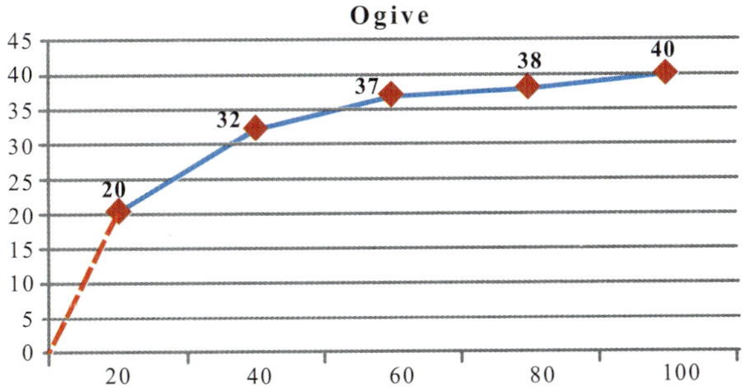

11. (a)

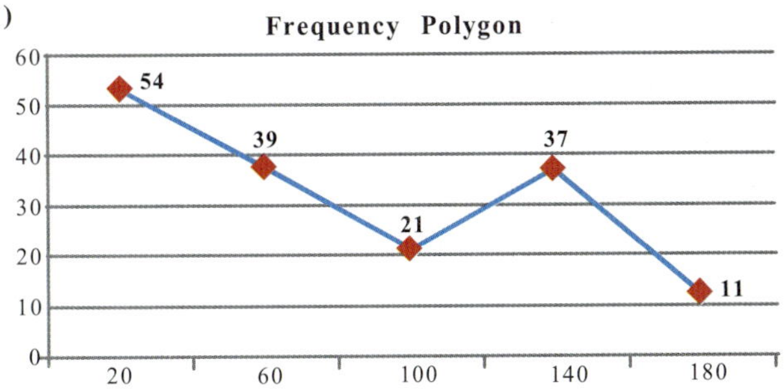

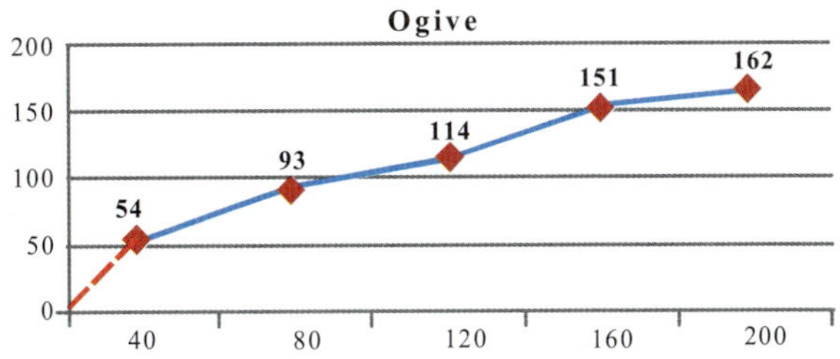

(b) None of these.

13. (a) and (b)

x	F	RF	CF
1	23	0.14	23
2	30	0.19	53
3	28	0.18	81
4	17	0.11	98
5	33	0.21	131
6	29	0.18	160

(c) **Frequency Histogram**

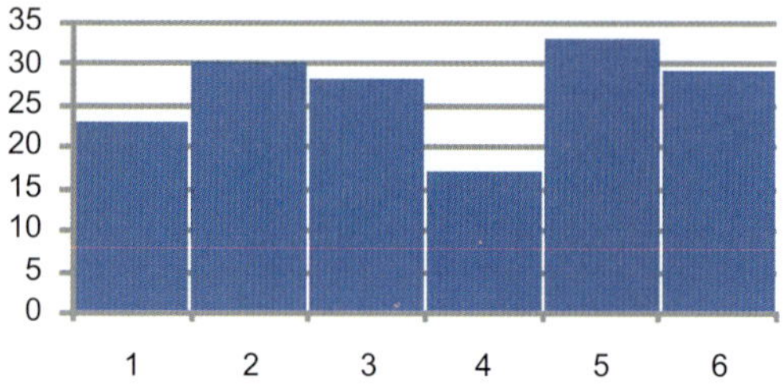

(d) **Frequency Polygon**

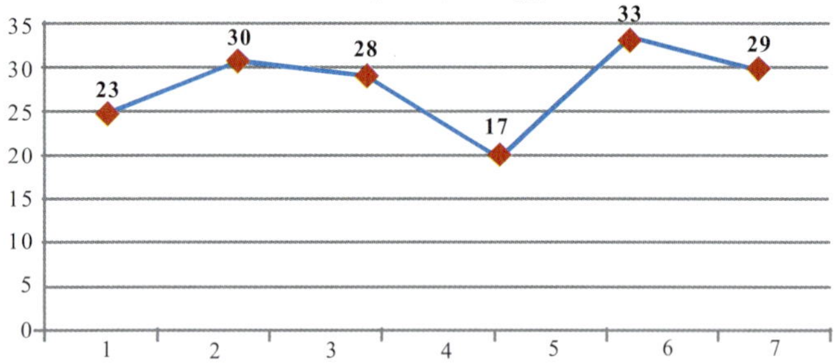

(e) **Relative Frequency Polygon**

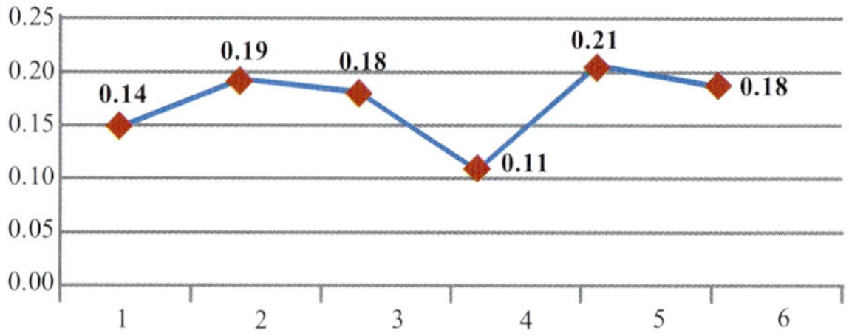

(f) **Cumulative Frequency Polygon (Ogive)**

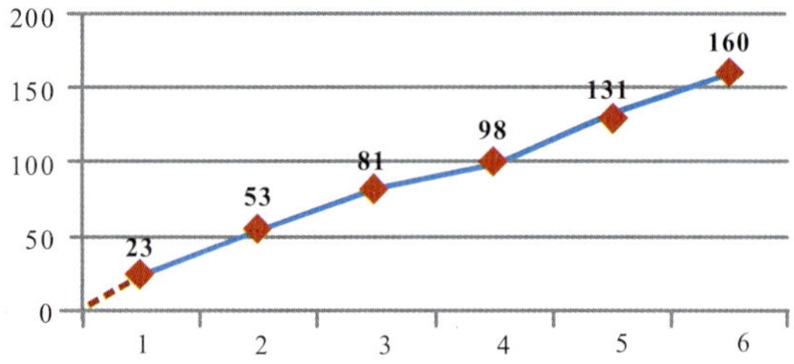

15.

0	1 1 1 1 1 1 1 2 5 6 7 7 7 8 8 9
1	0 1 2 3 5 5 6 7 8 9 9 9
2	1 2 2 2 3 7 8 8 8 9
3	0 0 0 1 1 2 4 6 7 9
4	0
5	
6	
7	
8	
9	
10	0 2 5
11	5 7
12	
13	3 8

17. **Stem and Leaf Plot**

4	1
5	
6	0 1 4 5 5 7 7 9
7	0 2 3 6 7 8 9
8	0 0 1 1 3 6 7 8 9 9
9	0 1 2 3 5 5 6 8 9
10	0 0 1 2 3 5 6 8 9
11	0 1 1 2 3 5 7 8
12	0 1 5 7 7 9
13	1 2 5 8 9
14	0
15	3

19. var = 522.2

21. (a)

x	F	RF
1	6	0.2
2	6	0.2
3	6	0.2
4	4	0.13
5	5	0.17
6	3	0.1

(b) Mean = 3.2 ; SD = 1.7

23. (a)

Classes	Frequency	Marked X
30,000-50,000	2	40000
50,000-70,000	9	60000
70,000-90,000	17	80000
90,000-110,000	15	100000
110,000-130,000	8	120000
130,000-150,000	4	140000
Total	55	

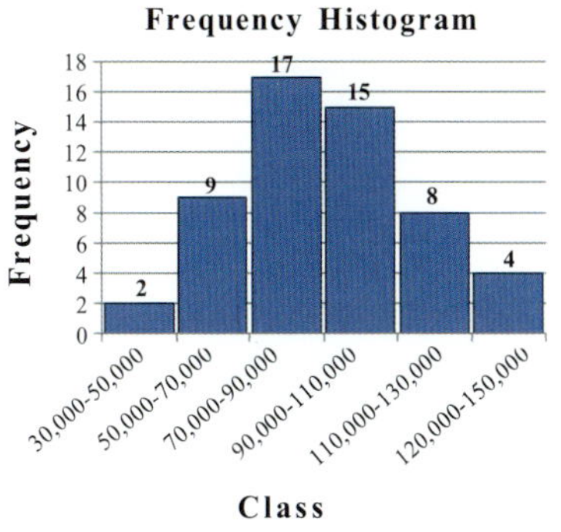

(b) Range = Maximum − Minimun = 141,367 − 30,273 = 111,094

Mean = (30,273 + ... + 141,367)/55 = 91,569

Standard Deviation = $\sqrt{\dfrac{(40,000-91,569)^2 \times 2 + \ldots + (140,000-91,569)^2 \times 4}{99}}$ =11293.42

Usual salary range = Mean ±2 × 11293.42 = (68982.367, 114156.05)

So the unusual salary range is lower than 68,982.367 or higher than 114,156.05.

(c) Since \$200,000 is higher than \$114,156.05, it is an unusual salary.

25. (a) Mean = 0.37 , SD = 0.29 ;

Range of usual values = (−0.21 , 0.95)

(b) z-score = −0.65

27. No (unusual)

29. (a) Mean IQ = 97.2

31. (a) Mean = 48.75 ; Variance = 314.6 ; SD = 17.7 ; Range of usual values = (13.27 , 84.23) ; Range rule of thumb = 14.3

All gasoline expense below 13.40 or above 84.20 are considered unusual.

(b) SD = 14.3

(c) $Z(43) = -0.33$

(d) Q1 = 35.7 , Median = 50.7 , Q3 = 63.3,

Range of outliers = All observations below −19.5 or above 118.5 are considered as outliers.

33. (a)

5 number summary

min	Q1	median	Q3	max
10	16.75	29.5	39.5	73

(b)

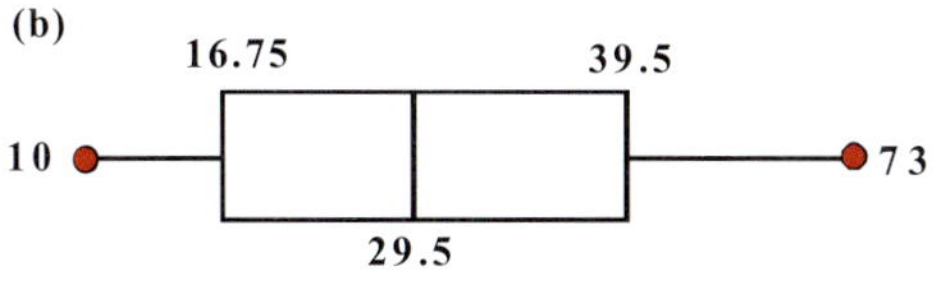

(c) $P_{58} = 32$

35. $P_{85} = 701$

37. (a) $P_{32} = 35$ **(b)** $P_{79} = 73$, $P_5 = 19$

(d) Mean = 50.9 , Median = 49 , Q1 = 30.5 , Q3 = 71 , Variance = 553.4519027, SD = 23.5 ,

Range of usual values = (3.9, 97.9) , Range of outliers = All data values below −48.8 or above 150.1 are considered as outliers.

39. (a)

Stem and Leaf Plot

0	3 5 5 6 7 7 7 8 8 8 8 8 8 8 9 9 9 9 9 9
1	0 0 0 0 1 2 2 2 2 2 3 3 4 5 5 5 5 5 6 6 6 6 7 8 9 9 9 9
2	0 0 1 3 4 4 5 7 8
3	5 6 8

(b) Mean = 14.5 , Variance = 59.5 , SD = 7.7

(c) Mean = 14.2 , Variance = 68.8 , SD = 8.3

(d) Range of usual values = (−0.9, 29.9) , Range of unusual values = All data values below −0.9 or above 29.9 are considered as unusual values.

(e) $P_{95} = 28$

(f)

5 number summary

min	Q1	median	Q3	max
3	9	12.5	19	38

41. **(a)**

5 number summary

min	Q1	median	Q3	max
9	21.5	29	40.5	62

(b)

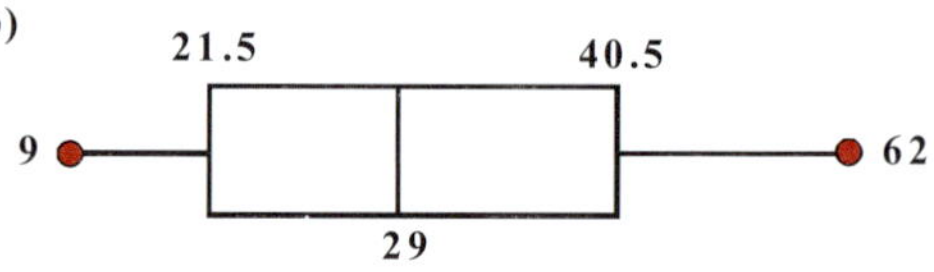

43. Mean = 22.6 , SD = 5.3

5 number summary

min	Q1	median	Q3	max
13	18	23	27	32

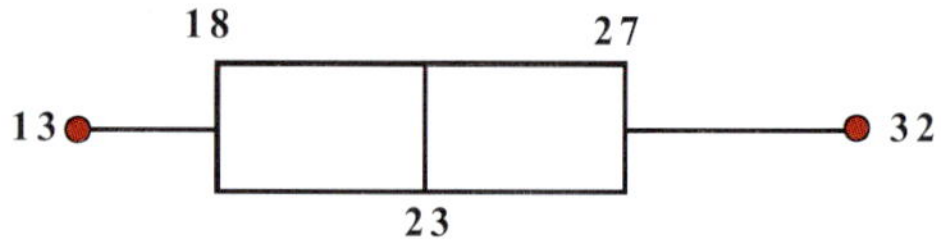

45. (\$1829 , \$3171) **47.** 68.3%

Chapter 3: Try it yourself

Section 3.2 A

1. $\left\{\begin{matrix}\phi, \{1\}, \{3\}, \{5\}, \{7\}, \{1,3\}, \{1,5\}, \{1,7\}, \\ \{3,5\}, \{3,7\}, \{5,7\}, \{1,3,5\}, \{1,3,7\}, \\ \{1,5,7\}, \{3,5,7\}, \{1,3,5,7\}\end{matrix}\right\}$

3. $\bar{A}$ = {27, 28, 29, 30}

Section 3.2 B

1. (a) {1, 2, 3, 4, 5, 6, 7, 8, 9, 11, 15, 16, 17, 18, 30} **(b)** {2, 11, 15}

(c) {1, 3, 16, 17, 18, 30} **(d)** {1, 3, 16, 17, 18, 30}

Section 3.3 A

1. 90 **3. (a)** 1220096908800 **(b)** 36^8 **5.** 35

Section 3.3 B

1. 36^6 **3.** 2^8 **5.**

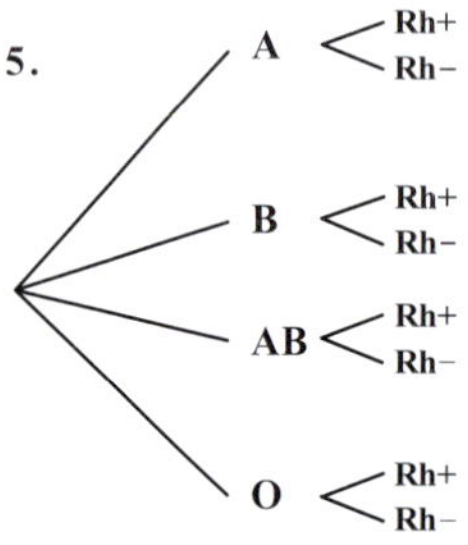

7.

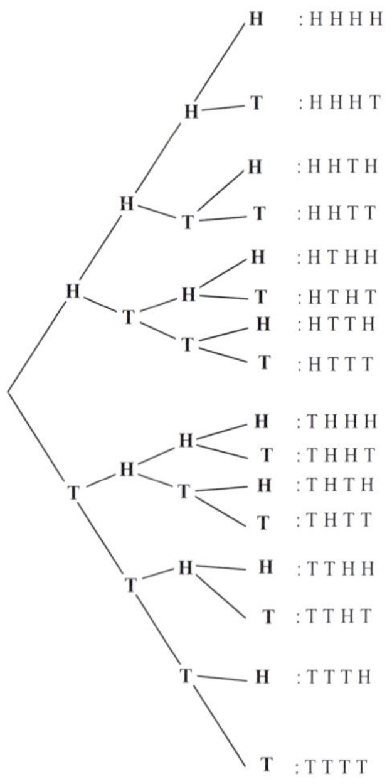

Section 3.3 C

1. 6!

Section 3.3 D

1. $\frac{8!}{5!}$ **3.** With repetition = 30240 ; Without repetition = 6!

1. $\frac{52!}{47!}$ **3.** 31

Section 3.4 B

1. (a) $\frac{58}{127}$ **(b)** $\frac{52}{58}$ **3. (a)** $\frac{C^{21}_{4}}{C^{28}_{4}}$ **(b)** $\frac{C^{7}_{4}}{C^{28}_{4}}$ **(c)** $\frac{C^{7}_{1} \cdot C^{21}_{3}}{C^{28}_{4}}$ **5.** $\frac{1}{15}$

Section 3.5 B

1. (a) $\frac{139}{200} = 0.695$ **(b)** $\frac{18}{25} = 0.72$ **3. (a)** $\frac{18}{47}$ **(b)** $\frac{32}{47}$

Section 3.6 C

1. 0.00018 **3. (a)** 0.00018 **(b)** 0.00018 **(c)** 0.0129 **(d)** 0.00048

5. $(0.03)^{15}$ **7.** 0.000048

Chapter 3: General Exercises

1. (a) $A \cup B = \{-3, 4, 5, 6, 7, 18, 24, 26, 120\}$ **(b)** $A \cap B = \{4, 5\}$ **3.** $\overline{E} = \{a, b, c, g, h\}$

5. (a) $A \cup B = \{1, 2, 3, 4, 5, 6, 7, 8\}$ **(b)** $A \cap B = \{1, 2, 3, 4\}$ **(c)** $A \cap B \cap C = \{4\}$

(d) $B \cap \overline{C} = \{1, 2, 3, 5, 7\}$ **(e)** $\overline{A} \cup C = \{-5, -4, -3, -2, -1, 0, 4, 5, 6, 7, 8, 9, 10\}$

(f) $\overline{A} \cup C \cap B = \{4, 6, 8\}$ **(g)** $A \cup \phi = A$ **(h)** $A \cap \phi = \phi$

7. 60 **9.** 16 **11. (a)** 100,000 **(b)** 30,240

13. 170,581,728,179,578,208,256 **15.** 12,157,665,459,056,928,801 **17.** 720

19. (a) 11,881376 **(b)** 7,893,600 **21. (a)** 216 **(b)** 120 **23. (a)** 10^{15}

25. (a) 1320 **(b)** 110 **27.** 120 **29.** 384 **31.** 120

33. (a) $C^{20}_{4} = 4845$ **(b)** $C^{19}_{3} = 969$ **35. (a)**

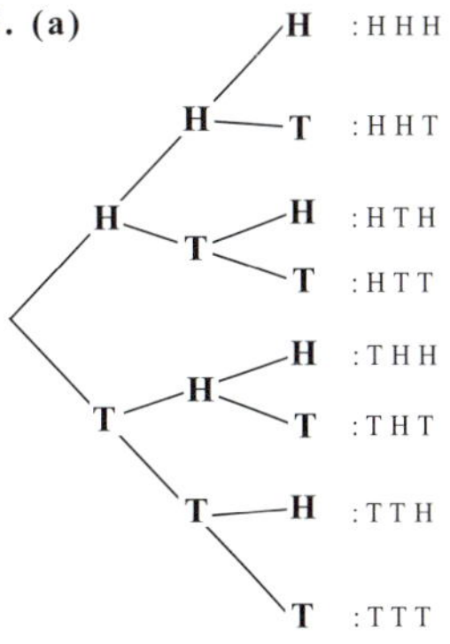

(b)

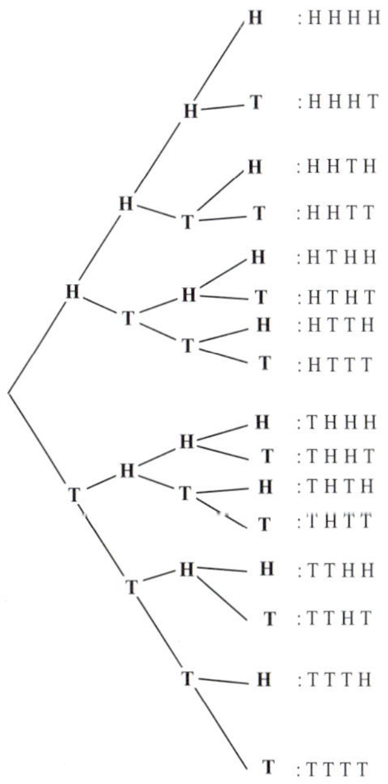

37.

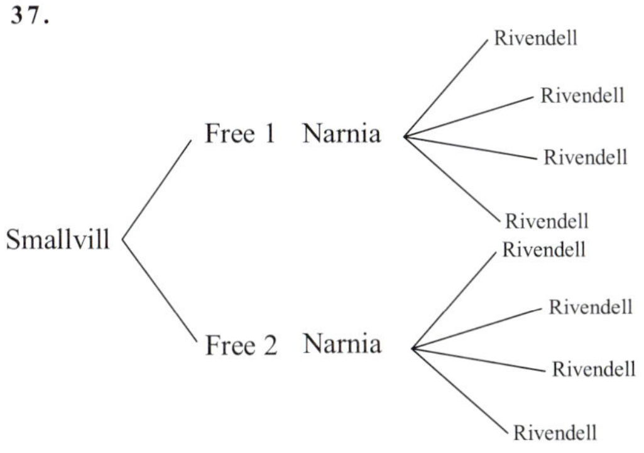

39. $\frac{2}{5} = 0.4$

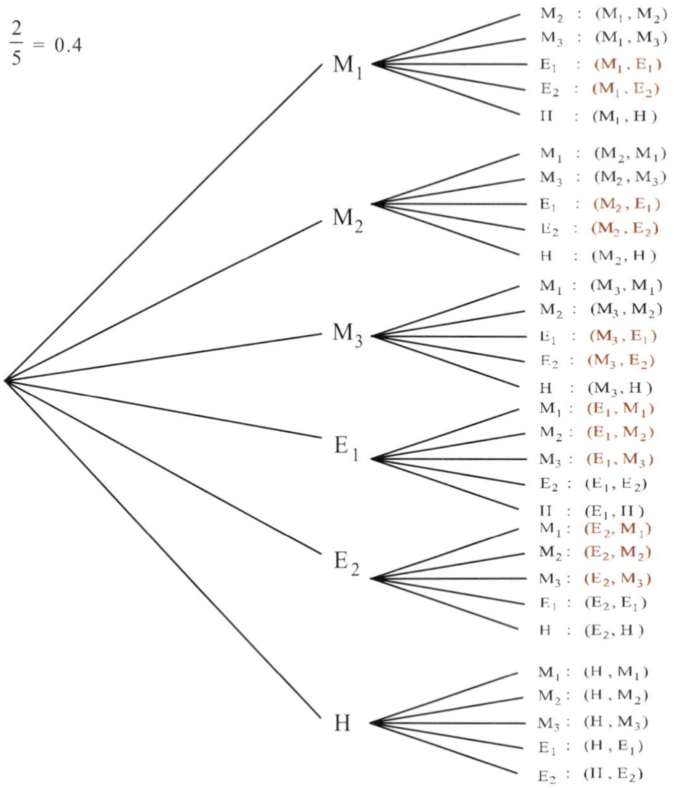

41. **(a)** 720 **(b)** 20 **(c)** 56 **(d)** 20 **(e)** 11,115,232,128,000

43. **(a)** 151,200

45. **(a)** 336 **(b)** 20 **(c)** 120 **(d)** 6

47. 30,240 **49.** 210

51. 495 **53.** 25

55. 9933

57. **(a)** 1 **(b)** ≈ 0.167 **(c)** 0.278

59. **(a)** $\left(\frac{1}{2}\right)$ **(b)** (1) **(c)** $\frac{1}{6}$

61. **(a)** $\frac{1}{4}$ **(b)** $\frac{1}{9}$ **(c)** $\frac{1}{4}$

63. $\frac{1}{2}$ **65.** 0.329

67. 0.0000000007 **69.** 0.104

71. $\frac{1}{12}$ **73.** 0.000004

75. 0.438 **77.** 0.17

79. 0.615 **81.** 0.667

83. **(a)** 0.399 **(b)** 0.998 **(c)** 1.000 **85.** 0.567 **87.** 0.055

89. **(a)** 0.0002 **(b)** 0.013 **(c)** 0.0005 **(d)** 0.017 **(e)** 0.013

91. **(a)** 0.0004 **(b)** 0.0005 **93.** 0.026 **95.** 0.080 **97.** 0.014 **99.** 0.0001

101. **(a)** 0.467 **(b)** 0.467 **(c)** 0.933 **103.** $\frac{1}{3}$ **105.** 567000 **107.** 38880

109. 127,400 **111.** 1,382,400,000 **113.** **(a)** 0.774 **(b)** 0.159 **(c)** 0 **115.** 0.999

117. 0.688 **119.** 0.5 **121.** **(a)** 0.385 **(b)** 0.047 **123.** ≈ 0.0000002

125. **(a)** 0.073 **(b)** 0.154 **(c)** 0.154 **(d)** 0.330 **127.** 0.861

129. **(a)** 120 **(b)** 0.008

131. **(a)** 0.431 **(b)** 0.770 **(c)** 0.856 **(d)** 0.603 **(e)** 0.517

133. **(a)** 0.0005 **(b)** 0.002 **(c)** 0.001 **(d)** 0.048 **(e)** 0.000004 **(f)** 0.000002

Chapter 4: Try it yourself

Section 4.3 D

1. **(a)**

X	0	1	2	3	4	5
$P(X = x)$	$\frac{1}{6}$	$\frac{5}{18}$	$\frac{2}{9}$	$\frac{1}{6}$	$\frac{1}{9}$	$\frac{1}{18}$

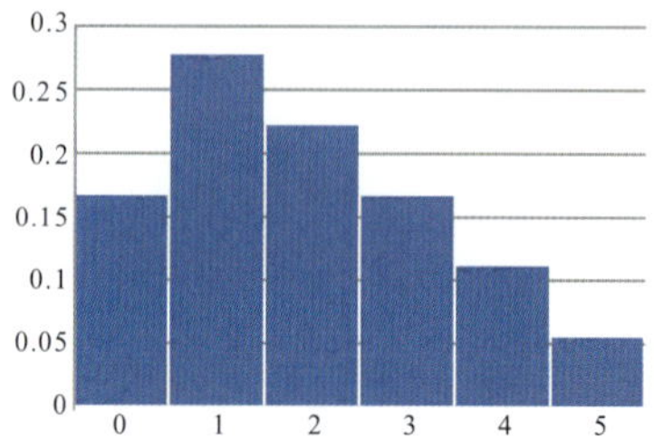

1. (b)

X	0	1	2	3	4	5
$P(X = x)$	$\frac{1}{36}$	$\frac{3}{36}$	$\frac{5}{36}$	$\frac{7}{36}$	$\frac{9}{36}$	$\frac{11}{36}$

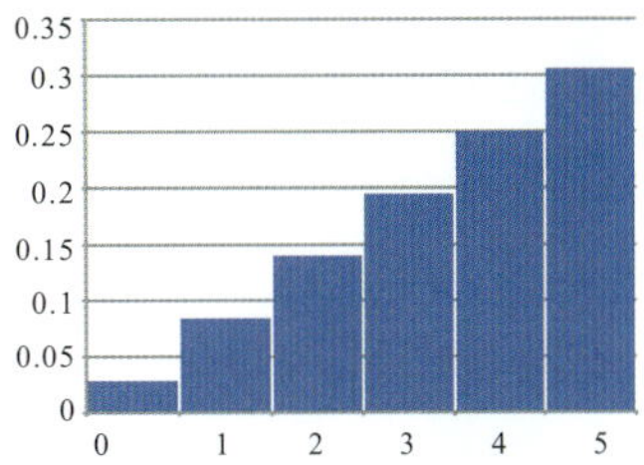

1. (c)

X	0	1	2	3	4	5
$P(X = x)$	$\frac{11}{36}$	$\frac{9}{36}$	$\frac{7}{36}$	$\frac{5}{36}$	$\frac{3}{36}$	$\frac{1}{36}$

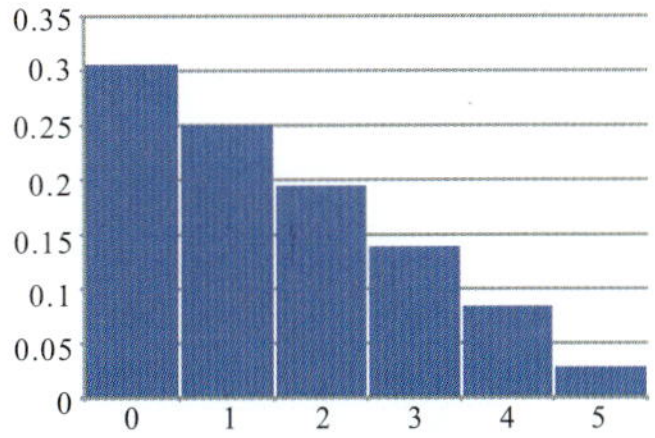

3.

X	0	1	2
$P(X = x)$	0.902	0.096	0.002

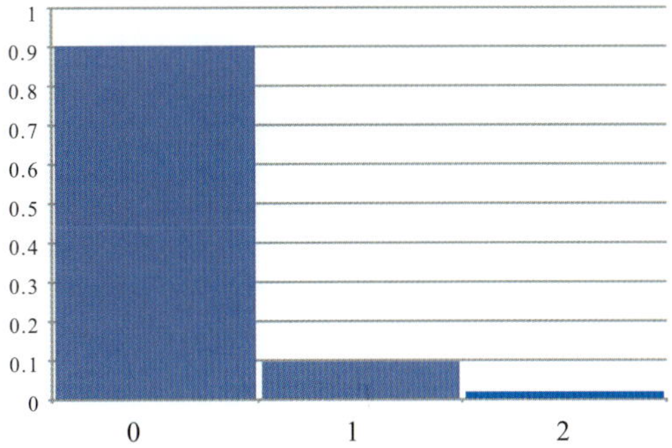

5. **(a)** 0.3417 **(b)** 0.9962

(c) 0.0004 **(d)** 0.0004

Section 4.4 B

1. $-\frac{17}{18}$; -94.4

3. $E(X) = 33.5$; $\text{Var}(X) = 635.25$; $\text{sd}(X) = 25.2$

5. $E(X) = 0.7$; $\text{Var}(X) = 2.61$; $\text{sd}(X) = 1.615$

Section 4.5 D

1. (a) 0.165 **(b)** 0.4612

3. 0.174

5. $P(X = 0) = 0.065$, $P(X = 1) = 0.1947$,

$P(X = 2) = 0.2726$, $P(X = 3) = 0.236$,

$P(X = 4) = 0.141$, $P(X = 5) = 0.0623$,

$P(X = 6) = 0.0208$, $P(X = 7) = 0.0053$,

$P(X = 8) = 0.00107$, $P(X = 9) = 0.00166$,

$P(X = 10) = 2 \times 10^{-5}$, $P(X = 11) = 1.8 \times 10^{-6}$,

$P(X = 12) = 1.21 \times 10^{-7}$, $P(X = 13) = 5.58 \times 10^{-9}$,

$P(X = 14) = 1.6 \times 10^{-10}$, $P(X = 15) = 2.13 \times 10^{-12}$

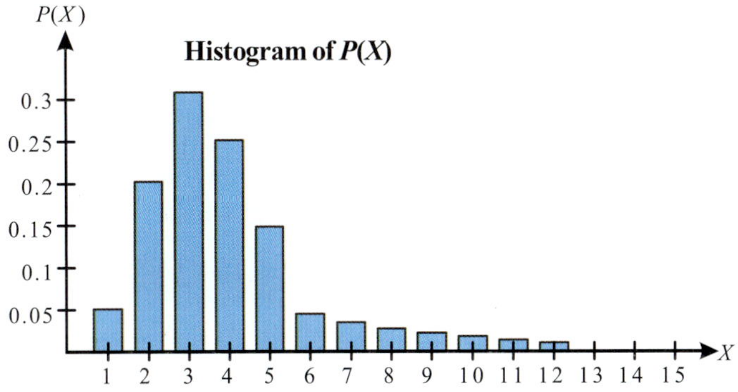

Section 4.6 C

1. (a) 0.0000 **(b)** 0.2745 **(c)** 0.8069 **(d)** 0.0433

3. 0.0011

Section 4.6 D

1. (a) (i) 0.7788 **(ii)** 0.2212 **(iii)** 0.9735 **(b)** 0.25, 0.25

1. **(a)**

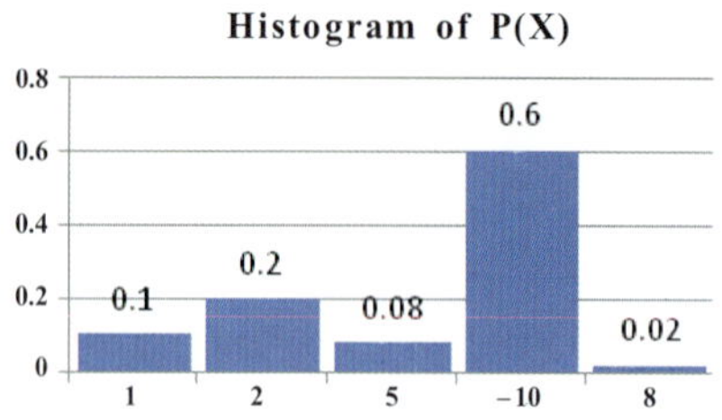

(b)

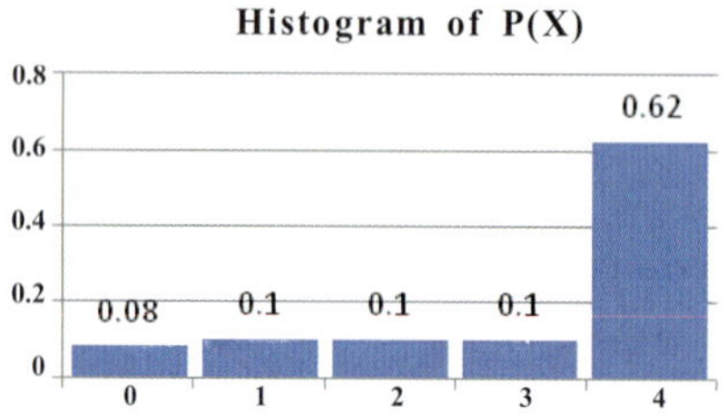

(c)

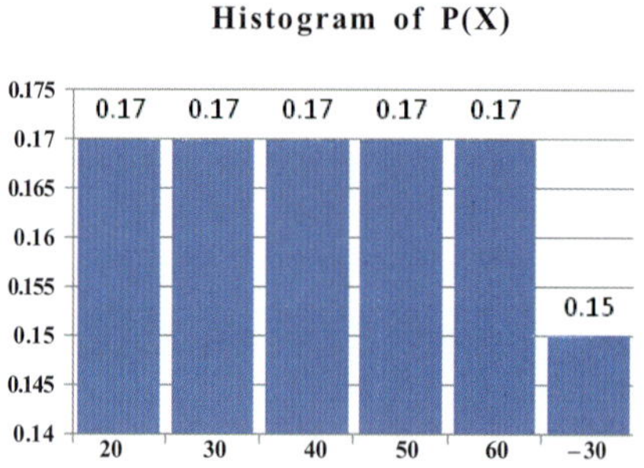

3. **(a)** Yes **(b)** 2.55

(c) Variance = 1.4075, Range of usual values = (0.177, 4.923)

5. **(a)** 1.74 **(b)** SD = 1.507

7. **(a)** $P(X = 3) = 0.11$ **(b)** $V(X) = 0.8124$, $SD(X) = 0.901$

9. **(a)** $E(X) = 21.915$ **(b)** $V(X) = 1.0178$, $SD(X) = 1.009$

11. **(a)** $a = \frac{1}{81}$

(b) $P(X < 3) = \frac{1}{9}$, $P(X \geq 3) = \frac{8}{9}$, and $P(0 < X < 5) = \frac{8}{27}$.

13. **(a)**

X	Frequency	P(X)
1	27	0.16875
2	36	0.225
3	44	0.275
4	33	0.20625
5	20	0.125

(b)

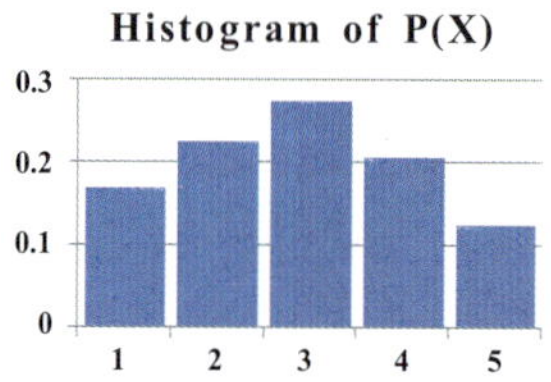

15. **(a)** $P(X) = 1$, $P(X = 0) = 0$, $P(X = 1) = \frac{1}{15}$, $P(X = 2) = \frac{2}{15}$, $P(X = 3) = \frac{3}{15}$, $P(X = 4) = \frac{4}{15}$, $P(X = 5) = \frac{5}{15}$

(b) $E(X) = 3.67$; $V(X) = 1.56$; and $\sigma = 1.25$

(c)

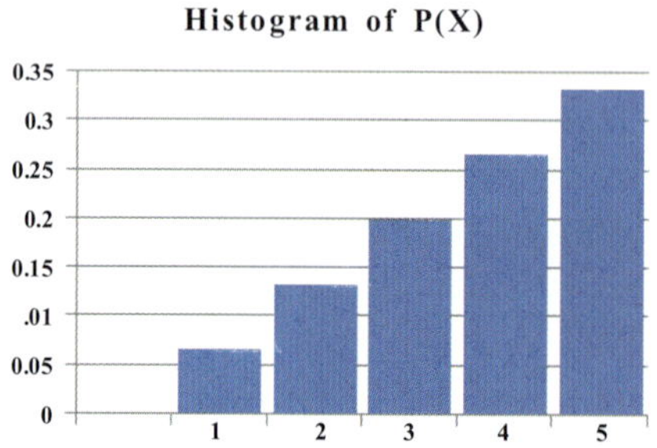

(d) $P(X = 1 \text{ or } X = 2) = 0.2$

(e) $P(X \leq 3) = 0.4$

(f) The sum of areas of all rectangles is equal to 1, and area of histogram represents the probability of variable at different points.

17. **(a)** $P(X \geq 3) = 0.9349$ **(b)** $P(X \leq 4) = 0.3971$

(c) $P(X < 5) = 0.3971$ **(d)** $P(X > 6) = 0.1738$

(e) $P(X = 2) = 0.0513$ **19.** $E(X) = 104$

21. $E(X) = 28.18$, $Var(X) = 178.51$

23.

X	P(X)
0	0.4666667
1	0.4666667
2	0.0666667

$E(X) = 0.6$; $Var(X) = 0.373$; $Sd(X) = 0.611$

25. $E(X) = -\frac{2}{3}$

27. $E(X) = 2$;

$Var(X) = 1$;

$SD(X) = 1$

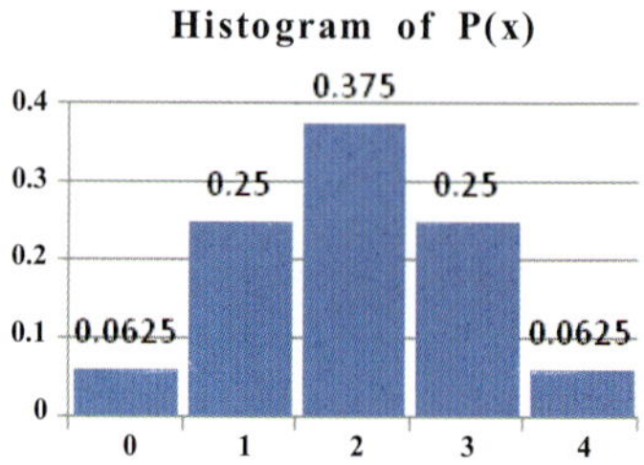

29. E(X) = 2; Var(X) = 1; SD (X) = 1;

31. E(X) = −4.869

33. Var(X) = 0.6248

35. E(X) = 0.636

37. E(X) = −8, 29.44; SD(X) = 29.44

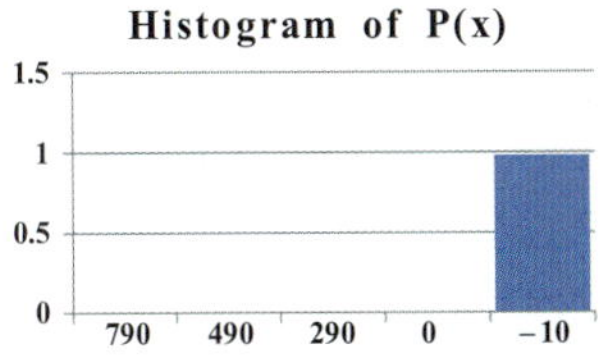

39. E(X) = −5.263

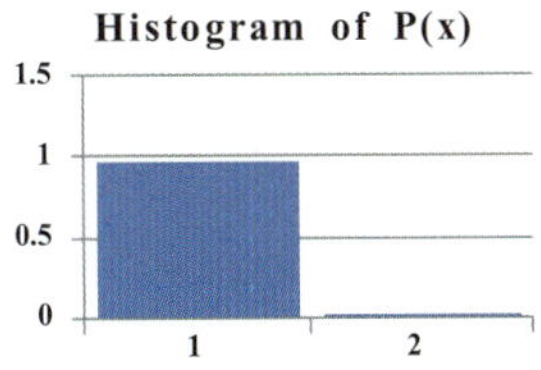

41.

X	0	1	2	3	4	5	6
P(X)	2.14E-05	0.09375	0.23438	0.3125	0.234375	0.09375	2.14E-05

E(X) = 3; SD (X) = 1.22;

43. P(X = 2) = 0.0042

45. P(X = 3) = 0.0833

47. P(X = 3) = 0.1865

49. **(a) i)** P(X ≥ 3) = 0.0401 **ii)** P(X ≤ 5) = 0.9999 **iii)** P(X = 0) = 0.4344

(b) E(X) = 0.8; SD(X) = 0.8579, Range of usual values(−0.9158, 2.5158)

51. **(a)** P(X ≤ 2) = 0.0065 **(b)** P(X ≥ 8) = 0.3953 **(c)** P(X = 1) = 0.0009

53. **(a) i)** P(X = 6) = 0.0162 **ii)** P(X ≥ 4) = 0.2241 **iii)** P(X ≤ 3) = 0.7759 **iv)** P(X > 5) = 0.0197

(b) P(X ≥ 7) = 0.0035 **(c)** mean = 2.5 ; SD(X) = 1.369

55. **(a) i)** P(X = 6) = 0.0097 **ii)** P(X ≥ 7) = 0.0024 **iii)** P(X ≤ 4) = 0.9522 **iv)** P(X > 8) = 0.000046

v) P(X < 4) = 0.8523 **(b)** mean = 2.1; Var(X) = 1.806

57. Average = 60.5 and SD = 7.29

59. Mean = 9.92, Variance = 3.77

61. Expected sale = 0.2; σ = 0.445; Usual range = (−0.7, 1.1)

63. μ = 44.64 ≈ 45; σ = 6.12

65. **(a)** P(X = 0) = 0.002479 **(b)** P(X < 14) = 0.996 **(c)** P(X > 12) = 0.0088 **(d)** P(X ≤ 9) = 0.916

67. **(a)** P(X = 0) = 0.0025 **(b)** P(X = 6) = 0.1606 **(c)** P(X ≥ 2) = 0.9826

69. **(a)** P(X = 8) = 0.0463 **(b)** P(X = 5 or 6) = 0.2989 **(c)** P(X ≥ 3) = 0.8264

71. **(a)** P(X = 0) = 0.0000 **(b)** P(X = 25) = 0.0131 **(c)** P(X ≥ 20) = 0.237 **(d)** P(X ≤ 9) = 0.6846

Chapter 5: Try it yourself

Section 5.3 C

1. Smallest to largest Figure No.-(1), (4), (2), (3)

3. 0.157

Section 5.4 D

1. 0.683

(3) 0.023

(5) 0.001

7. Symmetric about mean, area under the normal curve is 1 and
Area between mean ±1 sd = 0.683
Area between mean ±2 sd = 0.954
Area between mean ±3 sd = 0.997

9. Empirical Rule
$P(-1 < z < 1) = 68\%$
$P(-2 < z < 2) = 95\%$
$P(-3 < z < 3) = 99.7\%$

Section 5.5 C

1. 0.443

3. 0.125

5. 0.974

7. 0.025

9. 0.858

11. 0.127

13. 1.64

15. 1.133

17. −0.362

19. ±1.96

21. 0.675

23. −0.385

25. −0.675

Section 5.6 C

1. **(a)** 0.599 **(b)** 0.236 **(c)** 0.788 **(d)** 0.006 **(e)** 0.841 **(f)** 0.071

3. 89.05 **5.** 105.897 **7.** **(a)** 0.115 **(b)** 0.0603 **(c)** 0.0547

9. **(a)** **i)** 30.85% **ii)** 6.68% **iii)** 62.47% **iv)** 1.3% **(b)** 146.45

Section 5.7 A

1. 0.002 **3.** 0.5702 **5.** **(a)** 0.859 **(b)** 0.917 **(c)** 0.023 **(d)** 0.989

Chapter 5: General Exercises

1. **(a)** 0.9911 **(b)** 0.9950

3. **(a)** $c = 1.33$ **(b)** $c = 2.575$ **(c)** $c = 0.75$ **(d)** $c = 0.78$ **(e)** $c = -0.91$ **(f)** $c = -1.19$

5. **(a)** 0.1378 **(b)** 0.0367 **(c)** 0.3557 **(d)** 0.9243 **(e)** 0.0939 **(f)** 0.9812

7. **(a)** 0.67 **(b)** 0.36 **(c)** −0.1 **(d)** 1.48 **(e)** −1.175 **(f)** −0.67

(g) −0.25 **(h)** −0.84 **(i)** 0.95 **(j)** −2.05 **(k)** −1.96 **(l)** 2.05

9. 0.5 **11.** The height of Mr. Vartan is higher than Mr. Mardir with respect to their regions.

13. **(a)** $Q_3 = 162.56$ **(b)** $P_{30} = 127.79$ **(c)** $P_{97} = 197.54$ **(d)** $P_{15} = 112.94$ **(e)** $P_{12} = 108.93$

15. $\sigma = 2.59$ **17.** $x = 11.156$ **19.** $x = 8.2375$

21. **(a)** 0.7586 **(b)** 0.6372 **(c)** 0.1463 **(d)** $x = 1072.20$

23. **(a)** 61.25% **(b)** 25.79% **(c)** $Q_3 = 10.51$ **(d)** $Q_1 = 7.68$

25. 0.0275 **27.** 8.96% **29.** 61.2% **31.** 0.6363

33. **(a)** 0.3414 **(b)** 0.1898 **(c)** 0.9742 **35.** $x = 44.4$ **37.** 0.3138 **39.** $x = 1058.79$

41. $P_{98} = 1436.99$ **43.** **(a)** 0.7333 **(b)** 0.6915 **(c)** 0.6615

45. **(a)** 0.153 **(b)** 0.7154 **(c)** 0.1316 **47.** 0.3605 **49.** **(a)** 0.9192 **(b)** 0.7257

51. **(a)** 0.5328 **(b)** 0.8413 **(c)** 0.3085 **53.** **(a)** 0.1151 **(b)** 0.8849

55. **(a)** 0.9987 **(b)** 0.0013 **(c)** 0.9974 **57.** 0.1619 **59.** 0.9836 **61.** 0.9072

63. **(a)** 0.9942 **(b)** 0.2457 **(c)** 0.8201 **(d)** 0.2457 **(e)** 0.1636 **65.** 0.2043

67. **(a)** 0.0484 **(b)** 0.8208 **(c)** 0.0414

Chapter 6: Try it yourself

Section 6.3 B

1. **(a)** **i)** 5, 2.828 **ii)**

$\bar{x}$	1	2	3	4	5	6	7	8	9
p	0.04	0.08	0.12	0.16	0.2	0.16	0.12	0.08	0.04

iii) 5, 2 **3.** 26.5, 1.181 **5.** 0.000 **7.** 104.643, 107.357

Section 6.4 B

1. 0.021 **3.** **(a)** 0.072 **(b)** 0.975 **(c)** 0.835

Section 6.5 C

1. **(a)** (48.865, 52.595) **(b)** (108.843, 131.157) **(c)** (14.791, 18.809) **3.** (7.342, 8.658) **5.** 2 **7.** 267

Section 6.6 B

1. 3.169 **3.** ±1.948 **5.** ±2.764 **7.** (16.508, 19.492)

Section 6.7 B

1. (0.197, 0.363) **3.** (0.251, 0.389) **5.** (0.4177, 0.5822) **7.** 619

Section 6.8 D

1. **(a)** (1.541, 3.211) **(b)** (1.351, 3.831) **3.** (0.989, 2.627)

Chapter 6: General Exercises

1. **(a)** Sample space = (2, 2), (2, 3), (2, 6), (2, 12), (2, 15), (3, 2), (3, 3), (3, 6), (3, 12), (3, 15), (6, 2), (6, 3), (6, 6), (6, 12), (6, 15), (12, 2), (12, 3), (12, 6), (12, 12), (12, 15), (15, 2), (15, 3), (15, 6), (15, 12), (15, 15)

(b) Population mean = 7.6, Population standard deviation = 5.08 **(d)** 1.016

3. **(a)** Sample space = (1, 5, 8), (1, 5, 12), (1, 5, 15), (1, 5, 19), (1, 8, 12), (1, 8, 15), (1, 8, 19), (1, 12, 15), (1, 12, 19), (1, 15, 19), (5, 8, 12), (5, 8, 15), (5, 8, 19), (5, 12, 15), (5, 12, 19), (5, 15, 19), (8, 12, 15), (8, 12, 19), (8, 15, 19), (12, 15, 19)

Mean of all sample means $= \dfrac{200}{20} = 10$, Population mean $= \dfrac{60}{6} = 10$

(b) SE of sample mean = 2.708, Population standard deviation(σ) = 6.055

$$\text{SD}(\bar{X}) = \sigma_{\bar{X}} = \sigma\sqrt{\frac{N-n}{n(N-1)}} = 6.055\sqrt{\frac{6-3}{3(6-1)}} = 2.708$$

5. Mean of sampling distribution of $(\bar{X}) = 860$,

Standard deviation of sampling distribution of $(\bar{X}) = \dfrac{\sigma}{\sqrt{n}} = \dfrac{60}{\sqrt{25}} = 12$

7. Mean of sampling distribution of $(\bar{X}) = 18$,

Standard deviation of sampling distribution of $(\bar{X}) = \dfrac{\sigma}{\sqrt{n}} = \dfrac{4.2}{\sqrt{20}} = 0.94$

9. **(a)** 0.6306 **(b)** 0.9522 **11.** 0.9199

13. **(a)** 0.0000 **(b)** 0.2071 **(c)** 0.8618 **(d)** 0.0000 **15.** 66.43

17. **(a)** 0.0014 **(b)** 0.9324 **(c)** 0.0675 **19.** 0.9555

21. **(a)** 0.58 **(b)** 1.4935 **(c)** (72.007, 74.994) **23.** (98.965, 105.036) **25.** (68.08, 75.92)

27. (2911.137, 3106.863) **29.** (10.243, 11.931) **31.** (672.060, 693.940) **33.** 22 **35.** 973

37. 44 **39** (31.254, 34.347) **41.** **(a)** mean = 98.18 **(b)** (98.038, 98.322) **43.** (7.624, 9.776)

45. (5.282, 8.978) **47.** (9.588, 11.872) **49.** (10.955, 14.445), (8.616, 10.584) **51.** (4.196, 5.004)

53. (13.348, 17.044) **55.** (0.231, 0.369) **57** **(a)** (0.242, 0.258) **(b)** (0.242, 0.258) **59.** (0.281, 0.307)

61. (0.0537, 0.1098) **63.** (0.761, 0.839) **65.** (0.553, 0.807) **67.** (0.1096, 0.2126), (0.2462, 0.3760)

69. (0.519, 0.681) **71.** 1842 **73.** 908 **75.** 3238 **77.** (4.347, 6.956)

79. (19.582, 36.418) **81.** (10.612, 41.694) **83.** (0.269, 0.511) **85.** (1.581, 4.498), (1.780, 7.816)

87. **(a)** (5.680, 15.167) **(b)** (42.343, 145.766) **89.** (2.710, 7.191) **91.** (6.583, 30.555)

Chapter 7: Try it yourself

Section 7.3 B

1. $H_0 : \mu = 30$
$H_1 : \mu > 30$
Test statistics (z) = 8.081
CV = 1.645
Decision = Reject H_0,
p-value = 0.0000

90% CI for μ : (32.55, 33.85)

3. $H_0 : \mu = 15$
$H_1 : \mu \neq 15$
Test statistics (z) = −5.276
CV = −2.576
Decision = Reject H_0,
p-value = 0.0000

99% CI for μ : (12.92, 14.28)

5. $H_0 : \mu = 12$
$H_1 : \mu \neq 12$
Test statistics (z) = 3.514
CV = 1.959
Decision = Reject H_0,
p-value = 0.0004

95% CI for μ : (12.97, 15.43)

7. $H_0 : \mu = 1080$
$H_1 : \mu \neq 1080$
Test statistics (z) = −8.156
CV = −2.576
Decision = Reject H_0,
p-value = 0.0000

99% CI for μ : (668.15, 865.85)

Section 7.4 B

1. $H_0 : \mu = 8.6$
$H_1 : \mu > 8.6$
Test statistics (t) = −9.428 df = 49
CV = 1.677
Decision = Fail to reject H_0,
p-value = 1.000

3. $H_0 : \mu = 15$
$H_1 : \mu \neq 15$
Test statistics (t) = −3.478 df = 19
CV = −1.729
Decision = Reject H_0,
p-value = 0.003

5. $H_0 : \mu = 575$
$H_1 : \mu \neq 575$
Test statistics (t) = −2.4 df = 35
CV = −2.724
Decision = Fail to reject H_0,
p-value = 0.022

Section 7.5 B

1. $H_0 : p = 0.43$
$H_1 : p \neq 0.43$
Test statistics (z) = −0.383
CV = −1.645
Decision = Fail to reject H_0,
p-value = 0.702

3. $H_0 : p = 0.8$
$H_1 : p < 0.8$
Test statistics (z) = 1.945
CV = −2.326
Decision = Fail to reject H_0,
p-value = 0.974

5. $H_0 : p = 0.64$
$H_1 : p \neq 0.64$
Test statistics (z) = 0.031
CV = 2.576
Decision = Fail to reject H_0,
p-value = 0.975

7. $H_0 : p = 0.33$
$H_1 : p < 0.33$
Test statistics (z) = −0.150
CV = −1.645
Decision = Fail to reject H_0,
p-value = 0.440

9. $H_0 : p = 0.8$
$H_1 : p \neq 0.8$
Test statistics (z) = −3.558
CV = −2.578
Decision = Reject H_0,
p-value = 0.0004

Section 7.6 C

1. $H_0 : \sigma = 5$
$H_1 : \sigma \neq 5$
Test statistics (χ^2) = 56.448 df = 80

CV's: χ^2_L = 51.172 and χ^2_R = 116.32

Decision = Fail to reject H_0,

3. $H_0 : \sigma^2 = 324$
$H_1 : \sigma^2 < 324$
Test statistics (χ^2) = 38.329 df = 44

CV: χ^2_L = 29.788

Decision = Fail to reject H_0,

5. $H_0 : \sigma = 0.707$
$H_1 : \sigma < 0.707$
Test statistics (χ^2) = 1.029 df = 17

CV: χ^2_L = 8.672

Decision = Reject H_0,

7. $H_0 : \sigma = 1.5$
$H_1 : \sigma \neq 1.5$
Test statistics (χ^2) = 69.333 df = 39
CV's: χ_L^2 = 23.654 and χ_R^2 = 58.120
Decision = Reject H_0,

Chapter 7: General Exercises

1. (a) $H_0 : \mu = 2.7$
$H_1 : \mu > 2.7$
Test statistics (z) = 0
CV = 2.33
Fail to reject H_0,
P-value = 0.5
(b) 98% CI for μ : (2.56, 2.84)

3. (a) $H_0 : \mu = 3.99$
$H_1 : \mu \neq 3.99$
Test statistics (z) = 2.02
CV = 2.575
Fail to reject H_0,
P-value = 0.043
(b) 99% CI for μ : (3.98, 4.10)

5. (a) $H_0 : \mu = 0$
$H_1 : \mu > 0$
Test statistics (z) = 6.146
CV = 1.645
Reject H_0,
P-value = 0.0000
(b) 95% CI for μ : (0.272, 0.528)

7. (a) $H_0 : \mu = 200$
$H_1 : \mu < 200$
Test statistics (z) = −3.792
CV = −1.282
Reject H_0,
P-value = 0.0000
(b) 80% CI for μ : (178.46, 189.33)

9. (a) $H_0 : \mu = 40$
$H_1 : \mu \neq 40$
Test statistics (z) = 4.78
CV = 1.96
Reject H_0,
P-value = 0.0000
(b) 95% CI for μ : (41.3, 43.3)

11. (a) $H_0 : \mu = 7$
$H_1 : \mu > 7$
Test statistics (z) = 2.996
CV = 1.476
Reject H_0,
P-value = 0.0013
(b) 86% CI for μ : (7.41, 8.19)

13. (a) $H_0 : \mu = 6.9$
$H_1 : \mu \neq 6.9$
Test statistics (t) = −5.315 with $d.f.$ = 345
CV = −1.65
Reject H_0,
P-value = 0.0000
(b) 90% CI for μ : (6.376, 6.624)

15. (a) $H_0 : \mu = 30$
$H_1 : \mu < 30$
Test statistics (t) = −3.6091 with $d.f.$ = 34
CV = −2.032
Reject H_0,
P-value = 0.0005
(b) 95% CI for μ : (17.76, 26.58)

17. (a) $H_0 : \mu = 10$
$H_1 : \mu < 10$
Test statistics (t) = −3.733 with $d.f.$ = 17
CV = −2.57
Reject H_0,
P-value = 0.0008
(b) 98% CI for μ : (6.29, 9.31)

19. (a) $H_0 : \mu = 128.1$
$H_1 : \mu \neq 128.1$
Test statistics (t) = 0.85 with $d.f.$ = 21
CV = 1.72
Fail to reject H_0,
P-value = 0.404
(b) 90% CI for μ : (19.83, 23.17)

21. (a) $H_0 : \mu = 25$
$H_1 : \mu > 25$
Test statistics (t) = −0.565 with $d.f.$ = 7
CV = 1.89
Fail to reject H_0,
P-value = 0.7054
(b) 90% CI for μ : (20.65, 27.35)

23. (a) $H_0 : \mu = 6.8$
$H_1 : \mu \neq 6.8$
Test statistics (t) = −0.797 with $d.f.$ = 25
CV = −2.06
Fail to reject H_0,
P-value = 0.433
(b) 95% CI for μ : (6.44, 6.95)

25. (a) $H_0 : \mu = 260$
$H_1 : \mu > 260$
Test statistics (t) = 0.768 with $d.f.$ = 84
CV = 2.37
Fail to reject H_0,
P-value = 0.2222
(b) 98% CI for μ : (249.6, 280.4)

27. (a) $H_0 : \mu = 14$
$H_1 : \mu > 14$
Test statistics (t) = −6.274 with $d.f.$ = 10
CV = 1.812
Fail to reject H_0,
P-value ≈ 0.9999
(b) 90% CI for μ : (8.82, 11.14)

29. (a) $H_0 : \mu = 1.5$
$H_1 : \mu > 1.5$
Test statistics $(t) = -1.1048$ with $d.f. = 49$
CV = 1.299
Fail to reject H_0,
P-value = 0.8627

(b) 80% CI for μ : (1.39, 1.51)

31. (a) $H_0 : p = 0.2$
$H_1 : p < 0.2$
Test statistics $(Z) = -0.5$
CV = −1.645
Fail to reject H_0,
P-value = 0.309

(b) 90% CI for p : (0.114, 0.246)

33. (a) $H_0 : p = 0.6$
$H_1 : p < 0.6$
Test statistics $(Z) = 0.210$
CV = −2.326
Fail to reject H_0,
P-value = 0.583

(b) 98% CI for p : (0.422, 0.813)

35. (a) $H_0 : p = 0.55$
$H_1 : p > 0.55$
Test statistics $(Z) = -3.187$
CV = 1.645
Fail to reject H_0,
P-value = 0.9993

(b) 90% CI for p : (0.446, 0.517)

37. (a) $H_0 : p = 0.5$
$H_1 : p > 0.5$
Test statistics $(Z) = 0.6$
CV = 2.326
Fail to reject H_0,
P-value = 0.274

(b) 98% CI for p : (0.414, 0.646)

39. (a) $H_0 : p = 0.8$
$H_1 : p > 0.8$
Test statistics $(Z) = 9.879$
CV = 1.282
Reject H_0,
P-value = 0

(b) 80% CI for p : (0.907, 0.939)

41. (a) $H_0 : \sigma = 0.15$
$H_1 : \sigma < 0.15$
TS: $\chi^2 = 28.44$
CV: $\chi^2_L = 9.312$
Fail to reject H_0,

(b) 80% CI for σ : (0.165, 0.262)

43. (a) $H_0 : \sigma = 9$
$H_1 : \sigma > 9$
TS: $\chi^2 = 15.474$
CV: $\chi^2_R = 22.362$
Fail to reject H_0,

(b) 90% CI for σ : (7.487, 14.586)

45. (a) $H_0 : \sigma^2 = 0.32$
$H_1 : \sigma^2 < 0.32$
TS: $\chi^2 = 41.625$
CV: $\chi^2_L = 19.960$
Fail to reject H_0,

(b) 98% CI for σ^2 : (0.222, 0.667)

47. (a) $H_0 : \sigma^2 = 9$
$H_1 : \sigma^2 \neq 9$
TS: $\chi^2 = 15.357$
CV's: $\chi^2_L = 7.564$ and $\chi^2_R = 30.191$
Fail to reject H_0,

(b) 95% CI for σ^2 : (4.578, 18.272)

49. (a) $H_0 : \sigma = 20000$
$H_1 : \sigma > 20000$
TS: $\chi^2 = 14.641$
CV: $\chi^2_R = 30.579$
Fail to reject H_0,

(b) 98% CI for σ : (13839.22, 33465.061)

51. (a) $H_0 : \sigma = 0.6$
$H_1 : \sigma < 0.6$
TS: $\chi^2 = 46.875$
CV: $\chi^2_L = 18.493$
Fail to reject H_0,

(b) 98% CI for σ : (0.576, 1.062)

53. (a) $H_0 : \sigma = 0.47$
$H_1 : \sigma > 0.47$
TS: $\chi^2 = 29.311$
CV: $\chi^2_R = 30.578$
Fail to reject H_0,

(b) 98% CI for σ : (0.460, 1.113)

Chapter 8: Try it yourself

Section 8.2 F

1. (1.522, 3.078)

3. (−7.851, 20.851)

5. $H_0 : \mu_1 = \mu_2$
   $H_1 : \mu_1 \neq \mu_2$
   Test statistics (z) = −1.343
   CV = −1.96
   Decision = Fail to reject H_0,
   p-value = 0.179

Section 8.3 C

1. **(a)** $H_0 : \mu_1 = \mu_2$
   $H_1 : \mu_1 \neq \mu_2$
   Test statistics (z) = 1.103
   CV = 2.807
   Decision = Fail to reject H_0,
   p-value = 0.281

   (b) 99% CI for μ : (−10.041, 23.041)

3. **(a)** $H_0 : \mu_1 = \mu_2$
   $H_1 : \mu_1 > \mu_2$
   Test statistics (z) = 4.658
   CV = 1.3049
   Decision = Reject H_0,
   p-value = 0.0000

   (b) 80% CI for μ : (1.656, 2.944)

Section 8.4 B

1. (−3.202, −1.798)

3. $H_0 : \mu_d = 0$
   $H_1 : \mu_d \neq 0$
   Test statistics (t) = 1.561 df = 7
   CV = ±2.365
   Decision = Fail to reject H_0,
   p-value = 0.162

Section 8.5 C

1. $H_0 : p_1 = p_2$
   $H_1 : p_1 \neq p_2$
   Test statistics (t) = 2.967 df = 128
   CV = ±2.576
   Decision = Reject H_0,
   p-value = 0.003
   99% CI for p : (0.029, 0.403)

3. $H_0 : p_1 = p_2$
   $H_1 : p_1 < p_2$
   Test statistics (t) = −0.264 df = 44
   CV = −1.645
   Decision = Fail to reject H_0,
   p-value = 0.396
   90% CI for p : (−0.145, 0.105)

Section 8.6 B

1. $H_0 : \sigma_1^2 = \sigma_2^2$
   $H_1 : \sigma_1^2 > \sigma_2^2$
   Test statistics (F) = 4.291 with df (14, 7)
   CV = 3.529
   Decision = Reject H_0,

3. $H_0 : \sigma_1 = \sigma_2$
   $H_1 : \sigma_1 < \sigma_2$
   Test statistics (F) = 0.827 with df = (7, 7)
   Decision = Fail to reject H_0,

Chapter 8: General Exercises

1. **(a)** $H_0 : \mu_1 = \mu_2$
   $H_1 : \mu_1 \neq \mu_2$
   Test Statistics = 0.654
   CV = 1.96
   Decision = Fail to reject H_0,
   p-value = 0.513

   (b) 95% CI : (−0.599, 1.199)

3. **(a)** $H_0 : \mu_1 = \mu_2$
   $H_1 : \mu_1 < \mu_2$
   Test Statistics = −8.941
   CV = −2.326
   Decision = Reject H_0,
   p-value = 0.0000

   (b) 98% CI : (−11719.712, −6880.288)

5. **(a)** $H_0 : \mu_1 = \mu_2$
   $H_1 : \mu_1 > \mu_2$
   Test Statistics = 5.067
   CV = 2.326
   Decision = Reject H_0,
   p-value = 0.0000

   (b) 98% CI : (0.941, 2.539)

7. **(a)** $H_0: \mu_1 = \mu_2$
$H_1: \mu_1 \neq \mu_2$
Test Statistics = −11.201
CV = −1.96
Decision = Reject H_0,
p-value = 0.0000

(b) 95% CI : (−17.625, −12.375)

9. **(a)** $H_0: \mu_1 = \mu_2$
$H_1: \mu_1 \neq \mu_2$
Test Statistics = 0.31
CV = 2.576
Decision = Fail to reject H_0,
p-value = 0.757

(b) 99% CI : (−2.192, 2.792)

11. **(a)** $H_0: \mu_1 = \mu_2$
$H_1: \mu_1 \neq \mu_2$
Test Statistics = 0.161
CV = 2.048
Decision = Fail to reject H_0,
p-value = 0.873

(b) 95% CI : (−7.016, 8.216)

13. **(a)** $H_0: \mu_1 = \mu_2$
$H_1: \mu_1 < \mu_2$
Test Statistics = −0.746
CV = −2.552
Decision = Fail to reject H_0,
p-value = 0.233

(b) 98% CI : (−15.036, 8.236)

15. **(a)** $H_0: \mu_1 = \mu_2$
$H_1: \mu_1 \neq \mu_2$
Test Statistics = −1.297
CV = −1.993
Decision = Fail to reject H_0,
p-value = 0.199

(b) 95% CI : (−9889.853, −2089.853)

17. **(a)** $H_0: \mu_1 = \mu_2$
$H_1: \mu_1 \neq \mu_2$
Test Statistics = −1.608
CV = −2.845
Decision = Fail to reject H_0,
p-value = 0.124

(b) 99% CI : (−11.078, 3.078)

19. **(a)** $H_0: \mu_1 = \mu_2$
$H_1: \mu_1 \neq \mu_2$
Test Statistics = 1.857
CV = 1.668
Decision = Reject H_0,
p-value = 0.034

(b) 90% CI : (0.408, 7.592)

21. **(a)** $H_0: \mu_1 = \mu_2$
$H_1: \mu_1 < \mu_2$
Test Statistics = −0.835
CV = −2.528
Decision = Fail to reject H_0,
p-value = 0.207

(b) 98% CI : (−52.371, 26.371)

23. **(a)** $H_0: \mu_1 = \mu_2$
$H_1: \mu_1 \neq \mu_2$
Test Statistics = 8.454
CV = 2.074
Decision = Reject H_0,
p-value = 0.0000

(b) 95% CI : (6.893, 11.374)

25. **(a)** $H_0: \mu_1 = \mu_2$
$H_1: \mu_1 > \mu_2$
Test Statistics = 0.198
CV = 2.821
Decision = Fail to reject H_0,
p-value = 0.424

(b) 98% CI : (−3.305, 3.805)

27. **(a)** $H_0: \mu_1 = \mu_2$
$H_1: \mu_1 \neq \mu_2$
Test Statistics = 1.799
CV = 2.032
Decision = Fail to reject H_0,
p-value = 0.081

(b) 95% CI : (−2.598, 42.598)

29. **(a)** $H_0: \mu_1 = \mu_2$
$H_1: \mu_1 > \mu_2$
Test Statistics = 4.0002
CV = 2.576
Decision = Reject H_0,
p-value = 0.0000

(b) 99% CI : (0.02, 0.096)

31. **(a)** $H_0: \mu_1 = \mu_2$
$H_1: \mu_1 > \mu_2$
Test Statistics = 0.738
CV = 3.747
Decision = Fail to reject H_0,
p-value = 0.251

(b) 98% CI : (−516.109, 769.309)

33. **(a)** $H_0: \mu_1 = \mu_2$
$H_1: \mu_1 < \mu_2$
Test Statistics = −1.881
CV = −1.678
Decision = Reject H_0,
p-value = 0.033

(b) 90% CI : (−27.627, −1.573)

35. **(a)** $H_0: \mu_1 = \mu_2$
$H_1: \mu_1 \neq \mu_2$
Test Statistics = 1.188
CV = 2.681
Decision = Fail to reject H_0,
p-value = 0.258

(b) 98% CI : (−6.914, 17.914)

37. **(a)** $H_0: \mu_1 = \mu_2$
$H_1: \mu_1 \neq \mu_2$
Test Statistics = −13.901
CV = −2.042
Decision = Reject H_0,
p-value = 0.0000

(b) 95% CI : (−23970.546, −17829.454)

39. **(a)** $H_0: \mu_1 = \mu_2$
$H_1: \mu_1 < \mu_2$
Test Statistics = −1.718
CV = −1.833
Decision = Fail to reject H_0,
p-value = 0.060

(b) 90% CI : (−22.738, 0.738)

41. **(a)** $H_0: \mu_1 = \mu_2$
$H_1: \mu_1 \neq \mu_2$
Test Statistics = 2.256
CV = 2.571
Decision = Fail to reject H_0,
p-value = 0.074

(b) 95% CI : (−0.158, 2.425)

43. $H_0: \mu_1 = \mu_2$
$H_1: \mu_1 \neq \mu_2$
Test Statistics = −9
CV = −2.064
Decision = Reject H_0,
p-value = 0.0000
95% CI : (−5.532, −3.468)

45. **(a)** $H_0: \mu_1 = \mu_2$
$H_1: \mu_1 < \mu_2$
Test Statistics = −2.842
CV = −1.753
Decision = Reject H_0,
p-value = 0.006

(b) 90% CI : (−14.249, −3.376)

47. (a) $H_0 : \mu_1 = \mu_2$
$H_1 : \mu_1 > \mu_2$
Test Statistics = 4.154
CV = 2.326
Decision = Reject H_0,
p-value = 0.0000

(b) 98% CI : (0.022, 0.078)

49. (a) $H_0 : \mu_1 = \mu_2$
$H_1 : \mu_1 \neq \mu_2$
Test Statistics = −6.791
CV = −1.96
Decision = Reject H_0,
p-value = 0.0000

(b) 95% CI : (−0.152, −0.084)

51. (a) $H_0 : \mu_1 = \mu_2$
$H_1 : \mu_1 < \mu_2$
Test Statistics = −12.845
CV = −2.576
Decision = Reject H_0,
p-value = 0.0000

(b) 99% CI : (−0.264, −0.176)

53. (a) $H_0 : \mu_1 = \mu_2$
$H_1 : \mu_1 > \mu_2$
Test Statistics = 4.305
CV = 2.326
Decision = Reject H_0,
p-value = 0.0000

(b) 98% CI : (0.055, 0.185)

55. (a) $H_0 : \mu_1 = \mu_2$
$H_1 : \mu_1 < \mu_2$
Test Statistics = −1.613
CV = −1.645
Decision = Fail to reject H_0,
p-value = 0.053

(b) 90% CI : (−0.002, 0.0)

57. (a) $H_0 : \mu_1 = \mu_2$
$H_1 : \mu_1 \neq \mu_2$
Test Statistics = −1.613
CV = −1.96
Decision = Fail to reject H_0,
p-value = 0.540

(b) 95% CI : (−0.21, 0.11)

59. (a) $H_0 : \mu_1 = \mu_2$
$H_1 : \mu_1 \neq \mu_2$
Test Statistics = −0.324
CV = −2.576
Decision = Fail to reject H_0,
p-value = 0.746

(b) 99% CI : (−0.18, 0.14)

61. (a) $H_0 : \mu_1 = \mu_2$
$H_1 : \mu_1 \neq \mu_2$
Test Statistics = 0.314
CV = 2.576
Decision = Fail to reject H_0,
p-value = 0.753

(b) 99% CI : (−0.144, 0.184)

63. (a) $H_0 : \sigma_1 = \sigma_2$
$H_1 : \sigma_1 \neq \sigma_2$
Test Statistics = 1.159
CV's = $F_{(\text{right})}$ = 1.888 and $F_{(\text{Left})}$ = 0.516
Decision = Fail to reject H_0,

65. (a) $H_0 : \sigma_1 = \sigma_2$
$H_1 : \sigma_1 < \sigma_2$
Test Statistics = 0.227
CV: $F_{(\text{Left})}$ = 0.358
Decision = Reject H_0,
p-value = 0.000
90% CI(σ) : (17.785, 52.135)
CI(σ^2) : (316.309, 2718.077)

67. (a) $H_0 : \sigma_1 = \sigma_2$
$H_1 : \sigma_1 > \sigma_2$
Test Statistics = 0.948
CV: $F_{(\text{right})}$ = 3.313
Decision = Fail to reject H_0,
p-value = 0.545
90% CI(σ) : (245.869, 768.376)
CI(σ^2) : (60451.414, 590401.542)

69. (a) $H_0 : \sigma_1 = \sigma_2$
$H_1 : \sigma_1 \neq \sigma_2$
Test Statistics = 1.778
CV's = $F_{(\text{right})}$ = 3.532 and $F_{(\text{Left})}$ = 0.311
Decision = Fail to reject H_0,
p-value = 0.227
99% CI(σ) : (10.207, 34.395)
CI(σ^2) : (104.182, 1183.039)

71. (a) $H_0 : \sigma_1 = \sigma_2$
$H_1 : \sigma_1 < \sigma_2$
Test Statistics = 0.993
CV: $F_{(\text{Left})}$ = 0.291
Decision = Fail to reject H_0,
p-value = 0.496
90% CI(σ) : (0.0001, 0.0004)
CI(σ^2) : (1.4323E-08, 1.69306E-07)

73. (a) $H_0 : \sigma_1 = \sigma_2$
$H_1 : \sigma_1 > \sigma_2$
Test Statistics = 19.255
CV: $F_{(\text{right})}$ = 1.779
Decision = Reject H_0,
p-value =
90% CI(σ) : (355448998.6, 636195575.4)
CI(σ^2) : (1.26344E+17, 4.04745E+17)

Chapter 9: Try it yourself

Section 9.2 D

1. H_0 : Current money making pattern is the same as that for 2000.
Test Statistics = 1.333
CV = 5.992
Decision = Fail to reject H_0,
p-value = 0.513

3. H_0 : Number of cars sold in different months of year an uniformly distributed.
Test Statistics = 42.364
CV = 17.275
Decision = Reject H_0,
p-value = 0.000

Section 9.3 B

1. H_0 : Person's age group and political beliefs are independent.
Test Statistics = 27.767
CV = 13.277
Decision = Reject H_0,
p-value = 0.000

3. H_0 : Selection of a Premium or a Basic Cable TV programming is independent of the viewer's level of education.
Test Statistics = 155.007
CV = 9.210
Decision = Reject H_0,
p-value = 0.000

Section 9.4 A

1. H_0 : Proportion of married men is same in all four cities.
Test Statistics = 14.894
CV = 7.815
Decision = Reject H_0,
p-value = 0.002

3. H_0 : Proportion of teenage smokers in all three school districts are same.
Test Statistics = 0.712
CV = 5.992
Decision = Fail to reject H_0,
p-value = 0.701

Chapter 9: General Exercises

1. H_0 : Frequency of fatal accidents in different days of month are Uniformely distributed.
Test Statistics = 4.229
CV = 16.812
Decision = Fail to reject H_0,
p-value = 0.646

3. H_0 : $P_{16-25} = 0.20$, $P_{26-35} = 0.30$, $P_{36-45} = 0.20$, $P_{46-55} = 0.15$, and $P_{56\text{ and older}} = 0.15$
Test Statistics = 16.200
CV = 9.488
Decision = Reject H_0,
p-value = 0.003

5. H_0 : $P_{\text{seniors}} = 0.10$, $P_{\text{juniors}} = 0.25$, $P_{\text{sophomores}} = 0.30$ and $P_{\text{freshmen}} = 0.35$
Test Statistics = 90.783
CV = 11.345
Decision = Reject H_0,
p-value = 0.000

7. H_0 : The color distribution of cars solds is Uniform.
Test Statistics = 6.694
CV = 15.086
Decision = Fail to reject H_0,
p-value = 0.244

9. H_0 : Peoples travel uniformly on different days of the week.
Test Statistics = 147.003
CV = 12.592
Decision = Reject H_0,
p-value = 0.000

11. H_0 : Distribution of full-time students enrolled in 4-year degree-granting public institutions are uniformly distributed.
Test Statistics = 21.296
CV = 27.688
Decision = Fail to reject H_0,
p-value = 0.067

13. H_0 : $P_{\text{both Parants}} = 0.60$, $P_{\text{Mother}} = 0.27$, $P_{\text{Father}} = 0.10$ and $P_{\text{Others}} = 0.03$.
Test Statistics = 3.007
CV = 11.345
Decision = Fail to reject H_0,
p-value = 0.391

15. H_0 : The different digits in the given data set follows Uniform distribution.
Test Statistics = 4.100
CV = 16.919
Decision = Fail to reject H_0,
p-value = 0.905

17. H_0 : Psychological characteristics have a uniform distribution.
Test Statistics = 4.250
CV = 11.070
Decision = Fail to reject H_0,
p-value = 0.514

19. H_0 : The day of the week students would prefer to take their examinations are uniformly distributed.
Test Statistics = 15.286
CV = 9.488
Decision = Reject H_0,
p-value = 0.004

21. H_0 : The level of mathematics courses and gender are independent from each other.
Test Statistics = 99.367
CV = 9.488
Decision = Reject H_0,
p-value = 0.000

23. H_0 : the use of technology in the classroom and student pass rates are independent.
Test Statistics = 3.789
CV = 3.842
Decision = Fail to reject H_0,
p-value = 0.052

25. H_0 : The production of fuels is independent of the year.
Test Statistics = 0.004
CV = 6.251
Decision = Fail to reject H_0,
p-value = 1.000

27. H_0 : People's responses are independent of the year in which they were surveyed.
Test Statistics = 27.144
CV = 6.251
Decision = Reject H_0,
p-value = 0.000

29. H_0 : The distribution of planes flying and planes ordered are independent of the airline company.
Test Statistics = 8.642
CV = 9.210
Decision = Fail to reject H_0,
p-value = 0.013

31. H_0 : The age-group and preferred museum type are independent.
Test Statistics = 21.317
CV = 9.488
Decision = Reject H_0,
p-value = 0.000

33. H_0 : The proportions of survivals in the three Hospitals are equal.
Test Statistics = 1.738
CV = 9.210
Decision = Fail to reject H_0,
p-value = 0.419

35. H_0 : The proportions of parents who favor the 12-month school year are the same in all four counties.
Test Statistics = 20.174
CV = 9.348
Decision = Reject H_0,
p-value = 0.000

37. H_0 : The proportions of defective items in the three locations are equal.
Test Statistics = 1.018
CV = 9.210
Decision = Fail to reject H_0,
p-value = 0.601

39. H_0 : $P_{\text{small car}} = 0.25$, $P_{\text{Medium}} = 0.35$, $P_{\text{large Car}} = 0.10$, $P_{\text{BMW}} = 0.25$ and $P_{\text{special Car}} = 0.05$.
Test Statistics = 11.902
CV = 13.277
Decision = Fail to reject H_0,
p-value = 0.019

41. H_0 : The digits in lottery numbers are uniformly distributed.
Test Statistics = 5
Tab. Value = 21.666
Decision = Fail to reject H_0,
p-value = 0.834

43. H_0 : Job satisfaction is independent of pay categories.
Test Statistics = 2.854
CV = 4.605
Decision = Fail to reject H_0,
p-value = 0.240

45. H_0 : The credits earned are choice of major are independent.
Test Statistics = 2.236
CV = 9.488
Decision = Fail to reject H_0,
p-value = 0.692

47. H_0 : $P_{\text{sodium Cloride}} = 0.54$, $P_{\text{Magnesium}} = 0.03$ and $P_{\text{other compounds}} = 0.43$.
Test Statistics = 36.646
CV = 5.992
Decision = Reject H_0,
p-value = 0.000

Chapter 10: Try it yourself

Section 10.2 C

1. (a)

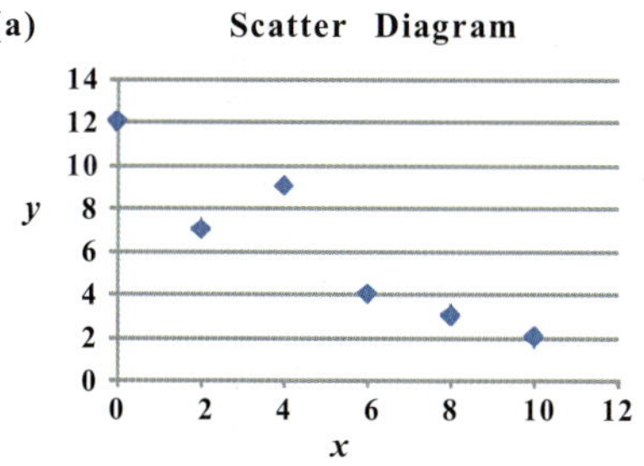

$r = -0.926$

(b)

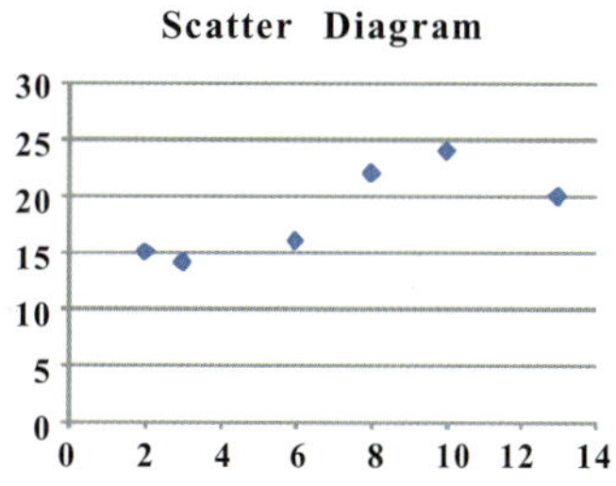

$r = 0.782$

(c)

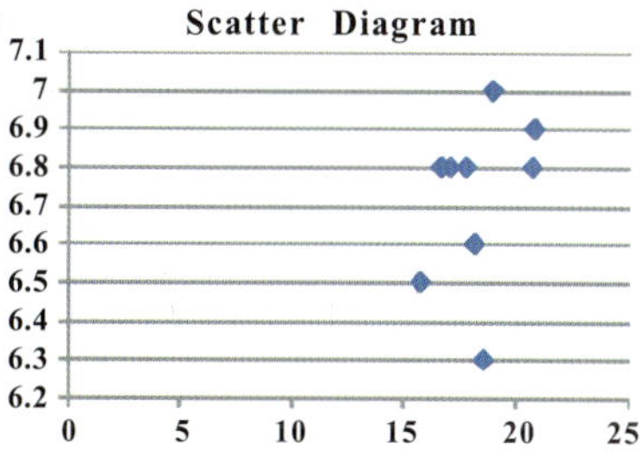

$r = 0.344$

3. $r = 0.64074$,
High Degree of Positive Correlation

5. **(a)** **Method 1:** $H_0 : \rho = 0$
$H_1 : \rho \neq 0$
Test statistics $(t) = 2.247$
CV = 2.262
Decision: Fail to reject H_0
p-value = 0.051
Method 2: Test statistics $(r) = 0.599$
CV for $(n = 11) = 0.602$
Decision: Fail to reject H_0

(b) CI : (−0.028, 1.227)

Section 10.3 D

1. $y = -1.063x + 141.298$
3. −19.109
5. **(a)** $y = 0.413x + 9.377$ **(b)** $\hat{y} = 50.632$

Section 10.4 D

1. **(a)** 2.427 **(b)** 2.427
3. (12.383, 44.391)

Chapter 10: General Exercises

1. **Scatter Diagram**

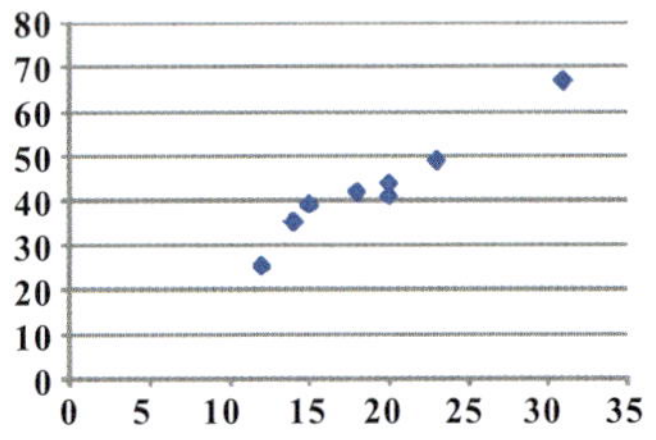

$r = 0.972$

3. **Scatter Diagram**

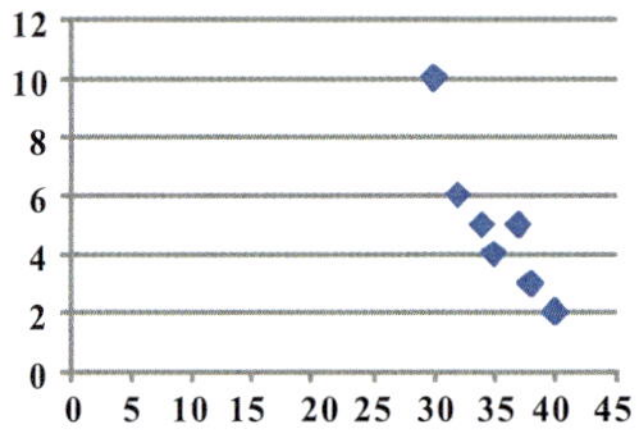

$r = -0.908$

5. **Scatter Diagram**

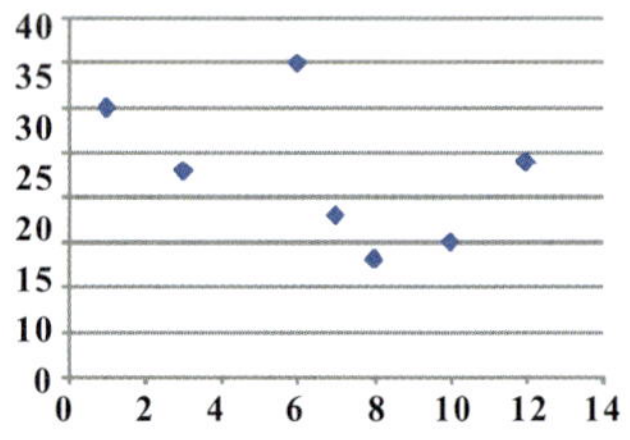

$r = -0.459$

7. $r = 0.855$

9. $r = -0.170$

11. **(a)** **Scatter Diagram**

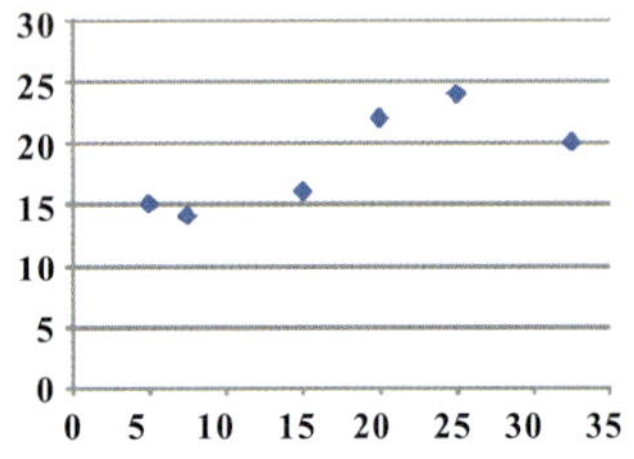

(b) $r = 0.782$, High degree of positive correlation

(c) **Method 1:** $H_0 : \rho = 0$
$H_1 : \rho \neq 0$
Test statistics $(t) = 2.506$
CV = 2.776
Decision: Fail to reject H_0
p-value = 0.066
Method 2: Test statistics $(r) = 0.782$
CV for $(n = 6) = 0.811$
Decision: Fail to reject H_0

(d) (0.341, 1.223)

13. **(a)** **Scatter Diagram**

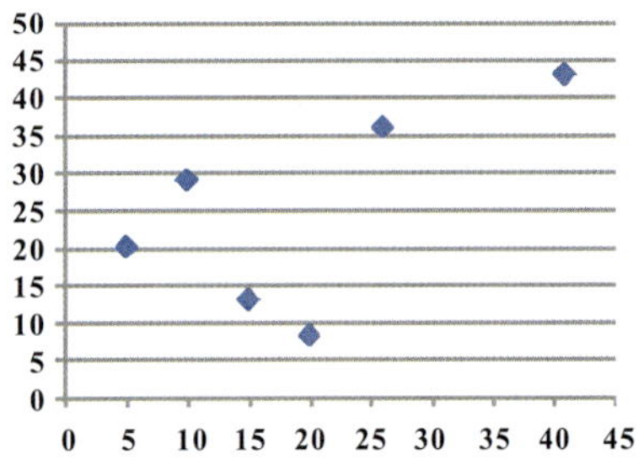

(b) $r = 0.618$, High degree of positive correlation

(c) **Method 1:** $H_0 : \rho = 0$
$H_1 : \rho \neq 0$
Test statistics $(t) = 1.573$
CV = 2.776
Decision: Fail to reject H_0
p-value = 0.191
Method 2: Test statistics $(r) = 0.618$
CV for $(n = 6) = 0.811$
Decision: Fail to reject H_0

(d) (−0.082, 1.319)

15. (a) Scatter Diagram

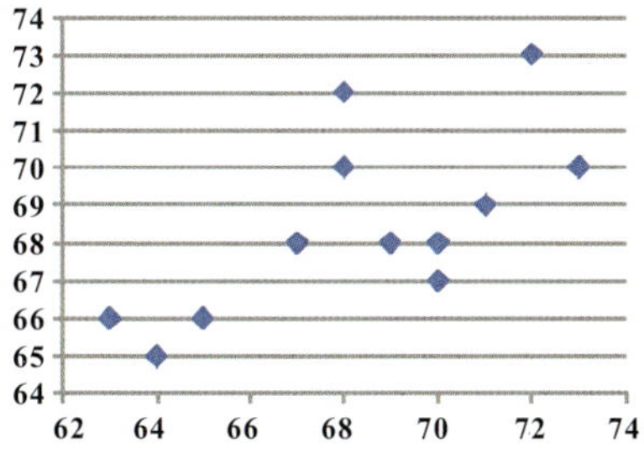

(b) $r = 0.724$, High degree of positive correlation

(c) **Method 1:** $H_0: \rho = 0$
$H_1: \rho \neq 0$
Test statistics $(t) = 3.482$
CV = 2.201
Decision: Reject H_0
p-value = 0.005

Method 2: Test statistics $(r) = 0.724$
CV for $(n = 13) = 0.553$
Decision: Reject H_0

(d) (0.434, 1.015)

17. $y = 0.929x - 1.125$

19. $y = 0.75x + 5.75$

21. −112.24

23. −34.782

25. (a) $y = 1.608x + 7.833$

(b)

y(est.)	30.349	38.392	46.433	41.608	49.649	43.217	31.958	39.999	38.392

(c)

Residuals	0.650	−2.392	1.567	−4.608	0.350	1.783	1.042	1.000	0.608

(d) $\sum e = 0.000$

27. (a) $y = 0.5x + 2.990$

(b)

y(est.)	7.991	6.991	9.491	7.491	8.491	9.991	5.991	4.991	8.991	6.491	5.491

(c)

Residuals	1.109	1.109	−0.791	1.309	0.809	−1.891	0.109	−1.891	0.109	5.491	−0.791

(d) $\sum e = 0.000$

29. (32.745, 50.755)

31. (86.007, 110.964)

33. (a) Scatter Diagram

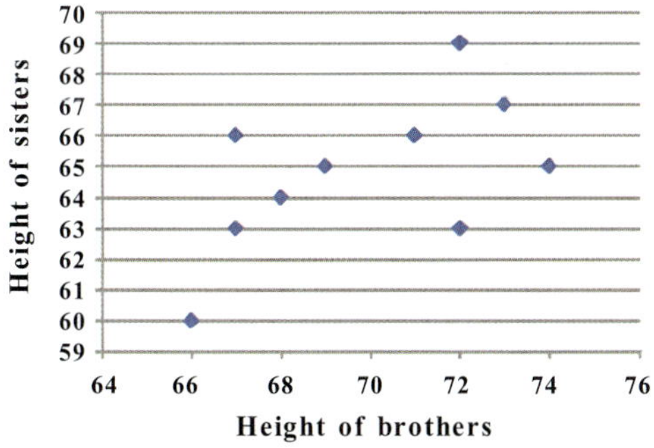

(b) $r = 0.570$

(c) **Method 1:** $H_0: \rho = 0$
$H_1: \rho \neq 0$
Test statistics $(t) = 2.082$
CV = 2.262
Decision: Fail to reject H_0
p-value = 0.067

Method 2: Test statistics $(r) = 0.570$
CV for $(n = 11) = 0.602$
Decision: Fail to reject H_0

(d) $y = 0.500x + 29.909$

(e) $\hat{y} = 64.909$

(f) (60.028, 69.790)

35. (a) Scatter Diagram

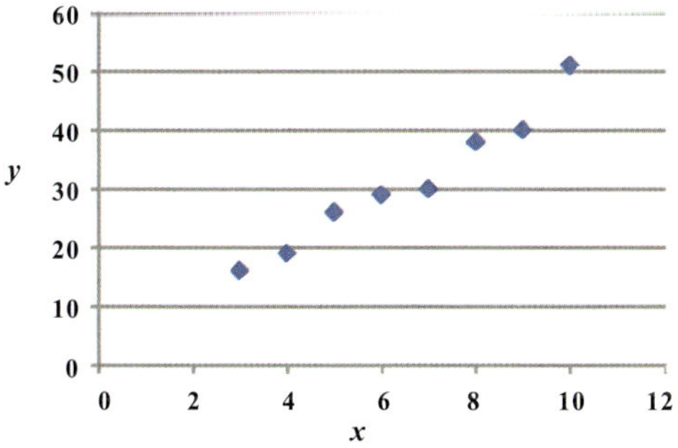

(b) $r = 0.979$

(c) **Method 1:** $H_0: \rho = 0$
$H_1: \rho \neq 0$
Test statistics $(t) = 11.960$
CV = 2.447
Decision: Reject H_0
p-value = 0.000

Method 2: Test statistics $(r) = 0.979$
CV for $(n = 8) = 0.707$
Decision: Reject H_0

(d) $y = 4.607x + 1.179$

(e) $\hat{y} = 70.286$

(f) (59.982, 80.589)

37. (a) Scatter Diagram

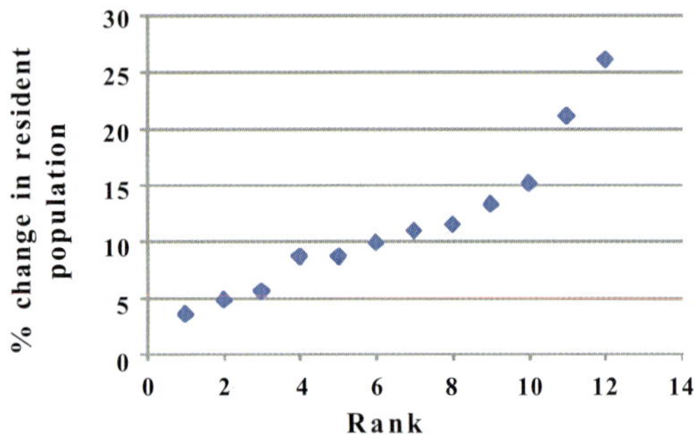

(b) $r = 0.939$

(c) Method 1: $H_0 : \rho = 0$
$H_1 : \rho \neq 0$
Test statistics $(t) = 8.676$
CV = 3.169
Decision: Reject H_0
p-value = 0.000

Method 2: Test statistics $(r) = 0.939$
CV for $(n = 12) = 0.708$
Decision: Reject H_0

(d) $y = 1.728x + 0.268$

(e) $\hat{y} = 26.188$ **(f)** (16.675, 35.701)

39. (a) Scatter Diagram

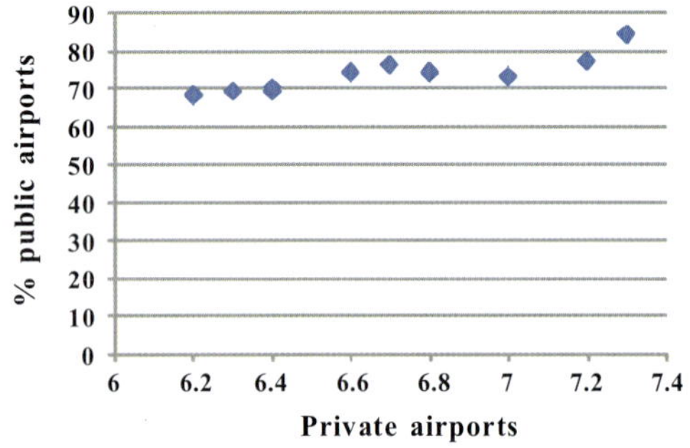

(b) $r = 0.903$

(c) Method 1: $H_0 : \rho = 0$
$H_1 : \rho \neq 0$
Test statistics $(t) = 6.638$
CV = 3.169
Decision: Reject H_0
p-value = 0.000

Method 2: Test statistics $(r) = 0.903$
CV for $(n = 12) = 0.708$
Decision: Reject H_0

(d) $y = 11.504x - 3.646$

(e) $\hat{y} = 37.769$ **(f)** (19.691, 55.849)

Appendix 1: General Exercises

1.

	A	B	C	D	E	F	G	H
1	12	18	2.00	144				
2	40	28	5.14	1600				
3	12	70	48.06	144				
4	5	50	40.50	25				
5	6	12	3.00	36				
6	7	98	84.50	49				
7	86	74	1.95	7396				
8	2	18	14.22	4				174.5762
9	34	45	2.69	1156				
10	4	36	28.44	16				
11	5	65	55.38	25				
12	6	82	70.44	36				
13	7	100	86.49	49				
14	665	17	24700.24	442225				
15	891	713	25143.05	452905				
16								

3. 2.2738 **5. (a)** 2.9993 **(b)** 1.4485

7.

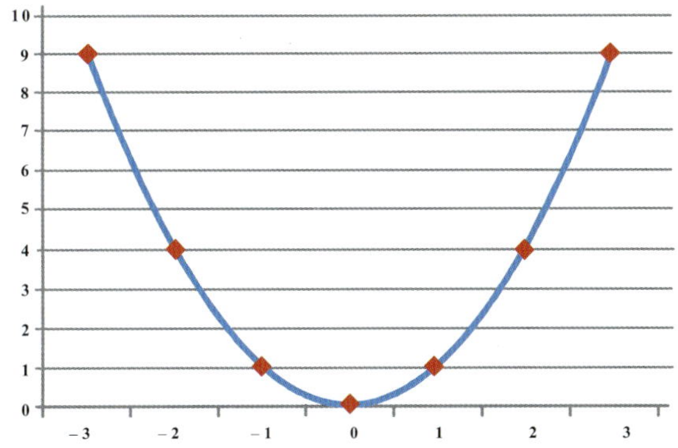

Appendix 2: General Exercises

1. Cal.V(F) = 10.294; T.V. = 3.682; Dec. = Reject H_0; P.V. = 0.002

3. Cal.V(F) = 6.373; T.V. = 4.419; Dec. = Reject H_0; P.V. = 0.007

5. Cal.V(F) = 2.434; T.V. = 4.256; Dec. = Fail to reject H_0; P.V. = 0.143

7. Cal.V(F) = 0.013; T.V. = 3.239; Dec. = Fail to reject H_0; P.V. = 0.998

9. Cal.V(F) = 0.889; T.V. = 6.701; Dec. = Fail to reject H_0; P.V. = 0.435

11. Cal.V(F) = 0.081; T.V. = 7.559; Dec. = Fail to reject H_0; P.V. = 0.923

13. Cal.V(F) = 0.954; T.V. = 3.885; Dec. = Fail to reject H_0; P.V. = 0.413

15. Cal.V(F) = 1.062; T.V. = 5.292; Dec. = Fail to reject H_0; P.V. = 0.393

17. Cal.V(F) = 8.053; T.V. = 2.866; Dec. = Reject H_0; P.V. = 0.0003

19. CF = 45227.111; SS = 356.889; SST = 29.556; SSV = 1.556; SSE = 325.778; Dec.= Fail to reject H_0

21. CF = 222778.125; SS = 2291.875; SST = 257.625; SSV = 1894.875; SSE = 139.375; Dec. = Reject H_0

23. CF = 2795.042; SS = 35.958; SST = 13.125; SSV = 9.708; SSE = 13.125; Dec. = Fail to reject H_0

25. CF = 92.881; SS = 1.429; SST = 0.286; SSV = 0.462; SSE = 0.682; Dec. = Fail to reject H_0

Appendix 3: General Exercises

1. $H_0 : \eta = 60$
$H_1 : \eta < 60$
TS : $Z = 1.114$
CV = −1.645
Decision: Fail to reject H_0
P.V. = 0.867

3. $H_0 : \eta = \$200$
$H_1 : \eta < \$200$
TS : $Z = -2.6517$
CV = −1.645
Decision: Reject H_0
P.V. = 0.000

5. $H_0 : \eta = 67$
$H_1 : \eta \neq 67$
TS : $Z = 0.348$
CV = 2.575
Decision: Fail to reject H_0
P.V. = 0.728

7. $H_0 : \eta = 65$
$H_1 : \eta < 65$
TS : $Z = 1.172$
CV = −2.053
Decision: Fail to reject H_0
P.V. = 0.568

9. $H_0 : \eta = 20$
$H_1 : \eta > 20$
$X = 8$, $n = 13$
Rejection Region: $X \geq 10$
Decision: Fail to reject H_0
P.V. = 0.291

11. $H_0 : \eta = 40$
$H_1 : \eta > 40$
$X = 3$, $n = 7$
Rejection Region: $X \geq 6$
Decision: Fail to reject H_0
P.V. = 0.773

13. $H_0: \eta = 50$
$H_1: \eta \neq 50$
$X = 4$, $n = 7$
Rejection Region: $X \leq 0$ or $X \geq 7$
Decision: Fail to reject H_0
P.V. $= 2P(X \geq 4) = 1.000$

15. $H_0: \eta_1 = \eta_2$
$H_1: \eta_1 \neq \eta_2$
$X = 2$, $n = 7$
Rejection Region: $X \leq 0$ or $X \geq 7$
Decision: Fail to reject H_0
P.V. $= 2P(X \leq 2) = 0.454$

Method 3- $X = 2$, $n = 7$, Rejection Region: $X \leq 0$, Decision: Fail to reject H_0

17. $H_0: \eta = 20$
$H_1: \eta > 20$
$T = \min(48, 43) = 43$
Rejection Region: $(T \leq 21)$
Decision: Fail to reject H_0
P-value $= P(T \leq 43) > P(T \leq 21) = 0.05$

19. $H_0: \eta = 40$
$H_1: \eta > 40$
$T = \min(12, 16) = 12$
Rejection Region: $(T \leq 4)$
Decision: Fail to reject H_0
P-value $= P(T \leq 12) > P(T \leq 4) = 0.05$

21. $H_0: \eta = 50$
$H_1: \eta \neq 50$
$T = \min(16, 12) = 12$
Rejection Region: $(T \leq 4)$
Decision: Fail to reject H_0
P-value $= P(T \leq 12) > P(T \leq 4) = 0.1$

23. $H_0: \eta_1 = \eta_2$
$H_1: \eta_1 < \eta_2$
$T = \min(8, 13) = 8$
Rejection Region: $(T \leq 2)$
Decision: Fail to reject H_0
P-value $= P(T \leq 8) > P(T \leq 2) = 0.05$

25. $H_0: \eta = 3$
$H_1: \eta \neq 3$
$T = \min(297, 28) = 28$
$Z = -3.619$
Decision: Reject H_0
P-value $= P(Z \leq -3.619) = 0.0003$

27. $H_0: \eta = 200$
$H_1: \eta < 200$
$T = \min(129, 399) = 129$
$Z = -2.524$
Decision: Reject H_0
P-value $= P(Z \leq -2.524) = 0.006$

29. $H_0: \eta = 575$
$H_1: \eta \neq 575$
$T = \min(419, 211) = 211$
$Z = -1.703$
Decision: Fail to reject H_0
P-value $= P(Z \leq -1.703) + P(Z \geq 1.703) = 0.089$

31. $H_0: \eta_1 = \eta_2$
$H_1: \eta_1 < \eta_2$
$T_X = 20.5$, $T_Y = 45.5$
$U_X = 10.5$, $U_Y = 17.5$ and $U = 10.5$
Rejection Region: $U \leq 4$
Decision: Fail to reject H_0
P-value $= P(U \leq 10.5) > P(U \leq 4) = 0.05$

33. $H_0: \eta_1 = \eta_2$
$H_1: \eta_1 > \eta_2$
$T_X = 36$, $T_Y = 42$
$U_X = 15$, $U_Y = 21$ and $U = 15$
Rejection Region: $U \leq 7$
Decision: Fail to reject H_0
P-value $= P(U \leq 15) > P(U \leq 7) = 0.05$

35. $H_0: \eta_1 = \eta_2$
$H_1: \eta_1 \neq \eta_2$
$T_X = 21.5$, $T_Y = 33.5$
$U_X = 6.5$, $U_Y = 18.5$ and $U = 6.5$
Rejection Region: $U \leq 4$
Decision: Fail to reject H_0
P-value $= P(U \leq 6.5) > P(U \leq 4) = 0.1$

37. $H_0: \eta_A = \eta_B$
$H_1: \eta_A < \eta_B$
$T_A = 111.5$, $T_B = 98.5$
$U_A = 56.5$, $U_B = 43.5$ and $U = 43.5$
TS : $Z = -0.49$
CV $= -1.64$
Decision: Fail to reject H_0
P-value $= P(Z \leq -0.49) = 0.312$

39. $H_0 : \eta_A = \eta_B$

$H_1 : \eta_A \neq \eta_B$

$T_A = 109,\ T_B = 101$

$U_A = 54,\ U_B = 46$ and $U = 46$

TS : $Z = -0.302$

CV = -1.96

Decision: Fail to reject H_0

P-value = $P(Z \leq -0.302) + P(Z \geq 0.302) = 0.7627$

41. $r_{sp} = -0.943$

43. $r_{sp} = -0.464$

45. $r_{sp} = 0.706$

CV = 0.591

Rejection Region = $r_{sp} < -0.591$ or $r_{sp} > 0.591$

Decision: Reject H_0

47. $r_{sp} = 0.992$

CV = 0.833

Rejection Region : $r_{sp} < -0.833$ or $r_{sp} > 0.833$

Decision: Reject H_0

49. TS = 4.909

CV = 2.576

Decision: Reject H_0

P-value = 0.000

51. TS = 5.901

CV = 2.576

Decision: Reject H_0

P-value = 0.000

INDEX

Index